T0293772

Herbal Medicines

Herbal Medicines
A Boon for Healthy Human Life

Edited by

Hifzur R. Siddique
Molecular Cancer Genetics & Translational Research
Lab, Section of Genetics, Department of Zoology,
Aligarh Muslim University, Aligarh, India

Maryam Sarwat
Amity Institute of Pharmacy, Amity University,
Noida, Uttar Pradesh, India

ACADEMIC PRESS
An imprint of Elsevier
elsevier.com/books-and-journals

Academic Press is an imprint of Elsevier

125 London Wall, London EC2Y 5AS, United Kingdom
525 B Street, Suite 1650, San Diego, CA 92101, United States
50 Hampshire Street, 5th Floor, Cambridge, MA 02139, United States
The Boulevard, Langford Lane, Kidlington, Oxford OX5 1GB, United Kingdom

Notices
Knowledge and best practice in this field are constantly changing. As new research and experience broaden our understanding, changes in research methods, professional practices, or medical treatment may become necessary.

Practitioners and researchers must always rely on their own experience and knowledge in evaluating and using any information, methods, compounds, or experiments described herein. In using such information or methods they should be mindful of their own safety and the safety of others, including parties for whom they have a professional responsibility.

To the fullest extent of the law, neither the Publisher nor the authors, contributors, or editors, assume any liability for any injury and/or damage to persons or property as a matter of products liability, negligence or otherwise, or from any use or operation of any methods, products, instructions, or ideas contained in the material herein.

British Library Cataloguing-in-Publication Data
A catalogue record for this book is available from the British Library

Library of Congress Cataloging-in-Publication Data
A catalog record for this book is available from the Library of Congress

ISBN: 978-0-323-90572-5

For Information on all Academic Press publications visit our website at https://www.elsevier.com/books-and-journals

Publisher: Stacy Masucci
Acquisitions Editor: Patricia Osborn
Editorial Project Manager: Sam W. Young
Production Project Manager: Maria Bernard
Cover Designer: Christian Bilbow

Typeset by Aptara, New Delhi, India

Working together
to grow libraries in
developing countries

www.elsevier.com • www.bookaid.org

Dedication

Dedicated to our parents whose love toward herbal medicines encouraged us to edit this book.

Contents

Contents

Contributors

Bhat M. Aalim, Academy of Scientific & Innovative Research (AcSIR), Ghaziabad-201002, UP, India; PK-PD and Toxicology Division, CSIR-Indian Institute of Integrative Medicine, Jammu, Jammu & Kashmir, India

Mohammad Afzal, Department of Zoology, Aligarh Muslim University, Aligarh, UP, India

Neeraj Agarwal, Department of Medicine, Cedars Sinai Medical Center, Los Angeles, CA, United States

Kirti Aggarwal, Amity Institute of Pharmacy, Amity University, Noida, Uttar Pradesh, India

Meenakshi Ahluwalia, Department of Pathology, Medical College of Georgia, Augusta University, Augusta, GA, United States

Sayeed Ahmad, Department of Pharmacognosy & Phytochemistry, School of Pharmaceutical Education & Research, Jamia Hamdard, New Delhi, India

Fahad Ahmed, Department of Basic Health, Yidirim Beyazit University, Üniversiteler Mah. İhsan Doğramacı Bulvarı Ankara Atatürk Eğitim Araştırma Hastanesi Yanı Bilkent Çankaya/Ankara, Turkey

Swalih P. Ahmed, Interdisciplinary Brain Research Centre (IBRC), J.N.Medical College, Faculty of Medicine, Aligarh Muslim University, Aligarh, UP, India

Ahmad Ali, University Department of Life Sciences, University of Mumbai, Mumbai, India

Asif Ali, Interdisciplinary Brain Research Centre (IBRC), J.N.Medical College, Faculty of Medicine, Aligarh Muslim University, Aligarh, UP, India

Sumera Ali, Department of Pharmaceutical Sciences, Chicago State University College of Pharmacy, Chicago, IL

Bader Alshehri, Department of Laboratory Sciences, College of Applied Medical Sciences, Majmaah University, Majmaah, Saudi Arabia

Abdel R. Al-Tawaha, Department of Biological Sciences, Al-Hussein Bin Talal University, Maan, Jordon

Arturo Analla, Department of Pharmaceutical Sciences, Chicago State University College of Pharmacy, Chicago, IL

Sajeeda Archoo, Academy of Scientific & Innovative Research (AcSIR), Ghaziabad-201002, UP, India; PK-PD and Toxicology Division, CSIR-Indian Institute of Integrative Medicine, Jammu, Jammu and Kashmir, India

Rajendra Awasthi, Amity Institute of Pharmacy, Amity University Uttar Pradesh, Noida, Uttar Pradesh, India

Omar Bagasra, Claflin University, 400 Magnolia Street, Orangeburg, SC 29115, USA

Samarpita Banerjee, Department of Botany, Shri Shikshayatan College, Kolkata, India

Jonathan M. Banks, Department of Periodontics, College of Dentistry, University of Illinois at Chicago, Chicago, Illinois, USA

Hina Bansal, Amity Institute of Biotechnology, Amity University Uttar Pradesh, Noida, Uttar Pradesh, India

Debora B. Barbosa, Department of Diagnosis and Surgery, São Paulo State University (UNESP), School of Dentistry, Araçatuba, São Paulo, Brazil

Parakh Basist, Department of Pharmacognosy & Phytochemistry, School of Pharmaceutical Education & Research, Jamia Hamdard, New Delhi, India

Gaber E-S Batiha, Department of Pharmacology and Therapeutics, Faculty of Veterinary Medicine, Damanhour University, Damanhour, AlBeheira, Egypt

Jessica Beamon, Department of Pharmaceutical Sciences, Chicago State University College of Pharmacy, Chicago, IL

Daniela A. Brandini, Department of Diagnosis and Surgery, São Paulo State University (UNESP), School of Dentistry, Araçatuba, São Paulo, Brazil

Ercan Bursal, Department of Biochemistry, Mus Alparslan University, Turkey

Anis A. Chaudhary, Department of Biology, College of Science, Al-Imam Mohammad Ibn Saud Islamic University (IMSIU), Riyadh, Kingdom of Saudi Arabia

Abhijit Dey, Department of Life Sciences, Presidency University, Kolkata

Neerupma Dhiman, Amity Institute of Pharmacy, Amity University Uttar Pradesh, Noida, Uttar Pradesh, India

Leonard C. D'Souza, Nitte (Deemed to be University), Nitte University Centre for Science Education and Research (NUCSER), Division of Environmental Health and Toxicology, Mangaluru, India

Charles E. Dowling II, Claflin University, 400 Magnolia Street, Orangeburg, SC 29115, USA

Berthon Eliche, Claflin University, 400 Magnolia Street, Orangeburg, SC 29115, USA

Mohammad Fareed, College of Medicine, Al-Imam Mohammad Ibn Saud Islamic University (IMSIU), Riyadh, Kingdom of Saudi Arabia

Homa Fatma, Molecular Cancer Genetics & Translational Research Lab, Section of Genetics, Department of Zoology, Aligarh Muslim University, Aligarh, Uttar Pradesh, India

Deepak Ganjewala, Amity Institute of Biotechnology, Amity University Uttar Pradesh, Noida, Uttar Pradesh, India

Mimosa Ghorai, Department of Life Sciences, Presidency University, Kolkata

Arabinda Ghosh, Department of Botany, Gauhati University, Guwahati, Assam, India

Suchhanda Ghosh, Department of Botany, Shri Shikshayatan College, Kolkata, India

Deepshikha Gupta, Amity Institute of Applied Sciences, Amity University, Noida, India

Meenakshi Gupta, Amity Institute of Pharmacy, Amity University, Noida, Uttar Pradesh, India

Mahmood A. Hashim, ITMO University, Faculty of Biotechnologies (BioTech), Saint Petersburg, Russia

Zhaozhao Hua, Department of Obstetrics, The Second Hospital, Guizhou University of Traditional Chinese Medicine, Guiyang, Guizhou Province, China

Brionna Hudson, Department of Pharmaceutical Sciences, Chicago State University College of Pharmacy, Chicago, IL

Chinnaswamy Jagannath, Department of Pathology and Genomic Medicine, Houston Methodist Research Institute, Houston, TX, United States

Christine Jeyaseelan, Amity Institute of Applied Sciences, Amity University, Noida, India

Zhigang Jiang, School of Public Health Sciences, Zunyi Medical University, Zunyi City, Guizhou Province, China

Mahsa Pourali Kahriz, Department of Field Crops, Facultury of Agriculture, Ankara University, Ankara

Parisa Pourali Kahriz, Department of Field Crops, Facultury of Agriculture, Ankara University, Ankara

Nandni Kakar, Department of Pharmaceutical Sciences, Chicago State University College of Pharmacy, Chicago, IL

Gurmanpreet Kaur, School of Public Health, University of Alabama at Birmingham, AL, United States

Arshad Khan, Department of Pathology and Genomic Medicine, Houston Methodist Research Institute, Houston, TX, United States

Faaeiza Khan, Department of Pharmaceutical Sciences, Chicago State University College of Pharmacy, Chicago, IL

Johra Khan, Department of Laboratory Sciences, College of Applied Medical Sciences, Majmaah University, Majmaah, Saudi Arabia

Khalid Mahmood Khawar, Department of Field Crops, Facultury of Agriculture, Ankara University, Ankara

Giriraj T. Kulkarni, Gokaraju Rangaraju College of Pharmacy, Hyderabad, Telangana, India

Munish Kumar, Department of Biochemistry, University of Allahabad, Prayagraj, Uttar Pradesh, India

Vinay Kumar, Department of Biotechnology, Modern College (Savitribai Phule Pune University), Pune, India

Jian Li, Department of Oncology, The Second Hospital, Guizhou University of Traditional Chinese Medicine, Guiyang, Guizhou Province, China

Dezhong Joshua Liao, Office of Research and Education Administration, The Second Hospital, Guizhou University of Traditional Chinese Medicine, Guiyang, Guizhou Province, China

Zhengqi Liu, Office of Research and Education Administration, The Second Hospital, Guizhou University of Traditional Chinese Medicine, Guiyang, Guizhou Province, China

Swati Madan, Amity Institute of Indian System of Medicine, Amity University, Noida, UP, India

Siphiwe G. Mahlangu, Centre for Bioprocess Engineering Research, Department of Chemical Engineering, University of Cape Town, Cape Town, South Africa

Abhishek Mishra, Department of Pathology and Genomic Medicine, Houston Methodist Research Institute, Houston, TX, United States

Anurag Mishra, Department of Biochemistry, University of Allahabad, Prayagraj, Uttar Pradesh, India

Ruchika Mittal, Amity Institute of Biotechnology, Amity University Uttar Pradesh, Noida, Uttar Pradesh, India

R'kia El Moudden, Department of Pharmaceutical Sciences, Chicago State University College of Pharmacy, Chicago, IL

Mariam Muradova, ITMO University, Faculty of Biotechnologies (BioTech), Saint Petersburg, Russia

Liudmila Nadtochii, ITMO University, Faculty of Biotechnologies (BioTech), Saint Petersburg, Russia

Shahid H. Naikoo, Academy of Scientific & Innovative Research (AcSIR), Ghaziabad-201002, UP, India; PK-PD and Toxicology Division, CSIR-Indian Institute of Integrative Medicine, Jammu, Jammu and Kashmir, India

Samapika Nandy, Department of Life Sciences, Presidency University, Kolkata, India

Afsar R. Naqvi, Department of Periodontics, College of Dentistry, University of Illinois at Chicago, Chicago, Illinois, USA

Fabrice Neiers, Centre des Sciences du Goût et de l'Alimentation (CSGA), Université de Bourgogne Franche-Comté, INRAE, CNRS, France

Potshangbam Nongdam, Department of Biotechnology, Manipur University, Imphal, Manipur, India

Pascaline Obika, Department of Pharmaceutical Sciences, Chicago State University College of Pharmacy, Chicago, IL

Jagdish G. Paithankar, Nitte (Deemed to be University), Nitte University Centre for Science Education and Research (NUCSER), Division of Environmental Health and Toxicology, Mangaluru, India

Devendra K. Pandey, Department of Biotechnology, Lovely Professional University, Punjab, India

Pragya Pandey, Amity Institute of Pharmacy, Amity University Uttar Pradesh, Noida, Uttar Pradesh, India

Additiya Paramanya, University Department of Life Sciences, University of Mumbai, Mumbai, India

Savan Patel, Department of Pharmaceutical Sciences, Chicago State University College of Pharmacy, Chicago, IL

Li Peng, Department of Cardiology, The Second Hospital, Guizhou University of Traditional Chinese Medicine, Guiyang, Guizhou Province, China

Payal Poojari, University Department of Life Sciences, University of Mumbai, Mumbai, India

Alena Proskura, ITMO University, Faculty of Biotechnologies (BioTech), Saint Petersburg, Russia

Melany Puglisi-Weening, Department of Pharmaceutical Sciences, Chicago State University College of Pharmacy, Chicago, IL

Sharma R. Raghu, Academy of Scientific & Innovative Research (AcSIR), Ghaziabad-201002, UP, India; PK-PD and Toxicology Division, CSIR-Indian Institute of Integrative Medicine, Jammu, Jammu & Kashmir, India

Anwita Revoori, Sharon High School, Sharon, MA, Unites States

Maryam Sarwat, Amity Institute of Pharmacy, Amity University, Noida, Uttar Pradesh, India

Meghna Saxena, Networking Key Services (NKS), Edinburgh, United Kingdom

Mehdi H. Shahi, Interdisciplinary Brain Research Centre (IBRC), J.N. Medical College, Faculty of Medicine, Aligarh Muslim University, Aligarh, UP, India

Mohd Shahid, Department of Pharmaceutical Sciences, Chicago State University College of Pharmacy, Chicago, IL

Anurag Sharma, Nitte (Deemed to be University), Nitte University Centre for Science Education and Research (NUCSER), Division of Environmental Health and Toxicology, Mangaluru, India

Bhupesh Sharma, Amity Institute of Pharmacy, Amity University Uttar Pradesh, Noida, Uttar Pradesh, India

Mahipal S. Shekhawat, Plant Biotechnology Unit, K.M. Government Institute for Post Graduate Studies and Reserach, Lawspet, Puducherry, India

Lubna Shihadeh, Department of Pharmaceutical Sciences, Chicago State University College of Pharmacy, Chicago, IL

Shoaib Shoaib, Department of Biochemistry, Faculty of Medicine, Aligarh Muslim University, Aligarh, India

Pallavi Shrivastava, Faculty of Pharmacy and Biochemistry, Catholic University of Santa Maria, Arequipa, Peru

Hifzur R. Siddique, Molecular Cancer Genetics & Translational Research Lab, Section of Genetics, Department of Zoology, Aligarh Muslim University, Aligarh, India

Deepti Singh, Molecular Cancer Genetics & Translational Research Lab, Section of Genetics, Department of Zoology, Aligarh Muslim University, Aligarh, India

Dipty Singh, Neuroendocrinology, ICMR- National Institute for Research in Reproductive Health, Mumbai, India

Vipul K. Singh, Department of Pathology and Genomic Medicine, Houston Methodist Research Institute, Houston, TX, United States

Gauri Srivastava, Amity Institute of Biotechnology, Amity University Uttar Pradesh, Noida, Uttar Pradesh, India

Siew L. Tai, Centre for Bioprocess Engineering Research, Department of Chemical Engineering, University of Cape Town, Cape Town, South Africa

Aline S. Takamiya, Department of Diagnosis and Surgery, São Paulo State University (UNESP), School of Dentistry, Araçatuba, São Paulo, Brazil

Sheikh A. Tasduq, Academy of Scientific & Innovative Research (AcSIR), Ghaziabad-201002, UP, India; PK-PD and Toxicology Division, CSIR-Indian Institute of Integrative Medicine, Jammu, Jammu & Kashmir, India

Pari Thakkar, Department of Periodontics, College of Dentistry, University of Illinois at Chicago, Chicago, Illinois, USA

Khanh Tran, Department of Molecular Medicine and Pathology, University of Auckland, Auckland, New Zealand

Champa Keeya Tudu, Department of Life Sciences, Presidency University, Kolkata

Begüm Ünlü, Department of Dermatology, Ağrı State Hospital, Ağrı, Turkey

Kumar Vaibhav, Department of Neurosurgery, Medical College of Georgia, Augusta University, Augusta, GA, United States

Sivakumar Vijayaraghavalu, Narayana Translational Research Centre, Narayana Medical College and Hospital, Nellore, Andhra Pradesh, India

Tao Xu, Department of Cardiology, The Second Hospital, Guizhou University of Traditional Chinese Medicine, Guiyang, Guizhou Province, China

Vanshika Yadav, Amity Institute of Pharmacy, Amity University, Noida, Uttar Pradesh, India

Khurram Yusuf, Kendriya Vidyalaya, Jabalpur, India

Nabiha Yusuf, School of Public Health, University of Alabama at Birmingham, AL, United States; Department of Dermatology, School of Medicine, University of Alabama at Birmingham, AL, United States

Kathy Zheng, Department of Periodontics, College of Dentistry, University of Illinois at Chicago, Chicago, Illinois, USA

About the Editors

Dr. Hifzur R. Siddique

Dr. Siddique is working as a Senior Assistant Professor at Aligarh Muslim University (AMU), India. He obtained his PhD in 2008 from the Indian Institute of Toxicology Research, India, and got post-doctoral trainings at the University of Wisconsin, University of Minnesota & University of S. California, USA. His research specializations are cancer biology, pharmacology, and toxicology. He is now the Programme Director of Molecular Cancer Genetics and Translational Research program at AMU. He is, so far, published more than 75 original research papers in journals like nature communications, hepatology, stem cells, clinical cancer research, cancer research, genes and cancer, seminars in cancer biology, seminars in cell and developmental biology, dalton transactions, co-ordination chemistry reviews, molecular cancer therapeutics, etc. He is the recipient of several academic awards, such as SBUR-Young Scientist Travel Award, USA; IABS-Outstanding Cancer Research Award, India; SPER-Innovative Researcher Award, India; "AEDS-Distinguish Scientist Award," India, etc. He is a member of several scientific societies/associations, such as the American Association of Cancer Research, National Academy Sciences, India, Indian Association for Cancer Research, etc. He regularly delivers invited talks, filed four patents, and 26 awards/honors to his credit. Further, he serves as an editorial member of five journals and reviewers of more than 50 reputed international journals. Multiple government agencies of India support his ongoing research.

Dr. Maryam Sarwat

Dr. Maryam Sarwat is an Indian Citizen, working as an Associate Professor in the Amity Institute of Pharmacy, Amity University, Noida. She has obtained her PhD in 2007 and completed postdoctoral research at ICGEB, New Delhi. She has received several research grants from various R & D agencies, such as DST, DBT, CCRUM, and SERB in India. She has presented her findings in scientific conferences in various countries like Germany, Czech Republic, France, etc. Dr. Sarwat has mentored six PhD and 15 masters' students. She has more than 55 international publications in reputed journals and three patents to her credit. She is the recipient of the prestigious "Scientist of the Year Award" in 2015. Dr. Sarwat has published two volumes of "Stress Signaling in Plants, Genomics and Proteomics Perspective" with

Springer Nature in 2013 and 2017. She has also authored two books with Elsevier entitled "Senescence Signaling and Control in Plants" in 2018 and "Saffron: The Age-Old Panacea in a New Light" in the early 2020. She has also published the title "Environment and Human Health" through King Abdul Aziz University Press in 2020. Her two volumes on "Ethnic Knowledge on Biodiversity, Nutrition and Health Security" are due for publication by Taylor and Francis in 2021. She has served various international journals as reviewer.

Foreword

Herbal Medicines: A Boon for Healthy Human Life

Global interest in herbal supplements and medicines is increasing every day, and according to some estimates, as much as 80% population worldwide might be using for some part of their healthcare. Herbal medicines are also getting significant attention in global health markets. At present, many researchers and health care personnel believe that herbal medicines may play an important role in health management. However, the major issue faced by scientists/clinicians is a lack of adequate research data, especially focusing on their safety, which hinders the regulatory aspects for such therapies. This hinders the clinical acceptance and approval of herbal agents. Indeed, significant additional research is required in this direction. In addition, dedicated books and journals can be very useful in compiling and informing the progress toward the potential use of herbal agents in health care. Considering global interest of general public as well as researchers and clinicians toward potential importance of herbal medicine/supplements, this book presents interesting chapters related to herbal medicines in the management of human health. These chapters summarize the current state and future directions, which should enlighten everyone who may be interested or has a stake in health care research or management. The book also covers mechanistic studies and studies focusing on identifying active compounds of herbal agents. The present book is a compilation of 31 chapters covering *neurological disorders, oral health, metabolic disorders, and overall health, dermatological diseases, reproductive diseases, infectious diseases, cardiovascular diseases, cancer. The last chapter is dedicated to the concern of discriminatory use of herbal medicine without scientific evidence* contributed by the leading experts from different parts of the world. Overall, I found the book entitled "*Herbal Medicines: A Boon for Healthy Human Life*" to be very interesting and quite unputdownable. I sincerely hope that it would evoke a great deal of interest amongst students, teachers, researchers, and clinicians. This book reflects the editors' deep scientific understanding and academic prowess. The editors Hifzur

R. Siddique and Maryam Sarwat complemented each other to bring about their very best on herbal medicine.

Prof. Nihal Ahmad, PhD

"Nelson M. Hagan Class of 1929" Endowed Professor

Vice-Chair for Research

Department of Dermatology | University of Wisconsin School of Medicine and Public Health

Co-Leader, Cancer Prevention and Control Program, Carbone Cancer Center

Address: 1111 Highland Avenue, WIMR-II; Room 7405 | Madison, WI 53705-2275

Phone: 608.263.5359 | Fax: 608.262.7137

Preface

Medicinal plants are important therapeutic aid for alleviating ailments of humankind since ancient times. From the beginning of human civilizations, there have been very strong traditional systems of medicine, such as Chinese, Ayurvedic, and the Unani were practiced. With advancements in modern science, our understanding of the composition of herbal drugs and their mechanism of action at the molecular level has increased immensely. These plants contain bioactive chemical substances, such as alkaloids, tannins, saponins, and other therapeutic potentials. A lot of research work has been carried out on some medicinal herbs, and they have been found to have definite action on the diseases related to the nervous, circulatory, respiratory, digestive, and urinary systems and the skin, vision, hearing, and sexual organs. Sustained research is being carried out worldwide to study their efficiency and mechanism of action, among many other aspects.

Our book has extensively highlighted various studies on herbal medicines against multiple diseases and the medicinal and pharmacological aspects of different herbal medicines and their products. The book has been divided into nine sections, with the last section covering the shortcomings of herbal medicine. Sections are divided into neurological disorders, oral health, metabolic disorders, and overall health, skin disorders, reproductive disorders, infectious diseases, cardiovascular disorders, and cancer. Each chapter summarizes the current state and future direction of the use of herbal medicines against multiple diseases from a translational point of view, making this reference a valuable source of information for a large audience, including researchers and healthcare providers interested in the field of herbal remedies. Discusses essential evidence-based information about herbal medicines provides an update to discoveries and recent advances on the use of herbal medicines to treat multiple human diseases. It covers all major medicinal compounds present in herbal medicines, including alkaloids, glycosides, polyphenols, and terpenes. In addition, detailed studies have been explained with the help of figures, images, and tables to make it even more exciting and descriptive. Even after our best efforts, we do feel that there may still be scope for improvement and addition in this work, especially given the rapidly evolving and ever-advancing nature of this field. Helpful criticism and suggestions from the readers are dearly welcome.

Finally, we are thankful to Elsevier for their keen interest and attention in bringing out this book in its present form.

Hifzur R. Siddique

Maryam Sarwat

Acknowledgments

We are grateful to our institutes Aligarh Muslim University, India, and Amity University, Noida, India, for providing infrastructure to complete this work smoothly. We also wish to express our most profound appreciation to our colleagues and friends for their expert insights, discussions, and communications throughout the process of book writing and editing that greatly assisted and enhanced the final outlook of our book. Dr. Hifzur R. Siddique is thankful to **Prof. Tariq Mansoor**, Vice Chancellor, for his constant encouragement to do work on herbal medicine, **Prof. Mohammed Saleem**, Rush University, USA; **Prof. Nihal Ahmad**, University of Wisconsin, USA; **Prof. Ajay P Singh**, University of South Alabama, USA; Prof. **D. Joshua Liao**, Three George University, China, **Prof. D Kar Chowdhuri**, Ex-Chief Scientist, Indian Institute of Toxicology Research, India for their support and critical comments, which help a lot to choose the topics for the book. Dr. Hifzur is also thankful to his research scholars, **Ms. Homa Fatma**, **Ms. Deepti Singh,** and **Mr. Mohammad Afsar Khan,** for their timely help.

Dr. Maryam is sincerely indebted to Prof. Ashok K. Chauhan, Founder President of Amity University, Dr. Atul Chauan, Chancellor AUUP, Prof. W. Selwamurthy, President of Amity Science, Technology and Innovation Foundation and Prof. B. C. Das, Dean of Amity Health and Allied Sciences, Amity University, Noida, for an all-time help and encouragement during the process of editing this book. Dr. Maryam owes special thanks to the Head of the Institute, Amity Institute of Pharmacy, Dr. Tanveer Naved, and the research scholars Ms. Shruti Rastogi, Ms. Meenakshi Gupta, Ms. Kumari Chandan, and Ms. Nidhi Sharma for assistance with comments and innovative ideas that greatly improved the manuscript. Dr. Maryam conveys special regards for her colleagues and friends Dr. Swati Madan, Dr. Shruti Chopra, and Dr. Ramanpreet Walia, for their unconditional love and support during the writing of this book and always.

This book would not have any essence of thoughts, concerns, and feel without having Ms. **Faria Zafar** (daughter of MS) and **Darakhsheed Rabbani** (sister of MS), Ms. **Zikra Siddique,** and Mr. **Daniyal Siddique** (children of HRS) on board. Dr. Siddique is especially indebted to his wife, Mrs. Jaheda Choudhury Siddique, for her time, support, and midnight tea/coffee during the book's preparation. The editors are also thankful to their colleagues, friends, and families for providing moral support while writing the book.

Finally, we would like to thank all the authors, reviewers, publisher, and their team members for sharing their pearls of wisdom with us during the writing and compilation of this book.

PART 1

Neurological disorders

Herbal drugs an alternative medicine for the treatment of neurodegenerative diseases: Preclinical and clinical trial review

Pallavi Shrivastava[a], Anwita Revoori[b], Kumar Vaibhav[c], Meenakshi Ahluwalia[d] and Meghna Saxena[e]

[a]Faculty of Pharmacy and Biochemistry, Catholic University of Santa Maria, Arequipa, Peru [b]Sharon High School, Sharon, MA, Unites States [c]Department of Neurosurgery, Medical College of Georgia, Augusta University, Augusta, GA, United States [d]Department of Pathology, Medical College of Georgia, Augusta University, Augusta, GA, United States [e]Networking Key Services (NKS), Edinburgh, United Kingdom

1.1 Introduction

Neurodegenerative diseases are major health problems in the elderly population and among them most common are Alzheimer's disease (AD) and Parkinson's disease (PD) [1,2]. AD is the sixth leading cause of death in the United States and a leading cause of cognitive impairment [1,3]. PD is the 14th cause of death in the United States and affects the motor cortex of mid brain that causes rigidity, slowness of movement, shaking, postural instability and difficulty with walking and gait disorders [2]. Both AD and PD develop dementia and impacts on a person's life significantly as it prohibits patients from following through with their daily routines [3,4]. Dementia is a term used to describe a loss of memory, language, problem-solving, and other thought skills that are serious enough to affect everyday life [5]. AD is a chronic, irreversible disease that causes impairment in intellectual functioning such as gradually destroying the ability to reason, remember, imagine, and learn [1,6]. Alzheimer's gradually damage the neurons in the brain and currently, the only few treatments are available to slow down the neural damage rather than completely halting it. On the other hand, PD is a neurodegenerative disorder that primarily affects dopamine-producing ("dopaminergic'') neurons in the substantia nigra region of the brain [2]. As a result of Parkinson's, patients may experience tremors in hands, bradykinesia, limb rigidity, or gait and balance problems

Herbal Medicines: A Boon for Healthy Human Life.
DOI: https://doi.org/10.1016/B978-0-323-90572-5.00031-7

A

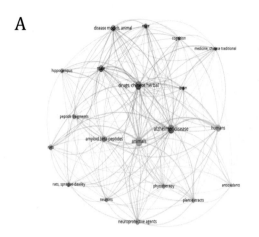

B

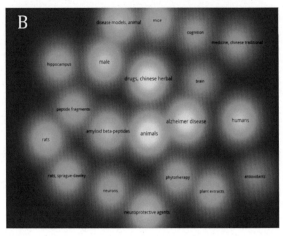

Figure 1.1
(A) Network analysis of timeline of herbal drugs published on AD in scientific publications in the past three decades. (B) Density plot of research published about AD in the past three decades (1990–2020) created by VOS viewer. *AD*, Alzheimer disease.

[2]. Similar to AD, PD has no cure, and treatment options vary as they include medications and surgery. Both of these diseases are generally more common in older people of age from 65 years and above as they experience these symptoms more often. The current treatments available for AD that are FDA approved are cholinesterase inhibitors and memantine [7,8]. Additionally, the current treatment available for people with PD is Levodopa which is considered as a gold standard drug for PD, but it develops tolerance [9]. Because these treatments do not provide the ultimate cure of the disease, there is an urgent need and demand for a more effective treatment for these diseases. As a result, many researchers have turned to herbal medicines to find any possible targets and treatments that may be more effective and less toxic than the current FDA-approved drugs. There have been many preclinical and clinical trials performed on herbal drugs, but some of the drugs used in preclinical trials unable to culminate into the clinical trials due to financial constraints (Figs. 1.1–1.3).

Herbal drugs have been extensively used for the treatment of other diseases like cancer, gastrointestinal disorders [10]. The benefit of herbal drugs is that they are natural and have less or no side effects as compared to Allopathic medicines. For AD, there have been 191 herbal drugs tested in preclinical and 52 clinical trials (1990–2020) held on herbal medicines around the world in the past three decades. Many of these trials occurred in China, Japan, and India, which is understandable as they have easy access to many exotic herbal plants necessary to explore the efficacy of herbal drugs. For PD, 107 herbal drugs have been screened in preclinical and clinical trials from 1900–2020. Many of these drugs that were able to go through the clinical trial phases had stopped at phase one, making it difficult for any of them to get FDA approval. Only one drug is FDA approved for the treatment of AD, that is Huperzine A, but it

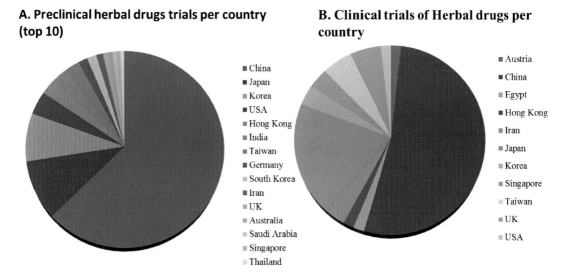

A. Preclinical herbal drugs trials per country (top 10)

- China
- Japan
- Korea
- USA
- Hong Kong
- India
- Taiwan
- Germany
- South Korea
- Iran
- UK
- Australia
- Saudi Arabia
- Singapore
- Thailand

B. Clinical trials of Herbal drugs per country

- Austria
- China
- Egypt
- Hong Kong
- Iran
- Japan
- Korea
- Singapore
- Taiwan
- UK
- USA

Figure 1.2

(A) Preclinical herbal drugs trials on AD in countries (top 10 countries presented based on number of preclinical publications). (B) Number of clinical trials performed by countries in the past three decades (1990–2020) on AD. *AD*, Alzheimer disease.

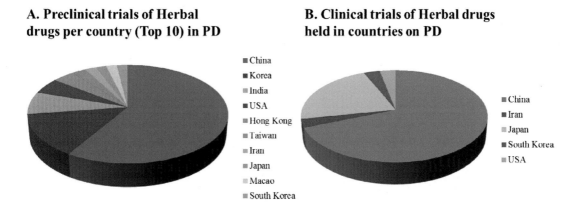

A. Preclinical trials of Herbal drugs per country (Top 10) in PD

- China
- Korea
- India
- USA
- Hong Kong
- Taiwan
- Iran
- Japan
- Macao
- South Korea

B. Clinical trials of Herbal drugs held in countries on PD

- China
- Iran
- Japan
- South Korea
- USA

Figure 1.3

(A) Preclinical herbal drugs trials on PD in countries (top 10 countries presented based on number of pre-clinical publications). (B) Number of clinical trials performed by countries in the past three decades (1990–2020) on PD. *PD*, Parkinson disease.

is only listed as a dietary supplement rather than a medicine for the treatment of AD [11,12]. In this chapter, we will discuss the top five herbal drugs published studies in preclinical and clinical models for AD and PD.

1.2 Herbal drugs formulations in preclinical trials

Several studies on herbal products, providing positive results on AD and PD as many formulations went under preclinical trials. This indicates the necessity of large and complete studies on herbal formulations. A few of the herbal drugs used in preclinical trials are mentioned here based on the number of published articles on Pubmed.

1.2.1 Effect of herbal drugs on Alzheimer's disease

1. **Danggui Shaoyao San (DSS):** There have been eight preclinical trials done using DSS drug. It is derived from China, and the main active principles within DSS are: Angelica Sinensis (Oliv.) Diels., Ligusticum chuanxiong Hort., Paeonia Lactiflora Pall., Poria cocos (Schw.) Wolf, Alisma orientalis (Sam.) Juzep., Atractylodes macrocephala Koidz [13]. In certain trials and through behavioral tests, APP/PS1 mice showed serious cognitive impairment. The cognitive deficits of APP/PS1 mice were markedly improved after treatment with DSS extract. Biochemical measurements revealed that APP/PS1 mice had higher TG, TC, LDL-c, and lower HDL-c than WT mice and DSS extract greatly slowed these changes. Low DHA content, low expression of iPLA2 and 15-LOX were found in the hippocampus and cortex of APP/PS1 mice, but DSS extract significantly reversed these improvements. DSS also improves memory deficiencies in APP/PS1 mice by increasing DHA content by upregulating iPLA2 and 15-LOX, which decreases oxidative stress and inflammation and thus improves DHA content [14]. Through multiple trials, it was found that DSS had significantly reduced symptoms [15].

2. **Ginkgo Biloba (EGb):** There have been eight preclinical trials done in AD using this EGb. As mentioned previously, it is derived from China and its main active principles include terpene lactones, ginkgo flavone glycoside [16]. In a specific study, TgCRND8 AD mice were administered Ginkgo Biloba to specifically overexpress human Alzheimer's amyloid precursor protein (APP) in neurons [17]. The Barnes Maze test revealed that mice treated for five months with EGb 761 had significantly increased cognitive function. It also avoided the absence of synaptic function proteins such as PSD-95, Munc18-1, and SNAP25. After five months of therapy with EGb 761, microglial inflammatory activity in the brain was blocked. EGb 761 therapy lowered NLRP3 protein levels in microglia that were colocalized with LC3-positive autophagosomes or autolysosomes [18]. By inhibiting secretase production and A aggregation, long-term therapy with EGb 761 can minimize a disease in the brain [18]. Therefore, with this data, the researchers were able to conclude that this drug improves the symptoms of AD.

3. **Huanglian-Jie-Du-Tang (HLJDT):** The HLJDT drug is also very commonly used and 11 preclinical trials were published. This drug is derived from China and its main components are berberine, baicalin, baicalein, and geniposide and can also include Rhizoma

Coptidis, Cortex Phellodendri, and Fructus Gardeniae without Radix Scutellariae [19]. The drug increased learning and memory in a 1–42 oligomer-induced AD model in a preclinical experiment involving mice [20]. Furthermore, the NMDA receptor-mediated glutamatergic transport and the adenosine/ATPase/AMPK cascade were discovered to be implicated in the fundamental mechanism of AD [21]. A modified formulation of Huanglian-Jie-Du-Tang was found to reduce memory impairments and β-amyloid plaques in a triple transgenic mouse model of AD [22]. In conclusion, the study proved that HLJDT may be a sufficient therapeutic drug to treat AD.

4. **Kaixin San (KXS):** This drug was very effective for treating AD which was found through the 11 preclinical trials conducted. KXS is derived from China and its main active components are Ginseng Radix et Rhizome (root and rhizome of P. ginseng), Polygalae Radix (root of P. tenuifolia), Acori Tatarinowii Rhizoma (rhizome of A. tatarinowii), and Poria (sclerotium of P. cocos) [23]. To determine the drug's effectiveness, brain tissue samples were collected and analyzed using high-throughput lipidomics based on UPLC-Q/TOF-MS in a study involving this drug. This research showed that high-throughput lipidomics can be used to recognize damaged pathways and lipid biomarkers as potential targets for determining the therapeutic effects of KXS [24]. The efficacy of effect of six class of Kaixin San formulas were studied on the pharmacological and preliminary mechanism of AD mice. The KXS significantly improved the learning and memory ability evaluation indicators, significantly increased BDNF and Ach in the hippocampus of AD model mice, reduced the Aβ, Tau protein, p-Tau protein in hippocampus of AD model mice, decreased the NT-proBNP and AchE in serum of AD mice, the effect is more significant [25]. Thus these pre-clinical suggests the significant role as a potential drug for treatment of AD.

5. **Yokukansan (YKS):** Yokukansan is a very effective drug and 18 preclinical trails have been published in the last 3 decades. It is found in Japan and its main active components are Atractylodis lanceae rhizoma, Poria, Cnidii rhizoma, Uncariae uncis cum ramulus, Angelicae radix, Bupleuri radix, and Glycyrrhizae radix [26]. YKS greatly improved the behavioral performance of both aged and 5xFAD mice in novel object detection and contextual fear conditioning exercises [27]. Increased neuroprotective signaling through protein kinase B/Akt was the typical mode of action in both aged and 5xFAD YKS-treated mice. The results show that YKS has beneficial effects in both 5xFAD mice and aged mice through Akt signaling, with several additional mechanisms possibly leading to its beneficial effects in the elderly. Further, ameliorative effects of yokukansan were found on learning and non-cognitive disturbances in the Tg2576 mouse model of AD [28]. The cholinergic involvement and synaptic dynamin 1 expression were studied in Yokukansan-mediated improvement of spatial memory in a rat model of early AD [29]. These studies suggest that YKS is useful in the treatment of memory defects in AD and could be a potential treatment for AD.

1.2.2 Preclinical trials of herbal drugs in Parkinson's disease

1. **Baicalein:** This drug was used in multiple preclinical trials to evaluate the effect of herbal drugs on PD. It is derived from China and its main active components are Baicalin and baicalein and it is found in Scutellaria species, including S. lateriflora and S. galericulata [30]. When baicalein was given, MPTP-induced motor impairment, dopaminergic neuron death, and proinflammatory cytokine elevation were all reversed [31]. Baicalein also inhibited NLRP3 and caspase-1 activation, which suppressed GSDMD-dependent pyroptosis [32]. Baicalein also inhibited the activation and proliferation of disease-associated proinflammatory microglia. These results indicate that baicalein, by blocking the NLRP3/caspase-1/GSDMD pathway, can reverse MPTP-induced neuroinflammation in mice [32]. Baicalein also attenuates α-synuclein aggregation, inflammasome activation, and autophagy in the MPP+-treated nigrostriatal dopaminergic system in vivo [33]. These findings suggest that baicalein may be useful in the treatment of PD.

2. **Bushen Huoxue Yin:** The drug is used for evaluating the effects of herbal drugs on PD. It is found in China and its main active ingredients are Rou Cong Rong (Herba Cistanches), Shan Yu Rou (Fructus Corni), Dang Gui (Radix Angelicae Sinensis), He Shou Wu (Radix Polygoni Multiflori), Chi Shao (Radix Paeoniae Rubra), Chuan Xiong (Rhizoma Chuanxiong), Wu Gong (Scolopendra), and Shi Chang Pu (Rhizoma Acori Graminei) [34]. The model group had NF-KB concentrations in brain tissue of 14.04 +/- 4.38 microg x L(-1), which were higher than the normal in a particular sample. However, there were no significant differences in NF-KB and NO content between the BSHXY and standard classes (P > 0.05). The suppression of NF-KB activation and decreased NO levels in the brain was believed to be the mechanism of action of BSHXY in the treatment of PD [35].

3. **Mucuna pruriens (Mp):** This drug was also frequently used in preclinical trials with PD. The drug's main active components include levodopa, coenzyme Q10, and nicotine [36]. Mp treatment recovered the number of TH-positive cells in both the SN region and the striatum while reducing the expression of iNOS and GFAP in the SN. Treatment with Mp significantly increased the levels of dopamine, DOPAC and homovanillic acid compared to MPTP intoxicated mice [37]. In PQ-induced PD, Mp treatment reduced iNOS expression, nitrite intake, and lipid peroxidation, indicating that it lowers nitric oxide levels. iNOS mRNA, protein expression, and immunoreactivity were all significantly decreased by Mp therapy, while TH immunoreactivity was significantly increased [38]. In conclusion, NO damage to dopaminergic neurons in the substantia nigra is prevented by Mp.

4. **Tenuigenin (TEN):** This is a very effective drug used in preclinical trials involving PD. It is derived from China and one of its main active components is Polygala tenuifolia. Tenuigenin significantly reduced dopaminergic neuron degeneration and inhibited NLRP3 inflammasome activity in the substantia nigra of MPTP mice [39]. Tenuigenin also reduced intracellular reactive oxygen species activity and inhibited NLRP3 inflammasome

activation, caspase-1 cleavage, and interleukin-1 secretion in BV2 microglia cells [39]. Tenuigenin protects dopaminergic neurons from inflammation in part by inhibiting the NLRP3 inflammasome in microglia. Tenuigenin also protects dopaminergic neurons from inflammation-mediated damage induced by the lipopolysaccharide (LPS) [40]. The neuroprotective effects of tenuigenin in a SH-SY5Y cell model with 6-OHDA-induced injury have been reported [41], these reports suggests that it may be used to treat PD clinically.

5. Triptolide (T10): The drug T10 has been frequently used in preclinical trials that evaluate the effect on PD. It is derived from China and one of its main components is Tripterygium wilfordii [42]. In a preclinical analysis, the function of mGlu5 in the anti-inflammatory effect of T10 in a LPS-induced PD model was investigated. Triptolide upregulates metabotropic glutamate receptor 5 to inhibit microglia activation in the LPS-induced model of PD [43]. T10 enhanced mRNA expression and protein stability, which was inhibited by LPS, and thereby increased mGlu5 expression. T10 also blocked receptor-mediated mitogen-activated protein kinase activity and reversed the LPS-induced decrease in mGlu5 membrane localization. Triptolide (T10) protects neurons by inhibiting microglia activation and has anti-inflammatory and immunosuppressive effects. Triptolide Inhibits Preformed Fibril-Induced Microglial Activation by Targeting the MicroRNA155-5p/SHIP1 pathway in PD [44]. Further, triptolide protects against 1-methyl-4-phenyl pyridinium-induced dopaminergic neurotoxicity in rats [45]. These studies suggest the active role of triptolide to protect dopaminergic neurons in preclinical toxin-induced model of PD.

1.2.3 Herbal drugs formulations in clinical trials

Despite several studies on herbal products, providing positive results on AD and PD, very few formulations went under clinical trials. This indicates the necessity of large and complete studies on herbal formulations in AD and PD. A few of the herbal drugs used in clinical trials have been discussed here:

1.2.3.1 Clinical trials on herbal drugs in Alzheimer's disease

1. **Panax Ginseng**: One of the most popular herbal medicines used in AD clinical trials has been this one. Rb1 is one of the most active ingredients, and the plant itself is from China. The ginseng group and the control group were randomly divided in one clinical trial (clinical trial: NCT00391833), and the ginseng group received Panax ginseng powder (4.5 g/d) for 12 weeks. On the MMSE and ADAS scales, there were no baseline variations between the classes. After ginseng therapy, the ADAS cognitive subscale and the MMSE score improved by up to 12 weeks (P = 0.029 and P = 0.009 vs baseline, respectively). The stronger ADAS and MMSE scores returned to the control group's rate after the ginseng was stopped. The study was pursued for phase-1 and phase-2 trials. This

clearly demonstrates that Panax Ginseng was clinically useful in improving AD patients' cognitive performance [6].

2. **Ginkgo Biloba**: This drug has also been commonly used in both clinical and preclinical trials and has shown to not be as effective. It is derived from China and its main active principles include terpene lactones, ginkgo flavone glycosides. In a clinical trial designed as a 12-week randomized, placebo-controlled, double-blind study, patients between the age of 75 years and older, were suffering from mild to moderate dementia were allocated into one of the three treatments: Ginkgo biloba (120 mg daily dose), donepezil (5 mg daily dose), or a placebo group (clinical trial: NCT00010803) [46]. One important thing to note was that in this study, they compare cholinesterase inhibitors to Ginkgo biloba in the treatment of AD, which may be a helpful addition to the discussion. Furthermore, the results indicate that there are no major variations in the efficacy of EGb 761 and donepezil in the treatment of mild to severe AD, suggesting that both medications should be used.

3. **Tiaobu Xinshen Recipe (TXR)**: In addition, TXR has also been seen to be very effective when treating AD. The drug is derived from China and its main active principles are: CM, Astragaloside IV, Astragalus membranaceus-polysaccharide, Radix Angelica Sinensis. In a study involving 88 AD patients with heart and kidney deficiencies, 47 patients were treated with TXR and 41 patients were treated with donepezil [8]. The two groups' MMSE and MoCA scores improved after treatment compared to before treatment (P0.05). However, there was no statistical difference in MMSE or MOCA scores between the two groups after therapy (P > 0.05). The experimental group's CM dementia syndrome score was significantly lower after treatment than the control group (P0.01)[8]. As a result, it proves that TXR could effectively improve cognitive impairment of MCI-AD patients, and especially those with other medical problems.

4. **Yishen Huazhuo Decoction (YHD)**: This drug has been found to have significant effects when used to treat AD. It is derived from China and its main active principles are: Yinyanghuo (Epimedium), Nvzhenzi (Fructus Ligustri Lucidi), Buguzhi (Psoralea fruit), Heshouwu (Radix Polygoni Multiflori), Huangqi (Radix Astragali), Chuanxiong(Ligusticum wallichi Franchat), and Shichangpu (Acorusgramineus). In a specific clinical trial (Chinese Clinical Trial Registry ChiCTR-TRC-12002846), YHD was used to compare its effects to patients with a treatment of donepezil 5 mg/day. At the end of the 24-week treatment period, both the YHD and DH groups increased their ADAS-cog and MMSE mean scores. The findings also showed that YHD outperformed DH in terms of improving ADAS-cog and MMSE mean scores [47]. The findings indicate that the Chinese herbal formula YHD is useful and effective for improving cognitive function in patients with mild AD, with the mechanism being that it reduces amyloid- (A) plaque deposition in the hippocampus.

5. **Yokukansan (YKS)**: This is the most commonly used herbal drug in both clinical and preclinical trials by a significant margin. It is derived from Japan and its main active components are: Atractylodis lanceae rhizoma, Poria, Cnidii rhizoma, Uncariae uncis

cum ramulus, Angelicae radix, Bupleuri radix, and Glycyrrhizae radix. In a specific study, the efficacy and safety of YKS in patients with AD in a nonblinded, randomized, parallel-group comparison was investigated. The study included patients with at least one symptom score of four or more on the neuropsychiatric inventory (NPI) subscales. In the YKS-treated group, the NPI total score increased significantly more than in the non-YKS-treated group [48]. The YKS-treated group improved significantly more than the non-YKS-treated group on the NPI subscales of agitation/aggression and irritability/lability, but there was no statistically significant improvement with YKS in the other subscales. Therefore, it was shown through the study that YKS was safe and effective in treating BPSD in Alzheimer's patients.

1.2.4 Clinical trials on herbal drugs in Parkinson's disease

1. **Bushen Huoxue Granule (BHG):** This has been found to be commonly used when conducting clinical trials on Parkinson's patients. It is derived from China and its main active components are: Astragali radix, Angelicae sinensis radix, Ligustici Chuanxiong Rhizoma, Cuscutae semen, Taxilli Herba, and Dipsaci Radix. In a study involving patients in between the ages of 50 and 80, 120 participants were moved to the BHG group or the placebo group. There were major statistical variations in mobility, physical well-being, stigma, cognition (P0.01), and body discomfort (P0.05). There was no statistical disparity in the measurements of social care, ADL, and connectivity between the data of these two classes of patients (P0.05). The total index of the PDQ-39 was found to be significantly different [49]. As a result, BHG is very impactful as it has significant effects in most aspects of PD patients life quality, especially in mobility, emotional well-being, stigma and cognition.

2. **Gulling Pa'an Capsule (GPC):** This drug has been significant for the use in PD as seen through multiple trials. It is derived from China and its main active principles include Radix Paeoniae (RP), Radix Cyathulae (RC), Rhizoma Chuanxiong (RCX), Cortex Lyci (CL), Radix Saposhnikoviae (RS), Cassia Twig (CT), Sargassum pallidum (SP), Polygonatum sibiricum (PG), Astragali Radix (AgR), Ramulus mori (RM), Silybum marianum (SM), and Orostachys fimbriata (OF). In a clinical trial involving 242 PD patients, were split into multiple groups based on their prior medications and current statistics [12]. After therapy, 1 out of 28 patients in group A had dramatically improved symptoms, and 11 had improved symptoms; the significantly improving rate was 3.6%, and the improving rate was 39.3%; the equivalent rates in group B were 0 (0/25) and 28.0% (7/25) respectively, showing a negligible disparity between the two groups [12]. Through the trial it was found that levodopa combined with GPC for treating PD patients is significantly more effective than that of levodopa alone, meaning that GPC is very effective in terms of treatment of PD patients.

3. **SQJZ:** This drug has found to have an impactful effect on PD patients treated with SQJZ. It is derived from China and its main components are Rehmannia glutinosa, Cornus officinalis, Ophiopogon japonicus, Poria cocos, Trichosanthes kirilowii, Cuscuta chinensis, Semen ziziphi spinosae, Schisandra chinensis, and Aurantii fructus immaturus. In a specific clinical trial, about 240 PD patients having a Hoehn and Yahr scale score ≤ 4 will be allocated into a SQJZ or a placebo in a 2:1 ratio [50]. The primary outcome will be calculated using the NMS scale, while secondary outcomes will be measured using the composite PD rating scale, PD sleep scale, Parkinson fatigue scale, constipation intensity instrument, and PD Questionnaire-39 [50]. The key effectiveness review will be focused on the intention-to-treat methodology which will use mixed-model repeated-measures tests. Through the findings, it was found that SQJZ is efficient and safe in treating NMS in Parkinson's patients.

4. **Xifeng Dingchan Pill (XFDCP):** This drug has been found to be somewhat effective in the use of treating PD. It is also derived from China, but its main active principles are unknown. In a clinical trial involving 320 patients, some were given XFDCP and others were in a control group and given Madopar and Piribedil [51]. Throughout the trial, the Unified PD Rating Scale scores, TCM symptom scores, quality of life, change of Madopar's dosage and the toxic, and adverse effects of Madopar were used to track the changes. The results of the study have proven that XFCDP is beneficial to developing a comprehensive therapy regimen, which can improve the life of a Parkinson's patient [51].

5. **Yokukansan (YKS):** Similar to AD, this drug was very common in clinical trials evaluating the effect of herbal drugs on PD. Once again, it is derived from China and its main active components are: Atractylodis lanceae rhizoma, Poria, Cnidii rhizoma, Uncariae uncis cum ramulus, Angelicae radix, Bupleuri radix, and Glycyrrhizae radix. Twenty-five PD patients were treated with YKS for 12 weeks in one of the clinical trials. After 12 weeks, the median NPI total score fell from 12 to 4.0 ($P = 0.00003$) [52]. The NPI sub-scales for hallucinations, fear, and apathy all demonstrated significant improvements. Bad symptoms (anxiety-apathy) decreased significantly ($p = 0.00391$), while optimistic symptoms (delusions-hallucinations-irritability) decreased significantly ($p = 0.01660$). The Hoehn and Yahr scales, as well as the UPDRS-III, showed no significant improvement. sK decreased from 4.26 0.30 mEq/L to 4.08 0.33 mEq/L, a marginal decline [52, 53]. In general, through this trial, it was found that YKS increased neuropsychiatric symptoms associated with PD, such as hallucinations, anxiety, and apathy, without causing serious side effects or worsening PD.

1.3 Conclusion

Many of the drugs involved in these trials were from Asian countries, mostly China and Japan because they have easy access to many of these drugs. Some of the drugs that were used in the preclinical trials were not able to pursue it to the clinical trial stage because of financial constraints. For the same reason, some of the clinical trial drugs were not able to complete it to further stages as it requires more financial inputs to complete them. These clinical trails thus unable to pass phase 4 trials stage that could lead to the drug being FDA approved. As a result, the only drug that is FDA approved is Huperzine A, which is approved as only a supplement to treat AD. For PD, there are few synthetic drugs that are available like levodopa, carbidopa, Pramipexole, Opicapone, but there no herbal drug that is currently available for PD that is approved by FDA. With more motivation from people all over the world to find a solution for such a damaging disease, more funding could be provided, which could lead to a better treatment and possibly even a cure that would benefit millions of people all over the world.

References

[1]   EM Reiman, RJ Caselli, Alzheimer's disease, Maturitas 31 (3) (1999) 185–200.
[2]   R Balestrino, AHV Schapira, Parkinson disease, Eur J Neurol 27 (1) (2020) 27–42.
[3]   X Liu, D Hou, F Lin, J Luo, J Xie, Y Wang, et al., The role of neurovascular unit damage in the occurrence and development of Alzheimer's disease, Rev Neurosci 30 (5) (2019) 477–484.
[4]   DF Drake, S Harkins, A Qutubuddin, Pain in Parkinson's disease: pathology to treatment, medication to deep brain stimulation, NeuroRehabilitation 20 (4) (2005) 335–341.
[5]   G Livingston, J Huntley, A Sommerlad, D Ames, C Ballard, S Banerjee, et al., Dementia prevention, intervention, and care: 2020 report of the Lancet commission, Lancet 396 (10248) (2020) 413–446.
[6]   S-T Lee, K Chu, J-Y Sim, J-H Heo, M Kim, Panax ginseng enhances cognitive performance in Alzheimer disease, Alzheimer Dis Assoc Disord 22 (3) (2008) 222–226.
[7]   M Mazza, A Capuano, P Bria, S Mazza, Ginkgo biloba and donepezil: a comparison in the treatment of Alzheimer's dementia in a randomized placebo-controlled double-blind study, Eur J Neurol 13 (9) (2006) 981–985.
[8]   Z-Y Lin, T-W Huang, J-S Huang, G-Y Zheng, Tiaobu Xinshen Recipe (TXR) improved mild cognitive impairment of Alzheimer's disease patients with Xin (Heart) and Shen (Kidney) deficiency, Chin J Integr Med 26 (1) (2020) 54–58.
[9]   Y Zhang, C Lin, L Zhang, Y Cui, Y Gu, J Guo, et al., Cognitive improvement during treatment for mild Alzheimer's disease with a Chinese herbal formula: a randomized controlled trial, PLoS One 10 (6) (2015) e0130353.
[10]   K Okahara, Y Ishida, Y Hayashi, T Inoue, K Tsuruta, K Takeuchi, H Yoshimuta, K Kiue, Y Ninomiya, J Kawano, K Yoshida, S Noda, S Tomita, M Fujimoto, J Hosomi, Y Mitsuyama, Effects of Yokukansan on behavioral and psychological symptoms of dementia in regular treatment for Alzheimer's disease, Prog Neuropsychopharmacol Biol Psychiatry 34 (3) (2010) 532–536, doi:10.1016/j.pnpbp.2010.02.013.
[11]   M Li, HM Yang, DX Luo, JZ Chen, HJ Shi, Multi-dimensional analysis on Parkinson's disease questionnaire-39 in Parkinson's patients treated with Bushen Huoxue Granule: a multicenter, randomized, double-blinded and placebo controlled trial, Complement Ther Med 29 (2016) 116–120.
[12]   G-H Zhao, Q-G Meng, X Yu, A multi-centered randomized double-blinded controlled clinical study on efficacy of gulling pa'an capsule in treating Parkinson's disease, Zhongguo Zhong xi yi jie he za zhi Zhongguo Zhongxiyi jiehe zazhi = Chinese J Integr Tradit West Med 29 (7) (2009) 590–594.
[13]   X. Fu, Q. Wang, Z. Wang, H Kuang, P Jiang, Danggui-Shaoyao-San: New hope for Alzheimer's disease, Aging Dis 7 (4) (2016) 502–513.

[14] TM Dmitrieva, TM Nikolaeva, KV Golubtsov, The functional plasticity of taste reception in fish, Nervn Sist 29 (1990) 173–185.

[15] Y You, X Liu, Y You, D Liu, C Zhang, Y Chen, et al., Traditional Chinese medicine Danggui Shaoyao San for the treatment of Alzheimer's disease: a protocol for systematic review, Medicine (Baltimore) 99 (15) (2020) e19669.

[16] VS Sierpina, B Wollschlaeger, M Blumenthal, Ginkgo biloba, Am Fam Physician 68 (5) (2003) 923–926.

[17] Y Hou, MA Aboukhatwa, D-L Lei, K Manaye, I Khan, Y Luo, Anti-depressant natural flavonols modulate BDNF and beta amyloid in neurons and hippocampus of double TgAD mice, Neuropharmacology 58 (6) (2010) 911–920.

[18] X Liu, W Hao, Y Qin, Y Decker, X Wang, M Burkart, et al., Long-term treatment with Ginkgo biloba extract EGb 761 improves symptoms and pathology in a transgenic mouse model of Alzheimer's disease, Brain Behav Immun 46 (2015) 121–131.

[19] Y Qi, Q Zhang, H Zhu, Huang-Lian Jie-Du decoction: a review on phytochemical, pharmacological and pharmacokinetic investigations, Chin Med 14 (2019) 57.

[20] X Qiu, G Chen, T Wang, Effects of huanglian jiedu decoction on free radicals metabolism and pathomorphism of the hippocampus in App/PS1 double transgenic mice, Zhongguo Zhong xi yi jie he za zhi Zhongguo Zhongxiyi jiehe zazhi = Chinese J Integr Tradit West Med 31 (10) (2011) 1379–1382.

[21] Y Liu, T Du, W Zhang, W Lu, Z Peng, S Huang, et al.P Tucci (Ed.), Modified huang-lian-jie-du decoction ameliorates Aβ synaptotoxicity in a murine model of alzheimer's disease, Oxid Med Cell Longev (2019) 8340192. https://doi.org/10.1155/2019/8340192.

[22] SSK Durairajan, A Iyaswamy, SG Shetty, AK Kammella, S Malampati, W Shang, et al., A modified formulation of Huanglian-Jie-Du-Tang reduces memory impairments and β-amyloid plaques in a triple transgenic mouse model of Alzheimer's disease, Sci Rep. 7 (1) (2017) 6238.

[23] X-J Wang, A-H Zhang, L Kong, J-B Yu, H-L Gao, Z-D Liu, et al., Rapid discovery of quality-markers from Kaixin San using chinmedomics analysis approach, Phytomedicine 54 (2019) 371–381.

[24] X Zhang, Q Li, C Lv, H Xu, X Liu, Z Sui, et al., Characterization of multiple constituents in Kai-Xin-San prescription and rat plasma after oral administration by liquid chromatography with quadrupole time-of-flight tandem mass spectrometry, J Sep Sci 38 (12) (2015) 2068–2075.

[25] M-H Li, J Zhang, R-Q Zhao, X-Z Dong, Y Hu, T Chen, et al., Effect of six class of Kaixin San formulas on pharmacological and preliminary mechanism of Alzheimer's disease mice], Zhongguo Zhong yao za zhi = Zhongguo zhongyao zazhi = China J Chinese Mater medica 41 (7) (2016) 1269–1274.

[26] H Kitagawa, M Munekage, K Ichikawa, I Fukudome, E Munekage, Y Takezaki, et al., Pharmacokinetics of active components of yokukansan, a traditional Japanese herbal medicine after a single oral administration to healthy japanese volunteers: a cross-over, randomized study, PLoS One 10 (7) (2015) e0131165.

[27] R Kaushik, E Morkovin, J Schneeberg, AD Confettura, MR Kreutz, O Senkov, A Dityatev, Traditional Japanese herbal medicine yokukansan targets distinct but overlapping mechanisms in aged mice and in the 5xFAD mouse model of Alzheimer's disease, Front Aging Neurosci 10 (2018) 411, doi:10.3389/fnagi.2018.00411.

[28] M Tabuchi, T Yamaguchi, S Iizuka, S Imamura, Y Ikarashi, Y Kase, Ameliorative effects of yokukansan, a traditional Japanese medicine, on learning and non-cognitive disturbances in the Tg2576 mouse model of Alzheimer's disease, J Ethnopharmacol 122 (1) (2009) 157–162.

[29] N Uchida, K Takasaki, Y Sakata, A Nogami, H Oishi, T Watanabe, et al., Cholinergic involvement and synaptic dynamin 1 expression in yokukansan-mediated improvement of spatial memory in a rat model of early Alzheimer's disease, Phytother Res 27 (7) (2013) 966–972.

[30] R Kyo, N Nakahata, I Sakakibara, M Kubo, Y Ohizumi, Baicalin and baicalein, constituents of an important medicinal plant, inhibit intracellular Ca2+ elevation by reducing phospholipase C activity in C6 rat glioma cells, J Pharm Pharmacol 50 (10) (1998) 1179–1182.

[31] F.-Q Li, T Wang, Z Pei, B Liu, J.-S Hong, Inhibition of microglial activation by the herbal flavonoid baicalein attenuates inflammation-mediated degeneration of dopaminergic neurons, J Neural Transm 112 (3) (2005) 331–347.

[32] W Rui, S Li, H Xiao, M Xiao, J Shi, Baicalein attenuates neuroinflammation by inhibiting NLRP3/caspase-1/GSDMD pathway in MPTP induced mice model of Parkinson's disease, Int J Neuropsychopharmacol 23 (11) (2020) 762–773.

[33] K-C Hung, H-J Huang, Y-T Wang, AM-Y Lin, Baicalein attenuates α-synuclein aggregation, inflammasome activation and autophagy in the MPP(+)-treated nigrostriatal dopaminergic system in vivo, J Ethnopharmacol 194 (2016) 522–529.

[34] M Xie, Y Yu, Z Zhu, L Deng, B Ren, M Zhang, Simultaneous determination of six main components in Bushen Huoxue prescription by HPLC-CAD, J Pharm Biomed Anal 201 (2021) 114087.

[35] S-D Li, Y Liu, M-H Yang, Effects of bushen huoxue yin on brain NF-kB and NO content in the Parkinson's disease model mouse, J Tradit Chinese Med = Chung i tsa chih ying wen pan 32 (1) (2012) 67–70.

[36] LR Lampariello, A Cortelazzo, R Guerranti, C Sticozzi, G Valacchi, The magic velvet bean of Mucuna pruriens, J Tradit Complement Med 2 (4) (2012) 331–339.

[37] SK Yadav, J Prakash, S Chouhan, S Westfall, M Verma, TD Singh, et al., Comparison of the neuroprotective potential of Mucuna pruriens seed extract with estrogen in 1-methyl-4-phenyl-1,2,3,6-tetrahydropyridine (MPTP)-induced PD mice model, Neurochem Int 65 (2014) 1–13.

[38] SK Yadav, SN Rai, SP Singh, Mucuna pruriens reduces inducible nitric oxide synthase expression in Parkinsonian mice model, J Chem Neuroanat 80 (2017) 1–10.

[39] Z Fan, Z Liang, H Yang, Y Pan, Y Zheng, X Wang, Tenuigenin protects dopaminergic neurons from inflammation via suppressing NLRP3 inflammasome activation in microglia, J Neuroinflammation 14 (1) (2017) 256.

[40] H-L Yuan, B Li, J Xu, Y Wang, Y He, Y Zheng, et al., Tenuigenin protects dopaminergic neurons from inflammation-mediated damage induced by the lipopolysaccharide, CNS Neurosci Ther 18 (7) (2012) 584–590.

[41] Z Liang, F Shi, Y Wang, L Lu, Z Zhang, X Wang, et al., Neuroprotective effects of tenuigenin in a SH-SY5Y cell model with 6-OHDA-induced injury, Neurosci Lett 497 (2) (2011) 104–109.

[42] B Wang, L Ma, X Tao, PE Lipsky, Triptolide, an active component of the Chinese herbal remedy Tripterygium wilfordii Hook F, inhibits production of nitric oxide by decreasing inducible nitric oxide synthase gene transcription, Arthritis Rheum 50 (9) (2004) 2303–2995.

[43] Y-Y Huang, Q Zhang, J-N Zhang, Y-N Zhang, L Gu, H-M Yang, et al., Triptolide up-regulates metabotropic glutamate receptor 5 to inhibit microglia activation in the lipopolysaccharide-induced model of Parkinson's disease, Brain Behav Immun 71 (2018) 93–107.

[44] Y Feng, C Zheng, Y Zhang, C Xing, W Cai, R Li, et al., Triptolide inhibits preformed fibril-induced microglial activation by targeting the microRNA155-5p/SHIP1 pathway, Oxid Med Cell Longev 2019 (2019) 6527638.

[45] J-P Gao, S Sun, W-W Li, Y-P Chen, D-F Cai, Triptolide protects against 1-methyl-4-phenyl pyridinium-induced dopaminergic neurotoxicity in rats: implication for immunosuppressive therapy in Parkinson's disease, Neurosci Bull 24 (3) (2008) 133–142.

[46] BE Snitz, ES O'Meara, MC Carlson, AM Arnold, DG Ives, SR Rapp, et al., Ginkgo biloba for preventing cognitive decline in older adults: a randomized trial, JAMA 302 (24) (2009) 2663–2670.

[47] Y Zhang, C Lin, L Zhang, Y Cui, Y Gu, J Guo, et al., Correction: cognitive improvement during treatment for mild Alzheimer's disease with a Chinese herbal formula: a randomized controlled trial, PLoS One 13 (6) (2018) e0199895.

[48] K Furukawa, N Tomita, D Uematsu, K Okahara, H Shimada, M Ikeda, et al., Randomized double-blind placebo-controlled multicenter trial of Yokukansan for neuropsychiatric symptoms in Alzheimer's disease, Geriatr Gerontol Int 17 (2) (2017) 211–218.

[49] M Li, M Yang, Y Liu, [Effects of Chinese herbal medicine Bushen Huoxue Granule on quality of life of patients with Parkinson disease: a randomized, double-blinded and placebo-controlled trial]. Zhong Xi Yi Jie He Xue Bao. 2012;10(3):310–7.

[50] J Shi, J Tian, T Li, B Qin, D Fan, J Ni, et al., Efficacy and safety of SQJZ herbal mixtures on nonmotor symptoms in Parkinson disease patients: protocol for a randomized, double-blind, placebo-controlled trial, Medicine (Baltimore) 96 (50) (2017) e8824 Dec.

[51]   J Zhang, Y Ma, X Shen, Evaluation on the efficacy and safety of Chinese herbal medication Xifeng
Dingchan Pill in treating Parkinson's disease: study protocol of a multicenter, open-label, randomized
active-controlled trial, J Integr Med 11 (4) (2013) 285–290.

[52]   T Hatano, N Hattori, T Kawanabe, Y Terayama, N Suzuki, Y Iwasaki, et al., An exploratory study of the
efficacy and safety of yokukansan for neuropsychiatric symptoms in patients with Parkinson's disease, J
Neural Transm 121 (3) (2014) 275–281 Available from. https://doi.org/10.1007/s00702-013-1105-y .

[53]   R Kaushik, A Dityatev, O Senkov, et al., Traditional Japanese Herbal Medicine Yokukansan Targets Distinct but Overlapping Mechanisms in Aged Mice and in the 5xFAD Mouse Model of Alzheimer's Disease,
Front Aging Neurosci. 10 (2018) 411, doi:10.3389/fnagi.2018.00411.

CHAPTER 2

Review on correlations between depression and nutritional status of elderly patients

Liudmila Nadtochii[a], Alena Proskura[a], Mariam Muradova[a], Mahmood A. Hashim[a] and Fabrice Neiers[b]

[a]ITMO University, Faculty of Biotechnologies (BioTech), Saint Petersburg, Russia [b]Centre des Sciences du Goût et de l'Alimentation (CSGA), Université de Bourgogne Franche-Comté, INRAE, CNRS, France

2.1 Introduction

In 2001, the World Health Organization (WHO), for the first time, recognized the global problem of mental health at the level of the world's population. Around 450 million people worldwide have a mental health disorder, with one in four people worldwide at different stages of life. WHO study (2001) involving 28 countries of the world demonstrated that in all countries participating in the study, mental disorders of the population were noted up to 36% of the prevalence during life [1].

A person in the modern world is amenable to psycho-emotional stress, resulting in an expressed cognitive impairment and mental disorders in developed countries. Abnormalities in mental health are of a wide range, including cognitive impairments, neurodegenerative or mental disorders. They can initially exhibit such manifestations as anxiety disorder, cognitive decline, attention-deficit/hyperactivity disorder (ADHD), depression, etc. [2].

In 2013, WHO stated that depression is one of the most common causes of disability worldwide (on average, 11% of all years lived with a disability worldwide), and women are more susceptible to it [3]. The risk of depression is increased when it is diagnosed among children (from 0.4% to 2.5%) and adolescents (from 0.4% to 8.3%) [4]. According to the research [5], children 9–13 years old, who initially did not tend depression, developed depressive disorder by 16 years in more than 7% for boys and almost 12% for girls.

Herbal Medicines: A Boon for Healthy Human Life.
DOI: https://doi.org/10.1016/B978-0-32-390572-5.00011-1

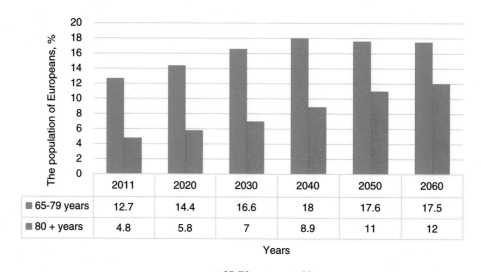

The population of Europeans, %

Years	2011	2020	2030	2040	2050	2060
■ 65-79 years	12.7	14.4	16.6	18	17.6	17.5
■ 80 + years	4.8	5.8	7	8.9	11	12

■ 65-79 years   ■ 80 + years

Figure 2.1
Europeans' population structure by 65--79 and 80+ age groups in EU-27 [7].

2.2 Correlations between depression and nutritional status of elderly patients

Recently there is an increase of the number of older people (people aged 65 and over). In 2019, there were approximately 703 million people aged 65 and over globally. This number is projected to increase to 1.5 billion by 2050 [6]. According to the European Commission data, in the next 40 years the number of Europeans' population aged 80 and older will twice in comparison between projections in 2020 and 2060 (Fig. 2.1) [7].

Elderly people have a greater susceptibility to developed mental health diseases [8]. The high degree of susceptibility to depression among the elderly population aggravates the already low level of life quality associated with increased rates of diseases such as metabolic syndrome/diabetes, cardiovascular diseases (CVD), and others [9–11].

Like physical health, mental health is a complex interplay of biological, psychological, and social factors. Depression is a prominent cause of disability worldwide, with a particularly high prevalence among the elderly, and it is linked to poor nutrition [13]. The existing dualism in this issue is associated with the fact that depression in old age is considered a psychological phenomenon with consequences for nutrition or a nutritional problem with psychological effects. If depression is the cause, general malnutrition will be the consequence, but the opposite is also true. However, if depression is associated with malnutrition, it will be a consequence, not a cause of nutritional deficiencies. Aging is associated with reduced chewing and a reduced salivary output that could provoke unpleasant sensory experience [14]. Indeed, saliva plays an important role in flavor perception, transporting for instance taste compounds up to the chemoreceptors [15]. Mental health diseases can reduce taste and smell sensitivity due to brain function alteration, consequently promoting malnutrition [16].

Table 2.1 shows the relationship between a deficiency of bioactive substances and the risk of mental health diseases, in particular depression, as well as the risk of premature aging and related noncommunicable diseases. Depending on the available data, it can be stated that the assessment of the deficiency of essential components of food, in particular vitamins, minerals, etc., can be successfully used to determine the degree of risk of elderly patients and their susceptibility to depression. Some of these vitamins can also impact on the redox status of saliva and as a result modulate the perception of food flavor [17]. Indeed, a study has demonstrated that the antioxidant capacity of saliva is one of the main physiological parameters explaining aroma release in the elderly [18]. Another study has reported a higher antioxidant capacity in the elderly suffering from hyposalivation and as a result, a lower release of aroma compounds [19]. This effect of the antioxidant capacity of saliva on aroma release has been correlated to the enzymatic activity of saliva on aroma compounds [20]. In several studies, a higher antioxidant capacity of saliva was associated to a higher metabolization of aroma compounds by salivary enzymes [18,20]. The decrease of aroma release could affect the pleasure of eating, which is one of the main drivers of food consumption [14].

2.3 Nutritional status of the elderly patients

Numerous studies indicate that the nutrition status of older age groups is characterized by a pronounced imbalance of nutrients in the diet, a deficiency of essential nutrients such as protein, vitamins, macro and microelements, dietary fiber. These factors can lead to the development and aggravation of nutritionally caused diseases, such as cardiovascular, endocrine, oncological, and diseases of the blood and hematopoietic organs, gastrointestinal tract, and metabolic disorders [73]. In addition, the mentioned defects in the nature and structure of nutrition result in the activation of the processes of premature aging and often too early disability and death [74]. Nutritional experts agree that the nature of diet affects the normal vital activity and functioning of practically all organs and systems in the human body [75–78]. The modern understanding of healthy (rational) nutrition of the older generation is formed in the following main positions: adequacy to the physiological needs of this age group, balance in chemical composition and energy value, as well as in the diversity of the diet [79,80]. Rational nutrition among the elderly contributes to the possible prevention of many nutritionally dependent diseases; however, as shown by recent studies, this age group demonstrates a lack of a conscious strategy to the formation of their diet and eating behavior [81,82]. It is also necessary to consider the growing influence of stress factors (technogenic, natural, climatic-ecological, psycho-emotional), which enhances the influence of negative effects on physiological and biochemical processes in organs and tissues, as well as changes in the hormonal and

Table 2.1: Role of certain bioactive substances in depressive disorders and premature aging among elderly patients.

Nutrients	Outcome	References
B1 (thiamine)	Low levels of thiamine were associated with a high prevalence of depressive symptoms among older Chinese adults; clinical signs of depression were revealed among 74 malnourished patients due to thiamine deficiency; symptoms of depression significantly improved in subjects with MDD after six weeks of thiamine supplementation compared with placebo; B1 deficiency can lead to mood disorders, anxiety, and depression as well as severe cardiovascular complications and heart failure; B1 can contribute to impairment of oxidative metabolism, neuroinflammation, endoplasmic reticulum stress, autophagy and neurodegeneration	[21–28]
B2 (riboflavin)	Depressed people have a higher prevalence of marginal riboflavin insufficiency; low levels of B2 are correlated to the reduced cognitive outcome, depression, aggression, and distinct alterations within the central nervous system	[29]
B6 (pyridoxal)	Marginal pyridoxal deficiency was significantly associated with the presence of depressive symptoms among 1371 elderly adults; higher intakes of vitamin B6 reduced the risk of depression within older women	[30–33]
B9 (folate)	Decreasing and low concentration of serum folate was associated with a higher risk of depressive symptoms in older Chinese adults; low B9 levels have been correlated with dementia and Alzheimer's disease (AD); patients with megaloblastic anemia have been exhibited to suffer from an impaired immune response, which can be treated with folic acid supplementation	[34–39]
B12 (cobalamin)	High total intakes of vitamin B12 are shown to be protective in depressive symptoms among older adults; decreasing and low concentration of cobalamin has been referred to severe depression, suicidal behaviors, reduced cognition, mental fatigue, low mood, mania, psychosis, and intense agitation	[40–42]
Vitamin C (ascorbic acid)	Ascorbic acid with antidepressants significantly reduced the overall Hamilton depression rating scale; mediates several beneficial effects on oxidative redox pathways and mitochondrial pathways on the immune system, inflame-aging, endothelial integrity, and lipoprotein metabolism	[43–45]
Vitamin D	In older people, vitamin D deficiency (less than 20 ng/mL) was linked to depression and impairment on two of four cognitive performance tests.; vitamin D supplementation can improve depression scores in people aged 60 and over; a positive association between vitamin D and neurological status was determined; reduced serum content of circulating 25 (OH) D form is a risk factor is linked to the development of cardiovascular disorders (arterial hypertension, hyperlipidemia, coronary artery disease), type 2 diabetes, and other ailments.	[46–50]

(continued on next page)

Table 2.1: Role of certain bioactive substances in depressive disorders and premature aging among elderly patients—cont'd

Nutrients	Outcome	References
Vitamin A	Vitamin A deficiency leads to poor or impaired immunity and an increase in common infections; actions of vitamin A defect include neurodegeneration, the physiological function of steroid and thyroid hormones; all-trans RA has protective effects in neurodegeneration	[51–53]
Vitamin E	Protects neuronal membranes from oxidation so that deficiency may affect the brain via increased inflammation; no convincing evidence from meta-analyses on randomized, placebo-controlled trials that vitamin E alone or in combination with other antioxidants prevents the progression of MCI or AD or improves cognitive function in older subjects; lower serum alpha-tocopherol levels in patients with major depression compared to controls;	[54–56]
Polyphenols	Verbal learning ameliorated in older adults with mild cognitive impairment (MCI) after consumption of Concord grape juice, blueberry juice, and flavanols; catechins play a crucial role in the neuroprotective functions; Apigenin (10 μM) inhibits oxidative stress and inflammation by suppressing the expression of NF-κB and increasing the level of Nrf2 and its subsequent antioxidant molecules; quercetin has anti-inflammatory effects; quercetin-rich onion facilitates cognitive function and prevents dementia in elderly people	[57–62]
Minerals	Consumption of minerals (potassium, calcium, magnesium, phosphorus, iron, zinc, copper) was significantly and negatively correlated with depressive symptoms among elderly Japanese women over 65 years old; higher intakes of magnesium, calcium, iron, and zinc were associated with lower prevalence of depressive status in Japanese employees; high concentrations of Sr and Ba were shown as a risk factor for depressive symptoms among 954 elderly subjects Magnesium shortage causes mitochondrial DNA damage, telomere shortening, cell-cycle arrest proteins activation, and premature aging; in an RCT, potassium in table salt was related to a 40% reduction in cardiovascular disease in elderly men compared to regulatable salt.	[63–66]
PUFA	PUFA at a dosage of more than 1.5 g/day promoted to reduce depressive symptoms in adults aged 60 years and older; omega-3 fatty acid deficiency is associated with melanoma and other cancers as well as cognitive dysfunction	[67–70]
Protein	Higher protein intake was associated with decreased risk of severely depressed mood in US men; In healthy older people, consuming a high-whey protein, leucine-enriched supplement resulted in a higher overall postprandial muscle protein synthesis rate than consuming a standard dairy product.	[71, 72]

immune systems [83]. Energy, protein, vitamin C, vitamin D, folate, iron, zinc, and fiber are considered essential nutrients for older adults. Numerous studies revealed that adherence to a high-quality diet, a relatively low dietary inflammatory index, fresh fruits, fish, and vegetables was associated with a lower risk of depressive symptoms.

2.3.1 Nutrients for the prevention of depressive disorders in the elderly people

2.3.1.1 Water-soluble vitamins

2.3.1.1.1 B1 (thiamine)

B1 (thiamine) is an essential component of glucose metabolism. B1 deficiency negatively affects the ability of the brain to oxidize glucose, which leads to the accumulation of reactive oxygen species and disruption of the blood-brain barrier, as well as leading to apathy. Sources of vitamin B1 are generally recognized as meat products (mainly pork), fish, legumes, nuts, and cereals.

2.3.1.1.2 B2 (riboflavin)

B2 (riboflavin) is essential for the oxidative processes of the body, ensuring the synthesis of monoamines and the methylation cycle. Although vitamin B2 deficiency is extremely rare and can affect human health due to prolonged restriction of vitamin B2 intake. However, among the elderly population (age ≥ 65 years), vitamin B2 deficiency occurs on average in about 20% of cases. Moreover, a low level of vitamin B2 is observed in the population prone to depression. Apparently, this is due to the role of vitamin B2 for the endogenous antioxidant - glutathione. Therefore, eating dairy, meat, fish products, and egg products, nuts, legumes, and greens will reduce vitamin B2 deficiency and associated diseases.

2.3.1.1.3 B6 (pyridoxal)

B6 (pyridoxal) is an essential component for the maintenance of the required levels of glycolysis, the methylation cycle, and the renewal of the brain's antioxidant glutathione. In addition, vitamin B6 deficiency has been linked to the prevalence and form of depression. Recognized sources of vitamin B6 are meat (mostly beef and poultry), fish products, fish, potatoes, and legumes.

2.3.1.1.4 B9 (folate) and B12 (cobalamin)

Several studies reported about the specific deficiency of nutrients in old age, in particular, B9 (folate) and B12 (cobalamin) [84–86]. Vital micronutrient deficiencies have been reported in older populations. They have been associated with social factors and nutritional statuses such as dietary restrictions, loss of appetite, drug abuse, and mental illness such as dementia and depression [87–89]. Low folate and cobalamin levels are more common in older adults with mental disorders but can also occur in healthy older adults [87,90]. Folate and cobalamin are involved in the re-methylation of homocysteine (Hcy) to form the amino acid methionine. It should be noted that methionine is the only endogenous source of homocysteine in our bodies. Accordingly, folate and/or cobalamin deficiency leads to hyperhomocysteinemia [91,92].

Scientists have identified high blood levels of Hcy in patients with depression and neurode-generative diseases (Alzheimer's disease and Parkinson's disease) [93,94]. The primary sources of folic acid are legumes (mostly lentils) and the green mass of plants. The main sources of vitamin B12 are dairy, meat, and fish products, and aquatic organisms.

2.3.1.1.5 Vitamin C (ascorbic acid)

Vitamin C is one of the most important water-soluble antioxidants found in the extracellular fluid of the body. The role of vitamin C in anxiety, stress, depression, fatigue, and mood state in humans was reviewed [95]. It has been hypothesized that oral vitamin C supplementation can improve mood and reduce stress and anxiety. Furthermore, they examined this propensity to have an antidepressant-like effect in a variety of ways:

- Vitamin C can activate the serotonin 1A (5-HT1A) receptor. This activation is a mechanism of action of many antidepressants, anxiolytic and antipsychotic drugs);
- Modulation of GABAergic systems (via activation of GABAA receptors and possible inhibition of GABAB receptors);
- Inhibition of N-methyl-D-aspartate (NMDA) receptors and L-arginine-nitric oxide (NO)-cyclic guanosine 3,5-monophosphate (cGMP) pathway—the blockade of NMDA receptor is associated with reduced levels of NO and cGMP, whereas reduction of NO levels within the hippocampus was shown to induce antidepressant-like effects;
- Blocking potassium (K+) channels whereas their inhibition plays a significant role in the treatment of depression;
- Induction of heme oxygenase one expression may be a link factor between inflammation, oxidative stress, and the biological as well functional changes in brain activity in depression; its decrease is associated with depressive symptoms;
- Vitamin C may play an antidepressant function also by its antioxidant properties.

2.3.1.2 *Fat-soluble vitamins*

2.3.1.2.6 Vitamin D

Vitamin D is known to be a fat-soluble vitamin. Its principal amount is produced in the skin during exposure to the sun when vitamin D3 is formed from 7-dehydrocholesterol under the action of ultraviolet radiation (wavelength 290–320 nm) [96]. Vitamin D is found in fatty fish (herring, salmon, tuna), fish liver oil, beef liver, cheese, and egg yolks. Even though the listed products are often found in the diet, the alimentary route of vitamin intake to the total level of content in the body is small [97].

The main reasons for vitamin D deficiency are the following [98]: low level of insolation, increased physiological need (pregnant women, lactating women, the elderly), strict vegetarianism, allergy to milk protein, intolerance lactose, overweight, inadequate food intake, malabsorption syndrome, nephrotic syndrome, drugs. Vitamin D deficiency occurs in almost all elderly people, especially those who have to constantly stay at home or in nursing homes [9]. At the same time, among these patients, so-called age-associated diseases are quite common, including diseases of the cardiovascular system, neurodegenerative diseases, type 2 diabetes mellitus, osteoporosis. Given the fact that vitamin D has an important effect on cell viability, it can be assumed that vitamin D deficiency can increase the rate of aging.

It was found that vitamin D plays a vital role in the processes of cell proliferation through proteins involved in the G1/S phase of the cell cycle [100]. Also, vitamin D affects the regulators of phosphorus-calcium metabolism, particularly morphogenetic proteins of fibroblasts-23 (FGF-23) and the Klotho factor [101]. Defects in these structures can cause premature phenotypic aging. Vitamin D induces the expression of FGF-23 through interaction with the calcitriol receptor (VDR) and reduces the concentration of systemic inflammation mediators such as interleukin-2 and tumor necrosis factor-alpha [102]. In addition, the immunosuppressive properties of vitamin D are confirmed by the inverse relationship between its concentration in blood plasma and the level of the inflammatory marker, C-reactive protein (CRP) [103]. Vitamin D deficiency is associated with many adverse neurological conditions, including depression, cognitive decline (in particular, the effect of vitamin D on the metabolism of the neurotransmitter dopamine), while treatment with vitamin D preparations is an effective means of preventing these conditions [104].

2.3.1.2.7 Vitamin A

Vitamin A plays several critical roles in the aging process, mainly immune function, and oxidative processes. The deficiency is associated with an impaired immune response to infection. All-trans retinoic acid (RA), a common form of active vitamin A, is involved in immune homeostasis through regulating cell homing and differentiation. It activates T-cell responses during infection or autoimmune disease [105]. The role of retinoids in preventing neuroinflammation in neurodegenerative processes has been reported in earlier studies [106]. Since retinoids significantly suppress IL-6 generation, retinoid therapy can be a helpful therapy against AD [107]. Retinoids have been found to suppress lipopolysaccharide or Aβ-induced production of tumor necrosis factor-alpha and inhibit the expression of inducible nitric oxide synthase in activated microglia inhibiting nuclear translocation of nuclear factor kappa B [108]. Retinol is usually made from the provitamin A carotenoids found in many colored fruits and vegetables or animal sources such as liver, egg yolks, or dairy products [109]. New studies focus on the effects of vitamin A on enhancing the T-cell response to cancer, infection, gut inflammation, and immune-mediated diseases in humans, including autoimmune diseases associated with aging [110].

2.3.1.2.8 Vitamin E

The brain is susceptible to oxidative stress, which is involved in the pathogenesis of neurodegenerative diseases such as AD. Several studies have found low levels of vitamin E in the cerebrospinal fluid of AD patients [111,112]. A sizeable placebo-controlled study in patients with mild neurological impairment showed that intake of 2000 IU of synthetic α-tocopherol per day for two years (equivalent to 900 mg per day of RRR-α-tocopherol) significantly slowed the progression of AD [113]. Vascular dementia is the most common type of dementia in the population after AD disease. A case-control study examining the risk factors for vascular dementia in older Japanese-American men has found that supplementation with vitamin E and vitamin C significantly reduced the risk of developing vascular dementia and other types of dementia, but not AD [114]. The major sources of vitamin E are vegetable oils, single pressing plant seed oils, overgrown wheat, sea buckthorn fruits, nuts, cereals, vegetables, fruits, meat, poultry, fish [115].

2.3.1.3 Polyphenols

One of the most common and abundant natural chemicals that exhibit biological and antioxidant activity is polyphenols. They are found in vegetables, fruits, grains, spices, as well as in wine, green and black tea, coffee, cocoa, etc. The presence of phytochemicals in fruits and vegetables reduces the risk of several major diseases, including cardiovascular disease, cancer, and neurodegenerative disorders. Consequently, people who consume more vegetables and fruits may have a lower risk of certain diseases caused by neuronal dysfunction [116,117]. Herbal medicine has long been used to treat neural symptoms. These properties have attracted the interest of researchers in polyphenols. Many studies have highlighted their potential role in the prevention and treatment of various pathological conditions associated with oxidative stress and inflammation, such as cancer, cardiovascular and neurodegenerative disorders, and cell damage caused by pollutants [118–120]. Numerous studies have identified the neuroprotective effects of medicinal mushrooms, which have antioxidant, anti-inflammatory, cholinesterase inhibiting, and neuroprotective properties [121–123]. The impact of *Hericium erinaceus* as an alternative medicine for treating depression was reviewed [124]. The authors noted the bioactive compounds extracted from the mycelia and fruiting bodies of *H. erinaceus* and the effect of these bioactive compounds to promote the expression of neurotrophic factors associated with cell proliferation, such as nerve growth factors. Based on the neurotrophic and neurogenic pathophysiology of depression, *H. erinaceus* can be a potential alternative medicine for the treatment of depression. Several studies using polyphenols, especially from red wine or green tea [125], have focused on their neuroprotective role in most neurodegenerative diseases, such as the neuroprotection of epigallocatechin gallate (EGCG) against neurotoxicity mediated by amyloid-beta [126]. However, polyphenols, such as flavan-3-ol, have a high affinity for salivary proteins especially salivary proline-rich proteins. These interactions could both affect their bioavailability but also lead to the aggregation of the mucosal pellicle [127] and *in fine* to the perception of astringency via a molecular singling pathway involving MUC1, a transmembrane protein with a sensing function [128]. Astringency is a negative flavor that could lead to food rejection. Thus, increasing the acceptability of phenolic compounds to increase them in the diet, or their use as additives, nutraceuticals, or pharmacological agents, while preventing their scavenging by salivary proteins, is considered promising for preventing neurodegenerative diseases [129].

2.3.1.4 Minerals

The need for such minerals as calcium, magnesium, potassium, iron in old age remains quite high [130,131]. Insufficient iron in the diet contributes to the development of iron deficiency anemia with age. Iron is necessary to form about 100 hemin and non-heme enzymes, the generation of interleukins, T-killers, T-suppressors, the synthesis of steroid hormones, DNA. Iron deficiency in the body affects the genetic, molecular, cellular, tissue, and organ levels. More than 10% of people over 65 have signs of anemia, and up to 50% of elderly patients with chronic diseases staying in nursing homes [132–134]. Food sources of iron are best for preventing nutrient-deficiency anemia, but supplementation is often necessary, especially for the elderly. Nonheme iron is abundant in cereals, egg yolk, and green leafy vegetables. However, it is not absorbed efficiently. Vitamin C is known to increase iron absorption and be grown in the diet by citrus fruit consumption [135]. Zinc is an essential trace element necessary for many biochemical and physiological processes associated with the normal functioning of the brain [136] and cellular metabolism [137]. Zinc can be found in red meat, oysters, and crabs. Interventional studies in humans and rodents have shown that zinc has antidepressant effects and improves mood [138,139]. Magnesium levels are essential for central nervous system (CNS) function and may play a role in Alzheimer's disease, diabetes, stroke, hypertension, migraines, and attention deficit hyperactivity disorder [140]. Several interventional studies have shown a beneficial role of magnesium supplementation in treating depression [141,142]; however other studies have revealed conflicting outcomes [143,144]. Magnesium deficiency has been shown to cause changes in the functioning of the central nervous system (CNS), especially in the glutamatergic transmission of the limbic system and the cerebral cortex, which is involved in the etiopathogenesis of depression [145]. A balanced diet that includes an adequate intake of foods containing zinc and other micronutrients can be an effective adjunct to antidepressants for relieving symptoms of depression.

2.3.1.5 Essential polyunsaturated fatty acids

Sufficient supply of the elderly with ω-3 PUFAs is of particularly important: long-chain acids - eicosapentaenoic (EPA) and docosahexaenoic (DHA). It is known that regular and adequate intake of EPA and DHA contributes to the prevention of cardiovascular disease. These acids are involved in the construction of cell membranes, myelin sheaths, activate normal stem cell division, the synthesis of regulatory proteins, supporting cognitive and mental functions in the elderly [70,146,147]. The effect of using two dietary supplements to beat depression and anxiety associated with menopause was studied [148]. They concluded that the supplementation of products enriched with unsaturated fatty acid in the blend was more beneficial to slow down the psychological menopause symptoms than natural estrogen-rich product consumption. Among fatty products, vegetable oils and marine fatty fish have preferred sources of phospholipids, phytosterols, monounsaturated and PUFA families ω-6 and ω-3 for the elderly in the treatment of neuropsychiatric disorders [149].

2.3.1.6 Protein

Recently, the issue of the protein content in the nutrition of people over 60 years old has acquired relevance. It is known that both an excess and a lack of protein in the diet can cause pathological changes in the body. Lack of protein leads to disruption of hormonal status, liver function, synthesis of enzymes involved in the exchange of macro-and micronutrients, antibodies, and progressive loss of muscle mass in sarcopenia. Excessive protein intake negatively affects the function of the kidneys, the digestive and nervous systems and promotes carcinogenesis [150]. Recent studies indicate that older people need more protein than young people, which is associated with age-related changes, primarily protein metabolism [151]. Older people also need more protein to compensate for the inflammatory and catabolic processes associated with chronic and acute diseases that usually occur with aging [152]. The effects of whey protein supplementation consumed either immediately pre or postrest time on skeletal muscle mass, muscular strength, and functional capacity in preconditioned older women were investigated [153]. They found that whey protein supplementation effectively promoted increases in skeletal muscle mass, muscular strength, and functional capacity in pre-conditioned older women, regardless of supplementation timing. Protein-rich supplements are commonly used in the older population, and increasingly so due to a growing recognition of the adverse health consequences of nutritional impairment. These impacts include weight loss and a reduction in appendicular muscle mass with associated functional impairment, which are symptoms of the anorexia of aging, defined by the loss of appetite and decreased energy intake associated with aging [154]. The age effects on energy intake, appetite, gastric emptying, blood glucose, and gut hormones in response to protein-rich drinks were established [155]. Authors concluded that aging reduces the responses of caloric beverages on hunger, the desire to eat, fullness, and energy intake, and protein-rich nutrition supplements may be an effective strategy to increase energy intake in undernourished older people; in addition, they reported that less suppression of appetite and energy intake by 'pure' whey protein drinks and small intestinal infusions of whey in healthy older men compared to younger men. Researchers concluded that the ingestion of a high whey protein, leucine-enriched supplement resulted in a more significant overall postprandial muscle protein synthesis rate in healthy older subjects than a conventional dairy product [156]. Protein needs are considered to increase with age,

Table 2.2: Major food sources of specific nutrients.

Nutrients	Dietary sources	References
B1 (thiamine)	Whole grains, brown rice, pork, poultry, soybeans, nuts, dried beans, peas	[157]
B2 (riboflavin)	Organ meats, poultry, fish, and eggs, dairy products (milk and cheese)	[158]
B6 (pyridoxal)	Fish, liver, starchy vegetables, legumes, nuts, bananas, avocados, egg yolks, whole grains	[159]
B9 (folate)	Dark green leafy vegetables, spinach, liver, asparagus	[160]
B12 (cobalamin)	Yeast, seaweed, liver, kidneys, caviar, eggs, cheese, milk, cottage cheese, meat, fish	[161]
Vitamin C (ascorbic acid)	Sea buckthorn, black currant, citrus, kiwi, strawberry, blueberry, cranberry, raspberry, bell pepper, fresh cabbage	[162]
Vitamin D	Mushrooms, liver, fish, fish (not only fatty fish), egg yolk	[163]
Polyphenols	Fruits, vegetables, cereals, berries, nuts, beans	[164]
Minerals	seafood, organ meats, legumes, nuts, iodized salt drinking water	[165]
PUFA	Chicken, eggs, fish, nuts	[146]
Protein	Fish, eggs, dairy, and fermented milk products, meat, beans	[166]

with protein consumption being associated with many positive outcomes. Protein-fortified products are often used to improve nutritional status and prevent age-related muscle mass loss in older adults.

Table 2.2 represents a summary of the available natural resources - sources of biologically active substances. The use of which in the daily diet will reduce the susceptibility to depression among the elderly.

2.4 Summary

The world scientific community is increasingly concerned about how mental health diseases, particularly depression, can be prevented, treated, and managed, including physiological aspects with greater attention to a balanced diet among elderly people.

In this integrated review, we examined the influence of the nutritional status of the elderly population, particularly the deficiency of essential nutrients (water- and fat-soluble vitamins, polyunsaturated fatty acids, polyphenols, minerals, and protein) on the risk of depression. The literature most strongly supports the role of foods containing essential components that can prevent the development of human depressive diseases due to their multiple bioactivities. In addition, studies considered a positive effect of a balanced diet on the quality of life of elderly people. It has been scientifically proven that dairy, meat, fish, and plant-based products are full-fledged sources of essential nutrients for preventing depressive diseases among elderly people and should be included in the daily diet of this population group.

References

[1] World Health OrganizationThe World health report: 2001: Mental health: new understanding, new hope. Rapport sur la santé dans le monde: 2001: La santé mentale: nouvelle conception, nouveaux espoirs, 2001.

[2] TM Liew, Subjective cognitive decline, anxiety symptoms, and the risk of mild cognitive impairment and dementia, Alz Res Therapy 12 (2020) 107.

[3] S Saxena, M Funk, D Chisholm, WHO's Mental Health Action Plan 2013–2020: what can psychiatrists do to facilitate its implementation? World Psychiatry 13 (2014) 107–109.

[4] B Birmaher, ND Ryan, DE Williamson, DA Brent, J Kaufman, RE Dahl, J Perel, B Nelson, Childhood and adolescent depression: a review of the past 10 years. Part I, J Am Acad Child Adolesc Psychiatry 35 (1996) 1427–1439.

[5] EJ Costello, S Mustillo, A Erkanli, G Keeler, A Angold, Prevalence and development of psychiatric disorders in childhood and adolescence, Arch Gen Psychiatry 60 (2003) 837–844.

[6] United NationsWorld Population Prospects 2019: Highlights, 2019. https://doi.org/10.18356/13bf5476-en.

[7] European Commission: Eurostat: Population structure and ageing. Available from: https://ec.europa.eu/eurostat/statistics_explained/index.php?title=Population_structure_and_ageing#Future_trends_in_population_ageing (accessed on: 21 September, 2021).

[8] TT Yoshikawa, Epidemiology and unique aspects of aging and infectious diseases, Clin Infect Dis 30 (2000) 931–933.

[9] M Bahall, G Legall, K Khan, Quality of life among patients with cardiac disease: the impact of comorbid depression, Health Qual Life Outcomes 18 (2020) 189.

[10] I Liguori, G Russo, F Curcio, et al., Depression and chronic heart failure in the elderly: an intriguing relationship, J Geriatr Cardiol 15 (2018) 451–459.

[11] C Vlachakis, K Dragoumani, S Raftopoulou, M Mantaiou, L Papageorgiou, S Champeris Tsaniras, V Megalooikonomou, D Vlachakis, Human emotions on the onset of cardiovascular and small vessel related diseases, In Vivo 32 (2018) 859–870.

[12] D Bolton, G Gillett, Biopsychosocial conditions of health and disease, The Biopsychosocial Model of Health and Disease, Springer International Publishing, Cham, 2019, pp. 109–145.

[13] Z Vafaei, H Mokhtari, Z Sadooghi, R Meamar, A Chitsaz, M Moeini, Malnutrition is associated with depression in rural elderly population, J Res Med Sci 18 (2013) S15–S19.

[14] C Muñoz-González, M Vandenberghe-Descamps, G Feron, F Canon, H Labouré, C Sulmont-Rossé, Association between salivary hypofunction and food consumption in the elderlies. A systematic literature review, J Nutr Health Aging 22 (2018) 407–419.

[15] C Muñoz-González, G Feron, F Canon, Main effects of human saliva on flavour perception and the potential contribution to food consumption, Proc Nutr Soc 77 (2018) 423–431.

[16] S Kremer, JHF Bult, J Mojet, JHA Kroeze, Food perception with age and its relationship to pleasantness, Chem Senses 32 (2007) 591–602.

[17] M Schwartz, F Neiers, G Feron, F Canon, The relationship between salivary redox, diet, and food flavor perception, Front Nutr 7 (2021) 612735.

[18] C Muñoz-González, G Feron, F Canon, Physiological and oral parameters contribute prediction of retronasal aroma release in an elderly cohort, Food Chem 342 (2021) 128355.

[19] C Muñoz-González, M Brulé, G Feron, F Canon, Does interindividual variability of saliva affect the release and metabolization of aroma compounds ex vivo? The particular case of elderly suffering or not from hyposalivation, J Texture Stud 50 (2019) 36–44.

[20] C Muñoz-González, G Feron, M Brulé, F Canon, Understanding the release and metabolism of aroma compounds using micro-volume saliva samples by ex vivo approaches, Food Chem 240 (2018) 275–285.

[21] MWP Carney, DG Williams, BF Sheffield, Thiamine and pyridoxine lack in newly-admitted psychiatric patients, Br J Psychiatry 135 (1979) 249–254.

[22] A Ghaleiha, H Davari, L Jahangard, M Haghighi, M Ahmadpanah, MA Seifrabie, H Bajoghli, E Holsboer-Trachsler, S Brand, Adjuvant thiamine improved standard treatment in patients with major depressive

disorder: results from a randomized, double-blind, and placebo-controlled clinical trial, Eur Arch Psychiatry Clin Neurosci 266 (2016) 695–702.

[23] K Mikkelsen, K Hallam, L Stojanovska, V Apostolopoulos, Yeast based spreads improve anxiety and stress, J Funct Foods 40 (2018) 471–476.

[24] K Mikkelsen, L Stojanovska, M Polenakovic, M Bosevski, V Apostolopoulos, Exercise and mental health, Maturitas 106 (2017) 48–56.

[25] K Mikkelsen, L Stojanovska, M Prakash, V Apostolopoulos, The effects of vitamin B on the immune/cytokine network and their involvement in depression, Maturitas 96 (2017) 58–71.

[26] K Mikkelsen, L Stojanovska, K Tangalakis, M Bosevski, V Apostolopoulos, Cognitive decline: a vitamin B perspective, Maturitas 93 (2016) 108–113.

[27] N Nemazannikova, K Mikkelsen, L Stojanovska, GL Blatch, V Apostolopoulos, Is there a link between vitamin B and multiple sclerosis? Med Chem 14 (2018) 170–180.

[28] G Zhang, H Ding, H Chen, X Ye, H Li, X Lin, Z Ke, Thiamine nutritional status and depressive symptoms are inversely associated among older Chinese adults, J Nutr 143 (2013) 53–58.

[29] M Naghashpour, R Amani, R Nutr, S Nematpour, MH Haghighizadeh, Riboflavin status and its association with serum hs-CRP levels among clinical nurses with depression, J Am Coll Nutr 30 (2011) 340–347.

[30] JE Digby, F Martinez, A Jefferson, N Ruparelia, J Chai, M Wamil, DR Greaves, RP Choudhury, Anti-inflammatory effects of nicotinic acid in human monocytes are mediated by GPR109A dependent mechanisms, Arterioscler Thromb Vasc Biol 32 (2012) 669–676.

[31] L Gougeon, H Payette, JA Morais, P Gaudreau, B Shatenstein, K Gray-Donald, Intakes of folate, vitamin B6 and B12 and risk of depression in community-dwelling older adults: the Quebec longitudinal study on nutrition and aging, Eur J Clin Nutr 70 (2016) 380–385.

[32] PS Lipszyc, GA Cremaschi, MZ Zubilete, MLA Bertolino, F Capani, AM Genaro, M.R Wald, Niacin modulates pro-inflammatory cytokine secretion. A potential mechanism involved in its anti-atherosclerotic effect, TOCMJ 7 (2013) 90–98.

[33] W-H Pan, Y-P Chang, W-T Yeh, Y-S Guei, B-F Lin, I-L Wei, FL Yang, Y-P Liaw, K-J Chen, WJ Chen, Co-occurrence of anemia, marginal vitamin B6, and folate status and depressive symptoms in older adults, J Geriatr Psychiatry Neurol 25 (2012) 170–178.

[34] EM Balk, Vitamin B6, B12, and folic acid supplementation and cognitive function: a systematic review of randomized trials, Arch Intern Med 167 (2007) 21.

[35] MM Corrada, CH Kawas, J Hallfrisch, D Muller, R Brookmeyer, Reduced risk of Alzheimer's disease with high folate intake: the Baltimore longitudinal study of aging, Alzheimers Dement 1 (2005) 11–18.

[36] J Du, M Zhu, H Bao, B Li, Y Dong, C Xiao, G.Y Zhang, I Henter, M Rudorfer, B Vitiello, The role of nutrients in protecting mitochondrial function and neurotransmitter signaling: implications for the treatment of depression, PTSD, and suicidal behaviors, Crit Rev Food Sci Nutr 56 (2016) 2560–2578.

[37] M Fava, D Mischoulon, Folate in depression: efficacy, safety, differences in formulations, and clinical issues, J Clin Psychiatry 70 (2009) 12–17.

[38] F Jernerén, AK Elshorbagy, A Oulhaj, SM Smith, H Refsum, AD Smith, Brain atrophy in cognitively impaired elderly: the importance of long-chain ω-3 fatty acids and B vitamin status in a randomized controlled trial, Am J Clin Nutr 102 (2015) 215–221.

[39] T-P Ng, L Feng, M Niti, E-H Kua, K-B Yap, Folate, vitamin B12, homocysteine, and depressive symptoms in a population sample of older chinese adults: folate and depression, J Am Geriatr Soc 57 (2009) 871–876.

[40] KA Skarupski, C Tangney, H Li, B Ouyang, DA Evans, MC Morris, Longitudinal association of vitamin B-6, folate, and vitamin B-12 with depressive symptoms among older adults over time, Am J Clin Nutr 92 (2010) 330–335.

[41] CG-C Viviana Loria-Kohen, P-M Samara, A-S Blanca, Estudio piloto sobre el efecto de la suplementación con ácido fólico en, Nutr Hosp 28 (2013) 807–815.

[42] JG Walker, PJ Batterham, AJ Mackinnon, AF Jorm, I Hickie, M Fenech, M Kljakovic, D Crisp, H Christensen, Oral folic acid and vitamin B-12 supplementation to prevent cognitive decline in community-dwelling older adults with depressive symptoms—the beyond ageing project: a randomized controlled trial, Am J Clin Nutr 95 (2012) 194–203.

[43] SM Aburawi, Effect of ascorbic acid on mental depression drug therapy: clinical study, J Psychol Psychother 4 (2014) 1.

[44] KA Naidu, Vitamin C in human health and disease is still a mystery? An overview, Nutr J 2 (2003) 7.

[45] SJ Padayatty, A Katz, Y Wang, et al., Vitamin C as an antioxidant: evaluation of its role in disease prevention, J Am Coll Nutr 22 (2003) 18–35.

[46] NM Alavi, S Khademalhoseini, Z Vakili, F Assarian, Effect of vitamin D supplementation on depression in elderly patients: a randomized clinical trial, Clin Nutr 38 (2019) 2065–2070.

[47] C Di Somma, E Scarano, L Barrea, VV Zhukouskaya, S Savastano, C Mele, M Scacchi, G Aimaretti, A Colao, P Marzullo, Vitamin D and neurological diseases: an endocrine view, Int J Mol Sci 18 (2017) 2482.

[48] SE Judd, V Tangpricha, Vitamin D deficiency and risk for cardiovascular disease, Am J Med Sci 338 (2009) 40–44.

[49] CH Wilkins, YI Sheline, CM Roe, SJ Birge, JC Morris, Vitamin D deficiency is associated with low mood and worse cognitive performance in older adults, Am J Geriatr Psychiatry 14 (2006) 1032–1040.

[50] MN Wu, F He, QR Tang, J Chen, X Gu, YJ Zhai, FD Li, T Zhang, XY Wang, J Lin, Association between depressive symptoms and supplemental intake of calcium and vitamin d in older adults, J Nutr Health Aging 24 (2020) 107–112.

[51] MS Chapman, Vitamin A: History, Current Uses, and Controversies, Semin Cutan Med Surg 31 (2012) 11–16.

[52] H-P Lee, G Casadesus, X Zhu, H Lee, G Perry, MA Smith, K Gustaw-Rothenberg, A Lerner, All-trans retinoic acid as a novel therapeutic strategy for Alzheimer's disease, Expert Rev Neurother 9 (2009) 1615–1621.

[53] J Watson, M Lee, MN Garcia-Casal, Consequences of inadequate intakes of vitamin A, vitamin B12, vitamin D, calcium, iron, and folate in older persons, Curr Geriatr Rep 7 (2018) 103–113.

[54] N Farina, D Llewellyn, MGEKN Isaac, N Tabet, Vitamin E for Alzheimer's dementia and mild cognitive impairment, Cochrane Database Syst Rev 4 (2017) CD002854.

[55] AJ Owen, MJ Batterham, YC Probst, BFS Grenyer, LC Tapsell, Low plasma vitamin E levels in major depression: diet or disease? Eur J Clin Nutr 59 (2005) 304–306.

[56] CK Sen, S Khanna, S Roy, Tocotrienol: the natural vitamin e to defend the nervous system? Ann N Y Acad Sci 1031 (2004) 127–142.

[57] F Ali, Naz F Rahul, S Jyoti, YH Siddique, Health functionality of apigenin: a review, Int J Food Prop 20 (2017) 1197–1238.

[58] G Desideri, C Kwik-Uribe, D Grassi, et al., Benefits in cognitive function, blood pressure, and insulin resistance through cocoa flavanol consumption in elderly subjects with mild cognitive impairment: the cocoa, cognition, and aging (CoCoA) study, Hypertension 60 (2012) 794–801.

[59] R Krikorian, MD Shidler, TA Nash, W Kalt, MR Vinqvist-Tymchuk, B Shukitt-Hale, JA Joseph, Blueberry supplementation improves memory in older adults, J Agric Food Chem 58 (2010) 3996–4000.

[60] R Krikorian, TA Nash, MD Shidler, B Shukitt-Hale, JA Joseph, Concord grape juice supplementation improves memory function in older adults with mild cognitive impairment, Br J Nutr 103 (2010) 730–734.

[61] M Nishimura, T Ohkawara, T Nakagawa, T Muro, Y Sato, H Satoh, M Kobori, J Nishihira, A randomized, double-blind, placebo-controlled study evaluating the effects of quercetin-rich onion on cognitive function in elderly subjects, FFHD 7 (2017) 353.

[62] M Russo, S Moccia, C Spagnuolo, I Tedesco, GL Russo, Roles of flavonoids against coronavirus infection, Chem Biol Interact 328 (2020) 109211.

[63] H-Y Chang, Y-W Hu, C-SJ Yue, Y-W Wen, W-T Yeh, L-S Hsu, S-Y Tsai, W-H Pan, Effect of potassium-enriched salt on cardiovascular mortality and medical expenses of elderly men, Am J Clin Nutr 83 (2006) 1289–1296.

[64] J Lv, Y Li, W Ren, et al., Increased depression risk for elderly women with high blood levels of strontium and barium, Environ Chem Lett 19 (2021) 1787–1796.

[65] T Miki, T Kochi, M Eguchi, et al., Dietary intake of minerals in relation to depressive symptoms in Japanese employees: The Furukawa Nutrition and Health Study, Nutrition 31 (2015) 686–690.

[66] T Thi Thu Nguyen, S Miyagi, H Tsujiguchi, Y Kambayashi, A Hara, H Nakamura, K Suzuki, Y Yamada, Y Shimizu, H Nakamura, Association between lower intake of minerals and depressive symptoms among elderly Japanese women but not men: findings from Shika Study, Nutrients 11 (2019) 389.

[67] M Hashimoto, S Kato, Y Tanabe, M Katakura, AA Mamun, M Ohno, S Hossain, K Onoda, S Yamaguchi, O Shido, Beneficial effects of dietary docosahexaenoic acid intervention on cognitive function and mental health of the oldest elderly in Japanese care facilities and nursing homes: effect of n-3 PUFA on the oldest elderly, Geriatr Gerontol Int 17 (2017) 330–337.

[68] JC McCann, BN Ames, Is docosahexaenoic acid, an n−3 long-chain polyunsaturated fatty acid, required for development of normal brain function? An overview of evidence from cognitive and behavioral tests in humans and animals, Am J Clin Nutr 82 (2005) 281–295.

[69] M Rondanelli, A Opizzi, M Faliva, M Mozzoni, N Antoniello, R Cazzola, R Savarè, R Cerutti, E Grossi, B Cestaro, Effects of a diet integration with an oily emulsion of DHA-phospholipids containing melatonin and tryptophan in elderly patients suffering from mild cognitive impairment, Nutr Neurosci 15 (2012) 46–54.

[70] Y Tajalizadekhoob, F Sharifi, H Fakhrzadeh, M Mirarefin, M Ghaderpanahi, Z Badamchizade, S Azimipour, The effect of low-dose omega 3 fatty acids on the treatment of mild to moderate depression in the elderly: a double-blind, randomized, placebo-controlled study, Eur Arch Psychiatry Clin Neurosci 261 (2011) 539–549.

[71] YC Luiking, NE Deutz, RG Memelink, S Verlaan, RR Wolfe, Postprandial muscle protein synthesis is higher after a high whey protein, leucine-enriched supplement than after a dairy-like product in healthy older people: a randomized controlled trial, Nutr J 13 (2014) 9.

[72] AR Wolfe, C Arroyo, SH Tedders, Y Li, Q Dai, J Zhang, Dietary protein and protein-rich food in relation to severely depressed mood: a 10year follow-up of a national cohort, Prog Neuropsychopharmacol Biol Psychiatry 35 (2011) 232–238.

[73] B Saka, O Kaya, GB Ozturk, N Erten, MA Karan, Malnutrition in the elderly and its relationship with other geriatric syndromes, Clin Nutr 29 (2010) 745–748.

[74] M Hickson, Malnutrition and ageing, Postgrad Med J 82 (2006) 2–8.

[75] C Cannella, C Savina, LM Donini, Nutrition, longevity and behavior, Arch Gerontol Geriatr 49 (2009) 19–27.

[76] L Fontana, L Partridge, Promoting health and longevity through diet: from model organisms to humans, Cell 161 (2015) 106–118.

[77] S Lim, Eating a balanced diet: a healthy life through a balanced diet in the age of longevity, J Obes Metab Syndr 27 (2018) 39–45.

[78] S Maggini, A Pierre, PC Calder, Immune function and micronutrient requirements change over the life course, Nutrients 10 (2018) 1531.

[79] T Ahmed, N Haboubi, Assessment and management of nutrition in older people and its importance to health, Clin Interv Aging 5 (2010) 207–216.

[80] R Mõttus, G McNeill, X Jia, LCA Craig, JM Starr, IJ Deary, The associations between personality, diet and body mass index in older people, Health Psychol 32 (2013) 353–360.

[81] A Drewnowski, WJ Evans, Nutrition, physical activity, and quality of life in older adults: summary, J Gerontol A 56 (2001) 89–94.

[82] E Tourlouki, A-L Matalas, DB Panagiotakos, Dietary habits and cardiovascular disease risk in middle-aged and elderly populations: a review of evidence, Clin Interv Aging 4 (2009) 319–330.

[83] LC Hawkley, JT Cacioppo, Stress and the aging immune system, Brain Behav Immun 18 (2004) 114–119.

[84] V Lerner, M Kanevsky, T Dwolatzky, T Rouach, R Kamin, C Miodownik, Vitamin B12 and folate serum levels in newly admitted psychiatric patients, Clin Nutr 25 (2006) 60–67.

[85] GI Papakostas, T Petersen, BD Lebowitz, D Mischoulon, JL Ryan, AA Nierenberg, T Bottiglieri, JE Alpert, JF Rosenbaum, M Fava, The relationship between serum folate, vitamin B12, and homocysteine levels in major depressive disorder and the timing of improvement with fluoxetine, Int J Neuropsychopharmacol 8 (2005) 523–528.

[86] MJ Taylor, SM Carney, GM Goodwin, JR Geddes, Folate for depressive disorders: systematic review and meta-analysis of randomized controlled trials, J Psychopharmacol 18 (2004) 251–256.

[87] KE D'Anci, IH Rosenberg, Folate and brain function in the elderly, Curr Opin Clin Nutr Metab Care 7 (2004) 659–664.

[88] HH Keller, Malnutrition in institutionalized elderly: how and why? J Am Geriatr Soc 41 (1993) 1212–1218.

[89] M Pirlich, H Lochs, Nutrition in the elderly, Best Pract Res Clin Gastroenterol 15 (2001) 869–884.

[90] J Selhub, LC Bagley, J Miller, IH Rosenberg, B vitamins, homocysteine, and neurocognitive function in the elderly, Am J Clin Nutr 71 (2000) 614S–620S.

[91] RV Banerjee, RG Matthews, Cobalamin-dependent methionine synthase, FASEB J 4 (1990) 1450–1459.

[92] T Bottiglieri, Folate, vitamin B_{12}, and S-adenosylmethionine., Psychiatr Clin North Am 36 (2013) 1–13.

[93] M Cordaro, R Siracusa, R Fusco, S Cuzzocrea, R Di Paola, D Impellizzeri, Involvements of hyperhomocysteinemia in neurological disorders, Metabolites 11 (2021) 37.

[94] A Rozycka, PP Jagodzinski, W Kozubski, M Lianeri, J Dorszewska, Homocysteine level and mechanisms of injury in Parkinson's disease as related to MTHFR, MTR, and MTHFD1 genes polymorphisms and L-dopa treatment, Curr Genomics 14 (2013) 534–542.

[95] Joanna Kocot, Dorota Luchowska-Kocot, Małgorzata Kiełczykowska, Irena Musik, Jacek Kurzepa, Does vitamin C influence neurodegenerative diseases and psychiatric disorders? Nutrients 9 (2017) 659.

[96] M Holick, J MacLaughlin, M Clark, S Holick, J Potts, R Anderson, I Blank, J Parrish, P Elias, Photosynthesis of previtamin D3 in human skin and the physiologic consequences, Science 210 (1980) 203–205.

[97] S-W Chang, H-C Lee, Vitamin D and health - the missing vitamin in humans, Pediatr Neonatol 60 (2019) 237–244.

[98] LU Gani, CH How, PILL series. Vitamin D deficiency, Singapore Med J 56 (2015) 433–436.

[99] FM Gloth, Vitamin D deficiency in homebound elderly persons, JAMA 274 (1995) 1683.

[100] KK Deeb, DL Trump, CS Johnson, Vitamin D signalling pathways in cancer: potential for anticancer therapeutics, Nat Rev Cancer 7 (2007) 684–700.

[101] B Lanske, MS Razzaque, Mineral metabolism and aging: the fibroblast growth factor 23 enigma, Curr Opin Nephrol Hypertens 16 (2007) 311–318.

[102] JM Lemire, Immunomodulatory role of 1,25-dihydroxyvitamin D3, J Cell Biochem 49 (1992) 26–31.

[103] P Oelzner, A Müller, F Deschner, M Hüller, K Abendroth, G Hein, G Stein, Relationship between disease activity and serum levels of vitamin D metabolites and PTH in rheumatoid arthritis, Calcif Tissue Int 62 (1998) 193–198.

[104] JP Kesby, DW Eyles, THJ Burne, JJ McGrath, The effects of vitamin D on brain development and adult brain function, Mol Cell Endocrinol 347 (2011) 121–127.

[105] Z Huang, Y Liu, G Qi, D Brand, SG Zheng, Role of vitamin A in the immune system, J Clin Med 9 (2018) 258.

[106] S Kuenzli, C Tran, J-H Saurat, Retinoid receptors in inflammatory responses: a potential target for pharmacology, Curr Drug Targets Inflamm Allergy 3 (2004) 355–360.

[107] K Shudo, H Fukasawa, M Nakagomi, N Yamagata, Towards retinoid therapy for Alzheimer's disease, Curr Alzheimer Res 6 (2009) 302–311.

[108] C Kaur, V Sivakumar, ST Dheen, EA Ling, Insulin-like growth factor I and II expression and modulation in amoeboid microglial cells by lipopolysaccharide and retinoic acid, Neuroscience 138 (2006) 1233–1244.

[109] C Gilbert, What is vitamin A and why do we need it? Community Eye Health 26 (2013) 65.

[110] M Oliveira L de, FME Teixeira, MN Sato, Impact of retinoic acid on immune cells and inflammatory diseases, Mediators Inflamm 2018 (2018) 1–17.

[111] FJ Jiménez-Jiménez, F de Bustos, JA Molina, et al., Cerebrospinal fluid levels of alpha-tocopherol (vitamin E) in Alzheimer's disease, J Neural Transm (Vienna) 104 (1997) 703–710.

[112] K Kontush, S Schekatolina, Vitamin E in neurodegenerative disorders: Alzheimer's disease, Ann N Y Acad Sci 1031 (2004) 249–262.

[113] M Sano, C Ernesto, RG Thomas, et al., A controlled trial of selegiline, alpha-tocopherol, or both as treatment for Alzheimer's disease. The Alzheimer's Disease Cooperative Study, N Engl J Med 336 (1997) 1216–1222.

[114] R Brigelius-Flohé, FJ Kelly, JT Salonen, J Neuzil, J-M Zingg, A Azzi, The European perspective on vitamin E: current knowledge and future research, Am J Clin Nutr 76 (2002) 703–716.

[115] A Trela, R Szymańska, Less widespread plant oils as a good source of vitamin E, Food Chem 296 (2019) 160–166.

[116] V Lobo, A Patil, A Phatak, N Chandra, Free radicals, antioxidants and functional foods: impact on human health, Pharmacogn Rev 4 (2010) 118.

[117] A Selvam, Inventory of vegetable crude drug samples housed in botanical survey of India, Howrah, Pharmacogn Rev 2 (2008) 61.

[118] B Salehi, E Azzini, P Zucca, et al., Plant-derived bioactives and oxidative stress-related disorders: a key trend towards healthy aging and longevity promotion, Appl Sci 10 (2020) 947.

[119] M Sharifi-Rad, NV Anil Kumar, P Zucca, et al., Lifestyle, oxidative stress, and antioxidants: back and forth in the pathophysiology of chronic diseases, Front Physiol 11 (2020) 694.

[120] B Uttara, AV Singh, P Zamboni, RT Mahajan, Oxidative stress and neurodegenerative diseases: a review of upstream and downstream antioxidant therapeutic options, Curr Neuropharmacol 7 (2009) 65–74.

[121] X Chen, P Ciarletta, H-H Dai, Physical principles of morphogenesis in mushrooms, Phys Rev E 103 (2021) 022412.

[122] SY Lew, SL Teoh, SH Lim, LW Lim, KH Wong, Discovering the potentials of medicinal mushrooms in combating depression – a review, MRMC 20 (2020) 1518–1531.

[123] B Muszyńska, M Łojewski, J Rojowski, W Opoka, K Sułkowska-Ziaja, Natural products of relevance in the prevention and supportive treatment of depression, Psychiatr Pol 49 (2015) 435–453.

[124] PS Chong, M-L Fung, KH Wong, LW Lim, Therapeutic potential of Hericium erinaceus for depressive disorder, IJMS 21 (2019) 163.

[125] S Mandel, T Amit, L Reznichenko, O Weinreb, MB Youdim, Green tea catechins as brain-permeable, natural iron chelators-antioxidants for the treatment of neurodegenerative disorders, Mol Nutr Food Res 50 (2006) 229–234.

[126] X Zhang, M Wu, F Lu, N Luo, Z-P He, H Yang, Involvement of $\alpha 7$ nAChR signaling cascade in epigallo-catechin gallate suppression of β-amyloid-induced apoptotic cortical neuronal insults, Mol Neurobiol 49 (2014) 66–77.

[127] S Ployon, M Morzel, C Belloir, A Bonnotte, E Bourillot, L Briand, E Lesniewska, J Lherminier, E Aybeke, F Canon, Mechanisms of astringency: structural alteration of the oral mucosal pellicle by dietary tannins and protective effect of bPRPs, Food Chem 253 (2018) 79–87.

[128] F Canon, C Belloir, E Bourillot, et al., Perspectives on astringency sensation: an alternative hypothesis on the molecular origin of astringency, J Agric Food Chem 69 (2021) 3822–3826.

[129] M-C Canivenc-Lavier, F Neiers, L Briand, Plant polyphenols, chemoreception, taste receptors and taste management, Curr Opin Clin Nutr Metab Care 22 (2019) 472–478.

[130] BN Ames, Low micronutrient intake may accelerate the degenerative diseases of aging through allocation of scarce micronutrients by triage, Proc Natl Acad Sci 103 (2006) 17589–17594.

[131] RJ Wood, PM Suter, RM Russell, Mineral requirements of elderly people, Am J Clin Nutr 62 (1995) 493–505.

[132] E Andrès, L Federici, K Serraj, G Kaltenbach, Update of nutrient-deficiency anemia in elderly patients, Eur J Intern Med 19 (2008) 488–493.

[133] European Food Safety Authority (EFSA), Dietary Reference Values for nutrients Summary report, Dietary Reference Values for nutrients Summary report, EFS3, 2017. https://doi.org/10.2903/sp.efsa.2017.e15121.

[134] KV Patel, Epidemiology of anemia in older adults, Semin Hematol 45 (2008) 210–217.

[135] M Alleyne, MK Horne, JL Miller, Individualized treatment for iron-deficiency anemia in adults, Am J Med 121 (2008) 943–948.

[136] CT Chasapis, AC Loutsidou, CA Spiliopoulou, ME Stefanidou, Zinc and human health: an update, Arch Toxicol 86 (2012) 521–534.

[137] T Kambe, T Tsuji, A Hashimoto, N Itsumura, The physiological, biochemical, and molecular roles of zinc transporters in zinc homeostasis and metabolism, Physiol Rev 95 (2015) 749–784.

[138] J Lai, A Moxey, G Nowak, K Vashum, K Bailey, M McEvoy, The efficacy of zinc supplementation in depression: systematic review of randomised controlled trials, J Affect Disord 136 (2012) e31–e39.

[139] T Sawada, K Yokoi, Effect of zinc supplementation on mood states in young women: a pilot study, Eur J Clin Nutr 64 (2010) 331–333.

[140] U Gröber, J Schmidt, K Kisters, Magnesium in prevention and therapy, Nutrients 7 (2015) 8199–8226.

[141] A Rajizadeh, H Mozaffari-Khosravi, M Yassini-Ardakani, A Dehghani, Effect of magnesium supplementation on depression status in depressed patients with magnesium deficiency: a randomized, double-blind, placebo-controlled trial, Nutrition 35 (2017) 56–60.

[142] EK Tarleton, B Littenberg, CD MacLean, AG Kennedy, C Daley, Role of magnesium supplementation in the treatment of depression: a randomized clinical trial, PLoS One 12 (2017) e0180067.

[143] L Barragán-Rodríguez, M Rodríguez-Morán, F Guerrero-Romero, Efficacy and safety of oral magnesium supplementation in the treatment of depression in the elderly with type 2 diabetes: a randomized, equivalent trial, Magnes Res 21 (2008) 218–223.

[144] FE Fard, M Mirghafourvand, S Mohammad-Alizadeh Charandabi, A Farshbaf-Khalili, Y Javadzadeh, H Asgharian, Effects of zinc and magnesium supplements on postpartum depression and anxiety: a randomized controlled clinical trial, Women Health 57 (2017) 1115–1128.

[145] D Peng, F Shi, G Li, D Fralick, T Shen, M Qiu, J Liu, K Jiang, D Shen, Y Fang, Surface vulnerability of cerebral cortex to major depressive disorder, PLoS One 10 (2015) e0120704.

[146] D Cutuli, Functional and structural benefits induced by omega-3 polyunsaturated fatty acids during aging, Curr Neuropharmacol 15 (2017) 534–542.

[147] Food and Agriculture Organization of the United NationsFats and Fatty Acids in Human Nutrition: Report of an Expert Consultation, Food and Agriculture Organization of the United Nations, Rome, 2010.

[148] S Fouad, SM El Shebini, M Abdel-Moaty, NH Ahmed, AMS Hussein, HA Essa, ST Tapozada, Menopause anxiety and depression; how food can help? Open Access Maced J Med Sci 9 (2021) 64–71.

[149] A Reimers, H Ljung, The emerging role of omega-3 fatty acids as a therapeutic option in neuropsychiatric disorders, Ther Adv Psychopharmacol 9 (2019) 204512531985890.

[150] I Delimaris, Adverse effects associated with protein intake above the recommended dietary allowance for adults, ISRN Nutr 2013 (2013) 126929–126929.

[151] C Nowson, S O'Connell, Protein requirements and recommendations for older people: a review, Nutrients 7 (2015) 6874–6899.

[152] J Bauer, G Biolo, T Cederholm, et al., Evidence-based recommendations for optimal dietary protein intake in older people: a position paper from the PROT-AGE Study Group, J Am Med Dir Assoc 14 (2013) 542–559.

[153] H Nabuco, C Tomeleri, P Sugihara Junior, et al., Effects of whey protein supplementation pre- or post-resistance training on muscle mass, muscular strength, and functional capacity in pre-conditioned older women: a randomized clinical trial, Nutrients 10 (2018) 563.

[154] JE Morley, AJ Silver, Anorexia in the elderly, Neurobiol Aging 9 (1988) 9–16.

[155] C Giezenaar, K Lange, T Hausken, KL Jones, M Horowitz, I Chapman, S Soenen, Effects of age on acute appetite-related responses to whey-protein drinks, including energy intake, gastric emptying, blood glucose, and plasma gut hormone concentrations—a randomized controlled trial, Nutrients 12 (2020) 1008.

[156] YC Luiking, NE Deutz, RG Memelink, S Verlaan, RR Wolfe, Postprandial muscle protein synthesis is higher after a high whey protein, leucine-enriched supplement than after a dairy-like product in healthy older people: a randomized controlled trial, Nutr J 13 (2014) 9.

[157] K Osiezagha, S Ali, C Freeman, NC Barker, S Jabeen, S Maitra, Y Olagbemiro, W Richie, RK Bailey, Thiamine deficiency and delirium, Innov Clin Neurosci 10 (2013) 26–32.

[158] JT Pinto, J Zempleni, Riboflavin, Adv Nutr 7 (2016) 973–975.

[159] PJ Stover, MS Field, Vitamin B-6, Adv Nutr 6 (2015) 132–133.

[160] AC Ross, B Caballero, RJ Cousins, KL Tucker, TR Ziegler, Modern Nutrition in Health and Disease, Wolters Kluwer Health, 2012.

[161] F Watanabe, Vitamin B12 sources and bioavailability, Exp Biol Med (Maywood) 232 (2007) 1266–1274.

[162] JF Gregory, Ascorbic acid bioavailability in foods and supplements, Nutr Rev 51 (1993) 301–303.

[163] C Lamberg-Allardt, Vitamin D in foods and as supplements, Prog Biophys Mol Biol 92 (2006) 33–38.

[164] C Manach, A Scalbert, C Morand, C Rémésy, L Jiménez, Polyphenols: food sources and bioavailability, Am J Clin Nutr 79 (2004) 727–747.

[165] RL Bailey, VL Fulgoni, DR Keast, JT Dwyer, Dietary supplement use is associated with higher intakes of minerals from food sources, Am J Clin Nutr 94 (2011) 1376–1381.

[166] JR Hoffman, MJ Falvo, Protein – which is best? J Sports Sci Med 3 (2004) 118–130.

CHAPTER 3

Herbal remedies against Huntington's disease: Preclinical evidences and future directions

Samarpita Banerjee[a], Champa Keeya Tudu[j], Samapika Nandy[j],
Devendra K. Pandey[b], Mimosa Ghorai[j], Mahipal S. Shekhawat[c],
Arabinda Ghosh[d], Potshangbam Nongdam[e], Abdel R. Al-Tawaha[f],
Ercan Bursal[g], Gaber E-S Batiha[h], Suchhanda Ghosh[a], Vinay Kumar[i] and
Abhijit Dey, PhD[j]

[a]Department of Botany, Shri Shikshayatan College, Kolkata, India [b]Department of Biotechnology, Lovely Professional University, Punjab, India [c]Plant Biotechnology Unit, K.M. Government Institute for Post Graduate Studies and Reserach, Lawspet, Puducherry, India [d]Department of Botany, Gauhati University, Guwahati, Assam, India [e]Department of Biotechnology, Manipur University, Imphal, Manipur, India [f]Department of Biological Sciences, Al-Hussein Bin Talal University, Maan, Jordon [g]Department of Biochemistry, Mus Alparslan University, Turkey [h]Department of Pharmacology and Therapeutics, Faculty of Veterinary Medicine, Damanhour University, Damanhour, AlBeheira, Egypt [i]Department of Biotechnology, Modern College (Savitribai Phule Pune University), Pune, India [j]Department of Life Sciences, Presidency University, Kolkata, India

3.1 Introduction

Huntington's disease (HD) is an autosomal dominant, devastating progressive neurodegenerative genetic disorder with a distinct phenotype distinguished by the gradual development of involuntary dystonia and chorea, cognitive deterioration, incoordination, neuropsychiatric problems and behavioral difficulties [1,2]. These phenotypes result in neuronal impairment which ultimately leads to death in selective regions of the brain, the principal targets being the striatum and cerebral cortex [1]. Huntingtin (mHtt), the protein product of the mutant gene responsible for HD (the gene, linked to a polymorphic DNA marker maps to human chromosome 4), occurs due to an expanded repeat of CAG which in turn leads to a polyglutamine strand of indefinite length at the N-terminus [2–4]. This protein has a molecular weight of 350-kilodalton and after mutation, it decreases the production of Brain-derived neurotrophic factor (BDNF) in the cortex [5]. The mutant protein, mHtt disturbs vesicle trafficking and decontrols autophagy in the nerve cells, hence leading to cellular death [6]. The onset of the disease depends on the length of the CAG repeat: the longer the repeat, the earlier the occurrence. Though the average age of showing symptoms is 30-50 years, there have been cases where 20 year olds have been diagnosed with HD (known as juvenile Huntington's disease, or HD). In the latter case, the CAG repeat often crosses 55 [7]. In the former case, patients affected with HD have been found to have 30–70 CAG repeats [8] (Fig. 3.1).

Besides the effects of HD discussed above, proteins get abnormally processed and aggregated, and cellular toxicity includes both cell-cell interaction as well as cell autonomous mechanisms. Neuroimaging studies and predictive genetic tests reveal that the onset of Huntington's disease starts many years prior to the revelation of symptoms. Therefore, HD is a potential model for developing therapeutic solutions to prevent its onset and not only delay its progression [4]. The prevalent neuropsychiatric symptoms, though not dependent on the motor and cognitive facets of HD, are managed by conventional therapeutics due to lack of proper treatment for gradual neuronal impairment [2,9]. Chorea, the abnormal and sudden jerky movements is treated by blocking dopamine receptor or lessening agents [7,10]. However, Tetrabenzine (TBZ), a dopamine-depleting medicine known to cure chorea has unique dosage instructions, side effects and drug interactions, hence limiting its usage [11]. Though some neuroleptic agents like olanzapine and aripiprazole have been proved to be efficacious for treating chorea and psychosis, limited treatment strategies have been discovered to overcome the cognitive, motor and psychiatric problems of HD. However, medical practitioners can use education and symptomatic strategies for patients and families suffering from HD [12,13]. A recent study has shown that there has been an upsurge of significant developments for understanding the macroscopic changes in the brain structure and cellular pathology, in therapeutic relief targets and clinical tests, in the last decade. The most effective of these are potential therapies targeted at decreasing the levels of mutant huntingtin, one of them being antisense oligonucleotide therapy with sufficient positive results from clinical trials still pending [14].

Approximately 3–7 people out of 100,000 are affected by HD and 20 out of 100,000 are carriers, on a worldwide basis [10]. Recent studies reveal that the prevalence of HD was lowest among the Asians, and highest among the Caucasians. There has been an increase in the rates for the past 50 years in North America, Western Europe and Australia [15]. The gradual development of the disease results in daily life discrepancy, requiring full-time care and eventually death. The primary cause of death of HD patients is pneumonia, succeeded by death [7]. The rate of suicide amidst HD patients has been reported to be much more than the general population, even though this disparity was attributed to rational suicide and not linked to mental illnesses [16].

Various experimental models were used to understand the molecular mechanisms of HD and how to treat it, effectively and therapeutically. 3-Nitopropionic acid is known to procure similar morphological, behavioral and biochemical changes to those affected by HD.

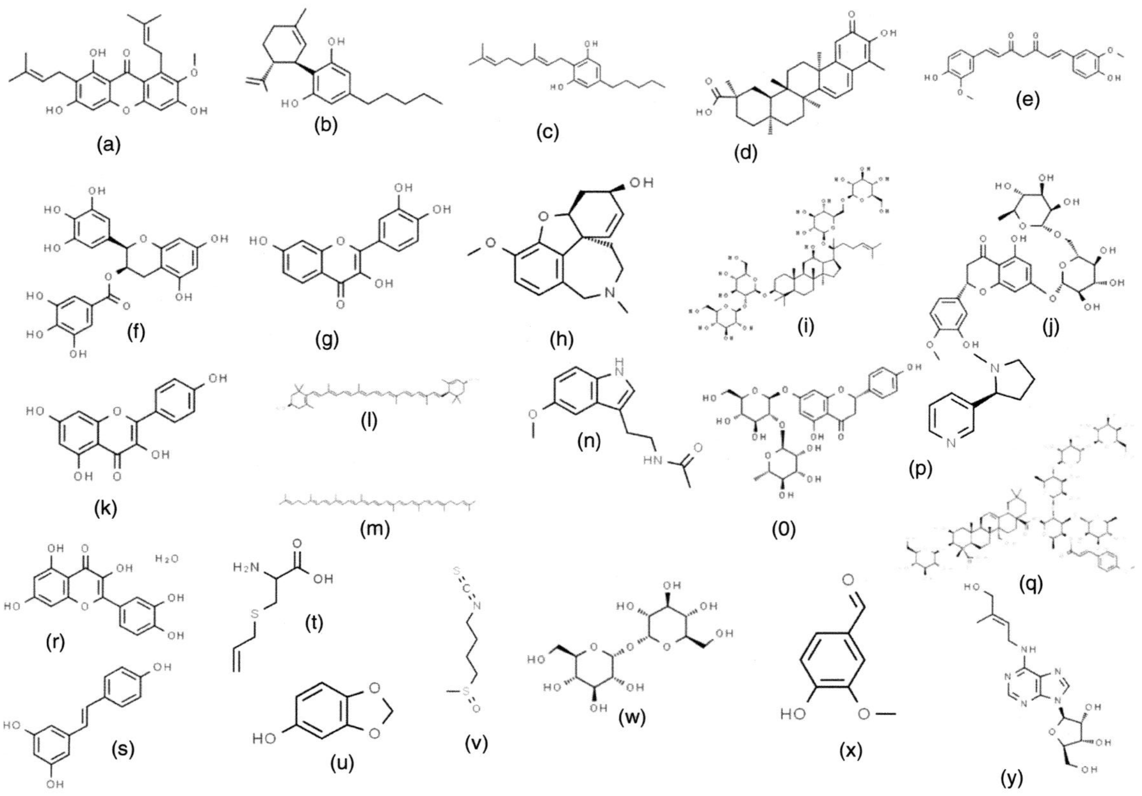

Figure 3.1

Chemical structures of some anti-HD natural compounds: (A) α-Mangostin, (B) cannabidiol, (C) cannabigerol, (D) celastrol, (E) curcumin, (F) EGCG, (G) fisetin, (H) galantamine, (I) ginsenoside Rb1, (J) hesperidin, (K) kaempferol, (L) lutein, (M) lycopene, (N) melatonin, (O) naringin, (P) nicotine, (Q) onjisaponin B, (R) quercetin, (S) resveratrol, (T) S-allylcysteine, (U) sesamol, (V) sulforaphane, (W) trehalose, (X) vanillin, and (Y) zeatin riboside (figures obtained from www.chemspider.com).

Therefore, it works as a suitable phenotypic model to mimic HD which further may help in the development of new therapies [17]. Transgenic animal models like mice, rats, *Drosophila* and *Caenorhabditis elegans* expressing mHtt are used for similar purposes because they exhibit HD-like symptoms [4,18]. However, a large number of genetic mouse models fail to replicate the striking degeneration found in patient brains [18].

Statistical studies conducted by WHO reveal that for primary healthcare, nearly 80% of the population in Asian and African countries uses traditional medicines [19]. Traditional medicines have been preferred over synthetic drugs due to their negligible side effects, easy availability, lower cost and therapeutic efficiency. Added to these, monoherbal and polyherbal formulations can be isolated which may lead to the characterization of active constituents [20,21]. Huntington's disease characterized by mitochondrial dysfunction and oxidative stress has been reported to be extensively cured by using a large number of medicinal plants [22]. However, further studies need to be conducted where we can investigate the toxicology, efficiency and tolerance of the herbal formulations against HD [23].

In this paper, we will discuss about the recent studies that suggest how various herbal formulations, plant extracts, fractions and isolated compounds showed effects on various neurotoxic and transgenic HD models (Tables 3.1–3.3).   Fig. 3.1 presents the structures of the bioactive phytochemicals reported against HD.

3.2 Anti-HD activity of crude/semi-purified plant fractions/extract

3.2.1 Bacopa monnieri (L.) Wettst. (Plantaginaceae)

The Indian subcontinent has been familiar with the usage of the herb *Bacopa monnieri* (BM) traditionally. Locally known as Brahmi, it has been used against memory loss, cognitive impairments as well as various other neurological disorders like Alzheimer's Disease, Parkinson's Disease, Epilepsy, Autism and Encephalomyelitis [24–28]. BM has been pharmacologically useful due to its anti-inflammatory, anti-oxidative and anti-apoptotic properties [29,30]. The effect of *Bacopa monnieri* alcoholic leaf extract was studied in a prepubertal mice brain. Pre-treating prepubertal mice brain with alcoholic leaf extract of *Bacopa monnieri* has been found to be neuroprotective against 3-NPA-induced exidative stress, both in vitro and in vivo by lowering oxidative stress and preventing cytotoxicity. Similar results were found when dopaminergic N27 cells were pre-treated with BME [31]. *Bacopa monnieri* ethanolic extract prevented 3-NP induced mitochondrial dysfunctions in the striatum by reviving the activities of ETC enzymes (cytochrome c oxidase, succinate-ubiquinone oxidoreductase, NADH: cytochrome c reductase and NADH: ubiquinone oxidoreductase). It also enhances the anti-oxidant status of the mitochondria in the striatum by restoring the activities of significant anti-oxidant enzymes like thioredoxin reductase, glutathione peroxidase, glutathione reductase and superoxide dismutase [31].

Table 3.1: Anti-HD activity of crude/semipurified plant fractions/extract.

Source plant (parts/crude or semipurified extracts)	Family	In-vitro/in-vivo models	Mode of action	References
Bacopa monnieri alcoholic leaf extract	Plantaginaceae	Striatum mitochondria, 3-NP; N27 cell lines, 3-NP	Free radical scavenging, antioxidant; lowers oxidative stress and prevents mitochondrial dysfunction	[31]
B. Monnieri ethanol extract			Lowers cytotoxicity and prevents mitochondrial dysfunction	[24]
Calendula officinalis L. flowers	Asteraceae	Rats, 3-NP	Anti-inflammatory, antioxidant, and estrogenic prevented significant behavioral alterations and neuronal loss in the striatum	[35]
Celastrus paniculatus Wild. standardized extract	Celastraceae	Rats,3-NP	Antioxidant, neuroprotective; ameliorates behavioral parameters	[38]
Centella asiatica L. aq. extract and leaf powder	Apiaceae	Brain mitochondria, 3-NP	Free radical scavenging, antioxidant; elevated levels of thiols and glutathione	[42]
Convolvulus pluricaulis Choisy hydromethanolic extract and its fractions: butanol, aqueous and ethyl acetate	Convolvulaceae	3-NP	Neuroprotective, antioxidant; improves memory, body weight, grip strength and locomotor activity	[45]
Cannabis sativa L. [Sativex((Δ(9)-THC) or CBD]	Cannabaceae	Malonate-lesioned rats, 3-NP	Antioxidant, neuroprotective; altered the increase of proinflammatory markers	[49]

(continued on next page)

Table 3.1: Anti-HD activity of crude/semipurified plant fractions/extract—cont'd

Source plant (parts/crude or semipurified extracts)	Family	In-vitro/in-vivo models	Mode of action	References
Garcinia kola Heckel aq. extract	Clusiaceae	Malnourished mice, 3-NP	Neuroprotective effect on cerebellum and hippocampus	[55]
Gastrodia elata Blume	Orchidaceae	PC12 cell, mHtt	Enhances proteasome activity, A(2A)-R, PKA; prevents mHtt accumulation	[61]
Ginkgo biloba L. (extract [EGb 761])	Ginkgoaceae	Rats, 3-NP	Antioxidant, antiapoptotic; increases energy metabolism, Bcl-xl in the striatum; downregulates GAPDH and Bax	[67]
Olea europaea L. (extra-virgin olive oil)	Oleaceae	Rats, 3-NP	Antioxidant; lowers lipid peroxidation and prevents GSH depletion	[70]
Panax ginseng C.A. Mey. (GTS)	Araliaceae	Rats, 3-NP	Downregulates behavioral impairments, inhibits intracellular Ca^{2+} upsurge and striatal neurodegeneration.	[73]
Psoralea corylifolia L. seed extracts	Fabaceae	PC12 cells, 3-NP	Induced mitochondrial respiration, prevented cell death and lowered ATP levels	[76]
Withania sominfera L. (Dunal) root extract	Solanaceae	Rats, 3-NP	Antioxidant, neuroprotective; prevents mitochondrial dysfunctions	[80]

Table 3.2: Anti-HD activity of plant derived natural compounds.

Compound	Source plant	In vitro/in vivo models	Mode of action	References
Alpha-Mangostin	*Garcinia mangostana* L. fruit (Clusiaceae)	CGNs, 3-NP	Neuroprotective, antioxidant; lowers ROS production	[86]
Astragalan	*Astragalus membranaceus* Moench(Fabaceae)	*C. elegans*, Mutant polyQ	Increases lifespan of adults and daf-2 and age-1; regulates DAF-16/FOXO; downregulates polyQ aggregation	[89]
Cannabidiol	*Cannabis sativa* L.(Cannabaceae)	Rats, 3-NP	Antioxidant; prevented neurodegeneration	[49]
Cannabigerol		R6/2 Mice,3-NP	Neuroprotective; reduced microgliosis, striatal mHtt and upsurge of proinflammmatory markers.	[94]
Celastrol	*Tripterygium wilfordii* Hook root bark (Celastraceae)	PC12 cells, 3-NP, mutant polyQ and HeLa	Modulates hsp gene; induces hsp70 within dopaminergic neurons; downregulates nuclear factor kB, tumor necrosis factor-α, striatal lesion volume, and astrogliosis	[96,97]
C-SLNs(Curcumin encapsulated solid lipid nanoparticles)	*Curcuma longa* L. root (Zingiberaceae)	Rats, 3-NP	Increases neuromotor coordination; prevents mitochondrial impairments	[101]
EGCG[(-)-epigallocatechingallate]	*Camellia sinensis* (L.) Green tea leaves (Theaceae)	HD flies overexpressing protein Httex1; Poly-Q mediated htt protein, HD yeast	Upregulates motor function; prevents mutant Httex1 protein accumulation and cytotoxicity; decreases photoreceptor degeneration	[106]
Fisetin	Various plants	*Drosophila* and PC12 cells expressing mutant Httex1; R6/2 mice	Upregulates ERK activation; decreases mHtt levels	[109]

(continued on next page)

Table 3.2: Anti-HD activity of plant derived natural compounds—cont'd

Compound	Source plant	In vitro/in vivo models	Mode of action	References
Galantamine	*Galanthus* sp. (Amaryllidaceae)	Rats, 3-NP	Neuroprotective and antiapoptotic; lessens striatal lesions; modulates nAChR	[112]
Ginsenosides (Rg5, Rc and Rb1)	*Panax ginseng* C.A. Meyer (Araliaceae)	MSN from YAC128 mouse	Prevents glutamate-induced Ca^{2+} responses	[115]
Hesperidin	Various citrus fruits	Rats, 3-NP	Anti-inflammatory and antioxidant; prevents alterations in PPI response and locomotor activity; increases striatal, hippocampal and cortical MDA levels	[67]
Kaempferol	Various plants	Rats, 3-NP	Prevents neurodegeneration, oxidative stress and striatal lesions; reduced mortality rates and motor deficits	[122]
Lutein	Various plants	Rats, 3-NP	Antioxidant; ameliorated behavioral alterations, mitochondrial enzymes complex activities and body weight	[124]
Lycopene	Tomatoes and other red vegetables and fruits	Rats, 3-NP	Modulates NO; improves memory, biochemical, cellular and behavioral alterations	[80,126]
Melatonin	Various plants	Rats,3-NP	Lowers oxidative stress	[129]
Naringin	Mainly citrus fruits	Rats,3-NP	Reduced oxidative stress and mitochondrial and behavioral impairments	[153]
Nicotine	*Nicotiana tabacum* L. (Solanaceae)	Rats,3-NP	Neuroprotective; prevented depletion of GSH and DA in the striatum	[133]
Onjisaponin B	Radix Polygalae (Polygalaceae)	PC12 cells, mHtt	Induce autophagy *via* AMPK-mTOR pathway; reduces toxicity	[135]
Quercetin	Various plants	Rats, 3-NP	Antioxidant; restores SOD, thiol, and catalase activities; prevents mitochondrial dysfunctions	[136]

(continued on next page)

Compound	Source	Model	Effect	Ref
Quercetin with fish oil		Rats, 3-NP	Antioxidant; reduces up-regulated AChE activities and mitochondrial dysfunctions	[205]
Resveratrol	Red grapes and others	PC12 cells and *Drosophila* expressing mutant Httex1; R6/2 mouse model	Activates ERK	[109]
S-allylcysteine	*Allium sativum* L. (Amaryllidaceae)	Rats, 3-NP	Antioxidant, neuroprotective; down-regulates lipid peroxidation and mitochondrial dysfunctions	[144]
(-)schisandrin B	*Schisandra chinensis* (Turcz.) Baill.	PC12 cells, 3-NP	Neuroprotective, antiapoptotic, and anti-necrotic	[148]
Sesamol	*Sesamum indicum* L. (Pedaliaceae)	Rats, 3-NP, H_2O_2-induced human SH-SY5Y cells	Antioxidant, upregulated free radical scavenging activity	[152,153]
Sulforaphane	Various vegetables	Mice, 3-NP	Reduces 3-NP induced striatal toxicity: activates the Keap1-Nrf2-ARE pathway and prevents the MAPKs and NF-kB pathways	[157]
Trehalose	Various plants	PC12 cells and COS-7 expressing mHtt	Enhances autophagy and the clearance of mHtt	[160]
Vanillin	*Vanilla planifolia* (Orchidaceae)	3-NP	Neuroprotective; enhanced learning and memory; reduced weight loss, neuronal, behavioral and biochemical changes	[162]
Zeatin riboside (a cytokinin)	Various plants	PC12 cells expressing mHtt	Modulates A(2A)-R signaling; prevents mHtt induced protein aggregations	[163]

Table 3.3: Anti-HD activity of herbal formulations.

Herbal formulations	Mode of action	References
B307	Decreases inflammation, apoptosis, and oxidative stress	[172]
CLMT	Declines HD symptoms	[166]
YGS	Declines HD symptoms	[166]

3.2.2 Calendula officinalis L. (Asteraceae)

Calendula officinalis, commonly known as marigold is commonly grown in warm temperate regions of the world. It is commonly used for treating nephrotixicity, jaundice, helps in healing wounds and purifying blood and can potentially cure cancer [32-34]. *C. officinalis* is known to be an antioxidant as well as an inhibitor for AChE [32]. Adult female Wistar rats were pretreated with *C. officinalis* flower extract (100 and 200 mg/kg for 7 days) with incorporation of 3-NP (15 mg/kg, i.p. for the following 7 days). The flower extract acted as an anti-inflammatory, anti-oxidant and estrogenic agent by preventing significant behavioral changes and neuronal loss in the striatum of the brains of adult female Wistar rats [35].

3.2.3 Celastrus paniculatus Wild. (Celastraceae)

Celastrus paniculatus Wild. (CP) has been used in Ayurveda since time immemorial. The Jyotishmati Oil, extracted from the seeds of this plant is known to be a useful nervine tonic, altering cognitive impairments and enhancing learning abilities and memory [36]. Apart from these, the plant shows other pharmacological properties like healing wounds and being a hypolipidemic, anti-arthritic, anti-nociceptive and anti-oxidant agent [37]. The effect of ethanol extract of macerated CP seeds (CPEE) against 3-NP induced HD-like symptoms in male Wistar rats was studied. Various extracts of CPEE were tested, but the aqueous fraction (AF) at 18mg/kg was the most successful in altering the 3-NP induced behavioral impairments due to its strong anti-oxidant property [38].

3.2.4 Centella asiatica (L.) Urb. (Apiaceae)

Centella asiatica commonly known as Gotu kola and native to South-East Asia, is now becoming prevalent in the West [39,40]. This herb is known for healing wounds and also for improving memory, cognition and treating anxiety [40]. *C. asiatica* acts as an anti-oxidant, prevents formation of amyloid plaque in AD and decreases neurotoxicity in PD [39]. Even though recent reports do not strongly suggest the usage of this herb for cognitive impairments, *C. asiatica* can be useful in anger management and improving alertness [41]. The aqueous extract and leaf powder of this plant blocked the 3-NP induced mitochondrial dysfunctions and oxidative stress in the brain regions of male mice and up-regulated the free radical scavenging activity, because of their anti-oxidant property [42].

3.2.5 Convolvulus pluricaulis *Choisy (Convulvulaceae)*

Convolvulus pluricaulis, commonly known as Shankhapushpi is a perennial herb known to be therapeutically significant. All the plant parts have been suggested to be pharmacologically important and has been used in Ayurvedic medicines against epilepsy, anxiety, sleeplessness, depression, amnesia, stress, hallucinations and many other CNS disorders. The active chemicals responsible for its biological significance are steroids, flavonoids, coumarins, alkaloids, glycosides, fatty acids, proteins and carbohydrates [43]. *C. pluricaulis* has been reported to possess neuroprotective, anti-apoptotic and anti-oxidant properties [44]. The effect of hydromethanolic extract (100 and 200 mg/kg) of this herb as well as its fractions such as aqueous (50 and 100 mg/kg), butanol (25 and 50 mg/kg) and ethyl acetate (15 and 30 mg/kg) against 3-NP challenge was studied. The extracts, supplied orally protected the rat brains against 3-NP induced HD-like symptoms and also improved memory, body weight, grip strength, and locomotor activity [45].

3.2.6 Cannabis sativa *L. (Cannabaceae)*

Cannabis sativa, a herbaceous species originating from Central Asia has been used in folk medicine since past times. Besides being rich in phytochemicals, this herb is also a rich source of woody and cellulosic fibers. The active phytochemicals are hemp phytochemicals including phenolic compounds, cannabinoids and terpenes [46]. The Cannabinoid (CB) system is known to cure various neurodegenerative disorders like multiple sclerosis, cerebral ischemia and AD. A closer look reveals that 2-arachidonoyl glycerol and anandamide, the endogenous CB molecules target CB_1 and CB_2, the G-protein coupled CB receptors. The phytocannabinoids isolated from *C. sativa* also follow similar mechanisms [47]. The most studied *C. sativa* extracts are delta-9-tetrahydrocannabinol (Δ9-THC) and cannabidiol (CBD) and they have been reported to be anti-inflammatory, neuroprotective and have immunomodulatory advantages [48]. Sativex, which consists of the abovementioned compounds in a 1:1 ratio, have been found to reverse the GABA deficiency, the down-regulated CB(1) and IGF-1 expression and Nissl-stained neuronal cells caused by the 3-NP induced HD-like symptoms in rat models and altered the increase of proinflammatory markers in malonate-lesioned rats [49].

3.2.7 Garcinia kola *Heckel (Clusiaceae)*

Garcinia kola, commonly known as bitter kola is also referred to as the "wonder plant" because nearly every part of this plant has been deemed to be of medical importance. This important tree species is found in tropical forests of Central and West Africa [50]. The phytochemicals that make *G. kola* pharmacologically important are cyanogenic glycosides, tannins, alkaloids, saponins and flavonoids. Bitter kola is known to possess anti-inflammatory, antiviral, antidiabetic, antihepatotoxic, and bronchodilator properties [51]. Garcinoic acid (GA), isolated from *G. kola* is responsible for the anti-inflammatory activities of this tree species [52]. When exposed to atrazine, kolaviron had a protective effect on human dopaminergic SH-SY5Y cells and PC12 cells [53,54]. Adult malnourished mice, when pretreated with 3-NP at 20 mg/kg b.w. followed by *G. kola* extracts at 200 mg/kg for 7 days showed an absence of neuronal cell degeneration [55].

3.2.8 Gastrodia elata *Blume (Orchidaceae)*

Gastrodia elata Blume (Orchidaceae) commonly known as Tian ma in China, is primarily distributed in countries of East Asia such as India, Japan, Korea and China. It has been prevalent in the Chinese herbal medicine since ancient times and is known to possess a number of biological activities such as hypnotic, sedative, anticonvulsive, antidepressant, antiepileptic, antianxiety, antivertigo, antipsychotic, anti-inflammatory, antioxidative, antivirus, antiaging, antitumor, modulating circulatory system, and memory improving effects [56]. 25 components have been isolated and identified from fresh and steamed *G. elata* [57]. Gastrodin and Gastrol B, the two phenolics isolated from the rhizome showed protective effects against H_2O_2-induced impairment in PC12 cells [58]. The plant is known to be effective against Alzheimer's disease, because it prevents amyloid beta-peptide induced toxicity [59] while it is also effective against Parkinson's disease by preventing apoptosis in dopaminergic neurons in vitro [60]. *G. elata* prevents mHtt accumulation by enhancing A(2A)-R, PKA and proteasome activity in PC12 cell lines, transiently transfected with mHtt [61].

3.2.9 Ginkgo biloba L. *(Ginkgoaceae)*

Ginkgo biloba, commonly known as maidenhair tree has been used to treat dementia, depression and age-related disorders in traditional Chinese medicine [62]. The leaf extract has been treated as a beneficial supplement since ages and it consists of numerous components including terpene lactones and flavanol glycosides [63]. This plant is also known to improve cognition and can also treat patients with Alzheimer's disease [64]. EGb 761, the *G. biloba* leaf extract 761 has been suggested to be effective against Parkinson's disease [65,66]. EGb 761 showed its anti-apoptotic and anti-oxidative properties against 3-NP-induced HD-like symptoms in rats by decreasing GADPH and Bax levels, while increasing the levels of Bcl-xl in the striatum. Hence, it worked as a neuroprotective agent [67].

3.2.10 Olea europaea L. *(Oleaceae)*

Olea europaea, the source of olive oil is pharmacologically important, mainly due to the varied amount of phytochemicals isolated from the plant: isochromans, iridoids, secoiridoids, flavonoids, biphenols, triterpenes, and benzoic acid derivatives. The varied amount of properties that makes olive oil medicinally important are antioxidant, antidiabetic, anti-inflammatory, anticonvulsant, antiviral, antimicrobial, antinociceptive, antihyperglycemic, anticancer, antihypertensive, gastroprotective, analgesic and wound healing effects [68]. The leaf extract is the source of the compound oleuropein which gives the olive fruits a unique taste. Oleuropein, a phenolic compound and its other hydrolyzed products function as antioxidants [69]. VOO (extra virgin olive oil) prevents 3-NP induced neurotoxicity in rat brains showing HD-like symptoms. It functions as an anti-oxidant by lowering lipid peroxidation and preventing GSH depletion [70].

3.2.11 Panax ginseng C.A. Mey. (Araliaceae)

Panax ginseng, commonly found in China and Korea, is a perennial herb [71]. It is also known as Asian or Korean ginseng and is known to have a lot of beneficial properties such as antioxidant, anti-inflammatory and anticancer effects. The primary active phytochemicals are ginsenosides [72]. The Korean red Ginseng saponins inhibited the neuronal degeneration in the striatum of rat brains induced by 3-NP. The saponins also down-regulated the intracellular Ca^{2+} levels and behavioral alterations, and also increased the survival rate in model animals [73].

3.2.12 Psoralea corylifolia L. (Fabaceae)

Psoralea corylifolia, commonly known as babchi, have been used in traditional medicines since time immemorial. It is a popular herb and has anti-oxidant, anti-inflammatory, antimicrobial and chemoprotective properties [74]. Its bioactive compounds, mainly isolated from fruits and seeds belongs to meroterpenes, coumarins and flavonoids groups [75]. The seed extracts of this herb were administered against 3-NP induced mitochondrial dysfunctions in cultured PC12 (pheochromocytoma) cells. It had a commendable protective property, by stimulating mitochondrial respiration, lowering the mitochondrial MP (membrane potential), ATP levels and preventing 3-NP induced cell death [76].

3.2.13 Withania somnifera L. (Dunal) (Solanaceae)

Withania somnifera, commonly known as Ashwagandha, is considered as the Indian ginseng. It can be used medicinally, to treat various CNS disorders, specifically Alzheimer's disease, Parkinson's disease, cerebral ischemia, epilepsy, tardive dyskinesia, stress and even for managing drug addiction [77]. This plant possesses various significant pharmacological properties such as anti-oxidant, anti-inflammatory, antistress, antitumor, hemepoetic, immunomodulatory and rejuvenating effects. Toxicity studies reveal that Ashwagandha is safe for use [78]. Ashwagandha consists of a group of bioactive compounds known as withanolides [79]. The root extract of *W. somnifera* acted as a neuroprotective agent against the 3-NP challenge in rat brains by behaving as an antioxidant and preventing mitochondrial dysfunctions and behavioral alterations [80].

3.3 Anti-HD activity of plant derived natural compounds

Biologically active plant compounds can be effective against HD, which are depicted in this section on the basis of various in vitro and/or in vivo studies.

3.3.1 α-Mangostin

Mangosteen or *Garcinia mangostana* L., the source plant for α-mangostin is a tree 6–25 meters long and is thought to have originated from Southeast Asia [81]. The pericarp of the fruit consists of mangosteen xanthones which are actually a family of tricyclic isoprenylated polyphenols, known to show various immunomodulatory, anticancer, antiacne and anti-inflammatory activities in in vitro and in vivo model systems through varied mechanisms of action [82–85]. α-Mangostin ameliorated neuronal death and production of ROS (reactive oxygen species) in a concentration-dependent manner against 3-NP induced primary cultures of CGNs (cerebellar granule neurons) [86].

3.3.2 Astragalan

Astragalus membranaceus, the primary source of Astragalan has been widely used in Chinese traditional medicine since time immemorial. It is used as antioxidant, antidiabetic, anticancer, antitumor, anti-atherosclerosis, immune stimulant, hepatoprotectant, tonic, diuretic, and expectorant [87,88]. It downregulated abnormal polyglutamine (polyQ) aggregation in *C. elegans* via regulating the DAF-16/FOXO (abnormal dauer formation/16-forkhead box O) transcription factor [89].

3.3.3 Cannabidiol

Cannabidiol (CBD), a major non-psychoactive constituent of glandular hairs of *Cannabis sativa* (Cannabaceae) is known to be pharmacologically beneficial [90]. Some of the significant pharmacological properties that it possesses are antipsychotic, anti-inflammatory, anticonvulsive, neuroprotective, hypnotic and sedative [91]. CBD is an extremely promising therapeutic agent because it lacks any unwanted psychotropic effect [92]. CBD acted as a neuroprotective and antioxidant, preventing atrophy on the striatal neurons that project to the substantia nigra against 3-NP induced rat models exhibiting HD-like symptoms, functioning independent of adenosine signaling and transient receptor potential cation channel subfamily V member(TRPV1) receptors [49].

3.3.4 Cannabigerol

Cannabigerol (CBG), a chemical phenotype in *Cannabis sativa* L. (Cannabaceae), is the direct precursor of CBD [93]. CBG reduced microgliosis and the upregulated proinflammatory markers by 3-NP induced in vivo models of HD. It also acted as a neuroprotective agent which normalized the series of genes linked to HD and the expressions of certain genes like BDNF, IGF-1, and PPAR-gamma in R6/2 mice [94].

3.3.5 Celastrol

Celastrol, extracted from the roots of *Tripterygium wilfordi* Hook. f. (Celastraceae) also known as "Thunder of God Vine" has been used since ancient times as an anti-inflammatory agent [64]. It also works as an anticancer, antioxidant and antiapoptotic agent [95,96]. Celastrol modulates the hsp (heat shock protein) gene to enhance the the viability of cells expressing polyQ protein [97]. It also reduces nuclear factor kB, tumor necrosis factor-α, striatal lesion volume and astrogliosis [96].

3.3.6 Curcumin

Curcumin responsible for the yellow color of turmeric, was derived from the rhizome of the plant *Curcuma longa* (Zingiberaceae) and has been used in Ayurveda since 1900 BC for pains, wounds, sprains, aches, gastrointestinal, pulmonary, skin, and liver disorders [98]. It is also known to possess anti-inflammatory, antioxidant and anticancer properties [99]. Curcumin is also known to have therapeutic effects against severe neurological disorders such as Alzheimer's disease and Parkinson's disease [100]. C-SLNs (Curcumin encapsulated solid lipid particles) enhanced the activities of cytochrome levels and mitochondrial complexes, restored SOD activity and glutathione levels and decreased lipid peroxidation, mitochondrial dysfunctions, ROS and protein carbonyls in 3-NP induced rat models of HD [101].

3.3.7 (-)-epigallocatechin-gallate

Epigallocatechin gallate (EGCG), a plant flavonoid of green tea (*Camellia sinensis* L. Kuntze [Theaceae]) is known to possess anticancer, antitumor and antiapoptotic properties [102–105]. Due to its protective effects against ßA-induced neuronal apoptosis, EGCG can also be beneficial against Alzheimer's [105]. It can dose-dependently down-regulate polyQ-mediated htt protein accumulation, prevent mutant htt exon 1 (Httex1) accumulation and enhance motor functions in models showing HD-like symptoms [106].

3.3.8 Fisetin

Fisetin, a natural flavonoid (3,3',4',7-tetrahydroxyflavone) found in various plants is known to be pharmacologically beneficial [107]. It is known to possess anticancer, antioxidant and anti-inflammatory properties [107,108]. Fisetin reduces the impairments caused by mHtt aggregation in all the HD models (PC12 cells and *Drosophila* expressing mutant Httex1 and R6/2 mouse model) probably by ERK activation [109].

3.3.9 Galantamine

Galantamine, extracted from *Galanthus sp.* (Amaryllidaceae) is an alkaloid that is an AChE inhibitor [110]. It is also a potential agent against Alzheimer's disease [111]. Galantamine acted as an anti-apoptotic agent, reduced striatal lesion volumes and down-regulated neurode-generation by modulating nAChR in Lewis rats induced by 3-NP [112].

3.3.10 Ginsenosides

Ginsenosides, a special group of triterpenoid saponins extracted from *Panax ginseng* (Araliaceae) can be divied into two types based on their aglycones skeleton (dammarene and oleanane type). They can be potential anti-inflammatory, antihypertensive, anti-allergic, antidiabetic, antistress, antiatherosclerotic, and immunomodulatory agents and are active against various CNS disorders like PD < AD < depression, cerebral ischemia, and others [113,114]. Ginsenosides Rg5, Rb1, and Rc extracted from the roots of *P. ginseng* prevented glutamate-induced Ca^{2+} responses in cultures of MSNs (medium spiny striatal neuronal cultures) from the YAC128 mouse model showing HD-like symptoms [115].

3.3.11 Hesperidin

Hesperidin, a biflavonoid extracted mainly from the rinds of citrus fruits possesses antihyper-lipidemic, anti-inflammatory, antihypertensive, anticancer, antifertility, antioxidant, analgesic, diuretic, and Ca^{2+} channel blocker properties [116–118]. It can also be a potential therapeutic agent against various CNS disorders like AD, PD, and epilepsy [119]. Hesperidin performed as an antioxidant and anti-inflammatory agent by increasing cortical, hippocampal, and striatal MDA (malondialdehyde) levels and preventing changes in locomotor activity and PPI response against 3-NP induced rat models of HD [67].

3.3.12 Kaempferol

Kaempferol(3,5,7-trihydroxy-2-(4-hydroxyphenol)-4H-1-benzopyran-4-one) found in various fruits and vegetables is a flavonoid known to have anti-inflammatory, antioxidant, anticancer, antimicrobial, antiosteoporotic, antidiabetic, neuroprotective, estrogenic, cardioprotective, analgesic, and anxiolytic activities [120,121]. It acts as a neuroprotective and antioxidant, prevents striatal lesions, mortality rates, and motor deficits [122].

3.3.13 Lutein

Lutein, generally found in various dark-green vegetables is a xanthophyllous carotenoid that has a number of pharmacological properties such as anti-inflammatory, antioxidant, anti-cancer, immunomodulatory, cardioprotective, and oculoprotective effects [123]. Administration of lutein daily in 3-NP induced rats ameliorated neurobehavioral alterations, mitochondrial enzymes complex activities, body weight, and reduced oxidative stress [124].

3.3.14 Lycopene

Lycopene, extracted from *Solanum lycopersicum* L. (Solanaceae) is known to be a very potent antioxidant and has therapeutic effects against cardiovascular diseases and cancer [125]. Lycopene treated 3-NP rats showed restored levels of glutathione system and improved memory and cognitive functions, and also improved biochemical, cellular and behavioral alterations by modulating NO pathways. It further acted as a neuroprotective agent [80,126].

3.3.15 Melatonin

Melatonin, extracted from various plants is known to have antioxidant, oncostatic, bone formation and protection, detoxification of free radicals, cardioprotective, neuroprotective and anti-psychotic properties [127]. The levels of melatonin is found to be disturbed in various neurological disorders like AD and PD, which suggests its involvement in the pathophysiology of these neurological disorders [128]. It acted as a neuroprotective and antioxidant agent by preventing the alterations induced by 3-NP in rats [129].

3.3.16 Naringin

Naringin (4',5,7-trihydroxyflavanone-7-rhamnoglucoside), mainly obtained from citrus fruits has been suggested to possess numerous biologically important properties such as antiapoptotic, antioxidant and anti-inflammatory effects [130]. It acted as a neuroprotective agent by preventing mitochondrial impairments and behavioral changes in 3-NP induced rats, via modulating NO pathways [131].

3.3.17 Nicotine

Nicotine extracted from *Nicotiana tabacum* and other plants belonging to the Solanaceae family has been proved to be therapeutic for the treatment of AD, PD, Tourette's syndrome, attention deficit disorder, sleep apnea and ulcerative colitis [132]. It protects the neuronal cells from the effects of 3-NP and prevents depletion of glutathione (GSH) and dopamine (DA) levels in rat models of HD [133].

3.3.18 Onjisaponin B

Onjisaponin B is extracted from the ethanol extracts of Radix Polygalae (Polygalaceae). It is effective against AD since it suppresses Aß production without blocking the ß-site of amyloid precursor protein cleaving enzyme1 (BACE1) and gamma-secretase activities [134]. Onjisaponin B induced autophagy via the AMPK-mTOR pathway and enhanced the removal of mHtt in PC12 cells expressing mHtt [135].

3.3.19 Quercetin

Quercetin, a polyphenol extracted from various plants and vegetables is known to possess a wide range of biological properties such as antiviral, anti-inflammatory and anti-carcinogenic effects [134]. Administration of Quercetin (25mg/kg b.w. orally for 21 days) in 3-NP induced animals restored SOD and catalase activities, ATP levels, reduced mitochondrial dysfunctions and motor deficits [136].

3.3.20 Resveratrol

Resveratrol (3,4',5-trihydroxy-trans-stilbene), a common phytoalexin is found in many edible plant materials like peanuts, red wine and grape skins. It acts as an antioxidant, anticancer, increases high-density lipoprotein cholesterol, enhances NO production, and inhibits platelet accumulation and a cardioprotective agent [137]. It is known to be anti-apoptotic and to protect neurons from oxidative toxicity and damage, and hence can be used against neurological disorders like Parkinson's disease (PD) and Alzheimer's Disease (AD) [138]. Administration of Resveratrol in PC12 and *Drosophila* expressing mHtt and R6/2 mouse model of HD stimulates the activation of ERKm [109].

3.3.21 S-allylcysteine

S-allylcysteine is an amino acid derived from garlic (*Allium sativum* L., Amaryllidaceae) [139]. It is an antioxidant, anti-cancer, anti-diabetic, neuroprotective, cardioprotective and anti-atherosclerotic agent [140–143]. S-allylcysteine acts as a neuroprotective and antioxidant, prevents mitochondrial dysfunctions and down-regulates lipid peroxidation against 3-NP induced rats [144,145].

3.3.22 (-)schisandrin B

(-)schisandrin B, an enantiomer of schisandrin B is present in *Schisandra chinensis* (Turcz.) Baill. (Schisandraceae) and is known to have therapeutic effects against vascular diseases [146]. It is also known to have therapeutic effects against neurological disorders due to its anti-inflammatory, antioxidant and anti-apoptotic properties [147]. (-) schisandrin B acted as an anti-apoptotic and anti-necrotic agent against 3-NP induced PC12 cells in rats, probably by blocking activated C-jun N-terminal kinase (JNK)- mediated pyruvate dehydrogenase(PDH) inhibition [148].

3.3.23 Sesamol

Sesamol extracted from *Sesamum indicum* L. (Pedaliaceae) has a wide range of biological properties like anti-diabetic, anti-oxidant and antimutagenic effects [149–151]. It can be effective against AD, PD and HD [152]. Pretreatment of sesamol (5, 10, and 20 mg/kg) upregulated free-radical scavenging activity in 3-NP induced rats [153]. SH-SY5Y cells pretreated with 1 μM Sesamol attenuated cell death caused by 400 μM H_2O_2. It also activated SIRT1-SIRT3-FOXO3a expression, inhibited pro-apoptotic protein BAX and enhanced anti-apoptotic protein, BCL-2 [152].

3.3.24 Sulforaphane

Sulforaphane, extracted from various vegetables, especially broccoli is known to be anti-metastatic and anti-apoptotic [154,155]. It was effective against HD, by preventing mHtt aggregation and cytotoxicity and up-regulated proteasome and autophagy activities in HD cell models [156]. It also acts as a neuroprotective agent against 3-NP induced mouse model of HD by attenuating apoptosis, mRNA or protein expression of inflammatory mediators, neuronal death, microglial activation and by reversing other alterations in the striatum of mice brain [157].

3.3.25 Trehalose

Trehalose is a naturally occurring sugar consisting of two D-glucose units linked together by a α,α-1,1 bond and unlike many sugars, it makes itself a popular constituent of health, food, beauty and pharmaceutical products [158]. Trehalose modulates autophagy in numerous diseases like cancer, neurodegenerative disorders, aging, metabolic disorders and other infectious diseases [159]. It also enhanced autophagy against mHtt aggregation in PC12 cells and COS-7 expressing mHtt [160].

3.3.26 Vanillin

Vanillin, naturally extracted from *Vanilla planifolia* (Orchidaceae) is a phenolic aldehyde (4-hydroxy-3-methoxybenzaldehyde) [161]. It acts as a therapeutic agent against HD by modulating the vanilloid receptor [162].

3.3.27 Zeatin riboside

Zeatin riboside, a cytokinin has therapeutic potential against several neurodegenerative disorders and can also be used as a mammalian immunomodulatory agent [163]. It modulates A(2A)-R signaling and reverses mHtt-induced protein aggregations in PC12 cells expressing mHtt [163].

3.4 Anti-HD activity of herbal formulations

Herbal formulations are advantageous over single herbs because they can regulate multiple targets [156]. Various Chinese herbal medicines possess a wide variety of pharmacological properties and can have therapeutic effects against various human diseases [164]. Herbal formulations like B307, YGS (Yi-Gan San), and CLMT (Chaihu-Jia-Longgu-Muli Tan) are known to be effective against HD [165,166]. YGS can effectively treat AD (by attenuating beta amyloid-induced cytotoxicity), schizophrenia, cognitive impairments, dementia and various other neurological disorders [167–169]. Saponins isolated from CLMT were effective against chronic mild-stress apoptosis and depression in vivo [170,171]. Oral incorporation of B307 prevents cardiac failure and is effective against HD by reducing inflammation, oxidative stress and apoptosis [172]. Even though herbal formulations are popular and are accepted widely in clinical practices, an interactive approach between medical practitioners and patients and evidence-based practices should be taken into account to deal with the treatment of uncertain phytochemical compositions against various mental disorders [173].

3.5 Discussion

Toxin models of HD can be induced by 3-nitropropionic acid (3-NP) or malonate as they both inhibit succinate dehydrogenase, complex II of the mitochondrial respiratory chain [174]. Hence, various experiments have been performed on 3-NP induced models of HD to understand the therapeutic effects of various aforementioned phytochemicals.

A primary approach to therapeutically cure HD can be by the induction of autophagy, which eliminates the toxic or aggregated mHtt proteins by delivering them to lysosomes for degradation. It is suggested that improper engulfment of cytosolic components by autophagosomes is responsible for their slower turnover, accumulation and functional decay inside HD cells [175,176]. Rapamycin induces autophagy via the mTOR inhibitor, thereby clearing mHtt fragments in HD models [177]. Among the aforementioned anti-HD phytochemicals, trehalose and onjisaponin B can induce autophagy against mHtt, onjisaponin B exhibiting autophagy via the AMPK-mTOR signaling pathway [135,160]. Fisetin decreases mHtt levels [109]. Another possible approach can be by reducing oxidative stress since oxidative damage and metabolic dysfunction are responsible for the pathogenesis of HD [178]. Oxidative damage to proteins, DNA and lipids may contribute to the development of HD characteristic CAG repeat [179]. Modulation of Nrf-2/ARE pathway, or up-regulation of PGC-1α expression increases the activity of the anti-oxidant enzymes [180]. CBD, curcumin, hesperidin, α-mangostin, lutein, kaempferol, resveratrol, melatonin and sesamol have anti-oxidant and free radical scavenging properties. Disruption of PPI (prepulse inhibition) is prevalent in HD [181]. Mitochondria, playing a role of "sensors" of the neurochemical environment, plays a signifiant role in the degeneration of striatum in HD [182]. Several natural compounds like hesperidin, S-allylcysteine, quercetin, EGCG, curcumin, naringin and lutein alters mitochondrial dysfunctions. It was suggested in another study that the usage of advanced spectroscopy, magnetic resonance technologies and non-invasive magnetic-resonance imaging (MRI) are useful in assessing 3-NP induced cerebral alterations [163]. Various plants are known to be natural sources of AChE inhibitors which makes them suitable for treating various neurological disorders, including HD due its effectiveness against dementia and cognitive impairment [183–185].

Plants and their natural products have been used in curing a wide range of neurological disorders including AD and PD [185]. Zeaxanthin, lycopene and lutein have been suggested to be therapeutic against AD [186,187]. Curcumin, resveratrol, S-allylcysteine, EGCG and kaempferol has been suggested to be effective against PD [20]. The aforementioned anti-HD herbs and herbal compounds are also therapeutic against a wide range of other neurological disorders. Crude plant extracts of *W. somnifera* (reported against anxiety, coginitve and various neurological disorders) [188], *P. ginseng* (neuroprotective) [189], *G. elata* (reported against CNS disorders: AD, PD, epilepsy) [190], *G. biloba* (reported against AD and various CNS disorders) [191], *C. asiatica* (reported for managing anxiety and cognitive impairments) [192] and *B. monnieri* (reported for treating various neurological disorders) [193] and plant compounds like trehalose (reported against PD) [194], celastrol (reported against AD, PD and ischemia) [195], naringin (against various neurological disorders) [196], and fistein (neurotrophic effects) [109] are some examples.

New approaches in drug delivery consider the bioavailability of phytochemicals, so that they get delivered to the target tissue [197]. A few examples are: α-Mangostin had low bioavailability with negligible oral absorption. A soft capsule with vegetable oil as the dispersion matrix can enhance the bioavailability of mangostin [198]. SEDDS (self-emulsifying drug delivery system) based on VESIsorb® formulation technology incorporating cannabidiol made the bioavailability of CBD easier [199]. Nanoemulsion formulation increased the bioavailability of fistein. However, this when incorporated intraperitoneally, shows a 24-fold increase [107]. Piperine increases bioavailability of EGCG in mice [200]. For increasing the bioavailability of ginsenosides, various micro-/nano-sized delivery systems such as polymeric particles, emulsions and vesicular systems can be used [201]. Lycopene loaded whey protein isolate nanoparticles(LYC-WPI-NPs) enhances the bioavailability of lycopene [202] and encapsulation of sesamol in phosphatidyl choline micelles increases the bioavailability of sesamol [203]. A novel water-soluble ternary nanoparticle of α-linoleic acid, ß-lactoglobulin and amylase which encapsulates naringin was developed to enhance the bioavailability of naringin [204]. Since a number of factors, including these govern the bioavailability and absorption of bioactive phytoconstituents, further research needs to be done in order to enhance the bioavailability of the anti-HD compounds, thereby increasing their efficacy against Huntington's disease.

3.6 Conclusion

We tried providing an insight into the current research trends of certain botanical derived products that possess anti-HD properties, in this review article. Apart from HD, these botanical products have therapeutic effects against various neurological disorders. A single compound acting on one or a few targets cannot cure HD because of the complex nature of the disorder. Instead, herbal formulations and extracts, containing more than one compound can be more efficient in acting on multiple targets. However, that being said, we need to keep in mind that there are a number of factors (some of them namely: poor bioactivity, biotransformation, bioavailability, poor quality, metabolism and dose dependent relationships of natural compounds) that are responsible for the decelerating research and drug development from botanical sources. Hence, detailed research should be done to explore and evaluate the therapeutic efficacy of the already discovered plant compounds and to discover novel plant-based bioactive components. Dosage, mechanism of action as well as toxicological studies of these phytochemicals should also be considered.

References

[1]   SE Browne, MF Beal, Oxidative damage in Huntington's disease pathogenesis, Antioxid Redox Signal 8 (2006) 2061–2073.
[2]   FO Walker, Huntington's disease, Lancet North Am Ed 369 (2007) 218–228.
[3]   JF Gusella, NS Wexler, PM Conneally, SL Naylor, MA Anderson, RE Tanzi, AB Young, A polymorphic DNA marker genetically linked to Huntington's disease, Nature 306 (1983) 234–238.
[4]   CA Ross, SJ Tabrizi, Huntington's disease: from molecular pathogenesis to clinical treatment, Lancet Neurol 10 (2011) 83–98.
[5]   C Zuccato, A Ciammola, D Rigamonti, B.R Leavitt, D Goffredo, L Conti, T Timmusk, Loss of huntingtin-mediated BDNF gene transcription in Huntington's disease, Science 293 (2001) 493–498.
[6]   JG Arabit, R Elhaj, SE Schriner, EA Sevrioukov, M Jafari, Rhodiola rosea improves lifespan, locomotion, and neurodegeneration in a Drosophila melanogaster model of Huntington's disease, Biomed Res Int (2018) 6726874.
[7]   RA Roos, Huntington's disease: a clinical review, Orphanet J Rare Dis 5 (2010) 40.
[8]   RG Snell, J.C MacMillan, J.P Cheadle, I Fenton, L.P Lazarou, P Davies, D.J Shaw, Relationship between trinucleotide repeat expansion and phenotypic variation in Huntington's disease, Nat Genet 4 (1993) 393–397.
[9]   J.S Paulsen, R.E Ready, J.M Hamilton, M.S Mega, J.L Cummings, Neuropsychiatric aspects of Huntington's disease, J Neurol Neurosurg Psychiatry 71 (2001) 310–314.

[10]   A Nayak, R Ansar, SK Verma, DM Bonifati, U Kishore, Huntington's disease: an immune perspective, Neurol Res Int 2011 (2011) 563784.

[11]   S Setter, J Neumiller, E Dobbins, L Wood, J Clark, C DuVall, A Santiago, Treatment of chorea associated with Huntington's disease: focus on tetrabenazine, Consult Pharmacis 24 (2009) 524–537.

[12]   S Frank, Treatment of Huntington's disease, Neurotherapeutics 11 (2014) 153–160.

[13]   A Killoran, KM Biglan, Current therapeutic options for Huntington's disease: good clinical practice versus evidence-based approaches? Mov Disord 29 (2014) 1404–1413.

[14]   P McColgan, SJ Tabrizi, Huntington's disease: a clinical review, Eur J Neurol 25 (2018) 24–34.

[15]   MD Rawlins, NS Wexler, AR Wexler, SJ Tabrizi, I Douglas, SJ Evans, L Smeeth, The prevalence of Huntington's disease, Neuroepidemiology 46 (2016) 144–153.

[16]   M Halpin, Accounts of suicidality in the Huntington disease community, Omega (Westport) 65 (2012) 317–334.

[17]   I Túnez, I Tasset, A Santamaría, 3-Nitropropionic acid as a tool to study the mechanisms involved in Huntington's disease: past, present and future, Molecules 15 (2010) 878–916.

[18]   XJ Li, S Li, Large animal models of Huntington's disease, in: H.H.P Nguyen, M Angela Cenci (Eds.), Behavioral Neurobiology of Huntington's Disease and Parkinson's Disease, Springer, Berlin, Heidelberg, 2013, pp. 149–160.

[19]   World Health Organization,Global status report on noncommunicable diseases 2014" (No. WHO/NMH/NVI/15.1), World Health Organization, 2014.

[20]   A Dey, JN De, Ethnomedicinal plants used by the tribals of Purulia district, West Bengal, India against gastrointestinal disorders. J Ethnopharmacol 143 (2012a) 68–80.

[21]   A Dey, JN De, Anti–snake venom botanicals used by the ethnic groups of Purulia District, West Bengal, India. J Herbs Spices Med Plants 18 (2012b) 152–165.

[22]   S Manoharan, GJ Guillemin, RS Abiramasundari, MM Essa, M Akbar, MD Akbar, The role of reactive oxygen species in the pathogenesis of Alzheimer's disease, Parkinson's disease, and Huntington's disease: a mini review, Oxid Med Cell Longev 2016 (2016) 8590578.

[23]   SC Man, SSK Durairajan, WF Kum, JH Lu, JD Huang, CF Cheng, M Li, Systematic review on the efficacy and safety of herbal medicines for Alzheimer's disease, J Alzheimers Dis 14 (2008) 209–223.

[24]   GK, Shinomol, and MS, Bharath, Neuromodulatory propensity of *Bacopa monnieri* leaf extract against 3-nitropropionic acid-induced oxidative stress: in vitro and in vivo evidences. Neurotox Res 22(2012a), 102–114.

[25]   N Uabundit, J Wattanathorn, S Mucimapura, K Ingkaninan, Cognitive enhancement and neuroprotective effects of *Bacopa monnieri* in Alzheimer's disease model, J Ethnopharmacol 127 (2010) 26–31.

[26]   C Jain, A Gupta, A Tewari, V Sharma, V Kumar, A Mathur, S Sharma, Molecular docking studies of bacoside from *Bacopa monnieri* with LRRK2 receptor, Biologia (Bratisl) 68 (2013) 1068–1071.

[27]   T Sandhya, J Sowjanya, B Veeresh, *Bacopa monniera* (L.) Wettst ameliorates behavioral alterations and oxidative markers in sodium valproate induced autism in rats, Neurochem Res 37 (2012) 1121–1131.

[28]   K, Das, Role of triterpenoid glycosides on axonal protection in multiple sclerosis and experimental autoimmune encephalomyelitis (2019). Phd Theses.

[29]   K Rai, N Gupta, L Dharamdasani, P Nair, P Bodhankar, *Bacopa Monnieri*: a wonder drug changing fortune of people, Int J Appl Sci Biotechnol 5 (2017) 127–132.

[30]   S Jose, S Sowmya, TA Cinu, NA Aleykutty, S Thomas, EB Souto, Surface modified PLGA nanoparticles for brain targeting of Bacoside-A, Eur J Pharm Sci 63 (2014) 29–35.

[31]   Shinomol, GK, and Bharath, MS (2012b). Pretreatment with *Bacopa monnieri* extract offsets 3-nitropropionic acid induced mitochondrial oxidative stress and dysfunctions in the striatum of prepubertal mouse brain. Can J Physiol Pharmacol 90, 595-606.

[32]   IG Mekinić, F Burčul, I Blaževieć, D Skroza, D Kerum, V Katalinić, Antioxidative/acetylcholinesterase inhibitory activity of some Asteraceae plants, Nat Prod Commun 8 (2013) 471–474.

[33]   PK Verma, R Raina, M Sultana, M Singh, P Kumar, Total antioxidant and oxidant status of plasma and renal tissue of cisplatin-induced nephrotoxic rats: protection by floral extracts of *Calendula officinalis* Linn, Ren Fail 38 (2016) 142–150.

[34] D Cruceriu, O Balacescu, E Rakosy, *Calendula officinalis*: potential roles in cancer treatment and palliative care, Integr Cancer Ther 17 (2018) 1068–1078.

[35] BD Shivasharan, P Nagakannan, BS Thippeswamy, VP Veerapur, P Bansal, MK Unnikrishnan, Protective effect of *Calendula officinalis* Linn. flowers against 3-nitropropionic acid induced experimental Huntington's disease in rats, Drug Chem Toxicol 36 (2013) 466–473.

[36] KA Deodhar, NW Shinde, Comparative study of pet ether extracts of all parts of *C. paniculatus* Wild, J Med Plant Res 3 (2015) 141–143.

[37] YA Kulkarni, S Agarwal, MS Garud, Effect of Jyotishmati (*Celastrus paniculatus*) seeds in animal models of pain and inflammation, J Ayurveda Integr Med 6 (2015) 82–88.

[38] J Malik, M Karan, R Dogra, Ameliorating effect of *Celastrus paniculatus* standardized extract and its fractions on 3-nitropropionic acid induced neuronal damage in rats: possible antioxidant mechanism, Pharmaceut Biol 55 (2017) 980–990.

[39] IE Orhan, Centella asiatica (L.) Urban: from traditional medicine to modern medicine with neuroprotective potential, Evid Based Complement Altern Med 2012 (2012) 946259.

[40] KJ Gohil, JA Patel, AK Gajjar, Pharmacological review on *Centella asiatica*: a potential herbal cure-all, Indian J Pharm Sci 72 (2010) 546–556.

[41] P Puttarak, P Dilokthornsakul, S Saokaew, T Dhippayom, C Kongkaew, R Sruamsiri, N Chaiyakunapruk, Effects of *Centella asiatica* (L.) Urb. on cognitive function and mood related outcomes: a systematic review and meta-analysis, Sci Rep 7 (2017) 1–12.

[42] GK Shinomol, Prophylactic neuroprotective property of *Centella asiatica* against 3-nitropropionic acid induced oxidative stress and mitochondrial dysfunctions in brain regions of prepubertal mice, Neurotoxicology 29 (2008) 948–957.

[43] P Agarwa, B Sharma, A Fatima, SK Jain, An update on Ayurvedic herb *Convolvulus pluricaulis* Choisy, Asian Pac J Trop Biomed 4 (2014) 245–252.

[44] P Rachitha, K Krupashree, GV Jayashree, HK Kandikattu, N Amruta, N Gopalan, F Khanum, Chemical composition, antioxidant potential, macromolecule damage and neuroprotective activity of *Convolvulus pluricaulis*, J Tradit Complement Med 8 (2018) 483–496.

[45] J Malik, S Choudhary, P Kumar, Protective effect of *Convolvulus pluricaulis* standardized extract and its fractions against 3-nitropropionic acid-induced neurotoxicity in rats, Pharm Biol 53 (2015) 1448–1457.

[46] CM Andre, JF Hausman, G Guerriero, *Cannabis sativa*: the plant of the thousand and one molecules, Front Plant Sci 7 (2016) 19.

[47] A Gowran, J Noonan, VA Campbell, The multiplicity of action of cannabinoids: implications for treating neurodegeneration, CNS Neurosci Ther 17 (2011) 637–644.

[48] J Maroon, J Bost, Review of the neurological benefits of phytocannabinoids, Surg Neurol Int 9 (2018) 91.

[49] O Sagredo, MR Pazos, V Satta, JA Ramos, RG Pertwee, J Fernández-Ruiz, Neuroprotective effects of phytocannabinoid-based medicines in experimental models of Huntington's disease, J Neurosci 89 (2011) 1509–1518.

[50] JC Onyekwelu, B Stimm, Garcinia kola, in: B Stimm, A Roloff, UM Lang, H Weisgerber (Eds.), Enzyklopädie der Holzgewächse: Handbuch and Atlas der Dendrologie, WILEY-VCH, Rudolf-Diestel-Str. 3, 86899 Landsberg, 2019, pp. 1–16.

[51] A Moneim, E Sulieman, *Garcinia Kola* (Bitter Kola): Chemical Composition, in: AA Mariod (Ed.), Wild Fruits: Composition, Nutritional Value and Products editor, Springer, Cham, 2019, pp. 285–299.

[52] M Wallert, J Bauer, J Kluge, L Schmölz, YC Chen, M Ziegler, H Pein, The vitamin E derivative garcinoic acid from *Garcinia kola* nut seeds attenuates the inflammatory response, Redox Biol 24 (2019) 101166.

[53] SO, Abarikwu, EO, Farombi, and AB, Pant, Biflavanone-kolaviron protects human dopaminergic SH-SY5Y cells against atrazine induced toxic insult. Toxicol in Vitro 25(2011a), 848–858.

[54] SO, Abarikwu, EO, Farombi, MP, Kashyap, and AB, Pant, Kolaviron protects apoptotic cell death in PC12 cells exposed to atrazine. Free Radic Res 45(2011b), 1061–1073.

[55] SA Ajayi, DA Ofusori, GB Ojo, OA Ayoka, TA Abayomi, AA Tijani, The microstructural effects of aqueous extract of *Garcinia kola* (Linn) on the hippocampus and cerebellum of malnourished mice, Asian Pac J Trop Biomed 1 (2011) 261–265.

[56] HD Zhan, HY Zhou, YP Sui, XL Du, WH Wang, L Dai, TL Jiang, The rhizome of *Gastrodia elata* Blume–an ethnopharmacological review, J Ethnopharmacol 189 (2016) 361–385.

[57] Y Li, XQ Liu, SS Liu, DH Liu, X Wang, ZM Wang, Transformation mechanisms of chemical ingredients in steaming process of *Gastrodia elata* Blume, Molecules 24 (2019) 3159.

[58] ZC Zhang, G Su, J Li, H Wu, XD Xie, Two new neuroprotective phenolic compounds from *Gastrodia elata*, J Asian Nat Prod Res 15 (2013) 619–623.

[59] GB Huang, T Zhao, SS Muna, HM Jin, JI Park, KS Jo, EO Park, Therapeutic potential of *Gastrodia elata* Blume for the treatment of Alzheimer's disease, Neural Regen Res 8 (2013) 1061.

[60] H Kumar, IS Kim, SV More, BW Kim, YY Bahk, DK Choi, Gastrodin protects apoptotic dopaminergic neurons in a toxin-induced Parkinson's disease model, Evid Based Complement Altern Med 2013 (2013) 1–13.

[61] CL Huang, JM Yang, KC Wang, YC Lee, YL Lin, YC Yang, NK Huang, *Gastrodia elata* prevents huntingtin aggregations through activation of the adenosine A2A receptor and ubiquitin proteasome system, J Ethnopharmacol 138 (2011) 162–168.

[62] AR Gaby, *Ginkgo biloba* extract: a review, Altern Med Rev 1 (1996) 236–242.

[63] N Mei, X Guo, Z Ren, D Kobayashi, K Wada, L Guo, Review of *Ginkgo biloba*-induced toxicity, from experimental studies to human case reports, J. Environ. Sci. Health C 35 (2017) 1–28.

[64] G Yang, Y Wang, J Sun, K Zhang, J Liu, *Ginkgo biloba* for mild cognitive impairment and Alzheimer's disease: a systematic review and meta-analysis of randomized controlled trials, Curr Top Med Chem 16 (2016) 520–528.

[65] P Rojas, S Montes, N Serrano-García, J Rojas-Castañeda, Effect of EGb761 supplementation on the content of copper in mouse brain in an animal model of Parkinson's disease, Nutrition 25 (2009) 482–485.

[66] P Rojas, P Montes, C Rojas, N Serrano-García, JC Rojas-Castañeda, Effect of a phytopharmaceutical medicine, *Ginko biloba* extract 761, in an animal model of Parkinson's disease: Therapeutic perspectives, Nutrition 28 (2012) 1081–1088.

[67] ET Menze, MG Tadros, AM Abdel-Tawab, AE Khalifa, Potential neuroprotective effects of hesperidin on 3-nitropropionic acid-induced neurotoxicity in rats, Neurotoxicology 33 (2012) 1265–1275.

[68] MA Hashmi, A Khan, M Hanif, U Farooq, S Perveen, Traditional uses, phytochemistry, and pharmacology of *Olea europaea* (olive), Evid Based Complement Altern Med (2015) 541591.

[69] N Acar-Tek, D Ağagündüz, Olive leaf (*Olea europaea* L. folium): potential effects on glycemia and lipidemia, Ann Nutr Metab 76 (2020) 63–68.

[70] I Tasset, AJ Pontes, AJ Hinojosa, R De la Torre, I Túnez, Olive oil reduces oxidative damage in a 3-nitropropionic acid-induced Huntington's disease-like rat model, Nutr Neurosci 14 (2011) 106–111.

[71] JT Coon, E Ernst, Panax ginseng, Drug Saf 25 (2002) 323–344.

[72] DS Kiefer, T Pantuso, Panax ginseng, Am Acad Fam Physician 68 (2003) 1539–1542.

[73] JH Kim, S Kim, IS Yoon, JH Lee, BJ Jang, SM Jeong, HC Kim, Protective effects of ginseng saponins on 3-nitropropionic acid-induced striatal degeneration in rats, Neuropharmacology 48 (2005) 743–756.

[74] PS Khushboo, VM Jadhav, VJ Kadam, NS Sathe, *Psoralea corylifolia* Linn.—"Kushtanashini", Pharmacogn Rev 4 (2010) 69–76.

[75] F Alam, GN Khan, MHHB Asad, *Psoralea corylifolia* L: Ethnobotanical, biological, and chemical aspects: a review, Phytother Res 32 (2018) 597–615.

[76] AR Im, SW Chae, G Jun Zhang, MY Lee, Neuroprotective effects of *Psoralea corylifolia* Linn seed extracts on mitochondrial dysfunction induced by 3-nitropropionic acid, BioMed Central Complement Altern Med 14 (2014) 370.

[77] SK Kulkarni, A Dhir, *Withania somnifera*: an Indian ginseng, Prog. Neuropsychopharmacol. Biological Psychiatry 32 (2008) 1093–1105.

[78]   LC Mishra, BB Singh, S Dagenais, Scientific basis for the therapeutic use of *Withania somnifera* (ashwagandha): a review, Altern Med Rev 5 (2000) 334–346.

[79]   G Singh, PK Sharma, R Dudhe, S Singh, Biological activities of *Withania somnifera*, Ann Biol Res 1 (2010) 56–63.

[80]   P Kumar, A Kumar, Possible neuroprotective effect of *Withania somnifera* root extract against 3-nitropropionic acid-induced behavioral, biochemical, and mitochondrial dysfunction in an animal model of Huntington's disease, J Med Food 12 (2009) 591–600.

[81]   A Ghasemzadeh, HZ Jaafar, A Baghdadi, A Tayebi-Meigooni, Alpha-mangostin-rich extracts from mangosteen pericarp: optimization of green extraction protocol and evaluation of biological activity, Molecules 23 (2018) 1852.

[82]   MY Ibrahim, NM Hashim, AA Mariod, S Mohan, MA Abdulla, SI Abdelwahab, IA Arbab, α-Mangostin from *Garcinia mangostana* Linn: an updated review of its pharmacological properties, Arabian J Chem 9 (2016) 317–329.

[83]   F Gutierrez-Orozco, ML Failla, Biological activities and bioavailability of mangosteen xanthones: a critical review of the current evidence, Nutrients 5 (2013) 3163–3183.

[84]   R Asasutjarit, T Meesomboon, P Adulheem, P Kittiwisut, P Sookdee, W Samosornsuk, A Fuongfuchat, Physicochemical properties of alpha-mangostin loaded nanomeulsions prepared by ultrasonication technique, Heliyon 5 (2019) e02465.

[85]   P Kasemwattanaroj, P Moongkarndi, K Pattanapanyasat, S Mangmool, E Rodpai, J Samer, K Sukapirom, Immunomodulatory activities of α-mangostin on peripheral blood mononuclear cells, Nat Prod Commun 8 (2013) 1257–1260.

[86]   J Pedraza-Chaverrí, LM Reyes-Fermín, EG Nolasco-Amaya, M Orozco-Ibarra, ON Medina-Campos, O González-Cuahutencos, R Mata, ROS scavenging capacity and neuroprotective effect of α-mangostin against 3-nitropropionic acid in cerebellar granule neurons, Exp Toxicol Pathol 61 (2009) 491–501.

[87]   J Fu, Z Wang, L Huang, S Zheng, D Wang, S Chen, S Yang, Review of the botanical characteristics, phytochemistry, and pharmacology of *Astragalus membranaceus* (Huangqi), Phytother Res 28 (2014) 1275–1283.

[88]   M Jin, K Zhao, Q Huang, P Shang, Structural features and biological activities of the polysaccharides from *Astragalus membranaceus*, Int J Biol Macromol 64 (2014) 257–266.

[89]   H Zhang, N Pan, S Xiong, S Zou, H Li, L Xiao, Z Huang, Inhibition of polyglutamine-mediated proteotoxicity by *Astragalus membranaceus* polysaccharide through the DAF-16/FOXO transcription factor in *Caenorhabditis elegans*, Biochemistry 441 (2012) 417–424.

[90]   R Mechoulam, M Peters, E Murillo-Rodriguez, LO Hanuš, Cannabidiol–recent advances, Chem Biodivers 4 (2007) 1678–1692.

[91]   C Scuderi, DD Filippis, T Iuvone, A Blasio, A Steardo, G Esposito, Cannabidiol in medicine: a review of its therapeutic potential in CNS disorders, Phytother Res 23 (2009) 597–602.

[92]   T Iuvone, G Esposito, D De Filippis, C Scuderi, L Steardo, Cannabidiol: a promising drug for neurodegenerative disorders? CNS Neurosci Ther 15 (2009) 65–75.

[93]   EPM De Meijer, KM Hammond, The inheritance of chemical phenotype in *Cannabis sativa* L.(II): cannabigerol predominant plants, Euphytica 145 (2005) 189–198.

[94]   S Valdeolivas, C Navarrete, I Cantarero, ML Bellido, E Muñoz, O Sagredo, Neuroprotective properties of cannabigerol in Huntington's disease: studies in R6/2 mice and 3-nitropropionate-lesioned mice, Neurotherapeutics 12 (2015) 185–199.

[95]   M Nagase, J Oto, S Sugiyama, K Yube, Y Takaishi, N Sakato, Apoptosis induction in HL-60 cells and inhibition of topoisomerase II by triterpene celastrol, Biosci. Biotechnol. Biochem. 67 (2003) 1883–1887.

[96]   C Cleren, NY Calingasan, J Chen, MF Beal, Celastrol protects against MPTP-and 3-nitropropionic acid-induced neurotoxicity, J Neurochem 94 (2005) 995–1004.

[97]   YQ Zhang, KD Sarge, Celastrol inhibits polyglutamine aggregation and toxicity though induction of the heat shock response, J Mol Med 85 (2007) 1421–1428.

[98] BB Aggarwal, C Sundaram, N Malani, H Ichikawa, Curcumin: The Indian Solid Gold, in: BB Aggarwal, Surh YJ, S Shishodia (Eds.), The Molecular Targets and Therapeutic Uses of Curcumin in Health and Disease, Springer Science & Business Media, US, 2007, pp. 1–75.

[99] E Kunchandy, MNA Rao, Oxygen radical scavenging activity of curcumin, Int J Toxicol 58 (1990) 237–240.

[100] A Goel, AB Kunnumakkara, BB Aggarwal, Curcumin as "Curecumin": from kitchen to clinic, Biochem Pharmacol 75 (2008) 787–809.

[101] R Sandhir, A Yadav, A Mehrotra, A Sunkaria, A Singh, S Sharma, Curcumin nanoparticles attenuate neurochemical and neurobehavioral deficits in experimental model of Huntington's disease, NeuroMol Med 16 (2014) 106–118.

[102] ME Waltner-Law, XL Wang, BK Law, RK Hall, M Nawano, DK Granner, Epigallocatechin gallate, a constituent of green tea, represses hepatic glucose production, J Biol Chem 277 (2002) 34933–34940.

[103] M Suganuma, S Okabe, M Oniyama, Y Tada, H Ito, H Fujiki, Wide distribution of [3H](-)-epigallocatechin gallate, a cancer preventive tea polyphenol, in mouse tissue, Carcinogenesis 19 (1998) 1771–1776.

[104] S Yoshizawa, T Horiuchi, H Fujiki, T Yoshida, T Okuda, T Sugimura, Antitumor promoting activity of (−)-epigallocatechin gallate, the main constituent of "Tannin" in green tea, Phytother Res 1 (1987) 44–47.

[105] YT Choi, CH Jung, SR Lee, JH Bae, WK Baek, MH Suh, SI Suh, The green tea polyphenol (−)-epigallocatechin gallate attenuates β-amyloid-induced neurotoxicity in cultured hippocampal neurons, Life Sci 70 (2001) 603–614.

[106] DE Ehrnhoefer, M Duennwald, P Markovic, JL Wacker, S Engemann, M Roark, PJ Muchowski, Green tea (−)-epigallocatechin-gallate modulates early events in huntingtin misfolding and reduces toxicity in Huntington's disease models, Hum Mol Genet 15 (2006) 2743–2751.

[107] H Ragelle, S Crauste-Manciet, J Seguin, D Brossard, D Scherman, P Arnaud, GG Chabot, Nanoemulsion formulation of fisetin improves bioavailability and antitumour activity in mice, Int J Pharm 427 (2012) 452–459.

[108] HH Park, S Lee, JM Oh, MS Lee, KH Yoon, BH Park, SH Kim, Anti-inflammatory activity of fisetin in human mast cells (HMC-1), Pharmacol Res 55 (2007) 31–37.

[109] P Maher, R Dargusch, L Bodai, PE Gerard, JM Purcell, JL Marsh, ERK activation by the polyphenols fisetin and resveratrol provides neuroprotection in multiple models of Huntington's disease, Hum Mol Genet 20 (2011) 261–270.

[110] J Marco-Contelles, M do Carmo Carreiras, C Rodríguez, M Villarroya, AG Garcia, Synthesis and pharmacology of galantamine, Chem Rev 106 (2006) 116–133.

[111] D Prvulovic, H Hampel, J Pantel, Galantamine for Alzheimer's disease, Exp Opin Drug Metab Toxicol 6 (2010) 345–354.

[112] JE Park, ST Lee, WS Im, K Chu, M Kim, Galantamine reduces striatal degeneration in 3-nitropropionic acid model of Huntington's disease, Neurosci Lett 448 (2008) 143–147.

[113] LP Christensen, Ginsenosides: chemistry, biosynthesis, analysis, and potential health effects, Adv Food Nutr Res 55 (2008) 1–99.

[114] HJ Kim, P Kim, C.Y Shin, A comprehensive review of the therapeutic and pharmacological effects of ginseng and ginsenosides in central nervous system, J Ginseng Res 37 (2013) 8.

[115] J Wu, HK Jeong, SE Bulin, SW Kwon, JH Park, I Bezprozvanny, Ginsenosides protect striatal neurons in a cellular model of Huntington's disease, J Neurosci Res 87 (2009) 1904–1912.

[116] Galati, EM, Monforte, MT, Kirjavainen, S, Forestieri, AM, Trovato, A, and Tripodo, MM (1994). Biological effects of hesperidin, a citrus flavonoid.(Note I): antiinflammatory and analgesic activity. *Farmaco (Societa Chimica Italiana: 1989)*40, 709.

[117] A Garg, S Garg, LJD Zaneveld, AK Singla, Chemistry and pharmacology of the citrus bioflavonoid hesperidin, Phytother Res 15 (2001) 655–669.

[118] PK Wilmsen, DS Spada, M Salvador, Antioxidant activity of the flavonoid hesperidin in chemical and biological systems, J Agric Food Chem 53 (2005) 4757–4761.

[119]   J Kim, MB Wie, M Ahn, A Tanaka, H Matsuda, T Shin, Benefits of hesperidin in central nervous system disorders: a review, Anat Cell Biol 52 (2019) 369–377.

[120]   JM Calderon-Montano, E Burgos-Morón, C Pérez-Guerrero, M López-Lázaro, A review on the dietary flavonoid kaempferol, Mini Rev Med Chem 11 (2011) 298–344.

[121]   AY Chen, YC Chen, A review of the dietary flavonoid, kaempferol on human health and cancer chemoprevention, Food Chem 138 (2013) 2099–2107.

[122]   R Lagoa, C Lopez-Sanchez, AK Samhan-Arias, CM Gañan, V Garcia-Martinez, C Gutierrez-Merino, Kaempferol protects against rat striatal degeneration induced by 3-nitropropionic acid, J Neurochem 111 (2009) 473–487.

[123]   T Madaan, AN Choudhary, S Gyenwalee, S Thomas, H Mishra, M Tariq, S Talegaonkar, Lutein, a versatile phyto-nutraceutical: an insight on pharmacology, therapeutic indications, challenges and recent advances in drug delivery, PharmaNutrition 5 (2017) 64–75.

[124]   Y Binawade, A Jagtap, Neuroprotective effect of lutein against 3-nitropropionic acid–induced huntington's disease–like symptoms: possible behavioral, biochemical, and cellular alterations, J Med Food 16 (2013) 934–943.

[125]   D Heber, QY Lu, Overview of mechanisms of action of lycopene, Exp Biol Med 227 (2002) 920–923.

[126]   P Kumar, H Kalonia, A Kumar, Lycopene modulates nitric oxide pathways against 3-nitropropionic acid-induced neurotoxicity, Life Sci 85 (2009) 711–718.

[127]   S Tordjman, S Chokron, R Delorme, A Charrier, E Bellissant, N Jaafari, C Fougerou, Melatonin: pharmacology, functions and therapeutic benefits, Curr Neuropharmacol 15 (2017) 434–443.

[128]   BS Alghamdi, The neuroprotective role of melatonin in neurological disorders, J Neurosci Res 96 (2018) 1136–1149.

[129]   I Túnez, P Montilla, M Del Carmen Munoz, M Feijóo, M Salcedo, Protective effect of melatonin on 3-nitropropionic acid-induced oxidative stress in synaptosomes in an animal model of Huntington's disease, J Pineal Res 37 (2004) 252–256.

[130]   S Bharti, N Rani, B Krishnamurthy, DS Arya, Preclinical evidence for the pharmacological actions of naringin: a review, Planta Med 80 (2014) 437–451.

[131]   P Kumar, A Kumar, Protective effect of hesperidin and naringin against 3-nitropropionic acid induced Huntington's like symptoms in rats: possible role of nitric oxide, Behav Brain Res 206 (2010) 38–46.

[132]   NL Benowitz, Pharmacology of nicotine: addiction and therapeutics, Annu Rev Pharmacol 36 (1996) 597–613.

[133]   M Tariq, HA Khan, I Elfaki, S Al Deeb, K Al Moutaery, Neuroprotective effect of nicotine against 3-nitropropionic acid (3-NP)-induced experimental Huntington's disease in rats, Brain Res Bull 67 (2005) 161–168.

[134]   Y Li, J Yao, C Han, J Yang, MT Chaudhry, S Wang, Y Yin, Quercetin, inflammation and immunity, Nutrients 8 (2016) 167.

[135]   AG Wu, VKW Wong, SW Xu, WK Chan, CI Ng, L Liu, BYK Law, Onjisaponin B derived from Radix Polygalae enhances autophagy and accelerates the degradation of mutant α-synuclein and huntingtin in PC-12 cells, Int J Mol Sci 14 (2013) 22618–22641.

[136]   R Sandhir, A Mehrotra, Quercetin supplementation is effective in improving mitochondrial dysfunctions induced by 3-nitropropionic acid: implications in Huntington's disease, Biochim. Biophys. Acta Mol Basis Dis 1832 (2013) 421–430.

[137]   KP Bhat, JW Kosmeder, JM Pezzuto, Biological effects of resveratrol, Antioxid Redox Signal 3 (2001) 1041–1064.

[138]   S Andrade, MJ Ramalho, MDC Pereira, JA Loureiro, Resveratrol brain delivery for neurological disorders prevention and treatment, Front Pharmacol 9 (2018) 1261.

[139]   Y Kodera, A Suzuki, O Imada, S Kasuga, I Sumioka, A Kanezawa, K Maeshige, Physical, chemical, and biological properties of S-allylcysteine, an amino acid derived from garlic, J Agric Food Chem 50 (2002) 622–632.

[140]   M Padmanabhan, PSM Prince, Effects of pharmacological amounts of S-allylcysteine on lipids in normal and isoproterenol-induced myocardial infarction in rats, J Sci Food Agric 86 (2006) 772–777.

[141] YD Wen, YZ Zhu, The pharmacological effects of S-propargyl-cysteine, a novel endogenous H 2 S-producing compound, in: PK Moore, M Whiteman (Eds.), Chemistry, Biochemistry and Pharmacology of Hydrogen Sulfide, Springer, Cham, 2015, pp. 325–336.

[142] S Sundaresan, P Subramanian, S-allylcysteine inhibits circulatory lipid peroxidation and promotes antioxidants in N-nitrosodiethylamine-induced carcinogenesis, Pol J Pharmacol Pharm 55 (2003) 37–42.

[143] VS Uddandrao, P Brahmanaidu, G Saravanan, Therapeutical perspectives of S-allylcysteine: effect on diabetes and other disorders in animal models, Cardiovasc Hematol Agents Med Chem 15 (2017) 71–77.

[144] MN Herrera-Mundo, D Silva-Adaya, PD Maldonado, S Galván-Arzate, L Andrés-Martínez, V Pérez-De La Cruz, A Santamaría, S-Allylcysteine prevents the rat from 3-nitropropionic acid-induced hyperactivity, early markers of oxidative stress and mitochondrial dysfunction, Neurosci Res 56 (2006) 39–44.

[145] Pérez-De La Cruz, González-Cortés V, Pedraza-Chaverrí C, Maldonado J, P D, L Andrés-Martínez, A Santamaría, Protective effect of S-allylcysteine on 3-nitropropionic acid-induced lipid peroxidation and mitochondrial dysfunction in rat brain synaptosomes, Brain Res Bull 68 (2006) 379–383.

[146] JN Chun, SY Kim, EJ Park, EJ Kwon, DJ Bae, IS Kim, I So, Schisandrin B suppresses TGFβ1-induced stress fiber formation by inhibiting myosin light chain phosphorylation, J Ethnopharmacol 152 (2014) 364–371.

[147] DQ Xin, ZM Hu, HJ Huo, XJ Yang, D Han, WH Xing, QH Qiu, Schisandrin B attenuates the inflammatory response, oxidative stress and apoptosis induced by traumatic spinal cord injury via inhibition of p53 signaling in adult rats, Mol Med Rep 16 (2017) 533–538.

[148] PY Lam, KM Ko, Schisandrin B as a hormetic agent for preventing age-related neurodegenerative diseases, Oxid Med Cell Longev 2012 (2012) 1–9.

[149] A Kuhad, K Chopra, Effect of sesamol on diabetes-associated cognitive decline in rats, Exp Brain Res 185 (2008) 411–420.

[150] M Uchida, S Nakajin, S Toyoshima, M Shinoda, Antioxidative effect of sesamol and related compounds on lipid peroxidation, Biol Pharm Bull 19 (1996) 623–626.

[151] IP Kaur, A Saini, Sesamol exhibits antimutagenic activity against oxygen species mediated mutagenicity, Mutat Res Genet Toxicol Environ 470 (2000) 71–76.

[152] W Ruankham, W Suwanjang, P Wongchitrat, V Prachayasittikul, S Prachayasittikul, K Phopin, Sesamin and sesamol attenuate H_2O_2-induced oxidative stress on human neuronal cells via the SIRT1-SIRT3-FOXO3a signaling pathway, Nutr Neurosci 24 (2) (2021) 90–101.

[153] P Kumar, H Kalonia, A Kumar, Protective effect of sesamol against 3-nitropropionic acid-induced cognitive dysfunction and altered glutathione redox balance in rats, Basic Clin Pharmacol Toxicol 107 (2010) 577–582.

[154] C Zhou, EJ Poulton, F Grün, TK Bammler, B Blumberg, KE Thummel, DL Eaton, The dietary isothiocyanate sulforaphane is an antagonist of the human steroid and xenobiotic nuclear receptor, Mol Pharmacol 71 (2007) 220–229.

[155] S Kanematsu, N Uehara, H Miki, K Yoshizawa, A Kawanaka, T Yuri, A Tsubura, Autophagy inhibition enhances sulforaphane-induced apoptosis in human breast cancer cells, Anticancer Res 30 (2010) 3381–3390.

[156] Y Liu, CL Hettinger, D Zhang, K Rezvani, X Wang, H Wang, Sulforaphane enhances proteasomal and autophagic activities in mice and is a potential therapeutic reagent for Huntington's disease, J Neurochem 129 (2014) 539–547.

[157] M Jang, I.H Cho, Sulforaphane ameliorates 3-nitropropionic acid-induced striatal toxicity by activating the Keap1-Nrf2-ARE pathway and inhibiting the MAPKs and NF-κB pathways, Mol Neurobiol 53 (2016) 2619–2635.

[158] AB Richards, S Krakowka, LB Dexter, H Schmid, APM Wolterbeek, DH Waalkens-Berendsen, M Kurimoto, Trehalose: a review of properties, history of use and human tolerance, and results of multiple safety studies, Food Chem Toxicol 40 (2002) 871–898.

[159] K Hosseinpour-Moghaddam, M Caraglia, A Sahebkar, Autophagy induction by trehalose: molecular mechanisms and therapeutic impacts, J Cell Physiol 233 (2018) 6524–6543.

[160]   S Sarkar, JE Davies, Z Huang, A Tunnacliffe, DC Rubinsztein, Trehalose, a novel mTOR-independent autophagy enhancer, accelerates the clearance of mutant huntingtin and α-synuclein, J Biol Chem 282 (2007) 5641–5652.

[161]   NJ Walton, MJ Mayer, A Narbad, Vanillin, Phytochemistry 63 (2003) 505–515.

[162]   S Gupta, B Sharma, Pharmacological benefits of agomelatine and vanillin in experimental model of Huntington's disease, Pharmacol Biochem Behav 122 (2014) 122–135.

[163]   YC Lee, YC Yang, CL Huang, TY Kuo, JH Lin, DM Yang, NK Huang, When cytokinin, a plant hormone, meets the adenosine A 2A receptor: a novel neuroprotectant and lead for treating neurodegenerative disorders? PLoS One 7 (2012) e38865.

[164]   BYK Law, SWF Mok, AG Wu, CWK Lam, MXY Yu, VKW Wong, New potential pharmacological functions of Chinese herbal medicines via regulation of autophagy, Molecules 21 (2016) 359.

[165]   CL Lin, SE Wang, CH Hsu, SJ Sheu, CH Wu, Oral treatment with herbal formula B307 alleviates cardiac failure in aging R6/2 mice with Huntington's disease via suppressing oxidative stress, inflammation, and apoptosis, Clin Interv Aging 10 (2015) 1173–1187.

[166]   T Satoh, T Takahashi, K Iwasaki, H Tago, T Seki, N Yaegashi, H Arai, Traditional Chinese medicine on four patients with Huntington's disease, Mov Disord 24 (2009) 453–455.

[167]   ZJ Zhang, QR Tan, XC Zhen, Y Tong, The potential benefits of herbal medicines for schizophrenia: from empirical observations to clinical trials, Clin Trials Psychopharmacol 16 (2010) 311–335.

[168]   SK Dubey, G Singhvi, KV Krishna, T Agnihotri, RN Saha, G Gupta, Herbal medicines in neurodegenerative disorders: an evolutionary approach through novel drug delivery system, J Environ Pathol Toxicol Oncol 37 (2018) 199–208.

[169]   BH May, M Lit, CC Xue, AW Yang, AL Zhang, MD Owens, DF Story, Herbal medicine for dementia: a systematic review, Phytother Res 23 (2009) 447–459.

[170]   W Zhu, S Ma, R Qu, D Kang, Antidepressant-like effect of saponins extracted from Chaihu-jia-longgu-muli-tang and its possible mechanism, Life Sci 79 (2006) 749–756.

[171]   Y Liu, S Ma, R Qu, SCLM, total saponins extracted from Chaihu-jia-longgu-muli-tang, reduces chronic mild stress-induced apoptosis in the hippocampus in mice, Pharm Biol 48 (2010) 840–848.

[172]   CL Lin, SE Wang, CH Hsu, SJ Sheu, CH Wu, Oral treatment with herbal formula B307 alleviates cardiac failure in aging R6/2 mice with Huntington's disease via suppressing oxidative stress, inflammation, and apoptosis, Clin Interv Aging 10 (2015) 1173–1187.

[173]   DM Gardner, Evidence-based decisions about herbal products for treating mental disorders, J Psychiatry Neurosci 27 (2002) 324–333.

[174]   M Gu, MT Gash, VM Mann, F Javoy-Agid, J.M Cooper, A.H.V Schapira, Mitochondrial defect in Huntington's disease caudate nucleus, Ann Neurol 39 (1996) 385–389.

[175]   M Martinez-Vicente, Z Talloczy, E Wong, G Tang, H Koga, S Kaushik, AM Cuervo, Cargo recognition failure is responsible for inefficient autophagy in Huntington's disease, Nat Neurosci 13 (2010) 567–576.

[176]   DD Martin, S Ladha, DE Ehrnhoefer, MR Hayden, Autophagy in Huntington disease and huntingtin in autophagy, Trends Neurosci 38 (2015) 26–35.

[177]   A Williams, S Sarkar, P Cuddon, EK Ttofi, S Saiki, FH Siddiqi, CJ O'kane, Novel targets for Huntington's disease in an mTOR-independent autophagy pathway, Nat Chem Biol 4 (2008) 295–305.

[178]   SE Browne, AC Bowling, U Macgarvey, MJ Baik, SC Berger, MM Muquit, MF Beal, Oxidative damage and metabolic dysfunction in Huntington's disease: selective vulnerability of the basal ganglia, Ann Neurol 41 (1997) 646–653.

[179]   SE Browne, MF Beal, Oxidative damage in Huntington's disease pathogenesis, Antioxid Redox Signal 8 (2006) 2061–2073.

[180]   A Johri, MF Beal, Mitochondrial dysfunction in neurodegenerative diseases, Pharmacol Exp Ther 342 (2012) 619–630.

[181] AE Khalifa, Neural monoaminergic mediation of the effect of St. John's wort extract on prepulse inhibition of the acoustic startle response in rats, J Psychopharmacol 19 (2005) 467–472.

[182] M Damiano, L Galvan, N Déglon, E Brouillet, Mitochondria in Huntington's disease, Biochim. Biophys. Acta Mol Basis Dis 1802 (2010) 52–61.

[183] PK Mukherjee, V Kumar, M Mal, PJ Houghton, Acetylcholinesterase inhibitors from plants, Phytomedicine 14 (2007) 289–300.

[184] JJ Vattakatuchery, R Kurien, Acetylcholinesterase inhibitors in cognitive impairment in Huntington's disease: a brief review, World J Psychiatry 3 (2013) 62.

[185] PJ Houghton, MJ Howes, Natural products and derivatives affecting neurotransmission relevant to Alzheimer's and Parkinson's disease, Neurosignals 14 (2005) 6–22.

[186] JA Mares-Perlman, AE Millen, TL Ficek, SE Hankinson, The body of evidence to support a protective role for lutein and zeaxanthin in delaying chronic disease. Overview., J Nutr 132 (2002) 518–524.

[187] JY Min, KB Min, Serum lycopene, lutein and zeaxanthin, and the risk of Alzheimer's disease mortality in older adults, Dement Geriatr Cogn Disord 37 (2014) 246–256.

[188] SK Verma, A Kumar, Therapeutic uses of *Withania somnifera* (ashwagandha) with a note on withanolides and its pharmacological actions, Asian J Pharm Clin Res 4 (2011) 1–4.

[189] IH Cho, Effects of Panax ginseng in neurodegenerative diseases, J Ginseng Res 36 (2012) 342.

[190] M Matias, S Silvestre, A Falcão, G Alves, Gastrodia elata and epilepsy: Rationale and therapeutic potential, Phytomedicine 23 (2016) 1511–1526.

[191] SK Singh, S Srivastav, RJ Castellani, G Plascencia-Villa, G Perry, Neuroprotective and antioxidant effect of *Ginkgo biloba* extract against AD and other neurological disorders, Neurotherapeutics 16 (2019) 666–674.

[192] U Jana, TK Sur, LN Maity, PK Debnath, D Bhattacharyya, A clinical study on the management of generalized anxiety disorder with *Centella asiatica*, Nepal Med Coll J 12 (2010) 8–11.

[193] R Jeyasri, P Muthuramalingam, V Suba, M Ramesh, JT Chen, Bacopa monnieri and their bioactive compounds inferred multi-target treatment strategy for neurological diseases: a cheminformatics and system pharmacology approach, Biomolecules 10 (2020) 536.

[194] M Khalifeh, GE Barreto, A Sahebkar, Trehalose as a promising therapeutic candidate for the treatment of Parkinson's disease, Br J Pharmacol 176 (2019) 1173–1189.

[195] Y Li, D He, X Zhang, Z Liu, X Zhang, L Dong, Y Chen, Protective effect of celastrol in rat cerebral ischemia model: down-regulating p-JNK, pc-Jun and NF-κB, Brain Res 1464 (2012) 8–13.

[196] S Ahmed, H Khan, M Aschner, MM Hasan, ST Hassan, Therapeutic potential of naringin in neurological disorders, Food Chem Toxicol 132 (2019) 110646.

[197] F Aqil, R Munagala, J Jeyabalan, MV Vadhanam, Bioavailability of phytochemicals and its enhancement by drug delivery systems, Cancer Lett 334 (2013) 133–141.

[198] Y Zhao, G Tang, Q Tang, J Zhang, Y Hou, E Cai, S Wang, A method of effectively improved α-mangostin bioavailability, Eur J Drug Metab Pharmacokinet 41 (2016) 605–613.

[199] K Knaub, T Sartorius, T Dharsono, R Wacker, M Wilhelm, C Schön, A novel self-emulsifying drug delivery system (SEDDS) based on VESIsorb® formulation technology improving the oral bioavailability of cannabidiol in healthy subjects, Molecules 24 (2019) 2967.

[200] JD Lambert, J Hong, DH Kim, VM Mishin, CS Yang, Piperine enhances the bioavailability of the tea polyphenol (−)-epigallocatechin-3-gallate in mice, J Nutr 134 (2004) 1948–1952.

[201] H Kim, JH Lee, JE Kim, YS Kim, CH Ryu, HJ Lee, J Lee, Micro-/nano-sized delivery systems of ginsenosides for improved systemic bioavailability, J Ginseng Res 42 (2018) 361–369.

[202] A Jain, G Sharma, G Ghoshal, P Kesharwani, B Singh, US Shivhare, OP Katare, Lycopene loaded whey protein isolate nanoparticles: An innovative endeavor for enhanced bioavailability of lycopene and anti-cancer activity, Int J Pharm 546 (2018) 97–105.

[203] PS Yashaswini, NK Kurrey, SA Singh, Encapsulation of sesamol in phosphatidyl choline micelles: Enhanced bioavailability and anti-inflammatory activity, Food Chem 228 (2017) 330–337.

[204]   T Feng, K Wang, F Liu, R Ye, X Zhu, H Zhuang, Z Xu, Structural characterization and bioavailability of ternary nanoparticles consisting of amylose, α-linoleic acid and β-lactoglobulin complexed with naringin, Int J Biol Macromol 99 (2017) 365–374.

[205]   DJM Kollareth, M Muralidhara, Neuroprotective efficacy of a combination of fish oil and *Bacopa monnieri* against 3-nitropropionic acid induced oxidative stress in rats, Int J Neurol Res 3 (2017) 349–357.

CHAPTER 4

Alternate medicinal approach for the treatment of depression/mood disorder

Swalih P. Ahmed[a], Mehdi H. Shahi[a], Mohammad Afzal[b] and Asif Ali[a]

[a]*Interdisciplinary Brain Research Centre (IBRC), J.N.Medical College, Faculty of Medicine, Aligarh Muslim University, Aligarh, UP, India* [b]*Department of Zoology, Aligarh Muslim University, Aligarh, UP, India*

4.1 Introduction

Depression is one of the most prevalent psychiatric disorders, and this causes about 25% of women and 12% of men to suffer at least one time during their lifetime [1]. Depression is the second most common psychological disorder; this disease increased in developed and developing countries [2]. Recent data shows that nearly 264 million people worldwide affect the Depression [54]. Depressive disorders are prevalent, with approximately 11.3 of all adults affected during every year [3]. Depression disorders show more frequently in women than in men [4].

The most commonly used depression treatments are antidepressants, psychotherapy and a combination of drugs and psychotherapy. Different type of depression treatment available but current psychiatric drugs cause many induce side effects, such as sexual dysfunction, metabolic syndrome, dependency, weight gain, etc. [5]. Globally, Complementary Alternative Medicine (CAM) has been widely used globally, but CAM in depression data is limited. Various antidepressant drugs, including herbal medicine, are suppressing the reuptake of neurotransmitters in the brain. Some herbal medicines act as sensitization of serotonin receptors or inhibit monoamine oxidases, leading to antidepression [6]. This study aimed to evaluate the use of herbal medicine in Depression.

4.2 Role of herbal medicine in depression/mood disorder

There are several types of herbal remedies are used to treat depression, some of which are listed below:

Herbal Medicines: A Boon for Healthy Human Life.
DOI: https://doi.org/10.1016/B978-0-323-90572-5.00008-1

1. Saffron

Saffron is the most expensive spice, and its use a traditional food additive; recent studies on Saffron indicate several therapeutic effects. It is used for depression treatment without side effects in Persian traditional medicine. Saffron treatment was adequate for the inhibiting of serotonin by nerve cells (neurons) to help treat mild to moderate depression [3]. The petroleum ether and dichloro- methane fractions are the active parts of corms of C. sativus; they produce antidepressant-like effects. These results show the efficacy of Crocus sativus in the treatment of mild to moderate depression [7]. Crocin and Safranal are active ingredients in Saffron. These play an essential role in the modulate the different neurotransmitter systems in the brain, including serotonin, dopamine, glutamate, and norepinephrine and helped in reducing depression [8,9]. The saffron dose was found to be effectively similar to fluoxetine and imipramine (antidepressant) without side effect [3,10] (Fig. 4.1).

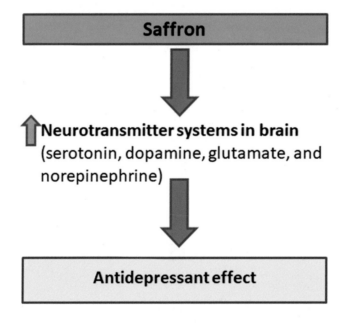

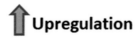

Figure 4.1
Role of Saffron in mood disorder/depression: Saffron enhances the various neurotranmittance like serotonin, dopamine, glutamate and nor-epinephrin in the brain. These neurotransmittance further showed antidepressant effect.

2. Rhodiola (Rhodiola Rosea)

Rhodiola (Rhodiola Rosea) is a perennial plant used to improve physical endurance and mental performance. It is also known as roseroot or golden root. This traditional medicine is seen in Asia and Eastern Europe. R. Rosea shows an ability to increase endogenous β-endorphin, which helps depression treatment. Rhodiola extract versus a conventional antidepressant for mild to moderate depression. Henceforth, R. rosea is better tolerated than sertraline, suggesting its potential as a treatment alternative for patients [11]. Rhodiola rosea influence the levels and activity of biogenic monoamines such as serotonin, dopamine, and norepinephrine in the brain and Rhodiola rosea root extracts inhibited activity on MAO-A (Monoamine oxidase-A) and MAO-B (Monoamine oxidase-B). It enhanced the permeability of the blood-brain barrier [12]. R. rosea effect on neuropeptide-Y (NPY) leads to upregulation of heat shock protein Hsp-70 this cause down-regulation of stress-induced JNK protein (suppressing glucocorticoid receptors and increasing cortisol), NPY also mediates monoamine and serotonin receptors involved in stress-induced depression. Activation of serotonin receptors leads to the release of many neuro-transmitters, including glutamate, acetylcholine, GABA, NA and, DA as well as many hormones including oxytocin, vasopressin, cortisol, corticotropin, prolactin, substance P, and others. R. rosea also mediated stress response, homeostasis of HPA axis activity regulation, and modulation of G- protein-coupled receptor (GPCR) signaling pathways and other molecular networks involved in depression. Therefore, Rhodiola is helpful in depression treatment [13] (Fig. 4.2).

3. Chinese herbal medicine

Traditional Chinese medicine improves mood and keeps healthy. Chinese herbal medicines are used for thousands of years in several Asian countries to cure various psychiatric disorders. Chinese herbal medicine (CHM) has great potential to assist in the development of therapy for depression. Herb, ginseng has been frequently used in the western world. The 20(S)-protopanaxadiol compound is isolated from ginseng, which acts as suitable antidepressant activities in rodent paradigms via upregulating the levels of N.E. and 5-HT. Paeonialactiflora Pall (Ranunculaceae) plant root portion is an essential ingredient of various Chinese medicines used to treat depression. Albiziajulibrissin is also called mimosa or silk tree, is widely used for depression treatment, mainly in Asia. Silk tree induces BDNF expression, this help in depression treatment. Perillafrutescens is traditional Chinese medicine, this medicine used for hundreds of years to cure various psychiatric disorders, mainly depression. Essential oil of Perillafrutescens (EOPF) reduces the depressive-like symptoms reported in rodent models [6].

4. Xiao Yao San

Xiao Yao San (XYS) is a traditional Chinese medicine used to treat depression [16]. It composed of eight herbs; this first documented in the medical book of Taiping Huimin Heji

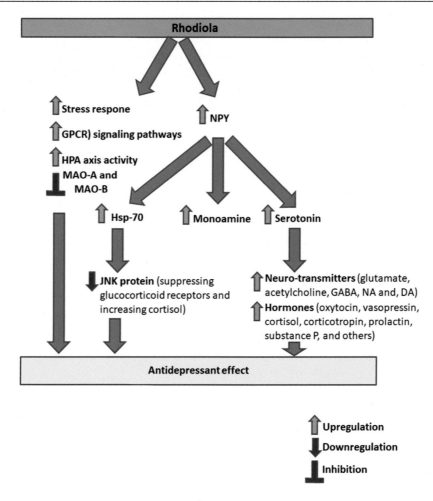

Figure 4.2
Role of Rhodiola in mood disorder/depression: Rhodiola increase the expression of Hsp-70, monoamine and serotonin. High expression of Hsp-70 downregulates the expression of JNK protein and further suppressing glucocorticoid receptor and increasing cortisol. High serotonin further increases the neurotransmitters like glutamate, acetyltcholin, GABA, and hormone like oxytocin, vasopression, cortisol, corticotrophin, prolactin, and others.

Jufang, which was written in the Song Dynasty (960–1127 AD). XYS help with depression treatment. It composed of eight herbs: Radix Bupleuri (Bupleurumchinense DC.), Radix Angelicae Sinensis (Angelica sinensis (Oliv.) Diel), Radix Paeoniae Alba (Paeonialactiflora Pall.), Rhizoma Atractylodis Macrocephalae (Atractylodesmacrocephala Koidz.), Poria (Poriacocos (Schw.) Wolf), HerbaMenthae (Menthahaplocalyx Briq.), Rhizoma Zingiberis Recens (Zingiber officinale Rosc.) and Radix Glycyrrhizae (Glycyrrhizauralensis Fisch). These eight herbs interact with each other and perform the pharmacological functions of

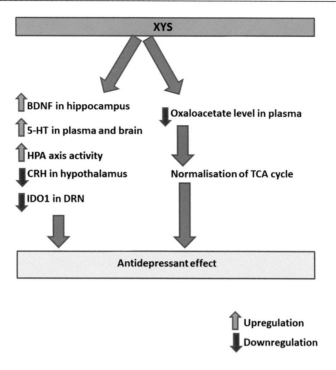

Figure 4.3

Role of Xia Yao San (XYS) in mood disorder/depression: XYS increases the 5-HT in plasma and brain, enhance the HPA axis activity and also decreases the CRH in hypothalamus. It also decreases the oxaloacetate level in plasma and showed the antidepressant effect.

the formula; this helps depression treatment [6]. XYS increase the expression of brain-derived neurotrophic factor (BDNF) in hippocampus, upregulate 5-HT (serotonin) levels in plasma and brain, downregulate corticotropin-releasing hormone(CRH) expression in the hypothalamus. XYS decreased the protein level of IDO1 (Indoleamine-2,3-dioxygenase 1) in the DRN (Dorsal raphe nucleus), increased 5-HT expression and decrease the microglia number to improve the IFN-α (interferon-gamma) induced depression in mice [14]. XYS downregulate oxaloacetate level in plasma; this leads to normalization of TCA cycle and improves the depression symptoms [15] (Fig. 4.3).

5. Banxia Xiexin decoction

Banxia Xiexin decoction is a traditional Chinese medicine used for different diseases treatment. BanxiaXiexin decoction comprises seven herbs such as Banxia, Ganjiang, Scutellaria, Coptis, ginseng, liquorice, and jujube. The biological mechanism of Banxia Xiexin decoction is to help depression treatment by regulating cytokines and inflammatory mediators through HIF-1α. Banxia Xiexin decoction directly participated in drug and lipid metabolism and regulating the functions of the nervous system, immune system, digestive system and other systems of the body; this help for other diseases treated with the same treatment [16].

6. Sihogayonggolmoryeo-tang (SGYMT)

Sihogayonggolmoryeo-tang (Chai-Hu-Jia-Long-Gu-Mu-Li-Tang) is one of the traditional medicine; this medicine was first introduced in the classical Chinese text "Treatise on Cold Damage Diseases. SGYMT comprises eleven herbs, including Bupleuri Radix, Pinelliae Rhizoma, Cinnamomi Ramulus, Poria, Scutellariae Radix, JujubaeFructus, Ginseng Radix or Codonopsis Radix, Ostreae Concha, Fossilia Ossis Mastodi, Zingiberis Rhizoma Recens, and RheiRhizoma. In recent studies SGYMT was recommended to treat the major depressive disorder (MDD) and PSD [17]. SGYMT prevents the collapse of the hypothalamopituitary-adrenal (HPA) axis, including dysfunction in the glucocorticoid negative feedback system leading to antidepressant effects [18].

7. Shuganjieyu capsule

Shuganjieyu capsule, is a mixture of extract of St. John's wort herb and Acanthopanax senticosus; this is the only pure herbal pharmaceutical product for the treatment of depression approved by the China Food and Drug Administration (CFDA).Efficacy and safety of St. John's wort herb on depression is evaluated in both clinical trials and systematic reviews, Acanthopanax senticosus shows some antidepressant effect and preliminary studies indicate antidepressant activity. Shuganjieyu capsule shows a vital role to reduce the depression symptoms and help to upregulation of hippocampal neuron generation, survival, and neogenesis, downregulate the protein levels of caspase-3, and reverse neurocyte apoptosis, This provides the same efficacy of fluoxetine in depression treatment [19] (Fig. 4.4).

8. Gan Mai Da Zao decoction

Gan Mai Da Zao (GMDZ) decoction is a Chinese herbal medicine is widely used for depression in East Asia. GMDZ is also known as Ganmck daecko tang in Korean and kambakutaisoto in Japanese [20]. It is also called Zhong Jing Da Zao Tang or Da Zao Tang [21]. It was first documented in the Chinese medical book Jin Gui Yao Lue by Zhang Zong Jing (AD 152–219). GMDZ decoction is a mixture of three herbal extracts: Glycyrrhiza, Fructus Tritici, and Jujube [20]. GMDZ upregulate horizontal movement and vertical movement, increase sugar consumption and amount of food consumed, upregulate norepinephrine (NE), 5-hydroxytryptamine receptors (5-HT); this leads to antidepressant effects [21] (Fig. 4.5).

9. Humuluslupulus L

Humuluslupulus L is a prolonged standing tradition medicine use for different treatment in Europe. Humuluslupulus L plant belongs Cannabaceae family [22]. The female inflorescences

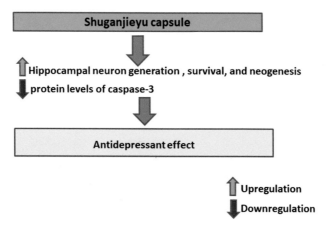

Figure 4.4
Role of Shuganjieyu capsule in mood disorder/depression: Shuganjieyu Increases the hippocampal neuron generation, survival, and neurogenesis. It decreases the caspase 3 and acts as antidepressant.

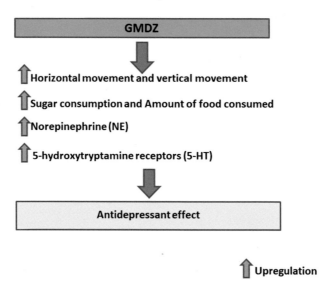

Figure 4.5
Role of Gan Man Da Zao (GMDZ) in mood disorder/depression: It increases the norepinephrin (NE) and 5-hydroxytryptamine receptor (5-HT) and further promote increased sugar consumption, horizontal and vertical movement. All these effect showed antidepressant characteristic.

of the Humuluslupulus Lare known as Hops, the common name of the strobiles (cones) commonly used for a herbal medicinal product primarily for sleep disturbances [23]. Humuluslupulus L significantly reduces the self-reported depression, anxiety and stress symptoms [11].

10. Berberine

Berberine, a natural isoquinoline alkaloid, is widely used in traditional Chinese medicine. It is isolated from several herbal species, including Berberis Hydrastiscanadensis (golden seal), Tinosporacordifolia, Phellodendronanureses (Amur cork tree), Coptis Chinensis (Chinese gold thread), Argemonemexicana (prickly poppy), Xanthorhiza simplicissima, (yellow root), and Eschscholzia california (Californian poppy) [24]. Berberine act as a variant type of action in depression such as inhibition of monoamine oxidase activity, $\alpha2$ autoreceptors, NOX and ROS, Increase of N.E., 5-HT and DA levels, Antagonism of D2 receptors and agonism of D1 receptors, involvement with substance P, sigma receptor l-arginine- NO-cGMP pathway, tumor necrosis factor β, interleukin-6, interleukin- 1-beta, IDO, kynurenine levels, reduction of lipid peroxide and superoxide dismutase levels, induction of NGF secretion, Activation of phosphoinositide 3-kinase/protein kinase/nuclear factor-E2-related factor 2-mediated regulation, BDNF-cAMP response element-binding protein and eEF2 pathways protect the gastrointestinal tract, decrease of plasma corticosterone levels and fluctuations in gonadal hormone levels [25]. The bioavailability of berberine is a severe problem for application and development [26] (Fig. 4.6).

11. Kaixinjieyu

Kaixinjieyu (K.J.) is an effective Chinese herbal medicine used to treat vascular Depression (V.D.). Vascular Depression (V.D.) is a subtype of late-life depression associated with vascular diseases or cerebrovascular risk factors mainly age, stroke, hypertension, hyperlipidemia, myocardial infarction, and diabetes. The effects of KJ act as an antidepressant by mediated the up-regulation of neurogenesis and the tight junction of the brain-blood barrier, and balance of the fibrinolytic system; this leads to restoration of NVU function [27].

12. Melissa officinalis L

Melissa officinalis L or Lemon Balm (Lamiaceae) is a herbal medicine use in European countries. It contains volatile oil with citral. It effectively treats nervous system disorders, reducing excitability, anxiety, and stress [28]. MO reduced the level of corticosterone in serum and decreased GABA-T levels in the hippocampus. These effects help depression treatment [29]. Melissa officinalis is the source of flavonoids. It is used for the treatment of depression [30] (Fig. 4.7).

13. Lavandulaangustifolia Mill

Lavandulaangustifolia Mill or Lavender (Lamiaceae) is an aromatic and evergreen subshrub natively grown in the Mediterranean. It is helpful for the treatment of mood disturbance such as restlessness or insomnia, nervous stomach irritation and nervous intestinal discomfort [28]. This leads to the increased strength of the nervous system, and it helps in reducing depression and nervous exhaustion [31]. Lavender acted as the antidepressant activity and showed fewer

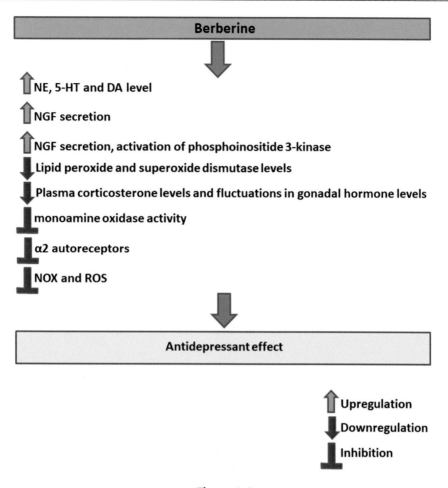

Figure 4.6

Role of Berberine in mood disorder/depression: berberine increases the expression of NE, 5-HT, DA level, decrease the lipid peroxide, superoxide dismutase and inhibit the NOX and ROS. All these effects lead to antidepressant property of berberine.

side effects compared to fluoxetine [32]. Flavonoid compounds are present in Lavender. This effective on benzodiazepine receptors and also present linalool in lavender. This leads to upregulation of noradrenaline and dopamine levels. These effects help depression treatment [33] (Fig. 4.8).

14.   Echium amoenum (Boraginaceae)

Echiumamoenumis a sizeable annual plant of the Boraginaceae family. This commonly shows parts of Europe and in northern parts of Iran. The flower borage is used as a medicinal herb in various countries [34]. E. amoenum flower extract showed a significant superiority over

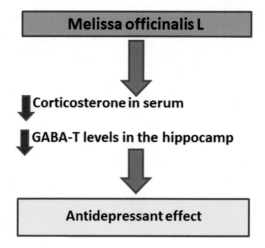

Figure 4.7
Role of Melissa officinalis L in mood disorder/depression: Melissa officinalis decrease the corticosteroid in serum, GABA-T in hippocampus and further causes antidepressant effect.

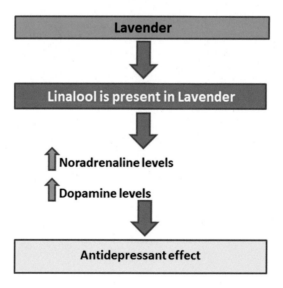

Figure 4.8
Role of lavender in mood disorder/depression: lavender increase the noradrenalin and dopamine and further promote antidepressant effect.

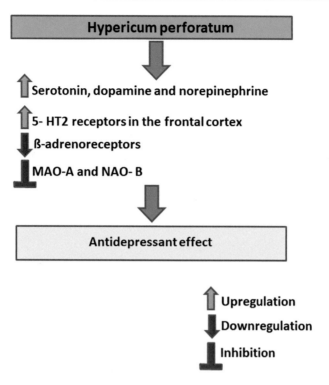

Figure 4.9

Role of Hypericum perforatum in mood disorder/depression: hypericum increases the serotonin, dopamine, norepinephrine, and 5-HT2 receptor in the frontal cortex. It also downregulates β-adrenoreceptor and inhibit the MAO-A and NAO-B. These effects are responsible for antidepressant effect.

placebo in reducing depressive symptoms but most commonly reported side effects such as headache, somnolence, vomiting, dry mouth, constipation and blurred vision. E. amoenum contains flavonoids, saponins and unsaturated sterols; these act as antidepressor [35]. However, the borage flower is used as a herbal medicine in various countries as an antifebrile and antidepressant [34] (Fig. 4.9).

15. Hypericum perforatum

Hypericum perforatum is a medicinal plant that shows antidepressant properties [36]. This plant is popularly called St. John's wort (SJW) [37]. St. John's wort has been well-studied chemical composition and documented pharmacological activities, including antidepressant, antiviral, and antibacterial effects [38]. Hypericum extracts used for depression treatment have fewer side effects than standard antidepressants [37]. St John's wort shows a more favourable short-term safety profile than do standard antidepressants, and this factor is essential in patients continuing to take medication [38]. H. perforatum acts as inhibition of MAO-A and NAO- B, inhibition of synaptosomal uptake of serotonin, dopamine and norepinephrine, and downregulation of beta-receptors and up-regulation of 5- HT2 receptors in the frontal cortex [39]. Hypericum inhibited serotonin uptake in post-synaptic receptors, down-regulated ß-adrenoreceptors, and decrease the binding activity of a benzodiazepine drug to benzodiazepine-binding sites in GABAA receptors. Hypericum also affects on serotonin, dopamine, norepinephrine, ß- adrenoreceptors, interleukins, GABA, MAO, and the HPA axis [40]. St John's wort upregulated gene transcription of GAD in the BST, CREB in the hippocampus and POMC in the pituitary [41]. St. John's wort antidepressant activity help to treatment of depression.

16. Curcumin

Curcumin is the principal curcuminoid found in turmeric (Curcuma longa). It shows antiinflammatory and antioxidant properties. Curcumin is a spice frequently used in dietary regimens of the Asian population, mainly in India and China Curcumin shows a similar effect to conventional antidepressants like fluoxetine and imipramine. Curcumin upregulate serotonin and dopamine, increase the levels of brain-derived neurotrophic factor (BDNF) [42], Inhibit monoamine oxidase A and B enzymes, upregulate the neurotransmitter levels in the brain, inhibit the inflammatory cytokines like nuclear factor-kappa B, NLRP3 inflammasome, and interleukin-1B [43], inhibit the monoamine oxidase (MAO), decrease the inflammatory cytokines IL-1β and TNFα levels, and decrease the salivary cortisol concentrations [44]. These effects help depression treatment (Fig. 4.10).

17. Ginsenosides

Ginseng is a famous herbal medicine in China. This medicine was used for thousands of years in East Asian countries, such as Japan, China, and Korea. [45].Ginseng derived from roots of some plants of the species Panax sp. major commercial ginsengs are Panaxginseng Meyer (P. ginseng; Korean ginseng), Panax quinquefolius(American ginseng), Panax notoginseng (Chinese ginseng), Panax japonicum (Japanese ginseng) and Panax viet-namensis (Vietnamese ginseng) [46]. Many active components are contained in ginseng, such as ginsenosides, polysaccharides (glycans), amino acids, minerals, volatile oils, and flavonoids. Ginsenosides are generally classified into three groups: protopanaxadiol (PPD-type), protopanaxatriol (PPT-type) and oleanolic acid (OA-type, Ro), Based on the chemical structure classified in different types such as Rb1, Rb2, Rc, Rd, Rg3, Rh3, Re, Rf, Rg1, Rg2 and Rh1. Ginsenoside binding to SIRT1. SIRT1 is attracted as a potential target in depression. Ginsenosides Rh2, Rg1, Rk1, and PPD shows significant antidepressant effect. Ginsenosid Rh2 related to the brain-derived neurotrophic factor/tropomyosin-related kinase B and SIRT1/NF-κB signaling pathways). Ginsenoside Rg1 upregulates the SIRT1-MAPK signaling pathway which further reduces the NF-κB transcriptional activity. Ginsenoside also

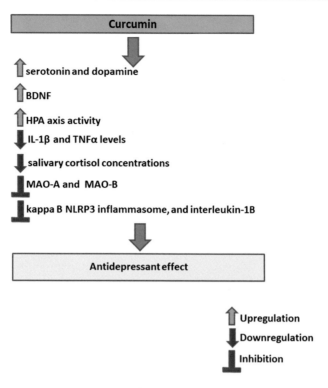

Figure 4.10

Role of Curcumin in mood disorder/depression: curcumin increases the serotonin, dopamine, BDNF, and HPA-axis activity. It also reduces the IL-1β, TNFα and inhibits the MAO-A, MAO-B, and interleukin-1B. All these effects lead to antidepressant.

downregulates Akt and mTOR [47]. Therefore, Ginsenosides are very useful for depression treatment (Fig. 4.11).

18.   Lion's mane mushroom (Hericium erinaceus)

Hericium erinaceus, a well-known edible mushroom this are used as traditional medicine in several Asian countries for treatment of various diseases [48]. Hericium erinaceus induce the various neurotrophic factors expression and monoamines and modulatation of inflammatory response. Mycelia and fruiting bodies found in H. erinaceus this are stimulate neurotrophic factor NGF through the JNK pathway. H. erinaceus inhibit the monoamine oxidase (MAO), upregulate the monoaminergic transmitters such as serotonin, norepinephrine, and dopamine [49], down regulate the pro-inflammatory factors IL-6, TNF-α and NF-κB, and the upregulation of BDNF [50].This leads to antidepressant activity. It showed the anti-oxidative and anti-inflammatory nature and also used for cognitive impairment. Although, H. erinaceous antidepressant activity is needed to validate and also compared with conventional

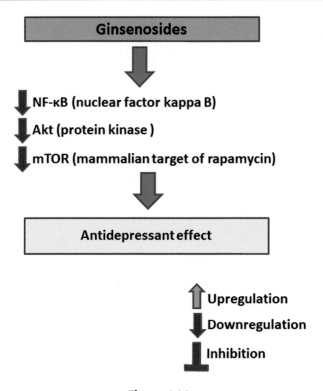

Figure 4.11
Role of Ginsenosides in mood disorder/depression: ginsenosides decreases the NF-$\kappa\beta$, Akt, and mTOR and showed the antidepressant nature.

antidepressants. Henceforth, it was suggested that H. erinaceus might be used as a potential alternative medicine for the treatment of depression [49] (Fig. 4.12).

19.   **Flavanones**

It was reported that dietary flavanones have the ability of neuroprotective and antidepressant effect. Among the dietary flavanones Naringenin (4,5,7-trihydroxy flavanone, NAR) is a natural flavanone, present in the peel of citrus fruits. It was reported that NAR could show antioxidative [51] and antidepressive (L.-T. [52]) (L. T. [52]). Another study investigated the role of NAR in CUMS Wistar rat model and shown that NAR inhibited the behavioral abnormalities in rats due to CUMS. It was also revealed that NAR repaired that morphological anomaly in the hippocampal CA1 region and cortex of Wistar rats due to CUMS. This study concluded that CUMS caused low expression of BDNF, Sonic hedgehog cell signaling and its downstream targets genes Nkx2.2 and Pax6 expression in the hippocampus of the Wistar rats. Intriguingly, all these down-regulated factors due to CUMS were upregulated in pre-NAR treated rats (Fig. 4.13) [53]. This study may suggest that NAR can be used as one of the potential antidepressants (Table 4.1).

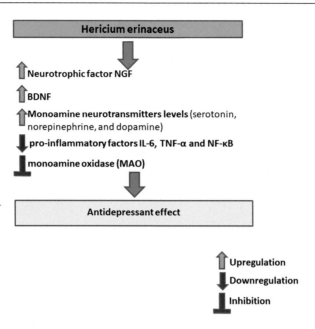

Figure 4.12
Role of Hericium erinaceus in mood disorder/depression: Hericium erinaceus upregulates neurotrophin factors NGF, BDNF, monoamine neurotransmitters, and decreases the proinflammatory factors IL-6, TNF-α, NF-$\kappa\beta$ and inhibit the monoamine oxidase.

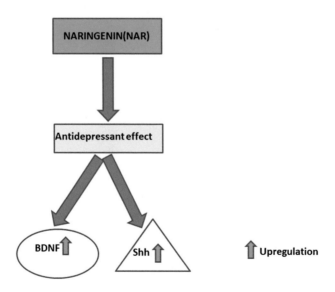

Figure 4.13
Role of Naringenin in mood disorder/depression: Naringenin showed antidepressant effect by increasing brain derived neurotrophic factor and sonic hedgehog cell signaling.

Table 4.1: Role of herbal medicine in depression/mood disorder.

S. No.	Herbal medicine	Effect on mediators	Reference
1.	Saffron	↑Neurotransmitter systems in brain (serotonin, dopamine, glutamate, and norepinephrine)	[8,9]
2.	Rhodiola	↑Stress response ↑GPCR) signaling pathways ↑HPA axis activity ↑NPY ↑Hsp-70 ↑Serotonin ↑Neurotransmitters (glutamate, acetylcholine, GABA, NA, and DA) ↑Hormones (oxytocin, vasopressin, cortisol, corticotropin, prolactin, substance P, and others) ↑Monoamine ↓MAO-A and MAO-B ↓JNK protein (suppressing glucocorticoid receptors and increasing cortisol)	[12,13]
3.	XYS	↑BDNF in hippocampus ↑5-HT in plasma and brain ↑HPA axis activity ↓CRH in hypothalamus ↓IDO1 in DRN ↓Oxaloacetate level in plasma	[14,15]
4.	Shuganjieyu capsule	↑Hippocampal neuron generation, survival, and neogenesis ↓protein levels of caspase-3	[19]
5.	GMDZ	↑Horizontal movement and vertical movement ↑Sugar consumption and Amount of food consumed ↑Norepinephrine (NE) ↑5-Hydroxytryptamine receptors (5-HT)	[21]
6.	Berberine	↑NE, 5-HT and DA level ↑NGF secretion ↑NGF secretion, activation of phosphoinositide 3-kinase ↓Lipid peroxide and superoxide dismutase levels ↓Plasma corticosterone levels and fluctuations in gonadal hormone levels ↓monoamine oxidase activity ↓α2 autoreceptors ↓NOX and ROS	[25]
7.	Melissa officinalis L	↓Corticosterone in serum ↓GABA-T levels in the hippocamp	[29]
8.	Lavender	↑Noradrenaline levels ↑Dopamine levels	[33]
9.	Hypericum perforatum	↑Serotonin, dopamine and norepinephrine ↑5- HT2 receptors in the frontal cortex ↓ß-Adrenoreceptors ↓MAO-A and NAO- B	[39,40,41]

(continued on next page)

Table 4.1: Role of herbal medicine in depression/mood disorder—cont'd

S. No.	Herbal medicine	Effect on mediators	Reference
10.	Curcumin	↑serotonin and dopamine ↑BDNF ↑HPA axis activity ↓IL-1β and TNFα levels ↓salivary cortisol concentrations ↓MAO-A and MAO-B ↓ kappa B NLRP3 inflammasome and interleukin-1B	[42, 43, 44]
11.	Ginsenosides	↓NF-κB (nuclear factor kappa B) ↓Akt (protein kinase) ↓mTOR (mammalian target of rapamycin)	[47]
12.	Hericium erinaceus	↑Neurotrophic factor NGF ↑Neurotrophic factor NGF ↑Monoamine neurotransmitters levels (serotonin, norepinephrine, and dopamine) ↓proinflammatory factors IL-6, TNF-α and NF-κB ↓monoamine oxidase (MAO)	[49,50]
13.	Naringenin (NAR)	↑BDNF ↑Shh	[53]

4.3 Conclusion

Depressive disorders are prevalent mental disorder. This causes 264 million people worldwide to affect the depression. Different type of depression treatment available but current psychiatric drugs cause many induce side effects, such as sexual dysfunction, metabolic syndrome, dependency, weight gain, etc. Herbal medicines can solve this problem to some extent; with minimum side effects. However, we need more scientific studies and investigation to explore the role of herbal medicine in the treatment of depression.

References

[1] JM Rubio, JC Markowitz, A Alegría, G Pérez-Fuentes, S-M Liu, K-H Lin, C Blanco, Epidemiology of chronic and nonchronic major depressive disorder: results from the national epidemiologic survey on alcohol and related conditions, Depress Anxiety 28 (2011) 622–631. https://doi.org/10.1002/da.20864.

[2] L Andrade, JJ Caraveo-Anduaga, P Berglund, RV Bijl, R De Graaf, W Vollebergh, E Dragomirecka, R Kohn, M Keller, RC Kessler, N Kawakami, C Kiliç, D Offord, TB Ustun, H-U Wittchen, The epidemiology of major depressive episodes: results from the International Consortium of Psychiatric Epidemiology (ICPE) Surveys, Int J Methods Psychiatr Res 12 (2003) 3–21. https://doi.org/10.1002/mpr.138.

[3] AA Noorbala, S Akhondzadeh, N Tahmacebi-Pour, AH Jamshidi, Hydro-alcoholic extract of Crocus sativus L. versus fluoxetine in the treatment of mild to moderate depression: a double-blind, randomized pilot trial, J Ethnopharmacol 97 (2005) 281–284. https://doi.org/10.1016/j.jep.2004.11.004.

[4] ME Medina-Mora, G Borges, C Benjet, C Lara, P Berglund, Psychiatric disorders in Mexico: lifetime prevalence in a nationally representative sample, Br J Psychiatry 190 (2007) 521–528. https://doi.org/10.1192/bjp.bp.106.025841.

[5] K Gao, DE Kemp, E Fein, Z Wang, Y Fang, SJ Ganocy, JR Calabrese, Number needed to treat to harm for discontinuation due to adverse events in the treatment of bipolar depression, major depressive disorder,

and generalized anxiety disorder with atypical antipsychotics, J Clin Psychiatry 72 (2011) 1063–1071. https://doi.org/10.4088/JCP.09r05535gre.

[6]   L Liu, C Liu, Y Wang, P Wang, Y Li, B Li, Herbal medicine for anxiety, depression and insomnia, Curr Neuropharmacol 13 (2015) 481–493. https://doi.org/10.2174/1570159x1304150831122734.

[7]   Y Wang, T Han, Y Zhu, CJ Zheng, QL Ming, K Rahman, L.P Qin, Antidepressant properties of bioactive fractions from the extract of crocus sativus L, J Nat Med 64 (2010) 24–30. https://doi.org/10.1007/s11418-009-0360-6.

[8]   F Jalali, SF Hashemi, The Effect of Saffron on Depression among Recovered Consumers of Methamphetamine Living with HIV/AIDS, Subst Use Misuse 53 (2018) 1951–1957. https://doi.org/10.1080/10826084.2018.1447583.

[9]   N Shahmansouri, M Farokhnia, SH Abbasi, SE Kassaian, Noorbala Tafti, Gougol AA, Yekehtaz A, Forghani H, Mahmoodian S, Saroukhani M, Arjmandi-Beglar S, Akhondzadeh AK, A randomized, double-blind, clinical trial comparing the efficacy and safety of Crocus sativus L. with fluoxetine for improving mild to moderate depression in post percutaneous coronary intervention patients, J Affect Disord 155 (2014) 216–222. https://doi.org/10.1016/j.jad.2013.11.003.

[10]   S Akhondzadeh, H Fallah-Pour, K Afkham, AH Jamshidi, F Khalighi-Cigaroudi, Comparison of Crocus sativus L. and imipramine in the treatment of mild to moderate depression: A pilot double-blind randomized trial [ISRCTN45683816], BMC Complement Altern Med 4 (2004) 1–5. https://doi.org/10.1186/1472-6882-4-12.

[11]   JJ Mao, SX Xie, J Zee, I Soeller, QS Li, K Rockwell, JD Amsterdam, Rhodiola rosea versus sertraline for major depressive disorder: A randomized placebo-controlled trial, Phytomedicine 22 (2015) 394–399. https://doi.org/10.1016/j.phymed.2015.01.010.

[12]   L Gao, C Wu, Y Liao, J Wang, Antidepressants effects of Rhodiola capsule combined with sertraline for major depressive disorder: a randomized double-blind placebo-controlled clinical trial, J Affect Disord 265 (2020) 99–103. https://doi.org/10.1016/j.jad.2020.01.065.

[13]   JD Amsterdam, AG Panossian, Rhodiola rosea L. as a putative botanical antidepressant, Phytomedicine 23 (2016) 770–783. https://doi.org/10.1016/j.phymed.2016.02.009.

[14]   M Wang, W Huang, T Gao, X Zhao, Z Lv, Effects of Xiao Yao San on interferon-α-induced depression in mice, Brain Res Bull 139 (2018) 197–202. https://doi.org/10.1016/j.brainresbull.2017.12.001.

[15]   X Liu, C Liu, J Tian, X Gao, K Li, G Du, X Qin, Plasma metabolomics of depressed patients and treatment with Xiaoyaosan based on mass spectrometry technique, J Ethnopharmacol 246 (2020). https://doi.org/10.1016/j.jep.2019.112219.

[16]   Y Yu, G Zhang, T Han, Analysis of the pharmacological mechanism of Banxia Xiexin decoction in treating depression and ulcerative colitis based on a biological network module, BMC Complement Med Ther 3 (2020) 1–13. https://doi.org/10.1186/s12906-020-02988-3.

[17]   C Kwon, B Lee, S Chung, JW Kim, A Shin, Y Choi, Y Yun, J Leem, Herbal medicine Sihogayonggolmoryeo-tang or Chai-Hu-Jia-Long-Gu-Mu-Li-Tang for the treatment of post-stroke depression, Medicine (Baltimore) 97 (2018) 1–6. https://doi.org/10.1097/MD.0000000000012384.1.

[18]   B Lee, Herbal Medicine (Sihogayonggolmoryeo-Tang or Chai-Hu-Jia-Long-Gu-Mu-Li-Tang) for treating hypertension: a systematic review and meta-analysis, Evid Based Complement Altern Med 2020 (2020) 1–13. https://doi.org/10.1155/2020/9101864.

[19]   X Zhang, D Kang, L Zhang, L Peng, Shuganjieyu capsule for major depressive disorder (MDD) in adults: a systematic review, Aging Ment Heal 18 (2014) 941–953. https://doi.org/10.1080/13607863.2014.899975.

[20]   JH Jun, TY Choi, JA Lee, KJ Yun, MS Lee, Herbal medicine (Gan Mai da Zao decoction) for depression: A systematic review and meta-analysis of randomized controlled trials, Maturitas 79 (2014) 370–380. https://doi.org/10.1016/j.maturitas.2014.08.008.

[21]   SR Kim, HW Lee, JH Jun, BS Ko, Effects of herbal medicine (Gan Mai Da Zao decoction) on several types of neuropsychiatric disorders in an animal model: A systematic review - Herbal medicine for animal studies of neuropsychiatric diseases, J Pharmacopuncture 20 (2017) 5–9. https://doi.org/10.3831/KPI.2017.20.005.

[22]   I Kyrou, A Christou, D Panagiotakos, C Stefanaki, K Skenderi, K Katsana, C Tsigos, Effects of a hops (Humulus lupulus L.) dry extract supplement on self-reported depression, anxiety and stress levels in

apparently healthy young adults: a randomized, placebo-controlled, double-blind, crossover pilot study, Hormones 16 (2017) 171–180.

[23] P Zanoli, M Zavatti, Pharmacognostic and pharmacological profile of Humulus lupulus L, J Ethnopharmacol 116 (2008) 383–396. https://doi.org/10.1016/j.jep.2008.01.011.

[24] AFG Cicero, A Baggioni, Berberine and Its Role in Chronic Disease, Adv Exp Med Biol 928 (2016) 27–45. https://doi.org/10.1007/978-3-319-41334-1_2.

[25] J Fan, K Zhang, R Cui, Y Jin, B Li, Pharmacological effects of berberine on mood disorders, J Cell Mol Med (2019) 21–28. https://doi.org/10.1111/jcmm.13930.

[26] D Xiao, Z Liu, S Zhang, M Zhou, F He, M Zou, J Peng, X Xie, Y Liu, D Peng, Berberine derivatives with different pharmacological activities via structural modifications, Mini Rev Med Chem 18 (2018) 1424–1441. https://doi.org/10.2174/1389557517666170321103139.

[27] J Pan, X Lei, J Wang, S Huang, Y Wang, Y Zhang, W Chen, D Li, Effects of Kaixinjieyu, a Chinese herbal medicine preparation, on neurovascular unit dysfunction in rats with vascular depression, BMC Complement Altern Med 15 (2015) 1–14. https://doi.org/10.1186/s12906-015-0808-z.

[28] M Araj-khodaei, AA Noorbala, R Yarani, F Emadi, E Emaratkar, A double-blind, randomized pilot study for comparison of Melissa officinalis L. and Lavandula angustifolia Mill. with Fluoxetine for the treatment of depression, BMC Complement Med Ther 5 (2020) 1–9. https://doi.org/10.1186/s12906-020-03003-5.

[29] H Haybar, AZ Javid, MH Haghighizadeh, E Valizadeh, SM Mohaghegh, A Mohammadzadeh, The effects of Melissa officinalis supplementation on depression, anxiety, stress, and sleep disorder in patients with chronic stable angina, Clin Nutr ESPEN 26 (2018) 47–52. https://doi.org/10.1016/j.clnesp.2018.04.015.

[30] J Ghazizadeh, S Hamedeyazdan, M Torbati, F Farajdokht, A Fakhari, J Mahmoudi, M Araj-Khodaei, S Sadigh-Eteghad, Melissa officinalis L. hydro-alcoholic extract inhibits anxiety and depression through prevention of central oxidative stress and apoptosis, Exp Physiol 105 (2020) 707–720. https://doi.org/10.1113/EP088254.

[31] G Buchbauer, L Jirovetz, W Jäger, H Dietrich, C Plank, Aromatherapy: evidence for sedative effects of the essential oil of lavender after inhalation, Z Naturforsch C 46 (1991) 1067–1072. https://doi.org/10.1515/znc-1991-11-1223.

[32] T Itai, H Amayasu, M Kuribayashi, N Kawamura, M Okada, A Momose, T Tateyama, K Narumi, W Uematsu, S Kaneko, Psychological effects of aromatherapy on chronic hemodialysis patients, Psychiatry Clin Neurosci 54 (2000) 393–397. https://doi.org/10.1046/j.1440-1819.2000.00727.x.

[33] MR Bazrafshan, M Jokar, N Shokrpour, H Delam, The effect of lavender herbal tea on the anxiety and depression of the elderly: a randomized clinical trial, Complement Ther Med 50 (2020). https://doi.org/10.1016/j.ctim.2020.102393.

[34] M Abolhassani, Antiviral activity of borage (Echium amoenum), Arch Med Sci 6 (2010) 366–369. https://doi.org/10.5114/aoms.2010.14256.

[35] Mehdi Sayyah, Mohammad Sayyah, M Kamalinejad, A preliminary randomized double blind clinical trial on the efficacy of aqueous extract of Echium amoenum in the treatment of mild to moderate major depression, Prog Neuropsychopharmacol Biol Psychiatry 30 (2006) 166–169. https://doi.org/10.1016/j.pnpbp.2005.10.005.

[36] R Rahimi, S Nikfar, M Abdollahi, Efficacy and tolerability of Hypericum perforatum in major depressive disorder in comparison with selective serotonin reuptake inhibitors: a meta-analysis, Prog Neuropsychopharmacol Biol Psychiatry 33 (2009) 118–127. https://doi.org/10.1016/j.pnpbp.2008.10.018.

[37] K Linde, MM Berner, L Kriston, St John's wort for major depression, Cochrane database Syst Rev 2008 (2008) CD000448. https://doi.org/10.1002/14651858.CD000448.pub3.

[38] J Barnes, LA Anderson, J.D Phillipson, St John's wort (Hypericum perforatum L.): a review of its chemistry, pharmacology and clinical properties, J Pharm Pharmacol 53 (2001) 583–600. https://doi.org/10.1211/0022357011775910.

[39] A Eatemadnia, S Ansari, P Abedi, S Najar, The effect of Hypericum perforatum on postmenopausal symptoms and depression: a randomized controlled trial, Complement Ther Med 45 (2019) 109–113. https://doi.org/10.1016/j.ctim.2019.05.028.

[40] AL Miller, St. John's Wort (Hypericum perforatum): clinical effects on depression and other conditions, Altern Med Rev 3 (1998) 18–26.

[41] V Butterweck, Mechanism of action of St John's wort in depression what is known? CNS Drugs 17 (2003) 539–562.

[42] L Fusar-Poli, L Vozza, A Gabbiadini, A Vanella, I Concas, S Tinacci, A Petralia, MS Signorelli, E Aguglia, Curcumin for depression: a meta-analysis, Crit Rev Food Sci Nutr 60 (2020) 2643–2653. https://doi.org/10.1080/10408398.2019.1653260.

[43] QX Ng, SSH Koh, HW Chan, CYX Ho, Clinical use of curcumin in depression: a meta-analysis, J Am Med Dir Assoc 18 (2017) 503–508. https://doi.org/10.1016/j.jamda.2016.12.071.

[44] FN Kaufmann, M Gazal, CR Bastos, MP Kaster, G Ghisleni, Curcumin in depressive disorders: an overview of potential mechanisms, preclinical and clinical findings, Eur J Pharmacol 784 (2016) 192–198. https://doi.org/10.1016/j.ejphar.2016.05.026.

[45] X Huang, N Li, Y Pu, T Zhang, B Wang, Neuroprotective effects of ginseng phytochemicals: recent perspectives, Molecules 24 (2019) 1–20. https://doi.org/10.3390/molecules24162939.

[46] Y Kim, SH Cho, The effect of ginsenosides on depression in preclinical studies: a systematic review and meta-analysis, J Ginseng Res 45 (2021) 420–432. https://doi.org/10.1016/j.jgr.2020.08.006.

[47] T Lou, Q Huang, H Su, D Zhao, X Li, Targeting Sirtuin 1 signaling pathway by ginsenosides, J Ethnopharmacol. 268 (2021) 1–39. https://doi.org/10.1016/j.jep.2020.113657.

[48] M Nagano, K Shimizu, R Kondo, C Hayashi, D Sato, K Kitagawa, K Ohnuki, Reduction of depression and anxiety by 4 weeks Hericium erinaceus intake, Biomed Res 31 (2010) 231–237. https://doi.org/10.2220/biomedres.31.231.

[49] PS Chong, ML Fung, KH Wong, LW Lim, Therapeutic potential of Hericium erinaceus for depressive disorder, Int J Mol Sci 21 (2020). https://doi.org/10.3390/ijms21010163.

[50] F Limanaqi, F Biagioni, CL Busceti, M Polzella, C Fabrizi, F Fornai, Potential antidepressant effects of scutellaria baicalensis, hericium erinaceus and rhodiola rosea, Antioxidants 9 (2020). https://doi.org/10.3390/antiox9030234.

[51] HJ Heo, DO Kim, SC Shin, MJ Kim, BG Kim, DH Shin, Effect of antioxidant flavanone, naringenin, from Citrus junos on neuroprotection, J Agric Food Chem 52 (2004) 1520–1525. https://doi.org/10.1021/jf035079g.

[52] L-T Yi, J Li, H-C Li, D-X Su, X-B Quan, X-C He, X-H Wang, Antidepressant-like behavioral, neurochemical and neuroendocrine effects of naringenin in the mouse repeated tail suspension test, Prog Neuropsychopharmacol Biol Psychiatry 39 (2012) 175–181. https://doi.org/10.1016/j.pnpbp.2012.06.009.

[53] M Tayyab, S Farheen, MPM Mariyath, N Khanam, Mobarak Hossain, MMH Shahi, Antidepressant and neuroprotective effects of naringenin via Sonic Hedgehog-GLI1 cell signaling pathway in a rat model of chronic unpredictable mild stress, NeuroMolecular Med 21 (2019) 250–261. https://doi.org/10.1007/s12017-019-08538-6.

[54] GBD, Global, regional, and national incidence, prevalence, and years lived with disability for 310 diseases and injuries, 1990-2015: a systematic analysis for the Global Burden of Disease Study 2015, Lancet (London, England) 388 (2016) 1545–1602. https://doi.org/10.1016/S0140-6736(16)31678-6.

PART 2

Oral health

CHAPTER 5

Herbal medicines: A boon for healthy human life

Jonathan M. Banks[a], Daniela A. Brandini[b], Debora B. Barbosa[b], Aline S. Takamiya[b], Pari Thakkar[a], Kathy Zheng[a] and Afsar R. Naqvi[a]

[a]Department of Periodontics, College of Dentistry, University of Illinois at Chicago, Chicago, Illinois, USA [b]Department of Diagnosis and Surgery, São Paulo State University (UNESP), School of Dentistry, Araçatuba, São Paulo, Brazil

5.1 Introduction

5.1.1 Plant products and their benefits

Plants and their derived extracts have been used for their believed health benefits for millennia. While plants have had positive and negative effects, with some acting as therapies and others as toxins, they were used to help people find relief throughout human history [1]. With hundreds of thousands of plant species on earth that are safe for human consumption, the possibilities for plant-based products and therapies remain largely understudied and unrealized. Commonly known and often used plant products such as ginger are widely renowned for their anti-inflammatory and pain-relieving abilities [2]. Essential oils and natural products like cinnamon are touted for their potency as multi-compound containing therapies, and they are considered to have an advantage over single-compound synthetic drugs because of their ability to combat pathogens in the body through a variety of mechanisms and avenues [3]. These oils are plant derivatives with often very powerful compounds that help aid plants in fending off microbes during their lifetime. These same compounds yield effective antimicrobial results when used in traditional medicine; in fact, approximately 60% of essential oils have antifungal compounds, and about 35% of them feature antibacterial activity, which can be used to fight disease-causing bacterium throughout the body [4]. To fight diseases in the oral cavity, flowering plants from the *Rosa* genus have been used to treat mild cases of oral inflammation, and essential oils from *Syzygium aromaticum* have been used to relieve oral pain associated with dental caries [5].

It has been estimated that approximately 80% of the population in developing countries still relies on traditional medicines [6]. For example, Ayurveda, the ancient Indian system of treating overall health, entails the prescription of specific herbs and minerals in order to help

Herbal Medicines: A Boon for Healthy Human Life.
DOI: https://doi.org/10.1016/B978-0-323-90572-5.00015-9

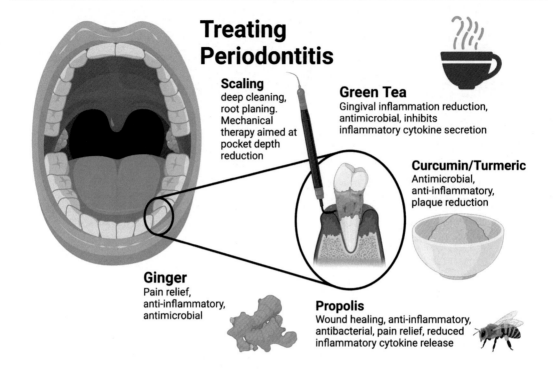

Figure 5.1
Treating periodontitis. Effective procedural and herbal treatments for periodontal inflammation and diseased periodontal tissue, including the benefits for each therapy. *Figure created with BioRender.com, 2021.*

cure various diseases, which can be used as an adjunct for oral health maintenance [7]. It has been found that several of these herbs have antibacterial, antiulcer genic, wound healing, anti-inflammatory, antimicrobial, and antioxidant properties, thus positively impacting oral health (Fig. 5.1, Table 5.1). These antibacterial properties are due to the potential bioactive compounds found in plants, thus aiding in the bacterial load present in the mouth. By reducing the bacterial load, plaque, dental caries, and ulcer formation are prevented. However, very few of these plant-based products are used on a day-to-day basis in clinical dental practice.

Outside of the body, plant products have been sought after for their antimicrobial capabilities in the field of food preservation [8,9]. However, within the body, the antimicrobial effects of plant products have been used for a variety of therapeutic purposes, and due to surge in studies examining the active compounds within these extracts, the molecular mechanisms behind these therapies are continually being brought to light (Table 5.1). In Brazil, *Hypericum connatum* extracts have been used to treat herpesvirus infections in gingivostomatitis and ulcers in the mouth [10]. Historically, Miswak, a derivative of the Arak plant *Salvadora persica*, is used to resist the formation of tooth caries by making it into chewing sticks [11]. The compound Plumbagin, derived from the roots of South African plants of the genus *Plumbago*, has been

Table 5.1: Plant products, their derivatives, and their effects on disease.

Plant product and derivatives	Observed diseases treated and product effects	Sources
Rosa genus flowering plants	Treat oral inflammation	[5]
Syzygium aromaticum essential oils	Dental caries pain relief	[5]
Hypericum connatum	Treat gingivostomatitis and ulcers arising from herpesvirus infections	[10]
Miswak, *Salvadora persica* derivative, mustard green	Preventative dental caries treatment, plaque reduction in toothpaste, anticariogenic, mixed in sealer cement for endodontic treatment	[11,38,40,136]
Plumbagin, *Plumbago* genus root derivative	Plasmid curing in *E. coli*, effective against resistant bacteria	[12,13,186,187]
Drynaria fortune	Treats inflammation, antibacterial. Effective against *S. mutans*, *P. gingivalis*, other prevalent species	[14]
PM014	Anti-inflammatory in pulmonary tissue, reduce inflammation caused by radiation, reduce inflammatory cytokine production by inhibiting NF-κB pathway	[16]
Licorice, *Glycyrrhiza glabra*, also known as Yashtimadhu Extracts: Isoliquiritigenin/ISL, licoricidin/LC, and licorisoflavan A/LIA	Reduce proinflammatory cytokine secretion, antioxidant, antiplatelet aggregation, plaque reduction in mouthwash, anticariogenic with minimal oral microbiome perturbation. Deemed Generally Recognized as Safe by the US Food and Drug Administration, and capable of reducing lipopolysaccharide-induced inflammatory cytokine production and secretion	[18,19,38,40,41,229]
Phloretin	Antioxidant, reduced anti-inflammatory cytokine secretion, treatment for prolonged inflammation	[20]
Silymarin	Antioxidant, reduced anti-inflammatory cytokine secretion, treatment for prolonged inflammation	[20]
Hesperetin	Antioxidant, reduced anti-inflammatory cytokine secretion, treatment for prolonged inflammation	[20]
Resveratrol	Antioxidant, reduced anti-inflammatory cytokine secretion, treatment for prolonged inflammation	[20]
Garlic, *Allium sativum*	Antimicrobial against *S. mutans*, evidenced by zone of inhibition, anti-cariogenic, antifungal treatment in rinse for denture stomatitis (DS), antifungal management	[21,40,51,54]
Ginger, *Zingiber officinale*	Antimicrobial against *S. mutans*, evidenced by zone of inhibition, anti-cariogenic, antifungal management for denture stomatitis (DS)	[21,40,54]
Lemon	Antimicrobial against *S. mutans*, evidenced by zone of inhibition	[21]
Honey	Antimicrobial against *S. mutans*, evidenced by zone of inhibition	[21]
Sage, *Salvia officinalis*	Anticariogenic (affecting *S. mutans*)	[40]

(*continued on next page*)

Table 5.1: Plant products, their derivatives, and their effects on disease—cont'd

Plant product and derivatives	Observed diseases treated and product effects	Sources
Guava, *Psidium guajava*	Anticariogenic (affecting *S. mutans*)	[40]
Oregano, *Origanum dubium*	Anticariogenic (affecting *S. mutans*)	[40]
Cinnamon, *Cinnamomum cassia*	Anticariogenic (affecting *S. mutans*)	[40]
Peppermint, *Mentha piperita*	Anticariogenic (affecting *S. mutans*)	[40]
Mint, *Mentha arvensis*	Anticariogenic (affecting *S. mutans*)	[40]
Mango, *Mangifera indica*	Plaque reduction in mouthwash, anticariogenic (affecting *S. mutans*)	[38,40]
Lemon, *Citrus limon*	Anticariogenic (affecting *S. mutans*)	[40]
Rosemary, *Rosmarinus officinalis*	Anticariogenic (affecting *S. mutans*)	[40]
Aloe, *Aloe vera*, active substances: acemannan, anthraquinone, anthracene and cinnamonic acid	Plaque reduction in mouthwash, anticariogenic (affecting *S. mutans*), antibacterial, antifungal, antivirus, anti-inflammatory properties, natural healer, promotes cell proliferation and growth, preservation of periodontal ligament cells	[38,40,131,170, 173,185]
Cashew, *Anacardium occidentale*	Anticariogenic (affecting *S. mutans*), plaque and gingival inflammation reduction	[40,62]
Pomegranate, *Punica granatum*	Plaque removal in toothpaste, anticariogenic (affecting *S. mutans*), antifungal management for denture stomatitis (DS)	[38,40, 54]
Melaleuca alternifolia	Plaque reduction in mouthwash, Antifungal treatment, management for denture stomatitis (DS)	[38,53,54]
Zataria multiflora	Antifungal treatment for denture stomatitis (DS)	[52,54]
Propolis	Antifungal management for denture stomatitis (DS), periodontal disease treatment, wound healing, anti-inflammatory, antibacterial, reduction in dental plaque, antiviral, antioxidant properties, intracanal medication to prevent pain, storage media for avulsed teeth, reduced inflammatory cytokine expression	[54,93–95,97,98,159, 160,174,179–181]
Curcumin, Curcuminoid, Turmeric, *Curcuma longa*	Antifungal management for denture stomatitis (DS), periodontal disease reduction, subgingival irrigator in mouthwash, toothpaste, plaque and gingival inflammation reduction, antioxidant, anti-inflammatory, antimicrobial, reduction in NF-κB activity, inhibiting inflammatory cytokine secretion pathway	[54,58,62,69–72, 193]
Green tea, *Camellia sinensis*	Antifungal management for denture stomatitis (DS), plaque and gingival inflammation reduction, protective antioxidants, decrease in pro-inflammatory cytokines, root canal irrigator, antibacterial components, anticarcinogenic	[54,62,104,105, 143,146,147]

(continued on next page)

Table 5.1: Plant products, their derivatives, and their effects on disease—cont'd

Plant product and derivatives	Observed diseases treated and product effects	Sources
Ricinus communis	Antifungal management for denture stomatitis (DS)	[54]
Lawsonia inermis	Antifungal management for denture stomatitis (DS)	[54]
Uncaria tomentosa	Antifungal management for denture stomatitis (DS)	[54]
Satureja hortensis	Antifungal management for denture stomatitis (DS)	[54]
Pelargonium graveolens	Antifungal management for denture stomatitis (DS)	[54]
Artemisia Sieberi	Antifungal management for denture stomatitis (DS)	[54]
Chamomile, *Matricaria chamomilla*	Plaque reduction in toothpaste, mouthwash, anti-inflammatory polyphenols reduce inflammatory cytokine levels	[38,194]
Neem, *Azadirachta indica*	Plaque reduction in toothpaste, mouthwash, gingival inflammation reduction, antibacterial, antifungal, pain relief, biofilm bacterial reduction as irrigator	[38,62,150–153,155]
Lippia sidiodes	Plaque reduction in toothpaste	[38]
Juniper	Plaque reduction in mouthwash	[38]
Schinus terebinthifolius	Plaque and gingival inflammation reduction	[62]
Protium heptaphyllum	Antti-inflammatory, pain relief, wound repair, reduced periodontal disease	[74,76]
Resveratrol	Anti-inflammatory, antioxidant, biofilm formation inhibitor, decreases inflammatory cytokine levels	[80,87]
Triphala	Anti-inflammatory, antioxidant, antimicrobial, antiseptic, treats halitosis, oral ulcers, prevents oral caries, reduces plaque accumulation	[113–119,123,127]
Moringa oleifera	Biofilm inhibitor for *E. faecalis*	[135]
Grape seed extract, *Vitis vinifera*	Irrigating solution for root canals, antimicrobial, antioxidant, immunomodulatory affecter	[142,165,167]
Chitosan	Antimicrobial effects on *C. albicans* and *E. faecalis*, interacts with bacterial cell membrane, alters membrane permeability	[168,169]
Amla, *Emblica officinalis*	Antimicrobial effect as a component in sealer cement	[136]
Nutmeg, *Myristica fragrans*	Antimicrobial effect as a component in sealer cement, especially against *Bacteroides fragilis*	[136]
Psidium spp.	Intracanal medication against bacteria such as *C. albicans, E. faecalis*. Biofilm inhibitor, antimicrobial	[176]
Coconut, *Cocos nucifera L*	Water can be used as blood plasma substitute to replace electrolytes, sugars, fluids	[183]
Copaiba oil, *Copaifera* spp.	Antimicrobial essential oil, effective against *Streptococcus mutans, Streptococcus sanguinis*	[188,189]
Meadowsweet, *Filipendula ulmaria*	Anti-inflammatory polyphenols reduce inflammatory cytokine levels	[194]
Willow bark, *Salix alba L*	Anti-inflammatory polyphenols reduce inflammatory cytokine levels	[194]
Inula viscosa	Anti-inflammatory, decreased p65/RelA levels, inhibited NF-κB pathway	[203,204]

Compiled list of documented herbal medicines, including their beneficial properties.

found to perform plasmid curing to remove multidrug-resistant (MDR) plasmids from *E. coli*, effectively wiping out the resistance of the bacteria by decreasing the number of plasmids copied and by decreasing the loss of fitness associated with plasmid loss by eliminating the typical toxic effect [12,13]. *Drynaria fortunei*, a plant historically used in traditional Korean medicine for treating inflammation due to its antibacterial effects, has been found remarkably effective as an antibacterial agent due to its polyphenols and flavonoids [14]. It is worth noting that, in other plant-based compounds, flavonoids have served to relieve oxidative stress in plants, and they have been associated with anti-inflammatory and antiviral effects in various plant species [15]. Lastly, PM014 is an extract containing seven key natural components from the Chung-Sang-Bo-Ha-Tang mixture, a mixture used for its anti-inflammatory potency in diseases involving inflamed pulmonary tissue [16].

These medicines have the advantage of multiple versatile compounds with active sites possessing various functions and abilities, allowing these herbs to provide relief for a multitude of ailments. However, a major setback in clinical use of plan-derived medicines is lack of knowledge on their mode of action. Advancing research in this field can provide the molecular basis of how plant derivatives can enhance human health to combat plethora of microbes. In this chapter, we will summarize recent advances in our understanding of plant-based products in improving oral health.

5.1.2 Impact on oral health

The oral cavity in humans is home to several microbes that can cause disease, including viruses, fungi, Archaea, protozoa and Eubacteria [17]. These oral microbes cause very common human diseases, including periodontal disease and dental caries, the two most common infectious diseases amongst humans. Periodontitis is a gum infection that causes inflammation in the surrounding areas around the tooth by causing damage to the bone and soft tissue that provide a foundation for the tooth. Gingivitis is a milder and a reversible form of periodontitis.

Therefore, oral diseases have recently become one of the most important problems in developing countries and is at the forefront of public health tasks to address [7]. A large majority of these oral diseases are caused by bacterial infections. For several years now, chemical agents have been the primary form of treatment for oral hygiene management [17]. Examples of these chemical agents include toothpaste, mouth rinses, antibiotics, and synthetic antimicrobials. However, now oral microbes are displaying resistance to these same chemical agents. Therefore, several antibiotics, for example, have decreased clinical efficacy, more so towards individuals with HIV/AIDS, alongside their pre-existing negative side-effects and ability to alter oral microbiota. In light of these findings, conventional methods of treating bacterial infections have now been more receptive to incorporating plant-based products, and can have similar positive impact on oral health [7,17].

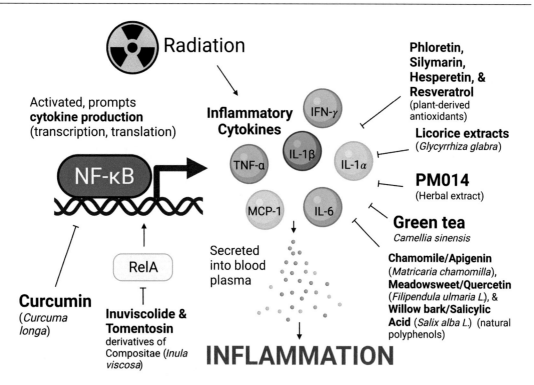

Figure 5.2
NF-kappa B inflammatory cytokines pathway and herbal medicine. Figure depicts plant-based derivatives and their influence on NF-kappa B activation, inflammatory cytokine secretion, and inflammation. *Figure created with BioRender.com, 2021.*

5.1.3 Pathways related to boosting immunity

Various studies have indicated that herbal medicines have the capability to operate at multiple levels and pathways, resulting in a much broader anti-inflammatory effect in the oral cavity than typical, pathway-oriented drugs. Furthermore, natural drugs are traditionally less expensive, and they possess fewer or no side effects, making them more tolerable and affordable for widespread use. Studies conducted on periodontal diseases have indicated that various herbal medicines have the capability to reduce the oral cavity's inflammatory response.

When tested using an *in vitro* macrophage model and *ex vivo* human whole-blood model, licorice extracts were found to greatly reduce lipopolysaccharide-induced proinflammatory cytokine secretion by macrophages and whole blood [18]. These were achieved by inhibiting the AP-1 and NF-κB pathways, resulting in the reduced secretions of proinflammatory cytokines, IL-1β, IL−6 and TNF-α (Fig. 5.2). In oral cancers, isoliquiritigenin (ISL), a flavonoid derived from licorice, has promising properties as a cancer chemotherapeutic agent [19]. Researchers found that ISL could mediate rapamycin (mTOR)-dependent autophagic and cell apoptosis in adenoid cystic carcinoma (ACC) through mTOR-dependent pathway. Not only is ISL safe for human consumption, but it also has anti-inflammatory, antioxidant, and antiplatelet aggregation properties (Table 5.1). This signaling pathway plays a vital role in regulating autophagy, which results in large degradations of cellular macromolecules and organelles. It is known that ISL suppresses the mTOR signaling pathway, however, many of the mechanisms behind ISL and the cytotoxicity it exerts are still unclear. Several other signal transduction pathways, like the p53-dependent pathway, have shown conflicting results as to whether ISL can induce apoptosis. A previous study has demonstrated that it can in HepG2 cells, but others have stated that it failed to affect the transcription activity of p53 [19].

Phloretin, Silymarin, Hesperetin, and Resveratrol were investigated to confirm whether they exhibit the ability to produce inflammatory mediators through innately-activated leukocytes [20]. These plant products are routinely used as antioxidants, which are critical for counteracting free radicals and oxidative stress. Peripheral blood mononuclear cells (PBMC) pretreated with each antioxidant exhibit inhibition of the Lipopolysaccharide (LPS)-induced cytokine secretion, which are the main pathogenic factor of immune diseases like periodontitis. As such, the production of key proinflammatory cytokines, TNF-α, IL-1α, IL-1β, IL-6, and IFN-γ, were inhibited (Fig. 5.2).

In addition to their anti-inflammatory capabilities, herbal medicines, like garlic, ginger, lemon, and honey, all have antimicrobial effects against *Streptococcus mutans*: the initiator of dental caries (Fig. 5.1, Table 5.1) [21]. Known for their antioxidative and anti-inflammatory properties, these plant derivatives have been used in traditional medicines throughout history worldwide. Researchers tested the efficiency of each commercially available product in isolation and in combination against *S. mutans* at varying concentrations. When the extracts were compared, the garlic in isolation showed the greatest antimicrobial activity with a mean zone of inhibition (34.9 $\pm$ 0.58 mm). This is largely due to the presence of allicin, which is the prominent antibacterial agent in garlic. Allicin is very effective against multiple strains of Gram-negative and positive bacteria. Honey in isolation showed the least antimicrobial activity (0.5 $\pm$ 0.6 mm). Zone of inhibition of 27.6 $\pm$ 0.43 mm, was observed with a combination of lemon and garlic had the greatest anti-microbicidal activity. Among the combinations, lemon and garlic combination showed the greatest zone of inhibition (27.6 $\pm$ 0.43 mm) [21].

5.2 Therapeutic use of plant products on oral diseases

According to the World Health Organization (WHO), in a document from its publication about The World Medicines Situation in 2011, therapeutic products including natural extracts, teas, vitamins and others similar were used by about 80% of worldwide population [22]. In 2019, WHO published the Global Report on Traditional and Complementary Medicine describing how many countries regulate the use of herbal medicines [23].

A healthy human body presents approximately 3.8×10^{13} cells of microorganisms, which living together are known as a microbiome [24,25]. The microbiome consists of a community of profuse bacteria, viruses, archaea, and fungi coexisting in a symbiotic relationship with the host. There exists a dynamic balance between them: a balance that, when disturbed by systemic disease, diet, or the prolonged use of antibiotics, can lead to the derangement of the microbial homeostasis. This phenomenon is known as dysbiosis [26]. When it occurs, typical microorganism-host interactions cease, and the body may become more susceptible to disease [26].

Mira et al. highlighted the importance of understanding the mechanisms involved in oral dysbiosis and promote oral symbiosis [27,28]. Through research and discovery, researchers have observed how plant-based compounds have influenced microbial activities within the oral cavity. These findings illuminate the potential uses of herbal medicine in prevalent oral diseases.

5.2.1 Caries and gingivitis

Dental caries, commonly known as cavities, are a major global public health concern, since they are included among the most prevalent diseases worldwide [29,30]. Data from the World Health Organization and others confirms the widespread prevalence of the disease. The data is appalling: 2.4 billion people, 35% of the global population, presented untreated caries in permanent teeth, making dental caries the most prevalent condition worldwide in 2010 [29,31]. Kassebaum et al. also reported that untreated caries in deciduous teeth were present in 9% of the global population [31]. With such prevalence, disease prevention must be at the center of managing dental caries. Fortunately, preventive oral health intervention in the past years has been on the rise [29,32]. In addition, the risks factors involved in dental caries must be properly assessed as well as factors affecting the oral microbiota [28,32].

The occurrence and development of oral diseases including dental caries are closely related to oral microorganisms. Caries results from acid production by acidogenic bacteria, which reproduce and form a biofilm when inhabiting a sugary environment [32]. However, similar oral environments have yielded different results as far as caries development, suggesting other factors may be at play [28]. These factors contribute to the microbial imbalance and environmental changes that alter the ecology and favor the growth and metabolism of key bacteria involved with the formation of caries [32].

Thus, in response to dental caries, a number of antimicrobial agents have been incorporated into oral care products to target the development and maturation of bacterial biofilms [29,33]. Despite this, Devine et al. in 2015 understood the need for therapies that maintained healthy microbial levels, leaving healthy, beneficial bacteria alive while killing pathogenic species [33]. Although the study's conclusions about the use of chlorhexidine for anticaries purpose have been inconsistent [34], their potent inhibitory activity against dental biofilms has been also proved [35]. At low and high concentrations, chlorhexidine can cause bacterial cytoplasmic leakage, cytoplasm condensation, and denaturation [29]. Though they can be effective antibacterial treatments, one concern about conventional chemical agents is their side effects. Besides the perturbation of the oral microbial balance, the long-term use of chlorhexidine gluconate in mouthwashes may lead to not only oral tissue discoloration but also oral mucosal ulcerations and paresthesia [36,37]. Because of their ability to regulate oral microbial activity to prevent caries while maintaining symbiosis, herbal medicines provide a clever alternative to typical chemical agents for dental practices.

While investigating the efficacy of herbal toothpastes and mouth rinses in reducing dental biofilm or gingivitis when compared to conventional chemical ones, Janakiram et al. conducted a systematic review and meta-analysis in 2020 by examining 493 randomized controlled trials [38]. Toothpastes with a variety of herbs and plant extracts were observed. For 712 adults polled from 11 studies, herbal toothpaste was superior in removing dental biofilm during a four-week period, and at long-term use (12 weeks) herbal and nonherbal toothpastes produced similar supra-gingival plaque (biofilm) scores. A small difference was found between herbal and nonherbal mouth rinses in reducing dental biofilm or gingivitis independently of the study duration. In this case, all trials used non-herbal mouth rinses based on chlorhexidine. This chemical agent presents strong evidence of anti-biofilm and anti-gingivitis effectiveness acting against a variety of microorganisms including bacteria, viruses, fungi, and yeasts. Chlorhexidine in mouth rinses is retained in the mouth for 12 h, and presents a sufficient substantivity, the persistence of active compounds' effectiveness, to inhibit or kill oral microorganisms [38,39]. On the other hand, the substantivity of oral herbal mouth rinses is unknown. According to the Janakiram review, the results were statistically significant, though the authors stopped short of recommending herbal mouth rinses over chlorhexidine rinses to combat dental biofilm formation and gingivitis.

Several herbal agents with anticariogenic properties are listed in a recent review published by researches from Iran [40]. Almost 30 anti-cariogenic in vitro studies involving numerous plants were selected from 2014 to 2020. Among them are well known plants such as sage (*Salvia officinalis*), ginger (*Zingiber officinale*), guava (*Psidium guajava*), licorice (Glycyrrhiza glabra), oregano (*Origanum dubium*), aromatic cinnamon (*Cinnamomum cassia*), peppermint (*Mentha piperita*), garlic (*Allium sativum*), mint (*Mentha arvensis*), mango (*Mangifera indica*), lemon (*Citrus limon*), rosemary (*Rosmarinus officinalis*), mustard (*Salvadora persica*), aloe (*Aloe vera*), cashew (*Anacardium occidentale*), and pomegranate (*Punica granatum*). Sixteen clinical trials using "herbal dentistry" for anticariogenic application were identified in this review. Mouthwashes (8 trials), lollipops [3], chewing stick or gum [3], gel [1], and herbal extract [1] were selected from 2015 to 2019. These clinical trials were particularly aimed on growth inhibition of *Streptococcus mutans* or rate of its eradication in biologic samples such as saliva. Particularly a clinical trial cited in this review demonstrated an interesting finding after giving liquorice lollipop with glycyrrhizol A for 37 preschool children [41]. Through quantitative PCR & 16S rRNA gene analysis before and after licorice-lollipop consumption by the children, *S. mutans* population was reduced without interfering in the oral microbiota diversity.

Results obtained by our own lab illustrated that the addition of pomegranate peel extract in non-alcohol mouthwashes containing a polyphosphate and lower concentrations of fluoride (patent deposit number: BR 10 2019 027251 1) improves its anticaries effect, in both in vitro and in situ studies. When compared to a commercial non-alcohol and low fluoride mouthwash, the mouthwash was not as effective against dual-species biofilm of *Streptococcus mutans* and *Candida albicans*. On the other hand, macrophages produced more cytokines after being in contact with that commercial mouthwash than the mouthwash containing pomegranate peel extract. The cytotoxicity was also statistically higher for the commercial mouthwash tested (Barbosa et al. Unpublished Results).

The revolutionary potential of plant-based remedies lies in their ability to target cariogenic bacteria while preserving the overall balance of the oral microbiome in a manner that promotes oral health. Added anti-inflammatory and antioxidant effects were observed in different plant extracts. These promising findings point to a bright future in plant-based caries treatment and prevention.

5.2.2 Denture stomatitis and candidiasis

Denture stomatitis (DS) is an oral *Candida*-associated inflammation [42,43], and it is one of the clinical types of the oral candidiasis [44]. DS is the term widely used in the literature to describe an inflammatory change [45] of palatal mucosa underlying complete and partial removable dentures [46]. Its prevalence can vary from 15% to over 70% among denture-wearing populations, according to a review of many studies with various population samples conducted by Gendreau and Loewy [47]. Although microbial biofilm remains the most relevant etiological factor of DS, a number of individual and denture factors predispose its occurrence and seriousness [45]. Denture overnight use, poor denture hygiene, denture age, and denture faults are factors present with a significant impact on DS [45,46,48], and patients should be informed of these risk factors by their clinician [49]. Factors related to the host (patient), including salivary flow, tobacco use, dietary factors, and chronic diseases like diabetes and other immunocompromised conditions can affect both the onset and the severity of DS [44,45,49,50], and these factors can make it challenging to control the disease.

One of the recommendations to reduce the odds of chronic candidiasis is to improve oral hygiene by the regular and frequent use of denture cleansers [49]. Also, antifungal therapy presents a considerable importance in the management strategy for individuals with associated diseases [49]. A systematic review and meta-analyses to assess the efficacy and effectiveness of interventions for the management of DS, with searches performed in 2016, revealed no evidence to provide any guidance regarding methods for preventing DS. Although, in general, medications containing nystatin and immersion of dentures in disinfectants seemed to be more effective than inactive controls, the current evidence from the randomized clinical trials selected in this review does not clarify whether one or more specific therapies is more effective for treating DS.

A systematic review and meta-analysis compared the efficacy of antifungals with alternative methods including disinfection agents, antiseptic mouthwashes, and natural products with antimicrobial properties. These treatments for denture stomatitis were reviewed by researchers from Montreal, Canada in 2014 [51]. They included all relevant randomized controlled trials (RCT) with at least 7 days as a period of treatment. In comparing chlorhexidine digluconate to amphotericin B, both were found effective in clinical and microbiological outcomes. However, 14 days after those treatments, recurrence of DS was noted. There were no differences between the use of Listerine® mouthwash and nystatin oral suspension or between hexetidine mouth rinse and fluconazole capsules. This review also mentioned two RCT which used natural substances [52,53]. Similar clinical positive results were observed when essential oil of *Melaleuca alternifolia* or nystatin was incorporated in a tissue conditioner over a period of 12 days [53]. Gels of 0.1% *Zataria multiflora* essential oil or 2%miconazole was applied on the internal surface of the upper denture four times per day for two weeks, and the follow up of two weeks was carried out [52]. Both treatments significantly reduced the palatal erythema when compared to the pre-treatment conditions. Although miconazole gel was more efficient in reducing the number of yeast colonies, both groups presented similar efficacy by day 21. The authors also cited a RCT study where patients used nystatin mouthwash or garlic aqueous solution for 4 weeks, and both produced statistically significant reduction in the length and width of erythema in palatal area. In that study an interesting finding was observed: patients reported greater satisfaction using the garlic rinse rather than the nystatin. The authors highlighted the side effects of the overuse of antifungals, including not only systemic effects such as gastrointestinal disturbances, renal and liver toxicity, hypersensitivity, and unwanted interactions with other drugs, but local problems like unpleasant tastes and microbial resistance were also mentioned. Additionally, the high relapse of DS after treatment with antifungals was observed from 2 weeks to a few months after the conclusion of the treatment. For those reasons, a combination of treatments may be required to combat DS, including antifungal therapy, dentures disinfection, removing dentures overnight, and denture adjustment, realigning, or reconstructing.

Thus, despite the high efficacy of nystatin, miconazole, amphotericin B, chlorhexidine and other conventional antifungals, their side effects, the high relapse incidence, and the emergence of resistant pathogens owing to its overuse have brought a worldwide concern [54]. The use of natural products has been emerging as an alternative of candidiasis management with clinical therapeutic effects and minimum adverse reactions. A systematic review and meta-analysis made by Shui et al. investigated whether or not herbal medicines or phytotherapy are as efficient as conventional antifungal therapies for DS management [54]. Almost a hundred relevant RCT trials were initially collected from electronic databases, but only 19 were included in the systematic review. Among the trials, 8 were from Iran, 7 from Brazil, 2 from Thailand, 1 from India and 1 from Chile. The potential natural compounds candidates considered to the management of DS were propolis, curcuminoid, garlic, *Camellia sinensis, Ricinus communis, Lawsonia inermis, Zingiber officinale, Uncaria tomentosa, Satureja hortensis, Pelargonium graveolens, Artemisia Sieberi, Melaleuca alternifolia, Zataria multiflora,* and *Punica granatum*. Natural compounds were like typical antifungals (miconazole and nystatin) clinically and microbiologically, and participants indicated fewer side effects and more satisfaction. The studies prescribed the use of both antifungals and natural compounds for 2 - 4 times/day during 7 – 30 days. Notably, the typical antifungals garnered the lowest satisfaction and highest reported side effects. When 1.25% *Punica granatum* gel and 2% miconazole were used, all patients manifested adverse reactions (mostly nausea and gastric disorders) after using miconazole while no complaint was made for the *P. granatum* gel. Propolis extract (20 mg/g) was approved by 95% of the patients versus 70% in miconazole group. Patients related desirable taste to 1% *Artemisia sieberi* mouthwash while 500,000 UI nystatin was undesirable. Garlic mouthwash (40 mg/mL) produced more satisfaction by the patients than nystatin (100,000 UI), and 75% related more side effects for nystatin. Also, ginger (20 mL) mouthwash yield 86.7% patient satisfaction in contrast to 13.3% for those using nystatin (500,000 UI). Natural compounds ultimately demonstrated varying levels of efficacy; some showed better, similar, or lower effectiveness compared with typical disinfectants (sodium hypochlorite, chlorhexidine, or Listerine®). Side effects such as burning sensation by using Listerine®, and staining and burning sensation on the tongues produced by chlorhexidine were noted by the patients. Only two studies were eligible for quantified meta-analyses, and they compared propolis and miconazole. There was no statistically significant difference between them for clinical and microbiological parameters evaluated. Although the meta-analysis did not provide conclusive evidence that herbal medicine may absolutely meet clinical demands without using conventional antifungal agents, the authors highlighted the promise and potential that herbal medicines present related to the treatment of DS due to their comparable efficacy and minimal side effects.

In conclusion, considering the fact that natural compounds present not only antimicrobial properties, but also anti-inflammatory effects [54–56], satisfactory biocompatibility [54], and virtually no side effects [54], plant-based treatments for mitigating DS collectively show tremendous promise and upside. Further, broader, and more comprehensive RCT experiments are warranted to continue exploring their potential as effective medicines and preventative options. Furthermore, the severity and chronic nature of DS, as is the case with many oral diseases, can be traced back to the condition of the oral microbiome and the relationships between oral microorganisms and the host. As such, herbal medicines used in treating the disease can only reach maximum efficacy by preserving the microbial balance within the oral cavity.

5.2.3 Periodontal diseases

There are many diseases that affect the periodontium; however, infectious diseases such as gingivitis and periodontitis are the primary focus for dental surgeons due to their wide incidence and prevalence. These pathologies are interconnected, with periodontitis being an evolution of gingivitis. Periodontitis represents a public health disease with its high incidence and negative impact on chewing ability, aesthetics, halitosis, and ultimately one's quality of life [57]. The disease is characterized as an inflammatory disease that leads to progressive destruction of the tooth's supportive tissues. This destruction is mediated by the host's immune response, and it is associated with dysbiosis due to dental plaque biofilm formation [57,58].

For the control and treatment of these diseases, the microbiome is often targeted. The main bacterial species implicated in periodontal disease are *Aggregatibacter actinomycetemcomitans, Dialister pneumosintes, Porphyromonas gingivalis, Prevotella intermedia, Prevotella nigrescens, Fusobacterium nucleatum, Tannerella forsythia*, and *Treponema denticola* [59]. Periodontitis is sometimes associated with other gram-negative and gram-positive microorganisms, mainly strict anaerobes, and these bacteria can modulate the immune response and alter the result of the treatment instituted [60]. As with any therapy, the first step is to change the patient behavior to promote the successful removal of supragingival dental biofilm and to minimize risk factors. Secondly, the dentist works to control subgingival biofilm and calculus formation, and thereafter treatment focuses on treating areas of the dentition that are not responding adequately to the second step of therapy. Following these therapies, periodontal stability is maintained and preventative treatment is administered [57]. For the steps involving patient behavior and periodontal health maintenance, dentifrices, toothpastes, powders, gel, or mouth rinses are used to keep the teeth and oral cavity clean while promoting fresh breath. The mechanical effectiveness of tooth brushing with water does not achieve complete biofilm reduction in most individuals [61], and the function of oral hygiene products that aid in plaque removal thus remain important.

The effectiveness of herbal oral care products in reducing dental plaque and gingivitis were review by Janakiram et al. [38]. In various toothpastes, chamomile, *Salvadora persica, Punica granatum, Azadirachta indica* and *Lippia sidiodes* were analyzed for plaque-reducing efficacy. A significantly greater reduction in plaque was observed for herbal toothpastes compared to nonfluoride toothpastes, but the difference disappeared when compared with fluoride toothpastes. The mouth rinses included in the study had chlorhexidine and natural products such as essential oils, licorice, *Melaleuca*, chamomile, *A. indica*, mango, *Aloe vera* and juniper, and the findings significantly favored the herbal mouth rinses compared to the nonherbal rinses. Santi et al. concluded the *Camelia sinensis, A. indica, Anacardium occidentale,*

Schinus terebinthifolius and *Curcuma longa* showed better results than chlorhexidine in dental plaque and gingival inflammation reduction [62].

Root planing, or scaling, is often the first choice for periodontal treatment; however, in more severe cases with periodontal pockets greater than 6 mm or in situations where the clinical picture cannot be improved with non-surgical mechanical therapy, the prescription of antibiotics [63] and/or local adjuvant treatments such as antimicrobial irrigators like chlorhexidine [64] as well as surgical periodontal therapies [65] may have additional benefits to assist scaling and root planing in the reduction or elimination of subgingival biofilm [57]. The evidence of non-responsive patients to conventional periodontal therapy highlights the need for therapeutic alternatives to treat periodontitis [66]. Systemic antibiotic therapy should only be performed in specific cases [63], and a typical approach involves using amoxicillin and metronidazole, followed by metronidazole alone and azithromycin [63]. However, for the later parts of treatment, natural products yielded a better reduction of probing pocket depth when compared with scaling and root planing alone or scaling and root planing plus placebo in adults with periodontitis [66]. In the selected literature *Clonorchis sinensis* (10, 50, 120 mg/mg), Propolis (0.02 mg/mL), *Centella asiatica*, *P. granatum* (0.3 mg/ mg), *Sonchus asper* (80 mg/mL), *Garcinia mangostana L.* (40 and 800 mg/mg), Olive oil (140 mg/mL), *C. longa* (0.01 and 10 mg/mL), Blue-green algae, Spirulina (40 mg/mg), *Cymbopogon citratus* (20 mg/mg), *Ginkgo biloba* (0.01 mg/mg), *Quercus brantii* (200 mg/mg), *Coriandrum sativum* (10 mg/mg), *Achyranthes aspera* (40 mg/mg), *Emblica officinalis* (100 mg/mg), *Aloe vera* (NI and 10 mg/mg), Grape seed (20 mg/mg), and Curry leaf (10 mg/mg) were used, and this review concluded that adjunctive natural products provided superior reduction of probing pocket depth when compared to scaling and root planing alone or scaling and root planing plus placebo in a follow-up of 3–6 months.

Among many natural products, some are gaining more attention in the treatment of periodontal diseases, such as Curcumin, *Protium heptaphyllum*, Resveratrol, Propolis, *Camellia sinensis* and Triphala (Fig. 5.1). Curcumin is a yellow pigment found in the spice turmeric and a main functional constituent of the rhizomes of *Curcuma longa* L [67]. Curcumin has been used clinically a subgingival irrigator in combination with SRP [68], in the form of mouthwash [69], toothpaste [70], and carried to resorbable membranes [71]. It has antioxidant, anti-inflammatory, and antimicrobial properties, in the gingiva and alveolar bone [72], and it resists periodontal pathogenic bacteria [73], such as *Porphyromonas gingivalis* (Pg) and *Aggregatibacter actinomycetemcomitans* (Aa). Thus, it can be a potent agent in the treatment and prevention of periodontal diseases. Curcumin, in conjunction with scaling and root planing, has shown clinical benefits in periodontal disease reduction at 3 months both as an irrigator or photosensitizer, in comparison with scaling and root planing alone [58].

The Brazilian flora contains the oil-resin obtained from the *Protium heptaphyllum* tree, widely used in folk medicine. It is currently being studied for its anti-inflammatory and analgesic properties and wound repair [74]. This resin is rich in a mixture of triterpenes, identified as α-amyrin, β-amyrin, rosin, maniladiol, α-amirone, β-amirone and lupenone, in addition to an essential oil rich in mono- and sesquiterpenes [75]. Oral administration of a solution with this oily resin in rats with periodontitis induced by ligature placement revealed that resin extracted from the *P. heptaphyllum* tree slows acute inflammation [76] and reduces the evolution of periodontal disease, decreasing the speed of alveolar bone resorption. The inflammatory modulation of periodontitis in the acute phase was significant in decreasing the levels of TNF-α, an important inducer of osteoclastic activity in the alveolar bone [76].

Resveratrol is a polyphenol stilbene found in red wine [77], peanuts [78], apples, various vegetables and berries [79], and has been investigated for prophylaxis and the therapeutic treatment of periodontal disease based on its anti-inflammatory and antioxidant properties [80]. It inhibits the biofilm formation of some pathogenic bacteria, including *E. coli, Pseudomonas aeruginosa, Fusobacterium nucleatum, Vibrio cholerae, Staphylococcus aureus* [81], *Actinobacillus actinomycetemcomitans,* and *Porphyromonas gingivalis* [82], and it has a potent nonstick property that influences bacterial biofilm formation [81].

In addition, resveratrol is able to decrease period ontal degradation, promoting immunomodulatory effects on the host response [83] by increasing osteoblastic proliferation and differentiation [84–86], decreasing proinflammatory cytokine levels (TNF-α and IL-6), cells exhibiting neutrophil and macrophage marker (MPO), cell proliferation marker (Ki67), and pre-osteoblastic marker (RUNX2); and increase the number of CD31-, F4/80-, and osteocalcin-(OCN) positive cells [87]. Most studies evaluating the effects of resveratrol on periodontitis were experimental in animals [84,86], with intraperitoneal [85], subcutaneous [86], oral [88], and gavage administration [80,84,89] and with doses ranging from 10 mg/kg(84) to 25 mg/kg [80]. The duration of treatment also varied from a single dose of 0.001% (w/w) of body weight [85] to seven [80] and 30 days [84]. In humans, the effects of daily intake of 480 mg capsules of resveratrol [90] showed efficacy in decreasing the average depth of the periodontal pockets of diabetic patients with chronic periodontitis.

Propolis is a resinous compound collected by bees and has been used for many years to treat influenza, wound healing and periodontitis [91,92]. There are a variety of chemical compounds in propolis, including flavonoids, terpenoids, phenolics, artepilin C, baccarat and ursolic acid [93,94] that have anti-inflammatory and antibacterial effects [95], especially against periodontopathic bacteria, such as *P. gingivalis* [94,96]. In addition, propolis has therapeutic effects with reduction of dental plaque [97,98], microbial activity stabilization, and improvement of the clinical parameters of periodontitis such as probing depth; level of clinical insertion, and bleeding in the probe [75,97,99–101].

Propolis, for the treatment of periodontal disease, has already been studied in the form of irrigation with 20% hydroalcoholic propolis solution and the application of 3 mL of the solution twice a week for 2 weeks [102], as a mouthwash in different concentrations, (1%, 2.5%, 5%,

10%) [96,97] and as an ointment with Propolis extracted with 0.01 mg / ethanol mL in CMC ointment, three applications per month for 3 months [99]. Overall, propolis is safe to use and can improve the results of periodontal disease treatment, reducing probing pocket depth compared with treatment with a placebo [103].

Green tea, *Camellia sinensis* has been explored due to the antioxidant action of polyphenols that has the ability to protect against various oral diseases, such as dental caries, gingivitis, periodontitis, halitosis and oral malignancy; and for preventing oral oxidative stress, inflammation, as well as reducing dentin erosion and abrasion [104]. There are different ways of applying green tea for the treatment of periodontitis, such as tea bags for infusion [105], strips of hydroxypropylcellulose (HPC), gel [106,107], and toothpaste [108] all associated with the mechanical treatment of periodontitis.

Some studies indicate that periodontal disease is inversely related to the consumption of green tea; that is, the higher the consumption of green tea, the better the periodontal health [109]. It is suggested that the supportive role in maintaining periodontal health is due to factors that restrict the development and colonization of periodontopathic bacteria, such as *Porphyromonas gingivalis, Prevotella intermedia*, and *Prevotella nigrescens* [110,111]. It can also decrease gingival oxidative stress and manifestation of pro-inflammatory cytokines [105]. However, a recent systematic review suggested that it is uncertain whether or not green tea can be an efficient adjunct to conventional nonsurgical periodontal treatment for chronic periodontitis in reducing the probing depth and loss of clinical insertion [112].

Triphala is a combination of three medicinal plants: family *Amalaki Phyllanthus emblica* (syn. *Emblica officinalis*) *Phyllanthaceae*, family *Haritaki (Terminalia chebula) Combretaceae* and family *Bahera (Terminalia bellirica) Combretaceae*. They are known for their anti-inflammatory and antioxidant action that slow down the oxidation process and protect cells from damage caused by free radicals [113,114]; in addition to antimicrobial and antiseptic properties [115–119]. Triphala is generally used as a mouthwash (10% triphala) and has shown an efficacy similar to chlorhexidine when compared to the plaque index and Streptococcus reduction [120–124]. It was also able to neutralize the salivary pH without altering the taste and color of the teeth [125], being an economical and easily available alternative, with limited side effects in the periodontal tissues [126]. Following these findings, Triphala is deemed as a safe and effective treatment for halitosis, oral ulcers, and dental caries prevention [127]. Further, it can be used to supplement preventive treatment of periodontal diseases by reducing the accumulation of plaque and gingival inflammation [123].

5.2.4 Endodontic infection and wound healing

It is well established that endodontic infection is caused by multiple microbial species, mainly anaerobic and facultative strains. Infected and inflamed pulp, if left untreated, can lead to periradicular disease around the root of the tooth, and it can ultimately evolve into an abscess and cellulitis [128,129]. The main goal of endodontic treatment is root canal disinfection, eliminating the bacterial population with chemomechanical instruments to create a favorable environment for healing [130].

Enterococcus faecalis is one of the most prevalent microorganisms associated with failed root canal treatment since this microorganism has the ability to survive canal preparation, irrigation, and even intracanal medication in almost 77% of periradicular lesions [131,132]. This characteristic pathogenicity is associated with the capacity to penetrate dentinal tubules and accessory canals, survive without other bacteria, form an anaerobic biofilm, and resist acidic and alkaline environments [133–135]. The growing increase of resistant pathogens in endodontic infections, and the side effects induced by synthetic drugs, has prompted clinicians and researchers to seek out natural compound's alternatives [136,137].

In nature, the plant kingdom is a very rich source of natural compounds that are usable in medical and dental practice [138]. Bioactive molecules present in medicinal plants, such as flavonoids, phenolic compounds, and alkaloids have become a safe alternative for combating resistant infections and promoting tissue repair [137,139]. For many years, it has been stated that herbal extracts and natural products have advantages such as potent antimicrobial effects, excellent biocompatibility, cost-effectiveness, antioxidant and anti-inflammatory properties, and availability due to them being renewable in nature [132,140,141].

5.2.4.1 Root canal irrigator

The success of a root canal treatment is directly associated with the complete eradication of microorganisms, debris such as components of the bacteria wall, lipopolysaccharides, and other toxins. In this sense, irrigating solutions are essential for penetrating into a complex root canal system, flushing out debris, lubricating the files, and cleaning the root canal and dissolute necrotic pulp tissue without damaging the peri radicular region, ultimately allowing for healing [142–144].

Several natural compounds have been cited for use in canal preparation as irrigating solution such as green tea, propolis, grape seed extract, chitosan, *Aloe vera*, Turmeric extract, *Morinda citrifolia*, *Pinus elliottii*, berberine, and *Moringa oleifera*. These compounds have shown attractive properties such as easy availability, antimicrobial effect, low cost, and biocompatibility [137,145].

5.2.4.1.1 Green tea (Camellia sinensis)

A potent antibacterial effect, similar to chlorhexidine, against *E. faecalis* has been observed when green tea (25mg/50ml water) was used as irrigating solution [143]. This effect is resulting from the destruction of the bacteria cell wall through binding free radicals and hydroxyl groups. Furthermore, polyphenols from green tea have an excellent antioxidant effect with the combination of aromatic rings and hydroxyl. It has been stated that epigallocatechin-3-gallate, the major component of green tea has also anti-inflammatory, anticarcinogenic and antimicrobial properties [146,147].

5.2.4.1.2 Neem (Azadirachta indica)

Neem is derived from a tree known as Margosa or Indian neem [148,149]. Antibacterial, antifungal and analgesic properties have been attributed to an active compound azadirachtin which is a strong inhibitor of cellular membrane synthesis [150–153]. According to Prabhakar et al., the substances responsible for therapeutics benefits are alkaloids, glycosides, terpenoids, steroids, and tannins [154]. Arévalo-Híjar et al. has evaluated the effects of *Moringa oleifera* (Moringa) and *Azadirachta indica* (Neem) methanolic extracts against *E. faecalis* and reported that 25 μg/ml was more efficient than chlorhexidine in inhibiting *E. faecalis* biofilm for 24 and 48 h of treatment [135]. Another study also observed a significant reduction in bacterial load of anaerobic biofilm in infected root canals after exposure to neem as irrigating solution [150,155]. The methanolic extracts of neem were effective against *E. faecalis, S. mutans,* and *S. aureus* even in low concentrations (3 mg) [156].

5.2.4.1.3 Propolis

Produced by honeybees, propolis is a natural resinous mixture derived from plants. Its composition depends on the geographical region and local flora [157,158]. It has been reported that propolis has potent antibacterial, antifungal, antiviral, anti-inflammatory, and antioxidant properties. Flavonoids, aromatic and phenolic compounds are responsible for antimicrobial and antioxidant characteristics [159,160]. Ehsani et al. observed that ethanolic propolis extract was more effective against *E. faecalis* compared to chlorhexidine [131]. In a study done by Jaiswal et al., propolis and chitosan combined with chlorhexidine were as efficient as hypochlorite in reducing *E. faecalis*, showing propolis to be a good alternative as an irrigating agent for root canals [144]. Many other studies have reported the antimicrobial effect of propolis against *E. faecalis* and *E. coli* in concentrations of 400–1600 μg/mL [161,162]. Jolly M et al. have tested the anti-inflammatory potential of propolis as an irrigating agent and observed that initially, propolis extracts induced a significant inflammatory response. However, as time progressed, the inflammation was less significantly reduced. Researchers concluded that the reduction of inflammation in the initial time induced by propolis could be important for the healing process of periapical tissues [163]. As exhibited by Silva et al. aqueous and alcoholic propolis extract were biocompatible, inducing a lower irritating response *in vivo*. According to the authors, this response could be attributed to the ability of propolis to inhibit phospholipase A2 and consequently the inflammation [164].

5.2.4.1.4 Grape seed extract (Vitis vinifera)

The literature has reported that grape seed extracts have many desirable properties as irrigating root canal solutions such as antimicrobial activity, biocompatibility to dental tissues, antioxidant capabilities, and immunomodulatory effects [142,165]. D'aviz et al. investigated the antimicrobial activity of grape seed extracts (6.5%) as gel and solution, sodium hypochlorite, and chlorhexidine during root canal instrumentation against *E. faecalis*. They reported that grape seed extracts and chlorhexidine exhibited similar results in bacterial load reduction; however, both yielded lower values compared to hypochlorite. Grape seed extract gel showed better results than the extract solution. Moreover, the antimicrobial effect was attributed to the positive charges of the extract that interact with negative charges on bacterial cell walls, causing osmotic imbalance and bacterial death [142]. Another experiment involved extracted human mandibular molars which were inoculated with *E. faecalis*, and researchers observed a reduction in bacterial population upon adding 50% grape seed extract at a level comparable to sodium hypochlorite in high concentrations [166]. Fiallos et al. on the other hand, observed a more potent effect of grape seed extract than chlorhexidine against *E. faecalis* biofilm. The mechanism of grape seed action was attributed to phenolic compounds that were capable of disrupting the cell wall and plasma membrane of bacteria [165] Furthermore, grape seed extract could be considered as an alternative for irrigating solution stead of sodium hypochlorite once that is capable to increase mechanical properties of dentin, mainly in cases of the thin root structure [167].

5.2.4.1.5 Chitosan

Chitosan is a natural compound obtained from crustaceans, primarily crab and shrimp shells, and it is a product of the deacetylation of chitin. Chitosan is a biocompatible and biodegradable polysaccharide with a good antimicrobial effect and low production costs [168]. Yada et al. evaluated the activity of chitosan on *Candida albicans* and *E. faecalis* biofilm and the biocompatibility. They observed that chitosan caused a significant reduction of colony-forming units for both *C. albicans* and *E. faecalis*. This effect could be explained by two mechanisms [1]: chitosan possessed positive charges which interacted with the negative charges on bacteria cell walls, altering cell permeability, leading to leakage of intercellular components and consequent cell death, and [2] chitosan causes DNA damage, inhibiting the synthesis of mRNA and proteins [169]. In a study conducted by Jaiswal et al., chitosan and chlorhexidine showed similar antibacterial activity to sodium hypochlorite (5%) against *E faecalis*. Moreover, other mechanisms of action have been attributed to chitosan such as the ability to chelating metal ions and inhibiting essential enzymes which can lead the bacteria cell to death [144].

5.2.4.2 Sealer cement

Although the goal of root canal treatment is abolishing bacterial infection, even after root instrumentation, it is possible that a few microorganisms remain in the dentin tubules and accessory canals. Thus, the use of sealers with the ability to seal the root canal and dentin tubes is critical, and these sealers must possess antimicrobial activity to eliminate residual microorganisms and prevent recontamination [136].

Recently, methanolic herbal extracts from *Emblica officinalis* (Amla), *Myristica fragrans* (Nutmeg) and *Salvadora persica* (Miswak) were added to three different sealers (Endomethasone, AH plus, and Apexit plus) and were examined for antimicrobial effect against *S. aureus, S. β haemolyticus, E. faecalis, E. coli, P. aeruginosa, Peptostreptococcus sp., B. fragilis, L. casei*, and *V. parvula* biofilms. Endomethasone sealer (zinc-oxide-eugenol-based) combined with *Myristica fragrans* exhibited the highest antimicrobial effect against *Bacteroides fragilis* and the minimum effect against *E. faecalis*. When AH plus (epoxy resin-based) was mixed with *Salvadora persica* (Miswak) it was observed maximum efficacy against *L. casei* and minimum for *E. coli*. Apexit plus (calcium-hydroxide-based) mixed with *Salvadora persica* (Miswak) showed antibacterial activity only for *Peptostreptococcus* strains. It has been stated that *Myristica fragrans* have compounds such as tannis, saponins, essential oils, phenolic products and flavonoids which are probably responsible for its high antimicrobial effect. These active biomolecules act on the bacterial enzymes, cell wall and cell adhesion [136].

5.2.4.3 Intracanal medication

Although chemomechanical instrumentation can reduce the bacterial load in the root canal system, the complete elimination of microorganisms is difficult to achieve. Thus, it is recommended the use of intracanal medication to supplement the elimination of persistent bacteria [158]. Among the intracanal medication used in dental clinics, calcium hydroxide is the most preferred due to its antimicrobial effect, inhibition of root resorption, pulp healing induction, low cytotoxicity, and stimulation of hard tissue formation [142,158]. However, the action of calcium hydroxide at the deeper layer of dentinal tubes is difficult, allowing for the survival of unwanted microorganisms. Thus, phytotherapeutic substances such as *Aloe vera*, mushroom, lemon, propolis, and *Psidium cattleianum* have become more popular for intracanal medication [142,170].

Aloe vera belongs to *Asphodelaceae* (*Liliaceae*) family [148]. It is a plant that adapts well to dry and hot weather. It has been stated that aloe vera has antibacterial, antifungal, antivirus, and anti-inflammatory properties. Varshini et al. demonstrated that *Aloe vera* showed better antimicrobial activity against *E. faecalis* than calcium hydroxide [170]. Acemannan, anthraquinone, anthracene and cinnamonic acid were the active subtances responsible for the biological effect of aloe vera [131,170]. Kurian et al. evaluated the efficacy of calcium hydroxide, extracts of mushroom and *Aloe vera* as an intracanal medicament on *E. faecalis* using human teeth and observed that mushroom extract (40 mg/mL) and *Aloe vera* (60 mg/mL) were able to significantly reduce *E. faecalis* biofilm [171]. The antimicrobial effect induced by *Aloe vera* has been related to its capacity to disrupting cell membrane, inhibition of protein synthesis and cellular respiration leading to cell leakage and death [171,172]. Furthermore, *Aloe vera* is considered a natural healer, since it is biocompatible and stimulates cell migration, proliferation, and growth [173].

Recently, Chinese propolis was tested in a clinical trial as intracanal medication to prevent acute exacerbation of pain known as "flare-up". It was observed that Chinese propolis was similar to calcium hydroxide in postoperative endodontic pain, however, propolis induced fewer side effects and could be preferably indicated as a medicament in necrotic cases [174]. Madhubala et al. also evaluated propolis as an intracanal medication, comparing it to a tri-antibiotic mixture. The results showed that propolis exhibited a 100% reduction in colony counts of *E. faecalis* by the second day of treatment, making it more effective than the tri-antibiotic mixture. The mechanism of antimicrobial action attributed to propolis is related to flavonoids, esters of caffeic and gallangin. The active compounds of propolis can inhibit RNA polymerase, leading to bacterial cell death [175].

Psidium spp. belongs to the *Myrtaceae* family and is found in America. Ethanolic extracts of *Psidium catleianum* have been implicated as intracanal medicators against *C. albicans* and *E. faecalis*. Research has shown that *Psidium catleianum* is the fastest and most effective inhibitor of *E. faecalis* and *C. albicans* biofilm. *Psidium catleianum* contains flavonoids and tannins that form protein complexes that are toxic for yeasts and bacteria, resulting in high antimicrobial activity [176].

5.2.4.4 Storage media for avulsed teeth

When immediate reimplantation is not possible in cases of the dental avulsion, the use of a storage media capable to maintain the viability of the periodontal ligament cells and minimize the risks of resorption is determinant for the successful replantation [177,178]. Several natural compounds have been studied as an alternative for storage media such as propolis, green tea, aloe vera, coconut water, castor oil, and many others, and they have been shown to be more effective than synthetic compounds [179].

Recently, it was stated that Thai propolis extract at 0.625mg/mL was biocompatible for periodontal ligament cells, and it induced periostin expression similarly to fresh periodontal ligament cells for up to 12 h, suggesting that Thai propolis could be a storage media for avulsed teeth [180]. Furthermore, Brazilian propolis at 10µg/mL also showed to be biocompatible to human periodontal ligament cells; however, the calcium deposition and mineralized nodules showed to be smaller than that induced by milk. It has been reported that Brazilian propolis reduced IL-6, IL-1β and IL-8 expression compared to milk [181]. Mori et al. evaluated the effect of Brazilian propolis as a storage media for avulsed teeth *in vivo* and observed that propolis was able to preserve the cementum layer and root surface for 6 hours of extra alveolar period. The period of 6 hours would be necessary for the incorporation of active principles of propolis for the antimicrobial and anti-inflammatory effect, allowing the periodontal ligament to heal [182].

Coconut (*Cocos nucifera L*) is a fruit of coconut palm. Coconut water is known as a blood plasma substitute since it can replace electrolytes, sugars, and fluids in the body after physical exercises [183]. It has been reported that coconut water is composed of potassium, calcium, and magnesium, sodium, chloride, phosphate, proteins, amino acids, vitamins, and minerals. In a recent study, coconut water induced better cell viability than propolis, milk, and Hank's balanced salt solution. This result was attributed to the presence of glucose, fructose, and essential amino acids, which maintain the ideal osmolarity for cell viability [183].

An *in vivo* study has also tested a powered coconut water formula in beagle dogs that had teeth replanted after storage in this compound with two different osmolalities (250 and 372 mOsm/kg). Coconut water with 250 mOsm/kg has shown to induce a better response, exhibiting normal periodontium in most areas of the periodontal ligament of avulsed teeth and lower levels of external root reabsorption. Thus, similarly to immediate replantation, coconut water is being considered as an alternative for storage media [184].

Aloe vera has also been reported as a medicinal plant with desirable properties such as anti-inflammatory, antioxidant, and antimicrobial effects. Aloe vera at 10, 30 and 50% can preserve human periodontal ligament cells *in vitro* at least for up to 9 hours. It has been stated that aloe vera at these three concentrations showed ideal osmolarity for cells, and essential nutrients present in the plant extract were responsible for the positive results of cell viability [185].

5.3 Molecular mechanisms of action of natural products against oral pathogens

One major advantage of natural products in contrast to synthetic drugs is their increased efficacy against pathogens that have developed drug resistance. Due to a combination of factors, including the widespread misuse and overuse of antibiotic drugs in humans and animals, bacterial resistance is a global problem with increasingly problematic implications. When bacteria develop resistance, the genetic information relevant to that resistance is typically stored in a plasmid, a smaller loop of coiled genetic code that can be then transferred between members of the same bacterial species, allowing the resistance to spread quickly through a population. To confront bacteria armed with a multitude of advantageous plasmids, multi-pronged therapies with various antibacterial mechanisms are needed, thus making plant-based products a promising candidate.

To fight plasmid-equipped bacteria, one method is to use "plasmid curing," which is when a plasmid is removed from a bacterial population without killing the bacteria, causing any pathogenic and antimicrobial resistant bacteria to once again be susceptible to typical general antibiotics [13]. Essentially undoing bacterial resistance when done properly, plasmid curing has the potential to be a powerful tool in the arsenal of researchers and healthcare providers to fight bacteria, including oral pathogens. This method has been conducted with success in limited *in vivo* trials, and very little experimental data is available. Some known plasmid curing mechanisms require certain compounds to integrate into the DNA of the desired bacterial plasmid. In doing so, the compound and can cause plasmid breakage, dysregulate the

DNA supercoiling, or inhibit conjugation, with either option ultimately resulting in a decreased plasmid count in the bacterial population as successful plasmid replication is inhibited [13].

Researchers have discovered that plant-derived compounds have acted as plasmid curing agents, acting particularly effectively against multidrug-resistant bacteria (MDR). As previously mentioned, Plumbagin, the *Plumbago* derivative with the chemical name 5-hydroxy-2-methyl-1,4- naphthoquinone, is a yellow dye that is obtained from the root of these tropical South African plants [12,13]. Two species, *Plumbago zeylanica* and *Plumbago auriculata*, were tested with isolates at below inhibitory concentrations, and they were deployed on MDR isolated samples of *E. coli*. For *P. auriculata,* the derivative was also tested for *Pseudomonas aeruginosa, Proteus vulgaris* and *Klebsiella pneumoniae. P. zeylanica* was only mildly effective, with only 14% of the *E. coli* population cured of the plasmid [186]. In contrast, the samples tested with the extract of the *P. auriculata* root effectively cured plasmids for each bacterial species, suggesting a potency that allowed for effective treatment in multiple bacterial species [187]. For even better efficacy, these plasmid curing mechanisms can be used in conjunction, involving extracts from various plant species. This combination of derivatives allows the various compounds to work in tandem, providing an effective plasmid curing effect and resulting in more effective antibiotic treatments for pathogenic bacteria that would otherwise exhibit resistance.

The *Copaifera* species native to tropical Latin America and Western Africa can be extracted to produce copaiba oil, a natural essential oil [4]. This oil is like other essential oils in that it is believed to possess antimicrobial compounds that can combat pathogenic bacteria. Specifically, copaiba oil features hydrophobic compounds within the oil. Functional groups within the compounds allow for hydrophobic interactions with the bacterial cell membrane, and these interactions cause a change in cellular membrane permeability structure. This can lead to cellular imbalance, ultimately resulting in bacterial death [4]. These hydrophobic compounds within the oil can be remarkably effective, acting on bacteria related to both dental caries formation and the onset of periodontal disease. Bacteria such as *Streptococcus mutans* and *Streptococcus sanguinis* have been affected by the copaiba oil extracts, as found in both *in vitro* zone of inhibition analysis and *in vivo* studies involving dog animal models [188,189]. These promising results further support the potential use of copaiba oil as an effective natural product to treat a wide range of oral diseases.

The hydrophobic compounds present in copaiba oil extracts perturbed the bacterial membrane of pathogenic bacterial species, and other plant-based compounds have mechanisms involving different compound types with a similar target: the cellular membrane of the bacteria. Plant-based phenolic compounds contain an active hydroxyl group, and this group allows the phenols to engage in hydrogen bonding with bacterial membranes, causing membrane disruption in a manner that can lead to ruptured membranes, inhibited membrane transport, failed pH gradient maintenance, and improper ATP level regulation [9]. These can lead to cell death, and the phenolic compounds can accomplish this cell death through multiple avenues in different bacterial populations. Because of alkyl groups present in certain polyphenolic compounds within plant derivatives, varied interactions with bacterial membranes can have an enhanced bactericidal impact [9]. Essentially, the molecular structure of these compounds upsets the membrane structure and cytoplasmic balance in bacterial membranes, resulting in unstable bacteria that often die. For example, the *Drynaria fortune* plant that has been used as a Korean herbal medicine was found to be effective as an antibacterial agent against various bacterial species, including *Streptococcus mutans, Streptococcus sanguinis, Streptococcus sobrinus, Streptococcus ratti, Streptococcus cricetid, Streptococcus anginosus, Streptococcus gordonii, Prevotella intermedia, and Porphylomonas gingivalis* [14]. The polyphenols present in the isolated extracts are believed to inhibit bacterial growth and glucosyltransferase activity, an enzymatic process involved in plaque formation, and the flavonoids from *D. fortunei* are thought to interact with proteins within the bacterial cell wall in an antibacterial manner.

5.4 Plant products and immunity boosting

5.4.1 Antioxidative functions

Antioxidants have been found to either prevent or greatly delay oxidation of substrates when they are at lower concentrations than the actual substrate at hand [190]. Antioxidants can either be synthesized in vivo or ingested through one's diet. The latter is sourced from plants to provide exogenous antioxidants. Almost all the world's plants that have been found to have medicinal properties (approximately T of the world's plants) are also known for their antioxidative functions. This finding is most likely due to plants' ability to synthesize both secondary metabolites (e.g., phenolic compounds) and non-enzymatic antioxidants (e.g., ascorbic acid and glutathione).

Exogenous antioxidants from plants first gained attention when ascorbic acid was found in plants [190]. Ascorbic acid, more commonly known as vitamin C, is a powerful antioxidant in diets that has a pro-oxidant activity [191]. This activity is due to its antioxidant functions of creating reactive free radicals, which are then able to cause cytotoxic effects. It has been shown that this pro-oxidant activity of ascorbic acid is due to its hypothesized dose-dependent Fenton mechanism. Several in vitro, animal, and ex vivo studies have concluded that vitamin C is indeed an effective agent of cytotoxicity against oral neoplastic cells, while causing no harm towards healthy cells. However, despite this conclusion, there is a lack of sufficient clinical as well as *in vivo* and human trials to fully evaluate the antitumor activity of ascorbic acid as a treatment against oral neoplasms.

Soon after the discovery of ascorbic acid in plants, it was also found that oxidative stress, being a cause of the progression and development of several diseases, has been combated by supplementing or boosting exogenous or endogenous antioxidants, respectively, in one's body [190]. These methods will help address the negative effects of reactive oxygen species (ROS) and its induced damage.

Overall, very few *in vitro* studies' conclusions have been confirmed or even attempted to be analyzed in *in vivo* studies. Therefore, the applicability of these conclusions to *in vivo* systems is still unknown. That being said, several phytochemicals in *in vitro* assays have been found to have antioxidant properties. Yet, only a few have been found to have positive impacts in *in vivo* studies, potentially due to the interference with other phytopharmacological processes (e.g., metabolism, distribution, storage and excretion, and absorption).

5.4.2 Activation of anti-inflammatory pathways

Several studies have shown that the anti-inflammatory impact of several naturally occurring plant products may be due to their macrophage regulation [192]. Macrophages originate from monocytes and can be found in all parts of the body, as they are integral to the host immune system. Specifically, macrophages are key components of an inflammatory response and aid in antigen presentation, immunomodulation, tissue repair and phagocytosis. Due to changes in their environment, macrophages are further differentiated into either M1 (having proinflammatory properties) or M2 (having anti-inflammatory properties), thus allowing macrophages to be either positive or negative regulators of inflammation. Regulation of this polarization is integral to several processes related to inflammatory processes, perhaps those that are found in the oral cavity. One form of regulation occurs with certain signaling pathways. A certain pathway named the phosphoinositide 3-kinase/protein kinase B (PI3K/AKT) cascade is activated in immune cells once a Toll-like receptor (TLR4) binds to a lipopolysaccharide (LPS), a necessary part of bacteria. A mammalian target of rapamycin (mTORC1) and protein kinase B (AKT or PKB) are both downstream targets of this PI3K/AKT pathway. The pathway activation is directly related to the inhibition of apoptosis and instead, promotion of cell cycle progression while preventing autophagy. Autophagy, being directly involved in several anti-inflammatory processes, plays a role in macrophage physiology as well. The PI3K/AKT/mTORC1 pathway inhibition activates autophagy and safeguards macrophages from inflammation. It has been found that PI3K/AKT signal transduction can be suppressed by several anti-inflammatory plant extracts, which can then in turn play an important role in the immune system. Therefore, these plant-based products have the potential to become a large part of therapeutic strategies used in the oral cavity against several inflammatory oral diseases, (e.g., periodontitis).

Another study has shown that PM014, a herbal extract, is able to reduce radiation-induced pulmonary fibrosis due to regulation of the NF-κB and TGF-b1/NOX4 pathways [16]. Increased radiation-induced NF-κB activity, due to PM014 administration, as well as reduced p65 translocation, DNA damage, the epithelial-mesenchymal transition and ROS production was observed. Therefore, PM014 has the potential to increase efficacy of radiotherapy, via radiation-induced lung fibrosis. However, further *in vivo* studies are needed to fully understand the functionality of these plant-based compounds and the molecular mechanisms by which they function and have anti-inflammatory effects via certain anti-inflammatory pathways.

5.4.3 Inhibition of cytokine secretions and nuclear factor-κB

The ability of plant compounds to influence human immune regulation is a testament to the broad scope of possibilities in plant-based medicine. Researchers have discovered natural compounds that are anti-inflammatory in nature, and they combat inflammation by inhibiting the secretion of pro-inflammatory cytokines (Fig. 5.2). One such compound is Curcumin, a natural compound obtained from the *Curcuma longa* plant, specifically from the plant's rhizome [193]. Researchers tested Curcumin *in vivo* with a murine model and found that, in cases of acute kidney injury (AKI), the addition of Curcumin prompted a decrease in the transcription of mRNA involved with the secretion of pro-inflammatory cytokines such as interleukin-1β (IL-1β), interleukin-6 (IL-6), tumor necrosis factor-α (TNF-α), and monocyte chemoattractant protein-1 (MCP-1). Additionally, the introduction of Curcumin extract led to lowered transcription and translation of Macrophage-inducible C-type lectin as well as decreased activation of nuclear factor-κB (NF-κB), influencing the related NF-κB pathway [193] via these mechanisms, the plant-based compound is believed to have potentially significant immunoregulatory capabilities that could prove useful in the treatment of inflammatory diseases like AKI. This example of a plant-based product acting as an anti-inflammatory agent suggests that other plant-derived compounds could similarly act on inflammatory diseases within the oral cavity by mediating the host immune response and suppressing pro-inflammatory cytokine secretion.

Pro-inflammatory cytokines like IL-1β, IL-6, and others are involved in the human immune response, and their production and secretion in the blood plasma can influence inflammation, including prolonged inflammation. In instances of unresolved inflammation, tissue damage can occur; this is prevalent in oral inflammatory diseases such as chronic periodontitis. By targeting the production of these pro-inflammatory agents, prolonged inflammation can be treated effectively. In one study mentioned previously, plant-derived antioxidants, many of which are commonly found in fruits and berries, were used to target cytokine production. Peripheral blood mononuclear cells (PBMCs) obtained from healthy donor samples were treated with Phloretin, Silymarin, Hesperetin, and Resveratrol, all plant-derivatives with antioxidant capabilities. The four were found to modify the blood cytokine profile significantly, with lowered production for cytokines IL-1β, IL-6, TNF-α, and even IL-1α, and IFN-γ [20]. The results of this study further suggested that the antioxidant compounds acted on the NF-κB pathway by measuring mRNA expression levels following antioxidant treatment.

The decrease in activation of IκB-α following antioxidant treatment, as measured by RNA isolation and RT-PCR analysis, supports the notion that the antioxidant compounds act on the NF-κB pathway, inhibiting key elements of the pathway upstream of IκB-α mRNA synthesis and resulting in the decrease in various inflammatory cytokine levels [20].

Another example of herbal medicines with immunoregulatory effects is PM014, one of the several herbal medicines commonly used in Asian countries historically as mentioned previously. These medicines have the advantage of multiple versatile compounds with active sites possessing various functions and abilities, allowing these herbs to provide relief for a multitude of ailments. An *in vivo* murine model was used to examine the healing effects of PM014 on mice after being exposed to radiation [16]. Following radiation treatment, a regular treatment for lung cancer, PM014 was found to mitigate some of the symptoms from the common side effect of radiation therapy: pulmonary fibrosis. Interestingly, PM014 was found to protect against some of the harmful side effects that were typically displayed during radiation therapy for lung cancer. Radiation is known to increase inflammatory cytokine secretion as a part of the NF-κB pathway, and due to the symptom mitigation and prevention exhibited by PM104, the evidence suggests a possible inhibitory effect by PM014, suppressing proinflammatory cytokine levels by downregulating the NF-κB pathway. Once again, PM014 provides another example of herbal, plant-derived compounds acting on the inflammatory pathways involved in immunity by downregulating inflammatory cytokines and eliminating their inflammatory symptoms, this time in epithelial pulmonary tissue.

Natural polyphenols have also recently been shown to have anti-inflammatory effects, specifically those from chamomile *(Matricaria chamomilla)*, meadowsweet *(Filipendula ulmaria* L.), and willow bark *(Salix alba* L.) [194]. The principal anti-inflammatory polyphenols found in these chamomile, meadowsweet, and willow bark are apigenin, quercetin and salicylic acid respectively. It was found that quercetin and apigenin, the latter with a concentration ≥10 μM, can reduce levels of proinflammatory cytokines TNF-α and IL-6. IL-1β was not found to be significantly decreased by any of these three polyphenols. Comparatively, salicylic acid had the least effective anti-inflammatory properties, with only a high concentration of 25 μM being sufficient to significantly decrease IL-6 levels. However, this finding was quite surprising, given the fact that this polyphenol is resemblant of acetylsalicylic acid, otherwise known as aspirin (a very widely-used anti-inflammatory drug) [195]. Acetylsalicylic acid mechanistically inhibits cyclooxygenase enzymes, including COX-1 and COX-2 [196]. However, salicylate relies on the inhibition of the NF-κB transcription factor [197]. Additionally, salicylate has been found to be less effective at preventing inflammation *in vitro*, compared to *in vivo*, which pushes for additional animal models to be used to address this question [194,195]. These isolated polyphenols demonstrated greater anti-inflammatory effects, compared to their herbal extract counterparts, but both did show a significant reduction in proinflammatory cytokine levels [194]. Compared to the other two, willow bark showed the largest reduction in IL-6 and TNF-α, and higher concentrations of this extract significantly reduced IL-1β levels. Willow bark has been shown to have lower polyphenol amounts, but it instead has a higher amount of flavonoids (compared to that of meadowsweet) as well as higher salicin levels, which have anti-inflammatory properties [198–200]. The different results in relation to IL-1β levels may be due to a different mechanism employed by this cytokine, of binding to two different types of receptors on the cell surface [201,202]. Therefore, these results may be due to different binding abilities of some cells over others, as macrophages produce this cytokine but only monocytes may have the receptor present on their surface [201]. Thus, autocrine effects of IL-1β on macrophages specifically may be minimal, especially at low concentrations of these polyphenols.

In addition, it has previously been found that the plant *Inula viscosa* (Compositae) has been shown to have anti-inflammatory properties via both *in vivo* and *in vitro* studies, due to the presence of secondary metabolites in its leaves [203,204]. It was previously observed that purified sesquiterpene lactone compounds, Inuviscolide (Inv) and Tomentosin (Tom), from this plant decreased the levels of p65/RelA, a component of the transcription factor NF-κB via a dose-dependent mechanism [205]. NF-κB has recently been found to be a master regulator of immune responses, including the regulation of several cytokines including IL-1β, IL-2, and IFNγ [206–208]. A recent study was aimed at analyzing the effects of Inv and Tom on pro-inflammatory cytokines released from human peripheral blood mononuclear cells (PBMCs) as a result of stimulation using phorbol myristate acetate (PMA) or lipopolysaccharide (LPS), using ELISA assay [209]. Prior to this experiment, the mechanisms of action behind these effects were also unknown, so they were also studied. This *in vitro* experiment explored whether the downregulation of NF-κB also played a vital role in the anti-inflammatory activities of plants. In addition, the protein levels of both NF-κB and STAT1, both being transcription factors heavily involved in cytokine production, using Western blot analysis. Both lactone compounds were found to decrease IL-2, IL-1β, and IFNγ production. TNFα production was slightly increased, and IL-6 secretion experienced no change. Although the levels of TNFα were greater than normal, this increased concentration was not found to significantly affect the viability of human PBMCs in this study. Western blot analysis found that there was a decrease in the protein levels of two components: p65/RelA subunit of nuclear factor-kB (NF-κB) along with the signal transducer and activator of transcription 1 (STAT1), all of which was achieved via proteasomal degradation. The expression level of STAT1 can be downregulated by both proteasomal degradation as well as decreased secretion level of IFNγ. The nuclear translocation of p65 and phosphorylation of STAT1 were not explicitly addressed in this study, as it was assumed that these processes and the amount of each active compound were also decreased due to a decrease in their production levels. There were no other significant effects observed in levels of the NF-κB component, the signal transducer and activator of transcription 3 (STAT3) or p50 (NF-κB). Proteasomal degradation of p65/RelA as well as STAT1 by Inv and Tom has never been found prior to this study [204]. Thus, this study was able to provide a new understanding of the mechanisms behind the previously observed anti-inflammatory benefits of the *Inula viscosa* plant.

One implication of these experimental results is the possible control of inflammation by Inv and Tom when the primary causes of inflammation can be mediated by IL-2, IFNγ, and IL-1 levels. Furthermore, NF-κB doesn't equally affect various cytokine types, so this specification may explain differing mechanisms and activity levels [210–212]. Finally, researchers were able to conclude that in the cases of overstimulation of cytokine secretion, these agents of Inv and Tom could possibly be used as or supplementary to anti-inflammatory treatments, especially given that these compounds are very effective at low concentrations, not toxic to normal human cells, and have the ability to target NF-κB and STAT1 [209]. Moreover, chronic inflammation has been shown to cause increased ROS levels, which has the potential to damage cells or even encourage the development of cancer [213–216]. Given these findings, plant-based products have the great potential to act as a prophylactic agent in which pro-inflammatory cytokine-supported chronic inflammation presents itself before tumor formation. Therefore, these compounds could very well be used as anti-cancer agents, alongside anti-inflammatory agents as well. Future studies should focus on additional tests to measure decreased occupation of proinflammatory cytokine promoters by NF-κB, given that it has previously been found that Sesquiterpene Lactones have the ability to inhibit NF-κB DNA, due to direct alkylation of the p65 subunit [217,218].

5.4.4 Impact of proliferation of immune cells

As a part of the immune system, immune cells aid the body in fighting diseases and infections. The human immune response is especially crucial in response to oral diseases. MicroRNAs (miRNAs) exert large effects on the expression of these cells and the body's inflammatory response against bacterial pathogens [219]. These miRNAs target inflammatory regulators during the immune response through toll-like receptor signaling. Especially in periodontal diseases, miRNAs control the functions of macrophages, dendritic cells, T cells, neutrophils, and B cells. Phagocytic cells, neutrophils, and macrophages are periodontal innate immunity: the body's initial line of defense against oral pathogens. An antigen-specific immune response, B and T cells, are adaptive immunity: the body's second line of defense. MiRNAs act as the bridge between both types of immunity. The latter type of immunity relies on the recognition of an antigen on an infected cell's surface, resulting in an immune response that attacks it [219].

A promising area of new research involves studying the altered miRNA expression profile in the periodontal tissues of patients with periodontal diseases. Regulation of how the immune cells enter the oral epithelium will have large implications for how susceptible the patient is to oral diseases later.

Due to its routine exposure to the tooth adherent biofilm, the gingival interface is an especially vulnerable region for various inflammatory diseases. The makeup of the immune cell network in this mucosal site was previously not well understood. Researchers have since found that there is a large presence of T cells, granulocytes/neutrophils, a network of antigen presenting cells (APC), and minimal B and lymphoid cells (ILC) at the gingival interface [220]. To analyze how the levels of immune cells shift in the advent of periodontitis, researchers studied a group of severe chronic periodontitis patients. They found that there was a significant increase in the presence of inflammatory cells in the region. Consistent with the finding from health, the CD3+ T cells still had a large presence in disease relative to other immune cells. However, the overall quantity of T cells in the diseased interface were significantly larger, resulting in a 10-fold rise in inflammatory cells. While B cells (CD19+ cells) were almost nonexistent in the lymphocyte compartment during health, their presence became detectable in periodontitis. It should be noted that the DC Mac APC compartment (HLADR+CD19−) did not have any noticeable changes in the advent of the disease. The most significant difference was the large increase in neutrophils (CD15+ CD16+ cells) [220]. Neutrophils typically help mediate the body's microbial surveillance and innate response, though it has also been found that they use efferocytosis by tissue phagocytes and release anti-inflammatory molecules to aid in mediating inflammation. Further analysis on the T cell subsets in the gingival interface indicated that CD4+ helper T cells predominated, which was followed by CD8+ T cells and $\gamma\delta$ T cells. The terminally differentiated memory subset (rTEM) in both the CD4+ and CD8+ compartments were especially crucial. Such differentiated memory subsets can irreversibly differentiate into cells with specialized site protection. Other tissue memory cells are seen in other barrier sites, such as the lungs and intestines. Researchers have found using a parabiosis model and *in vivo* labeling with mice that tissue-resident memory CD4+ and CD8+ T cells provide immunity to region-specific pathogens in the skin and lungs [221]. Further development on tissue-resident memory T cells in the oral barrier can lead to the potential development of a mucosal vaccine [220].

In a study on apical periodontitis (AP), researchers found that in addition to conventional T cells, mucosal-associated invariant T (MAIT) cells composing of Vα7.2-Jα33/20/12 α-chain rearrangements were in the oral mucosal tissue [222]. Prior to the study, the role of MAIT cells in preventing AP was largely unknown. Researchers compared surgically resected AP and gingival tissues to determine that AP tissues had much greater quantities of Vα7.2-Jα33, Vα7.2-Jα20, Vα7.2-Jα12, Cα and tumor necrosis factor (TNF), interferon (IFN)-γ and interleukin (IL)-17A transcripts than healthy tissue. These all indicate a MAIT cell signature. Researchers also found that MAIT cells can contribute to cytokine expression in AP by producing proinflammatory cytokines TNF-α, IFN-γ, and IL-17 in response to antigenic stimulus. In addition, MAIT cells were found in an *in vivo* pulmonary bacterial infection experiment to destroy riboflavin-pathway microbe infected cells, inhibiting their cell growth [223]. Analyzing the microbiome and immune data further indicated that the abundance of bacteria was negatively correlated with the Vα7.2-Jα33, Cα, and IL-17A transcript expression in an AP infected oral tissue barrier [222]. These results indicate that the MAIT cells' defense mechanism was upregulated, and it has the potential to provide local defense in the AP region, resulting in the possible future development in immune sensing of polymicrobial-related oral diseases [222]. However, despite the presence of MAIT, AP was still able to progress due to an overwhelming bacterial load. More conventional T cells, such as Cα and CD3+, can help mediate the inflammation in AP in addition to MAIT cells.

5.4.5 Skewing of immune cell differentiation

Macrophages can split into the M1 group (activated macrophages that have pro-inflammatory capabilities) and the M2 group (alternatively activated macrophages that have anti-inflammatory reactions and can remodel tissue). Depending on the polarization of macrophages, these cells can help regulate the development of inflammatory diseases [224]. However, the targeting of macrophage polarization is still in its early stages of development. Researchers have found through a human study *in vitro* and studies using mouse models that chemokine (C-C motif) ligand 2 (CCL2) inhibitors and anti-colony-stimulating factor 1 (anti-CSF-1) antibodies have the ability to induce M2-like skewing of macrophage function to inhibit tumor growth [225]. CCL2 is commonly expressed in various cancerous tumors: cervical, breast, lung, etc. such a finding is especially promising regarding oral cancers.

There are many limitations that have hindered the progress of this area of research. For instance, the collection of fresh macrophages from humans is inconvenient. As such, imperfect cell lines must be utilized. In addition, there are distinct differences between the macrophages in humans and mice. In fact, the M2 macrophage Arg1 is present in mice, but not expressed by *in vitro* macrophages in humans.

As immune cells, such as neutrophils, monocytes, and macrophages, are the main lines of defense against periodontal pathogens, inflammaging (aging in terms of inflammation and immunology) worsens the progress of oral diseases such as periodontitis [226]. For instance, the levels of proinflammatory cytokines IL-1β and -6 and TNF-α and M1 macrophages are increased in patients with the disease. Older patients are unable to initiate a robust immune response in the periodontal area to respond to these abnormal stimuli. Overall, as the body and immune system ages, the activation of immune cells becomes irregular. This ultimately leads to a skewing of innate and adaptive immunity cells, resulting in the alteration of immune cells that lead to persistent inflammation. Researchers have found that aged mesenchymal stem cells (MSCs) direct hematopoiesis toward oncogenesis or the myeloid lineage [227]. This results in the large production of pro-inflammatory cytokines, which promotes the body's ability to skew macrophages to M1. Not only do aged MSCs lose their ability to suppress M1 polarization, but their altered features allow them to produce inducers for M1 differentiation: IFN-γ, IL-1 and DAMP, which activates the NF-κB signal. In contrast, young MSCs can mediate the macrophage polarization from M1 to M2. Researchers found using *in vitro* co-cultures that macrophages with non-aged MSCs showed the M2 markers Arg1 and IL-10 [228]. The cocultured macrophages with aged MSCs expressed M1-related markers, such as TNF-α. Migratory abilities were exhibited in the macrophages co-cultured with aged MSCs. As such, MSCs are a crucial regulator in inflammaging and in oral diseases caused by it. A more systemic approach to analyzing the relationship between MSCs and tissue-specific signaling cascades of receptors is needed [227].

5.5 Routinely used immune modulating plant products

Unlike typical, pathway-oriented drugs, herbal medicines have the upper hand of being more accessible to patients while also exhibiting anti-inflammatory characteristics. In fact, some are already commonly used in everyday life. Yashtimadhu, more commonly known as licorice, has been utilized across the world for over 4000 years as a sweetener and a flavoring agent in both the production of typical medicines and food. Provided that it is consumed in conservative quantities, licorice has been deemed Generally Recognized as Safe (GRAS) by the United States Food and Drug Administration (FDA) [229]. To test the role of licorice in reducing the oral cavity's inflammatory response in the advent of periodontitis, researchers used an *in vitro* macrophage model and *ex vivo* human whole-blood model. They determined that licorice extracts had the potential to reduce lipopolysaccharide-induced proinflammatory cytokine secretion [18], lipopolysaccharides (LPS) are the main pathogenic factor of periodontitis and its associated pathogens, and macrophages are the main target for LPS. To examine whether licorice could interfere with LPS signaling and decrease the production of proinflammatory molecules, researchers treated macrophages and whole blood with a custom-made CO_2-supercritical extract of *G. uralensis* before being stimulated by LPS [18]. Various concentrations of the licorice extract were added to the monocyte-derived macrophage cultures and whole blood samples. The results indicated that the licorice extract reduced the secretions of proinflammatory cytokines, IL-1β and −6 and TNF-α, that were induced by macrophages stimulated with periodontopathogen LPS. This conclusion was supported by a study analyzing the effect of roasted licorice extracts on LPS signaled inflammatory responses in murine macrophages [230]. The immunoblot analyses indicated that the licorice extract achieved this through reducing the activities of AP-1 and NF-κB [18]. Part of the Jun and Fos proto-oncogene families, AP-1 is a complex of heterodimer proteins that regulate the transcription of proinflammatory mediators, specifically, IL-1β, −6, and −8 and TNF-α. *Actinomycetemcomitans* LPS induces the phosphorylation of Jun on serines 63 and 73, and in doing so, LPS boosts AP-1 activity. Researchers found that the phosphorylation of Jun was greatly decreased by the licorice extract, meaning the extract also has a strong effect on the activity of AP-1

[18]. The extract was also found to decrease the expression of IL-1β, -6, and -8 and TNF-α by inhibiting the secretion of the NF-κB p65 subunit, which in turn decreases the activation of the inflammatory transcription factor NF-κB's pathway. By targeting the AP-1 and NF-κB early on in their pathways, the licorice extract was able to reduce the initial releasing of proinflammatory cytokines rather than inhibit them after they have been released (Fig. 5.2).

Follow-up studies have instead focused on two main components from the licorice extract: licoricidin (LC) and licorisoflavan A (LIA) [231]. In addition to analyzing licorice extract's effect on the cytokine response, researchers examined LC and LIA's ability to inhibit the secretion of matrix metalloproteinases (MMPS) by human monocyte-derived macrophages that were induced by *A. actinomycetemcomitans* LPS. Partially expressed through the response of cytokines, MMPS are a class of zinc-dependent endopeptidases that play a key role in the progression of periodontitis. Examining macrophages is vital because they are found in high numbers in active periodontal lesions. This also indicates that they are large targets for LPS, resulting in the production of MMPs and inflammatory mediators. Researchers found that LC and LIA greatly reduced the LPS-induced secretion of IL-6 and CCL5, but not IL-8 secretion [231]. LC and LIA also reduced the production of MMP-7,-8, and -9, limiting their overexpression in periodontitis patients. This prevents MMP-7 from activating other MMPs and hydrolyzing various cell surface proteins: inhibiting a cascade of activation. Regarding the NF-κB p65 and AP-1 pathways, LC and LIA prevents the cytosolic inhibitor of NF-κB, IκB, from being phosphorylated and degraded by the proteasome. By preventing the degradation of such inhibitors, NF-κB cannot relocate to the nucleus and regulate active transcription. This overall process results in less cytokine and MMP secretion as well.

5.6 Conclusions and future perspectives

Due to diligent researchers and recent developments in the field, the collective knowledge surrounding plant-based therapies is continuing to grow. With each new discovery comes opportunities for better treatment options and improved health outcomes. With the worldwide prevalence of oral diseases, each step towards more effective treatment and prevention can change lives across the globe. Even now, herbal medicines are used to ease pain and soothe ailments. Despite the progress that has been made in herbal medicine research and plant-based therapies, there is so much more to explore within the field.

By continuing to research and understand the mechanisms behind how plants combat inflammation and bacterial resistance, and by unlocking the immunoregulatory capabilities of plants, researchers and clinicians will uncover additional advantageous treatment alternatives for chronic diseases like periodontitis, gingivitis, and dental caries. Plant-based medicine and oral health are irrevocably linked, just as oral health is tied to overall health and wellbeing. With time, more scientists and healthcare providers will continue to push forward and embrace the benefits of plant-based medicine: affordability, accessibility, renewability, and efficacy. Herbal medicine is not separate from the rest of medicine; they are linked. As such, oral medicine can only benefit from continuing to explore all that herbal medicine has to offer for the sake of science and for the wellbeing of people worldwide.

References

[1] MM Cowan, Plant products as antimicrobial agents, Clin Microbiol Rev 12 (4) (1999) 564–582.
[2] RB Semwal, DK Semwal, S Combrinck, AM Viljoen, Gingerols and shogaols: important nutraceutical principles from ginger, Phytochemistry 117 (2015) 554–568.
[3] NG Vasconcelos, J Croda, S Simionatto, Antibacterial mechanisms of cinnamon and its constituents: a review, Microb Pathog 120 (2018) 198–203.
[4] AL Diefenbach, F Muniz, HJR Oballe, CK Rösing, Antimicrobial activity of copaiba oil (Copaifera ssp.) on oral pathogens: systematic review, Phytother Res 32 (4) (2018) 586–596.
[5] E Valiakos, M Marselos, N Sakellaridis, C Th, Skaltsa H. Ethnopharmacological approach to the herbal medicines of the "Antidotes" in Nikolaos Myrepsosó Dynameron, J Ethnopharmacol 163 (2015) 68–82.
[6] EA Palombo, Traditional medicinal plant extracts and natural products with activity against oral bacteria: potential application in the prevention and treatment of oral diseases, Evid Based Complement Alternat Med 2011 (2011) 680354.
[7] NA Torwane, S Hongal, P Goel, BR Chandrashekar, Role of Ayurveda in management of oral health, Pharmacogn Rev 8 (15) (2014) 16–21.
[8] M Pateiro, PES Munekata, AS Sant'Ana, R Domínguez, D Rodríguez-Lázaro, JM Lorenzo, Application of essential oils as antimicrobial agents against spoilage and pathogenic microorganisms in meat products, Int J Food Microbiol 337 (2021) 108966.
[9] LB Chibane, P Degraeve, H Ferhout, J Bouajila, N Oulahal, Plant antimicrobial polyphenols as potential natural food preservatives, J Sci Food Agric 99 (4) (2019) 1457–1474.
[10] M Akram, IM Tahir, SMA Shah, Z Mahmood, A Altaf, K Ahmad, et al., Antiviral potential of medicinal plants against HIV, HSV, influenza, hepatitis, and coxsackievirus: a systematic review, Phytother Res 32 (5) (2018) 811–822.
[11] AE-L Hesham, SA Alrumman, El-Latif Hesham A. Antibacterial activity of Miswak Salvadora persica extracts against isolated and genetically identified oral cavity pathogens, Technol Health Care 24 (2016) S841–S848.
[12] U Mabona, A Viljoen, E Shikanga, A Marston, S Van Vuuren, Antimicrobial activity of southern African medicinal plants with dermatological relevance: From an ethnopharmacological screening approach, to combination studies and the isolation of a bioactive compound, J Ethnopharmacol 148 (1) (2013) 45–55.
[13] MMC Buckner, ML Ciusa, LJV Piddock, Strategies to combat antimicrobial resistance: anti-plasmid and plasmid curing, FEMS Microbiol Rev 42 (6) (2018) 781–804.
[14] J-D Cha, E-K Jung, S-Mi Choi, K-Y Lee, S-W Kang, J-D Cha, et al., Antimicrobial activity of the chloroform fraction of Drynaria fortunei against oral pathogens, J Oral Sci 59 (1) (2017) 31–38.
[15] S Kumar, AK Pandey, Chemistry and Biological Activities of flavonoids: an overview, SciWorld J 2013 (2013) 162750.
[16] S-H Park, J-Y Kim, J-M Kim, BR Yoo, SY Han, YJ Jung, et al., PM014 attenuates radiation-induced pulmonary fibrosis via regulating NF-kB and TGF-b1/NOX4 pathways, Sci Rep 10 (1) (2020) 16112.
[17] KC Chinsembu, Plants and other natural products used in the management of oral infections and improvement of oral health, Acta Trop 154 (2016) 6–18.
[18] C Bodet, VD La, S Gafner, C Bergeron, D Grenier, A licorice extract reduces lipopolysaccharide-induced proinflammatory cytokine secretion by macrophages and whole blood, J Periodontol 79 (9) (2008) 1752–1761.

[19] G Chen, X Hu, W Zhang, N Xu, F-Q Wang, J Jia, et al., Mammalian target of rapamycin regulates isoliquiritigenin-induced autophagic and apoptotic cell death in adenoid cystic carcinoma cells, Apoptosis 17 (1) (2012) 90–101.

[20] JB Fordham, A Raza Naqvi, S Nares, Leukocyte production of inflammatory mediators is inhibited by the antioxidants phloretin, silymarin, hesperetin, and resveratrol, Mediators Inflamm 2014 (2014) e938712.

[21] K Mathai, S Anand, A Aravind, P Dinatius, AV Krishnan, M Mathai, Antimicrobial effect of ginger, garlic, honey, and lemon extracts on Streptococcus mutans, J Contemp Dent Pract 18 (11) (2017) 1004–1008.

[22] World Health Organization. The World Medicines Situation 2011. Traditional Medicines: Global Situation, Issues and Challenges. Geneva, Switzerland: World Health Organization. (WHO/EMP/MIE/2011.2.3). 2011.

[23] World Health OrganizationWHO Global Report on Traditional and Complementary Medicine 2019, World Health Organization, Geneva, Switzerland, 2019.

[24] R Sender, S Fuchs, R Milo, Revised estimates for the number of human and bacteria cells in the body, PLoS Biol 14 (8) (2016) e1002533.

[25] D Verma, PK Garg, AK Dubey, Insights into the human oral microbiome, Arch Microbiol 200 (4) (2018) 525–540.

[26] GN Belibasakis, N Bostanci, PD Marsh, Zaura E. Applications of the oral microbiome in personalized dentistry, Arch Oral Biol 104 (2019) 7–12.

[27] A Mira, A Artacho, A Camelo-Castillo, S Garcia-Esteban, A Simon-Soro, Salivary immune and metabolic marker analysis (SIMMA): a diagnostic test to predict caries risk, Diagnostics (Basel) 7 (3) (2017) 38.

[28] BT Rosier, PD Marsh, A Mira, Resilience of the oral microbiota in health: mechanisms that prevent dysbiosis, J Dent Res 97 (4) (2018) 371–380.

[29] C Chen, P Feng, J Slots, Herpesvirus-bacteria synergistic interaction in periodontitis, Periodontology 82 (1) (2020) 42–64.

[30] Yadav, K Yadav, S Prakash, Dental caries: a microbiological approach, J Clin Infect Dis Practice 2017 (2017) 01–15.

[31] NJ Kassebaum, E Bernabé, M Dahiya, B Bhandari, CJL Murray, W Marcenes, Global burden of untreated caries: a systematic review and metaregression, J Dent Res 94 (5) (2015) 650–658.

[32] MJY Yon, SS Gao, KJ Chen, D Duangthip, ECM Lo, CH Chu, Medical model in caries management, Dent J (Basel) 7 (2) (2019) 37.

[33] PD Marsh, DA Head, DA Devine, Ecological approaches to oral biofilms: control without killing, Caries Res 49 (Suppl 1) (2015) 46–54.

[34] T Walsh, JM Oliveira-Neto, D Moore, Chlorhexidine treatment for the prevention of dental caries in children and adolescents. Cochrane Database Syst Rev. 13 (4) (2015) CD008457.

[35] HM van Rijkom, GJ Truin, MA van 't Hof, A meta-analysis of clinical studies on the caries-inhibiting effect of chlorhexidine treatment, J Dent Res 75 (2) (1996) 790–795.

[36] L Flötra, P Gjermo, G Rölla, J Waerhaug, Side effects of chlorhexidine mouth washes, Scand J Dent Res 79 (2) (1971) 119–125.

[37] SR Parwani, RN Parwani, PJ Chitnis, HP Dadlani, SVS Prasad, Comparative evaluation of anti-plaque efficacy of herbal and 0.2% chlorhexidine gluconate mouthwash in a 4-day plaque re-growth study, J Indian Soc Periodontol 17 (1) (2013) 72–77.

[38] C Janakiram, R Venkitachalam, P Fontelo, TJ Iafolla, BA Dye, Effectiveness of herbal oral care products in reducing dental plaque & gingivitis - a systematic review and meta-analysis, BMC Complement Med Ther 20 (1) (2020) 43.

[39] EF Corbet, JO Tam, KY Zee, MC Wong, EC Lo, AW Mombelli, et al., Therapeutic effects of supervised chlorhexidine mouthrinses on untreated gingivitis, Oral Dis 3 (1) (1997) 9–18.

[40] ET Moghadam, M Yazdanian, E Tahmasebi, H Tebyanian, R Ranjbar, A Yazdanian, et al., Current herbal medicine as an alternative treatment in dentistry: In vitro, in vivo and clinical studies, Eur J Pharmacol 889 (2020) 173665.

[41] Y Chen, M Agnello, M Dinis, KC Chien, J Wang, W Hu, et al., Lollipop containing Glycyrrhiza uralensis extract reduces Streptococcus mutans colonization and maintains oral microbial diversity in Chinese preschool children, PLoS One 14 (8) (2019) e0221756.

[42] M Perić, M Pekmezović, J Marinković, R Živković, V Arsić Arsenijević, Laboratory-Based Investigation of Denture Sonication Method in Patients with Candida-Associated Denture Stomatitis, J Pros 28 (5) (2019) 580–586.

[43] MF Khiyani, M Ahmadi, J Barbeau, JS Feine, RF de Souza, WL Siqueira, et al., Salivary biomarkers in denture stomatitis: a systematic review, JDR Clin Trans Res 4 (4) (2019) 312–322.

[44] L Martorano-Fernandes, LM Dornelas-Figueira, RM Marcello-Machado, B Silva R de, MB Magno, LC Maia, et al., Oral candidiasis and denture stomatitis in diabetic patients: systematic review and meta-analysis, Braz Oral Res 34 (2020) e113.

[45] VE Hannah, L O'Donnell, D Robertson, G Ramage, Denture stomatitis: causes, cures and prevention, Prim Dent J 6 (4) (2017) 46–51.

[46] S Ercalik-Yalcinkaya, M Özcan, Association between oral mucosal lesions and hygiene habits in a population of removable prosthesis wearers, J Prosthodont 24 (4) (2015) 271–278.

[47] L Gendreau, ZG Loewy, Epidemiology and etiology of denture stomatitis, J Prosthodont 20 (4) (2011) 251–260.

[48] A Yarborough, L Cooper, I Duqum, G Mendonça, K McGraw, L Stoner, Evidence regarding the treatment of denture stomatitis, J Prosthodont 25 (4) (2016) 288–301.

[49] JB Hilgert, A Giordani JM do, RF de Souza, E Wendland, OP D'Avila, FN Hugo, Interventions for the management of denture stomatitis: a systematic review and meta-analysis, J Am Geriatr Soc 64 (12) (2016) 2539–2545.

[50] N Kawanishi, N Hoshi, T Adachi, N Ichigaya, K Kimoto, Positive effects of saliva on oral candidiasis: basic research on the analysis of salivary properties, J Clin Med 10 (4) (2021).

[51] E Emami, M Kabawat, PH Rompre, JS Feine, Linking evidence to treatment for denture stomatitis: a meta-analysis of randomized controlled trials, J Dent 42 (2) (2014) 99–106.

[52] M Amanlou, JM Beitollahi, S Abdollahzadeh, Tohidast-Ekrad Z. Miconazole gel compared with Zataria multiflora Boiss. gel in the treatment of denture stomatitis, Phytother Res 20 (11) (2006) 966–969.

[53] A Catalán, JG Pacheco, A Martínez, MA Mondaca, In vitro and in vivo activity of Melaleuca alternifolia mixed with tissue conditioner on Candida albicans, Oral Surg Oral Med Oral Pathol Oral Radiol Endod 105 (3) (2008) 327–332.

[54] Y Shui, J Li, X Lyu, Y Wang, Phytotherapy in the management of denture stomatitis: a systematic review and meta-analysis of randomized controlled trials, Phytother Res 35 (8) (2021) 4111–4126.

[55] MJ Balunas, AD Kinghorn, Drug discovery from medicinal plants, Life Sci 78 (5) (2005) 431–441.

[56] B Kouidhi, YMA Al Qurashi, K Chaieb, Drug resistance of bacterial dental biofilm and the potential use of natural compounds as alternative for prevention and treatment, Microb Pathog 80 (2015) 39–49.

[57] M Sanz, D Herrera, M Kebschull, I Chapple, S Jepsen, T Beglundh, et al., Treatment of stage I-III periodontitis-The EFP S3 level clinical practice guideline, J Clin Periodontol 47 (Suppl 22) (2020) 4–60.

[58] EQM Souza, TE da Rocha, LF Toro, IZ Guiati, OA Freire J de, E Ervolino, et al., Adjuvant effects of curcumin as a photoantimicrobial or irrigant in the non-surgical treatment of periodontitis: Systematic review and meta-analysis, Photodiagnosis Photodyn Ther 34 (2021) 102265.

[59] RG Ledder, P Gilbert, SA Huws, L Aarons, MP Ashley, PS Hull, et al., Molecular analysis of the subgingival microbiota in health and disease, Appl Environ Microbiol 73 (2) (2007) 516–523.

[60] M Puig-Silla, JM Montiel-Company, F Dasí-Fernández, JM Almerich-Silla, Prevalence of periodontal pathogens as predictor of the evolution of periodontal status, Odontology 105 (4) (2017) 467–476.

[61] M Yaacob, HV Worthington, SA Deacon, C Deery, AD Walmsley, PG Robinson, et al. Powered versus manual toothbrushing for oral health. Cochrane Database Syst Rev 6 (2014) CD002281.

[62] SS Santi, M Casarin, AP Grellmann, L Chambrone, FB Zanatta, Effect of herbal mouthrinses on dental plaque formation and gingival inflammation: a systematic review, Oral Dis 27 (2) (2021) 127–141.

[63] W Teughels, M Feres, V Oud, C Martín, P Matesanz, D Herrera, Adjunctive effect of systemic antimicrobials in periodontitis therapy: a systematic review and meta-analysis, J Clin Periodontol 47 (Suppl 22) (2020) 257–281.

[64] H Zhao, J Hu, L Zhao, Adjunctive subgingival application of Chlorhexidine gel in nonsurgical periodontal treatment for chronic periodontitis: a systematic review and meta-analysis, BMC Oral Health 20 (1) (2020) 34.

[65] NP Lang, GE Salvi, A Sculean, Nonsurgical therapy for teeth and implants-when and why? Periodontol 2000 79 (1) (2019) 15–21.

[66] ET de Sousa, JSM de Araújo, AC Pires, EJ Lira Dos Santos, Local delivery natural products to treat periodontitis: a systematic review and meta-analysis, Clin Oral Investig 25 (7) (2021) 4599–4619.

[67] A Jitoe-Masuda, A Fujimoto, T Masuda, Curcumin: from chemistry to chemistry-based functions, Curr Pharm Des 19 (11) (2013) 2084–2092.

[68] M Siddharth, P Singh, R Gupta, A Sinha, S Shree, K Sharma, A comparative evaluation of subgingivally delivered 2% curcumin and 0.2% chlorhexidine gel adjunctive to scaling and root planing in chronic periodontitis, J Contemp Dent Pract 21 (5) (2020) 494–499.

[69] D Anusha, PE Chaly, M Junaid, JE Nijesh, K Shivashankar, S Sivasamy, Efficacy of a mouthwash containing essential oils and curcumin as an adjunct to nonsurgical periodontal therapy among rheumatoid arthritis patients with chronic periodontitis: a randomized controlled trial, Indian J Dent Res 30 (4) (2019) 506–511.

[70] D Dave, P Patel, M Shah, S Dadawala, K Saraiya, A Sant, Comparative evaluation of efficacy of oral curcumin gel as an adjunct to scaling and root planing in the treatment of chronic periodontitis, Adv Human Biol 8 (2) (2018) 79–82.

[71] V Anitha, P Rajesh, M Shanmugam, BM Priya, S Prabhu, V Shivakumar, Comparative evaluation of natural curcumin and synthetic chlorhexidine in the management of chronic periodontitis as a local drug delivery: a clinical and microbiological study, Indian J Dent Res 26 (1) (2015) 53–56.

[72] LH Theodoro, ML Ferro-Alves, M Longo, MAA Nuernberg, RP Ferreira, A Andreati, et al., Curcumin photodynamic effect in the treatment of the induced periodontitis in rats, Lasers Med Sci 32 (8) (2017) 1783–1791.

[73] VM Kumbar, MR Peram, MS Kugaji, T Shah, SP Patil, UM Muddapur, et al., Effect of curcumin on growth, biofilm formation and virulence factor gene expression of Porphyromonas gingivalis, Odontology 109 (1) (2021) 18–28.

[74] FA Oliveira, GM Vieira-Júnior, MH Chaves, FRC Almeida, MG Florêncio, RCP Lima, et al., Gastroprotective and anti-inflammatory effects of resin from Protium heptaphyllum in mice and rats, Pharmacol Res 49 (2) (2004) 105–111.

[75] EM de Lima, DSP Cazelli, FE Pinto, RA Mazuco, IC Kalil, D Lenz, et al., Essential oil from the resin of protium heptaphyllum: chemical composition, cytotoxicity, antimicrobial activity, and antimutagenicity, Pharmacogn Mag 12 (Suppl 1) (2016) S42–S46.

[76] SA Holanda Pinto, LMS Pinto, GMA Cunha, MH Chaves, FA Santos, VS Rao, Anti-inflammatory effect of alpha, beta-Amyrin, a pentacyclic triterpene from Protium heptaphyllum in rat model of acute periodontitis, Inflammopharmacology 16 (1) (2008) 48–52.

[77] RF Pastor, P Restani, C Di Lorenzo, F Orgiu, P-L Teissedre, C Stockley, et al., Resveratrol, human health and winemaking perspectives, Crit Rev Food Sci Nutr 59 (8) (2019) 1237–1255.

[78] JM Sales, AVA Resurreccion, Resveratrol in peanuts, Crit Rev Food Sci Nutr 54 (6) (2014) 734–770.

[79] JA Baur, DA Sinclair, Therapeutic potential of resveratrol: the in vivo evidence, Nat Rev Drug Discov 5 (6) (2006) 493–506 Jun.

[80] Y-T Chin, G-Y Cheng, Y-J Shih, C-Y Lin, S-J Lin, H-Y Lai, et al., Therapeutic applications of resveratrol and its derivatives on periodontitis, Ann N Y Acad Sci 1403 (1) (2017) 101–108.

[81] J-W Zhou, T-T Chen, X-J Tan, J-Y Sheng, A-Q Jia, Can the quorum sensing inhibitor resveratrol function as an aminoglycoside antibiotic accelerant against Pseudomonas aeruginosa? Int J Antimicrob Agents 52 (1) (2018) 35–41.

[82]   DJ O'Connor, RWK Wong, ABM Rabie, Resveratrol inhibits periodontal pathogens in vitro, Phytother Res 25 (11) (2011) 1727–1731.

[83]   X Gao, YX Xu, N Janakiraman, RA Chapman, SC Gautam, Immunomodulatory activity of resveratrol: suppression of lymphocyte proliferation, development of cell-mediated cytotoxicity, and cytokine production, Biochem Pharmacol 62 (9) (2001) 1299–1308.

[84]   MG Corrêa, S Absy, H Tenenbaum, FV Ribeiro, FR Cirano, MZ Casati, et al., Resveratrol attenuates oxidative stress during experimental periodontitis in rats exposed to cigarette smoke inhalation, J Periodontal Res 54 (3) (2019) 225–232.

[85]   E Ikeda, Y Ikeda, Y Wang, N Fine, Z Sheikh, A Viniegra, et al., Resveratrol derivative-rich melinjo seed extract induces healing in a murine model of established periodontitis, J Periodontol 89 (5) (2018) 586–595.

[86]   G Bhattarai, SB Poudel, S-H Kook, J-C Lee, Resveratrol prevents alveolar bone loss in an experimental rat model of periodontitis, Acta Biomater 29 (2016) 398–408.

[87]   N Adhikari, Y Prasad Aryal, J-K Jung, J-H Ha, S-Y Choi, J-Y Kim, et al., Resveratrol enhances bone formation by modulating inflammation in the mouse periodontitis model, J Periodontal Res 56 (4) (2021) 735–745.

[88]   N Tamaki, R Cristina Orihuela-Campos, Y Inagaki, M Fukui, T Nagata, H-O Ito, Resveratrol improves oxidative stress and prevents the progression of periodontitis via the activation of the Sirt1/AMPK and the Nrf2/antioxidant defense pathways in a rat periodontitis model, Free Radic Biol Med 75 (2014) 222–229.

[89]   L Zhen, D Fan, Y Zhang, X Cao, L Wang, Resveratrol ameliorates experimental periodontitis in diabetic mice through negative regulation of TLR4 signaling, Acta Pharmacol Sin 36 (2) (2015) 221–228.

[90]   A Zare Javid, R Hormoznejad, HA Yousefimanesh, M Zakerkish, MH Haghighi-Zadeh, P Dehghan, et al., The impact of resveratrol supplementation on blood glucose, insulin, insulin resistance, triglyceride, and periodontal markers in type 2 diabetic patients with chronic periodontitis, Phytother Res 31 (1) (2017) 108–114.

[91]   GA Burdock, Review of the biological properties and toxicity of bee propolis (propolis), Food Chem Toxicol 36 (4) (1998) 347–363.

[92]   JM Sforcin, Biological properties and therapeutic applications of propolis, Phytother Res 30 (6) (2016) 894–905.

[93]   S Huang, C-P Zhang, K Wang, GQ Li, F-L Hu, Recent advances in the chemical composition of propolis, Molecules 19 (12) (2014) 19610–19632.

[94]   Y Yoshimasu, T Ikeda, N Sakai, A Yagi, S Hirayama, Y Morinaga, et al., Rapid bactericidal action of propolis against Porphyromonas gingivalis, J Dent Res 97 (8) (2018) 928–936.

[95]   A Jităreanu, G Tătărîngă, A-M Zbancioc, C Tuchiluş, U Stănescu, [Antimicrobial activity of some cinnamic acid derivatives], Rev Med Chir Soc Med Nat Iasi 115 (3) (2011) 965–971.

[96]   NN Sanghani, S Bm, S S, Health from the hive: propolis as an adjuvant in the treatment of chronic periodontitis - a clinicomicrobiologic study, J Clin Diagn Res 8 (9) (2014) ZC41–ZC44.

[97]   S Sparabombe, R Monterubbianesi, V Tosco, G Orilisi, A Hosein, L Ferrante, et al., Efficacy of an all-natural polyherbal mouthwash in patients with periodontitis: a single-blind randomized controlled trial, Front Physiol 10 (2019) 632.

[98]   T Piekarz, A Mertas, K Wiatrak, R Rój, P Kownacki, J Śmieszek-Wilczewska, et al., The influence of toothpaste containing Australian Melaleuca alternifolia Oil and ethanolic extract of polish propolis on oral hygiene and microbiome in patients requiring conservative procedures, Molecules 22 (11) (2017) 1957.

[99]   R Nakao, H Senpuku, M Ohnishi, H Takai, Y Ogata, Effect of topical administration of propolis in chronic periodontitis, Odontology 108 (4) (2020) 704–714.

[100] E Giammarinaro, S Marconcini, A Genovesi, G Poli, C Lorenzi, U Covani, Propolis as an adjuvant to non-surgical periodontal treatment: a clinical study with salivary anti-oxidant capacity assessment, Minerva Stomatol 67 (5) (2018) 183–188.

[101]   HM El-Sharkawy, MM Anees, TE Van Dyke, Propolis improves periodontal status and glycemic control in patients with type 2 diabetes mellitus and chronic periodontitis: a randomized clinical trial, J Periodontol 87 (12) (2016) 1418–1426.

[102]   A Coutinho, Honeybee propolis extract in periodontal treatment: a clinical and microbiological study of propolis in periodontal treatment, Indian J Dent Res 23 (2) (2012) 294.

[103]   N López-Valverde, B Pardal-Peláez, A López-Valverde, J Flores-Fraile, S Herrero-Hernández, B Macedo-de-Sousa, et al., Effectiveness of propolis in the treatment of periodontal disease: updated systematic review with meta-analysis, Antioxidants (Basel) 10 (2) (2021) 269.

[104]   Z Khurshid, MS Zafar, S Zohaib, S Najeeb, M Naseem, Green tea (Camellia Sinensis): chemistry and oral health, Open Dent J 10 (2016) 166–173.

[105]   A Chopra, BS Thomas, K Sivaraman, HK Prasad, SU Kamath, Green tea intake as an adjunct to mechanical periodontal therapy for the management of mild to moderate chronic periodontitis: a randomized controlled clinical trial, Oral Health Prev Dent 14 (4) (2016) 293–303.

[106]   SA Hattarki, SP Pushpa, K Bhat, Evaluation of the efficacy of green tea catechins as an adjunct to scaling and root planing in the management of chronic periodontitis using PCR analysis: a clinical and microbiological study, J Indian Soc Periodontol 17 (2) (2013) 204–209.

[107]   K Rattanasuwan, S Rassameemasmaung, V Sangalungkarn, C Komoltri, Clinical effect of locally delivered gel containing green tea extract as an adjunct to non-surgical periodontal treatment, Odontology 104 (1) (2016) 89–97.

[108]   TS Hrishi, PP Kundapur, A Naha, BS Thomas, S Kamath, GS Bhat, Effect of adjunctive use of green tea dentifrice in periodontitis patients - a randomized controlled pilot study, Int J Dent Hyg 14 (3) (2016) 178–183.

[109]   M Kushiyama, Y Shimazaki, M Murakami, Y Yamashita, Relationship between intake of green tea and periodontal disease, J Periodontol 80 (3) (2009) 372–377.

[110]   M Makimura, M Hirasawa, K Kobayashi, J Indo, S Sakanaka, T Taguchi, et al., Inhibitory effect of tea catechins on collagenase activity, J Periodontol 64 (7) (1993) 630–636.

[111]   S Sakanaka, Y Okada, Inhibitory effects of green tea polyphenols on the production of a virulence factor of the periodontal-disease-causing anaerobic bacterium Porphyromonas gingivalis, J Agric Food Chem 52 (6) (2004) 1688–1692.

[112]   JGA Melo, JP Sousa, RT Firmino, CC Matins, AF Granville-Garcia, CFW Nonaka, et al., Different applications forms of green tea (Camellia sinensis (L.) Kuntze) for the treatment of periodontitis: a systematic review and meta-analysis, J Periodontal Res 56 (3) (2021) 443–453.

[113]   T Vani, M Rajani, S Sarkar, CJ Shishoo, Antioxidant properties of the ayurvedic formulation triphala and its constituents, Pharm Biol. 35 (5) (1997) 313–317.

[114]   A Padmawar, U Bhadoriya, Phytochemical investigation and comparative evaluation of in vitro free radical scavenging activity of triphala & curcumin, Asian J Pharm Med Sci 1 (2011) 9–12.

[115]   DV Surya Prakash, N Sree Satya, S Avanigadda, M Vangalapati, Pharmacological review on Terminalia Chebula, Int J Res Pharm Biomed Sci 3 (2012) 679–683.

[116]   PA Dar, G Sofi, SA Parray, MA Jafri, Halelah Siyah (Terminalia Chebula Retz): in Unani system of medicine and modern pharmacology: a review, Int J Inst Pharm Life Sci 2 (2012) 138–149.

[117]   B Thomas, SY Shetty, A Vasudeva, V Shetty, Comparative evaluation of antimicrobial activity of triphala and commercially available toothpastes: an in-vitro study, Int J Public Health Dent 2 (2011) 8–12.

[118]   YS Biradar, S Jagatap, KR Khandelwal, SS Singhania, Exploring of antimicrobial activity of triphala mashi-an ayurvedic formulation, Evid Based Complement Alternat Med. 5 (1) (2008) 107–113.

[119]   A Desai, M Anil, S Debnath, A clinical trial to evaluate the effects of triphala as a mouthwash in comparison with chlorhexidine in chronic generalized periodontitis patient, Indian J Dent Adv 2 (2010) 243–247.

[120]   SU Baratakke, R Raju, S Kadanakuppe, NR Savanur, R Gubbihal, PS Kousalaya, Efficacy of triphala extract and chlorhexidine mouth rinse against plaque accumulation and gingival inflammation among female undergraduates: a randomized controlled trial, Indian J Dent Res 28 (1) (2017) 49–54.

[121]   SH Chainani, S Siddana, C Reddy, TH Manjunathappa, M Manjunath, S Rudraswamy, Antiplaque and antigingivitis efficacy of triphala and chlorhexidine mouthrinse among schoolchildren – a cross-over, double-blind, randomised controlled trial, Oral Health Prev Dent 12 (3) (2014) 209–217.

[122]   N Bajaj, S Tandon, The effect of Triphala and Chlorhexidine mouthwash on dental plaque, gingival inflammation, and microbial growth, Int J Ayurveda Res 2 (1) (2011) 29–36.

[123]   RS Naiktari, P Gaonkar, AN Gurav, SV Khiste, A randomized clinical trial to evaluate and compare the efficacy of triphala mouthwash with 0.2% chlorhexidine in hospitalized patients with periodontal diseases, J Periodontal Implant Sci 44 (3) (2014) 134–140.

[124]   D Gupta, RK Gupta, DJ Bhaskar, V Gupta, Comparative evaluation of terminalia chebula extract mouthwash and chlorhexidine mouthwash on plaque and gingival inflammation - 4-week randomised control trial, Oral Health Prev Dent 13 (1) (2015) 5–12.

[125]   D Gupta, DJ Bhaskar, RK Gupta, B Karim, V Gupta, H Punia, et al., Effect of Terminalia chebula extract and chlorhexidine on salivary pH and periodontal health: 2 weeks randomized control trial, Phytother Res 28 (7) (2014) 992–998.

[126]   AH AlJameel, SA Almalki, Effect of triphala mouthrinse on plaque and gingival inflammation: A systematic review and meta-analysis of randomized controlled trials, Int J Dent Hyg 18 (4) (2020) 344–351.

[127]   R Malhotra, V Grover, A Kapoor, D Saxena, Comparison of the effectiveness of a commercially available herbal mouthrinse with chlorhexidine gluconate at the clinical and patient level, J Indian Soc Periodontol 15 (4) (2011) 349–352.

[128]   C Yu, PV Abbott, An overview of the dental pulp: its functions and responses to injury, Aust Dent J 52 (1 Suppl) (2007) S4–16.

[129]   EM Almadi, AA Almohaimede, Natural products in endodontics, Saudi Med J 39 (2) (2018) 124–130.

[130]   JF Siqueira, IN Rôças, Microbiology and treatment of acute apical abscesses, Clin Microbiol Rev 26 (2) (2013) 255–273.

[131]   M Ehsani, M Amin Marashi, E Zabihi, M Issazadeh, S Khafri, A comparison between antibacterial activity of propolis and aloe vera on Enterococcus faecalis (an in vitro study), Int J Mol Cell Med 2 (3) (2013) 110–116.

[132]   BV Chaitanya, KV Somisetty, A Diwan, S Pasha, N Shetty, Y Reddy, et al., Comparison of antibacterial efficacy of turmeric extract, Morinda Citrifolia and 3% sodium hypochlorite on enterococcus faecalis: an in-vitro study, J Clin Diagn Res 10 (10) (2016) ZC55–ZC57.

[133]   V Zand, M Lotfi, MH Soroush, AA Abdollahi, M Sadeghi, A Mojadadi, Antibacterial efficacy of different concentrations of sodium hypochlorite gel and solution on Enterococcus faecalis biofilm, Iran Endod J 11 (4) (2016) 315–319.

[134]   P Babaji, K Jagtap, H Lau, N Bansal, S Thajuraj, P Sondhi, Comparative evaluation of antimicrobial effect of herbal root canal irrigants (Morinda citrifolia, Azadirachta indica, Aloe vera) with sodium hypochlorite: an in vitro study, J Int Soc Prev Community Dent 6 (3) (2016) 196–199.

[135]   L Arévalo-Híjar, A-L MÁ, S Caballero-García, N Gonzáles-Soto, J Del Valle-Mendoza, Antibacterial and cytotoxic effects of Moringa oleifera (Moringa) and Azadirachta indica (Neem) methanolic extracts against strains of Enterococcus faecalis, Int J Dent 2018 (2018) 1071676.

[136]   MT Devi, S Saha, AM Tripathi, K Dhinsa, SK Kalra, U Ghoshal, Evaluation of the antimicrobial efficacy of herbal extracts added to root canal sealers of different bases: an in vitro study, Int J Clin Pediatr Dent 12 (5) (2019) 398–404.

[137]   TS Vinothkumar, MI Rubin, L Balaji, D Kandaswamy, In vitro evaluation of five different herbal extracts as an antimicrobial endodontic irrigant using real time quantitative polymerase chain reaction, J Conserv Dent 16 (2) (2013) 167–170.

[138]   SD Caetano da Silva, MG Mendes de Souza, MJ Oliveira Cardoso, T da Silva Moraes, SR Ambrósio, RC Sola Veneziani, et al., Antibacterial activity of Pinus elliottii against anaerobic bacteria present in primary endodontic infections, Anaerobe 30 (2014) 146–152.

[139]   S Agarwal, P Tyagi, A Deshpande, S Yadav, V Jain, KS Rana, Comparison of antimicrobial efficacy of aqueous ozone, green tea, and normal saline as irrigants in pulpectomy procedures of primary teeth, J Indian Soc Pedod Prev Dent 38 (2) (2020) 164–170.

[140]   HA Yamani, EC Pang, N Mantri, MA Deighton, Antimicrobial activity of tulsi (Ocimum tenuiflorum) essential oil and their major constituents against three species of bacteria, Front Microbiol 7 (2016) 681.

[141]   P Dausage, RB Dhirawani, J Marya, V Dhirawani, V Kumar, A comparative study of ion diffusion from calcium hydroxide with various herbal pastes through dentin, Int J Clin Pediatr Dent 10 (1) (2017) 41–44.

[142]   FS D'avis, E Lodi, MA Souza, AP Farina, D Cecchin, Antibacterial efficacy of the grape seed extract as an irrigant for root canal preparation, Eur Endod J 5 (1) (2020) 35–39.

[143]   V Agrawal, S Kapoor, I Agrawal, Critical review on eliminating endodontic dental infections using herbal products, J Diet Suppl 14 (2) (2017) 229–240.

[144]   N Jaiswal, D-J Sinha, U-P Singh, K Singh, U-A Jandial, S Goel, Evaluation of antibacterial efficacy of Chitosan, Chlorhexidine, Propolis and Sodium hypochlorite on Enterococcus faecalis biofilm : an in vitro study, J Clin Exp Dent 9 (9) (2017) e1066–e1074.

[145]   H Kumar, An in vitro evaluation of the antimicrobial efficacy of Curcuma longa, Tachyspermum ammi, chlorhexidine gluconate, and calcium hydroxide on Enterococcus faecalis, J Conserv Dent 16 (2) (2013) 144–147.

[146]   S Bashir, BM Khan, M Babar, S Andleeb, M Hafeez, S Ali, et al., Assessment of bioautography and spot screening of TLC of green tea (Camellia) plant extracts as antibacterial and antioxidant agents, Indian J Pharm Sci 76 (4) (2014) 364–370.

[147]   JL Watson, M Vicario, A Wang, M Moreto, DM McKay, Immune cell activation and subsequent epithelial dysfunction by Staphylococcus enterotoxin B is attenuated by the green tea polyphenol (-)-epigallocatechin gallate, Cell Immunol 237 (1) (2005) 7–16.

[148]   N Venkateshbabu, S Anand, M Abarajithan, SO Sheriff, PS Jacob, N Sonia, Natural therapeutic options in endodontics - a review, Open Dent J 10 (2016) 214–226.

[149]   JA Hutter, M Salman, WB Stavinoha, N Satsangi, RF Williams, RT Streeper, et al., Antiinflammatory C-glucosyl chromone from Aloe barbadensis, J Nat Prod 59 (5) (1996) 541–543.

[150]   A Dutta, M Kundabala, Comparative anti-microbial efficacy of Azadirachta indica irrigant with standard endodontic irrigants: a preliminary study, J Conserv Dent 17 (2) (2014) 133–137.

[151]   T Lakshmi, V Krishnan, R Rajendran, N Madhusudhanan, Azadirachta indica: a herbal panacea in dentistry - an update, Pharmacogn Rev 9 (17) (2015) 41–44.

[152]   JR Lechien, CM Chiesa-Estomba, DR De Siati, M Horoi, SD Le Bon, A Rodriguez, et al., Olfactory and gustatory dysfunctions as a clinical presentation of mild-to-moderate forms of the coronavirus disease (COVID-19): a multicenter European study, Eur Arch Otorhinolaryngol 277 (8) (2020) 2251–2261.

[153]   DA Mahmoud, NM Hassanein, KA Youssef, MA Abou Zeid, Antifungal activity of different neem leaf extracts and the nimonol against some important human pathogens, Braz J Microbiol 42 (3) (2011) 1007–1016.

[154]   J Prabhakar, M Senthilkumar, MS Priya, K Mahalakshmi, PK Sehgal, VG Sukumaran, Evaluation of antimicrobial efficacy of herbal alternatives (Triphala and green tea polyphenols), MTAD, and 5% sodium hypochlorite against Enterococcus faecalis biofilm formed on tooth substrate: an in vitro study, J Endod 36 (1) (2010) 83–86.

[155]   A Dutta, M Kundabala, Antimicrobial efficacy of endodontic irrigants from Azadirachta indica: an in vitro study, Acta Odontol Scand 71 (6) (2013) 1594–1598.

[156]   KS Mistry, Z Sanghvi, G Parmar, S Shah, The antimicrobial activity of Azadirachta indica, Mimusops elengi, Tinospora cardifolia, Ocimum sanctum and 2% chlorhexidine gluconate on common endodontic pathogens: an in vitro study, Eur J Dent 8 (2) (2014) 172–177.

[157]   DS Dezmirean, C Paşca, AR Moise, O Bobiş, Plant sources responsible for the chemical composition and main bioactive properties of poplar-type propolis, Plants (Basel) 10 (1) (2020) 22.

[158] R Baranwal, V Duggi, A Avinash, A Dubey, S Pagaria, H Munot, Propolis: a smart supplement for an intracanal medicament, Int J Clin Pediatr Dent 10 (4) (2017) 324–329.

[159] JM Sforcin, Propolis and the immune system: a review, J Ethnopharmacol 113 (1) (2007) 1–14.

[160] K Salomão, AP Dantas, CM Borba, LC Campos, DG Machado, FR Aquino Neto, et al., Chemical composition and microbicidal activity of extracts from Brazilian and Bulgarian propolis, Lett Appl Microbiol 38 (2) (2004) 87–92.

[161] A Nara, Dhanu, P Chandra, L Anandakrishna, Dhananjaya, Comparative evaluation of antimicrobial efficacy of MTAD, 3% NaOCI and propolis against E Faecalis, Int J Clin Pediatr Dent 3 (1) (2010) 21–25.

[162] BJ Moncla, PW Guevara, JA Wallace, MC Marcucci, JE Nor, WA Bretz, The inhibitory activity of typified propolis against Enterococcus species, Z Naturforsch C J Biosci 67 (5–6) (2012) 249–256.

[163] M Jolly, N Singh, M Rathore, S Tandon, M Banerjee, Propolis and commonly used intracanal irrigants: comparative evaluation of antimicrobial potential, J Clin Pediatr Dent 37 (3) (2013) 243–249.

[164] FB Silva, JM Almeida, SMG Sousa, Natural medicaments in endodontics – a comparative study of the anti-inflammatory action, Braz Oral Res 18 (2) (2004) 174–179.

[165] NM Fiallos, D Cecchin, CO de Lima, R Hirata, EJNL Silva, LM Sassone, Antimicrobial effectiveness of grape seed extract against Enterococcus faecalis biofilm: a confocal laser scanning microscopy analysis, Aust Endod J 46 (2) (2020) 191–196.

[166] LT Soligo, E Lodi, AP Farina, MA Souza, MP Vidal C de, D Cecchin, Antibacterial efficacy of synthetic and natural-derived novel endodontic irrigant solutions, Braz Dent J 29 (5) (2018) 459–464.

[167] CMP Vidal, W Zhu, S Manohar, B Aydin, TA Keiderling, PB Messersmith, et al., Collagen-collagen interactions mediated by plant-derived proanthocyanidins: a spectroscopic and atomic force microscopy study, Acta Biomater 41 (2016) 110–118.

[168] P Yadav, S Chaudhary, RK Saxena, S Talwar, S Yadav, Evaluation of antimicrobial and antifungal efficacy of chitosan as endodontic irrigant against Enterococcus Faecalis and Candida Albicans biofilm formed on tooth substrate, J Clin Exp Dent 9 (3) (2017) e361–e367.

[169] M Shahini Shams Abadi, E Mirzaei, A Bazargani, A Gholipour, H Heidari, N Hadi, Antibacterial activity and mechanism of action of chitosan nanofibers against toxigenic Clostridioides (Clostridium) difficile Isolates, Ann Ig 32 (1) (2020) 72–80.

[170] R Varshini, A Subha, V Prabhakar, P Mathini, S Narayanan, K Minu, Antimicrobial efficacy of aloe vera, lemon, Ricinus communis, and calcium hydroxide as intracanal medicament against Enterococcus faecalis: a confocal microscopic study, J Pharm Bioallied Sci 11 (Suppl 2) (2019) S256–S259.

[171] B Kurian, D Swapna, R Nadig, M Ranjini, K Rashmi, S Bolar, Efficacy of calcium hydroxide, mushroom, and Aloe vera as an intracanal medicament against Enterococcus faecalis: an in vitro study, Endodontology. 28 (2) (2016) 137–142.

[172] S Yanakiev, Effects of cinnamon (Cinnamomum spp.) in dentistry: a review, Molecules 25 (18) (2020) 4184.

[173] S Rahman, P Carter, N Bhattarai, Aloe vera for tissue engineering applications, J Funct Biomater 8 (1) (2017) 6.

[174] J Shabbir, F Qazi, W Farooqui, S Ahmed, T Zehra, Z Khurshid, Effect of Chinese propolis as an intracanal medicament on post-operative endodontic pain: a double-blind randomized controlled trial, Int J Environ Res Public Health 17 (2) (2020) 445.

[175] MM Madhubala, N Srinivasan, S Ahamed, Comparative evaluation of propolis and triantibiotic mixture as an intracanal medicament against Enterococcus faecalis, J Endod 37 (9) (2011) 1287–1289.

[176] J Sangalli, EGJ Júnior, CRE Bueno, RC Jacinto, G Sivieri-Araújo, JEG Filho, et al., Antimicrobial activity of Psidium cattleianum associated with calcium hydroxide against Enterococcus faecalis and Candida albicans: an in vitro study, Clin Oral Investig 22 (6) (2018) 2273–2279.

[177] WR Poi, CK Sonoda, CM Martins, ME Melo, EP Pellizzer, MR de Mendonça, et al., Storage media for avulsed teeth: a literature review, Braz Dent J 24 (5) (2013) 437–445.

[178]   WR Poi, CK Sonoda, MF Amaral, AF Queiroz, AB França, DA Brandini, Histological evaluation of the repair process of replanted rat teeth after storage in resveratrol dissolved in dimethyl sulphoxide, Dent Traumatol 34 (2018) 254–263.

[179]   S Adnan, MM Lone, FR Khan, SM Hussain, SE Nagi, Which is the most recommended medium for the storage and transport of avulsed teeth? A systematic review, Dent Traumatol 34 (2) (2018) 59–70.

[180]   A Bunwanna, T Damrongrungruang, S Puasiri, N Kantrong, P Chailertvanitkul, Preservation of the viability and gene expression of human periodontal ligament cells by Thai propolis extract, Dent Traumatol 37 (1) (2021) 123–130.

[181]   A Yuan, S Wang, Z Li, C Huang, Psychological aspect of cancer: From stressor to cancer progression, Exp Ther Med 1 (1) (2010) 13–18.

[182]   GG Mori, DC Nunes, LR Castilho, IG de Moraes, WR Poi, Propolis as storage media for avulsed teeth: microscopic and morphometric analysis in rats, Dent Traumatol 26 (1) (2010) 80–85.

[183]   V Gopikrishna, T Thomas, D Kandaswamy, A quantitative analysis of coconut water: a new storage media for avulsed teeth, Oral Surg Oral Med Oral Pathol Oral Radiol Endod. 105 (2) (2008) e61–e65.

[184]   MVP Reis, CJ Soares, PBF Soares, AM Rocha, CCM Salgueiro, MHNR Sobral, et al., Replanted teeth stored in a newly developed powdered coconut water formula, Dent Traumatol 34 (2) (2018) 114–119.

[185]   S Badakhsh, T Eskandarian, T Esmaeilpour, The use of aloe vera extract as a novel storage media for the avulsed tooth, Iran J Med Sci 39 (4) (2014) 327–332.

[186]   AZ Beg, I Ahmad, Effect of Plumbago zeylanica extract and certain curing agents on multidrug resistant bacteria of clinical origin, World J Microbiol Biotechnol 16 (8–9) (2000) 841–844.

[187]   RB Patwardhan, PK Dhakephalkar, BA Chopade, DD Dhavale, RR Bhonde, Purification and characterization of an active principle, lawsone, responsible for the plasmid curing activity of Plumbago zeylanica root extracts, Front Microbiol 9 (2018) 2618.

[188]   C Simões, O Conde NC de, GN Venâncio, P Milério, M Bandeira, VF da Veiga Júnior, Antibacterial activity of copaiba oil gel on dental biofilm, Open Dent J 10 (1) (2016) 188–195.

[189]   FA Pieri, MC Mussi, JE Fiorini, JM Schneedorf, Clinical and microbiological effects of copaiba oil (Copaifera officinalis) on plaque-forming bacteria in dogs, Arq Bras Med Vet Zootec 62 (2010) 578–585.

[190]   DM Kasote, SS Katyare, MV Hegde, H Bae, Significance of antioxidant potential of plants and its relevance to therapeutic applications, Int J Biol Sci 11 (8) (2015) 982–991.

[191]   MC Putchala, P Ramani, HJ Sherlin, P Premkumar, A Natesan, Ascorbic acid and its pro-oxidant activity as a therapy for tumours of oral cavity – a systematic review, Arch Oral Biol 58 (6) (2013) 563–574.

[192]   A Merecz-Sadowska, P Sitarek, T Śliwiński, R Zajdel, Anti-Inflammatory activity of extracts and pure compounds derived from plants via modulation of signaling pathways, especially PI3K/AKT in macrophages, Int J Mol Sci 21 (24) (2020) 9605.

[193]   R-Z Tan, J Liu, Y-Y Zhang, H-L Wang, J-C Li, Y-H Liu, et al., Curcumin relieved cisplatin-induced kidney inflammation through inhibiting Mincle-maintained M1 macrophage phenotype, Phytomedicine 52 (2019) 284–294.

[194]   EM Drummond, N Harbourne, E Marete, D Martyn, J Jacquier, D O'Riordan, et al., Inhibition of proinflammatory biomarkers in THP1 macrophages by polyphenols derived from chamomile, meadowsweet and willow bark, Phytother Res 27 (4) (2013) 588–594.

[195]   SJ Preston, MH Arnold, EM Beller, PM Brooks, WW Buchanan, Comparative analgesic and anti-inflammatory properties of sodium salicylate and acetylsalicylic acid (aspirin) in rheumatoid arthritis, Br J Clin Pharmacol 27 (5) (1989) 607–611.

[196]   B Hinz, V Kraus, A Pahl, K Brune, Salicylate metabolites inhibit cyclooxygenase-2-dependent prostaglandin E(2) synthesis in murine macrophages, Biochem Biophys Res Commun 274 (1) (2000) 197–202.

[197]   E Kopp, S Ghosh, Inhibition of NF-kappa B by sodium salicylate and aspirin, Science 265 (5174) (1994) 956–959.

[198]   N Harbourne, E Marete, JC Jacquier, D O'Riordan, Effect of drying methods on the phenolic constituents of meadowsweet (Filipendula ulmaria) and willow (Salix alba), Food Sci Technol 42 (9) (2009) 1468–1473.

[199]   N Harbourne, JC Jacquier, D O'Riordan, Optimisation of the aqueous extraction conditions of phenols from meadowsweet (Filipendula ulmaria L.) for incorporation into beverages, Food Chem 116 (3) (2009) 722–727.

[200]   JR Vane, The fight against rheumatism: from willow bark to COX-1 sparing drugs, J Physiol Pharmacol 51 (4 Pt 1) (2000) 573–586.

[201]   JA Symons, PR Young, GW Duff, Soluble type II interleukin 1 (IL-1) receptor binds and blocks processing of IL-1 beta precursor and loses affinity for IL-1 receptor antagonist, Proc Natl Acad Sci U S A. 92 (5) (1995) 1714–1718.

[202]   C Janeway, Immunobiology: The Immune System in Health and Disease, Garland Science, New York; London, 2005.

[203]   S Máñez, MC Recio, I Gil, C Gómez, RM Giner, PG Waterman, et al., A glycosyl analogue of diacylglycerol and other antiinflammatory constituents from Inula viscosa, J Nat Prod 62 (4) (1999) 601–604.

[204]   V Hernández, M del Carmen Recio, S Máñez, JM Prieto, RM Giner, JL Ríos, A mechanistic approach to the in vivo anti-inflammatory activity of sesquiterpenoid compounds isolated from Inula viscosa, Planta Med 67 (8) (2001) 726–731.

[205]   S Rozenblat, S Grossman, M Bergman, H Gottlieb, Y Cohen, S Dovrat, Induction of G2/M arrest and apoptosis by sesquiterpene lactones in human melanoma cell lines, Biochem Pharmacol 75 (2) (2008) 369–382.

[206]   TD Gilmore, Introduction to NF-kappaB: players, pathways, perspectives, Oncogene 25 (51) (2006) 6680–6684.

[207]   J Hiscott, J Marois, J Garoufalis, M D'Addario, A Roulston, I Kwan, et al., Characterization of a functional NF-kappa B site in the human interleukin 1 beta promoter: evidence for a positive autoregulatory loop, Mol Cell Biol 13 (10) (1993) 6231–6240.

[208]   PG McCaffrey, J Jain, C Jamieson, R Sen, A Rao, A T cell nuclear factor resembling NF-AT binds to an NF-kappa B site and to the conserved lymphokine promoter sequence "cytokine-1, J Biol Chem 267 (3) (1992) 1864–1871.

[209]   G Abrham, S Dovrat, H Bessler, S Grossman, U Nir, M Bergman, Inhibition of inflammatory cytokine secretion by plant-derived compounds inuviscolide and tomentosin: the role of NFκB and STAT1~!2010-02-10~!2010-06-28~!2010-08-12~!, Open Pharmacol J 4 (2010) 36–44.

[210]   G Abrham, M Volpe, S Shpungin, U Nir, TMF/ARA160 downregulates proangiogenic genes and attenuates the progression of PC3 xenografts, Int J Cancer 125 (1) (2009) 43–53.

[211]   A Ryo, F Suizu, Y Yoshida, K Perrem, Y-C Liou, G Wulf, et al., Regulation of NF-kappaB signaling by Pin1-dependent prolyl isomerization and ubiquitin-mediated proteolysis of p65/RelA, Mol Cell 12 (6) (2003) 1413–1426.

[212]   T Tanaka, MJ Grusby, T Kaisho, PDLIM2-mediated termination of transcription factor NF-kappaB activation by intranuclear sequestration and degradation of the p65 subunit, Nat Immunol 8 (6) (2007) 584–591.

[213]   N Azad, Y Rojanasakul, V Vallyathan, Inflammation and lung cancer: roles of reactive oxygen/nitrogen species, J Toxicol Environ Health B Crit Rev 11 (1) (2008) 1–15.

[214]   H Clevers, At the crossroads of inflammation and cancer, Cell 118 (6) (2004) 671–674.

[215]   F Balkwill, KA Charles, A Mantovani, Smoldering and polarized inflammation in the initiation and promotion of malignant disease, Cancer Cell 7 (3) (2005) 211–217.

[216]   LM Coussens, Z Werb, Inflammatory cells and cancer: think different!, J Exp Med 193 (6) (2001) F23–F26.

[217]   AJ García-Piñeres, V Castro, G Mora, TJ Schmidt, E Strunck, HL Pahl, et al., Cysteine 38 in p65/NF-kappaB plays a crucial role in DNA binding inhibition by sesquiterpene lactones, J Biol Chem 276 (43) (2001) 39713–39720.

[218]   G Lyss, A Knorre, TJ Schmidt, HL Pahl, I Merfort, The anti-inflammatory sesquiterpene lactone helenalin inhibits the transcription factor NF-kappaB by directly targeting p65, J Biol Chem 273 (50) (1998) 33508–33516.

[219]   X Luan, X Zhou, A Naqvi, M Francis, D Foyle, S Nares, et al., MicroRNAs and immunity in periodontal health and disease, Int J Oral Sci 10 (3) (2018) 24.

[220]   N Dutzan, JE Konkel, T Greenwell-Wild, NM Moutsopoulos, Characterization of the human immune cell network at the gingival barrier, Mucosal Immunol 9 (5) (2016) 1163–1172.

[221]   JR Teijaro, D Turner, Q Pham, EJ Wherry, L Lefrançois, DL Farber, Cutting edge: Tissue-retentive lung memory CD4 T cells mediate optimal protection to respiratory virus infection, J Immunol 187 (11) (2011) 5510–5514.

[222]   H Davanian, RA Gaiser, M Silfverberg, LW Hugerth, MJ Sobkowiak, L Lu, et al., Mucosal-associated invariant T cells and oral microbiome in persistent apical periodontitis, Int J Oral Sci 11 (2) (2019) 16.

[223]   A Meierovics, WJ Yankelevich, SC Cowley, MAIT cells are critical for optimal mucosal immune responses during in vivo pulmonary bacterial infection, Proc Natl Acad Sci U S A. 110 (33) (2013) E3119–E3128.

[224]   YC Liu, XB Zou, YF Chai, YM Yao, Macrophage polarization in inflammatory diseases, Int J Biol Sci 10 (5) (2014) 520–529.

[225]   H Roca, ZS Varsos, S Sud, MJ Craig, C Ying, KJ Pienta, CCL2 and interleukin-6 promote survival of human CD11b+ peripheral blood mononuclear cells and induce M2-type macrophage polarization, J Biol Chem 284 (49) (2009) 34342–34354.

[226]   W Qi, Z Xinyi, D Yi, Effect of inflammaging on periodontitis, West China J Stomat 36 (1) (2018) 99–103.

[227]   BC Lee, KR Yu, Impact of mesenchymal stem cell senescence on inflammaging, BMB Rep 53 (2) (2020) 65–73.

[228]   Y Yin, RX Wu, XT He, XY Xu, J Wang, FM Chen, Influences of age-related changes in mesenchymal stem cells on macrophages during in-vitro culture, Stem Cell Res Ther 8 (1) (2017) 153.

[229]   P Sidhu, S Shankargouda, A Rath, P Hesarghatta Ramamurthy, B Fernandes, A Kumar Singh, Therapeutic benefits of liquorice in dentistry, J Ayurveda Integr Med 11 (1) (2020) 82–88.

[230]   JK Kim, S Oh, HS Kwon, YS Oh, SS Lim, HK Shin, Anti-inflammatory effect of roasted licorice extracts on lipopolysaccharide-induced inflammatory responses in murine macrophages, Biochem Biophys Res Commun 345 (3) (2006) 1215–1223.

[231]   VD La, S Tanabe, C Bergeron, S Gafner, D Grenier, Modulation of matrix metalloproteinase and cytokine production by licorice isolates licoricidin and licorisoflavan A: potential therapeutic approach for periodontitis, J Periodontol 82 (1) (2011) 122–128.

PART 3

Metabolic disorders and overall health

CHAPTER 6

Herbal medicines for the treatment of metabolic syndrome

Pascaline Obika[a], Jessica Beamon[a], Sumera Ali[a], Nandni Kakar[a],
Arturo Analla[a], R'kia El Moudden[a], Lubna Shihadeh[a], Savan Patel[a],
Brionna Hudson[a], Faaeiza Khan[a], Melany Puglisi-Weening[a], Parakh Basist[b],
Sayeed Ahmad[b] and Mohd Shahid[a]

[a]*Department of Pharmaceutical Sciences, Chicago State University College of Pharmacy, Chicago, IL*
[b]*Department of Pharmacognosy & Phytochemistry, School of Pharmaceutical Education & Research, Jamia Hamdard, New Delhi, India*

6.1 Introduction

Definition: Metabolic syndrome (MetS), also known as syndrome X, 'insulin resistance syndrome', and 'hypertriglyceridemic waist', is a clustering of obesity, hyperglycemia, insulin resistance, and dyslipidemia. MetS increases the risk of developing atherosclerotic cardiovascular disease (CVD) and is now considered one of the most important cardiovascular risk factors. The World Health Organization (WHO) developed the first definition of MetS in 1998 which was based on the absolute requirement of insulin resistance. According to WHO, MetS is defined as the presence of insulin resistance (could be defined as presence of impaired fasting glucose (IGT), glucose tolerance, elevated homeostatic model assessment of insulin resistance (HOMA-IR) or type 2 diabetes mellitus) in addition to two of additional risk factors. These can include obesity, hyperlipidemia, hypertension, and microalbuminuria [1]. This definition for the first time centrally connected different key disorders to MetS and provided a solid framework to diagnose MetS. However, due to inaccessibility to the required clinical tests, this definition was not readily applicable across the board. Later, other global organizations proposed other iterations of this definition while retaining insulin resistance as central to the pathophysiology. Four most popular definitions that are used clinically and for epidemiological studies are given below in Table 6.1 [2].

As mentioned above MetS is a multitude of factors, including obesity, elevated blood pressure, dyslipidemia, and insulin resistance leading to increased risk for the cardiovascular complications. The linkage of these factors creates the overarching syndrome characterized by cardio-metabolic disorders. According to the Centers for Disease Control and Prevention (CDC) [6], MetS is one of the leading causes of multiple chronic diseases and death in the

Table 6.1: Definitions of metabolic syndrome.

Program	WHO (1998)[1]	NCEP-ATM III (2005)[3]	IDF (2006)[4]	EGIR (1999)[5]
Criteria	Insulin resistance or glucose >6.1 mmol/L (110 mg/dl), 2 h glucose >7.8 mmol (140 mg/dl) (required), plus at least any two of the following	Any three or more of the following	Central obesity (Waist >94 cm (men) or >80 cm (women)), plus two or more of the following	Hyperinsulinemia, plus two of the following
Blood glucose	Insulin resistance	>5.6 mmol/L (100 mg/dl) or on drug treatment or on drug treatment	Fasting, >5.6 mmol/L (100 mg/dl)	Insulin resistance
HDL cholesterol	<0.9 mmol/L (35 mg/dl) in men, <1.0 mmol/L (40 mg/dl) in women	<1.0 mmol/L (40 mg/dl) in men, <1.3 mmol/L (50 mg/dl) in women or on drug treatment	<1.0 mmol/L (40 mg/dl) in men, <1.3 mmol/L> (50 mg/dl) in women or on drug treatment	<39 mg/dl
Triglycerides	>1.7 mmol/L (150 mg/dl)	>1.7 mmol/L (150 mg/dl) or on drug treatment	>1.7 mmol/L (150 mg/dl) or on drug treatment	>177 mg/dl
Body weight	Waist/hip ratio >0.9 (men) or >0.85 (women) or BMI >30 kg/m2	Waist >102 cm (men) or >88 cm (women)	Obesity required	Waist ≥ 94 cm (men) or ≥ 0.80 cm (women)
Blood pressure	>140/90 mmHg	>130/85 mmHg or on drug treatment	>130/85 mmHg or on drug treatment	>140/90 mmHg or on drug treatment
Other	Microalbuminuria			

EGIR, European Group for the Study of Insulin Resistance; IDF, International Diabetes Federation; NCEP-ATM III, (National Cholesterol Education Program) Adult Treatment Plan 3.

modern world. It originates from an imbalance of calorie intake and energy expenditure and is caused by multitude of factors but mostly triggered by the genetic makeup of individuals. Other factors, such as a sedentary lifestyle, limited physical activity, quality, food composition, and gut microbes' composition also play critical roles in developing metabolic disorders. These factors increase the risk of developing chronic kidney diseases, arthritis, cardiovascular diseases, cancer, and schizophrenia. Environmental factors that can contribute to metabolic syndrome are smoking, increase in age, low socioeconomic status, and most importantly weight classification. Although a large population is susceptible to MetS, individuals that classify as obese are at the highest risk of having the syndrome. Initially seen in the western world, the rise of MetS and its key components has been witnessed globally, likely due to the spread of the Western lifestyle. Amongst the age group 20-85 years the prevalence of MetS was estimated to be 61.6% in the obese group, 33.2% in the overweight group, and 8.6% in the normal-weight group when analyzed in 12047 individuals in US [7], indicating a strong link between weight gain and MetS. Interestingly, MetS was highly associated with socioeconomic status in this group. Patients with MetS were less educated, less physically active, and had a lower income and a higher prevalence of smoking than their no-MetS counterparts [7].

6.2 Epidemiology of metabolic syndrome

National Health and Nutrition Examination Survey (NHNES) data estimates that among the US adults prevalence of MetS substantially increased from 25% during the period of 1988-1994 to 35% during 2007-2012 [8]. During this period, the prevalence of MetS increased both in men (33.4% from 25.6%) and women (34.9% from 25%). In particular, the largest increases were observed in non-Hispanic African black men (55%) and non-Hispanic white women (44%). Age was highly associated with the rise in the incidences of MetS. The MetS occurence markedly increased from 10% among those aged 18-29 years to close to 70% among women aged 70 or older in 2007-2012 [8]. As expected, rise in MetS incidence occurred in conjunction with an increase in the rate of obesity and type 2 diabetes. According to US National Center for Health Statistics, during 1988– 2010, average BMI increased by 0.37% per year in both men and women and waist circumference increased by 0.37 and 0.27% per year in men and women, respectively [9]. Whereas, type 2 diabetes was present in about 30.2 million adults aged 18 years or older in 2017 [9].

MetS is no longer considered a Western disease. Its prevalence has dramatically increased in developing world in the recent times. A meta-analysis of published studies in India until August 2019 revealed that prevalence of MetS in Indian adult population was 30% with the highest burden observed among the adults aged 50-59. The females and adults living in urban areas had higher prevalence than their male and rural counterparts [10]. Moreover, according to a cross-sectional study the prevalence of MetS in China between 2013 and 2014 was estimated to be 14.39% with the highest prevalence observed among the adults aged 50-59 years [11]. Likewise, prevalence of obesity in Chinese urban areas increased from 20 to 29% between 1992 and 2002, whereas MetS incidences increased from 8 to 10.6% [12]. Assuming the same rate of increase, the prevalence of MetS in China in 2017 would be about 15.5%.

Incidences of obesity and diabetes—the two key components of MetS—are rising at an alrming rate both in adults and children worldwide. Presence of visceral adiposity and insulin resistance or hyperglycemia are the central features of MetS and their presence independently increased the risk of MetS [13]. These two parameters are the absolute requirement for the diagnosis of MetS. According to WHO, obesity is defined as the accumulation of abnormal or unnecessary fat that poses a health risk. Obesity is a complicated, multifactorial, and mostly preventable disease, which has risen to epidemic proportions, with over 4 million individuals dying each year because of being overweight or obese in 2017 [14]. The current most generally utilized measures for classifying obesity is by the body mass index (BMI). It is classified from underweight (<18.5 kg/m2) to extreme (≥40 kg/m2) as described in the table below [14]. In children, body weight classifications differ from adults because there is a variability in the body composition as the child grows and develops. The variation is primarily based on sex as differences in sexual development and maturation impact body mass. According to WHO, overweight is described in US young people as age- and sex-specific BMI ≥85th and <95th percentile, and weight ≥95th percentile is considered obese [15]. The recent rise in childhood obesity is of great concerns amongst the policy makers and family physicians [16]. Globally, a total of 603.7 million adults and 107.7 million children in 195 countries were obese in 2015. Since 1980, prevalence of obesity doubled in 73 countries and substantially increased in many other countries. From 1990 through 2015, high BMI-related global death rate increased by 28.3%, from 41.9 deaths per 100,000 population in 1990 to 53.7 deaths per 100,000 population in 2015 [16].

Weight classification in adult

Weight status	BMI (kg/m^2)
Underweight	less than 18.5
Normal	18.5–24.9
Overweight	25–29.9
Obese Class I	30–34.9
Obese Class II	35–39.9
Obese Class III	Over 40

In contrast to the observation of high incidence of MetS in obese individuals, there exist a small group of obese people who are metabolically healthy and accordingly termed as "metabolically healthy obese (MHO)" [17]. These individuals appear to have high insulin sensitivity and normal lipids levels. Epidemiological survey in the US population has revealed a relatively higher presence of MHO in overweight and obese individuals. More importantly, however, was the presence of an unexpectedly high number (23.5%) of insulin resistant individuals in population with normal weight [17]. These data suggest that obesity doesn't necessarily translate to MetS and vice versa.

As for diabetes, WHO and IDF statistics indicate that 8.5-8.8% of adults globally aged 18 years and older had diabetes and it is expected to increase to 10.4% (642 m) by 2040. Diabetes became the direct cause of 1.6 million deaths in 2016. The data also shows that between 2000 and 2016, there was a 5% increase in premature mortality from diabetes. It was observed that there was a decrease in premature mortality rate due to diabetes in high-income countries from 2000 to 2010 but then increased in 2010-2016. There was an increase in premature mortality rate across both periods in lower-middle-income countries [18]. Over 50% of all diabetics lived in Southeast Asia and Western Pacific region. However, highest prevalence (11.5%) was observed in North America and Caribbean region [18]. According to the 2020 National Diabetes Statistics Report, 34.2 million (10.5%) people in the US have diabetes, that includes 7.3 million people who remain undiagnosed. Moreover, the estimated annual number of newly diagnosed cases in the United States among children and adolescents with type 1 diabetes were 18,291 and children and adolescents age 10 to 19 years with type 2 diabetes were 5,758 [3].

Since there are no uniform criteria to measure MetS incidences across the nations it is harder to measure the accurate prevalence rate of MetS globally. Additionally, many other important risk factors (other than discussed above) are not taken into consideration [13]. These may include age, family history, gender, and lifestyle (e.g., smoking, alcohol consumption). As a result, there is a paucity of clear global data on MetS and the associated risk factors. However, the data from US and many other nations across the globe as discussed above suggest that MetS occurrence may be as high as 3-folds than diabetes and thus, MetS may conceivably be afflicting approximately one billion people in the world.

6.3 Pathogenesis of metabolic syndrome

The pathophysiological mechanism/s of MetS, a constellations of many individual disorders and a heterogenous syndrome, are complex. The pathogenesis of MetS is defined by the complex interplay between different pathological processes, including visceral adiposity, insulin resistance, dyslipidemia, and/or elevated blood pressure. Individuals with 2 or more of these factors present are at high risk of developing atherosclerotic cardiovascular disease. In contrast, patients who have only one of these factors, for example obesity, are not at as high a risk to develop CVD and do not necessarily manifest MetS [17]. Individual components of MetS may exist independently with their distinct pathologies. However, in MetS they may represent manifestations from a common origin that are intricately linked to each other. Insulin resistance and the visceral adiposity are the primary trigger for the central pathological mechanisms leading to MetS. A vast majority of data now suggest that visceral adiposity, insulin resistance, and subclinical chronic inflammation play a central role in the initiation and progression of MetS as well as its transition to CVD [19-24]. Each of these parameters are discusssed in details below.

6.4 Visceral adiposity

Recent studies have revealed important roles of adipocytes in regulation of metabolism which till recently were seen as an inert energy storage depot. While white adipocytes store energy in the form of fat, brown and beige (brown-like and inducible) adipocytes express high levels of UCP1 and dissipate stored energy in the form of heat [25]. Adipocytes are metabolically more active than previously thought. They secret over a dozen of hormones and local factors that affect energy metabolism of the body by altering appetite and regulating function of other metabolic organs. Moreover, these factors also affect immune cells migration into the adipose tissue and thereby indirectly affecting energy regulation and insulin sensitivity [26,27]. Adispose tissue immune cells such as monocytes/macrophages secret inflammatory cytokines and catecholamines depending upon the local environment that can directly or indirectly affect energy expenditure and insulin sensitivity. Leptin, a hormone secreted by white adipocytes suppresses appetite via a central action. It binds to and activate the leptin receptor in the hypothalamus to activate JAK/STAT pathway to inhibits pro-opiomelanocortin (POMC) and neuropeptide Y (NPY) neurons to reduce appetite. In consistence to this, deficiency of leptin has been demonstrated to cause severe obesity in mice and rat models suggesting an indispensable role of leptin in energy regulation [28,29]. High leptin levels and leptin resistance have been observed in obese and diabetic individuals with metabolic syndrome. This may be due to leptin receptor mutations or inhibition of the downstream signal. Taken together, high leptin level is now considered as an independent risk factor for transition of obesity into cardiovascular disease [30,31].

In contrast, adiponectin—another important hormone secreted by adipocytes—produces the opposite effects to leptin. Adiponectin increases energy metabolism by inducing beige fat formation and increasing insulin sensitivity in adipocytes. It increases mitochondrial content in adipocytes and decreases adipocytes size and numbers, and utilization of free fatty acids (FFA) [32,33]. Adiponectin levels are negatively corelated with visceral fat accumulation [34]; its levels reportedly decrease in obesity [35]. Lower levels of adiponectin were associated with diabetes mellitus, coronary artery disease and hypertension, hallmarks of MetS [36,37]. Furthermore, in a mouse model of polycystic ovary syndrome, adiponectin overexpression protected mice against the disorders of MetS, including abdominal adiposity and insulin resistance, demonstrating a positive effect of adiponectin on energy regulation [33].

In addition to adipocytes, the immune cells infiltrated in the adipose tissue play a crucial role in energy metabolism and insulin sensitivity. Specifically, the quantity and the type of macrophages residing in the fat or those recruited from bone marrow can have a major impact on energy metabolism. Classically activated macrophages (CAM) have been demonstrated to inhibit UCP1 expression and energy dissipation by adipocytes. They promote inflammation and impair insulin sensitivity [38–41]. In contrast, alternatively activated macrophages (AAM) promote white to beige fat transformation by inducing the expression of UCP1 and thermogenic genes. Further, AAM inhibit inflammation and improve insulin sensitivity and restrict obesity [42–44]. Proinflammatory CAM have been reported to contribute to hepatic steatosis and adipogenesis, whereas AAMs inhibit these phenomena [38,41].

Recently, innate lymphoid type 2 cells (ILC2) were found to be critical for the maintenance of energy expenditure in healthy mice and humans. Adipose tissue ILC2s produce methionine-enkephalin peptides that act directly on surrounding adipocytes to drive thermogenic gene UCP1 expression and induce beige fat formation in mice [43,44]. In contrast, ILC2s levels in adipose tissues were decreased both in humans and mice, indicating that ILC2s, in addition to macrophages, regulate adipose tissue function and metabolic homeostasis and dysregulated ILC2 responsiveness leads to obesity [43,44]. Furthermore, pharmacological administration of cytokines IL-4 or IL-33 has been found to restrict high fat diet-induced weight gain and insulin resistance by promoting alternative activation of macrophages and recruitment of ILC2s [42–44]. These studies provide a new direction in the therapeutic field for MetS.

Another important factor that was recently shown to play an important role in the pathogenesis of metabolic syndrome is Endoplasmic reticulum stress or ER stress. Proteins are synthesized in ribosomes and matured in the endoplasmic reticulum (ER). However, if ER functions are compromised, defective proteins with incomplete folding begin to accumulate in the cells. This triggered a defensive mechanism called the unfolded protein response (UPR) or ER stress. Factors, including activating transcription factor 6 (ATF6), RNA-dependent protein kinase (PKR)-like ER kinase (PERK), and inositol-requiring enzyme 1 alpha (IRE1α) are required for UPR response to transduce to the outside of the ER [45]. It has been reported that deficiency of IRE1α in mice prevented HFD-induced obesity, insulin resistance, and hepatic steatosis which could be partly due to alternative activation of adipose tissue macrophages which in turn promotes browning of white fat and thereby increasing energy expenditure [39,46].

Accumulating evidence now suggests that altered gut microbiome diversity plays an important role in obesity and metabolic syndrome. The microbial metabolites short chain fatty acids act locally in the GI tract and regulate production of GI hormones such as cholecystokinin, glucagon-like peptide 1, peptide tyrosine-tyrosine, and leptin to increase fatty acid metabolism and energy homeostasis. Modifying gut's microbiota through fecal transplantation in rodents alters secretion of these hormones to subsequently affect the metabolism. For example, replenishment of one specific gut microbe *Akkermansia muciniphila*—that resides in the mucus layer of GI tract and degrade mucin—by feeding mice with prebiotic attenuated the metabolic syndrome in obese mice. This effect was associated with increased production of endocannabinoids in the intestine that inhibit inflammation and repair mucosal barrier [47].

It should be noted though that gut microbiome is extremely sensitive to dietary composition. David et al. demonstrated that gut microbiome responds differentially to plant- versus animal-based diet. Human volunteers were placed on either an animal-based diet (meats, eggs, and cheeses) or a plant-based diet (grains, legumes, fruits, and vegetables) for 5 days. There was a rapid growth of bile-tolerant microbes (*Alistipes*, *Bilophila*, and *Bacteroides*) while a decrease in the abundance of fiber-fermenting bacteria in individuals received animal-based diet [48]. Dietary fat content and type can also dramatically impact gut microbes' composition. For example, feeding rats with a lard-based diet triggered proliferation of the proinflammatory *Bilophila wadsworthia*, whereas a fish oil–based diet induced growth of *Lactobacillus* and *Akkermansia muciniphila*-that negatively regulates obesity and type 2 diabetes [49–52]. To confirm whether these observations have any implications in terms of metabolic syndrome, a recent proof-of-concept double-blind, placebo-controlled study was conducted in obese humans. The study demonstrated that oral supplementation of *Akkermansia muciniphila* for 3 months in the diet of overweight/obese insulin-resistant individuals significantly reduced plasma insulin and total cholesterol levels resulting in improved insulin sensitivity [53]. Collectively, these studies highlight an important link between gut microbiome and metabolic syndrome and that altering microbiota can provide a therapeutic solution to metabolic syndrome and its associated disorders.

6.5 Insulin resistance

Most human cells utilize glucose as the primary substrate for their normal function. Cellular glucose uptake requires insulin. Although almost all types of cells express insulin receptors, the role of insulin in glucose metabolism that has the most impact on energy homeostasis occurs in skeletal muscle, liver, and white adipocytes. In skeletal muscle, insulin increases glucose utilization by enhancing glucose transport into the cells and its conversion into glycogen for long-term storage. In liver, insulin increases glucose turnover by upregulating glycogen synthesis and lipogenesis and decreasing gluconeogenesis. Likewise, in WAT, insulin increases glucose transport, and promotes lipogenesis while suppressing lipolysis. The vast disparity in insulin effects in different organs is largely due to distinct downstream effector pathways, although the proximal mechanisms are strikingly similar.

When normal plasma insulin levels are unable to elicit an integrated glucose-lowering response in the target tissues, it is defined as insulin resistance. An integrated response involves increased glucose uptake/utilization, enhanced glucose storage (glycogen), decreased endogenous glucose production, and suppression of lipolysis [54–56]. This refractoriness causes a compensatory increase in insulin secretion to effectively reduce the plasma glucose, resulting in increased circulating insulin levels [57,58]. Thus, when higher circulating insulin levels are required to achieve the desired glucose-lowering response, a subject is considered insulin resistant. Insulin resistance is the underlying cause for many metabolic disorders and can lead to type 2 diabetes, obesity, atherosclerosis, high blood pressure etc. Various underlying molecular mechanisms have been highlted recently that play a central role in the pathophysiology of insulin resistance. This includes but not limited to defective insulin receptor sensitivity due to mutations or altered downstream signaling transduction pathway, chronic inflammation, oxidative stress, ER stress, and mitochondrial dysfunction.

It is now believed that insulin resistance-mediated increase in circulating FFA due to enhanced lipolysis in adipose tissue and liver plays a key role in the pathogenesis of MetS. Under insulin sensitive condition, insulin inhibits lipolysis, however, in the setting of insulin resistance, lipolysis is upregulated causing larger release of FFA into the circulation. Elevated levels of FFAs induce a negative feedback mechanism in skeletal muscles to inhibit diacylglycerol (DAG)/protein kinase (PKC) pathway, inhibiting glucose uptake in the muscle cells [19,59]. In contrast, FFA activate other isoforms of PKC in the liver that promotes gluconeogenesis and lipogenesis resulting in hepatosteatosis (fatty liver disease). Due to increased flux of FFA into the liver, production of VLDL triglycerides and apolipoprotein B increases, further contributing to hepatosteatosis [60,61]. To compensate for the loss of insulin sensitivity, insulin secretion increases in an attempt to achieve normal glucose lowering response, causing a hyperinsulinemic state. Visceral fat depots play a predominant role in the installation of insulin resistance more than the other fat depots. FFAs from visceral adipose tissue are directly drained into the liver and inhibit glucose uptake and promote gluconeogenesis contributing to insulin resistance [19,59,62]. Furthermore, visceral adipose depot contributes to insulin resistance through the proinflammatory cytokines released by adipocytes and the infiltrated immune cells. FFA directly induce the production of inflammatory cytokines such as tumor necrosis factor (TNF-α), interleukin (IL) IL-6, and IL-1β by activating toll-like receptors (TLRs) and NF-κB pathways [63–66].

An important factor that is not only a key component of MetS but also directly linked to insulin resistance is endothelial dysfunction. Endothelial dysfunction can lead to the development of hypertension, atherosclerosis, and ischemic disease [67]. Insulin elicits an endothelial-dependent vasodilatory effect in the systemic vasculature mediated by endothelial nitric oxide (NO) pathway. This pathway is impaired in insulin resistance due to reduced bioavailability of NO, leading to the development of hypertension [68]. One of the most important mechanism for this is inhibition of endothelial nitric oxide synthase (eNOS) phosphorylation at S1177 site [69]. Additionally, elevated FFA levels directly impair endothelium-dependent vasodilation by inhibiting (eNOS) activity and (NO) production in endothelial cells. This effect is partly mediated by the activation NOD-like" receptor family pyrin domain containing 3 (NLRP3) inflammasome [63,70]. Finally, increase in the visceral adipose tissue mass, as seen in obesity, results in large production of plasminogen activator inhibitor (PAI-1) and heparin-binding epidermal growth factor like growth factor resulting in a prothrombotic and proliferative state, all contributing to endothelial dysfunction [71,72].

6.6 Inflammation

Obesity is now considered a chronic inflammatory disease that alters metabolic homeostasis in the body via a local and remote action in different metabolic tissues [73]. Inflammation contributes to insulin resistance in metabolic tissues via pro-inflammatory cytokines in an autocrine and paracrine manners [74]. However, a role of inflammation in installation of insulin resistance is complex. A line of investigation has revealed that adipose tissue inflammation is not required for the initiation of insulin resistance [75,76]. Nevertheless, installation of visceral adiposity and insulin resistance culminates in a final common pathway that leads to chronic inflammation. Inflammation level, as indicated by the levels of circulating pro-inflammatory cytokines, increases as MetS worsens. The activity levels of several pro-inflammatory markers, including TNF-α, IL-6, IL-18, CRP, IL-1β, and PAI-1 increase in MetS [77–79]. Inhibition of inflammation, for example by neutralizing TNF-α, genetic silencing of Jun N-terminal kinase (JNK), or inhibition of inhibitor of nuclear factor kappa-B kinase subunit epsilon (IKKε) improved insulin resistance and overall metabolic state in many animal models and obese individuals, suggesting a pathogenic role of inflammation in insulin resistance [23,80–82].

Tumor necrosis factor-α. Circulating TNF-α levels have been shown to increase with obesity and insulin resistance [83]. Interestingly, elevated levels were normalized with weight loss, indicating a strong relevance of TNF-α levels in determining the severity of MetS [84]. TNF-α has been shown to directly contribute to the different components of MetS, including insulin resistance, elevated FFA levels, increased glucose levels. It inactivates insulin receptor IRS by inducing its phosphorylation in the adipose tissue and smooth muscle cells, increases circulating FFA levels by inducing lipolysis, aggravates inflammation and inhibits energy expenditure by suppressing adiponectin production [84]. However, attempt to treat MetS with neutralizing antibody has produced differential results. Lo et. al. observed that neutralization of TNF-α in patients with MetS improves inflammatory markers and total adiponectin levels without impacting insulin sensitivity, which could be due to opposing effects on inflammatory cytokines and body composition indexes [85].

C-reactive protein (CRP). Elevated levels of CRP have been associated with insulin resistance [86], high BMI [87], and hyperglycemia [88]. In terms of MetS, its levels increase as the number of metabolic disorders associated with MetS increases. Many studies highlighted the strong correlation between CRP levels and risk of MetS Although CRP is a general inflammatory marker which is used to determine the systemic inflammatory status, it can predict the prognosis of future CVD events [89]. For this reason, CRP levels are considered as an independent predictor of adverse cardiovascular outcomes of MetS.

Interleukin 6 (IL-6). Interleukin 6 (IL-6) is produced by both adipose tissue and immune cells in humans. IL-6 has been shown to exert differential effect depending upon the stimulus and the surrounding environment. It has both proinflammatory and anti-inflammatory actions [78]. In MetS, its level has been shown to be elevated with increased BMI and insulin resistance [90,91]. Moreover, IL-6 promotes a prothrombotic state in circulation by upregulating adhesion molecule expression in endothelial cells and increasing fibrinogen levels [92].

Resistin. Resistin is a relatively recently discovered adipokines in humans. It is secreted by adipocytes and infiltrated immune inflammatory cells. Its levels reportedly increase with BMI, insulin resistance, and triglycerides levels [89]. Interestingly enough, resistin levels appears to have a heritable trait related to insulin resistance, suggesting an important link between this cytokine and MetS [93]. Although, no significant association was observed between MetS and differential genotype distribution of resistin gene polymorphism in a Han Chinese population [94]. Nonetheless, resistin levels were positively corelated with body weight, BMI, triglycerides, and endothelin-1 in patients with MetS, and thus its levels could be measured to predict the future occurrence of CVDs [95].

6.7 Current modern therapies for metabolic syndrome

Metabolic syndrome groups multiple metabolic defect disorders, which are treated based on the abnormality patient presents with [96]. For example, if a patient with metabolic syndrome reports elevated blood pressure, they would be categorized under hypertension. When considering treatment doctors will prescribe medication that correlates with this symptom Fig. 6.1. A list of modern medications used to treat different disorders of MetS are given in Table 6.2.

6.7.1 Limitation of current modern therapies

Although pharmacological interventions can greatly improve the patients' health outcome by reducing specific risk factors, there are many limitations that are associated with them, which includes the adverse effects associated with pharmacological interventions, drug development process is long, FDA regulations are strict and adherence with medication is difficult.

1. Primarily, the adverse effects associated with pharmacological intervention can reduce adherence as well as cause unwanted symptoms. Many of the different classes of drugs tend to have common side effects such as diarrhea, headache, and nausea, However, there are class specific side effects that are unique to each class of medications. For example, some of the hypertension medications can cause erectile function as well as nervousness. Anti-obesity drugs can cause incontinence and oily spotting, whereas dyslipidemia drugs can cause hepatic failure and rhabdomyolysis [97]. The major side effect that is caused by antidiabetic medications is hypoglycemia [98]. Patients can be hypersensitive to ingredients used to produce the drug, which may lead to an allergic reaction that can be severe leading to anaphylaxis.

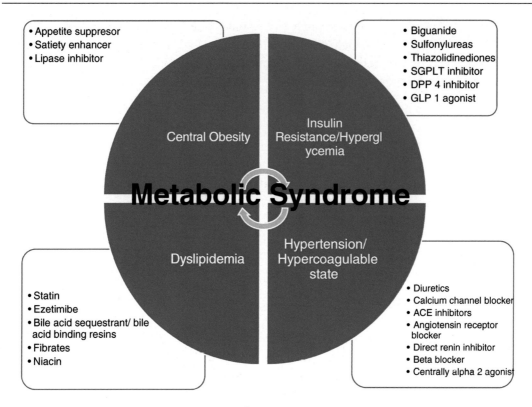

Figure 6.1
Current treatments for different components of metabolic syndrome.

2. Another dilemma with pharmacological therapy is that the drug development process can be long and tedious. On average drug development usually takes approximately 10 years, which consists of five primary steps. These five steps are discovery and development, pre-clinical research, clinical research, FDA review and FDA post-market safety monitoring. Lastly, once the drug is on the market, FDA continues to monitor it for safety. There are programs in place such as MedWatch, which the patients can use to report side effects [99].

3. Medication adherence is a tremendous issue that affects patients on poly drug regimen. In fact, it may affect approximately 40 to 50 percent of the patient population. It can lead to unnecessary deaths and medical costs. This can occur due to various reasons, which can be divided into patient related barriers and treatment related barriers. Patient related barriers include lack of motivation, depression, denial, cognitive impairment, drug or alcohol use, cultural issues, low educational level, and alternate belief systems. Whereas, treatment related barriers include complexity of treatment, fear of side effects, inconvenience, cost, and time. Patients with metabolic syndrome may potentially be on a multidrug regimen, which can result in forgetfulness to take medication especially if they

Table 6.2: Current modern drugs for metabolic syndrome with their major mechanism of action (MOA) (Please see Fig. 6.1).

Drug class	Drugs within the class	MOA
Central obesity		
Appetite suppressor	Phentermine	Dopamine agonist; partial norepinephrine agonist
	Topiramate ER	AMPA antagonist
	Diethylpropion	Norepinephrine reuptake inhibitor
	Naltrexone	Opioid receptor antagonist
	Bupropion	Dopamine and norepinephrine reuptake inhibitor
Satiety enhancer	Lorcaserin	Serotonin receptor agonist
	Liraglutide	GLP1 receptor agonist
Lipase inhibitor	Orlistat	Gastric and pancreatic lipase inhibitor
Insulin resistance/diabetes		
Biguanides	Metformin	Decrease hepatic production of glucose and absorption of glucose in intestines
Sulfonylureas	Glipizide	Stimulate pancreatic beta cells to secrete insulin
	Glimepiride	
	Glyburide	
Thiazolidinediones	Pioglitazone	PPARγ agonist
	Rosiglitazone	
SGLT2 inhibitor	Canagliflozin	Decreases glucose reabsorption and increase excretion of glucose by the kidney
	Dapagliflozin	
	Empagliflozin	
DPP4 inhibitor	Sitagliptin	Inhibits breakdown of incretins.
	Sitagliptin	
	Linagliptin	
	Alogliptin	
GLP1 agonist	Exenatide	Incretin analogs
	Liraglutide	
	Dulaglutide	
	Lixisenatide	
	Semaglutide	
Dyslipidemia		
Statins	Atorvastatin	HMG-CoA reductase
	Fluvastatin	
	Lovastatin	
	Pitavastatin	
	Pravastatin	
	Rosuvastatin	
	Simvastatin	
	Ezetimibe	Inhibits absorption of cholesterol in the small intestine
Bile acid sequestrants and bile acid resins	Cholestyramine	Binds cholesterol to facilitate excretion.
	Coleveselam	
	Colestipol	

(continued on next page)

Table 6.2: Current modern drugs for metabolic syndrome with their major mechanism of action (MOA) (Please see Fig. 6.1)—cont'd

Drug class	Drugs within the class	MOA
Fibrates	Fenofibrate	PPARα activators
	Gemfibrozil	
Niacin	Niacor	Decreases TG and LDL by decreasing the rate of vldl synthesis
	Niaspan	
	Slo-Niacin	
Hypertension		
Thiazide diuretics	Chlorothiazide	Inhibits sodium reabsorption In distal convoluted tubules
	Chlorthalidone	
	Hydrochlorothiazide	
	Indapamide	
	Metolazone	
Calcium channel blockers (DHP)	Amlodipine	Inhibition of calcium entrance into vascular smooth muscle and myocardial cells
	Felodipine	
	Isradipine	
	Nifedpine	
	Diltiazem	
	Verapamil	
ACE inhibitor	Benazepril	Inhibit angiotensin converting enzyme
	Captopril	
	Enalapril	
	Fosinopril	
	Lisinopril	
	Quinapril	
	Ramipril	
Angiotensin receptor blocker	Azlisartan	Inhibit angiotensin 2 receptor type 1
	Candesartan	
	Eprosartan	
	Irbesartan	
	Losartan	
	Olmesartan	
	Valsartan	
Nonselective beta blocker	Nadalol	Block beta 1 and beta 2 receptors
	Pindolol	
	Propranolol	
	Timolol	
Centrally acting alpha 2 adrenergic receptors	Clonidine	Stimulates alpha 2 adrenergic receptors
	Methyldopa	
Direct vasodilators	Hydralazine	Vasodilation of arterioles
	Minoxidil	

must be taken at different times. In addition, medication cost continues to rise due to high cost involved with manufacturing.

6.7.2 Herbal medicines

Definition. According to National Cancer Institute, herbal medicines are defined as a type of medicine that uses roots, stems, leaves, flowers, or seeds of plants to improve health, prevent disease, and treat illness. A more comprehensive definition is "a natural preparations synthesized from purely and originally herbal substances, not chemically altered, and are responsible for the therapeutic effects of the product" [100]. Herbal formulations can be processed into several different dosage forms to extract and deliver their active pharmaceutical ingredient (API); some of these forms include teas, syrups, essential oils, ointments, salves, rubs, capsules, or ground into a powder. These substances can be applied as a preventative or a treatment to several chronic or acute diseases. In laboratory settings several herbal medicines were shown to have beneficial therapeutic effects for diseases including cardiovascular disease, depression, inflammation, prostate problems, gastrointestinal problems and to treat vitamin deficiencies, to name a few [101]. For example, St. John's Wort has been used to treat depression, Ginseng for menopause and fatigue, Black cohosh for painful menstruation and uterine spasm, Echinacea for improving immune system, Gingko biloba for memory, garlic for hyperlipidemia, and Saw Palmetto for enlarged prostate etc. [102,103]. WHO reported in 2005 that up to 90% of the African population, along with 70% of the Indian population primarily used herbal medications and in the United States it was observed that 38% of adults used some form of over-the-counter herbal medications [104]. With a projected 15% year-over-year increase in international herbal medicine trade, herbal medicine therapy is, again, becoming an important and fast-growing field [105].

6.7.3 FDA requirement for the use of herbal medicines

When delving into herbal medicine, it is important to define the laws and regulations concerning the subject. In the United States, under the Dietary Supplement Health and Education Act (DSHEA) of 1994, herbal medicines are classified under dietary supplements which are defined as, "a large heterogeneous group of products intended to supplement the diet that are not better described as drugs, foods, or food additives. Supplements may contain, in whole or as a concentrate, metabolite, constitute, or extract, any combination of 1 or more vitamins, minerals, amino acids, herbs, or other botanicals and other substances used to increase total dietary intake, including enzymes, organ tissues and oils. They must be intended for ingestion; sold in the form of capsules, tablets, soft gels, gel caps, powders, or liquid; and not to be marketed as food items" [106]. Under the DSHEA act, herbal medicines do not have to meet the FDA premarket standards prior to being available to the public. The extent to which the FDA can regulate herbal medicines are limited to the following acts:

The Nutrition Labeling and Education Act (NLEA) of 1990

The FDA has the authority to regulate health claims on food labels, these products cannot claim to be a treatment, preventative, or a cure to certain diseases without prior approval. The statement "This product is not intended to diagnose, treat, cure or prevent any disease." must be printed on every label.

The FDA modernization Act of 1997

Any health claim can be made without prior approval of the FDA as long as the product is backed by a scientific governing body. The intent of the health claims needs to be made clear to the FDA prior to marketing. It limits what can be used on dietary labels to, health claims, nutrition content claims and structure claims.

Dietary Supplement and Non-Prescription Drug Protection Act (2006)

Requires manufacturers of over-the-counter drugs and herbal medicines to report all serious adverse events associated with the product.

As a result of these acts, the only time the FDA has the authority to remove a product from the market is if there has been well-documented evidence of a rising public health concern. However, "new ingredients" not known to the FDA prior to October 1994 may be subject to review, prior to them being released to the public [104]. Moreover, in 2007 the FDA published current good manufacturing practice (cGMP) guidelines. The aforementioned guidelines are intended to ensure the quality and safety of herbal products and dietary supplements.

6.7.4 Herbal medicines for metabolic syndrome

Metabolic syndrome is an intricate multisystem disease that includes cluster of diseases mainly obesity, hyperglycemia, hyperuicemia, dyslipidemia, and hypertension. Complementary and Alternative Medicine (CAM) exhibits the multi-target and less toxic effects, leads to more attraction and attention from the practitioners and researches in finding safe and effective alternatives to treat metabolic syndrome and associated disorders. Vast number of research studies have reported therapeutic potential of herbal medicines in MetS [107]. Due to involvement of complex mechanism in MetS, polytherapy is popularly and widely practiced as a treatment regimen for various MetS-related co-morbidities. Increase in number of medications prescribed can leads to lowering of drug adherence as well as rise in risk of adverse effects. Several limitations associated with modern drug therapy for MetS lead to increase demands of alternative treatments. Multiple natural medicine-based therapies for MetS with proven safety and efficacy are gradually emerging. Several recent studies involving *in vitro* and *in vivo* assessment reported anti-diabetic, anti-obesity, and anti-hypertensive effects of herbal medicines. List of various recent studies on herbal medicinal plants are listed in Table 6.3.

6.8 Possible mechanism/s underlying the protective effects of herbal medicines in metabolic syndrome

The pathophysiology of MetS is highly complicated and involves multiple interaction of different elements including genetic, behavioral, and environment. Currently, synthetic based

Table 6.3: List of herbal medicines exhibiting efficacy against metabolic syndrome and its key components.

Source	Extract	Dose	Model	Proposed mechanism	Inference	Reference
Allium cepa	High pressure-processed onion extracts and onion hot water extracts	8.1 g/kg	Sprague-Dawley rats fed with high cholesterol diet	• Reduction in LDL levels, total hepatic lipids and cholesterol levels • Increase in lipid as well as cholesterol excretion • HMG-CoA reductase activity inhibited	Hypolipidemic	(Jung et al., 2015) [108]
	Ethanolic peel extract	75 and 100 μg/ml 0.72%	3T3-L1 preadipocytes Sprague-Dawley rat supplemented with high-fat diet	• Down-regulation of activating protein mRNA levels • Up-regulation of carnitine palmitoyl transferase-1 α as well as fatty acid binding protein 4 • Lowering of body, retroperitoneal as well as mesenteric fat weights • Down-regulation of mRNA levels of PPAR γ in epididymal fat, C/EBPα (CCAAT/enhancer binding protein α) • CPT-1α as well as uncoupling protein-1 mRNA levels were up-regulated • Reduction in fatty acid synthase and acetyl-CoA carboxylase	Antiobesity	(Moon et al., 2013) [109]
	Aqueous extract	300 mg/kg	Alloxan monohydrate (200 mg/kg) induced diabetes in rabbits	• Reduction of blood glucose levels • Decrement in levels of SOD, CAT, GPx and GSH • Increment in MDA levels	Antidiabetic	(Ogunmodede et al., 2012) [110]
	Powder	100 g/kg	Wistar rats fed with high cholesterol diet	• Improvement in anti-oxidant status by regulating SOD and GPx enzyme • Reverse of impaired endothelium-dependent acetylcholine relaxation in mesenteric arteries	Hypotensive	(González-Peña et al., 2014) [111]

(continued on next page)

Table 6.3: List of herbal medicines exhibiting efficacy against metabolic syndrome and its key components—cont'd

Source	Extract	Dose	Model	Proposed mechanism	Inference	Reference
Aloe vera (Aloeaceae)	Aloe-sterol rich gel extract	200 µL/10 g	C57BL/6 mice fed with HFD HepG2 cell	• Suppression of gain of body, liver and fat weight • Increase in mRNA expression of brown adipose tissue • Increase in FGF21 expression leads to regulation of nutrient, energy homeostasis as well as brown adipose tissue	Antiobesity	(Tada et al., 2020) [112]
	Peptide and polypeptide fraction	0.450 mg/kg	Wistar rats injected with STZ	• Reduction in levels of fasting plasma glucose with increase insulin • Mitigating loss of intestinal permeability by increasing GLP-1 and decreasing DPP-IV and zonulin levels	Antidiabetic	
	Ethanolic gel extract	300 mg/kg	WNIN/GR-Ob mutant obese rats injected with STZ 35 mg/kg, i.p.	• Prominent reduction in TG, VLDL, TG to HDL ratio along with blood glucose levels and dipeptidyl peptidase IV • Increase in levels of serum insulin • Reduction in both homeostatic model assessment-insulin resistance and –β values • SEM analysis showed preservation of islets as well as increase in β-cell diameter	Antidiabetic	(Deora et al., 2021) [113]

(continued on next page)

Plant	Extract	Dose	Model	Effects	Activity	Reference
Azadirachta indica	Ethanolic seed and leave extract	500 mg/kg	Albino rats injected with alloxan monohydrate (120 mg/kg), i.p. *In vitro* assays	Decrease in glucose level	Antidiabetic	(Saleem et al., 2018) [114]
	Yogurt Aqueous leave extract	—		• Inhibition of α-amylase, α-glucosidase, and angiotensin-1 converting enzyme • Exhibits anti-oxidant activity	Antidiabetic and antihypertension	(Shori and Baba, 2013) [115]
	Chloroform	100 µg/200µL	Swiss mice were treated with STZ (3 mg/25g), i.p.	• Reduction in intestinal glucosidase activity • Upregulation of glucose-6-phosphate dehydrogenase as well as muscle glycogen • Insulin-producing cells regenerated • Plasma insulin and c-peptide levels increased	Antidiabetes	(Bhat et al., 2011) [116]
	Methanolic leaves extract	100 and 200 mg/kg	Wistar rat injected with N^ω-nitro-L-arginine methyl ester (40 mg/kg), p.o.	• Augmentation of NO bioavailability • Modulation of myeloperoxidase activity • Regulation of hydrogen peroxide, GSH, thiols, SOD, and MDA levels • Increase in GST and CAT and decrease in GPx levels • Normalization of blood pressure • Modulation in caspase-3 expression	Antihypertensive and renoprotective	(Omobowale et al., 2020) [117]
Bougainvillea spectabilis	Aqueous	100 µg/200µL	Swiss mice were treated with STZ (3 mg/25g), i.p.	• Reduction in intestinal glucosidase activity • Upregulation of glucose-6-phosphate dehydrogenase as well as muscle glycogen • Insulin-producing cells regenerated • Plasma insulin and c-peptide levels increased	Antidiabetes	(Bhat et al, 2011) [116]

(continued on next page)

Table 6.3: List of herbal medicines exhibiting efficacy against metabolic syndrome and its key components—cont'd

Source	Extract	Dose	Model	Proposed mechanism	Inference	Reference
Camellia simensis	Hot water extract	250 mg/5ml	Pancreatic β-cell line Clonal pancreatic β-cells Primary mouse pancreatic islets Sprague-Dawley rat fed with HFD	• Increment in insulin secretion • Significant increase in glucose uptake as well as insulin action in adipocyte cells • Inhibition of protein glycation, DPP-IV enzyme, starch digestion as well as glucose diffusion • Improvement in gain of body weight as well as energy intake • Regulation of insulin levels in both plasma and pancreas • Islet size and β-cell mass improved	Antidiabetic	(Ansari et al., 2020) [118]
	Strictinin isolated from leaves	100 and 130mg/kg	C57BL/6 mice fed with HFD	• Reduction in TG levels, cholesterol and glucose • Decrease in body weight gain and epididymal • Significant reduction in enlarged adipocytes	Antiobesity	(Chen et al., 2018) [119]
	Instant tea	5 mg/ml	C57BL/6 mice fed with HFD	• Decrease in gain of body weight as well as visceral fat • Increase in HDL-C levels and albumin to globin ratio • Lowering of LDL-C and leptin levels • Decrease in AST and ALT levels	Antiobesity	(Sun et al., 2019) [120]
	Tea extract	2 and 4 g/kg	Wistar rats fed with high sodium diet	• Significant decrease in serum concentration of TC and LDL • Reduction in insulin levels and homeostatic model assessment	Antihypertensive	(Stepien et al., 2018) [121]

(continued on next page)

Capsicum annuum	Methanolic seed extract	50, 100 and 200 μg/ml	Adipocyte (3T3-L1) cells	• Decrease in accumulation of lipid • Reduction in activity of glycerol-3-phosphate dehydrogenase • Reduction in expression of adipogenic transcription factor (C/EBP-α,-β and PPAR γ)	Antiobesity	(Jeon et al., 2010) [122]
	capsicoside G-rich fraction	100 mg/kg	C57BL/6 mice fed with HFD	• Reduction in body weight as well as food efficiency ratio • Lowering of weight of epididymal adipose tissue, adipocyte hypertrophy and liver fat deposition • Modulation of elevated expression of adipocyte differentiation regulators (PPARγ, C/EBPα, binding protein 1c of sterol regulator along with target genes	Antiobesity	(Sung et al., 2016) [123]
	Aqueous seed extract	200 mg/kg	C57BL/Ksj *db/db* mice	• Improvement in levels of fasting glucose, hemoglobin as well as insulin levels • Decrement in levels of oro-inflammatory cytokines (TNF-α and IL-6) and TG • Diminishing gluconeogenesis by regulating glucose 6-phosphatas and phosphoenolpyruvate carboxykinase enzymes • Increase in phosphorylation of FOXO1 and AMPK	Antidiabetic, anti-inflammatory, and antiobesity	(Kim et al., 2020) [124]

(continued on next page)

Table 6.3: List of herbal medicines exhibiting efficacy against metabolic syndrome and its key components—cont'd

Source	Extract	Dose	Model	Proposed mechanism	Inference	Reference
Crocus sativus	Ethanolic extract	40 and 80 mg/k	Male sprague Dawley rat fed with HFD for 12 weeks	• Significant reduction in food consumption • Body weight gain rate, total fat pad as well as epididymal fat to body ratio also reduces • Triacylglycerol and total cholesterol levels decreases • Improvement in atherogenic index	Antiobesity	(Mashmoul et al., 2014) [125]
	Stigma aqueous extract	40 and 80 mg/kg	Wistar albino rats treated with single STZ (60 mg/kg) dose, i.p.	• Increase in TNF-α level • Blood glucose, glycosylated serum protein as well as advanced glycation end products serum levels decreases • Reduction in cholesterol, triglyceride and LDL levels • Increment in HDL levels • Remarked upregulation of GSH, SOD, and CAT levels • Significant decrement of cognitive deficit, TNF-α and iNOS in hippocampus tissue	Antidiabetic	(Samarghandian et al., 2014) [126]
	Sequential (n-hexane, chloroform, methanol, and water) stigma powder extract	50 μL 20 mg/kg	Mouse myoblast (C2C12) cell line Type 2 diabetic KK-Aʸ mice	• PTP1B inhibition • Induction of ligand-independent activation of insulin signalling • Translocation of glucose transporter 4 causes enhancement of glucose uptake • Improvement in impaired tolerance of glucose	Antidiabetic	(Maeda et al., 2014) [127]

(continued on next page)

Plant	Extract	Dose	Model	Effects	Activity	Reference
Feronia limonia	Ethanolic and aqueous leaf extract	200 mg/kg	Swiss albino mice injected with STZ (500 mg/kg), i.p.	Reduction in blood glucose as well as glycosylated hemoglobin level; Lowering in TC, TG, and LDL levels	Antidiabetic and antihyperlipidemic	(Chowdary et al., 2018) [128]
	Hydroalcoholic bark extract	300 mg/kg	Wistar albino rats treated with alloxan (120 mg/kg), i.v.	Reduction in blood glucose level; Improvement in lipid profile by regulating levels of TC, TG, LDL, and HDL; Histopathological studies reveled reverse in disturbed pancreatic cells	Antidiabetic	(Hingwasia et al., 2018) [129]
Garcinia indica	Fruit extract	10 µg/ml; 40 mg/kg (extract) and 8 mg/kg (garcinol)	Mouse embryo fibroblasts (3T3L1); C57/BL6 mice fed with HFD	Inhibition of adipogenesis; Increment in uncoupling protein-1; Reduction in endoplasmic reticulum stress; Lowering of visceral fat accumulation	Antiobesity	(Majeed et al., 2020) [130]
		0.005%	3T3-L1 cells; Obese mice fed with HFD	• Regulation of TG content and glycerol released • Downregulation of PPARγ and C/EBPα protein expression • Decrement in body weight as well as adipocyte size • Promotion of fatty acid β-oxidation • Modulation of gut microbiota	Antiobesity	(Tung et al., 2021) [131]
	Polyphenol-enriched ethyl acetate fruit extract	800 µg/ml	3T3-L1 preadipocyte	• Inhibition of digestive enzyme • Prominent anti-oxidant effect • Inhibition of pancreatic lipase as well as α-amylase	Antiobesity	(Munjal et al., 2020) [132]
	Aqueous fruit extract	200 mg/kg	Wistar albino rats treated with STZ (90 mg/kg), i.p.	• Decrement in blood glucose (fasting and postprandial) • Restoration of erythrocyte GSH	Antidiabetic	(Kirana and Srinivasan, 2010) [133]

(continued on next page)

Table 6.3: List of herbal medicines exhibiting efficacy against metabolic syndrome and its key components—cont'd

Source	Extract	Dose	Model	Proposed mechanism	Inference	Reference
Hibiscus sabdariffa	Polyphenolic calyx extract	200 mg/kg	Sprague-Dawley rats fed with HFD and treated with STZ (35 mg/kg), i.p.	• Decrement in triacylglycerol, TC and LDL/HDL ratio • Reduction in elevated formation of advanced glycation end product (AGE) and lipid peroxidation • Recovering of weight loss • Immunohistological analysis showed inhibition of receptor of AGE and connective tissue growth factor expression	Antidiabetic	(C.-H. et al., 2011) [134]
	Polyphenol-rich calyces extract	100 mg/kg	Sprague-Dawley rat treated with STZ (60 mg/kg), i.p.	• Lowering of glucose level • Decrease in TG, TC and LDL levels • Increase in HDL levels • Modulation of SOD, CAT, and GSH levels • Significant reduction in levels of MDA and protein carbonyl • Improvement in thoracic aorta morphology	Antidiabetic	(Zainalabidin et al., 2018) [135]
	Aqueous extract	25, 50 and 100 mg/kg	Hamster rat fed with HFD	• Reduction of liver fat accumulation • Lowering of TC and TG • Regulation of lipid peroxides • Decrement in levels of ALT and AST	Antiobesity	(Huang et al., 2015) [136]
	Polyphenol-rich calyces extract	10 and 25 mg/kg	C57BL/6J mice treated with HFD	• Reduction in weight increase • Improvement of glucose tolerance and insulin sensitivity • Normalization of LDL/HDL ratio • Reduction in adipokines and pro-inflammatory expression • Reinforcement of gut integrity through up-regulation of mucus and protein expressions • Amelioration of gut microbiota changes	Antiobesity	(Diez-Echave et al., 2020) [137]
	Aqueous calyx extract	250 and 500 mg/kg	Rats fed with high-salt diet	Significant reduction in blood pressure both diastolic and systolic	Antihypertensive	(Abubakar et al., 2015) [138]

(continued on next page)

Hypericum perforatum	Leaves powder	6.25%	Wistar rat injected with STZ (60 mg/kg)	• Decrease in TC, LDL levels • Increase in HDL levels • Lowering of hepatic enzyme AST	Antidiabetic	(Ghosian Moghadam et al., 2017) [139]
	Hydro-alcoholic whole plant extract	100 and 200 mg/kg	Charles Foster rat fed with HFD and fructose	• Lowering of TC and LDL-C • Inhibition of weight gain • Normalization of dyslipidemia • Improvement in insulin sensitivity	Antiobesity and hypolipidemic	(Husain et al., 2011) [140]
	Infusion	1% (w/v)	Sprague-Dawley rat fed with high-fructose and fat diet	• Decrease in both body weight and abdominal fat mass • Reduction in TG, TC, LDL, and C-reactive protein levels • Inhibition of pancreatic lipase	Antiobesity	(Hernández-Saavedra et al., 2016) [141]
	Ethyl acetate leaf extract	50, 100 and 200 mg/kg	Wistar albino rats injected with STZ (40 mg/kg)	• Decrease in fasting glucose • Reduction in glucose level, TC, TG, and glucose-6-phosphate • Significant increase HDL-C, tissue glycogen content as well as glucose-6 phosphate dehydrogenase	Antidiabetic	(Arokiyaraj et al., 2011) [142]
	Hydro-alcoholic plant extract	150 mg/kg	New Zealand rabbit fed with cholesterol (1%)	• Decrease in levels of TG, TC, LDL-C, oxidized LDL and athersoclerosis index • Reduction in levels of apoB and ratio of apoB/apoA • Lowering of malondialdehyde and C-reactive protein • Histopathological analysis showed restriction of atherosclerotic lesions	Antiatherosclerosis, and cardioprotective	(Asgary et al., 2012) [143]

(continued on next page)

Table 6.3: List of herbal medicines exhibiting efficacy against metabolic syndrome and its key components—cont'd

Source	Extract	Dose	Model	Proposed mechanism	Inference	Reference
Ilex paraguariensis	Aqueous leave extract	0.5, 1 and 2 g/kg	C57BL/6j mice fed with high-fat diet	• Decrease in pre-adipocytes differentiation and lipid accumulation in adipocytes • Lowering of body weight gain and growth rate of adipose tissue • Serum TC, TG and glucose levels also reduced	Antiobesity and antidiabetic	(Kang et al., 2012) [144]
	Aqueous tea	1 g/kg	Swiss mice fed with HFD	• Improvement in blood glucose level as well as insulin response by restoring insulin substrate receptor-1 and AKT phosphorylation • Down-regulation of TNF-α, IL-6 as well as iNOS genes • Modulation of NF-κB	Antiobesity, antidiabetic and anti-inflammatory	(Arçari et al., 2011) [145]
	Ethanolic fraction	50 µg/ml	3T3-L1 adipocytes	• Inhibition of accumulation of TG • Modulation of expression of PPARγ2, leptin, C/EBPα and TNF-α	Antiobesity	(Gosmann et al., 2012) [146]
Olea europaea	Ethanolic leaf extract	200, 400 and 600 µg/ml; 300 and 500 mg/kg	Nerve growth factor-treated pheochromocytoma (PC12) cells; Male Wistar rat injected with single STZ (55 mg/kg) dose, i.p.	• Reverse of elevated glucose and caspase-3 activation • Decrease in cell damage • Amelioration of hyperalgesia and caspase-3 activation • Reduction in Bax/Bcl2 ratio	Diabetic neuropathic pain	(Kaeidi et al., 2011) [147]
	Methanolic leaf extract	10 and 25 mg/kg	C57BL/6j mice fed with HFD	• Lowering of body weight gain as well as basal glycaemia • Improvement in lipid profile • Reduction in insulin resistance • Amelioration of altered expression of PPARs, adiponectin as well as leptin receptor (adipogenic genes) • Reduction of TNF-α, IL-1β and IL-6 RNA expression • Counteracting of dysbiosis in colonic microbiota	Antiobesity, antidiabetic, and anti-inflammatory	(Vezza et al., 2019) [148]

(continued on next page)

Ethanolic leaves extract	200 and 400 mg/kg	Wistar rats fed with HFD followed by STZ (35 mg/kg), i.p.	• Significant decrease in body weight • Improvement in levels of glucose • Reduction in AST • Increase in insulin levels • Reduction in levels of IL-6, -1, TNF-α, and IFN-γ • Increase in IL-10 • Decrement in levels of leptin and resistin as well as increment in adiponectin	Antidiabetes	(Guex et al., 2019) [149]
Aqueous leaves extract	200 and 400 mg/kg	Wistar albino rats treated with STZ (60 mg/kg)	• Decrease in body weight gain • Lowering of levels of glucose, TG, TC, LDL-C, VLDL-C, creatine kinase, lactate dehydrogenase and MDA • Increase in levels of HDL-C, SOD, GSH and CAT • Up-regulation of IRS1 and IRA • Improvement in histopathological characteristics	Antidiabetes	(Al-Attar and Alsalmi, 2019) [150]
Leaves extract	—	Sprague-Dawley rats Human erythrocyte subjected to oxidative stress	• Increase in dieresis as well as saluresis • Significant reduction in systolic blood pressure • Dose-dependent decrease of erythrocyte membrane degradation	Antihypertensive, antioxidant, and diuretic	(Ghibu et al., 2015) [151]

(continued on next page)

Table 6.3: List of herbal medicines exhibiting efficacy against metabolic syndrome and its key components—cont'd

Source	Extract	Dose	Model	Proposed mechanism	Inference	Reference
Persea americana	Hydro-alcoholic fruit extract	100 mg/kg	Sprague-Dawley rat fed with HFD	• Reduction in body mass index, adiposity index, and total fat pad mass • Decrease in levels of TC, TG, and LDL • Reduction in mRNA expression of fatty acid synthase, leptin, and lipoprotein lipase • Increase in gene expression of fibroblast growth factor 21	Antiobesity and hypolipidemic	(Monika and Geetha, 2015) [152]
	Hydro-alcoholic fruit extract	100 mg/kg	Sprague-Dawley rat fed with HFD	• Reduction in body mass index, adiposity index, total fat pad mass • Decrease in levels of LDL and lipid peroxides • Increase in expression of mRNA of adiponectin, PPAR-γ as well as protein expression of PPAR-γ • Improvement of histopathological characters of adipose tissue, heart, and liver	Antiobesity and hypolipidemic	(Padmanabhan and Arumugam, 2014) [153]
	Hydro-alcoholic leaves extract	0.15 and 0,3 g/kg	Wistar rat treated with STZ (50 mg/kg)	• Lowering of blood glucose level • Regulation of glucose uptake in both liver and muscles • Activation of Protein kinase B/Akt • Restoration of intracellular energy	Antidiabetic	(Lima et al., 2012) [154]
	Aqueous leaves extract	50, 100, and 150 mg/kg	Sprague-Dawley rats administered with ethanol-sucrose- and epinephrine-induced hypertension	Decrement in blood pressure (both systolic and diastolic)	Antihypertensive and hypotensive	(Sokpe et al., 2020) [155]

(continued on next page)

Plant	Extract	Dose	Model	Effects	Activity	Reference
Phaseolus vulgaris (Fabaceae)	Aqueous pods extract	200 mg/kg	Wistar rat injected with STZ (45 mg/kg), i.p.	• Reduction in blood glucose level	Antidiabetes	(Almuaigel et al., 2017) [156]
	Ethanolic seed extract	100 mg/kg	Wistar albino rats fed with high-salt diet	• Normalization of lipid levels • Reduction of elevated blood pressure • Modulation of rennin-angiotensin system	Antihypertensive	(Jawaid et al., 2017) [157]
Rosmarinus officinalis	Leaves extract	30 µg/ml	Human primary omental preadipocyte as well as adipocytes	• Reduction of TG incorporation • Stimulation of lipolytic activity in adipocytes (differentiating pre as well as mature adipocytes) • Down-regulation of expression of cyclin D1, cyclin-dependent-kinase 4 and -kinase inhibitor 1A • Up-regulation of expression of wingless-type MMTV integration site family, GATA binding protein 3 and member 3A mRNA levels	Antiobesity	(Stefanon et al., 2015) [158]
	Leaves extract	500 mg/kg	Male C57BL/6J mice fed with HFD	• Reduction in weight of body as well as epididymal fat • Increase in faecal fat • Reduction in fasting glucose and cholesterol levels • Inhibition of pancreatic lipase along with activity of PPAR-γ agonist	Antiobesity and antidiabetic	(Ibarra et al., 2011) [159]
	Aqueous leaf extract	200 mg/kg	Albino rats injected with STZ (200 mg/kg), p.o.	• Reduction in fasting blood glucose, TC, and triacylglycerol level • Increase in protein, α-amylase, anti-oxidant enzymes, macromolecules (albumin, ferritin, and ceruloplasmin)	Antidiabetic	(Emam, 2012) [160]

(continued on next page)

Table 6.3: List of herbal medicines exhibiting efficacy against metabolic syndrome and its key components—cont'd

Source	Extract	Dose	Model	Proposed mechanism	Inference	Reference
	Aqueous leaf extract	200 mg/kg	Albino rats injected with STZ (200 mg/kg), i.p.	• Reduction in fasting blood glucose • Recovering in activities of AST, creatine phosphokinase and lactate dehydrogenase • Decrease in levels of TC, TG, LDL-C, VLDL-C and increase in HDL-C	Antidiabetic and antihypertensive	(S. Alnahdi, 2012) [161]
Silybum marianum	Aqueous leaf extract	5 g/kg	C57BL/6J mice fed with HFD for 16 weeks	• Less body and epididymal fat weigh increase as compared to toxic group • Fecal fat excretion increased • Reduction in fasting glycaemia as well as cholesterol levels • Modulation of pancreatic lipase along with PPAR-γ agonist activity	Antiobesity	(Ibarra et al., 2011) [159]
Vitis vinifera	Anthrocyanidin-rich skin extract	250 and 500 mg/kg	C57BL/6J mice fed with HFD	• Reduction in weight gain • Prevention of dyslipidemia and insulin resistance • Regulation of levels of adiponectin, leptin, and resistin • Modulation of insulin signaling protein, pAMK/AMPK ratio as well as GLUT4 expression • Restoration of anti-oxidative and anti-inflammatory action	Antidiabetic, antioxidation, and anti-inflammatory	(da Costa et al., 2017) [162]
	Petroleum ether, chloroform, and ethanol stem bark extracts	100 and 200 mg/kg	Albino Wistar rat injected with alloxan (80 mg/kg)	• Decrement in blood glucose level • Lowering of TC, TG, LDL-C, and VLDL-C levels as well as lipid peroxidation • Increase in HDL-C levels as well as SOD, CAT, and peroxidase levels	Antidiabetic	(Ahmed et al., 2012) [163]

(continued on next page)

Plant	Preparation	Dose	Model	Effects	Activity	Reference
	Aqueous and methanolic leaves extract	100, 200, and 400 mg/kg	Wistar rat fed with high cholesterol diet	• Improvement in body weight gain and glucose level • Decrease in TC, TG, LDL, and VLDL levels • Increase in HDL levels • Attenuation of elevated SGOT and SGPT levels • Improvement in histopathological characters of blood vessels • Normalization of atherogenic index	Antihypercholesterolemia	(Devi and Singh, 2017) [164]
	Polyphenols from skin extract	200 mg/kg	Wistar Kyoto rat treated with STZ (60 mg/kg), i.p.	• Decrease in blood glucose level • Decrease in systolic blood pressure • Improvement in augmentation of vascular endothelium-dependent relaxation (P1 purinergic) • Decreasing in vasoconstriction (P2 purinergic) • Increasing atrial negative inotropism (P1 purinergic)	Antihypertensive and antidiabetic	(Bomfim et al., 2019) [165]
Zingiber officinale	Extract	400 mg/kg	Swiss mice fed with high cholesterol diet	• Improvement in insulin sensitivity as well as decrease in TAG levels • Decrease in TNF-α and IL-1β levels • Lowering of oxidative damage as well as reactive species	Antiobesity, antioxidant, and anti-inflammatory	(Luciano et al., 2020) [166]
	Ethyl acetate rhizome extract	5 µg/ml	Mouse myoblasts (L6) cell and mouse pre-adipocyte (3T3-L1) cells	• Enhancement of glucose uptake • Increase expression of GLUT 4 in cell surface	Antidiabetic	(Rani et al., 2012) [167]
	Rhizomes	2 and 4%	Wistar albino rat fed with high cholesterol diet	• Inhibition of angiotensin-1-converting enzyme • Decrease in levels of TC, TG, VLDL-C, LDL-C as well as MDA levels • Increase in hdl-c levels	Antihypercholesterolemia and antihypertensive	(Akinyemi et al., 2014) [168]

chemical drugs are utilized for the treatment of several MetS that impose adverse effects and high cost. Despite high investments in research involving development of effective drugs for MetS, only few have been approved and used. Because of this, physicians and patients are looking towards and adopting CAM methods such as medicinal plants and related products for treatment [169]. The availability of wide range of extracts and phytoconstituents of herbs may provide a safe, effective, and economical alternative or additive to the modern medicines for MetS. Medicinal plants and their products act through multi-target mechanisms as shown in Fig. 6.2 and Table 6.3, and could be an ideal candidate for multigenic disease like MetS.

The metabolic role of CAM has been evaluated and confirmed in various *in vitro* as well as *in vivo* studies, demonstrating promising potential mechanism of action. The precise molecular mechanisms/pathways affected by CAM in metabolic disorder remain relatively unknown. However, a number of recent studies have provided molecular cues to their potential anti-metabolic syndrome action. For example, in hypothalamus, neuropeptide Y (NPY) neuron plays a major role in regulation of energy homeostasis in the body. NPY release leads to increase in food intake concomitant with a reduction in dietary fat oxidation and body thermogenic capacity. Also, it causes inhibition of sympathetic outflow in brown tissue and thus reducing metabolic rate and energy dissipation. Blockade of NPY by herbal medicines inhibits arcuato-paraventricular projection that reduces food intake. Kim and colleagues reported that ginseng suppressed intake of food through inhibition of NPY expression in hypothalamus [64]. Various studies have reported that CAM treatment increases low density lipid receptor (LDLR) stability in liver via modulating extracellular signal-regulated kinase (ERK) and JNK signaling pathways. This results in improved clearance of LDL and total cholesterol [170]. For example, Berberine, the major phytoconstituent present in various plant extracts reduced cholesterol levels by increasing expression of LDLR in liver [170]. Furthermore, various studies have reported the potential of CAM in protection of β-islet cells function through anti-oxidant action as well as inhibition of β-cell apoptosis, resulting in improved glucose utilization [171]. Treatment with Ginsenoside Re in HFD-fed rats activated insulin-signaling pathway, including IRS-1 and the downstream mediator phosphatidylinositol 3-kinase in 3T3-L1 adipocytes. Ginsenoside Re treatment also increased translocation of GLUT4, thereby enhancing glucose disposal [172]. These effects of Ginsenoside Re appeared to be mediated by inhibition of inflammatory pathways; JNK and NF-κB signaling [172]. Collectively, these data provide important mechanistic cues to the insulin sensitive effects and anti-obesity effects of herbal medicines.

Globally, humans adapted unique indigenous healing traditions defined by their environment, beliefs, and culture which triumphant the health requirements of specific communities over centuries. Significant increase in the use of traditional medicine encourages WHO to promote the incorporation of traditional, complementary, and alternative medicine into health systems of some countries and development of policies and regulations within the national health care systems [173].

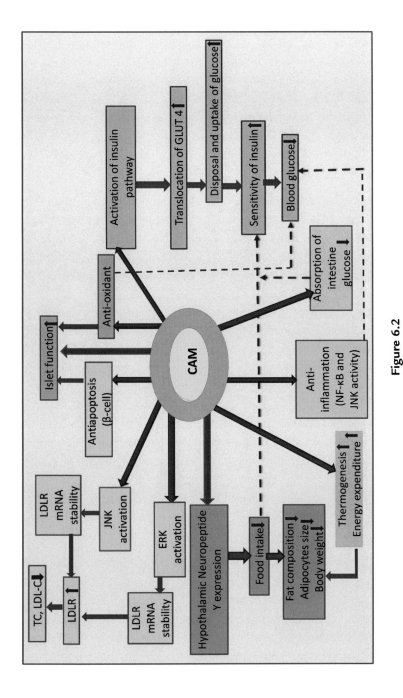

Figure 6.2

Common mechanism of action activated by different herbal formulations in management of metabolic syndrome. CAM, complementary and alternative medicines; ERK, extracellular signal regulated kinase; JNK - c, Jun N-terminal kinase; LDLR, low density lipoprotein receptor; TC, total cholesterol.

Despite the reported benefits as well as global use of CAM, there are limited scientific studies conducted to assess the safety as well as effectiveness of medicinal products and practices. Various studies reported adverse reactions of herbal medicines when utilized either alone or with conventional medicine. Despite, global diversity as well as adoption of CAM in different regions and cultures, no parallel growth in development of standards and methods at international level for evaluation has been done. Lack of national regulations and policies for CAM in different countries where they are available and used, makes it strenuous to completely regulate CAM products, practices and also practitioners due to divergence of categorizations and definitions of CAM therapies [174]. Further, availability of insufficient knowledge on sustaining and preserving procedures of plants and their use for medicinal purposes is a dormant threat to CAM sustenance.

Identification and authentication of substances along with quality of herbal medicine is considered as the core criteria [175]. However, due to complex formulations and lack of standardized analytical methods and procedures, the quality control evaluation of herbal products is challenging. Sensory and the phytochemical screening methods are employed for identification and authentication of raw and final herbal medicines in order to detect the species-based specific compounds and characteristics [176]. Herbal products are complex formulations, resulting from multi-processing steps, which leads to peculiar challenges to identification as well as authentication using organoleptic, microscopic and morphological characters and standard analytical assessments to detect specific target constituents [177]. Further, various factors affect the quality of raw material and end herbal product which requires careful consideration while opting analytical method for authentication and quality control evaluation [178]. For instance, secondary metabolites in medicinal plants are prone to variation in natural conditions, factors mainly including age, season, altitude, latitude, soil conditions, having direct impact on concentration of marker compounds assessed in variable batches of same plant material [179]. Involvement of complex processing in production will be reflected in the quality of end herbal product. Also, presence of numerous ingredients, with variable processing and extraction, hampers both authentication and quality control. Vast number of unreported ingredients used in herbal formulations may range from misleading labeling to serious poisoning or adverse drug reactions due to presence of toxic compounds.

CAM was considered safe by majority of users erroneously related to their natural sources. This misconception is considered as major reason for using herbal medicines. Considering the popularity of CAM among the different communities as well as ignoring toxicities lead to mandate the requirement of evaluation of safety, efficacy, and quality. Further, public educational program may be considered as beneficial for minimizing potential adverse effects. Also, more detailed and systematic research is warranted to evaluate the various parameters to ensure the safety and quality of raw material as well as herbal drug products. A plethora of research as well as clinical trials must be designed to address the issues and to evaluate the adverse effects related with CAM.

6.9 Efficacy in clinical trials

As indicated above an overwhelming number of animal studies have demonstrated strong efficacy of various herbal formulations against MetS and obesity. To prove clinical efficacy of herbal drugs in MetS and to translate the vast number preclinical findings into clinic, several clinical trials have been conducted in past two decades. A large number of clinical studies have shown their effectiveness in patients with varying degrees, whereas a number of other studies have demonstrated little or no effects. Promisingly enough though, a recent systematic meta-analysis of clinical trials of herbal medicines in obesity and MetS reported that herbal medicines containing green tea, *Phaseolus vulgaris, Garcinia cambogia, Nigella sativa*, puerh tea, *Irvingia gabonensis*, and *Caralluma fimbriata* and their active ingredients were effective in managing MetS [169]. In particular, consumption of *P. vulgaris, N. sativa*, or green tea caused a significant reduction in body weight, BMI, waist circumference, total cholesterol, and triglyceride, respectively. Moreover, a meta-analysis of clinical trials till November 2017 of the effect of ginger intake on MetS (14 studies with 473 subjects) revealed that ginger supplementation significantly decreased body weight, waist-to-hip ratio (WHR), hip ratio, fasting glucose and insulin resistance index (HOMA-IR), as well as significantly increased HDL-cholesterol levels[180].

In a randomized, double blind placebo controlled clinical trial. Forty-three adults aged 29-59 with a BMI >25 kg/m^2, or a waist circumference >94 cm (male), >80 cm (female) were treated with *Caralluma fimbriata* (an edible succulent) extract or placebo as 500 mg capsules for 12 weeks. After the completion of the protocol, the *C. fimbriata*-treated group lost 6.5 cm in waist circumference as compared to 2.6 cm in the placebo group. This effect was concomitant with improvement in waist/hip ratio in drug-treated group. It is to be noted that *C. fimbriata* treatment was also associated with concomitant decline in the sodium intake and the palatability of the food, which may have contributed to its observed positive effects [181]. Fenugreek (*Trigonella foenum-graceum*) is another herb, the hypoglycemic effect of which is well documented. Administration of hot-water-soaked fenugreek seeds (10g/day) in diabetic patients for 8 weeks significantly reduced fasting blood glucose, triglycerides and VLDL-cholesterol levels [182]. Similarly, the results from a double-blind controlled trial in patients with type 2 diabetes mellitus demonstrated that treatment with Alicor, a time released garlic powder tablet, significantly reduced the fasting blood glucose, fructosamine and triglycerides levels, improving overall metabolic control [183].

In another study conducted in Korean overweight and obese subjects with mild metabolic syndrome administration of a pterocarpan-high soybean leaf extract for 12 weeks significantly improved the homeostatic index of insulin resistance concomitant with reductions in the glycosylated hemoglobin (HbA1c), plasma glucose, FFA, total cholesterol, and non-HDL cholesterol levels without impacting the body weight and BMI [184]. The leaf extract also reduced systemic inflammation levels as determined by plasma levels of PAI-1, TNF-α, and IL-6. It is evident that this pterocarpan-rich herbal product clearly improved many key components of MetS, including insulin resistance, systemic inflammation, and hypercholesterolemia [184]. Moreover, administration of Kepar—a natural supplement containing extracts of *Curcuma longa*, silymarin, and guggul—in patients with MetS as add-on therapy resulted in significant reduction in body weight, BMI, waist circumference, and fasting glucose and total cholesterol levels [185], providing another evidence of potent efficacy of herbal medicines against MetS. A recent trial looked at the efficacy of the active constituent of *Curcuma longa*, curcuminoids, which is obtained from the rhizome of the plant in patients with nonalcoholic fatty liver disease (NAFLD). Administration of curcuminoid capsule along with piperine (to increase intestinal absorption of former) for 8 weeks significantly reduced body weight and the severity of NAFLD. The treatment also suppressed plasma levels of inflammatory cytokines, including TNF-α and MCP-1 [186]. The study provided the much-needed evidence for the presence of potent active molecules in the herbal products.

In contrast to the studies with positive outcome, there are a number of clinical trials that failed to translate preclinical findings into clinic with regard to effectiveness of herbal medicines in MetS. For example, a meta-analysis of randomized trials for the efficacy of almond against serum lipid profile in patients found no significant effect on LDL, HDL, or triglycerides levels [187]. Similarly, in a randomized, double-blind, placebo-controlled, crossover trial, *curcuminoids* administration (1 gm/day for 30 days) did not affect serum levels of total cholesterol, LDL, HDL, and inflammatory marker CRP protein. The treatment, however, significantly reduced the serum triglycerides levels [188]. Moreover, treatment with cinnamon capsule (*Cinnamomum cassia*), which has been shown to exert an insulin-sensitizing action in several animal models, in patients with type 2 diabetes for 3 months produced no significant effect on fasting blood glucose, total cholesterol, LDL, triglycerides, HBA1c, or insulin levels [189]. Therefore, high quality trials are still warranted to unequivocally demonstrate the clinical efficacy of the herbal drugs in MetS. In addition, the clinical efficacy of several potential herbal medicines has not been investigated extensively yet. Thus, there is a need to conduct large randomized controlled trials and systematic studies to determine the efficacy versus toxicity of botanicals in MetS. A list of clinical trials conducted recently is included in Table 6.4.

6.10 Limitations of herbal medicines

a. **Herb-drug interaction**. It is important to note that there is a common perception that herbal medicines are a safer/ healthier alternative to modern medicine and that is not always the case [190]. It is a largely under-researched area of study and combinations of certain herbs in different doses can increase the risk of allergies, adverse reactions, and often cross reactions with other pharmaceutical agents [191]. In the United States, from 2004-2013, herbal related drug induced liver injury increased from 7% to 20% [192].

Table 6.4: Clinical trials conducted on herbal based medicine and plant extracts for their potential in metabolic diseases (https://clinicaltrials.gov/).

Government identifier number	Trial year (phase)	Official Title	Extract used (dose)	Placebo	Study type	Total enrolments	Parameter assessed	Status
NCT04075032	2018-2019 (NA)	Effect of a pomegranate extract on metabolic and inflammatory markers, and the gut microbiota of polymedicated metabolic syndrome patients	Pomegranate extract (900 mg/day)	Yes (microcrystalline cellulose (900 mg/day)	Interventional (crossover)	60	Metabolic markers Inflammatory markers Modulation of the gut microbiota	Completed
NCT01190358	2010-2015 (NA)	Grape seed extract and postprandial oxidation and inflammation: A pilot study in people with the metabolic syndrome	Grape seed extract (300 mg)	Yes (Sugar pill containing maltodextrin)	Interventional (crossover)	12	Inflammatory response	Completed
NCT02764957	2016-2017 (3)	Effect of green coffee extract supplementation on anthropometric measurements, glycemic control, blood pressure, lipid profile	Green coffee extract (400 mg)	Yes (capsules containing starch)	Interventional (parallel)	43	Anthropometric measurements glycemic control Blood pressure Lipid profile	Completed
NCT03265184	2017 (NA)	Effects of green coffee extract supplementation on oxidative stress, systemic and vascular inflammation in patients with metabolic syndrome: A randomized clinical trial	Green coffee extract (400 mg)	Yes (capsules containing starch)	Interventional (parallel)	48	Oxidative stress Systemic and vascular inflammation	Completed

(continued on next page)

NCT number	Date	Title	Intervention	Placebo	Type	N	Outcomes	Status
NCT01889368	2012-2015 (NA)	Effect of a grape seed extract (GSE) on insulin resistan	Grape seed extract (300 mg)	Yes (maltodextrin)	Interventional (crossover)	19	Insulin resistance, Oxidative stress, Oxidized LDL levels	Completed
NCT01534910	2012-2014 (4)	Effect of aged garlic extract on atherosclerosis	Aged garlic extract (2400 mg)	Yes (sugar pill)	Interventional (parallel)	65	Tests of plaque (CT scan of the heart, carotid ultrasound)	Completed
NCT03973996	2019-2021 (NA)	Gut-level anti-inflammatory activities of green tea in metabolic syndrome	Green tea extract (1 g)	Yes (matched gummy confections)	Interventional (crossover)	40	Metabolic endotoxemia, Gut barrier function	Completed
NCT02065180	2014 (4)	The effect of a red rice and olive extract nutrition supplement on cholesterol	Red rice and olive extract	Yes	Interventional (parallel)	50	Cholesterol levels	Completed
NCT00327054	2006-2007 (2)	Effectiveness of Nigella Sativa (kalonji) seed in dyslipidemia: a randomized controlled trial	Nigella sativa seed	No	Interventional (parallel)	80	Cholesterol levels	Completed
NCT03479346	2018-2020 (NA)	Exploring the efficacy and safety of herbal medicine on Korean obese women with or without metabolic syndrome risk factors - a study protocol for a double-blind, randomized, multicenter, placebo-controlled clinical trial	GGT (3 g)	Yes	Interventional (parallel)	140	Obesity and metabolic biomarkers	Completed
NCT02651454	2015-2020 (4)	A pilot study exploring the efficacy and safety of herbak medicine on Korean obese women with metabolic syndrome risk factors - double blinded, randomized, multicenter, placebo-controlled clinical trial	Daesiho-tang (6 g) Jowiseungcheung-tang (6 g)	Yes	Interventional (parallel)	120	Obesity and metabolic biomarkers	Completed

Often, taking herbal medications alone is not associated with any major risk, the problem arises when the patient takes herbal medications in conjunction with their prescribed medications. For example, St. John's Wort is the most common herb-drug interaction known in the marketplace [193]. As a result, taking St John's Wort without consulting with a healthcare professional can lead to serious adverse effects. It has been shown to cause serotonin syndrome when used concurrently with serotonin reuptake inhibitors or with 5-HT receptor agonist [193]. In addition, hepatic CYP3A4 enzyme, a major drug metabolizing enzyme can be inhibited or induced by several herbal medicines including Goldenseal, Echinacea, St. John's wort, Ginseng, and Green tea [194–196]. This can alter the metabolism of several drugs such as cyclosporine, midazolam, buspirone, etc., causing toxicity or failure of the therapy.

b. **Herb-herb interactions.** Adverse reactions associated with herbal medicines in the past have been well documented [197]. The risk of adverse reactions associated with herbal medicines increases significantly in the presence of multiple herbal medications [198]. Many polyherbal Chinese medications have been shown to contain toxic compounds that increase the risk of cellular toxicity or genotoxicity [199]. A polyherbal medication 'Yoyo Bitters' highlights has been linked to increased liver toxicity, hypokalemia, and ultimately liver failure in a mouse model [200].

c. **Adulteration/contamination**. Adulteration is defined as altering/ adding non plant product components, using low-price varieties, low grade material from above-ground parts or processing waste, alternative species, or none at all to increase therapeutic efficacy of the herbal medicines [201]. The most common adulterated medicines are diabetes, calm/sleep, sexual dysfunction, pain relief, and rheumatism herbal medications. Herbal medicines lack the regulation and are usually marketed without any premarket safety or toxicology evaluations. A 2010 US Government Accountability Office report indicated that of the 40 dietary supplements tested, 93% of the samples contained the presence of lead, arsenic, mercury, cadmium, or pesticides [106].

d. **Variability**. As stated previously, herbal medications are not subject to the same scrutiny that prescription medications are. The probability of an untrained botanical may vary from lot to lot and are dependent on growing conditions, the manufacturer, and the time of year the plant is grown. Consequently, these factors result in varying quality and contents of each herbal formulation from batch to batch [202]. Ginseng is one of the most commonly utilized herbs on the market, totaling an estimated $2.1Billion [203]. A content analysis done on 507 ginseng plants from the global supply found that the concentrations of important therapeutic active ingredients were significantly varied. There was a 15- 36- fold difference in the concentrations of ginsenosides and these levels varied from location to location. Herbal medicine "echinacea" is another example of variability, 80% of 59 echinacea products tested contained substantially less echinacoside, or cichoric acid than what was stated on the label [204].

e. **Regulation**. This brings rise to the bigger issue that limits herbal medications, the lack of regulations outlined in the DSHEA act of 1994. The DSHEA act leaves the manufacturer in charge of the quality, safety, and efficacy on these batches of herbal medicines and therefore, there is no way of knowing if the quality from batch to batch will be consistent [205]. As a result, it is to no surprise that this type of regulation has led to inconsistency and unsafe herbal medications being marketed [206]. Thus, reclassifying herbal medications would provide a framework for the rapid development in the safety and quality of herbal medicines.

6.11 Modern approaches to investigate for the molecular mechanisms of herbal medicines

The identification of adulterated species in medicinal herbs is crucial because these products are usually cultivated from the wild or on unregulated farms [207]. Nevertheless, the detection of adulteration in herbal medicines is a complicated and challenging process that usually employs multiple methods to determine the undeclared substances. To further add to the problem, herbal formulations vary in chemical composition based on the sources they are obtained from. There is a need to develop novel tools or strategies that can help standardize the herbal products with a reasonable amount of time and efforts. Some of these processes include Gas Chromatography-Mass Spectrometry (GC-MS), Liquid Chromatography–Mass Spectrometry (LC-MS), Ultra High-Performance Liquid Gas Chromatography- Mass Spectroscopy (UPLC-TOP/MS), NMR, X-Ray powder diffractometry [208]. Recently, Hydrogen GC-MS has begun to gain popularity in the detection of adulterated substances in herbal medicines. The increased speed, lower gas separation temperatures, longer column life, greater availability, and fewer environmental concerns makes it a strong candidate for the detection of adulterants. However, substances that exhibit potential reactivity with Hydrogen must be evaluated early in the process or risk unwanted protonation or hydrogenation which will alter the lab results, using Helium or Nitrobenzene should be considered as an alternative [209]. With 75-80% of the world's population using herbal medicines, it is important to refine and master the process of detecting adulterated products. A recent study outlined a process that is in the development stages and has shown promise. Researchers used DNA barcode-based species-specific sequence characterized amplified regions (SCAR) markers to identify adulterated fungi species. Further, the SCAR markers were evaluated using a real-time polymerase chain reaction (PCR) in fungi species to detect taxonomic origin, the degree of contamination, and the amount of the adulterant in the samples. In essence, the success of this trial gives hope to the potential ease and reliability to the detection of adulterants using species-specific SCAR markers and real time PCR [210].

Another approach could be the use of microarray-based transcriptome analysis, a system-toxicology based and high throughput technique [211]. The technique can be used to detect the potential to induce toxicity for multiples individual compounds present in a formulation at an early stage of development. Different concentrations of the formulation can be employed to treat multiple cell lines and the data analyzed using the microarray-based transcriptome analysis. Several crucial molecular pathways can be investigated using this technique, such as, inflammation, senescence, apoptosis, cell stress, and cell growth. Thus, employing such novel system-based and high-throughput approaches, it is possible to classify the tested compounds based on their system toxicological profiles and draw the meaningful biological information that is relevant for specific organ system such as liver, heart etc. [212]

Unlike modern medicines which undergo stringent safety and therapeutic screening prior to approval for marketing, herbal drugs can be marketed without any clinical and safety studies. While herbal products are usually considered beneficial and self-prescribed, recent studies have highlighted important risks and adverse effects associated with their usages. The FDA's adverse event reporting system has revealed several toxic events such as liver failure, neural birth defect, etc. and critical drug interaction with prescription drugs associated with dietary supplements. Thus, it is imperative to invest effort into enhancing our understanding of pharmacological, toxicological, and safety profile of herbal drugs.

References

[1] A KG, PZ Zimmet, Definition, diagnosis and classification of diabetes mellitus and its complications. Part 1: diagnosis and classification of diabetes mellitus provisional report of a WHO consultation, Diabetic medicine: a journal of the British Diabetic Association 15 (1998) 539–553.

[2] KG Alberti, RH Eckel, SM Grundy, PZ Zimmet, JI Cleeman, KA Donato, et al., Harmonizing the metabolic syndrome: a joint interim statement of the International Diabetes Federation Task Force on Epidemiology and Prevention; National Heart, Lung, and Blood Institute; American Heart Association; World Heart Federation; International Atherosclerosis Society; and International Association for the Study of Obesity, Circulation 120 (2009) 1640–1645.

[3] Third Report of the National Cholesterol Education Program (NCEP), Expert panel on detection, evaluation, and treatment of high blood cholesterol in adults (adult treatment panel iii) final report, Circulation 106 (2002) 3143–3421.

[4] P Zimmet, D Magliano, Y Matsuzawa, G Alberti, J Shaw, The metabolic syndrome: a global public health problem and a new definition, Journal of atherosclerosis and thrombosis 12 (2005) 295–300.

[5] B Balkau, MA Charles, Comment on the provisional report from the WHO consultation. European Group for the Study of Insulin Resistance (EGIR), Diabetic medicine 16 (1999) 442–443.

[6] Prevention CfDCa. National Diabetes Statistics Report 2020, Estimates of Diabetes and Its Burden in the United Stated. 2020.

[7] TH Shi, B Wang, S Natarajan, The influence of metabolic syndrome in predicting mortality risk among us adults: importance of metabolic syndrome even in adults with normal weight, Preventing chronic disease 17 (2020) E36.

[8] JX Moore, N Chaudhary, T Akinyemiju, Metabolic syndrome prevalence by race/ethnicity and sex in the united states, national health and nutrition examination survey, 1988-2012, Preventing chronic disease 14 (2017) E24.

[9] LiverTox Clinical and Research Information on Drug-Induced Liver Injury Bethesda (MD), Bethesda, Maryland: National Institute of Diabetes and Digestive and Kidney Diseases, 2012.

[10]   Y Krishnamoorthy, S Rajaa, S Murali, T Rehman, J Sahoo, SS Kar, Prevalence of metabolic syndrome among adult population in India: a systematic review and meta-analysis, PLoS One 15 (2020) e0240971.

[11]   Y Lan, Z Mai, S Zhou, Y Liu, S Li, Z Zhao, et al., Prevalence of metabolic syndrome in China: an updated cross-sectional study, PLoS One 13 (2018) e0196012.

[12]   Y Wang, J Mi, XY Shan, W QJ, KY Ge, Is China facing an obesity epidemic and the consequences? The trends in obesity and chronic disease in China, International journal of obesity (2005) 31 (2007) 177–188.

[13]   PL Huang, A comprehensive definition for metabolic syndrome, Disease models & mechanisms 2 (2009) 231–237.

[14]   A Hruby, FB Hu, The epidemiology of obesity: a big picture, Pharmacoeconomics 33 (2015) 673–689.

[15]   SE Barlow, Expert committee recommendations regarding the prevention, assessment, and treatment of child and adolescent overweight and obesity: summary report, Pediatrics 120 (Suppl 4) (2007) S164–S192.

[16]   A Afshin, MH Forouzanfar, MB Reitsma, P Sur, K Estep, A Lee, et al., Health effects of overweight and obesity in 195 countries over 25 years, The New England journal of medicine 377 (2017) 13–27.

[17]   R.P.M.P Wildman, K Reynolds, et al., The obese without cardiometabolic risk factor clustering and the normal weight with cardiometabolic risk factor clustering: prevalence and correlates of 2 phenotypes among the US population (NHANES 1999–2004), Arch Int Med 168 (2008) 1617–1624.

[18]   K Ogurtsova, JD da Rocha Fernandes, Y Huang, U Linnenkamp, L Guariguata, NH Cho, et al., IDF Diabetes Atlas: global estimates for the prevalence of diabetes for 2015 and 2040, Diabetes research and clinical practice 128 (2017) 40–50.

[19]   Y Matsuzawa, T Funahashi, T Nakamura, The concept of metabolic syndrome: contribution of visceral fat accumulation and its molecular mechanism, Journal of atherosclerosis and thrombosis 18 (2011) 629–639.

[20]   S Guo, Insulin signaling, resistance, and the metabolic syndrome: insights from mouse models into disease mechanisms, The Journal of endocrinology 220 (2014) T1–t23.

[21]   B Isomaa, A major health hazard: the metabolic syndrome, Life sciences 73 (2003) 2395–2411.

[22]   GS Hotamisligil, P Peraldi, A Budavari, R Ellis, MF White, BM Spiegelman, IRS-1-mediated inhibition of insulin receptor tyrosine kinase activity in TNF-alpha- and obesity-induced insulin resistance, Science (New York, NY) 271 (1996) 665–668.

[23]   GS Hotamisligil, NS Shargill, BM Spiegelman, Adipose expression of tumor necrosis factor-alpha: direct role in obesity-linked insulin resistance, Science (New York, NY) 259 (1993) 87–91.

[24]   M S, Impact of the metabolic syndrome on mortality from coronary heart disease, and all causes in United States adults, Circulation 110 (2004) 1245–1250.

[25]   T Yoneshiro, S Aita, M Matsushita, T Kayahara, T Kameya, Y Kawai, et al., Recruited brown adipose tissue as an antiobesity agent in humans, The Journal of clinical investigation 123 (2013) 3404–3408.

[26]   A Chawla, KD Nguyen, YP Goh, Macrophage-mediated inflammation in metabolic disease, Nature reviews Immunology 11 (2011) 738–749.

[27]   KD Nguyen, Y Qiu, X Cui, YP Goh, J Mwangi, T David, et al., Alternatively activated macrophages produce catecholamines to sustain adaptive thermogenesis, Nature 480 (2011) 104–108.

[28]   A Stefanović, J Kotur-Stevuljević, S Spasić, N Bogavac-Stanojević, N Bujisić, The influence of obesity on the oxidative stress status and the concentration of leptin in type 2 diabetes mellitus patients, Diabetes research and clinical practice 79 (2008) 156–163.

[29]   NA Abdella, OA Mojiminiyi, MA Moussa, M Zaki, H Al Mohammedi, E Al Ozairi, et al., Plasma leptin concentration in patients with Type 2 diabetes: relationship to cardiovascular disease risk factors and insulin resistance, Diabetic medicine: a journal of the British Diabetic Association 22 (2005) 278–285.

[30]   SG Wannamethee, J Tchernova, P Whincup, GD Lowe, A Kelly, A Rumley, AM Wallace, et al., Plasma leptin: associations with metabolic, inflammatory and haemostatic risk factors for cardiovascular disease, Atherosclerosis 191 (2007) 418–426.

[31]   MP Reilly, N Iqbal, M Schutta, ML Wolfe, M Scally, AR Localio, et al., Plasma leptin levels are associated with coronary atherosclerosis in type 2 diabetes, The Journal of clinical endocrinology and metabolism 89 (2004) 3872–3878.

[32] MWA Adamczak, The adipose tissue as an endocrine organ, Semin Nephrol 33 (2013) 2–13.

[33] A Benrick, B Chanclón, P Micallef, Y Wu, L Hadi, JM Shelton, et al., Adiponectin protects against development of metabolic disturbances in a PCOS mouse model, in: Proceedings of the National Academy of Sciences of the United States of America, 114, 2017, pp. E7187–e7196.

[34] M Ryo, T Nakamura, S Kihara, M Kumada, S Shibazaki, M Takahashi, et al., Adiponectin as a biomarker of the metabolic syndrome, Circulation journal 68 (2004) 975–981.

[35] Y Arita, S Kihara, N Ouchi, M Takahashi, K Maeda, J Miyagawa, et al., Paradoxical decrease of an adipose-specific protein, adiponectin, in obesity, Biochemical and biophysical research communications 257 (1999) 79–83.

[36] GA Laughlin, E Barrett-Connor, S May, C Langenberg, Association of adiponectin with coronary heart disease and mortality: the Rancho Bernardo study, American journal of epidemiology 165 (2007) 164–174.

[37] Y Iwashima, T Katsuya, K Ishikawa, N Ouchi, M Ohishi, K Sugimoto, et al., Hypoadiponectinemia is an independent risk factor for hypertension, Hypertension (Dallas, Tex: 1979) 43 (2004) 1318–1323.

[38] M Shahid, AA Javed, D Chandra, HE Ramsey, D Shah, MF Khan, et al., IEX-1 deficiency induces browning of white adipose tissue and resists diet-induced obesity, Scientific reports 6 (2016) 24135.

[39] B Shan, X Wang, Y Wu, C Xu, Z Xia, J Dai, et al., The metabolic ER stress sensor IRE1α suppresses alternative activation of macrophages and impairs energy expenditure in obesity, Nature immunology 18 (2017) 519–529.

[40] F Zatterale, M Longo, J Naderi, GA Raciti, A Desiderio, C Miele, et al., Chronic adipose tissue inflammation linking obesity to insulin resistance and type 2 diabetes, . Frontiers in physiology 10 (2019) 1607.

[41] Y Jing, F Wu, D Li, L Yang, Q Li, R Li, Metformin improves obesity-associated inflammation by altering macrophages polarization, Molecular and cellular endocrinology 461 (2018) 256–264.

[42] D Wu, AB Molofsky, HE Liang, RR Ricardo-Gonzalez, HA Jouihan, JK Bando, et al., Eosinophils sustain adipose alternatively activated macrophages associated with glucose homeostasis, Science (New York, NY) 332 (2011) 243–247.

[43] Y Qiu, KD Nguyen, JI Odegaard, X Cui, X Tian, RM Locksley, et al., Eosinophils and type 2 cytokine signaling in macrophages orchestrate development of functional beige fat, Cell 157 (2014) 1292–1308.

[44] JR Brestoff, BS Kim, SA Saenz, RR Stine, LA Monticelli, GF Sonnenberg, et al., Group 2 innate lymphoid cells promote beiging of white adipose tissue and limit obesity, Nature 519 (2015) 242–246.

[45] D.A.P.K Mori, P Walter, 2014 Lasker Basic Medical Research awardees: The unfolded protein response, in: Proceedings of the National Academy of Sciences of the United States of America, 111, 2014, pp. 17696–17697.

[46] U Ozcan, Q Cao, E Yilmaz, AH Lee, NN Iwakoshi, E Ozdelen, et al., Endoplasmic reticulum stress links obesity, insulin action, and type 2 diabetes, Science (New York, NY) 306 (2004) 457–461.

[47] A Everard, C Belzer, L Geurts, JP Ouwerkerk, C Druart, LB Bindels, et al., Cross-talk between Akkermansia muciniphila and intestinal epithelium controls diet-induced obesity, in: Proceedings of the National Academy of Sciences of the United States of America, 110, 2013, pp. 9066–9071.

[48] LA David, CF Maurice, RN Carmody, DB Gootenberg, JE Button, BE Wolfe, et al., Diet rapidly and reproducibly alters the human gut microbiome, Nature 505 (2014) 559–563.

[49] X Zhang, D Shen, Z Fang, Z Jie, X Qiu, C Zhang, et al., Human gut microbiota changes reveal the progression of glucose intolerance, PLoS One 8 (2013) e71108.

[50] M Yassour, MY Lim, HS Yun, TL Tickle, J Sung, YM Song, et al., Sub-clinical detection of gut microbial biomarkers of obesity and type 2 diabetes, Genome medicine 8 (2016) 17.

[51] J Li, F Zhao, Y Wang, J Chen, J Tao, G Tian, et al., Gut microbiota dysbiosis contributes to the development of hypertension, Microbiome 5 (2017) 14.

[52] R Caesar, V Tremaroli, P Kovatcheva-Datchary, C PD, F Bäckhed, Crosstalk between gut microbiota and dietary lipids aggravates WAT inflammation through TLR signaling, Cell metabolism 22 (2015) 658–668.

[53]   C Depommier, A Everard, C Druart, H Plovier, M Van Hul, S Vieira-Silva, et al., Supplementation with Akkermansia muciniphila in overweight and obese human volunteers: a proof-of-concept exploratory study, Nature medicine 25 (2019) 1096–1103.

[54]   GM Reaven, Banting lecture 1988. Role of insulin resistance in human disease, Diabetes 37 (1988) 1595–1607.

[55]   BB Kahn, JS Flier, Obesity and insulin resistance, The Journal of clinical investigation 106 (2000) 473–481.

[56]   CR Kahn, Insulin resistance, insulin insensitivity, and insulin unresponsiveness: a necessary distinction, Metabolism: clinical and experimental 27 (1978) 1893–1902.

[57]   SE Kahn, The relative contributions of insulin resistance and beta-cell dysfunction to the pathophysiology of Type 2 diabetes, Diabetologia 46 (2003) 3–19.

[58]   MP Czech, Insulin action and resistance in obesity and type 2 diabetes, . Nature medicine 23 (2017) 804–814.

[59]   G Boden, GI Shulman, Free fatty acids in obesity and type 2 diabetes: defining their role in the development of insulin resistance and beta-cell dysfunction, European journal of clinical investigation 32 (Suppl 3) (2002) 14–23.

[60]   GF Lewis, KD Uffelman, LW Szeto, G Steiner, Effects of acute hyperinsulinemia on VLDL triglyceride and VLDL apoB production in normal weight and obese individuals, Diabetes 42 (1993) 833–842.

[61]   HN Ginsberg, YL Zhang, A Hernandez-Ono, Regulation of plasma triglycerides in insulin resistance and diabetes, Archives of medical research 36 (2005) 232–240.

[62]   P Arner, Not all fat is alike, Lancet (London, England) 351 (1998) 1301–1302.

[63]   JH Xing, R Li, YQ Gao, MY Wang, YZ Liu, J Hong, et al., NLRP3 inflammasome mediate palmitate-induced endothelial dysfunction, Life sciences 239 (2019) 116882.

[64]   F Kim, KA Tysseling, J Rice, M Pham, L Haji, BM Gallis, et al., Free fatty acid impairment of nitric oxide production in endothelial cells is mediated by IKKbeta, Arteriosclerosis, thrombosis, and vascular biology 25 (2005) 989–994.

[65]   VR Sopasakis, M Sandqvist, B Gustafson, A Hammarstedt, M Schmelz, X Yang, et al., High local concentrations and effects on differentiation implicate interleukin-6 as a paracrine regulator, Obesity research 12 (2004) 454–460.

[66]   F Hube, H Hauner, The role of TNF-alpha in human adipose tissue: prevention of weight gain at the expense of insulin resistance? Hormone and metabolic research = Hormon- und Stoffwechselforschung = Hormones et metabolisme 31 (1999) 626–631.

[67]   MA Gimbrone Jr., JN Topper, T Nagel, KR Anderson, G Garcia-Cardeña, Endothelial dysfunction, hemodynamic forces, and atherogenesis, Annals of the New York Academy of Sciences 902 (230-9) (2000) 239–240.

[68]   JE Tooke, MM Hannemann, Adverse endothelial function and the insulin resistance syndrome, Journal of internal medicine 247 (2000) 425–431.

[69]   D Fulton, JP Gratton, TJ McCabe, J Fontana, Y Fujio, K Walsh, et al., Regulation of endothelium-derived nitric oxide production by the protein kinase Akt, Nature 399 (1999) 597–601.

[70]   A Aljada, P Dandona, Effect of insulin on human aortic endothelial nitric oxide synthase, Metabolism: clinical and experimental 49 (2000) 147–150.

[71]   I Juhan-Vague, A MC, PE Morange, Hypofibrinolysis and increased PAI-1 are linked to atherothrombosis via insulin resistance and obesity, Annals of medicine 32 (Suppl 1) (2000) 78–84.

[72]   Y Ikeda, S Hama, K Kajimoto, T Okuno, H Tsuchiya, K Kogure, Quantitative comparison of adipocytokine gene expression during adipocyte maturation in non-obese and obese rats, Biological & pharmaceutical bulletin 34 (2011) 865–870.

[73]   YS Lee, J Wollam, JM Olefsky, An integrated view of immunometabolism, Cell 172 (2018) 22–40.

[74]   AR Saltiel, JM Olefsky, Inflammatory mechanisms linking obesity and metabolic disease, The Journal of clinical investigation 127 (2017) 1–4.

[75] JK Kim, O Gavrilova, Y Chen, ML Reitman, GI Shulman, Mechanism of insulin resistance in A-ZIP/F-1 fatless mice, The Journal of biological chemistry 275 (2000) 8456–8460.

[76] L Zhou, SY Park, L Xu, X Xia, J Ye, L Su, et al., Insulin resistance and white adipose tissue inflammation are uncoupled in energetically challenged Fsp27-deficient mice, Nature communications 6 (2015) 5949.

[77] MA Cornier, D Dabelea, TL Hernandez, RC Lindstrom, AJ Steig, NR Stob, et al., The metabolic syndrome, Endocrine reviews 29 (2008) 777–822.

[78] SK Fried, DA Bunkin, AS Greenberg, Omental and subcutaneous adipose tissues of obese subjects release interleukin-6: depot difference and regulation by glucocorticoid, The Journal of clinical endocrinology and metabolism 83 (1998) 847–850.

[79] I Mertens, A Verrijken, JJ Michiels, M Van der Planken, R JB, LF Van Gaal, Among inflammation and coagulation markers, PAI-1 is a true component of the metabolic syndrome, International journal of obesity (2005) 30 (2006) 1308–1314.

[80] J Hirosumi, G Tuncman, L Chang, CZ Görgün, KT Uysal, K Maeda, et al., A central role for JNK in obesity and insulin resistance, Nature 420 (2002) 333–336.

[81] EA Oral, SM Reilly, AV Gomez, R Meral, L Butz, N Ajluni, et al., Inhibition of IKK[and TBK1 improves glucose control in a subset of patients with type 2 diabetes, Cell metabolism 26 (2017) 157–170 e7.

[82] TL Stanley, MV Zanni, S Johnsen, S Rasheed, H Makimura, H Lee, et al., TNF-alpha antagonism with etanercept decreases glucose and increases the proportion of high molecular weight adiponectin in obese subjects with features of the metabolic syndrome, The Journal of clinical endocrinology and metabolism 96 (2011) E146–E150.

[83] C Tsigos, I Kyrou, E Chala, P Tsapogas, JC Stavridis, R SA, et al., Circulating tumor necrosis factor alpha concentrations are higher in abdominal versus peripheral obesity, Metabolism: clinical and experimental 48 (1999) 1332–1335.

[84] GS Hotamisligil, DL Murray, LN Choy, BM Spiegelman, Tumor necrosis factor alpha inhibits signaling from the insulin receptor, in: Proceedings of the National Academy of Sciences of the United States of America, 91, 1994, pp. 4854–4858.

[85] J Lo, LE Bernstein, B Canavan, M Torriani, MB Jackson, RS Ahima, et al., Effects of TNF-alpha neutralization on adipocytokines and skeletal muscle adiposity in the metabolic syndrome, American journal of physiology Endocrinology and metabolism 293 (2007) E102–E109.

[86] R Deepa, K Velmurugan, K Arvind, P Sivaram, C Sientay, S Uday, et al., Serum levels of interleukin 6, C-reactive protein, vascular cell adhesion molecule 1, and monocyte chemotactic protein 1 in relation to insulin resistance and glucose intolerance–the Chennai Urban Rural Epidemiology Study (CURES), Metabolism: clinical and experimental 55 (2006) 1232–1238.

[87] S Guldiken, M Demir, E Arikan, B Turgut, S Azcan, M Gerenli, et al., The levels of circulating markers of atherosclerosis and inflammation in subjects with different degrees of body mass index: Soluble CD40 ligand and high-sensitivity C-reactive protein, Thrombosis research 119 (2007) 79–84.

[88] AS González, DB Guerrero, MB Soto, SP Díaz, M Martinez-Olmos, O Vidal, Metabolic syndrome, insulin resistance and the inflammation markers C-reactive protein and ferritin, European journal of clinical nutrition 60 (2006) 802–809.

[89] PM Ridker, JE Buring, NR Cook, N Rifai, C-reactive protein, the metabolic syndrome, and risk of incident cardiovascular events: an 8-year follow-up of 14 719 initially healthy American women, Circulation 107 (2003) 391–397.

[90] AN Vgontzas, DA Papanicolaou, EO Bixler, A Kales, K Tyson, GP Chrousos, Elevation of plasma cytokines in disorders of excessive daytime sleepiness: role of sleep disturbance and obesity, The Journal of clinical endocrinology and metabolism 82 (1997) 1313–1316.

[91] AD Pradhan, JE Manson, N Rifai, JE Buring, PM Ridker, C-reactive protein, interleukin 6, and risk of developing type 2 diabetes mellitus, JAMA 286 (2001) 327–334.

[92] WB, The inflammatory syndrome: the role of adipose tissue cytokines in metabolic disorders linked to obesity, J Am Soc Nephrol 15 (2004) 2792–2800.

[93]   C Menzaghi, A Coco, L Salvemini, R Thompson, S De Cosmo, A Doria, et al., Heritability of serum resistin and its genetic correlation with insulin resistance-related features in nondiabetic Caucasians, The Journal of clinical endocrinology and metabolism 91 (2006) 2792–2795.

[94]   Y Fu, Y Yu, Y Wu, Y You, Y Zhang, C Kou, Association between two resistin gene polymorphisms and metabolic syndrome in Jilin, Northeast China: a case-control study, Disease markers 2017 (2017) 1638769.

[95]   SZA Samsamshariat, F Sakhaei, L Salehizadeh, M Keshvari, S Asgary, Relationship between resistin, endothelin-1, and flow-mediated dilation in patient with and without metabolic syndrome, Advanced biomedical research 8 (2019) 16.

[96]   T Binesh Marvasti, K Adeli, Pharmacological management of metabolic syndrome and its lipid complications. *Daru: journal of Faculty of Pharmacy*, Tehran University of Medical Sciences 18 (2010) 146–154.

[97]   S Dias, S Paredes, L Ribeiro, Drugs involved in dyslipidemia and obesity treatment: focus on adipose tissue, International journal of endocrinology 2018 (2018) 2637418.

[98]   S Matthaei, M Stumvoll, M Kellerer, HU Häring, Pathophysiology and pharmacological treatment of insulin resistance, Endocrine reviews 21 (2000) 585–618.

[99]   WR BMaB, The Complete German Commission E Monograph, American Botanical Council, Austin, 1998.

[100]   AH Alostad, DT Steinke, EI Schafheutle, International Comparison of Five Herbal Medicine Registration Systems to Inform Regulation Development: United Kingdom, Germany, United States of America, United Arab Emirates and Kingdom of Bahrain, Pharmaceutical medicine 32 (2018) 39–49.

[101]   A Vickers, C Zollman, R Lee, Herbal medicine, The Western journal of medicine 175 (2001) 125–128.

[102]   F Firenzuoli, L Gori, Herbal medicine today: clinical and research issues, Evidence-based complementary and alternative medicine: eCAM 4 (2007) 37–40.

[103]   E Mills, C Cooper, D Seely, I Kanfer, African herbal medicines in the treatment of HIV: hypoxis and sutherlandia. An overview of evidence and pharmacology, Nutrition journal 4 (2005) 19.

[104]   S Wachtel-Galor, IFF Benzie, Herbal medicine: an introduction to its history, usage, regulation, current trends, and research needs, in: IFF Benzie, S Wachtel-Galor (Eds.), Herbal Medicine: Biomolecular and Clinical Aspects, CRC Press/Taylor & Francis, Boca Raton (FL), 2011.

[105]   R Srirama, JU Santhosh Kumar, GS Seethapathy, SG Newmaster, S Ragupathy, KN Ganeshaiah, et al., Species adulteration in the herbal trade: causes, consequences and mitigation, Drug safety 40 (2017) 651–661.

[106]   RR Starr, Too little, too late: ineffective regulation of dietary supplements in the United States, American journal of public health 105 (2015) 478–485.

[107]   LM Kemppainen, TT Kemppainen, JA Reippainen, S ST, PH Vuolanto, Use of complementary and alternative medicine in Europe: Health-related and sociodemographic determinants, Scandinavian journal of public health 46 (2018) 448–455.

[108]   HJ Jung, JH Wee, KM Kim, HM Sung, HK Shin, Effect of onion (Allium cepa) ultra-high pressure processing and hot water extracts on the serum cholesterol level in high cholesterol-fed rats, Food Sci Biotech 24 (2015) 287–294.

[109]   J Moon, HJ Do, K OY, MJ Shin, Antiobesity effects of quercetin-rich onion peel extract on the differentiation of 3T3-L1 preadipocytes and the adipogenesis in high fat-fed rats, Food and chemical toxicology: an international journal published for the British Industrial Biological Research Association 58 (2013) 347–354.

[110]   OS Ogunmodede, LC Saalu, B Ogunlade, GG Akunna, AO Oyewopo, An evaluation of the hypoglycemic, antioxidant and hepatoprotective potentials of onion (Allium cepa L.) on alloxan-induced diabetic rabbits. International Journal of Pharmacology 8 (2012) 21–29.

[111]   D González-Peña, J Angulo, S Vallejo, C Colina-Coca, B de Ancos, CF Sánchez-Ferrer, et al., High-cholesterol diet enriched with onion affects endothelium-dependent relaxation and NADPH oxidase activity in mesenteric microvessels from Wistar rats, Nutrition & metabolism 11 (2014) 57.

[112]   A Tada, E Misawa, M Tanaka, M Saito, K Nabeshima, K Yamauchi, et al., Investigating anti-obesity effects by oral administration of aloe vera gel extract (AVGE): possible involvement in activation of brown adipose tissue (BAT), Journal of nutritional science and vitaminology 66 (2020) 176–184.

[113]   N Deora, MM Sunitha, M Satyavani, N Harishankar, MA Vijayalakshmi, K Venkataraman, et al., Alleviation of diabetes mellitus through the restoration of β-cell function and lipid metabolism by Aloe vera (L.) Burm. f. extract in obesogenic WNIN/GR-Ob rats, Journal of ethnopharmacology 272 (2021) 113921.

[114]   Saleem T., Mumtaz, U., Bashir, M.U., Qureshi, H.J., Saleem, A. Comparison of hypoglycemic effects of azadirachta indica seeds and leaves on alloxan induced diabetes in male albino rats. Pak J Med Health Sci. 12, 753–756.

[115]   AB Shori, AS Baba, Antioxidant activity and inhibition of key enzymes linked to type-2 diabetes and hypertension by Azadirachta indica-yogurt, J Saudi Chem Soc 17 (2013) 295–301.

[116]   M Bhat, SK Kothiwale, AR Tirmale, B SY, BN Joshi, Antidiabetic properties of Azardiracta indica and Bougainvillea spectabilis: in vivo studies in murine diabetes model, Evidence-based complementary and alternative medicine: eCAM 2011 (2011) 561625.

[117]   T Omobowale, A Oyagbemi, O Adejumobi, F Ugbor, E Asenuga, T Ajibade, et al., Antihypertensive effect of methanol leaf extract of Azadirachta indica is mediated through suppression of renal caspase 3 expressions on N ω -Nitro-l-arginine methyl ester induced hypertension, Pharmacog Res 12 (2020) 460–465.

[118]   P Ansari, PR Flatt, P Harriott, YHA Abdel-Wahab, Anti-hyperglycaemic and insulin-releasing effects of Camellia sinensis leaves and isolation and characterisation of active compounds, The British journal of nutrition 126 (8) (2020) 1–15.

[119]   TY Chen, MMC Wang, SK Hsieh, MH Hsieh, WY Chen, JTC Tzen, Pancreatic lipase inhibition of strictinin isolated from Pu'er tea (Cammelia sinensis) and its anti-obesity effects in C57BL6 mice, J Funct Foods 48 (2018) 1–8.

[120]   Y Sun, Y Wang, P Song, H Wang, N Xu, Y Wang, et al., Anti-obesity effects of instant fermented teas in vitro and in mice with high-fat-diet-induced obesity, Food & function 10 (2019) 3502–3513.

[121]   M Stepien, M Kujawska-Luczak, M Szulinska, M Kregielska-Narozna, D Skrypnik, J Suliburska, et al., Beneficial dose-independent influence of Camellia sinensis supplementation on lipid profile, glycemia, and insulin resistance in an NaCl-induced hypertensive rat model, Journal of physiology and pharmacology: an official journal of the Polish Physiological Society 69 (2) (2018) 69.

[122]   G Jeon, Y Choi, SM Lee, Y Kim, HS Jeong, J Lee, Anti-obesity activity of methanol extract from hot pepper (Capsicum annuum L.) seeds in 3T3-L1 adipocyte, Food Sci Biotech 19 (2010) 1123–1127.

[123]   J Sung, J HS, J Lee, Effect of the capsicoside G-rich fraction from pepper (Capsicum annuum L.) seeds on high-fat diet-induced obesity in mice, Phytotherapy research: PTR 30 (2016) 1848–1855.

[124]   HK Kim, J Jeong, EY Kang, GW Go, Red pepper (Capsicum annuum L.) seed extract improves glycemic control by inhibiting hepatic gluconeogenesis via phosphorylation of FOXO1 and AMPK in obese diabetic db/db mice, Nutrients (2020) 12.

[125]   M Mashmoul, A Azlan, BNM Yusof, H Khaza'ai, N Mohtarrudin, MT Boroushaki, Effects of saffron extract and crocin on anthropometrical, nutritional and lipid profile parameters of rats fed a high fat diet, J Func Food 8 (2014) 180–187.

[126]   S Samarghandian, M Azimi-Nezhad, F Samini, Ameliorative effect of saffron aqueous extract on hyperglycemia, hyperlipidemia, and oxidative stress on diabetic encephalopathy in streptozotocin induced experimental diabetes mellitus, BioMed research international 2014 (2014) 920857.

[127]   A Maeda, K Kai, M Ishii, T Ishii, M Akagawa, Safranal, a novel protein tyrosine phosphatase 1B inhibitor, activates insulin signaling in C2C12 myotubes and improves glucose tolerance in diabetic KK-Ay mice, Molecular nutrition & food research 58 (2014) 1177–1189.

[128]   P Ranadheer Chowdary DP, A Niventhi, M Vijey Aanandhi, Antidiabetic and antihyperlipidemic activities of aqueous and ethanolic leaf extracts of Feronia limonia, Drug Invention Today 10 (2018) 151–153.

[129]   N Hingwasia, S Khare, BK Dubey, A Joshi, S Dhakad, A Jain, Evaluation on antidiabetic activity of hydroalcoholic extract of bark of Feronia limonia, Asian J Pharm Pharmacol 4 (2018) 168–172.

[130]   M Majeed, S Majeed, K Nagabhushanam, L Lawrence, L Mundkur, Garcinia indica extract standardized for 20% Garcinol reduces adipogenesis and high fat diet-induced obesity in mice by alleviating endoplasmic reticulum stress, J Func Food (2020) 67.

[131]   Y-AS Y.-C.T., Nagabhushanam K., Ho C.-T., An-Chin Cheng, Min-Hsiung Pan. Coleus forskohlii and Garcinia indica extracts attenuated lipid accumulation by regulating energy metabolism and modulating gut microbiota in obese mice. Food research international (Ottawa, Ont). 142:110143.

[132]   K Munjal, S Ahmad, A Gupta, A Haye, S Amin, S Mir, Polyphenol-enriched fraction and the compounds isolated from Garcinia indica fruits ameliorate obesity through suppression of digestive enzymes and oxidative stress, Pharmacog Mag 16 (2020) 236–245.

[133]   H Kirana, B Srinivasan, Aqueous extract of garcinia indica choisy restores glutathione in type 2 diabetic rats, Journal of young pharmacists: JYP 2 (2010) 265–268.

[134]   CH Peng, CC Chyau, KC Chan, TH Chan, CJ Wang, CN Huang, Hibiscus sabdariffa polyphenolic extract inhibits hyperglycemia, hyperlipidemia, and glycation-oxidative stress while improving insulin resistance, Journal of agricultural and food chemistry 59 (2011) 9901–9909.

[135]   S Zainalabidin, SB Budin, NNM Anuar, NA Yusoff, NLM Yusof, Hibiscus sabdariffa Linn. improves the aortic damage in diabetic rats by acting as antioxidant, J App Pharm Sci 8 (2018) 108–114.

[136]   TW Huang, CL Chang, ES Kao, JH Lin, Effect of Hibiscus sabdariffa extract on high fat diet-induced obesity and liver damage in hamsters, Food & nutrition research 59 (2015) 29018.

[137]   P Diez-Echave, T Vezza, A Rodríguez-Nogales, AJ Ruiz-Malagón, L Hidalgo-García, J Garrido-Mesa, et al., The prebiotic properties of Hibiscus sabdariffa extract contribute to the beneficial effects in diet-induced obesity in mice, Food research international (Ottawa, Ont). 127 (2020) 108722.

[138]   MG Abubakar, AN Ukwuani, UU Mande, F Science, Antihypertensive activity of Hibiscus aqueous calyx extract in Albino rats Sabdariffa, Sky J Biochem Res 4 (2015) 16–20.

[139]   M Ghosian Moghadam, I Ansari, M Roghani, A Ghanem, N Mehdizade, The effect of oral administration of Hypericum perforatum on serum glucose and lipids, hepatic enzymes and lipid peroxidation in streptozotocin-induced diabetic rats, Gal Med J 6 (2017) 319–329.

[140]   GM Husain, SS Chatterjee, PN Singh, V Kumar, Hypolipidemic and antiobesity-like activity of standardised extract of Hypericum perforatum L. in rats, ISRN pharmacology 2011 (2011) 505247.

[141]   D Hernández-Saavedra, IF Pérez-Ramírez, M Ramos-Gómez, S Mendoza-Díaz, G Loarca-Piña, et al., Phytochemical characterization and effect of Calendula officinalis, Hypericum perforatum, and Salvia officinalis infusions on obesity-associated cardiovascular risk, Med Chem Res 25 (2016) 163–172.

[142]   S Arokiyaraj, R Balamurugan, P Augustian, Antihyperglycemic effect of Hypericum perforatum ethyl acetate extract on streptozotocin-induced diabetic rats, Asian Pacific journal of tropical biomedicine 1 (2011) 386–390.

[143]   S Asgary, A Solhpour, S Parkhideh, H Madani, P Mahzouni, N Kabiri, Effect of hydroalcoholic extract of Hypericum perforatum on selected traditional and novel biochemical factors of cardiovascular diseases and atherosclerotic lesions in hypercholesterolemic rabbits: a comparison between the extract and lovastatin, Journal of pharmacy & bioallied sciences 4 (2012) 212–218.

[144]   YR Kang, HY Lee, JH Kim, DI Moon, MY Seo, SH Park, et al., Anti-obesity and anti-diabetic effects of Yerba Mate (Ilex paraguariensis) in C57BL/6J mice fed a high-fat diet, Laboratory animal research 28 (2012) 23–29.

[145]   DP Arçari, W Bartchewsky Jr., TW dos Santos, KA Oliveira, CC DeOliveira, ÉM Gotardo, et al., Anti-inflammatory effects of yerba maté extract (Ilex paraguariensis) ameliorate insulin resistance in mice with high fat diet-induced obesity, Molecular and cellular endocrinology 335 (2011) 110–115.

[146]   G Gosmann, AG Barlette, T Dhamer, DP Arçari, JC Santos, ER de Camargo, et al., Phenolic compounds from maté (Ilex paraguariensis) inhibit adipogenesis in 3T3-L1 preadipocytes, Plant foods for human nutrition (Dordrecht, Netherlands) 67 (2012) 156–161.

[147]   A Kaeidi, S Esmaeili-Mahani, V Sheibani, M Abbasnejad, B Rasoulian, Z Hajializadeh, et al., Olea europaea L.) leaf extract attenuates early diabetic neuropathic pain through prevention of high glucose-induced apoptosis: in vitro and in vivo studies, Journal of ethnopharmacology 136 (2011) 188–196.

[148] T Vezza, A Rodríguez-Nogales, F Algieri, J Garrido-Mesa, M Romero, M Sánchez, et al., The metabolic and vascular protective effects of olive (Olea europaea L.) leaf extract in diet-induced obesity in mice are related to the amelioration of gut microbiota dysbiosis and to its immunomodulatory properties, Pharmacological research 150 (2019) 104487.

[149] CG Guex, FZ Reginato, PR de Jesus, JC Brondani, GHH Lopes, LF Bauermann, Antidiabetic effects of Olea europaea L. leaves in diabetic rats induced by high-fat diet and low-dose streptozotocin, Journal of ethnopharmacology 235 (2019) 1–7.

[150] AM Al-Attar, FA Alsalmi, Effect of Olea europaea leaves extract on streptozotocin induced diabetes in male albino rats, Saudi journal of biological sciences 26 (2019) 118–128.

[151] CM Steliana Ghibu, O Vostinaru, N Olah, C Mogosan, Adriana Muresan Diuretic, antihypertensive and antioxidant effect of olea europaea leaves extract, in rats, Archives of Cardiovascular Diseases Supplements 7 (2015) 184.

[152] P Monika, A Geetha, The modulating effect of Persea americana fruit extract on the level of expression of fatty acid synthase complex, lipoprotein lipase, fibroblast growth factor-21 and leptin–A biochemical study in rats subjected to experimental hyperlipidemia and obesity, Phytomedicine: international journal of phytotherapy and phytopharmacology 22 (2015) 939–945.

[153] M Padmanabhan, G Arumugam, Effect of Persea americana (avocado) fruit extract on the level of expression of adiponectin and PPAR-γ in rats subjected to experimental hyperlipidemia and obesity, Journal of complementary & integrative medicine 11 (2014) 107–119.

[154] CR Lima, CF Vasconcelos, JH Costa-Silva, CA Maranhão, J Costa, TM Batista, et al., Anti-diabetic activity of extract from Persea americana Mill. leaf via the activation of protein kinase B (PKB/Akt) in streptozotocin-induced diabetic rats, Journal of ethnopharmacology 141 (2012) 517–525.

[155] A Sokpe, MLK Mensah, GA Koffuor, KP Thomford, R Arthur, Y Jibira, et al., Hypotensive and antihypertensive properties and safety for use of Annona muricata and Persea americana and their combination products, Evidence-based complementary and alternative medicine: eCAM 2020 (2020) 8833828.

[156] MF Almuaigel, MA Seif, HW Albuali, O Alharbi, A Alhawash, Hypoglycemic and hypolipidemic effects of aqueous extract of phaseolus vulgaris pods in streptozotocin-diabetic rats, Biomedicine & pharmacotherapy = Biomedecine & pharmacotherapie 94 (2017) 742–746.

[157] T Jawaid, M Kamal, S Kumar, Antihypertensive effect of the alcoholic extract of seeds of Phaseolus vulgaris L. (Fabaceae) on high salt diet induced hypertension in male rats. Int J Pharm Sci Res 8 (2107) 3092–3097.

[158] B Stefanon, E Pomari, M Colitti, Effects of Rosmarinus officinalis extract on human primary omental preadipocytes and adipocytes, Experimental biology and medicine (Maywood, NJ) 240 (2015) 884–895.

[159] A Ibarra, J Cases, M Roller, A Chiralt-Boix, A Coussaert, C Ripoll, Carnosic acid-rich rosemary (Rosmarinus officinalis L.) leaf extract limits weight gain and improves cholesterol levels and glycaemia in mice on a high-fat diet, The British journal of nutrition 106 (2011) 1182–1189.

[160] E MA, Comparative evaluation of antidiabetic activity of Rosmarinus officinalis L. and Chamomile recutita in streptozotocin induced diabetic rats, Agric Biol J North Am 3 (2012) 247–252.

[161] A S, Effect of Rosmarinus Officinalis extract on some cardiac enzymes of streptozotocin-induced diabetic rats, J Health Sci 2 (2012) 33–37.

[162] GF da Costa, IB Santos, GF de Bem, VSC Cordeiro, CA da Costa, L de Carvalho, et al., The beneficial effect of anthocyanidin-rich Vitis vinifera L. grape skin extract on metabolic changes induced by high-fat diet in mice involves antiinflammatory and antioxidant actions, Phytotherapy research: PTR 31 (2017) 1621–1632.

[163] M Ahmed, SV Hegde, A Chavan, Evaluation of hepatoprotective activity of Vitis vinifera stem bark, J Pharm Res 5 (2012) 5228–5230.

[164] S Devi, R Singh, Evaluation of antioxidant and anti-hypercholesterolemic potential of Vitis vinifera leaves, Food Sci Hum Well 6 (2017) 131–136.

[165] GHS Bomfim, DC Musial, R Miranda-Ferreira, SR Nascimento, A Jurkiewicz, NH Jurkiewicz, et al., Antihypertensive effects of the Vitis vinifera grape skin (ACH09) extract consumption elicited by functional

improvement of P1 (A 1) and P2 (P2X1) purinergic receptors in diabetic and hypertensive rats, Pharm Nut 8 (2019) 100146.

[166] TF Luciano, CT De Souza, RA Pinho, SO Marques, GP Luiz, NDS Tramontin, et al., Effects of Zingiber officinale extract supplementation on metabolic and genotoxic parameters in diet-induced obesity in mice, The British journal of nutrition 126 (7) (2020) 1–12.

[167] MP Rani, MS Krishna, KP Padmakumari, KG Raghu, A Sundaresan, Zingiber officinale extract exhibits antidiabetic potential via modulating glucose uptake, protein glycation and inhibiting adipocyte differentiation: an in vitro study, Journal of the science of food and agriculture 92 (2012) 1948–1955.

[168] AJ Akinyemi, AO Ademiluyi, G Oboh, Inhibition of angiotensin-1-converting enzyme activity by two varieties of ginger (Zingiber officinale) in rats fed a high cholesterol diet, Journal of medicinal food 17 (2014) 317–323.

[169] M Payab, S Hasani-Ranjbar, N Shahbal, M Qorbani, A Aletaha, H Haghi-Aminjan, et al., Effect of the herbal medicines in obesity and metabolic syndrome: A systematic review and meta-analysis of clinical trials, Phytotherapy research: PTR 34 (2020) 526–545.

[170] HLS Lee, JH Park, KS Lee, Y Jang, HY Park, Berberine-induced LDLR up-regulation involves JNK pathway, Biochemical and biophysical research communications 362 (2007) 853–857.

[171] LL John Zeqi Luo, American ginseng stimulates insulin production and prevents apoptosis through regulation of uncoupling protein-2 in cultured beta cells, Evidence-based complementary and alternative medicine: eCAM 3 (2006) 365–372.

[172] ZZ XL, W Lv, Y Yang, H Gao, J Yang, Y Shen, et al., Ginsenoside Re reduces insulin resistance through inhibition of c-Jun NH2-terminal kinase and nuclear factor-kappaB, Mol Endocrinol 22 (2008) 186–195.

[173] Organization WH. WHO global report on traditional and complementary medicine. 2019;2021.

[174] N Gilbert, Regulations: herbal medicine rule book, Nature 480 (2011) S98–S99.

[175] M Heinrich, S Anagnostou, From pharmacognosia to DNA-based medicinal plant authentication - pharmacognosy through the centuries, Planta medica 83 (2017) 1110–1116.

[176] M Heinrich, Quality and safety of herbal medical products: regulation and the need for quality assurance along the value chains, British journal of clinical pharmacology 80 (2015) 62–66.

[177] M Heinrich, Ethnopharmacology in the 21st century - grand challenges, Frontiers in pharmacology 1 (2010) 8.

[178] IA Khan, T Smillie, Implementing a "quality by design" approach to assure the safety and integrity of botanical dietary supplements, Journal of natural products 75 (2012) 1665–1673.

[179] A Zhang, H Sun, G Yan, X Wang, Recent developments and emerging trends of mass spectrometry for herbal ingredients analysis, TrAC Trends Anal Chem 94 (2017) 70–76.

[180] N Maharlouei, R Tabrizi, KB Lankarani, A Rezaianzadeh, M Akbari, F Kolahdooz, et al., The effects of ginger intake on weight loss and metabolic profiles among overweight and obese subjects: a systematic review and meta-analysis of randomized controlled trials, Critical reviews in food science and nutrition 59 (2019) 1753–1766.

[181] KJ Astell, ML Mathai, AJ McAinch, CG Stathis, XQ Su, A pilot study investigating the effect of Caralluma fimbriata extract on the risk factors of metabolic syndrome in overweight and obese subjects: a randomised controlled clinical trial, Complementary therapies in medicine 21 (2013) 180–189.

[182] N Kassaian, L Azadbakht, B Forghani, M Amini, Effect of fenugreek seeds on blood glucose and lipid profiles in type 2 diabetic patients, International journal for vitamin and nutrition research Internationale Zeitschrift fur Vitamin- und Ernahrungsforschung Journal international de vitaminologie et de nutrition 79 (2009) 34–39.

[183] IA Sobenin, LV Nedosugova, LV Filatova, MI Balabolkin, TV Gorchakova, AN Orekhov, Metabolic effects of time-released garlic powder tablets in type 2 diabetes mellitus: the results of double-blinded placebo-controlled study, Acta diabetologica 45 (2008) 1–6.

[184] R Ryu, TS Jeong, YJ Kim, JY Choi, SJ Cho, EY Kwon, et al., Beneficial effects of pterocarpan-high soybean leaf extract on metabolic syndrome in overweight and obese Korean subjects: randomized controlled trial, Nutrients 8 (2016).

[185] AM Patti, K Al-Rasadi, N Katsiki, Y Banerjee, D Nikolic, L Vanella, et al., Effect of a natural supplement containing curcuma longa, guggul, and chlorogenic acid in patients with metabolic syndrome, Angiology 66 (2015) 856–861.

[186] M Saberi-Karimian, M Keshvari, M Ghayour-Mobarhan, L Salehizadeh, S Rahmani, B Behnam, et al., Effects of curcuminoids on inflammatory status in patients with non-alcoholic fatty liver disease: a randomized controlled trial, . Complementary therapies in medicine 49 (2020) 102322.

[187] OJ Phung, SS Makanji, CM White, CI Coleman, Almonds have a neutral effect on serum lipid profiles: a meta-analysis of randomized trials, Journal of the American Dietetic Association 109 (2009) 865–873.

[188] ASA Mohammadi 1, M Iranshahi, M Amini, R Khojasteh, M Ghayour-Mobarhan, GA Ferns, Effects of supplementation with curcuminoids on dyslipidemia in obese patients: a randomized crossover trial, Phytotherapy research: PTR 27 (2013) 374–379.

[189] SM Blevins, MJ Leyva, J Brown, J Wright, RH Scofield, CE Aston, Effect of cinnamon on glucose and lipid levels in non insulin-dependent type 2 diabetes, Diabetes care 30 (2007) 2236–2237.

[190] JP Cosyns, M Jadoul, JP Squifflet, W FX, C van Ypersele de Strihou, Urothelial lesions in Chinese-herb nephropathy, American journal of kidney diseases: the official journal of the National Kidney Foundation 33 (1999) 1011–1017.

[191] CC Falzon, A Balabanova, Phytotherapy: an introduction to herbal medicine, Primary care 44 (2017) 217–227.

[192] VJ Navarro, H Barnhart, HL Bonkovsky, T Davern, RJ Fontana, L Grant, et al., Liver injury from herbals and dietary supplements in the U.S. drug-induced liver injury network, Hepatology (Baltimore, Md) 60 (2014) 1399–1408.

[193] F Borrelli, AA Izzo, Herb-drug interactions with St John's wort (Hypericum perforatum): an update on clinical observations, The AAPS journal 11 (2009) 710–727.

[194] JC Gorski, SM Huang, A Pinto, MA Hamman, JK Hilligoss, NA Zaheer, et al., The effect of echinacea (Echinacea purpurea root) on cytochrome P450 activity in vivo, Clinical pharmacology and therapeutics 75 (2004) 89–100.

[195] HH Chow, IA Hakim, DR Vining, JA Crowell, CA Cordova, WM Chew, et al., Effects of repeated green tea catechin administration on human cytochrome P450 activity, Cancer epidemiology, biomarkers & prevention: a publication of the American Association for Cancer Research, cosponsored by the American Society of Preventive Oncology 15 (2006) 2473–2476.

[196] X Wu, Q Li, H Xin, A Yu, M Zhong, Effects of berberine on the blood concentration of cyclosporin A in renal transplanted recipients: clinical and pharmacokinetic study, European journal of clinical pharmacology 61 (2005) 567–572.

[197] JL Vanherweghem, JP Degaute, The policy of admission to the education in medicine and dentistry in the French-speaking community of Belgium, Acta clinica Belgica 53 (1998) 2–3.

[198] C Frenzel, R Teschke, Herbal hepatotoxicity: clinical characteristics and listing compilation, International journal of molecular sciences 17 (5) (2016) 17.

[199] IM Rietjens, MJ Martena, MG Boersma, W Spiegelenberg, GM Alink, Molecular mechanisms of toxicity of important food-borne phytotoxins, Molecular nutrition & food research 49 (2005) 131–158.

[200] TAOC John, M Ekor, Some phytochemical, safety, antimicrobial and hameatological studies of superb and seven keys to power blood purifier herbal tonics, Nig J Med Res (1997) 37–43.

[201] KS Lee, SM Yee, STR Zaidi, RP Patel, Q Yang, YM Al-Worafi, et al., Combating Sale of counterfeit and falsified medicines online: a losing battle, Frontiers in pharmacology 8 (2017) 268.

[202] MI Avigan, RP Mozersky, LB Seeff, Scientific and regulatory perspectives in herbal and dietary supplement associated hepatotoxicity in the United States, International journal of molecular sciences 17 (2016) 331.

[203] V Manzanilla, A Kool, L Nguyen Nhat, H Nong Van, H Le Thi Thu, HJ de Boer, Phylogenomics and barcoding of Panax: toward the identification of ginseng species, BMC evolutionary biology 18 (2018) 44.

[204] CM Gilroy, JF Steiner, T Byers, H Shapiro, W Georgian, Echinacea and truth in labeling, Archives of internal medicine 163 (2003) 699–704.

[205]   AH Alostad, DT Steinke, EI Schafheutle, Herbal medicine classification: policy recommendations, Frontiers in medicine 7 (2020) 31.

[206]   S Bent, Herbal medicine in the United States: review of efficacy, safety, and regulation: grand rounds at University of California, San Francisco Medical Center, Journal of general internal medicine 23 (2008) 854–859.

[207]   SH Ganie, Z Ali, S Das, PS Srivastava, MP Sharma, Genetic diversity and chemical profiling of different populations of Convolvulus pluricaulis (convolvulaceae): an important herb of ayurvedic medicine, 3 Biotech 5 (2015) 295–302.

[208]   YP Lin, YL Lee, CY Hung, C CF, Y Chen, Detection of adulterated drugs in traditional Chinese medicine and dietary supplements using hydrogen as a carrier gas, PLoS One 13 (2018) e0205371.

[209]   A Watanabe, C Watanabe, RR Freeman, N Teramae, H Ohtani, Hydrogenation reactions during pyrolysis-gas chromatography/mass spectrometry analysis of polymer samples using hydrogen carrier gas, Analytical chemistry 88 (2016) 5462–5468.

[210]   BC Moon, WJ Kim, I Park, GH Sung, P Noh, Establishment of a PCR assay for the detection and discrimination of authentic cordyceps and adulterant species in food and herbal medicines, Molecules (Basel, Switzerland) 23 (8) (2018) 23.

[211]   I Gonzalez-Suarez, A Sewer, P Walker, C Mathis, S Ellis, H Woodhouse, et al., Systems biology approach for evaluating the biological impact of environmental toxicants in vitro, Chemical research in toxicology 27 (2014) 367–376.

[212]   IMF Gonzalez-Suarez, J Hoenig, MC Peitsch, Mechanistic network models in safety and toxicity evaluation of nutraceuticals, in: RC G (Ed.), Nutraceuticals: Efficacy, Safety and Toxicity editor, Academic Press, Amsterdam, 2016, pp. 287–304.

CHAPTER 7

Pharmacological importance of the active molecule "guggulsterone" in overall human health

Meenakshi Gupta, Vanshika Yadav and Maryam Sarwat

Amity Institute of Pharmacy, Amity University, Noida, Uttar Pradesh, India

7.1 Introduction

Guggulsterone is the active component of guggulipid. Guggulipid is derived from the gum resin (Guggulu) of the tree *Commiphora mukul. Commiphora* species are flowering shrub or small trees which belong to the family Burseraceae [1]. These plants are found in the regions of Gujarat, Karnataka, Rajasthan, and Assam [1].

Guggulsterone is an analog of progesterone and exists in two forms, viz. E-isomer and Z-isomer (Fig. 7.1). It has been traditionally used as a highly valued and safe ayurvedic medicine. It is used in the treatment of various ailments like rheumatoid arthritis, inflammatory bowel disease, and bone resorption. The isomers have been demonstrated to exhibit their biological activities by binding to nuclear receptors [2]. Guggulsterone also shows its hypolipidemic activity and cardio-protective effects with collateral benefits against inflammation and obesity [3]. It exhibits anti-diabetic, anti-obesity, nodulocystic (acne), neuro-protective and anti-cancer activities (Fig. 7.2, Table 7.1) [1].

Further, guggulsterone is earning high attention in cancer treatment due to its known pharmacological effects which target multiple pathways and molecules [4]. It is also becoming a popular choice as a health supplement to manage weight loss [4]. It has also been therapeutically beneficial for diseases related to oxidative stresses such as myocardial ischemia and neurodegenerative diseases [5].

It has been reported that guggulsterone regulates the gene expression by exhibiting control over other molecular targets including transcription factors such as nuclear factor (NF-KB), signal transducer, the activator of transcription (STAT), steroid receptors [1]. Also, the Z-guggulsterone induces apoptosis in various cancer cell types such as prostate cancer, acute

Table 7.1: Medicinal properties of guggulsterone in curing various ailments and its mechanism of action.

S. No.	Condition	Type of study	Type of animals/cell line used	Mechanism of action	Administered dose	Observations	Reference
1	Hypolipidemia	In vivo	Albino rats	Inhibition of FXR activation Inhibition of dopamine beta-hydroxylase activity	0.5 gm daily for 12 weeks	Lowers serum cholesterol, phospholipid levels, and triglycerides levels	[9]
				Inhibition of chenodeoxycholic acid-activated nuclear farnesoid X-receptor	3 gm thrice a daily	Decreases in the lipid levels	[11]
				Inhibition of FXR receptor	Guggul ether extract (0.5 gm daily for 3 weeks)	Decreases the cholesterol levels and lipid levels also	[11]
2	Cardioprotective	In vivo	Albino rats	Increased level of phospholipase and cytosolic lipid peroxidase and reduction of cardiac glycogen	50mg/kg of body weight	Reversal in the isoproterenol-induced myocardial damage	[14]
3	Anti-inflammatory	In vivo	Lewis rats and rabbit	Inhibition of the LPS-induced upregulation of TNF and Cox-2; Attenuation of kappa-B binding activity of nuclear P50 and P65 levels	30 mg/kg of body weight	Reduction in several inflating cells, total protein, and inflammatory markers	[17]
4	Pancreatic cancer	In vitro	CD18/HPAF PC cells	Reduces the motility and suppress the invasion which leads to inhibition of FAK, disruption of cytoskeletal organization, and Src kinase signaling	Combined treatment of guggulsterone (20–30 µg/ml) with gemcitabine (10 µg/ml)	Inhibition of growth and apoptosis by downregulating the NFK-B activity	[28]
5	Head and neck cancer	In vivo	Albino rats	Induces apoptosis and cell cycle arrest	0.01–100 µM of guggulsterone with erlotinib (5 µM), cetuximab (4 µg/ml), and cisplatin (20 µM)	Reduction in apoptotic protein is noted	[31]
				Induces apoptosis and cell cycle arrest; Abrogates the effects of smokeless tobacco/ nicotine via PBk AKT pathway	Combined treatment of guggulsterone (50 mg/kg body weight)	Reduces the expression of anti-apoptotic proteins, Bcl-2, survivin, c-Myc, and cyclin-D1	[32,33]

(continued on next page)

#						
6	Breast cancer	In vitro	U937 cell, murine 4T1 cells, MDA-MB-231 cells.	Activation of NF-KB pathway; prevention of the MMP-9 expression and MAPK/AP-1 signaling pathway	Combined treatment of guggulsterone and bexarotene(10 µM) with doxorubicin (5 µM)	Inhibits tumor cells proliferation by inhibiting DNA synthesis, induces S-phase arrest and promotes apoptosis. Reduced cellular levels of BCRP by inducing its association and secretion with exosomes [36]
7	Lung cancer	In vitro	H1299 cells, A549 cells, and murine B-cell lines	NF-KB activation	50 µg	Inhibition in growth and apoptosis [37]
8	Prostate cancer	In vitro	PC-3 and LNCaP cells	ATP-lyase-regulated Akt inactivation; inactivation of AKT regulation ATP citrate lyase signalling, caspase-dependent apoptosis	20–30 µg	Retarded prostate cancer growth via inactivation of AKT regulation by ATP citrate lyase signalling in prostate cancer cells [39]
9	Thyroid	In vivo	Albino rats	Increase in iodine uptake by the thyroid gland; Increased level of thyroid peroxidase and protease; Increases in the oxygen consumption	1 mg/100 gm of body weight	Decrease in lipid peroxidation [44,45]

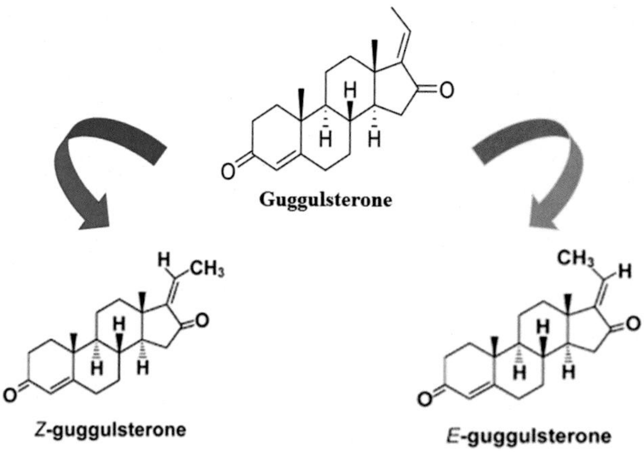

Figure 7.1
Structure of active compound guggulsterone with its isomers namely, Z and E guggulsterone.

myelocytic leukemia, colon cancer, oesophageal cancer, and gastric cancer [1]. It has been also used as a farnesoid X receptors (FXR) antagonist to treat obesity, lipid metabolism dysfunction, hypothyroidism, inflammation, and arthritis (Fig. 7.3) [6–8]. Differences in study design, methodological quality, statistical analysis, sample size, and subject population result in certain inconsistencies in the response to therapy, the data from *in vitro*, preclinical and clinical studies describe the therapeutic effects of guggulsterone. This chapter aims to examine in detail the properties of the compound so far reported in the literature from a pharmacological point of view.

7.2 Therapeutic effects of guggulsterone

For a very long time, guggul and its active constituents (like guggulsterone) are known for their cardioprotective, hypolipidemic, renal protective, gastroprotective, antioxidant, anti-inflammatory, and anticancer properties at different doses (Fig. 7.6).

7.2.1 Hypolipidemic activity

The hypolipidemic activity of guggulsterone has been well established through various *in vivo* and clinical studies. Several animal studies prove that the administration of guggulsterone reduces hyperlipidemia when animals are fed a high-fat diet. Treatment with guggulsterone decreases the liver cholesterol in mice fed with a cholesterol diet but is found ineffective in FXR null mice, providing shreds of evidence that guggulsterone acts on the principle of the inhibition of the FXR activation [9]. Rats fed with a high-fat diet show increased serum glucose, cholesterol, and triglycerides; treatment with guggulsterone (0.5 gm daily for 12 weeks) helps in reducing these levels. It lowers the serum cholesterol, phospholipid levels,

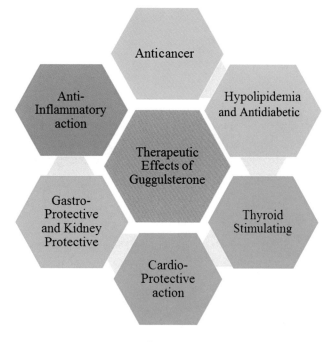

Figure 7.2
Therapeutic effects of guggulsterone.

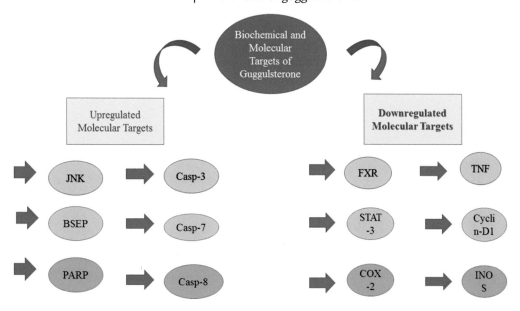

Figure 7.3
Biochemical and molecular targets of guggulsterone (upregulated and downregulated targets) which involves enzyme targets, protein kinase targets, transcription targets, and activation/ cleavage targets.

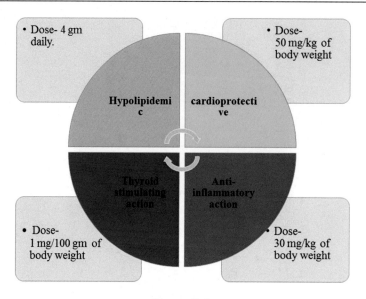

Figure 7.4
Doses of guggulsterone needed/required to correct the various diseases.

and triglycerides level and enhances glucose tolerance, fasting blood glucose, plasma insulin level, low-density proteins, very low-density lipoprotein, cholesterol, triglycerides, and expression of various genes involved in lipid metabolism [10]. Reduction in pancreatic beta-cell size, increase in adipocytes, and steatosis of the liver are also observed when rats are fed with high fat diet-fed; guggulsterone attenuates these effects. It also suppresses insulin secretion by mouse beta-cells via FXR activation and KATP channel inhibition by bile acids [10].

Out of the two isomers, Z guggulsterone in clinical trials proved to be the antagonists of the bile acid receptor (BAR; member of the intracellular receptor superfamily). It antagonizes the activity of BAR *in vitro* and cell structure systems on promoter genes and endogenous target genes as well. It mediates its effect through BAR, which in turn inhibits the FXR activation. The FXR agonist induces the transcription of the bile salt export pump (BSEP) which is a major hepatic bile acid transporter. Guggulsterone enhances the endogenous BSEP expression with a maximum induction by an FXR agonist alone. Therefore, inhibition of FXR activation is considered a mechanism for the hypolipidemic effects (Fig. 7.4) [9]. Another important mechanism for its hypolipidemic effect is the inhibition of dopamine beta-hydroxylase in the brain with some marked alteration in the heart by catecholamines [11]. Guggulsterone also shows its effect in high-fat diabetes-induced rats with improved PPAR gamma expression under *in vitro* and *in vivo* conditions [10]. Another animal study was conducted in rabbits where they were fed with hydrogenated vegetable oil to raise their cholesterol levels. Group of rabbits treated with guggul had normal serum cholesterol and lipid levels as compared to control, where the serum and lipids levels were found elevated [12].

A randomized controlled clinical trial conducted on 120 patients showed that treatment with guggul (2 gm twice daily) significantly reduce the serum lipid levels in non-obese patients [10].

7.2.2 Antidiabetic

Several studies revealed the anti-diabetic potential of guggulsterone. It shows the reduction in pancreatic beta-cell size and increase in adipocytes in high-fat-diet-induced diabetes in rats. It also inhibits the pre-adipocyte differentiation and has both hypo-glycaemic and hypo-lipidemic effects, that can help cure type 2 diabetes [10]. It also suppresses insulin secretion via FXR- activation and KATP channel inhibition stimulated by bile acids [13].

The administration of alcoholic extract of guggul at a dose of 200mg/kg for 60 continuous days reduces plasma glucose levels in streptozotocin-induced diabetic rats. This study shows that guggulsterone has a hypoglycaemic activity that can help cure type II disease. Guggulsterone prevents the iFN-gamma cells damage and reduces the iNOS and PGE2 production. It also prevents STAT activation, downregulates suppressors of cytokine signaling-3, and impairs secretion of insulin [1].

7.2.3 Antioxidant

Both guggulipid and guggulsterone act as potent anti-oxidant. Guggulipid is known for its dose-dependent decrease in the accumulation of LDL-derived cholesterol esters in mouse macrophages [10]. Guggulsterone isomers Z and E forms exhibit potent inhibitory activity against the production of nitric oxide-induced by bacterial lipopolysaccharides in macrophages with IC_{50} values of 1.1 and 3.3μM respectively [1]. It is generally accepted that overproduction of nitric oxide is associated with oxidative stress, which is involved in the pathogenesis of cardiovascular diseases or chronic inflammation, diabetes, rheumatoid arthritis, neurodegenerative diseases [4]. These findings indicate that guggulsterone is of therapeutic benefit in diseases associated with oxidative stress, such as myocardial ischemia and neurodegenerative diseases [4]. Guggulsterone reverses the myocardial damage and the induced metabolic changes in rats challenged with isoproterenol. This damage includes a marked increase in creatinine phosphokinase, phospholipase, and xanthine oxidase activities, enhanced levels of lipid peroxidases, and lowering of superoxide dismutase [10].

7.2.4 Cardioprotective benefits

Guggulsterone is shown to be an effective cardio-protective agent. Isoproterenol insult causes an increase in serum creatinine phosphokinase and glutamate pyruvate transaminase. Additionally, an increase in phospholipase, xanthine oxidase, and lipid peroxidases, and depletion of glycogen, phospholipids, and cholesterol are also observed. Guggulsterone reverses all the metabolic changes induced by isoproterenol hydrochloride (85mg/kg) in rats and showed a protective effect [14]. Treatment with guggulsterone at a dose of 50 mg/kg significantly protects cardiac damage as assessed by the reversal of blood and heart biochemical parameters in ischemic rats [14]. Guggulsterone also shows a reduction in post ischemic myocardial apoptosis and also the SiRNA- mediated silencing of endogenous FXR [15]. Accumulation of LDL is a major cause of the formation of atherosclerotic plaque; treatment with guggulsterone inhibits LDL oxidation and provides protection [16].

7.2.5 Anti-inflammatory

Guggulsterone is the only known antagonist of the Farnesoid X receptor, which is a bile acid receptor [17]. Different bile acids chenodeoxycholic acid acts as a natural ligand for FXR, whose expression is elevated in the liver and intestine [9]. When FXR binds to its ligand, it gets activated and reaches the cell nucleus, where it forms a heterodimer. This heterodimer binds to the hormone response elements on DNA and regulates various genes also which in turn inhibits the transcription of the CYP7A1 gene. This has been confirmed that guggulsterone has proinflammatory signals, together with transcription factor NF-kB [18]. The guggulsterone is a potent anti-inflammatory agent which inhibits LPS induced upregulation of tumor necrosis factor-alpha and cyclooxygenase-2. The guggulsterone suppressed the activation of NF-kB and exhibited the antiproliferative activity by inhibiting c-Myc and cyclin D1; pretreatment with it suppresses cyclooxygenase -2 protein production as well [19]. It strongly inhibits LPS or IL-1 beta-induced intracellular adhesion molecule 1 gene expression [19]. Guggulsterone also abolishes the increase of kappa-B binding activity of nuclear P50 and P65 levels along with kappa-B alpha depletion in cells. A study conducted on Lewis rats shows that endotoxin-induced uveitis in animals can be treated with guggulsterone (30 mg/kg). It reduces the number of inflating cells, total protein, and inflammatory markers and also prevents the expression of coX-2 proteins, NF-kappa B in the eye tissues of rats [20].

7.2.6 Arthritis

Guggulsterone has higher anti-inflammatory activities with the least side effects so it can be used in the treatment of diseases like osteoarthritis, rheumatoid arthritis, and gout. Ethanolic extract of guggulsterone shows cyclooxygenase enzyme inhibitory activities and also inhibits the PGE synthesis [17]. The Triphala guggul formulation is proved to be a potent inhibitor of hyaluronidase (which contributes significantly to cartilage degradation) and is involved in the treatment of arthritis and rheumatism [21]. In a clinical trial, it is found that guggulsterone capsules, administered to patients with osteoarthritis (OA), led to significant improvement without side effects [22]. The Triphala guggul is a combination of three powdered fruits, namely *amalaka* or T1 (Phyllanthus emblica), and *haritaki* or T2 (Terminalia chebula), and *bibhitaki* or T3 (Terminalia belerica) [23]. Two assays were done in which we tested the Triphala on the activity of hyaluronidase, an enzyme that destroys the hyaluronic acid, the backbone of the cartilage matrix [24]. Secondly, we tested the effects of Triphala on the gelatinase activity of collagenase type 2, as matrix metalloproteinase (MMPs)-1, -3, and -8 express collagenase activity which degrades cartilage matrix [25]. In mono-iodoacetate-induced osteoarthritis in rats, it acts as a chondroprotective with less damage to proteoglycan in the epiphyseal plate in 2 weeks of treatment. Guggulsterone (45 mg/kg) reduces pain and stiffness, relieves symptoms of osteoarthritis after one month of treatment, crucially improving the ill effects of the disease [26]. In a study where rabbits were challenged with arthritis, treatment with guggulsterone fraction led to a decrease in joint swelling and reduces the thickness of the joint [18].

7.2.7 Gastroprotective

Pre-treatment with guggulsterone is known to suppress the TNF- alpha-induced activation of IKB kinase and NF-kB signaling in MKN-45 cells [27]. Oral administration of guggulsterone at doses 50, 100 and 200 mg/kg in acetic acid-induced ulcer model reduce gastric lesions by 38%, 43%, and 55%, respectively when compared with vehicle controls. This is probably because of its antioxidant and anti-inflammatory properties through the downregulation of NF-Kb expression [27].

7.2.8 Anticancer effects

Guggulsterone shows its effects in various cancers like breast cancer, head and neck cancer, prostate cancer, lung cancer, leukemia, melanoma, gall bladder cancer, esophageal cancer, colon cancer, and brain tumors as well [14]. The use of guggulsterone in different types of cancer exemplifies its anti-proliferative, anti-metastatic, anti-apoptotic properties in many cell lines and animal models. Anticancer effects of guggulsterone are the result of suppression of constitutive NF-KB activation through inhibition of IKK, inhibition of DNA synthesis, induction of apoptosis, and inhibition of cell migration *via* AKT and STAT signaling (Fig. 7.5) [28].

7.2.8.1 Pancreatic cancer

Guggulsterone reduces the motility and suppresses the invasion in pancreatic cells via inhibition of FAK, disruption of the cytoskeletal organization, and Src kinase signaling [29]. It also mediates the FXR inhibition that has been reported in cell migration and invasion [30]. The combination of guggulsterone and gemcitabine *in vitro* results in more inhibition of growth and apoptosis by down-regulating the Nfk -B activity than the drugs/active constituent tested alone. Under *in vivo* conditions, the combination altered tumor growth inhibition through the same mechanism in tumor tissue [28]. Overall, this combined therapy is mediated by the

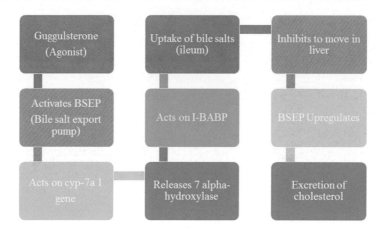

Figure 7.5
Representation of pathway of guggulsterone followed to eliminate the cholesterol by binding to its molecular target FXR by acting as agonists in case of hypercholesterolemia.

suppression of Akt and NF-kB activity and the modulation of anti-apoptotic and proapoptotic proteins [28].

7.2.8.2 Head and neck cancer

Guggulsterone induces apoptosis and cell cycle arrest and enhances the efficacy of erlotinib, cetuximab, and cisplatin which inhibit the invasion in head and neck squamous cell carcinoma cell lines [31]. It not only inhibits proliferation but also induces apoptosis by abrogating the effects of smokeless interleukin-C in tobacco/nicotine on the AKT pathway [32]. It also decreases the level of ST, and nicotine-induced secreted interleukin-6 in culture media of head and neck carcinoma cells [33].

7.2.8.3 Lung cancer

Guggulsterone suppresses the Nf-kB activation induced by the tumor necrosis factor in lung cancer cells. It significantly suppresses the Cox-2 and Cyclin 1 expression along with proliferation and apoptosis inhibition [34,35].

7.2.8.4 Breast cancer

In breast cancer, the Nf-KB pathway is highly activated; an important pathway for the regulation of many apoptotic genes that cause tumors. NF-kB has been shown to regulate the expression of several genes (ciap, suvivin, traf, bcl-2, and bcl-xl) whose products are involved in tumorigenesis. Guggulsterone inhibits NF-kB activation (by downregulating the expression of inducible nitric oxide synthetase), suppresses the expression of anti-apoptotic gene products,

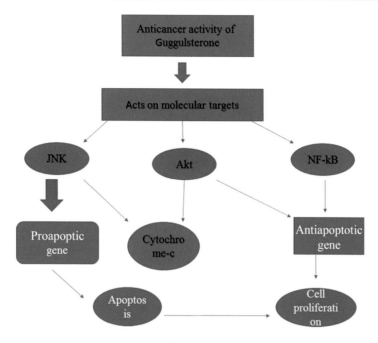

Figure 7.6
Representation of mechanism of action of anticancer activity of guggulsterone.

and enhances apoptosis. Guggulsterone suppresses the DNA binding of NF-kB induced by tumor necrosis factor (TNF), phorbol ester, okadaic acid, hydrogen peroxide, and interleukin-1. It inhibits the NF-kB activation through suppression of IkBα kinase, IkBα phosphorylation, and degradation, p65- phosphorylation, and nuclear translocation. It decreases the expression of the anti-apoptotic gene (IAP1, Bcl-2, surviving), proliferation (cyclin D1 and c-Myc), and metastasis (MMP-9, COX-2, and VEGF). Guggulsterone abrogated the expression of NF-kB-regulated gene products that inhibits apoptosis and promotes inflammation and tumor metastasis [36].

Active forms of guggulsterone are observed to prevent the MMP-9 expression and MAPK/AP-I signal pathway in breast cancer lines [37]. The Z form of guggulsterone reduces the b-catechin/TCF-4 complex and Wnt/b- catechin targeting genes like Cyclin D1, TCF-4, and C-Myc in breast cancer cells, indicating that the b-catechin signaling pathway is the target for guggul lipid-induced apoptosis and growth inhibition in human breast cancer [38].

7.2.8.5 Prostate cancer

Several *in vitro* studies have been done to show the anticancer activity of guggulsterone in prostate cancers. Treatment with guggulsterone results in efficient cytotoxic effects in human prostate cells (PC3) without affecting normal prostate epithelial cells [13]. It has been found that ATP lyase-regulated Akt inactivation is involved in guggulsterone-mediated prostate cancer growth inhibition. It retards prostate cancer growth *via* inactivation of Akt regulation by ATP citrate lyase signaling in human prostate cancer cells (PC-3 and LNCaP) [39]. Guggulsterone also induces caspase-dependent apoptosis mediated by Bax and Bak in prostate cancer cells [39].

7.2.8.6 Colon cancer

Treatment with guggulsterone decreases the Bcl-2, CIAP-1, and CIAP-2 levels and raised the level of P-Jnk [40]. It also inhibits angiogenesis by blocking STAT-3 and VEGF expression and reduction of MMP-2 and MMP-9 enzyme activity [41].

7.2.9 Pancreatitis

Acute pancreatitis is an inflammatory disease of the pancreas. Guggulsterone inhibits the infiltration of macrophages and neutrophils and also suppresses cytokine production in murine-induced pancreatitis. It attenuates the histological damage and decreases the serum lipase levels. Guggulsterone attenuates cerulein-induced acute pancreatitis with the inhibition of ERK and JNK activation [42].

7.2.10 Renal protective effects

Treatment with guggulsterone inhibits the production of proinflammatory molecules like coX-2, IL-6, and TNF-alpha cells and also shows inhibition of inflammatory responses in collecting duct cells which may lead to kidney injuries due to infection [43].

7.3 Conclusion and future perspectives

Guggulsterone is a natural product with low molecular weight and is the biologically active component that has been used for over 1000 years in Ayurveda to treat various diseases including cancer, urinary problems, liver disorders, inflammation, etc. This chapter suggests that guggulsterone may set up direct medicinal applications as a pharmaceutical agent or also serve as the synthesis of new substances for the treatment of human diseases. Further studies and clinical trials are also required to find out the specific intracellular site of action and targets to fully know the mechanism of various diseases like anti-inflammatory, anticancer, hypolipidemic, and other activities to further know its potential role as a therapeutic agent in the prevention and cure of diseases. There are various mechanisms of action of guggulsterone that have not been fully explained but with improved pharmacodynamics, it will serve more advanced benefits in the future.

References

[1]   R Shah, V Gulati, EA Palombo, Pharmacological properties of guggulsterones, the major active components of gum guggul, Phytother Res 26 (2012) 1594–1605.

[2]   T Yokota, The structure, biosynthesis and functions of brassinosteroids, Trends Plant Sci 2 (1997) 137–143.

[3]   BZ Yu, R Kaimal, S Bai, KA El Sayed, SA Tatulian, RJ Apitz, MK Jain, R Deng, OG Berg, Effect of guggulsterone and cembranoids of Commiphora mukul on pancreatic phospholipase A2: role in hypocholesterolemia, J Nat Prod 72 (2009) 24–38.

[4]   G Jain, P Das, Guggulsterone: mechanism of action and prospects of chemo prevention in prostate cancer, Int J Pharm Sci Res 11 (2020) 1527–1536.

[5]   G Saxena, SP Singh, R Pal, S Singh, R Pratap, C Nath, Gugulipid, an extract of Commiphora whightii with lipid-lowering properties, has protective effects against streptozotocin-induced memory deficits in mice, Pharmacol Biochem Behav 86 (2007) 797–805.

[6]   S Shishodia, KB Harikumar, S Dass, KG Ramawat, BB Aggarwal, The guggul for chronic diseases: ancient medicine, modern targets, Anticancer Res 28 (2008) 3464–3647.

[7]   T Yamada, K Sugimoto, Guggulsterone and its role in chronic diseases, Drug Discov Mother Nat 929 (2016) 329–361.

[8]   P Sarup, S Bala, S Kamboj, Pharmacology and phytochemistry of oleo-gum resin of Commiphora wightii (Guggulu), Scientifica 2015 (2015) 1–14.

[9]   NL Urizar, AB Liverman, DT Dodds, FV Silva, P Ordentlich, Y Yan, FJ Gonzalez, RA Heyman, DJ Mangelsdorf, DD Moore, A natural product that lowers cholesterol as an antagonist ligand for FXR, Science 296 (2002) 1703–1706.

[10] B Sharma, R Salunke, S Srivastava, C Majumder, P Roy, Effects of guggulsterone isolated from Commiphora mukul in high fat diet induced diabetic rats, Food Chem Toxicol 47 (2009) 2631–2639.

[11] M Srivastava, NK Kapoor, Guggulsterone induced changes in the levels of biogenic monoamines and dopamine ß-hydroxylase activity of rat tissues, J Biosci 10 (1986) 15–19.

[12] GV Satyavati, C Dwarakanath, SN Tripathi, Experimental studies on the hypocholesterolemic effect of Commiphora mukul Engl. (Guggul), Indian J Med Res 57 (1969) 1950–1962.

[13] M Dufer, K Horth, R Wagner, B Schittenhelm, S Prowald, TF Wagner, J Oberwinkler, R Lukowski, FJ Gonzalez, P Krippeit-Drews, G Drews, Bile acids acutely stimulate insulin secretion of mouse beta-cells via farnesoid X receptor activation and K(ATP) channel inhibition, Diabetes 61 (2012) 1479–1489.

[14] R Chander, F Rizvi, AK Khanna, R Pratap, Cardioprotective activity of synthetic guggulsterone (E and Z-isomers) in isoproterenol induced myocardial ischemia in rats: a comparative study, Indian J Clin Biochem 18 (2003) 71–79.

[15] J Pu, A Yuan, P Shan, E Gao, X Wang, Y Wang, WB Lau, W Koch, XL Ma, B He, Cardiomyocyte-expressed farnesoid-X-receptor is a novel apoptosis mediator and contributes to myocardial ischaemia/reperfusion injury, Eur Heart J 34 (2013) 1834–1845.

[16] WC Wang, YH Uen, ML Chang, KP Cheah, JS Li, WY Yu, KC Lee, CS Choy, CM Hu, Protective effect of guggulsterone against cardiomyocyte injury induced by doxorubicin in vitro, BMC Complement Altern Med 12 (2012) 1–10.

[17] JJ Song, SK Kwon, CG Cho, SW Park, SW Chae, Guggulsterone suppresses LPS induced inflammation of human middle ear epithelial cells, Int J Pediatr Otorhinolaryngol 74 (2010) 1384–1387.

[18] JN Sharma, Comparison of the anti-inflammatory activity of Commiphora mukul (an indigenous drug) with those of phenylbutazone and ibuprofen in experimental arthritis induced by mycobacterial adjuvant, Arzneimittel Forschung 27 (1977) 1455–1457.

[19] MR Meselhy, Inhibition of LPS induced NO production by the oleogum resin of commiphora wightii and its constituents, Phytochemistry 62 (2003) 213–218.

[20] NM Kalariya, M Shoeb, AB Reddy, M Zhang, FJ van Kuijk, KV Ramana, Prevention of endotoxin-induced uveitis in rats by plant sterol guggulsterone, Invest Ophthalmol Vis Sci 51 (2010) 5105–5113.

[21] VN Sumantran, AA Kulkarni, A Harsulkar, A Wele, SJ Koppikar, R Chandwaskar, V Gaire, M Dalvi, UV Wagh, Hyaluronidase and collagenase inhibitory activities of the herbal formulation Triphala guggulu, J Biosci 32 (2007) 755–761.

[22] BB Singh, LC Mishra, SP Vinjamury, N Aquilina, N Shepard, The effectiveness of Commiphora Mukul for osteoarthritis of the knee. An outcome study, Altern Ther Health Med 9 (2003) 74–79.

[23] RE Svoboda, Ayurvedic life health and longevity, J Biosci 32 (1992) 221–258.

[24] K Pavelka, Š Forejtová, M Olejarova, J Gatterova, L Šenolt, P Špaček, A Pavelkova, Hyaluronic acid levels may have predictive value for the progression of knee osteoarthritis, Osteoarthr Cartil 12 (2004) 277–283.

[25] G Murphy, MH Lee, What are the roles of metalloproteinases in cartilage and bone damage? Ann Rheum Dis 64 (2005) 44–47.

[26] M Hadipour Jahromy, R Mozaffari Kermani, F Nobakht, Influence of Commiphora Mukul resin on the knee articular cartilage of rats in experimental osteoarthritis induced by iodoacetate, Pak J Med Sci 25 (2009) 269–273.

[27] JM Kim, SH Kim, SH Ko, J Jung, J Chun, N Kim, HC Jung, JS Kim, The guggulsterone derivative GG-52 inhibits NF-κB signaling in gastric epithelial cells and ameliorates ethanol-induced gastric mucosal lesions in mice, Am J Physio Gastrointest Liver Physiol 304 (2013) 193–202.

[28] DW Ahn, JK Seo, SH Lee, JH Hwang, JK Lee, JK Ryu, YT Kim, YB Yoon, Enhanced antitumor effect of combination therapy with gemcitabine and guggulsterone in pancreatic cancer, Pancreas 41 (2012) 1048–1057.

[29] MA Macha, S Rachagani, S Gupta, P Pai, MP Ponnusamy, SK Batra, M Jain, Guggulsterone decreases proliferation and metastatic behavior of pancreatic cancer cells by modulating JAK/STAT and Src/FAK signalling, Cancer Lett 341 (2013) 166–177.

[30] JY Lee, KT Lee, JK Lee, KH Lee, KT Jang, JS Heo, SH Choi, Y Kim, JC Rhee, Farnesoid X receptor, overexpressed in pancreatic cancer with lymph node metastasis promotes cell migration and invasion, Brit J Cancer 104 (2011) 1027–1037.

[31] RJ Leeman Neill, SE Wheeler, SV Singh, SM Thomas, RR Seethala, DB Neill, JR Grandis, Guggulsterone enhances head and neck cancer therapies via inhibition of signal transducer and activator of transcription-3, Carcinogenesis 30 (2009) 1848–1856.

[32] MA Macha, A Matta, SS Chauhan, KM Siu, R Ralhan, Guggulsterone (GS) inhibits smokeless tobacco and nicotine-induced NF-κB and STAT3 pathways in head and neck cancer cells, Carcinogenesis 32 (2011) 368–380.

[33] MA Macha, A Matta, SS Chauhan, KM Siu, R Ralhan, Guggulsterone targets smokeless tobacco induced PI3K/Akt pathway in head and neck cancer cells, PLoS One 6 (2011) 14728.

[34] AA Bhat, KS Prabhu, S Kuttikrishnan, R Krishnankutty, J Babu, RM Mohammad, S Uddin, Potential therapeutic targets of guggulsterone in cancer, Nutr Metab 14 (2017) 1–11.

[35] S Shishodia, G Sethi, KS Ahn, BB Aggarwal, Guggulsterone inhibits tumor cell proliferation, induces S-phase arrest, and promotes apoptosis through activation of c-Jun N-terminal kinase, suppression of Akt pathway, and downregulation of antiapoptotic gene products, Biochem Pharmacol 74 (2007) 118–130.

[36] S Shishodia, BB Aggarwal, Guggulsterone inhibits NF-κB and IκBα kinase activation, suppresses expression of anti-apoptotic gene products and enhances apoptosis, J Biol Chem 279 (2004) 47148–47158.

[37] JN Kong, Q He, G Wang, S Dasgupta, MB Dinkins, G Zhu, A Kim, S Spassieva, E Bieberich, Guggulsterone and bexarotene induce secretion of exosome-associated breast cancer resistance protein and reduce doxorubicin resistance in MDA-MB-231 cells, Int J Cancer 137 (2015) 1610–1620.

[38] G Jiang, X Xiao, Y Zeng, K Nagabhushanam, M Majeed, D Xiao, Targeting beta-catenin signaling to induce apoptosis in human breast cancer cells by z-guggulsterone and Gugulipid extract of Ayurvedic medicine plant *Commiphora mukul*, BMC Complement Altern Med 13 (2013) 203.

[39] SV Singh, Y Zeng, D Xiao, VG Vogel, JB Nelson, R Dhir, YB Tripathi, Caspase-dependent apoptosis induction by guggulsterone, a constituent of Ayurvedic medicinal plant *Commiphora mukul*, in PC-3 human prostate cancer cells is mediated by Bax and Bak, Mol Cancer Ther 4 (2005) 1747–1754.

[40]   MJ An, JH Cheon, SW Kim, ES Kim, TI Kim, WH Kim, Guggulsterone induces apoptosis in colon cancer cells and inhibits tumor growth in murine colorectal cancer xenografts, Cancer Lett 279 (2009) 93–100.

[41]   ES Kim, SY Hong, HK Lee, SW Kim, MJ An, TI Kim, KR Lee, WH Kim, JH Cheon, Guggulsterone inhibits angiogenesis by blocking STAT3 and VEGF expression in colon cancer cells, Oncol Rep 20 (2008) 1321–1327.

[42]   Kim D.G., Bae G.S., Choi S.B., Jo I.J., Shi J.Y., Lee S.K., Kim M.J., Kim M.J., Jeong H.W., Choi C.M., Seo S.H., Choo G.C., Seo S.W., Song H.J., Park S.J. (2015a). Guggulsterone attenuates cerulein induced acute pancreatitis via inhibition of ERK and JNK activation. Int Immunopharmacol 26, 194–202.

[43]   Kim D.G., Bae G.S., Jo I.J., Choi S.B., Kim M.J., Jeong J.H., Kang D.G., Lee H.S., Song H.J., Park S.J. (2015b). Guggulsterone attenuated lipopolysaccharide-induced inflammatory responses in mouse inner medullary collecting duct-3 cells. Inflammation 39, 87–95.

[44]   YB Tripathi, OP Malhotra, SN Tripathi, Thyroid stimulating action of Z-guggulsterone obtained from *Commiphora Mukul*, Planta Med 50 (1984) 78–80.

[45]   YB Tripathi, P Tripathi, OP Malhotra, SN Tripathi, Thyroid stimulatory action of (Z)-guggulsterone: mechanism of action, Planta Med 54 (1988) 271–277.

CHAPTER 8

Herbal medicines for diabetes: Insights and recent advancement

Mohammad Fareed[a] and Anis A. Chaudhary[b]

[a]College of Medicine, Al-Imam Mohammad Ibn Saud Islamic University (IMSIU), Riyadh, Kingdom of Saudi Arabia [b]Department of Biology, College of Science, Al-Imam Mohammad Ibn Saud Islamic University (IMSIU), Riyadh, Kingdom of Saudi Arabia

8.1 Introduction

Diabetes mellitus (DM) is a noninfectious endocrine disorder, which is characterized by the disturbance in metabolism of carbohydrate and associated with hyperglycemia [1,2]. According to International Diabetes Federation survey in 2016, diabetes is a disorder that affects 415 million people in the world, and it may increase to 642 million by the year 2040 [3].

DM is linked with development of various serious diseases like micro vascular (nephropathy, retinopathy, nephropathy) and macro vascular (peripheral vascular disease and coronary heart diseases) [4]. Glucose blood levels are maintained by insulin which is a hormone released from the pancreas. When glucose level increases, insulin is produced from the pancreas and maintains the level of glucose. In diabetic patients, the production of insulin is absent or less which causes hyperglycemia [5]. The symptoms of DM are polydipsia, polyuria, polyphagia, fatigue, nausea, vomiting, impotence in men, slow healing wound, and blurred vision [6].

DM are three types type 1, type 2 and gestational DM. Type 1 DM is known as insulin dependent DM which is due to total loss of function of β cell of islets of Langerhans which are present in pancreas [7]. Type 2 DM is known as insulin nondependent DM which is due to insulin resistance among body cells and insulin is not able to metabolize glucose, eventually leading to hyperglycemia state in cells [8]. Type 2 DM mostly occur in obese persons and associated with high blood pressure and high cholesterol levels. The precise underlying mechanisms for the development of insulin resistance are still incomplete, however, increasing evidence has suggested that low-grade chronic inflammation, as well as multiple defects in intracellular insulin signaling and glucose uptake in skeletal and adipose tissues, plays a pivotal role in the manifestation of insulin resistance [9]. Gestational diabetes is a type of diabetes which is present with hyperglycemia in pregnant women. It usually appears in 2%–4% pregnancies in 2nd or 3rd trimester [10].

Herbal Medicines: A Boon for Healthy Human Life.
DOI: https://doi.org/10.1016/B978-0-323-90572-5.00026-3

Herbal medicines have been used for thousands of years in many ethnic cultures such as Chinese, Korean, Indian, and Mexican to treat and manage diabetes and its complications [11–14]. In the last Few decades, modern science has uncovered the benefits of using herbal medicines in the management of particular diabetic complications, such as vascular inflammation, nephropathy and retinopathy [15].

Many traditional herbs that have been comprehensively studied and show promising clinical applications in regard to diabetes contain pentacyclic triterpenoids as their active components. Some examples of these herbs are: huangqi (the root of *Astragalus membranaceus*), baical skullcap (the root of *Scutellaria baicalensis*), bitter melon (the fruit of *Momordica charantia*), fenugreek (the seed of *Trigonella foenum graecum*), ginseng (the root of *Panax ginseng*), green tea (the flower buds of *Camellia sinensis*), gymnema (the leaves of *Gymnema sylvestre, G. Montanum, G. Inodorum*), hawthorn (the fruit of *Crataegus monogyna*), licorice (the root of *Glycyrrhiza glabra*), oats (the seeds of *Avena sativa*), olive (the fruit of *Olea europaea*), yam (the rhizome of *Dioscorea opposita*), ashwagandha (the root of *Withania somnifera*), gotu kola (the leaves of *Centella asiatica*), and kudzu (the root of *Pueraria lobata*) [16].

Now a days, more than 400 plants are being used in different forms for their hypoglycemic effects in treating diabetes, with tall claims of efficacy by patients and practitioners [17]. Despite the presence of known antidiabetic medicines in the pharmaceutical market, remedies from medicinal plants are used with varying success by a good number of diabetic patients. Plant drugs and herbal formulations are frequently considered to be less toxic and freer of side effects. According to WHO recommendations, hypoglycemic agents of plant origin used in traditional medicine are important in the management of Diabetes [18]. The attributed antihyperglycemic effect of these plants is due to their ability to restore the function of pancreatic tissues and thereby increase the insulin output. The other mechanism may be by inhibiting the intestinal absorption of glucose or facilitation of metabolites in insulin-dependent pathways. These actions of herbal drugs protect the beta cells and iron out the excursions of blood glucose [19]. Another mechanism by which plant extracts help to contain the diabetic pathology is by acting as antioxidants. Free radicals generated by the metabolic process in the body leads to an oxidative stress, damaging different proteins. The antioxidant properties of herbal drugs may be able to help contain this damage. Very little biological knowledge on the specific modes of action of these herbal compounds are known so far, but most of the plants have been found to contain substances like glycosides, alkaloids, terpenoids and flavonoids, which are frequently quoted as having antidiabetic effects [20].

This chapter discusses the importance of various herbal plants and their extracts in the treatment of DM which are commonly used in Indian subcontinent and some other countries of Asia during past few decades. The active ingredient of these plants has antidiabetic potential which are documented from different studies has been explained. Additionally, the recent advancement for the use of these herbal plants in treatment of DM is explained by cellular and molecular mechanism for active compounds present in these herbal plants as evidenced by some *in vitro* and *in vivo* studies.

8.2 Herbal medicines of diabetes

According to world ethanobotanical 800 medicinal plants are used for the prevention of DM. Clinically proven that only 450 medicinal plants possess anti diabetic properties from which 109 medicinal plants have complete mode of action. In ancient time doctor and lay person used traditional medicinal plants with their active constituents and properties for the treatment of different diseases like heart diseases, cancer and diabetes. There is an extended history of traditional plants used for the control of diabetes in India and China. There are various books available such as Charaka Samhita and Susruta Samhita which explains phytopharmacology features of diabetes and its adverse effect [21].

Synthetic drugs which are used for treatment of diabetes are related to various adverse effect like sickness, vomiting, dysentery, alcohol flush, migraine, swelling, malignant anemia, and faintness. Herbal drugs are proved to be a far better choice over synthetic drugs due to less side effects and adverse effects. Herbal formulations are easily available without prescription. These herbal drugs are used for life threatening disease. These drugs also are used when chemical drugs are ineffective in treatment of disease. These are natural and safe drugs that is there is no toxic effects. Herbal drugs permanently cure person and treat the disease while synthetic drugs don't seem to be permanently cured the diseases [22]. Herbal formulations contain natural herbs and fruits and vegetables extract which are beneficial in treatment of different diseases with nonadverse effects. On the opposite hand chemical drugs are prepared synthetically and have side effect also. Herbal formulations are cheap as compared to all or any opathic medicines. Herbal formulations are Eco friendly. Herbal formulations are produced from natural products while all opathic medicines are produced from chemical and chemically modified natural products (Fig. 8.1). Herbal formulations are available without prescription while all opathic medicines are available with prescription [23,24].

8.3 Traditional herbal antidiabetic drugs

Currently the medicinal plants and herbs are getting used in extract forms for their antidiabetic activity. Various clinical studies confirmed that medicinal plants extracts show antidiabetic activity and restoring the action of pancreatic β cells (Table 8.1) [25].

8.3.1 Allium sativum

It is usually recognized as "garlic" and belongs to Liliaceae, which is a family of allium sativum [5]. Ethanolic extract of garlic (10 mL/kg/day) frequently shows hypoglycemic activity [26]. Extract of garlic was more efficient than anti diabetic drug glibenclamide [28]. Ester, ethanol, and petroleum ether extract was observed to point out an antidiabetic activity in STZ

Figure 8.1
Pictures of some herbal plants used for treatment of diabetes mellitus.

induced rats. Garlic shows various therapeutic effect like anti platelet, antibacterial, lowering the blood pressure and lowering the cholesterol level within the body [28].

8.3.2 *Aloe borbadensis*

It is referred to as Ghikanvar, which belongs to Liliaceae family. It looks sort of a cactus plant with green blade shaped leaves that are heavy narrowing, hairy and crammed with clear

Table 8.1: Herbal species used for treatment of diabetes mellitus.

S. No.	Plant species	Family	Common name	Part used	Active constituents	Mode of action	References
1.	*Allium sativum*	Lilliacae	Garlic	Petroleum	Allyproyl disulfide	Improve plasma lipid metabolism and plasma antioxidant activity	[6,27–29]
2.	*Aloe borbadensis*	Asphodelaceae	Ghikanwar	Leaf pulp	β-sitosterol, campesterol	Improvement in impaired glucose tolerance	[27,30]
3.	*Azadirachta indica*	Meliaceae	Neem	Leaf extracts	Azadirachitin nimbin	Glycogenolytic effect due to epinephrine	[7,27,31]
4.	*Brassica juncea*	Brassicaceae	Mustard	Aqueous	Sulforaphane	Increase activity of glycogen synthetase	[32]
5.	*Carica papaya*	Caricaceae	Papaya	Aqueous seed	Papain, chymopapin	Lowered fasting blood sugar, triglyceride,	[33]
6.	*Catharanthus roseus*	Apocaceae	Vinca	Hot water	Catharanthaine	Lowering of glycemia	[6,27]
7.	*Coriandrum sativum*	Apicaceae	Coriander	Seed extract	p-cymene	Increases the activity of β-cells and	[27]
8.	*Eugenia jambolana*	Myrataceae	Jamun	Pulp of fruit	Oleanolic acid, ellagic acid	Inhibited insulinase activity from liver and	[34]
9.	*Gymnema sylvestre*	Asclepidaceae	Gudmar	Dried leaves	Dihydroxy gymnemic	Increase the serum G peptide level which	[5]
10.	*Mangifera indica*	Anacardiaceae	Mango	Leaves extract	β-Carotene	Reduction in the intestinal absorption of	[36]
11.	*Momordica charantia*	Curcubitaceae	Bitter	Fresh green	Charantin, sterol	Activates PPARs α and γ and lower the	[37]
12.	*Ocimumsanctum*	Labi ateae	Tulsi	Entire herbs	Eugenol	Increased insulin release	[38]
13.	*Tinospora cardifolia*	Menispermaceae	Gulvel	Aqueous extract of root	Tinosporonetinosporic acid	Decrease of glycemia and brain lipids	[39,40–43]
14.	*Aeglemarmelos*	Rutaceae	Bael	Leaves extracts	Aegle marmelosine	Improve functional state of pancreatic β-cell	[32]
15.	*Allium cepa*	Lilliaceae	Onion	Dried powder	Dipropyl disulfide oxide	Stimulating the effects on glucose	[43,44]

viscid gel. Oral administration of aqueous extract of aloe vera during a dose of 150mg/kg of body weight significantly lowering the blood sugar level [26]. Burn plant gel consist various therapeutic effects like anti diabetic, antioxidant, increases the decrease level of glutathione by fourfold in diabetic rats [29].

8.3.3 Azadirachta indica

It is commonly named as 'neem' in India and it belongs to Meliaceae family. Its availability is predominent in India and Burma [6]. Its ethanolic and aqueous extract is observed as reduction in high blood sugar levels. It could be combined with allopathic drugs among type 2 diabetic patients whose diabetes is not maintained by solely allopathic drugs [26]. Large numbers of patients are treated by natural neem tablets at global level. Extract of neem improves the blood circulation by helping in enlargement of the blood vessels and predominantly helpful in the reduction of the blood sugar levels [30].

8.3.4 Brassica juncea

It is referred to as Rai which belongs to family Brassicaceae. It is widely used as spice in various food items. Aqueous seed extract has blood glucose lowering activity which was observed in alloxan induced diabetic rats. 250, 350, 450 mg/kg doses of extract show hypoglycemic activity [31].

8.3.5 Carica papaya

It is referred to as papaya which belongs to family caricaceae. Seed and leaves extract shows lowering of blood glucose level, lowering of lipid within the body and healing of wound activities in alloxan induced diabetic rats [32].

8.3.6 Catharanthus roseus

It is referred to as Vinca roseus which belongs to family apocynaceae. Methanolic extract of leaves and twigs shows decrease in blood glucose level within the alloxan induced diabetic rats. Oral administration of 500 mg/kg dose of leaves and twigs extract was beneficial in animals for lowering in blood glucose level [5]. The mechanism of action of catharanthus roseus is increases the synthesis of insulin from β cells of Langerhans [26].

8.3.7 Coriandrum sativum

It is mainly referred to as coriander which belongs to family Umbelliferae. It is widely used as spice in various food items. 200 mg/kg seed extract frequently increases the action of the

β cells of Langerhans and reduces serum sugar in alloxan induced diabetic rats and synthesis insulin from β cells of the pancreas. Extract of coriander shows blood glucose lowering property and insulin synthesizer [26].

8.3.8 Eugenia jambolana

It is referred to as jamun belongs to Myretaceae family. It contains dried seeds and mature fruits of Eugenia jambolana. It containsmalvidin3-laminaribiosidea and ferulic acid as active constituents. Extract of dried seeds (200 mg/kg) used for treatment of diabetic patients [33].

8.3.9 Gymnema sylvestre

It is commonly referred to as Gudmar which suggests "sugar destroying" and consists of Asclepidaceae family. Leaf extract of G.sylvestre (3.4/13.4 mg/kg) showed significant reduction of blood glucose level in streptozotocin induced rats. It is mostly utilized in Indian ayurvedic medicines for treatment of diabetes. The active constituents in G. Sylvester are alkaloids, flavonoids, saponins and carbohydrates [34].

8.3.10 Mangifera indica

Itis commonly referred to as mango and consists of family anacardiaceae. Antidiabetic activity shows by leaves extract (250 mg/kg), but oral administration of aqueous extract failed to change the blood sugar level in alloxan induced diabetic rats [35].

8.3.11 Momordica charantia

Itis commonly referred to as bitter melon (karela) and belongs to Cucurbitaceae family. The active constituents of balsam pear are momordic I and momordic II, cucurbitacin B. It is used within the treatment of diabetes. It consists lectin, which has insulin like activity. Lectin is a nonprotein, which is linked to insulin receptors. This lectin decreases the blood glucose level by acting on peripheral tissues [36]. Fruit extract of M. Charantia (200 mg/kg) shows hypoglycemic activity.

8.3.12 Ocimum sanctum

Itis referred to as tulsi and belongs to Labiateae family. It is widely found everywhere India. It is utilized in Indian ayurvedic medicines for treatment of varied diseases. Various animal studies proved that aqueous extract of Ocimum sanctum leaves (200 mg/kg) showed the hypoglycemic activity in streptozotocin induced rats. It's also used for treatment of viral infection, treatment of mycosis, reduces stress, treatment of tumor and treatment of peptic ulcer [37].

8.3.13 Tinospora cardifolia

It is well-known as guduchi and consists of the Menispermaceae family. The active constituents of T. Cardifolia are diterpene compounds which consists tinosporone, tinosporic acid, Syringen, berberine, and giloin [38]. Root extract of T. Cardifolia (50–200 mg/kg) shows decrease in blood and urine sugar in streptozotocin induced diabetic rats during oral administration for six weeks. It is mostly utilized in Indian ayurvedic medicines for treatment of diabetes. Root extract also forbid the reduction of body weight [39–42]. Various plant parts like roots, stem, leaves and fruits are extracted by maceration, infusion, percolation, decoction and soxhlet extraction generally. Mostly various solvents are used like ethanol, methanol and petroleum ether.

8.4 Herbal marketed formulations of diabetes mellitus

Currently, there are many polyherbal formulations in Indian market that are utilized in different form like Vati, Churna, Arkh, Quath for the treatment of diabetes [19]. These formulations may consist aqueous extract or powders of the different plants part which are used for the treatment of diabetes. These formulations are called as poly herbal formulation because they contain 3 to 25 herbs within the formula [39–41].

8.4.1 Aegle marmelos

Itis referred to as Bael and belongs to Rutaceae family. It's inherited to India and parts of plant like leaves, barks, roots, and fruits are utilized in the Ayurveda and in various medicines which is employed for cure of different diseases. Leaves of neem and tulsi together with leaves of A. Marmelos are dried, powdered, and administered thrice daily for 15 days. Animal studies proved that Aegle marmelos (100, 200, and 500 mg/kg) are used for treatment of different diseases like treatment of cancer, treatment of varied viral diseases, treatment of varied microbial diseases [31].

8.4.2 Allium cepa

It is locally referred to as onion or pyaz belongs to Liliaceae, a family of allium cepa. Antihyperglycemic activity shown by ether soluble part and ether insoluble a part of dried onion powder. It contains chemical ingredient allyl propyl disulfide which is understood as APDS and it inhibits the insulin destruction by the liver and provoke the production of insulin by the pancreas which reinforces the concentration of insulin and reduces the glucose levels within the blood. Crucial oil (100 mg/kg) collected from red onion frequently shows antihyperglycemic activity, antistatin, and antioxidant effects in alloxan induced diabetic rats. 300 mg/kg are most useful percentage in treatment of hyperglycemia and hyperlipidemia

[42]. Various clinical trails and animal research provided information that onion is used for treatment of asthma, treatment of diabetes, treatment of cancer, and treatment of different viral diseases [43].

8.5 Recent advancement in herbal medicines of diabetes

There are many herbal plants whose antidiabetic potential role has been established by many *in vitro* and *in vivo* studies by screening of their herbal extracts [44]. With the recent advancement in medical and biomedical sciences since past three decades, the constituent and active compounds of these herbal are monitored for their efficacy and mechanism in treatment of diabetes.

Some of the active compounds present in known anti diabetic herbal plants and their mechanism of action are discussed under following sections:

8.5.1 Curcumin

Curcumin is usually found in rhizomes of Zingiberaceae plants like, turmeric commonly referred to as turmeric [45]. Curcumin significantly improved the glucose intolerance, hyperglycemia, hyperlipidemia, hypoinsulinemia, and reduce the glucose intolerance. Curcumin can modulate the activity of cell signaling molecules like, expression of various transcription factors (like TNF-α) levels and control the free carboxylic acid level in plasma [46]. Similarly, curcuminoids protect β cell islets of pancreas from oxidative stress by inducing the modulatory subunit of γ-glutamyl-cysteine ligase; expression of NADPH:quinone oxidoreductase-1; and haem oxygenase-1 at both the mRNA and protein levels. Curcumin also modulates the diabetic neuropathy by inhibiting the event of diabetic cataracts by normalizing the lipid peroxides [44]. Curcumin normalizes the extent of cytokine (e.g., tumor necrosis factor-α and interleukin-1β), stimulated translocation of nuclear factor kappa-light-chain-enhancer of activated B cells (NF-κb) and increase β-cell mass and reduce cellular oxidative stress in islet cells of pancreas. Curcumin also improve the extent of anti-oxidant enzymes like superoxide dismutase, catalase and glutathione peroxidase; hepatic glucose regulating enzymes (glucose-6 phosphatase and phosphoenolpyruvate carboxykinase); hepatic lipid controlling metabolism enzymes (fatty acid synthase, acyl-coa: cholesterol acyltransferase and 3-hydroxy-3-methylglutaryl coenzyme reductase), and malondialdehyde [47].

8.5.2 Catechins

Various flavonoids including Catechins are extensively studied for their potential as therapeutic agents within the management of diabetes [48]. Epigallocatechin gallate (EGCG) is sort of catechins which is a naturally occurring phenolic compound found in tea, cocoa and palms often conjugated with gallic acid. It improves glucose uptake, while catechin glycosides suppressed the active glucose transportation [49]. Hence, catechins are found to be effective in plasma glucose control and play vital role in normalized the lipid and carbohydrate metabolism. The drinking of oolong tea rich in EGCG (386 mg), significantly reduced glucose and fructosamine level in plasma. EGCG mimics the action of insulin to stimulate phosphorylation of tyrosine of the insulin receptor and the substrate of insulin receptor. Thus, EGCG significantly reduced expression of gluconeogenic enzyme phosphoenolpyruvate carboxykinase gene, which could be a key enzyme of gluconeogenesis. Moreover, EGCG normalized the cytokine-induced β-cell damage and prevents β-cell mass degeneration from diabetic progression [50].

8.5.3 Ginsenosides

Ginsenosides are chemical constituents of a perennial plant species ginseng (Panax ginseng), which utilized in traditional medicine in China, Japan, Korea, and other countries [44]. Ginsenosides exhibited anti-hyperglycemic and anti-obesity activity through the peroxisome proliferator-activated receptor (PPAR) mediated pathway. PPAR-γ mRNA expression in patients of DM was significantly increased after two weeks treatment of ginsenosides. Ginsenoside Rg3 stimulate insulin secretion, which is expounded with ATP sensitive K channel for lowering the plasma glucose level. Ginsenoside Rg3 epimers exhibited differential action than Rg3, and among these 20(S)-Rg3 epimer showed higher level discharge of insulin and activation of AMPK. 20(S)-Ginsenoside Rg3 administration normalized the increased blood sugar level and glycosylated hemoglobin and improved renal dysfunction of diabetic rats [51].

8.5.4 Gingerols

Gingerol is that the active constituents derived from the rhizomes of ginger (Zingiber officinale). 6-gingerol is reported to control adipocyte function through inhibiting adiponectin expression in mouse adipocytes by inhibiting TNF-α and thus, inhibiting c-Jun N-terminal kinases (JNK) phosphorylation [44]. Gingerol showed structural homology with capsaicin and may need a role in inhibiting inflammatory activities of fat in dyslipidenia condition. Supplementation of 6-gingerol to high fat fed mice significantly reduced increase of body weight. Also, reduction in adiposity is related to a modification in metabolism of cholesterol and oxidation of carboxylic acid [52].

8.5.5 Stevosides

Stevioside is documented for intense non-caloric sweetness and found in leaf of Stevia rebaudiana. Intake of stevioside and related compounds (rebaudioside, steviol, and isosteviol) reduced the glucose level in plasma during glucose tolerance test followed by overnight fasting in fit individuals [53]. Phosphoenolpyruvate carboxykinase (PEPCK) gene is over expressed in diabetic rat with very low insulin level and stevioside controlled the expression of mrna and protein of phosphoenolpyruvate carboxykinase (PEPCK) gene [54], prevent gluconeogenesis process [55]. Injection of stevioside alongside glucose in Goto-Kakizaki (GK) rats normalizes the blood sugar level in response to glucose tolerance by inducing insulin secretion and suppresses glucagon level within the plasma [56]. Stevioside exhibited hypoglycemic activity on diabetic rat by regulating the expression of PEPCK, which control production of glucose from non-carbohydrate molecules within the liver during gluconeogensis [55].

8.5.6 Berberine

Berberine, an active compound of Berberis spp., Coptidis spp., Hydrastis xanthorhiza, Mahonia spp., cork tree, T. Cordifolia,and many other species, has been commonly utilized for the adjuvant administration of type 2 diabetes since an extended time [57]. Berberine treatment improves hyperglycemic conditions by increasing AMP-activated protein kinase (AMPK) action in adipocytes and L6 myotubes. It also improves glucose uptake by increasing translocation of glucose transporter type 4 protein (GLUT-4) in muscle cells by regulating phosphatidylinositol 3-kinase, and inhibit lipid deposition in adipocytes [58]. Berberine controls the glucose and lipid metabolism as demonstrated by several studies in several animal model. Significant reduction (up to 20%) in glycated hemoglobin (HBA1c), postprandial blood sugar, fasting blood sugar and plasma triglycerides were recorded in persons treated with berberine [59]. Supported a meta-analysis review, berberine was found beneficial for controlling blood sugar level within the treatment of diabetic mellitus. Berberine also found effective as like conventional oral hypoglycaemic drugs [60]. Various mechanism of action of berberine in treatment of diabetes and its complications are reported beside the activity of berberine derivatives and its combination with other molecules [58]. Berberine is additionally reported to control non-coding rnas to exert its therapeutic action in diabetes [61]. Intervention of berberine with improved lifestyle cared-for decrease the extent of fasting plasma glucose post-prandial glucose and glycated hemoglobin than lifestyle improvement alone or with the treatment of placebo demonstrated no significant difference between berberine and oral hypoglycemic drugs [62].

8.5.7 Capsaicin

Capsaicin is that the potent anti-diabetic compound found in chili peppers (e.g., Capsicum annum), which is employed as a spice everywhere around the globe. Capsaicin inhibits adipogenesis and lowers the intracellular triglycerides in adipocytes. Capsaicin suppresses hyperlipidemia induced by sub-clinical grade inflammation by improving adipokine release. Dietary capsaicin significantly reduced fasting glucose and triglyceride levels, and improves insulin level [44]. Similarly, in adipocyte tissue expression and discharge of interleukin-6 and monocyte chemo attractant protein-1 (MCP-1) are suppressed by capsaicin. Capsaicin showed anti-inflammatory activity by specifically binding to PPARs, which inhibit the production of TNF-α and inactivate NF-κb [51]. Additionally, capsaicin drastically reduced JNK activation, induced by carboxylic acid, thus normalize the pro-inflammatory effects of fatty acids [63]. Capsaicin inhibited activation of macrophage to release pro-inflammatory cytokines [51]. Hence, capsaicin has been considered as a beneficial natural product for management of diabetic hyperlipidemia.

8.6 Conclusion

It is very well-established fact that, DM is a metabolic disorder affecting many individuals of all age groups around the globe. The prevalence of DM has been increased in past few decades and is expected to elevate in future due to change in lifestyle and dietary habits among human population. The prevention, treatment, and management are still a great challenge for the health professionals and researchers. There are many well established pharmaceutically synthesized drugs in modern medicine which have a very high anti-diabetic potential. But those drugs have many adverse effects leading to severe hypoglycemia and other complications if consumed for many years. There are many herbal drugs having anti-diabetic importance and without any adverse effects as discussed in this chapter. These drugs have some active compounds which help in the regulation of blood glucose as evidenced by many laboratory studies. The mode of action of these herbal compounds at cellular and molecular level for hyperglycemia is well established in cellular and animal models. There are many studies for human trials of these herbal drugs which significantly support their anti-diabetic potential, but these studies are very scarce. So, it is concluded that in future, more studies should be done by clinical researchers among DM patients to establish the more efficacy of the active compounds present in these herbs. It is also recommended to health professionals for more use of these herbal drugs considering evidenced based medicine approach to minimize the adverse effects of synthesized pharmaceutical drugs among DM patients.

References

[1]   A Kumar, MK Goel, RB Jain, P Khanna, V Chaudhary, India towards diabetes control: key issues, Australas Med J 6 (10) (2013) 524–531.

[2]   MA Rahimi, Review: Anti Diabetic medicinal plants used for diabetes mellitus, Bull Env Pharmacol Life Sci (4) (2015) 163–180.

[3]   Arogyaworld.org. Available from: FACT SHEET: Diabetes in India. http://www.arogyaworld.org/wpcontent/uploads/2010/10/arogyaworldINDIAdiabetes/factsheets/. [Accessed 5 March 2021].

[4]   MU Rao, M Sreenivasulu, B Chengaiah, KJ Reddy, CM Chetty, Herbal medicines for diabetes mellitus. A review, Int J PharmTech Res (2) (2010) 1883–1892.

[5]   R Bordoloi, KNB Dutta, A review on herbs used in the treatment of diabetes mellitus, J Pharm Chem Biol Sci 2 (2014) 86–92.

[6]   Org.uk. Available from: International Diabetes Federation Atlas. Seventh Edition. 2015. https://www.diabetesatlas.org/upload/resources/previous/files/7/IDF%20Diabetes%20Atlas%207th.pdf. [Accessed 5 March 2021].

[7]   S Verma, M Gupta, H Popli, G Aggarwal, Diabetes mellitus treatment using herbal drugs, Int J Phytomedicine 10 (1) (2018) 1–10.

[8]   M Fareed, N Salam, AT Khoja, MA Mahmoud, M Ahamed, Life style related risk factors of type 2 diabetes mellitus and its increased prevalence in Saudi Arabia: a brief review, Int J Med Res Health Sci 6 (3) (2017) 125–132.

[9]   G Sesti, Pathophysiology of insulin resistance, Best Pract Res Clin Endocrinol Metab 20 (2006) 665–679.

[10]   WA Wannes, B Marzouk, Research progress of Tunisian medicinal plants used for acute diabetes, J Acute Dis 5 (5) (2016) 357–363.

[11]   WL Li, HC Zheng, J Bukuru, N De Kimpe, Natural medicines used in the traditional Chinese medical system for therapy of diabetes mellitus, J Ethnopharmacol 92 (1) (2004) 1–21.

[12]   A Andrade-Cetto, M Heinrich, Mexican plants with hypoglycaemic effect used in the treatment of diabetes, J Ethnopharmacol 99 (3) (2005) 325–348.

[13]   PK Mukherjee, K Maiti, K Mukherjee, PJ Houghton, Leads from Indian medicinal plants with hypoglycemic potentials, J Ethnopharmacol 106 (1) (2006) 1–28.

[14]   T-Y Jeong, B-K Park, J-H Cho, Y-I Kim, Y-C Ahn, C-G Son, A prospective study on the safety of herbal medicines, used alone or with conventional medicines, J Ethnopharmacol 143 (3) (2012) 884–888.

[15]   KH Alqahtania, A Kama, KH Wonga, Z Abdelhaka, V Razmovski-Naumovskia, K Chana, et al., The pentacyclic triterpenoids in herbal medicines and their pharmacological activities in diabetes and diabetic complications, Curr Med Chem 20 (7) (2013).

[16]   EA Omar, A Kam, A Alqahtani, KM Li, V Razmovski-Naumovski, S Nammi, et al., Herbal medicines and nutraceuticals for diabetic vascular complications: mechanisms of action and bioactive phytochemicals, Curr Pharm Des 16 (34) (2010) 3776–3807.

[17]   N Malviya, S Jain, S Malviya, Antidiabetic potential of medicinal plants, Acta Pol Pharm 67 (2) (2010) 113–118.

[18]   WHO Expert Committee on Diabetes Mellitus, World Health Organization. WHO expert committee on diabetes mellitus: second report of the WHO expert committee. World Health Organization; 1980. http://apps.who.int/iris/bitstream/handle/10665/41399/WHO_TRS_646.pdf?sequence=1 [Accessed 5 March 2021].

[19]   RV Jayakumar, Herbal medicines for type-2 diabetes, Int J Diabetes Dev Ctries 30 (3) (2010) 111.

[20]   D Loew, M Kaszkin, Approaching the problem of bioequivalence of herbal medicinal products, Phytother Res 16 (8) (2002) 705–711.

[21]   PK Prabhakar, M Doble, Mechanism of action of natural products used in the treatment of diabetes mellitus, Chin J Integr Med (17) (2011) 563–574.

[22]   S Verma, M Gupta, H Popli, G Aggarwal, Treatment using herbal drug, Int J Phytomedicine 10 (1) (2018) 1–10.

[23]   K Kumar, V Fateh, B Verma, S Pandey, Some herbal drugs used for treatment of diabetes, Int J Res Dev Pharm Life Sci 3 (5) (2014) 1116–1120.

[24]   S Wachtel-Galor, IFF Benzie, Herbal medicine: an introduction to its history, usage, regulation, current trends, and research needs, in: IFF Benzie, S Wachtel-Galor (Eds.), Herbal Medicine: Biomolecular and Clinical Aspects, CRC Press/Taylor & Francis, Boca RatonFL, 2011.

[25]   R Gupta, KG Bajpai, S Johri, M Saxenaa, An overview of Indian novel traditional medicinal plants with anti-diabetic potentials, Complement Altern Med 5 (1) (2008) 1–17.

[26]   R Malvi, S Jain, S Khatri, A Patel, S Mishra, A review on antidiabetic medicinal plants and marketed herbal formulations, Int J Pharm Biol Arch 2 (2011) 1344–1355.

[27]   G Gebreyohannes, M Gebreyohannes, Medicinal values of garlic: a review, Int J Med Med Sci 5 (9) (2013) 401–408.

[28] MS Lakshmi, KSS Rani, UKT Reddy, A review on and the herbal plants used for its treatment, Asian J Pharm Clin Res 5 (4) (2012) 15–21.

[29] R Bordoloi, KN Dutta, A review on herbs used in the treatment of diabetes mellitus, J Pharm Chem Biol Sci 2 (2) (2014) 86–92.

[30] RMA Review, A review: anti diabetic medicinal plants used for diabetes mellitus, Bull Environ Pharmacol Life. Sciences 4 (2) (2015) 163–180.

[31] P Giovannini, MR Jayne, E Howes, S E, Medicinal plants used in the traditional management of diabetes and its sequelae in Central America: a review, J Ethnopharmacol 184 (2016) 58–71.

[32] CP Dwivedi, S Daspaul, Antidiabetic herbal drugs and polyherbal formulation used for diabetes: a review, J Phytopharmacol 2 (3) (2013) 44–51.

[33] Y Khan, I Aziz, B Bihari, H Kumar, M Roy, VK Verma, A review- phytomedicines used in treatment of diabetes, Asian J Pharm Res 4 (3) (2014) 135–154.

[34] MS Lakshmi, KSS Rani, UKT Reddy, A review on diabetes mellitus and the herbal plants used for its treatment, Asian J Pharm Clin Res 5 (4) (2012) 15–21.

[35] D Kumar, N Trivedi, RK Dixit, Herbal medicines used in the traditional indian medicinal system as a therapeutic treatment option for diabetes management: a review, World J Pharm Pharm Sci 4 (4) (2015) 368–385.

[36] JC Ozougwu, Anti-diabetic effects of Allium cepa (onions) aqueous extracts on alloxan-induced diabetic Rattus novergicus, J Med Plant Res 5 (7) (2011) 1134–1139.

[37] AE Nugroho, M Andrie, NK Warditiani, E Siswanto, S Pramono, E Lukitaningsih, Antidiabetic and antihiperlipidemic effect of Andrographis paniculata (Burm. F.) Nees and andrographolide in high-fructose-fat-fed rats, Indian J Pharmacol 44 (3) (2012) 377–381.

[38] BA Khan, A Abraham, S Leelamma, Hypoglycemic action of Murraya koeingii (curry leaf) and Brassica juncea (mustard): mechanism of action, Indian J Biochem Biophys 32 (2) (1995) 106–108.

[39] KJ Abesundara, MK matsuit, Alpha-glucosidase inhibitory activity of some Sri Lanka plant extracts one of which Cassia auriculata exerts a strong antihyperglycemic effect in rats comparable to the therapeutic drug acarbose, J Agric Food Chem 52 (9) (2004) 2541–2545.

[40] DS Narayan, VJ Patra, SC Dinda, Diabetes and Indian traditional medicines an overview, Int J Pharm Pharm Sci 4 (3) (2012) 45–53.

[41] E Jarald, SB Joshi, DC Jain, Diabetes and herbal medicines, Iran J Pharmacol Ther 7 (1) (2008) 97–106.

[42] Available from: BGR 34 for Diabetes. https://www.ayurtimes.com/bgr-34-fordiabetes. [Accessed 5 March 2021].

[43] IO Rosalie, E EL, Antidiabetic potentials of common herbal plants and plant products: a glance, Int J Herb. Med 4 (4) (2016) 90–97.

[44] AK Jugran, S Rawat, HP Devkota, ID Bhatt, Rawal RS. Diabetes and plant-derived natural products: From ethnopharmacological approaches to their potential for modern drug discovery and development, Phytother Res 35 (1) (2021) 223–245.

[45] H Li, A Sureda, HP Devkota, V Pittalà, D Barreca, AS Silva, et al., Curcumin, the golden spice in treating cardiovascular diseases, Biotechnol Adv 38 (2020) 107343.

[46] MF El-Azab, FM Attia, AM El-Mowafy, Novel role of curcumin combined with bone marrow transplantation in reversing experimental diabetes: Effects on pancreatic islet regeneration, oxidative stress, and inflammatory cytokines, Eur J Pharmacol 658 (1) (2011) 41–48.

[47] VO Gutierres, CM Pinheiro, RP Assis, RC Vendramini, MT Pepato, IL Brunetti, Curcumin-supplemented yoghurt improves physiological and biochemical markers of experimental diabetes, Br J Nutr 108 (3) (2012) 440–448.

[48] J Chen, S Mangelinckx, A Adams, ZT Wang, WL Li, N De Kimpe, Natural flavonoids as potential herbal medication for the treatment of DM and its complications, Nat Prod Commun 10 (1) (2015) 187–200.

[49] K Johnston, P Sharp, M Clifford, L Morgan, Dietary polyphenols decrease glucose uptake by human intestinal Caco-2 cells, FEBS Lett 579 (7) (2005) 1653–1657.

[50] U Ullmann, J Haller, JP Decourt, N Girault, J Girault, AS Richard Caudron, et al., A single ascending dose study of epigallocatechin gallate in healthy volunteers, J Int Med Res 31 (2) (2003) 88–101.

[51] KS Kang, N Yamabe, HY Kim, JH Park, T Yokozawa, Therapeutic potential of 20(S)-ginsenoside Rg3 against streptozotocin induced diabetic renal damage in rats, Eur J Pharmacol 591 (1–3) (2008) 266–272.

[52] JH Beattie, F Nicol, MJ Gordon, MD Reid, L Cantlay, GW Horgan, et al., Ginger phytochemicals mitigate the obesogenic effects of a high-fat diet in mice: a proteomic and biomarker network analysis, Mol Nutr Food Res 55 (2) (2011) 203–213.

[53] V Ilic, S Vukmirovic, NC Stilinovic, M Arsenovic, B Milijasevic, Insight into anti-diabetic effect of low dose of stevioside, Biomed Pharmacother 90 (2017) 216–221.

[54] TH Chen, SC Chen, P Chan, YL Chu, HY Yang, JT Cheng, Mechanism of the hypoglycemic effect of stevioside, a glycoside of Stevia rebaudiana, Planta Medica 71 (2) (2005) 108–113.

[55] EB Ferreira, RN de Assis, MA da Costa, WA do Prado, L de Araujo Funari Ferri, RB Bazotte, Comparative effects of Stevia rebaudiana leaves and stevioside on glycaemia and hepatic gluconeogenesis, Planta Medica 72 (8) (2006) 691–696.

[56] PB Jeppesen, S Gregersen, KK Alstrup, K Hermansen, Stevioside induces antihyperglycaemic, insulinotropic and glucagonostatic effects in vivo: studies in the diabetic Goto-Kakizaki (GK) rats, Phytomedicine 9 (1) (2002) 9–14.

[57] AWK Yeung, IE Orhan, BB Aggarwal, M Battino, TBA Belwal, AG Atanasov, Berberine, a popular dietary supplement for human and animal health: Quantitative research literature analysis-a review, Anim Sci Pap Rep 38 (1) (2020) 5–19.

[58] YS Lee, WS Kim, KH Kim, MJ Yoon, HJ Cho, Y Shen, JB Kim, Berberine, a natural plant product, activates AMP- activated protein kinase with beneficial metabolic effects in diabetic and insulin-resistant states, Diabetes 55 (8) (2006) 2256–2264.

[59] Y Zhang, X Li, D Zou, W Liu, J Yang, N Zhu, G Ning, Treatment of type 2 diabetes and dyslipidemia with the natural plant alkaloid berberine, J Clin Endocrinol Metab 93 (7) (2008) 2559–2565.

[60] W Chang, L Chen, GM Hatch, Berberine as a therapy for type 2 diabetes and its complications: From mechanism of action to clinical studies, Biochem Cell Biol 93 (5) (2015) 479–486.

[61] W Chang, Non-coding RNAs and Berberine: A new mechanism of its anti-diabetic activities, Eur J Pharmacol 795 (2017) 8–12.

[62] J Lan, Y Zhao, F Dong, Z Yan, W Zheng, J Fan, G Sun, Meta-analysis of the effect and safety of berberine in the treatment of type 2 DM, hyperlipemia and hypertension, J Ethnopharmacol 161 (2015) 69–81.

[63] SE Choi, TH Kim, SA Yi, YC Hwang, WS Hwang, SJ Choe, et al., Capsaicin attenuates palmitate-induced expression of macrophage inflammatory protein 1 and interleukin 8 by increasing palmitate oxidation and reducing c-Jun activation in THP-1 (human acute monocytic leukemia cell) cells, Nutr Res 31 (6) (2011) 468–478.

CHAPTER 9

Diabetes – a metabolic disorder: Herbal medicines on rescue

Ahmad Ali[a], Johra Khan[b] and Bader Alshehri[b]

[a]University Department of Life Sciences, University of Mumbai, Mumbai, India [b]Department of Laboratory Sciences, College of Applied Medical Sciences, Majmaah University, Majmaah, Saudi Arabia

9.1 Introduction

Diabetes mellitus (DM) is a metabolic disorder caused due to the inability of the pancreas to produce insulin or the inability of cells to effectively utilize the produced insulin. The former condition is characterized as type 1 DM whereas the latter as type 2 DM. In addition, it can be clinically manifested in the form of gestational diabetes, neonatal diabetes or drug- induced diabetes. Recent times have witnessed an increased occurrence of type 2 DM, whereas the type 1 DM is a less common auto-immune disorder commonly occurring in children. Several factors like ageing, sedentary lifestyle, unhealthy diet schedule or genetic predisposition may progressively increase the chances of development of diabetes [1,2]. The obvious clinical symptoms of diabetes include high blood sugar level and presence of sugar in urine. The natural defences of the human body try to correct these symptoms by releasing more fluids in the blood stream to normalize the blood sugar level. Hence the patients experience excessive thirst and dehydration, and frequently have liquids in the form of water or juice. This, in turn, causes polyuria [3,4].

The statistical observations by World Health Organization (WHO) have indicated a drastic increase in diabetic patients in the near future. They are expected to reach approximately 366 million by the year 2030 and 629 million by 2045 [5,6]. Although the management of DM seems simple, considering its characteristics, the contribution of several unclear factors often complicates the pathology of the disease. Moreover, the sustained hyperglycemic condition may damage blood vessels and other vital body organs. In addition, the loss of structure and function of proteins can lead to several degenerative disorders and ageing [7–9]. Hence, the increased diagnosis of DM globally is a matter of great concern and requires immediate attention.

Diabetes cannot be treated completely. However, several regimens are available for effective management of the disease. Conventional therapy includes the use of synthetic

Herbal Medicines: A Boon for Healthy Human Life.
DOI: https://doi.org/10.1016/B978-0-323-90572-5.00013-5

223

hypoglycemic agents that inhibit the digestion of complex sugars or enhance the uptake of glucose by adipose and muscle tissues [10]. Synthetic insulin and agents that stimulate the pancreas to produce more insulin are also available commercially [11]. However, all these agents consequently promote the development of complications of DM mainly by inducing oxidative stress, DNA damage and increased inflammatory responses [12,13].

9.2 Herbal plants as a remedy for diabetes (Fig. 9.1)

Due to the severe complications of conventional anti-diabetic drugs, alternative remedies are sought by patients. The traditional systems like that of Ayurveda, Unani and Chinese medicines use a variety of herbs for managing the complications of diabetes. These systems advice modification of normal diet in diabetic patients and prescribe herbs that naturally regulate blood sugar levels. For instance, the intake of fibers can regulate blood sugar and insulin concentrations. Hence, the use of psyllium husk, barley and other high fiber foods are promoted, as the first step in natural systems of medicine, to manage diabetes [14]. In addition, the intake of green leafy vegetables is highly encouraged [15].

On ingestion, food or medicine is metabolized in the digestive system. When the herbs exhibiting anti-diabetic effect are metabolized, the biocomponents of plant interact with several regulatory enzymes and thus modulate their function. This results in useful outcomes due to activation or inhibition of various metabolic biochemical pathways involved in regulating the blood glucose levels. Several factors are known to influence the blood sugar level in diabetic patients. These factors are represented in Fig. 9.1. The conventional anti-diabetic drugs are often designed to work against one or more of these factors and manage diabetes. Similarly, several herbs are also identified and reported in the literature that can modulate the regulatory factors of diabetes and result in the effective management of the same. For instance, several herbal plants can potentially inhibit or substantially reduce the metabolism of carbohydrates [16,17].

Since enzymes like Q-glucosidase and Q-amylase are necessary for catabolism of glycogen to glucose, its inhibition can significantly control the release of glucose in blood cells. Similarly, a variety of herbs can stimulate the production of insulin and aid in proliferation of the pancreatic beta cell [18,19]. Hence, they can be used as natural phytoproducts or medicines for the management of diabetes and related disorders. A list of various herbs and their specific mode of action as an anti-diabetic agent is indicated in Table 9.1. Other advantages of phytomedicines are their easy digestibility, cost- effectiveness, non-toxicity and negligible side effects, if any [17,20].

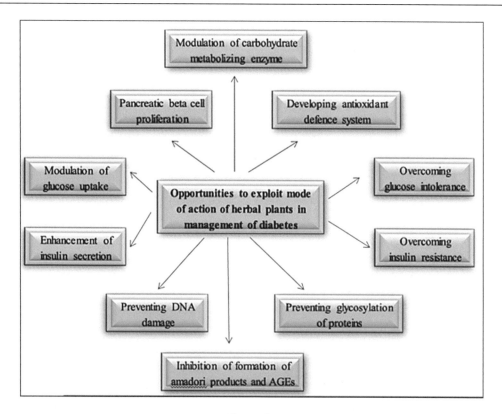

Figure 9.1
Factors influencing blood sugar levels.

9.3 Herbal plants exhibiting multiple mode of action in management of diabetes (Fig. 9.2)

In addition to the plants listed in Table 9.1, several individual studies have collectively implicated multiple modes of action of several herbal plants on diabetes. The most effective and commonly available herbs exhibiting anti-diabetic effect and their mechanisms of action are discussed below. Fig. 9.2 represents few examples of plants with anti-diabetic activity.

9.3.1 Gymnema sylvestre *(gurmar)*

The plant *Gymnema sylvestre* is commonly known as "gurmar" which can be literally translated to "sugar destroyer." It has been used in the Ayurvedic system of medicine since ages. However, very few clinical studies have proved the anti-diabetic potential of this plant. A past study reported that gymnemic acid IV obtained from crude methanol extract using the leaves of *G. sylvestre* could promote the release of insulin and manage obesity in streptozotocin

Table 9.1: Mode of action of herbal plants exhibiting antidiabetic activity.

Sr. No.	Plant (common name)	Plant part/extract/component	Mode of action	Effect	Experimental model	Reference
			Herbs that modulate glucose uptake			
1.	*Coptis chinensis* (Huang lian)	Polysaccharide obtained by aqueous extraction and alcohol precipitation method	Regulation of JNK/IRS1/PI3K pathway and improvement in Glut4 expression	Initiation of antioxidative mechanisms and increased glucose uptake by muscle and adipose tissues	Streptozotocin induced diabetic Male Wistar rats	[21]
2.	*Azadirachta indica* (Neem)	Fresh leaves	Blocking the action of epinephrine on glucose metabolism	Increase in peripheral glucose utilization to reduce blood sugar level	Streptozotocin induced diabetic rat	[22,23]
			Herbs that modulate carbohydrate metabolizing enzymes			
3.	*Ocimun sanctum* (Holy Basil)	Ethanolic leaf extract	Decrease in Serum Glutamate Oxaloacetate Transaminases (SGOT) and Serum Glutamate Pyruvate Transaminases (SGPT)	Lipid metabolism and glucose homeostatis	Alloxan induced diabetic model rats	[24]
4.	*Mangifera indica* (Mango)	Magniferin in ripe mango	Inhibition of α-amylase and α-glucosidase	Slowing of glucose metabolism to reduce post-prandial glucose level	In silico docking analysis with Autodock software	[25,26]
5.	*Glycyrrhiza foetida* (Licorice)	Amorfrutins extracted from roots	Activation of peroxide proliferator-activated receptor-γ (PPAR-γ)	Lipid metabolism and glucose homeostatis	Diet-induced obese and genetically manipulated diabetic mouse models	[22,27]
			Herbs that enhance insulin secretion and pancreatic beta cell proliferation			
6.	*Anoectochilus roxburghii* (Jewelled orchid)	n- butanol extract of whole plant	Regeneration of pancreatic β- cells	Enhancement of insulin production and hence reduction in blood glucose level	Streptozotocin induced diabetic rat	[28]
7.	*Carthamus tinctorius* (safflower)	Serotonin derivatives extracted in Hydro-alcoholic extract of dried flowers	Increase in size of islets of Langerhans	Reversal of β- cell cytotoxicity and hence increase in insulin production	Alloxan induced diabetic male white Wistar rats	[29]

(continued on next page)

8.	*Capsicum annuum* (Capsicum)	Capsaicin solution prepared in 2% gum acacia solution	Inhibition of CYP2C9 and CYP3A4	Enhancement in activity of gliclazide leading to reduced blood glucose levels, increased insulin levels and enhanced β-cell function	Alloxan induced diabetic model rats and rabbits	[30]
			Herbs that overcome insulin resistance and glucose intolerance			
9.	*Vaccinium corymbosum* (Blueberry)	Fruit	Improvement in the homeostatic model assessment-estimated insulin resistance (HOMA-IR model)	Increased insulin sensitivity	Diet-induced obese and genetically manipulated diabetic mouse models	[28]
10.	Litchi chinensis (Lychee)	Pavetannin B2 and procyanidin A2 in ethanol extract of seed	Inhibition of PTP1b	Increased insulin sensitivity	In vitro enzyme inhibitory assays	[31]
			Herbs that develop antioxidant defence system			
11.	Capparis decidua (Caper berry)	Seed extracts	Modulation of superoxide dismutase and catalase	Reduction in oxidative stress and increased glucose uptake by muscle and adipose tissues	In vitro assays	[32]
			Herbs that inhibit formation of advanced glycation end products			
12.	Seriphium plumosum (bankrupt bush)	Acetone extract	Anti-glycation activity	Decreases the oxidation of glucose and amadori products to prevent or delay the formation of advanced glycation end-products	In vitro assays	[33]
13.	Mentha arvensis (wild mint)	Methanolic extract of leaves	High anti-glycation activity	Inhibition of α-amylase and α-glucosidase activity leading to decrease in blood glucose levels	Starch induced diabetic Wistar rats	[29]

Figure 9.2
Few examples of common plants exhibiting antidiabetic activity are (A) gurmar plant, (B) fenugreek seeds, (C) garlic, (D) bitter melon, (E) aloe vera, (F) green tea, (G) ginseng plant, (H) ginseng roots, (I) onion, (J) cinnamon, and (K) black seeds.

induced diabetic mice [34]. Gymnemic acid was later found to interact with glyceraldehyde-3-phosphate dehydrogenase which is a key enzyme in glycolysis pathway [34,35]. The extracts of *G. sylvestre* leaves are also reported to play an important role in regeneration of pancreatic β-cells [36,37]. An interesting study reported that the similarity in atomic arrangements of gymnemic acids to sugars allows them to attach competitively to the external layer of intestine which prevents the absorption of sugar molecules by intestine. In addition, a decrease in blood HbA1c levels and improved glucose metabolism have been reported on supplementation of *G. sylvestre* formulations. Moreover, several studies have reported the antioxidant activities of solvent extract fractions of *G. sylvestre* leaves [38,39].

9.3.2 Trigonella foenum graecum (fenugreek)

Another common herb which has proved to be immensely beneficial in management of DM is fenugreek seeds. It is a pungent aromatic spice widely used as a flavoring agent in Indian recipes. The alkaloids (gonelline, nicotinic acid, and coumarin) present in the fenugreek seed is reported to prevent all catabolic processes in the metabolism of sugars. Besides, they aid in elevating the levels of glucose-6-phosphate dehydrogenase which enhance glucose uptake by adipose and muscle tissue and reduce the blood sugar levels considerably. The seeds are also rich in fiber. This indirectly helps in delaying the intestinal absorption of glucose [40]. There also exist strong evidence of decrease in fasting sugar levels and improvement in dyslipidemia on intake of fenugreek seeds for 8 weeks [41,42]. Moreover, both human and animal studies have confirmed the glucose stimulated insulin release on supplementation of 4-hydroxy-iso-leucine- a novel amino acid of fenugreek seeds. Other compounds of fenugreek seeds like diosgenin, galactomannan, trigoneosides have been reported to be beneficial in reverting insulin resistance [43,44].

9.3.3 Allium sativum (garlic)

Allium sativum commonly known as garlic is an irreplaceable kitchen ingredient in Indian homes. It is used for the treatment of numerous diseases. In fact, it is a common ingredient in the formulations of the tradition medicinal systems. The clinical trial studies have reported a decrease in cholesterol levels on intake of garlic with no effect on blood sugar levels [45]. However, the garlic oil has been reported to improve GLUT4 expression, and hence insulin and glucose tolerance, in streptozotocin induced diabetic rats [46,47]. Also, allicin- a sulfur-containing component of garlic has been reported to show significant hypoglycemic activity by exhibiting increased hepatic metabolism, insulin secretion and decreasing insulin resistance [48,49].

9.3.4 Momordica charantia (bitter melon)

Bitter melon is mainly known for its antioxidant and anti-diabetic activities. Its efficacy has been evaluated in alloxan-induced diabetic mice. The study reported that charantin - the active ingredient of bitter melon stimulates glucose utilization of adipocytes and skeletal muscle [50]. It is also responsible for down-regulation of MAPKs and NF-κB that induces insulin signaling. The extracts of this plant were found to up-regulate the peroxisome proliferator-activated receptor gamma (PPAR-γ gene) expression. Moreover, it acts as a negative regulator of protein-tyrosine phosphatase 1B [51,52]. The modulation of above genes allows translocation of GLUT4 to the cell membrane which enhances glucose uptake and stimulates insulin secretion from pancreatic beta cells respectively. Hence, bitter melon is also considered an insulin-mimetic agent. In fact, a polypeptide called p-insulin has been identified in alcohol extracts of fruit as well as seeds of this plant [53]. Overall, the efficacy of bitter melon as an anti-diabetic agent is due to combined effects of enhanced glucose uptake and decrease in hepatic glucose production with simultaneous increase in the rate of its conversion into glycogen. The decrease in glucose synthesis results due to inhibition of key gluconeogenic enzymes like glucose-6-phosphatase and fructose-1, 6-biphosphatase [21,54]. Besides, it is also reported to have beneficial effects on diabetes- related complications like neuropathy, dyslipidemia, renal diseases as well as ophthalmologic complications [51].

9.3.5 Aloe barbadensis *(aloe vera)*

Aloe vera is typically recognized for its efficacy as a skin ailment. Lately, it is increasingly being studied for its effectiveness against more complicated diseases. Few studies have reported lowering of fasting blood sugar levels on intake of Aloe vera gel due to stimulation of β-cells of pancreas. In addition, a decrease in insulin resistance has also been reported with the use of aloe vera gel. Other studies have also reported a decrease in total cholesterol and triglyceride levels in diabetic rats. Hence, it may also be useful in the management of diabetes related complications [55,56]. Interestingly, most of the studies have reported indirect effect of aloe vera gel on diabetes management through anti-oxidation, prevention of formation of serum malondialdehyde and advanced glycation end products and anti- inflammatory responses. Among the active ingredients isolated from aloe vera, only glucomannan (a water-soluble fiber present in very small concentration) is identified as an antidiabetic agent [57,58].

9.3.6 Panex quiquefolius *(ginseng)*

Ginseng roots are believed to be loaded with components which are beneficial for overall health. Hence, it is used in several traditional formulations for the treatment of diseases as well as boosting immunity [59]. Although the mode of action of ginseng and its active ingredients, collectively known as ginsenosides (saponin compounds), is not well understood in management of diabetes, they are suggested to be involved in stimulating the insulin signaling pathway and insulin sensitivity. The roots of ginseng are commonly used; however, anti-diabetic activity of every part of this plant is reported in literature. Unlike the roots of ginseng, the berries and leaves are known to improve glucose tolerance in streptozotocin and high fat diet induced diabetic rats [60,61]. All plant parts are also effective in treating diabetes related complications like hyperlipidemia and hypercholesterolemia [62]. A study reported the positive effect of ginseng extracts on protein kinase B and insulin receptor substrate-1 which in turn stimulates the pancreatic β-cells to produce more insulin. Besides, the translocation of GLUT 4 is significantly increased on uptake of ginseng extracts. It also helps in alleviating the complications of diabetes by reducing oxidative stress and preventing inflammations [63]. However, the formulations containing higher concentration of ginseng extract should be taken only after prescription. This is because ginseng roots act as a natural blood thinner. Hence, it helps in overcoming common health issues like weakness, headache and anxiety (which also occurs in diabetes), but at the same time may interfere with certain medications like warfarin [64].

9.3.7 Cinnamomum zeylanicum *(cinnamon)*

The bark of *Cinnamomum zeylanicum* is a rich source of flavonoids, glycosides, anthraquinones, terpenoids, coumarins and tannins. These compounds are responsible for a number of health benefits. They are commonly used as flavoring agents in food and drinks around the world. The phenolic compounds of *C. zeylanicum* are particularly of interest to the researchers since they have shown insulin mimetic (polyphenol type A) and insulin signaling effects in diabetic mice. Several studies have reported the competitive and reversible inhibitory effect of *C. zeylanicum* on yeast and mammalian α-glucosidase [65]. They are also reported to effectively suppress the rise in blood sugar levels after meal in streptozotocin-induced diabetic rats. A recent study also suggested significant lowering of HbA1c, fasting blood sugar, cholesterol and lipid levels in diabetic human patients on intake of 1g of cinnamon for 12 weeks [66]. A study carried out to verify the anti-diabetic effect of cinnamon used diabetic mice models which were not given insulin treatment. Instead, they were injected with water extract of cinnamon. The study confirmed the upregulation of mitochondrial uncoupling protein and enhancement of translocation of GLUT4 in the muscle and adipose tissues in cinnamon extract treated mice models [67].

9.3.8 Allium cepa *(Onion)*

Onion is undoubtedly the most widely used food ingredient in the world. The organic sulfur compounds (allyl propyl disulfide and S-methyl cysteine sulfoxide) are believed to be the active hypoglycemic components of onion. The antidiabetic activity of Egyptian onion extract has been reported as early as 1965 [68]. Another study has reported that treatment of streptozotocin induced diabetic mice with ripe onion juice for 7 days significantly reduced the blood sugar levels [69]. Moreover, the onion skin extracts have shown a dose dependent antioxidant as well as anti-α-amylase activity in vitro [70]. This characteristic is especially helpful in formulation of drug preparations, and clinical trial studies on effect of doses on different phase and pathophysiology of a disease. Although the exact mechanism of onion in management of diabetes is not clear, several studies have reported that the indirect mechanisms like anti oxidation and prevention of formation of glycated proteins may be responsible for the anti-diabetic activity of onion [71].

9.3.9 Camellia sinensis (Green tea)

Camellia sinensis or green tea is increasingly becoming a popular drink due to its proven anti-oxidative and anti-inflammatory properties. The polyphenols of green tea like catechin, epicatechin, epigallocatechin, and their gallates, tannin, and caffeine have shown several highly beneficial pharmacological properties including anti-cancer activity. A novel vitamin extracted from green tea i.e., pyrroloquinoline quinone and epigallo-catechin-3-gallate are identified as the main constituent responsible for anti-diabetic activity of green tea. Extensive studies have been carried out on the anti-diabetic activity of green tea. These studies have indicated that green tea can regulate glucose homeostasis by stimulating cAMP/PKA, AGEs/RAGE/TGF-β1 signaling pathway and GLUT4 translocation via PI3K/AKT. Moreover, it can stimulate glucose uptake by ameliorating TNF-α induced insulin resistance and inhibiting p38 and JNK pathways in HepG2 cells. In addition, they also exhibit α-glucosidase and α- amylase inhibitory activities along with anti-oxidation potential, anti-inflammatory activity, and lipid and cholesterol lowering ability [72].

9.3.10 Nigella sativa (Black seed)

Nigella sativa, commonly known as black seed, is most widely used in Ayurvedic and Unani formulations of medicines due to the range of health benefits they show. Most interestingly, the studies on black seed have indicated its efficacy on almost all aspects of the diabetes. Powdered seed supplements have been reported to reduce fasting and postprandial glucose levels, lipids, triglycerides, bad cholesterol as well as glycosylated hemoglobin [73]. In addition, they decrease insulin resistance, inhibit gluconeogenesis pathways and activate gluconeogenic enzymes (like glucose-6-phosphatase and fructose 1, 6-bisphosphatase), and promote the proliferation of islets cells. Studies on diabetic mice models have indicated the effectiveness of black seed extracts on glucose tolerance and management of diabetes related complications [74]. Their efficacy is reported on the functioning of liver, kidney and intestinal enzymes that promote the management of diabetes. Besides, the black seeds can actively reverse oxidative stress and initiate anti-inflammatory responses to help in regeneration and protection of delicate beta cells [33,75].

9.4 Conclusions

Whether we use modern or traditional sources of medicines, we are entirely dependent on natural (mostly plant based) products for the treatment of diseases. The importance of traditional systems of medicines that utilize herbal compounds in its crude form can be comprehended based on the fact that over 80% population in developing countries still rely on them. Moreover, the modern medicines are nothing, but purified or synthetic compounds originally identified in the natural flora. Hence, considering the unusual complications associated with diabetes, it is not unusual to witness an increased interest in herbal remedies to manage it. The extensive literature on antidiabetic activity of plants further proves the efficacy of a variety of herbs that can be used for preparing formulations that show negligible or no side effects as compared to existing antidiabetic drugs. However, the vast literature on ethnopharmacological aspects also points to the lack of, or very few, certified and regulatory clinical trial studies of common as well as rare herbs.

References

[1] P Zimmet, K Alberti, J Shaw, Global and societal implications of the diabetes epidemic, Nature 414 (6865) (2001) 782–787.

[2] J Khan, Magnesium deficiency and its correlation with insulin resistance in obese females in Majmaah, Am J Physiol 237 (2018) E214–EE23.

[3] KF Hanssen, Blood glucose control and microvascular and macrovascular complications in diabetes, Diabetes 46 (Supplement 2) (1997) S101–S1S3.

[4] J Khan, MA Alaidarous, A Naseem, Dyslipidemia relationship with socioeconomic status in East Champaran population, Int J Pharm Res Allied Sci 9 (2) (2020) 130–138.

[5] YM Farag, MR Gaballa, Diabesity: an overview of a rising epidemic, Nephrol Dial Transplant 26 (1) (2011) 28–35.

[6] B Tripathy, HB Chandalia, RSSDI Textbook of Diabetes Mellitus. JAYPEE Brothers, India (2012).

[7] X Luo, J Wu, S Jing, L-J Yan, Hyperglycemic stress and carbon stress in diabetic glucotoxicity, Aging Dis 7 (1) (2016) 90.

[8] M Piero, G Nzaro, J Njagi, Diabetes mellitus-a devastating metabolic disorder, Asian J Biomed Pharm Sci 5 (40) (2015) 1.

[9] N Sajjad, MM Mir, J Khan, IA Rather, EA Bhat, Recognition of TRAIP with TRAFs: current understanding and associated diseases, Int J Biochem Cell Biol 115 (2019) 105589.

[10] A Chaudhury, C Duvoor, VS Reddy Dendi, S Kraleti, A Chada, R Ravilla, et al., Clinical review of antidiabetic drugs: implications for type 2 diabetes mellitus management, Front Endocrinol 8 (2017) 6.

[11] Donner T., Sarkar S. Insulin–pharmacology, therapeutic regimens, and principles of intensive insulin therapy. 2015.

[12] E Shim, J Babu, Glycated albumin produced in diabetic hyperglycemia promotes monocyte secretion of inflammatory cytokines and bacterial adherence to epithelial cells, J Periodontal Res 50 (2) (2015) 197–204.

[13] SM Son, Reactive oxygen and nitrogen species in pathogenesis of vascular complications of diabetes, Diabetes Metab J 36 (3) (2012) 190.

[14] M Galisteo, R Morón, L Rivera, R Romero, A Anguera, A Zarzuelo, Plantago ovata husks-supplemented diet ameliorates metabolic alterations in obese Zucker rats through activation of AMP-activated protein kinase. Comparative study with other dietary fibers, Clin Nutr 29 (2) (2010) 261–267.

[15] PY Wang, JC Fang, ZH Gao, C Zhang, SY Xie, Higher intake of fruits, vegetables or their fiber reduces the risk of type 2 diabetes: a meta-analysis, J Diabetes Investig 7 (1) (2016) 56–69.

[16] MF Mahomoodally, AH Subratty, A Gurib-Fakim, MI Choudhary, KS Nahar, Traditional medicinal herbs and food plants have the potential to inhibit key carbohydrate hydrolyzing enzymes in vitro and reduce postprandial blood glucose peaks in vivo, Sci World J 2012 (2012) 1–9, doi:10.1100/2012/285284.

[17]   SS Abd El-Karim, MM Anwar, YM Syam, MA Nael, HF Ali, MA Motaleb, Rational design and synthesis of new tetralin-sulfonamide derivatives as potent anti-diabetics and DPP-4 inhibitors: 2-D & 3-D QSAR, in vivo radiolabeling and bio distribution studies, Bioorg Chem 81 (2018) 481–493.

[18]   YS Oh, Plant-derived compounds targeting pancreatic beta cells for the treatment of diabetes, Evid Based Complement Alternat Med 2015 (2015) 1–12. https://doi.org/10.1155/2015/629863.

[19]   F Hajimoosayi, S Jahanian Sadatmahalleh, A Kazemnejad, R Pirjani, Effect of ginger on the blood glucose level of women with gestational diabetes mellitus (GDM) with impaired glucose tolerance test (GTT): a randomized double-blind placebo-controlled trial, BMC Complement Med Ther 20 (2020) 1–7.

[20]   O Said, S Fulder, K Khalil, H Azaizeh, E Kassis, B Saad, Maintaining a physiological blood glucose level with 'glucolevel', a combination of four anti-diabetes plants used in the traditional Arab herbal medicine, Evid Based Complement Alternat Med 5 (4) (2008) 421–428.

[21]   S Jiang, Y Wang, D Ren, J Li, G Yuan, L An, et al., Antidiabetic mechanism of Coptis chinensis polysaccharide through its antioxidant property involving the JNK pathway, Pharm Biol 53 (7) (2015) 1022–1029.

[22]   SN Ahamed, J Khan, S Rahamathulla, T Venkateswarulu, S Krupanidhu, Metabolic and molecular effects of edible oils on PPAR modulators in rabbit liver, Adv Life Sci 8 (2) (2021) 154–159.

[23]   R Chattopadhyay, R Chattopadhyay, A Nandy, G Poddar, S Maitra, The effect of fresh leaves of Azadirachta indica on glucose uptake and glycogen content in the isolated rat hemi diaphragm, Bull Calcutta Sch Trop Med 35 (1987) 8–12.

[24]   C Egbuna, CG Awuchi, G Kushwaha, M Rudrapal, KC Patrick-Iwuanyanwu, O Singh, et al., Bioactive compounds effective against type 2 diabetes mellitus: a systematic review, Curr Top Med Chem 2021 (2021) 1–29, doi:10.2174/1568026621666210509161059. PMID:33966619.

[25]   J Khan, PK Deb, S Priya, KD Medina, R Devi, SG Walode, et al., Dietary flavonoids: cardioprotective potential with antioxidant effects and their pharmacokinetic, toxicological and therapeutic concerns, Molecules 26 (2021) 4021, doi:https://doi.org/10.3390/molecules26134021.

[26]   V Sekar, S Chakraborty, S Mani, V Sali, H Vasanthi, Mangiferin from Mangifera indica fruits reduces postprandial glucose level by inhibiting α-glucosidase and α-amylase activity, S Afr J Bot 120 (2019) 129–134.

[27]   A Lavecchia, C Di Giovanni, Amorfrutins are efficient modulators of peroxisome proliferator-activated receptor gamma (PPARγ) with potent antidiabetic and anticancer properties: a patent evaluation of WO2014177593 A1, Expert Opin Ther Pat 25 (11) (2015) 1341–1347.

[28]   D Li, Y Zhang, Y Liu, R Sun, M Xia, Purified anthocyanin supplementation reduces dyslipidemia, enhances antioxidant capacity, and prevents insulin resistance in diabetic patients, J Nutr 145 (4) (2015) 742–748.

[29]   SB Agawane, VS Gupta, MJ Kulkarni, AK Bhattacharya, SS Koratkar, Chemo-biological evaluation of antidiabetic activity of Mentha arvensis L. and its role in inhibition of advanced glycation end products, J Ayurveda Integr Med 10 (3) (2019) 166–170.

[30]   U Lagisetty, H Mohammed, S Ramaiah, Effect of capsaicin on pharmacodynamic and pharmacokinetics of gliclazide in animal models with diabetes, Pharmacogn Res 10 (4) (2018) 437.

[31]   S-A Choi, JE Lee, MJ Kyung, JH Youn, JB Oh, WK Whang, Anti-diabetic functional food with wasted litchi seed and standard of quality control, Appl Biol Chem 60 (2) (2017) 197–204.

[32]   M Zia-Ul-Haq, S Ćavar, M Qayum, I Imran, Feo Vd, Compositional studies: antioxidant and antidiabetic activities of Capparis decidua (Forsk.) Edgew, Int J Mol Sci 12 (12) (2011) 8846–8861.

[33]   R Pandey, D Kumar, A Ahmad, Nigella sativa seed extracts prevent the glycation of protein and DNA, Curr Perspect Med Aromat Plants 1 (1) (2018) 1–7.

[34]   Y Sugihara, H Nojima, H Matsuda, T Murakami, M Yoshikawa, I Kimura, Antihyperglycemic effects of gymnemic acid IV, a compound derived from Gymnema sylvestre leaves in streptozotocin-diabetic mice, J Asian Nat Prod Res 2 (4) (2000) 321–327.

[35]   S Ishijima, T Takashima, T Ikemura, Y Izutani, Gymnemic acid interacts with mammalian glycerol-3-phosphate dehydrogenase, Mol Cell Biochem 310 (1) (2008) 203–208.

[36]   VR Aralelimath, SB Bhise, Anti-diabetic effects of gymnema sylvester extract on streptozotocin induced diabetic rats and possible B-cell protective and regenerative evaluations, Dig J Nanomater Biostructures 7 (1) (2012) 135–142.

[37]   OO Olaokun, LJ McGaw, IJ van Rensburg, JN Eloff, V Naidoo, Antidiabetic activity of the ethyl acetate fraction of Ficus lutea (Moraceae) leaf extract: comparison of an in vitro assay with an in vivo obese mouse model, BMC Complement Altern Med 16 (1) (2016) 1–12.

[38]   ABA Ahmed, N Komalavalli, M Muthukumar, J Benjamini, A Rao, M Rao, Pharmacological activities, phytochemical investigations and in vitro studies of Gymnema sylvestre R. Br.—a historical review, Comprehensive Bioactive Natural Products Vol 1 Potential and Challenges. Studium Press Llc, USA (2009) 75–99.

[39]   Z Wang, J Wang, P Chan, Treating type 2 diabetes mellitus with traditional Chinese and Indian medicinal herbs, Evid Based Complement Altern Med 2013 (2013) 1–17. https://doi.org/10.1155/2013/343594.

[40]   M Ranade, N Mudgalkar, A simple dietary addition of fenugreek seed leads to the reduction in blood glucose levels: a parallel group, randomized single-blind trial, Ayu 38 (1-2) (2017) 24.

[41]   N Kassaian, L Azadbakht, B Forghani, M Amini, Effect of fenugreek seeds on blood glucose and lipid profiles in type 2 diabetic patients, Int J Vitam Nutr Res 79 (1) (2009) 34–39.

[42]   Y Belaïd-Nouira, H Bakhta, Z Haouas, I Flehi-Slim, F Neffati, MF Najjar, et al., Fenugreek seeds, a hepatoprotector forage crop against chronic AlCl 3 toxicity, BMC Vet Res 9 (1) (2013) 1–9.

[43]   S Fuller, JM Stephens, Diosgenin, 4-hydroxyisoleucine, and fiber from fenugreek: mechanisms of actions and potential effects on metabolic syndrome, Adv Nutr 6 (2) (2015) 189–197.

[44]   AD Kandhare, SL Bodhankar, V Mohan, PA Thakurdesai, Prophylactic efficacy and possible mechanisms of oligosaccharides based standardized fenugreek seed extract on high-fat diet-induced insulin resistance in C57BL/6 mice, J Appl Pharm Sci 5 (2015) 35–45.

[45]   J Wang, X Zhang, H Lan, W Wang, Effect of garlic supplement in the management of type 2 diabetes mellitus (T2DM): a meta-analysis of randomized controlled trials, Food Nutr Res 61 (1) (2017) 1377571.

[46]   A Ali, R Sharma, S Sivakami, Role of natural compounds in the prevention of DNA and proteins damage by glycation, Bionano Front 7 (2014) 25–30.

[47]   N Eser, A Yoldas, A Turk, A Kalaycı Yigin, A Yalcin, M Cicek, Ameliorative effects of garlic oil on FNDC5 and irisin sensitivity in liver of streptozotocin-induced diabetic rats, J Pharm Pharmacol 73 (6) (2021) 824–834.

[48]   SJ Showande, TO Fakeye, M Kajula, J Hokkanen, A Tolonen, Potential inhibition of major human cytochrome P450 isoenzymes by selected tropical medicinal herbs—Implication for herb–drug interactions, Food Sci Nutr 7 (1) (2019) 44–55.

[49]   S Ramalingam, M Packirisamy, M Karuppiah, G Vasu, R Gopalakrishnan, K Gothandam, et al., Effect of β-sitosterol on glucose homeostasis by sensitization of insulin resistance via enhanced protein expression of PPRγ and glucose transporter 4 in high fat diet and streptozotocin-induced diabetic rats, Cytotechnology 72 (3) (2020) 357–366.

[50]   II Abo-Ghanema, RM Saleh, Some Physiological effects of Momordica charantia and Trigonella foenumgraecum extracts in diabetic rats as compared with Cidophage®, Int J Anim Vet Sci 6 (4) (2012) 222–230.

[51]   S Jia, M Shen, F Zhang, J Xie, Recent advances in Momordica charantia: functional components and biological activities, Int J Mol Sci 18 (12) (2017) 2555.

[52]   A Mishra, S Gautam, S Pal, A Mishra, AK Rawat, R Maurya, et al., Effect of Momordica charantia fruits on streptozotocin-induced diabetes mellitus and its associated complications, Int J Pharm Pharm Sci 7 (2015) 356–363.

[53]   A Paul, SS Raychaudhuri, Medicinal uses and molecular identification of two Momordica charantia varieties-a review, Electron J Biol 6 (2) (2010) 43–51.

[54]   M Abdollah, A Zuki, A Goh, A Rezaeizadeh, M Noordin, The effects of Momordica charantia on the liver in streptozotocin-induced diabetes in neonatal rats, Afr J Biotechnol 9 (31) (2010) 5004–5012.

[55]   M Yimam, L Brownell, Q Jia, Aloesin as a medical food ingredient for systemic oxidative stress of diabetes, World J Diabetes 6 (9) (2015) 1097.

[56] L-W Qi, E-H Liu, C Chu, Y-B Peng, H-X Cai, P Li, Anti-diabetic agents from natural products—an update from 2004 to 2009, Curr Top Med Chem 10 (4) (2010) 434–457.

[57] A Surjushe, R Vasani, D Saple, Aloe vera: a short review, Indian J Dermatol 53 (4) (2008) 163.

[58] K Kim, MH Chung, S Park, J Cha, JH Baek, S-Y Lee, et al., ER stress attenuation by aloe-derived polysaccharides in the protection of pancreatic β-cells from free fatty acid-induced lipotoxicity, Biochem Biophys Res Commun 500 (3) (2018) 797–803.

[59] EJ Esra'Shishtar, A Jenkins, V Vuksan, Effects of Korean White Ginseng (Panax Ginseng CA Meyer) on vascular and glycemic health in type 2 diabetes: results of a randomized, double blind, placebo-controlled, multiple-crossover, acute dose escalation trial, Clin Nutr Res 3 (2) (2014) 89.

[60] Z Shakib, N Shahraki, BM Razavi, H Hosseinzadeh, Aloe vera as an herbal medicine in the treatment of metabolic syndrome: a review, Phytother Res 33 (10) (2019) 2649–2660.

[61] AMH Abo-Youssef, BAS Messiha, Beneficial effects of Aloe vera in treatment of diabetes: comparative in vivo and in vitro studies, Bull Fac Pharm, Cairo Univ 51 (1) (2013) 7–11.

[62] H Mollazadeh, D Mahdian, H Hosseinzadeh, Medicinal plants in treatment of hypertriglyceridemia: a review based on their mechanisms and effectiveness, Phytomedicine 53 (2019) 43–52.

[63] M Sedighi, M Bahmani, S Asgary, F Beyranvand, M Rafieian-Kopaei, A review of plant-based compounds and medicinal plants effective on atherosclerosis, J Res Med Sci 30 (2017) 22.

[64] D Mahdian, K Abbaszadeh-Goudarzi, A Raoofi, G Dadashizadeh, M Abroudi, E Zarepour, et al., Effect of Boswellia species on the metabolic syndrome: a review, Iran J Basic Med Sci 23 (11) (2020) 1374.

[65] S Asgary, A Sahebkar, N Goli-Malekabadi, Ameliorative effects of Nigella sativa on dyslipidemia, J Endocrinol Investig 38 (10) (2015) 1039–1046.

[66] T Farkhondeh, S Samarghandian, AM Pourbagher-Shahri, Hypolipidemic effects of Rosmarinus officinalis L, J Cell Physiol 234 (9) (2019) 14680–14688.

[67] A Rezaei, A Farzadfard, A Amirahmadi, M Alemi, M Khademi, Diabetes mellitus and its management with medicinal plants: a perspective based on Iranian research, J Ethnopharmacol 175 (2015) 567–616.

[68] N Tran, B Pham, L Le, Bioactive compounds in anti-diabetic plants: from herbal medicine to modern drug discovery, Biology 9 (9) (2020) 252.

[69] YA Selim, MI Sakeran, Effect of time distillation on chemical constituents and anti-diabetic activity of the essential oil from dark green parts of Egyptian Allium ampeloprasum L, J Essent Oil Bear Plants 17 (5) (2014) 838–846.

[70] JD Teshika, AM Zakariyyah, T Zaynab, G Zengin, KR Rengasamy, SK Pandian, et al., Traditional and modern uses of onion bulb (Allium cepa L.): a systematic review, Crit Rev Food Sci Nutr 59 (sup1) (2019) S39–S70.

[71] M Thomson, ZM Al-Amin, KK Al-Qattan, LH Shaban, M Ali, Anti-diabetic and hypolipidaemic properties of garlic (Allium sativum) in streptozotocin-induced diabetic rats, Int J Diabetes Metab 15 (3) (2007) 108–115.

[72] SA Zaahkouk, SZ Rashid, A Mattar, Anti–diabetic properties of water and ethanolic extracts of Balanites aegyptiaca fruits flesh in senile diabetic rats., Egypt J Hosp Med 10 (1) (2003) 90–108.

[73] A Najmi, M Nasiruddin, R Khan, SF Haque, Therapeutic effect of Nigella sativa in patients of poor glycemic control, Asian J Pharm Clin Res 5 (3) (2012) 224–228.

[74] MR Mahmoodi, M Mohammadizadeh, Therapeutic potentials of Nigella sativa preparations and its constituents in the management of diabetes and its complications in experimental animals and patients with diabetes mellitus: a systematic review, Complement Ther Med (2020) 102391.

[75] AS Rao, S Hegde, LM Pacioretty, J DeBenedetto, JG Babish, Nigella sativa and Trigonella foenumgraecum supplemented chapatis safely improve HbA1c, body weight, waist circumference, blood lipids, and fatty liver in overweight and diabetic subjects: a twelve-week safety and efficacy study, J Medi Food 23 (9) (2020) 905–919.

CHAPTER 10

Rheumatoid arthritis and alternative medicine

Anis A. Chaudhary[a] and Mohammad Fareed[b]

[a]Department of Biology, College of Science, Al-Imam Mohammad Ibn Saud Islamic University (IMSIU), Riyadh, Kingdom of Saudi Arabia [b]College of Medicine, Al-Imam Mohammad Ibn Saud Islamic University (IMSIU), Riyadh, Saudi Arabia

10.1 Introduction

Rheumatoid arthritis (RA) may be a disease of unknown etiology characterized by a chronic polyarthritis during which the main target tissue is that the synovial lining of the joints, bursa and tendons sheath leading to varying degrees of joint deformities and associated muscle wasting. It can occur at any age, but peaks between 30 and 50 years [1]. Despite rheumatoid arthritis is traditionally considered as a systemic, inflammatory, autoimmune disorder but it differs from organ specific autoimmune disease entities in several reports, which means that your immune system attacks on your healthy cells by mistake, producing inflammation in the affected body parts [2]. RA commonly attacks in the joints area, usually many joints directly affected. RA commonly affects joints within the hands, wrists, and knees. In a joint with RA, the liner of the joint becomes inflamed, causing damage to joint tissue [3]. Damage of the tissue can cause long-lasting or chronic pain, wobbliness and malformation (abnormality). RA represents a chronic and systemic inflammatory disease, the most common form of arthritis, which is unknown etiology and is marked by synovial hyperplasia with local incursion of bone and gristle leading to joint destruction [4]. The signs and symptoms of RA may diverge in severity, stages of increased disease activity, called flares, alternate with periods of relative remission; when the swelling and pain diminish or disappear. Over the time, RA can cause joints to deform and shift out of place and other signs and symptoms of RA, is stiffness, pain or aching, swelling and tenderness in more than one joint, weight loss, fever, fatigue or tiredness, and disability is common and significant cause [5]. RA can also affect other tissues throughout the body and cause complications in organs like the lungs, heart, and eyes. About 40% of the people with RA have also experience signs and symptoms that don't involve the joints [6]. Rheumatoid arthritis can affect many nonjoint structures, as skin, eyes, lungs, heart, kidneys, salivary glands, nerve tissue, bone marrow and blood vessels. The etiology of RA is multifactorial autoimmune disease, like many autoimmune diseases, with 50 percent of RA

Herbal Medicines: A Boon for Healthy Human Life.
DOI: https://doi.org/10.1016/B978-0-323-90572-5.00025-1

risk attributable to genetic factors [7]. Human leukocyte antigen-DR4 [8], and DRB1, and a variety of alleles called the shared epitope and genetic associations for RA [9–10]. Many studies based on genome have found some additional genetic reasons that increase the risk of RA and other autoimmune diseases as including STAT4 gene and CD40 locus [11].

It has been reported that RA affects around 1% of the adult population and more women afflicted than men, most of its damage within the initial year, early diagnosis and aggressive therapy is precarious. Left untreated RA may reduce life expectancy, as much as 18 years [12]. The parts attacked are the linings of the joints in RA (places within the body where two bones connect). The reasons of this are complex and not fully understood. Smoking is the foremost environmental cause for RA, particularly in those with a genetic susceptibility [13]. Although infections may uncover an autoimmune response, no specific pathogen has been proven to cause RA [14]. Considerable advances in recent years in both the clinical and basic research aspects of atrophic arthritis are made. Clinical progress of RA has shown significant development and authentication for clinical trials and, consequently, innovative trial designs. Some basic research has found clues to the pathogenic events connecting RA, and advances have facilitated the development of new classes of therapeutics [15]. Enhanced thoughtful of molecular pathogenesis has enabled development of innovative biological agents that focus on specific parts of the system. These treatments and management have changed the course and face of RA and outcomes for patients and society. New researches have arisen of how environmental factors interrelate with susceptibility genes and the immune system in the pathogenesis of a major subgroup of RA [16]. At present, the present modalities for treating arthritis are symptomatic and haven't been exposed to either block or reverse the cartilage deprivation and joint destruction. Generally, patients trust on the conventional treatment that consist of NSAIDs and medicines that target local inflammation although recently the introduction of disease modifying anti-rheumatic drugs (DMARDs) has led to important gain in our generally facility to treat RA patients [17]. The values of DMARDs also are limited due to its side effects and expensive to use than traditional NSAIDs. This has been resulted in heightened interest toward the use of complementary and alternative medicine (CAM) therapies, like acupuncture and extracts of medicinal herbs. According to some reports, 60–90% of disappointed arthritis patients are expected to search the choice of CAM therapy [18].

10.2 Medicinal herbs in arthritis

Historically, there are a spread of herbal medicines that are utilized in the treatment of rheumatological conditions, including conditions resembling what we now know as RA. Along with the increasing application of clinical test procedures for the assessment of traditional medicines for RA and other diseases, has been increasing consideration to the methodical assessment of the pre-modern and classical medical literature [19]. Many studies have focused on drug discovery from active compounds found in the herbal products used as traditional medicines [20–22]; identification of instances of long-term traditional use of herbal products for certain diseases or symptoms [23–26]. At the level of individual ingredient, there are 15 top herbs frequently used in the conventional formulas, have also been reported to validate anti-inflammatory activity for RA drug discovery. It has been suggested that long-standing traditional use could be considered as a foundation of evidence [27], for example, in Chinese medicine "Lei Gong Teng" (thunder god vine or "three-wing nut") has been used for centuries to treat inflammatory tissue swelling [28], preparations of Salix species, devil's claw (Harpagophytum procumbens) and nettle (Urtica dioica) are popular antirheumatic remedies in Europe, as *Withania Somnifera, Zingiber officinalis, Capsicum frutescens, Mentha piperita, Arnica montana, Curcuma longa,* and *Tanacetum parthenium* etc (Fig. 10.1).

10.2.1 Withania Somnifera *(ashwagandha)*

It is a medicinal plant of Indian origin with wide pharmacological actions. The effect of the ashwagandha has demonstrated on the adjuvant induced arthritis and it showed amelioration to the level of significance [29]. In another set of experiments, the plant has demonstrated to be adaptogenic, which allow the coping power to increase in the stress conditions, [30]. The use of *W. somnifera* are alterative, diuretic, aphrodisiac, tranquilizer, deobstruent, sedative, and curative in nature.

10.2.1.1 Chemical constituents

The methanol, hexane and ether extract from both leaves and roots of ashwagandha has been reported. Alkaloid percentage in roots ranges from 0.13% to 0.31%. The pharmacological activity of the root extract is accredited to the alkaloids and steroidals lactones. The total alkaloid content has been reported to vary between 0.13 and 0.3 within the roots of Indian types, though much high yields (up to 4.3%) have been recorded elsewhere. Many biochemical heterogeneous alkaloids, including choline, tropanol, pseudotopanol, cuscokygrene, 3-tigioyloxytropana, isopelletierine, and several other steroidal lactories [31]. Twelve alkaloids, 35 withanolides, and several sitoindosides have been isolated from the roots of the plant have been studied.

A sitoindoside may be a biologically active constituent referred to as withanolide containing a glucose molecule at carbon 27. Indian ginseng's pharmacological activity has been accredited to 2 more common withanolides, withaferin-A and withanolide-D. Withaferin-A is therapeutically active with anolide present in leaves. In addition to alkaloids, the roots are reported to contain starch, reducing sugars, glycosides, dulcitol, withancil, an acid and a neutral compound [31]. The amino acids reported from the roots include amino acid, glycine, tyrosine, alanine, glutaminic acid, and cysteine.

Figure 10.1

Picture of herbal plants use in Rheumatoid Arthritis.

10.2.2 Calotropis procera *(Ait, Arka)*

It has showed anti-inflammatory property when tested different models of arthritis especially adjuvant model [32]. Furthermore, the anti-inflammatory and analgesic activities of the root of the same plant were also deciphered by Paula (2005) [33]. In the latest studies, by Kumar et al. (2007) [34] has been shown to decrease oxidative stress in the arthritic rats.

10.2.2.1 Chemical constituents

In leaves mudarine is isolated as principal active constituent. Besides a yellow bitter acid, resin and three toxic glycosides calotropin calotoxin and uscharin, a powerful bacteriolytic enzyme in fluid, a very toxic glycoside calactin (as defense mechanism on insect or grasshopper attack in increased concentration), calotropin F I, calotropin F II, calotropin D I, calotrapin D II and nontoxic proteolytic enzyme calotopin (2%–3%) had been identified. This calotopin is more proteolytic than papain, and bromelain and coagulates milk, digests meat, gelatin and casein. whole plant contains a- and b-amyrin, b-amyrin, teraxasterol, gigantin, giganteol, isogiganteol, b-sitosterol, and a wax [35].

10.2.3 Butea frondosa *(Palash)*

Butea frondosa has been evaluated for ocular anti-inflammatory activity in rabbits [36]. Tumor necrosis factor alpha was seen to reduce in response to herbal preparation by Mines et al. (1995) [37] along with COX-2 inhibition. The pharmacological actions of *B. frondosa* show potential healing value for the treatment of inflammatory and other diseases related to activated mast cells play a role.

10.2.3.1 Chemical constituents

The bark exudes a red juice that dries to make the gum butea. The gum contains leucocyanidin, tetramer, procyanidin, acid, and mucilaginous material.

Flower - Triterpene, several flavonoids butein, butin, isobutrin, coreopsin, isocoreopsin (butin 7-glucoside), sulphurein, monospermoside (butein 3-e-D-glucoside) and isomonospermoside, chalcones, aurones, isobutyine, palasitrin, $3',4',7$- trihydroxyflavone. Myricyl alcohol, stearic, palmitic, arachidic and lignoceric acids, glucose, fructose, histidine, amino acid, alanine, and phenylalanine. Gum - tannins, mucilaginous material, pyrocatechin [38].

Seed - Oil, lypolytic and proteolytic enzymes, plant proteinase and polypeptidase. A nitrogenous acidic compound, alongside palasonin is present in seeds. It also contains monospermoside (butein3-e-D-glucoside) and somonospermoside. allophanic acid, several flavonoids, fatty acids like myristic, palmitic, stearic, arachidic, behenic, lignoceric, oleic, linoleic and linolenic, monospermin.

Stem - 3-Z-hydroxyeuph-25-ene and 2,14-dihydroxy-11,12-dimethyl-8-oxo-octadec-11-enylcyclohexane, Stigmasterol-e-D-glucopyranoside and nonacosanoic acid, flavonoid 8-C-prenylquercetin 7,4'-di- O-methyl-3-O-α-L-rhamnopyranosyl (1-4)-α-L-rhamnopyranoside. 3-hydroxy-9 methoxypterocarpan. In addition to stigmasterol-3-α-L- arabinopyranoside, four compounds isolated from the stem of dhak are characterized as 3-methoxy-8,9-methylenedioxypterocarp-6-ene, 21- methylene-22-hydroxy-24-oxooctacosanoic acid, 4-pentacosanylphenol, and pentacosanyl-β-D-glucopyranoside.

10.2.4 Ocimum sanctum (tulsi)

It is offered higher protection against carrageenin induced paw edema in rats and acetic acid induced writhing in mice [39]. Fixed oil of *O. sanctum* and linolenic acid were found to possess significant anti-inflammatory activity [40]. The anti-inflammatory and antipyretic role of the plant was also demonstrated by [41].

10.2.4.1 Chemical constituents

Main constituents of basil are tannins, alkaloids and essential oil. A range of biologically active compounds has been isolated from the leaves including ursolic acid, apigenin, and luteolin. Oil consists of eugenol, ursolic acid, rosmarinic acid, thymol, linalool, methyl chavicol, citral, and beta-caryophyllene. Three new compounds, ocimumosides A (6-deoxy 6-amino-α-D glucopyranosyl) and B (α-D glucopyranosyl (1"-6')-O-β-D-galctopyranosyl) have been isolated from an extract of the leaves of holy basil (Ocimum sanctum), together with eight main substances apigenin, apigenin-7-O-β-d-glucopyranoside, apigenin-7-O-β-d-glucuronic acid, apigenin-7-O-β-d-glucuronic acid 6"-methyl ester, luteolin-7-O-β-d-glucuronic acid 6"-methyl ester, luteolin-7-O-β-d-glucopyranoside, luteolin-5-O-β-d-glucopyranoside, and 4-allyl-1-O-β-d-gluco pyronosy l-2-hydroxybenzene, and two known cerebrosides [42].

10.2.5 Pongamia pinnata (kanji)

It is showed significant anti-inflammatory activity in carrageenan and PGE1 induced edema models. The anti-inflammatory activity of the leaves of the plant were demonstrated in all models of inflammation, acute or chronic [43].

10.2.5.1 Chemical constituents

The phytochemical investigation of the methanolic extract of the seeds of *P. pinata* have many different secondary metabolites such as β-sitosterol, β-sitosteryl acetate, β-sitosteryl galactoside, stigmasterol, stigmasteryl galactoside, and sucrose. β-sitosterol has been extracted from different parts of *P. pinata* like flowers, heart wood, leaves and seeds, from the seeds of *P. pinnata* eight fatty acids have been extracted among them three were saturated and five unsaturated. Out of all detected fatty acids oleate had the very best relative percentage of occurrence (44.24%). However, the share of monounsaturated fatty acid within the seed oil of P. pinnata reported in literature [44] was 8.36%. Among saturated fatty acids, stearic acid has detected in the highest percentage that is, 29.64%, while palmic acid was present in the text highest amount i.e. 18.58% in the seeds of P. pinnata. Apart from oleic acid, all the other unsaturated fatty acids occurred in small amounts such as heptadecylenic (2.21%) and linoleic acids (1.77%).

10.2.6 *Azadirachta indica (neem)*

A. indiaca exerted significant anti-inflammatory activity in the cotton pellet granuloma assay in rats [45]. A study was done to highlight the immunomodulating role of the neem and it was done both experimentally and clinically, [46]. The mode of the action has also been investigated [47] describing the antinociceptory role of neem.

10.2.6.1 *Chemical constituents*

Oil from *A. indica*, contains namely nimbin and nimbidol, the bitter substances; a paraffin alcohol, sugiol, and a new oxophenol, nimbiol, has been isolated from trunk bark. Tetranor-triterpenoids, epoxyazadiradione, azadiradione, and azadirone have been isolated from seed oil and a new meliacin, nimbolide, from leaves. The seed oil yielded meliantriol; deacetylnimbin has been reported from seed and bark. Quercetin and b-sitosterol from leaves, cyclolucalenol, methylenecycloartanol, and b-sitosterol from wood oil of West African tree, a new tetranortriterpenoid, meldenin from seed oil [48].

10.2.7 **Camellia sinensis** *(green tea)*

Green tea is one among the foremost commonly consumed beverages within the world with no reported side effects. The established pharmacological properties of tea are attributed to its high content of polyphenols/catechins, mainly epigallocatechin-3-gallate (EGCG). Since the production of *C. sinensis* has not satisfied demand, *C. hawkesii* has been reported as a substitute for *C. sinensis*. Comparison of the chemical constituents between *C. hawkesii* and *C. sinensis* based on thin-layer chromatography has shown that amino acid, alkaloid, sterol, and organic acid contents are similar [49].

10.2.7.1 *Chemical constituents*

The chemical constituents of *C. sinensis* have been reported amino acids, steak acid, D-mannitol, ergosterol, mycose, uracil, adenosine, adenine, cholesterolpalmitate, palmitic acid, and 5a-8a-epidioxy-5a-ergosta-6, 22 dien-3pß-ol 48-l. Ergosterol, p-mannitol, tearic acid, mycose, uridine, uracil, adenine, adenosine, and 13 amino acids found in both the cultured broth and mycelium of *C. sinensis*. The D-mannitol contents in *C. sinensis* were 8.7% and

7.8%; and the ergosterol contents were 1.2% and 1-1%, respectively. The contents of proteins, total amino acids, and alkaloids in both species were also similar [50].

10.2.8 Uncaria tomentosa, Uncaria guianensis *(cat's claw)*

Cat's claw may be a Peruvian vine with medicinal properties that are well-documented in medicinal literature. The Extract of cat's claw has been shown to own antioxidant, anti-inflammatory and immunomodulating properties [51]. Extraction or preparations from the two species *U. tomentosa* or *U. guianensis* are interchangeably wont to treat various inflammatory and non-inflammatory conditions within the Peruvian medicinal system.

10.2.8.1 Chemical constituents

The chemical conformation of the aqueous extract of *U. tomentosa* vine includes oxindole alkaloids tannins, quinovic acid, glycosides, sterols and flavonoids. The additional group of chemicals called quinovic acid glycosides have reported antiviral and anti-inflammatory actions. Antioxidant chemicals (tannins, catechins and procyanidins) also as plant sterols (beta-sitosterol, stigmasterol, and campesterol) account for the plant's anti-inflammatory properties. A class of compounds referred to as carboxyl alkyl esters found in cat's claw has been documented with immunostimulant, anti-inflammatory, anticancerous, and cell-repairing properties.

10.2.9 Tripterygium wilfordii *Hook F*

T. wilfordii Hook F (TwHF) may be a perennial vine-like plant that grows in Southern China and Taiwan and is additionally referred to as "Thunder God Vine." The medicinal extract springs from the basis and has been used for the treatment of varied autoimmune and inflammatory diseases including RA, systemic LE, nephritis, psoriasis, and asthma for several centuries [52].

10.2.9.1 Chemical constituents

The major constituent of thunder god vine roots has been identified as the diterpenoid triptolide and other constituents, important as triptolide, include sesquiterpenes (for example, dihydroagarofurans, alkaloids), diterpenes (for example, tripdiolide, tripchlorolide) and triterpenes. *T. wilfordii* is a complex chemical constituent, some trace constituents have powerful bioactivity. The most active compounds of *T. wilfordii* are triptolide and celastrol, and also toxic components. Around 300 compounds have been isolated so far and identified from the *T. species*, including sesquiterpenoids, sesquiterpene alkaloids, triterpenes and diterpenes [53]. About 100 compounds among them have been confirmed to be biological active, and the order of their bioactivity from strong to weak is diterpenes, alkaloids, and triterpenoids.

10.2.10 Curcuma longa *(turmeric)*

Turmeric may be a widely used spice and coloring/flavoring agent that comes from the basis of the plant *C. longa*, a member of the Zingaberacea family [54]. The FDA classified turmeric among constituents "generally accepted as safe." In Ayurveda, turmeric has been used for several medicinal conditions as rhinitis, cold, wound healing and skin infections, liver and tract diseases and as a "blood purifier" [55]. Turmeric was originated to be effective even when given by several routes including topical, oral or by inhalation, hooked into the proposed use.

10.2.10.1 Chemical constituents

Major constituents are ar-turmerone, and ar-curcumene in essential oil. Some of the other compounds are a-and b-pinene, sabinene, myrcene, a-terpinene, limonene, p-cymene, perillyl alcohol, eugenol, turmerone, eugenol methyl ether, iso-eugenol, and iso-eugenol methyl ether. [56]. Dry rhizomes of *C. longa* produce 5.8% essential oil contain b-hellandrene, d-sabinene, cineole, borneol, Zingiberene and sesquiterpene ketones (50%). Fresh rhizomes yield 0.24% oil containing Zingiberine [57]. The recent analysis of the essential oil, turmerone (29.3%), ar-turmerine (23.6%) and sabinone (0.6%) have been recognized in the ketonic fraction besides p-cymene, iso-caryophyllene, trans- b-farnesene, d-curcumene, b-bisbolene and b-sesquiphellandrene. The main active constituents of turmeric are the flavonoid curcumin (diferuloylmethane) and various volatile oils, including tumerone, atlantone, and zingiberone. Other constituents include sugars, proteins, and resins. The most active constituent is curcumin, which covers 0.3%–5.4% of raw turmeric.

10.2.11 Colchicum autumnale (Saffron)

C. autumnale, commonly referred to as meadow saffron, autumn crocus or naked lady, may be a flower which resembles truth crocuses, but flowering in autumn. The name "naked lady" comes from the very fact that the flowers emerge from the bottom long after the leaves have died back, commonly cultivated in temperate areas.

10.2.11.1 Chemical constituents

C. autumanle have colchicine, a useful drug with a narrow therapeutic index. Colchicine has been approved by the US FDA for the treatment of gout and familial brucellosis. Colchicineis also used in plant breeding to produce polyploidy strains. Twelve know compounds, compounds, colchicine, 2-demethylcolchicine, 2-demethyldemecolcine, 2-demethylcolchifoline, 2-demethyl-lumicolchicine, lumicolchicine, demecolcine, luteolin, apigenin, n-hentriacontane, n-triacontanol, and sitosterol were isolated from corms and flowers of *C. autumnale* L. The contents of colchicine were 0.015% and 0.04% in corms and flowers of *C. autumnale*, and the content of demecolcine in corms was 0.043% [58].

10.2.12 Boswellia sacra *(gum resin)*

Boswellia gum resin, derived from the Boswellia serrata tree, is a traditional Ayurvedic remedy that is used for a variety of inflammatory diseases, such as rheumatoid arthritis, OA, and cervical spondylitis [59]. Boswellic acids have been shown to possess anti-inflammatory and anti-arthritic activity in a variety of animal experimental models as well as human studies [60]. The effectiveness of B. sacra extract has been reported on 260 RA patients using a range of different clinical approaches. Compared to placebo, boswellia produced a big reduction in joint pain and swelling and morning stiffness, and therefore the patients' general health and well-being improved. Overall, boswellia was found to be effective in reducing the symptoms of atrophic arthritis in 50% to 60% of the patients.

10.2.12.1 Chemical constituents

Oil of *B. serrata* have higher content of oxygenated monoterpenes (89.9%): 2,3-Epoxycarene (51.8%), 1,5-isopropyl-2-methyl bicycle, hex-3-en-2-ol (31.3%), 4-Terpinenylacetate (3.9%) and p-Menth-1-en-2-on (2.6%), while the oil of B. serrata contained (46.2%) phenylpropan derivates and oxygenated monoterpenes: 1-(2,4-Dimethylphenyl) ethanol (20.3%), α-campholenal (13.4%) and α-terpineol (12.4%). Oil also has hexane and bicycle heptane.

10.2.13 Tanacetum parthenium *(feverfew)*

T. parthenium, a small bush up to around 46 cm (18) in high, with citrus-scented leaves, covered by flowers reminiscent of daisies [61]. It spreads rapidly, and that they will cover a good area after a couple of years. The plant has been used as a herbal treatment to scale back fever and to treat headaches, arthritis and digestive problems.

10.2.13.1 Chemical constituents

Fourteen constituents around 90.5% has been reported in stem oil: six monoterpene hydrocarbons (31.5%), five oxygenated monoterpenes (39.7%), two sesquiterpene hydrocarbons (15.8%) and one oxygenated sesquiterpenes (3.5%). The major components were α-pinene (50.0%), E-β-farnesene (13.8%), bicyclogermacrene (11.0%), spathulenol (5.0%), β-eudesmol (4.2%), and limonene (3.1%) [62].

10.2.14 Capsicum frutescens *(pepper)*

Plant of C. frutescens is annual or short-lived perennial. Fruit typically grows a straw and matures to a bright red, but also can be other colors. C. frutescens features a smaller sort of subspecies, likely due to the shortage of human breeding compared to other capsicum species.

10.2.14.1 Chemical constituents

C. frutescens extracts are extensively utilized in food industry as natural flavoring but also as coloring agent for several foods like spicy culinary, meat products, cheese food coatings, popcorn oil, and cheeses [63]. In addition, it has used dried or fresh in various pharmacological preparations [64].

10.2.15 Arnica Montana *(Arnica)*

A. montana may be a plant that grows in mountainous areas, and may be found within the northern us, Canada and therefore the European Alps. The flowers and roots have used conventionally to establish topical remedies for relief of the pain. Arnica is great for bruises, soreness, sprains, swelling and for the relief of arthritis, muscle, and joint pain.

10.2.15.1 Chemical constituents

This oil has a high content of hexadecanoic acid (29.6%) and carvacrol (15.2%). Hydrocarbons such as alkanes and fatty acids classified as "others" constituted 47.5% of the essential oil, followed by oxygenated monoterpenes and oxygenated sesquiterpenes, with 17.6% and 17.1% [65].

10.3 Herbal marketed formulations for rheumatoid arthritis

Currently there are many poly herbal formulations in Indian market are utilized in different form like Vati, Churna, Arkh, Quath. for the treatment of RA. These formulations may consist aqueous extract or powders of the different plants part which are used for the treatment of arthritis. These formulations are called as poly herbal formulation because they contain 3 to 25 herbs within the formula [66–68].

10.4 Conclusion

RA is a chronic, autoimmune and inflammatory disease that effects on joints and connective tissue. It requires lifelong pharmacological management, although they have greatly improved the general course of RA, carry some notable side effects as well. Therefore, RA patients often demand alternative medicine, they seek additional sources of relief and/or less side effects. Alternative medicine has a very broad category, including mind-body therapies, energy medicine, herbs and any other modalities, as well as a synopsis of what may be recommended as safe and possibly effective without any adverse effects as discussed in this chapter. Many investigations on alternative medicine have shown consistent beneficial outcomes for the treatment of RA. In fact, 30%–60% of rheumatic patients use some form of alternative medicine, mostly used in early stages of the disease, in conjunction with mainstream treatments. Many alternative therapies may represent an opportunity to improve the quality of life of our patients, and in the future may be integrated in the management of RA patients. So, it is concluded that in future, more studies should be done by clinical researchers among RA patients to establish the more efficacy of the active compounds present in these herbs. It is also recommended to health professionals for more use of these herbal drugs considering evidenced based medicine approach to minimize the adverse effects of synthesized pharmaceutical drugs among RA patients.

References

[1]   J Bathon, C Tehlirian, et al., Rheumatoid arthritis clinical and laboratory manifestations, in: JH Klippel, JH Stone, LJ Crofford, et al. (Eds.), Primer on the Rheumatic Diseases, 13th ed., Springer, New York, NY, 2008, pp. 114–121.

[2]   M Bax, J van Heemst, TW Huizinga, RE Toes, Genetics of rheumatoid arthritis: what have we learned? Immunogenetics 63 (8) (2011) 459–466.

[3]   DK Modi, VS Chopra, U Bhau, Rheumatoid arthritis and periodontitis: biological links and the emergence of dual-purpose therapies, Indian J Dent Res 20 (1) (2009) 86–90.

[4]   SE Gabriel, Cardiovascular disease in rheumatoid arthritis, Arthritis Rheum 60 (2009) 2851–2852.

[5]   DL Scott, F Wolfe, TW Huizinga, Rheumatoid arthritis, Lancet 376 (2010) 1094–1108.

[6]   A Rojas-Villarraga, OD Ortega-Hernandez, LF Gomez, Al Pardo, SL Guzman, CA Ferreira, et al., Risk factors associated with different stages of atherosclerosis in Colombian patients with rheumatoid arthritis, Semin Arthritis Rheum 38 (2008) 71–82.

[7]   AJ MacGregor, H Snieder, AS Rigby, et al., Characterizing the quantitative genetic contribution to rheumatoid arthritis using data from twins, Arthritis Rheum 43 (1) (2000) 30–37.

[8]   G Orozco, A Barton, Update on the genetic risk factors for rheumatoid arthritis, Expert Rev Clin Immunol 6 (1) (2010) 61–75.

[9]   A Balsa, A Cabezón, G Orozco, et al., Influence of HLA DRB1 alleles in the susceptibility of rheumatoid arthritis and the regulation of antibodies against citrullinated proteins and rheumatoid factor, Arthritis Res Ther 12 (2) (2010) R62.

[10] A McClure, M Lunt, S Eyre, et al., Investigating the viability of genetic screening/testing for RA susceptibility using combinations of five confirmed risk loci, Rheumatology (Oxford) 48 (11) (2009) 1369–1374.

[11] G Orozco, A Barton, Update on the genetic risk factors for rheumatoid arthritis, Expert Rev Clin Immunol 6 (1) (2010) 61–75.

[12] MJ Amador, RA Rodriguez, OG Montoya, How does age at onset influence the outcome of autoimmune diseases? Autoimmune Dis 2012 (2012) 251730.

[13] SY Bang, KH Lee, SK Cho, et al., Smoking increases rheumatoid arthritis susceptibility in individuals carrying the HLA-DRB1 shared epitope, regardless of rheumatoid factor or anti-cyclic citrullinated peptide antibody status, Arthritis Rheum 62 (2) (2010) 369–377.

[14] RL Wilder, LJ Crofford, Do infectious agents cause rheumatoid arthritis? Clin Orthop Relat Res (265) (1991) 36–41.

[15] S Josef, S Günter, Therapeutic strategies for rheumatoid arthritis, Nat Rev Drug Discov 2 (2003) 473–488.

[16] A Klareskog, Stephen, rheumatoid arthritis, Lancet North Am Ed 373 (9664) (2009) 659–672.

[17] T Pincus, T Sokka, H Kautiainen, Patients seen for standard rheumatoid arthritis care have significantly better articular, radiographic, laboratory, and functional status in 2000 than in 1985, Arthritis Rheum 52 (2005) 1009–1019.

[18] LW Engel, SE Straus, Development of therapeutics: opportunities within complementary and alternative medicine, Nat Rev 1 (2002) 229–237.

[19] EDS Buenz, B Bauer, P Elkin, J Riddle, T Motley, Techniques: bioprospecting historical herbal texts by hunting for new leads in old tomes, Trends Pharmacol Sci 25 (9) (2004) 494–498.

[20] M Adams, C Berset, M Kessler, M Hamburger, Medicinal herbs for the treatment of rheumatic disorders-a survey of European herbals from the 16th and 17th century, J Ethnopharmacol 121 (3) (2009) 343–359.

[21] EJ Buenz, BA Bauer, HE Johnson, et al., Searching historical herbal texts for potential new drugs, Brit Med J 333 (7582) (2006) 1314–1315.

[22] F Watkins, B Pendry, O Corcoran, A Sanchez, Anglo-Saxon pharmacopoeia revisited: a potential treasure in drug discovery, Drug Discov Today 16 (23-24) (2011) 1069–1075.

[23] PK Mukherjee, K Maiti, K Mukherjee, PJ Houghton, Leads from Indian medicinal plants with hypoglycemic potentials, J Ethnopharmacol 106 (1) (2006) 1–28.

[24] JL Shergis, L Wu, BH May, et al., Natural products for chronic cough, Chronic Respir Dis 12 (3) (2015) 204–211.

[25] L Zhang, Y Li, XF Guo, et al., Text mining of the classical medical literature for medicines that show potential in diabetic nephropathy, Evid Based Complement Alt Med 2014 (2014) 12.

[26] HW Bae, SY Lee, SJ Kim, HK Shin, BT Choi, JU Baek, Selecting effective herbal medicines for attention-deficit/hyperactivity disorder via text mining of donguibogam, Evid Based Complement Alt Med 2019 (2019) 9.

[27] A Helmstädter, C Staiger, Traditional use of medicinal agents: a valid source of evidence, Drug Discov Today 19 (1) (2014) 4–7.

[28] MacPherson. Lei Gong Teng' (thunder god vine or 'three-wing nut') has been used for centuries to treat inflammatory tissue swelling. Eur J Oriental Med 1 (1994) 17–29.

[29] VH Begum, J Sadique, Effect of withania-somnifera on glycosaminoglycan synthesis in carrageenin-induced air pouch granuloma, Bioch Med Metabolic Bio 38 (3) (1988) 272–277.

[30] P RA Kaur, S Mathur, M Sharma, M Tiwari, KK Sdvastava, RA Chandra, Biologically active constituent of withania somnifera (ashwagandha) with anti-stress activity, Indian J Clin Biochem 16 (2) (2001) 195–198.

[31] P Kaur, M Sharma, S Mathur, et al., Effect of 1-oxo-5beta, 6beta-epoxy-witha-2-ene-27-ethoxy-olide isolated from the roots of *Withania somnifera* on stress indices in Wistar rats, J Altern Complement Med 9 (2003) 897–907.

[32] VL Kumar, S Arya, Medicinal uses and pharmacological properties of Calotropis procera, in: JN Govil (Ed.), Recent Progress in Medicinal Plants, 11 editors, Studium Press, Texas, 2006, pp. 373–388.

[33] MS Paula, RL Silvane, GM Samara, MA Marcelo, CAP Manoel, DT Cleverson, et al., Antinociceptive activity of Calotropis procera latex in mice, J Ethnopharma 99 (2005) 125–129.

[34] Kumar VL, Kumar V, inventors. Extraction of latex of Calotropis procera and a process for the preparation thereof. # 737/DEL/2004 and PCT/IN 2005/000106 Indian Patent Application.

[35] AM Moustafa, SH Ahmed, ZI Nabil, AA Hussein, MA Omran, Extraction and phytochemical investigation of Calotropis procera: effect of plant extracts on the activity of diverse muscles, Pharm Biol 48 (2010) 1080190.

[36] SA Mengi, SG Deshpande, Anti-inflammatory activity of Butea frondosa leaves, Fitoterapia 70 (1999) 521–522.

[37] SH Mishra, MS Lavhale, Evaluation of free radical scavenging activity of Butea monosperma Lam, Indian J Exp Biol 45 (2007) 376–384.

[38] BO Victor, OW Caleb, OS Ayodele, BO Samuel, Studies on the Antiinflammatory and analgesic properties of Tithonia diversifolia leaf extract, J Ethnopharmacol 90 (2004) 317–321.

[39] S Singh, DK Majumdar, MR Yadav, Chemical and pharmacological studies on fixed oil of *O. sanctum*, Indian J Exp Biol 34 (1996) 1212–1215.

[40] PK Mediratta, KK Sharma, Differential effects of benzodiazepines on immune responses in non-stressed and stressed animals, Indian J Med Sci 56 (1) (2002) 9–15.

[41] S Godhwani, JL Godhwani, DS Vyas, *Ocimum sanctum*: an experimental study evaluating its anti-inflammatory, analgesic and antipyretic activity in animals, J Ethnopharmacol 21 (2) (1987) 153–163.

[42] S Chinnasamy, G Balakrishnan, SV Kontham, SL Baddireddi, B Arun, Potential anti-inflammatory properties of crude alcoholic extract of *Ocimum basilicum* L. in Human peripheral blood mononuclear cells, J Health Sci 53 (4) (2007) 500–505.

[43] K Srinivasan, S Muruganandan, J Lal, S Chandra, SK Tandan, VR Prakash, Evaluation of anti-inflammatory activity of Pongamia pinnata leaves in rats, J Ethnopharmacol 78 (23) (2001) 151–157.

[44] T Toshiyuki, I Munekazu, Y Kaoru, F Yuko, M Mizuo, Flavonoids in root bark of Pongamia pinnata, Phytochemistry 31 (1992) 993–998.

[45] S Chattopadhyay, LH Ang, P Puente, XW Deng, N Wei, *Arabidopsis* bZIP protein HY5 directly interacts with light-responsive promoters in mediating light control of gene expression, Plant Cell 10 (1998) 673–683.

[46] S Ahmed, M Bamofleh, A Munshi, Cultivation of neem (*Azadirachta indica*) in Saudi Arabia, Econ Bot 43 (1989) 35–38.

[47] A Khanna, Neem gains honour as Indias wonder tree, Down Earth. Soc. Environ. Commun. 1 (1992) 5–11.

[48] B Khillare, TG Shrivastav, Spermicidal activity of *Azadirachta indica* (neem) leaf extract, Contraception 68 (2003) 225–229.

[49] YH Chen, X Qi, Y Zhu, RC Zhang, Determination of heavy metals in ordinary herbal pieces, Chinese Tradit Patent Med 25 (2003) 405–407.

[50] TH Hsu, LH Shiao, C Hsieh, A comparison of the chemical composition and bioactive ingredients of the Chinese medicinal mushroom Dong Chong Xia Cao, its counterfeit and mimic, and fermented mycelium of Cordyceps sinensis, Food Chem 78 (2002) 463–469.

[51] C Akesson, H Lindgren, RW Pero, T Leanderson, F Ivars, An extract of *Uncaria tomentosa* inhibiting cell division and NF-kappa B activity without inducing cell death, Int Immunopharmacol 3 (2003) 1889–1900.

[52] WZ Gu, SR Brandwein, S Banerjee, Inhibition of type II collagen-induced arthritis in mice by an immuno-suppressive extract of *Tripterygium wilfordii* Hook f, J Rheumatol 19 (1992) 682–699.

[53] V Kantayos, Y Paisooksantivatana, Antioxidant activity and selected chemical components of 10 *Zingiber* spp. in Thailand, J Dev Sus Agr 7 (2012) 89–96.

[54] S Mamta, S Jyoti, Phytochemical screening of *Acorus calamus* and *Lantana camara*, Int Res J Plant 3 (5) (2012) 324–326.

[55] DE Okwu, Phytochemicals and vitamin contents of indigenous species of South Eastern Nigeria, J Sustain Agric Environ 6 (2004) 30–34.

[56] KPM Dhamayanthi, TJ Zachariah, Studies on karyology and essential oil constituents in two cultivars of ginger, J Cytol Genet 33 (2) (1998) 195–199.

[57] MA Sukari, NWM Sharif, ALC Yap, SW Tang, BK Neoh, M Rahmani, et al., Chemical constituents variations of essential oils from rhizomes of four Zingiberaceae species, Med J Adv Sci 12 (2008) 638–644.

[58] F Alali, K Tawaha, RM Qasaymeh, Determination of colchicine in Colchicum steveni and C. hierosolymitanum (Colchicaceae): comparison between two analytical methods, Phytochem Anal 15 (2004) 27–29.

[59] A Poutaraud, P Girardin, Influence of chemical characteristics of soil on mineral and alkaloid seed contents of Colchicum autumnale, Environ Exp Bot (2004) 1–8.

[60] FI Abdullaev, LR Negrete, HC Ortega, M Hernandez, IP Lopez, Miranda RP Aguirre JJE. Use of in vitro assays to assess the potential antigenotoxic and cytotoxic effects of saffron (Crocus sativus L.), Tox in Vitro 17 (2003) 731–736.

[61] A Pareek, M Suthar, S Garvendra, BansalV Feverfew, (*Tanacetum parthenium* L.): A systematic review, Pharmacogn Rev 5 (9) (2011) 103–110.

[62] S Ali, S Hajar, O Khodamali, Composition and antibacterial activity of essential oils from leaf, stem and root of Chrysanthemum parthenium (L.) Bernh. from Iran, Nat Prod Commun 4 (6) (2009) 859–860.

[63] JS Pruthi, Chemistry and quality control of capsicums and capsicum products, in: AK De (Ed.), The Genus Capsicum, Taylor & Francis, London, 2003, pp. 25–70.

[64] Sukrasno, N Yeoman, MM Sukrasno, Yeoman Phenylpropanoid metabolism during growth and development of Capsicum frutescens fruits, Phytochemistry 32 (1993) 839–844.

[65] N Tabanca, B Demirci, SL Crockett, KH Başer, DE Wedge, Chemical composition and antifungal activity of Arnica longifolia, Aster hesperius, and Chrysothamnus nauseosus essential oils, J Agric Food Chem 55 (21) (2007) 8430–8435.

[66] KJ Abesundara, M,K matsuit, Alpha-glucosidase inhibitory activity of some Sri Lanka plant extracts one of which Cassia auriculata exerts a strong antihyperglycemic effect in rats comparable to the therapeutic drug acarbose, J Agri Food Chem 52 (9) (2004) 2541–2545.

[67] DS Narayan, VJ Patra, SC Dinda, Diabetes and Indian traditional medicines an overview, Int J Pharm Pharm Sci. 4 (2012) 45–53.

[68] E Jarald, SB Joshi, DC Jain, Diabetes and herbal medicines, Iran J Pharm Ther 7 (1) (2008) 97–106.

CHAPTER 11

Traditional nutritional and health practices to tackle the lifestyle diseases

Kirti Aggarwal[a], Swati Madan[b] and Maryam Sarwat[a]

[a]Amity Institute of Pharmacy, Amity University, Noida, Uttar Pradesh, India [b]Amity Institute of Indian System of Medicine, Amity University, Noida, UP, India

11.1 Introduction

Traditional nutrition/diets are more close to "healthy eating." It doesn't involve processed food, which is stripped of flavor, nutrients, and calories. Such type of diet includes organic ingredients mainly derived from nature. It contains seasonal and regionally produced foods, which are healthy to eat and contain high nutrients, vitamins, minerals probiotics, etc. Several studies have demonstrated the nutritional values of traditional diets. One of the well-known traditional eating patterns is the *"Sattvic"* diet. It is a pure vegetarian diet, which includes recurring fresh fruit, abundant amount of fresh vegetables, whole grains and pulses, nutritious sprouts, dried nuts, seeds, honey, fresh herbs, milk, and dairy products. These foods raise *"sattva"* or our consciousness levels. The person who follows *"Sattvic"* diet is calm, peaceful, tranquil, harmonious, and full of energy, enthusiasm, health, hope, aspirations, creativity, and balanced personality.

Typically, an ideal human life is totally dependent on adaptations. Those who adapt themselves to the surrounding environment are able to live happily in their habitat. Adaptation is necessary for flora, fauna, and humans. With increasing urbanization, people tend to adapt healthy lifestyle practices to fight against day-to-day life threatening problems. Unhealthy lifestyle practices and the disturbed equilibrium between the consumption of carbohydrate-rich food nutrient-deficient food leads to various acute and chronic diseases. Maintaining stability, balance, or equilibrium within the body tissues is essential for proper functioning of the human body. There are many physical and physiological factors that disturb this balance of life. Some are genetic factors while some are lifestyle disorders like: obesity, inappropriate diet, malnutrition, tobacco smoking, and alcohol consumption which speeds up the ailing process.

Herbal Medicines: A Boon for Healthy Human Life.
DOI: https://doi.org/10.1016/B978-0-323-90572-5.00003-2
253

Being self-aware is the only method to have a healthy life. Having positive attitude, balanced diet, exercising and performing physical therapies can be helpful in activating the immunity of an individual. For a healthy body, one has to be physically, mentally, and emotionally fit. "Early to bed and early to rise" is the sole logo of a balanced life. Thus, traditional nutrition and health practices can be a boom to battle against lifestyle diseases.

11.2 Impact of lifestyle on an individual's health

According to the studies, lifestyle plays a huge role on physical and mental health of a human being. Having realistic and idealistic lifestyle healthify our body, speeding up the mechanism of action of various bodily fluids to work against foreign particles. Food is the main power source to build up organic compounds that provide energy to perform many metabolic activities like muscle contraction, nerve impulse propagation, and chemical synthesis. Poor lifestyle practices like alcohol consumption, smoking, overeating, utilization of junk food, unbalanced diet and having negative atmosphere can be responsible for chronic diseases. Over the years the dietary habits of Indians have changed tremendously (Fig. 11.1).

The type of lifestyle determines the health of an individual. People who exercise daily, do yoga and meditate regularly are considered to have a healthy lifestyle with minimum chances to get ill. Problem solving and stress management techniques, both are very important in the present world, they can provide mental peace.

The relationship of lifestyle and health should be considered to a higher extent as they are directly proportional and interdependent to each other. For example, when a person catches common cold, not only the nose but the whole body gets affected and the lifestyle gets disturbed. Application of personal health practices like maintaining proper hygiene, drinking required amount of water to balance the body fluid, washing hands, using handkerchief, cleaning our environment, etc., helps a lot in the personal development as well.

11.3 Lifestyle diseases: an emerging epidemic which needs urgent attention

A lifestyle is mainly a pattern that we follow; the way we sleep, eat, do our work and perform physical activities, whether we smoke or consume alcohol. The unhealthy lifestyle practices can cause metabolic diseases which badly affect our life. Being consistent towards healthy living should be the main purpose of life. With the increase in industrialization, the lifestyle diseases are appearing to increase in frequency.

It totally depends on an individual that what type of lifestyle he/she acquires; the easy one or the right one. There are various factors that influence susceptibility to many diseases like

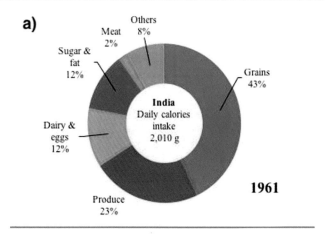

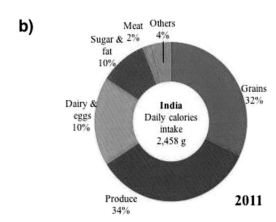

Figure 11.1
The change in food habits in Indian population over the course of 50 years; (A) year 1961 and (B) year 2011.

smoking tobacco, drinking alcohol, drug abuse, lack of exercising, physical inactivity, especially later in life. Nowadays, more and more people are seen smoking and drinking whether young or adult despite knowing the extent of complications they cause. Our nutritious and balanced meals are giving way to fast food and junk food, fresh fruits and vegetables are being rejected in favor of processed and packaged food and soft drinks are replacing milk. On the other hand, people prefer to commute by vehicles instead of walking, even for shorter distances.

Lifestyle disorders are accountable to kill approximate 35 million people each year. Some of them are chronic in nature that cannot be transferred from one person to person. These are the result of a combination of multiple factors that includes high calorie intake, genetics, physiology of body, surrounding environment and socio-economic behaviors. The main types of NCDs are cardiovascular and chronic respiratory diseases including cancer. NCDs such as stroke, diabetes, obesity and certain forms of cancer are associated to lifestyle choices which often lead to lifestyle disorders. With increasing modernization, the lack of curative treatment for various acute and chronic diseases brings our attention towards indigenous traditional knowledge which helps in identifying the nutritional novelties. The Food and Agriculture Organization of the United Nations (FAOSTAT) analyzed the calorie intake patterns of INDIA over a period of 50 years (1961–2011). It analyzed the average person's consumption in the six main food groups: grain, dairy and eggs, plant produce, meat, sugar and fat and others (pulses, alcoholic beverages). Some herbs have persuasive ingredients that have the potential to bring back the body to a state of natural equilibrium. Different herbs have a tendency to act on different parts of body. Herbal products are admired complementary or alternative goods for people. These are the supplements that are intended to enhance the diet containing one or more active dietary ingredients (including vitamins, minerals, herbs or other botanicals, amino acids, and other substances) or their constituents [1]. Awareness on the traditional food system can contribute to the healthy society with much more nutritional benefits. Food is an important part of Indian culture; it is like a festival where the food is prepared differently in all the four directions that are north, south, east and west in traditional and western ways. It helps in maintaining physical, mental, social and spiritual harmony and is the key to continuous good health.

11.3.1 Cardiovascular diseases

Ischemic heart disease, stroke, peripheral arterial disease, and congenital heart disease are a group of disorders that are considered as cardiovascular diseases. Difficulty in catching up breath, chest pain, nausea, back pain, vomiting, cold sweats, and paleness are some of the symptoms of these diseases.

11.3.1.1 Major risk factors

High blood pressure, abnormal blood lipids, tobacco use, physical inactivity, obesity, unhealthy diet, diabetes, heavy alcohol use, and genetic factors. Although genetic factors can contribute to the development of heart disease, unhealthy lifestyle plays a major role. Some of the unhealthy lifestyles that lead to heart diseases are:

- Inactive lifestyle
- Unhealthy diet (high in fat proteins, trans fats, sugary foods, and sodium)
- Smoking
- Excessive drinking
- Keeping body in a high-stress environment without proper stress management techniques
- Not managing your diabetes

Another heart disease that is caused due to alcohol abuse is alcoholic cardiomyopathy. Long-term alcohol abuse weakens the heart muscles, affecting the ability of heart to pump blood, leading to disturbed body functioning. This can lead to heart failure and various other life-threatening health problems.

11.3.2 Diabetes

Diabetes is a metabolic disorder that affects the way the body uses food as a source of nutrition and physical growth.

11.3.2.1 Major risk factors

Unhealthy diet, physical inactivity, obesity or overweight, high blood pressure, high cholesterol, heavy alcohol use, psychological stress, high consumption of sugar, and low consumption of fiber.

The National Institute of Diabetes and Digestive and Kidney Diseases estimated that people who have type II diabetes and especially the ones who have reached middle age are more likely to have heart disease or at least experience a stroke in comparison to those who don't have diabetes. Such people experience multiple heart attacks because they have insulin resistance or high blood glucose levels. The reason behind it is the static relationship between glucose and blood vessel health. Untreated high blood glucose levels can enlarge the amount of plaque that forms readily within the walls of the blood vessels, this resists or totally stops the blood flow toward the heart [2].

11.3.3 Cancer

Cancer affects different parts of the body and is characterized by a rapid growth of abnormal cells in a particular part of the body which can further invade other parts of the body as well, in an uncontrolled manner. Every year, around seven million people die due to cancer and 30% of these diseases are attributed to various unhealthy lifestyle choices [3].

11.3.3.1 Major risk factors

Hormone therapies, weight and physical activity, obesity, poor food habits low intake of fiber, unhealthy diet, and insufficient physical activity.

Early diagnosis is a key for cancer treatment. But symptoms of cancer are so common that people ignore it often and confuse them with any other easily curable disease. Irregular indigestion during food intake, persistent back pain, pelvic pain, bloating, or muscles spasms are some early signs of cancer.

11.3.4 *Chronic respiratory diseases*

Chronic respiratory diseases (CRDs), mostly under-diagnosed conditions are a potent cause of death, with 90% of the deaths taking place in low-income countries globally. Chronic obstructive pulmonary disease (COPD) and asthma are the two main types of chronic respiratory diseases.

11.3.4.1 *Major risk factors*

Cigarette smoke, dust and chemicals, environmental tobacco smoke, air pollution, infections, genetic, age.

The respiratory system bears the main impact of the atmospheric pollution. A person can readily choose or reject a food, a drink or a luxury, but there is no choice of the air we breathe. Whatever present in the atmosphere is going to enter into our respiratory system and all the noxious material, gases, fumes and chemical particles are inhaled along with it. In particular, children may also suffer from recurrent respiratory tract infections, wheezing, bronchitis, and asthma. Chronic exposures can cause more serious and chronic respiratory problems and disability.

11.3.5 *Polycystic ovarian disorder*

Polycystic ovarian disorder (PCOD) is a wide-range disorder with many symptoms, affecting women in their reproductive age, developing small cysts in their ovaries. Overall, these cysts are not harmful but lead to hormonal imbalance. The symptoms of PCOD include abnormal facial hair growth, baldness, acne, weight gain, irregular or absence of menstrual cycle and increased levels of male hormones.

11.3.5.1 *Major risk factors*

Genetics, sedentary lifestyle, environmental problems, nutritional deficiencies, metabolic disorders, chronic inflammation, and poor immunity.

In India, one in every four young women is said to have PCOD making its worldwide prevalence from 2.2% to 26%. One should try to control and prevent this disorder as long as they can, as, this can lead to long-term health problems like diabetes, heart diseases, infertility, gynecological cancers, hypertension, depression, and gestational diabetes [4].

11.4 *Occupational lifestyle diseases*

People are exposed to various types of diseases due to their way of living and occupational habits. They are often preventable and can be controlled with changes in eating habits, lifestyle, and environment. Lifestyle diseases are primarily based on an inappropriate relationship of people with their working environment and lifestyle they follow. The onset of action of these lifestyle diseases is dangerous and they take years to grow, and once encountered do not lend themselves easily to cure.

The main contributing factors of occupational lifestyle diseases include smoke, dust, fumes, noise, wrong body posture, extended working hours and disturbed biological clock. The diet and lifestyle of people belonging to different age groups partly determine the rates of developing cancer, and the basis for this assumption build up by results of studies showing that particular part of population who migrate from one country to another generally get hold of the cancer rates of the new host country [5].

11.4.1 Obesity

Obesity is a condition of abnormal or disproportionate fat accumulation in adipose tissue, to the extent that health is affected [6]. Although, it is a common problem globally, but it is quite underestimated in many countries. Obesity is presently a widespread epidemic in developed as well as underdeveloped countries, prevalent in all age groups. In 1995, there were an estimated 200 million obese adults, compared to 2000 where the number of obese adults increased to over 300 million. In developing countries, it is well estimated that, over 115 million people are suffering from obesity-related problems [7].

A rapid increase in obesity has also been reported in children [8]. Infrequently, obesity may be a sign of other medical complications like hypothyroidism, Cushing's syndrome and certain hypothalamic disorders. Another most common consequence of obesity is increased insulin secretion. The co-occurrence of hyperinsulinemia (with normal or elevated blood glucose levels) and obesity, can be due to "insulin resistance." This is very well accepted as the common underlying mechanism for a number of disease states namely, hypertension, dyslipidemia, and cardiovascular diseases (Fig. 11.2).

11.4.2 Deformities caused due to poor body posture

Poor body postures can be related to several muscular and physiological issues, which could contribute to generate positional or structural deviations from the normal body structuring, if the individual remains in inappropriate positions for a long time [9]. Mainly, spine is the major root cause of postural deviation, although other parts of the body have also been implicated in postural mis-alignment and are of high concern [10] (Fig. 11.3). The best body posture can be observed in children, however, as they grow older, they develop some habits like change in gait, adopting abnormal sitting and standing positions that predispose them to poor posture [11]. Poor body posture can occur in any stage of life and its occurrence is on the rise despite of the fact that immense efforts are done to educate people on the effects of poor posture, still the condition remains a major problem.

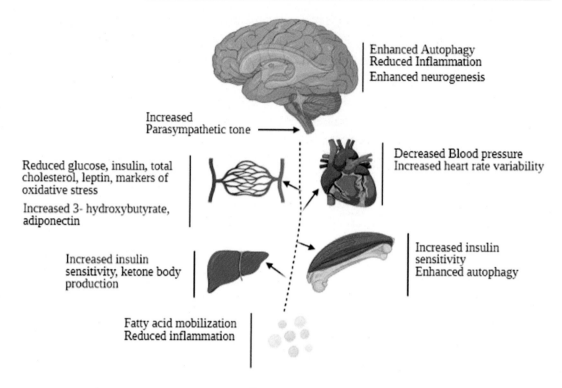

Figure 11.2

Obesity has a detrimental effect on various body organs; brain: enhanced autophagy, reduced inflammation and enhanced neurogenesis. Heart: decreased blood pressure, variable heart rate. Blood vessels: reduced glucose, insulin, total cholesterol, leptin, markers of oxidative stress, increased 3-hydroxybutyrate, adiponectin. Liver: increased insulin sensitivity, ketone body formation. Muscles: increased insulin sensitivity, enhanced autophagy. Fatty acid globules: reduced inflammation, mobilization of fatty acid.

Some of the postural deformities are kyphosis (an abnormal curvature of spine at front), scoliosis (sideway curvature of the spine), knock-knees (inward curvature of legs at knee), etc. These postural deformities reduce the efficiency of an individual to stand for longer time and cause many other health related issues like poor digestion, back pain, nerve constriction, misalignment of spine and joint imbalance [11]. All these body deformities can easily be corrected or prevented if steps are taken early in life. There are many types of corrective *asanas* and exercises which are helpful in correcting these postural deformities.

11.4.3 Depression

Physical health is of utmost importance as it is also the foundation of mental health. People who are not physically fit are always at an increased risk of developing chronic mental

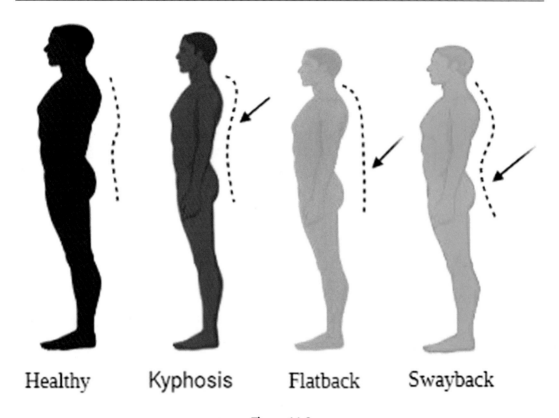

Healthy   Kyphosis   Flatback   Swayback

Figure 11.3
Deformities caused due to bad body postures. Healthy: normal body posture; kyphosis: excessive outward curvature of the spine; flatback: lower spine loses some of its curvature; swayback: hips and pelvis are tilted forward.

disabilities, such as depression. Those who follow unhealthy lifestyle practices find it more difficult to overcome depressive episodes compared to healthier people.

Negative lifestyle factors that can contribute to a depressive episode include, drug abuse, alcohol consumption, overworking, poor diet, consuming too much caffeine or sugar, lack of physical activity, poor sleep pattern, lack of leisure time, absence of recreational activities, etc.

Synthetic chemicals such as food additives and preservatives, pesticides, hormones and drugs, genetically modified foods, and industrial byproducts are bombarding our bodies.

The term "environmental causes" mainly describe various contributions made by the environment to depressive disorders. These are present in our environment in the form of air, water and food pollution, which is harming our mental health in a drastic manner. Various other nonchemical sources are also there in our surroundings like noise and electrical pollution, natural disasters, and other terrible environmental events. Other events, like childhood abuse, long-term stress at home or work, coping with the loss of a loved one, or traumatic events can also contribute adversely to mental problems.

11.5 Lifestyle affecting pregnancy and lactation

Over nutrition and lifestyle disorders begins early, even at intrauterine developmental stage. An example is the misconception that "an expectant mother has to eat for two" herself and the baby. The truth is a balanced diet of about 3000 calories per day is sufficient for her.

Excessive nutrition intake may result in enormous weight gain, greater risk of labor induction, cesarean section, higher birth weight and other complications during pregnancy and/or delivery. Higher calorie intake develops poor dietary habits and perhaps even metabolic characteristics that can have lifelong consequences. Also, poor lifestyle practices can cause mentally retarded or still born baby.

Lactating women must consume high amount of vitamin A, C, E, and certain B. Meat is considered to be the major source for saturated fat and cholesterol, it is also the most common source for easily ingestible pathogens and a rich supply of arachidonic acid, a pioneer of the immunosuppressive agent eicosanoid PGE2 [12]. In order to prevent neural tube defects, it is necessary to intake folates in required quantities prior to conception [13]. The amount of folate in vegetarian diet is quite high in comparison to a diet consisting of fast food [14]. To fight against lifestyle diseases, the expecting mother should take diet rich in calcium, phosphorus, magnesium, iron, zinc, potassium, selenium, copper, chromium, manganese, and molybdenum, etc. [15]. Prenatal vitamin-mineral formulas are advised in to meet their requirement.

There are different nutritional requirements throughout the life cycle. Over nutrition is the predominant problem in the developed countries, it has led to obesity and various chronic lifestyle diseases. Secondly, renewed focus on vegetables, fruits, whole grain, and legumes can help prevent weight gain problems and chronic illnesses, including cardiovascular diseases. National Academy of Sciences has proposed dietary reference intake (DRI) that are specific for various stages of life. Such guidelines are not meant for people who are ill or chronically ill or those who are at high risk of diseases due to age, genetic or lifestyle factors.

11.6 Infertility

There is a direct effect of nutritional and lifestyle factors on fertility, as it gives rise to many diseases including polycystic ovary syndrome, endometriosis, uterine fibroids and many more.

In females, obesity is considered to be the main reason of increasing rate of infertility [16]. Celiac disease, an immune mediated disease gets triggered due to uptake of gluten. It can impair fertility in women by causing amenorrhea, including malabsorption of nutrients needed for organogenesis and resulting in spontaneous abortion. In affected individuals' fertility can be improved by gluten free diet [17].

Infertility in females can also be caused due to physical inactivity or lack of exercise. Due to industrialization, we are prone to sedentary lifestyle because of white collar jobs depriving our body of various required endothermic and exothermic reactions, making it difficult to stay fit.

11.7 Ways to treat lifestyle disorders

Prevention of chronic diseases can be done through diet management and healthy lifestyle practices. Healthy lifestyle is like a natural medicine, it can be considered as a self-made antibiotic that has the potential to eliminate unwanted xenobiotics, brings inner happiness and release happy hormones which can bring positivity and harmony. Changing a person's natural habitat is nearly an impossible task but one can strive for that and by persistent efforts, it can be achieved. An important way of controlling non-communicable diseases is by controlling the risk factors associated with it. In other words, a number of communicable diseases can be prevented by controlling the behavioral or lifestyle habits associated with those diseases. There are a number of low-cost solutions that can be implemented by the government and other involved groups to reduce the common modifiable risk factors. Monitoring the trends of noncommunicable diseases and their associated risks is crucial for making policies and guidelines.

According to the World Health Organization (WHO) report, at least 80% of all heart diseases, stroke, type 2 diabetes, and up to 40% of cancer could be cured if people prefer to consume healthy nutritious food, engage themselves in physical activities and stop smoking tobacco. To encourage physical activities in young people, the Centers for Disease Control and Prevention (CDC) and the Youth Compendium of Physical Activities has prepared a list of 196 common activities in which youth takes part. The energy cost related to each activity is provided. It is a ready reference and will be helpful for researches, educationists, policy makers, fitness trainers, etc. Overall, 77.6% of men and 78.4% of women consume less than the minimum recommended five daily servings of fruits and vegetables [18]. "Neglecting one's healthcare can have multiple effects on the surrounding environment too such as family, workplace, and social network," said the study [19].

11.7.1 Ayurveda dietetics - a healthy outside, starts from healthy inside

The ayurvedic culture was originated in 6000 BC in the form of verbal communication with only some traces of the written material. It is composed of eight branches in total namely, *Ashtanga Ayurveda* (Fig. 11.4). Ayurvedic dietetics emphasizes on the nature of food articles,

Figure 11.4
Ashtanga Ayurveda: Eight branches of Ayurveda. *Vajeekarana* (aphrodisiac therapy), *Salya chikitsa* (surgery), *Urdhyanga chikitsa* (treatment of eyes, nose, throat, head, and related diseases), *Rasayana* (geriatrics), *Visha chikitsa* (toxicology), *Graha chitiksa* (psychiatry), *Kaya chikitsa* (internal medicine), and *Bala chikitsa* (pediatrics treatment).

wholesome/unwholesome diet *(pathya-apthaya Ahar)*, quantity *(matra)*, intake timings *(kaal)*, disciples of eating *(Ahar vidhi vidhan)*, etc. Acharya Charak in his teachings quoted the fact that our wellbeing is dependent on the food *(ahar)* we eat. Ayurveda believes in a balanced diet, pertaining to *panchbhautik* composition having *tridoshic* impacts on our body. *Tridoshic* refers to "having a balance of *vata, pitta,* and *kapha* doshas" on our body constitution. An equilibrium of the 'biological humor' and 'body tissues' is considered as a healthy state and this normalcy can be achieved by indulging in healthy food only. Food which maintains the symmetry of balanced body tissue *(samadhatu)* and regains the equilibrium of imbalanced tissue *(visham dhatu)* is called *pathya* or "wholesome food," whereas the diet which dislodge the vitiated biological humors *(doshas)* from the body but do not expel them out of the body are regarded as "unwholesome." Apart from this, a set of fairly satisfactory dietetic codes called *aharvidhivisheshaytana* has been identified and prescribed in Ayurveda. When the food is taken judiciously and according to code of dietetics then only the benefits of food can be achieved. Any aberration in the diet and its preparation style leads to poor health. *Aharvidhivisheshaytana* are certain dietary customs responsible for wholesome and unwholesome effect of food. Acharya Kashyapa so eloquently said, "Health is dependent on food, and food seeks out suitable methods." Ayurveda has also put a heavy focus on consuming the right amount of food in order to preserve good health. The quantity of food consumed is the amount that is digested and metabolized in a timely manner without disrupting the *dosha* balance (biological humors). The standard calculation of a person's food intake is based on their digestive ability, which varies depending on the season and age of the person. Dietic deficiency (*hinamatra*) causes affliction of the body, mind, intellect, senses, and *dhatus* excellence, while *atimatra* causes vitiation of all three doshas, vitiation of digestive capacity leading to indigestion. According to Acharya Vagbhata, the quantity of food one can consume is calculated not only by *agni* (digestive capacity), but also by *ahardravya* (nature of food substances). Therefore, *guru ahar dravya* should be consumed to half of one's digestive capacity, while *laghu ahar dravya* should be consumed until one is satisfied.

Ayurveda places a strong emphasis on *kala* (time factor) because it influences other factors such as quantity (*matra*), metabolism (*ahar parinaman*), power (*bala*), and even digestive capacity (*agni*). It has two aspects: *awasthik kala* (age and diseased/healthy state) and *nityaga kala* (season and time of day). Both of these factors have an effect on digestive capacity (*jatharagni*), which affects food digestion and metabolism. In Ayurvedic texts, it is mentioned that during the winter season, healthy people's digestive power increases and they are capable of digesting any heavy food, but during the spring season, *kapha* becomes irritated, so light and easily digestible food should be consumed. During the summer, a sweet, cold, liquid diet should be consumed, while during the rainy season, due to weakened digestive capacity, old barley, wheat, and red (*shali*) rice should be consumed. Similarly, food should be consumed at the right time of day, taking into account the digestive ability. Reasonable timings for meal intake, such as eating before the previous meal is digested (*adhyashan*), eating compatible and incompatible foods together (*samshan*), eating at an incorrect period (*vishamashan*), and incompatible food (*viruddhashan*), contribute to *agni* vitiation, resulting in the development of *ama*, the root cause of all diseases. While explaining the general principles of eating, Acharya Charak emphasizes the importance of eating hot, emollient food because it tastes good, stimulates digestive capacity, prevents flatulence, and increases body power. Meals should be eaten only after the previous meal has been digested, as ingested food easily vitiates all of the body's humors (*dosha*).

Ayurvedic cooking involves the use of herbs that can be helpful in boosting up the immunity and healing power if a person catches any type of acute or chronic ailment. It helps in cleaning accumulated toxins and rejuvenates the body, as each dish is cooked and spiced to achieve maximum digestibility. Ayurvedic foods are appetizing, full of flavor and aromatizing which offer good health when served in an inspiring atmosphere.

The effective cooking can improve the quality of food, and the "positive vibrations" of the cook and the surrounding atmosphere produce "happy hormones" in the body. Thus, converting the food into "natural medicines" with intensified beneficial properties. Traditional food system plays an essential part in maintaining the wellness and health of native people. Yet, evidence abounds showing that the traditional food base and knowledge of indigenous people are being eroded as people are occupying western culture. This has resulted in the usage of smaller number of species, decreased dietary miscellany due to household food insecurity and consequently poor health status. Thus, the knowledge of the traditional food system is imperative, it can change the current scenario. Indian traditional foods are also recognized as 'functional foods' because of the presence of body healing chemicals, antioxidants, dietary fibers and probiotics, which help in not only in supporting immunity and maintaining blood sugar level, but also, in curing various other diseases as well.

11.7.2 Yoga: a solution to lifestyle diseases

Yoga can be considered as the best lifestyle alignment tool. It has the capability to prevent, manage, and rehabilitate the most prevalent lifestyle disorders [20]. Yogic lifestyle, diet, attitudes, and various yogic practices help humans in strengthening and developing overall health, which enables them to withstand the stressful situations in a better manner.

Surya namaskar or sun salutation is one such *asana* that works on different parts of the body and strengthen them, ideally done facing the early morning sun which helps in absorption of vitamin D, which in return strengthen our bones and also helps to clear our vision. This process works by detoxifying the body and facilitating blood circulation, preventing backache, joint pain and relieve us from various types of stresses, improves the digestive system, etc.

Various yoga postures that can help prevent deformities in our body are:

1. Kyphosis: tadasana, shavasana, bhujangasana
2. Round shoulder: chakra asana, dhanurasana, ushtt asana
3. Lordosis: pachimotan asana, hal asana, Forward bending
4. Scoliosis: ardh chakra asana, trikon asana, chin-ups

11.7.3 Pranayam *for respiratory disorders*

Pranayam is a part of *ashtanga yoga*, a procedure of regulating mind by controlling the *vayu* (air) during inspiration and expiration of respiration. It has psychological as well as physiological benefits [21]. All the types of *pranayam* consist of physiological entities involved in respiration thereby, significantly improving the lung volumes and capacities. *Pranayam* helps in blood circulation within the body, thus helps in preventing cold, cough, asthma and pneumonia. *Pranayam* has proved beneficial in hypertension, obesity, and infertility as well.

During *pranayam,* there is slow and prolonged inspiration as well as expiration. This stretches elastin and collagen fibers of lung parenchyma and lungs get inflated to their total capacity. *Pranayam* aids in strengthening of respiratory muscles, release of surfactant & prostaglandins, stimulation of stretch receptors, release of undue tension, and adaptation of regulatory mechanisms and acclimatization of chemo receptors. It is beneficial for the improvement of lung volumes and capacities in healthy and diseased (of restrictive & obstructive respiratory diseases) individuals [22].

11.7.4 Meditation and stress

Among all the above approaches to cure lifestyle diseases, meditation has become one of the most popular way to relieve stress among people of all age-groups. This is an age-old practice, which can help combat unattended and unnoticed diseases.

The body is affected through meditation in exactly the opposite ways that stress does, by triggering the body's relaxation response in a positive manner. It helps in bringing back body to a relaxed state, ensuring it to repair itself and prevent new damage from the physical effects of stress. It can restore the 'peace of mind and body' by suppressing the stress-induced thoughts that keep your body's stress response triggered [23]. Throughout the day, when we acknowledge stress, our bodies have the tendency to automatically react in ways that prepare us to either fight or flight. This is your body's stress response reflux. In some cases when there is an extreme danger, this physical response seems to be very helpful. However, a prolonged state of such agitation can cause physical damage to every part of the body [24].

After learning to calm your body and mind, your physical and emotional stress can melt away. This will definitely make you feel better, refreshed, and ready to go and face the challenges of that comes your way with a healthy attitude. With regular practice over weeks or months, one can experience even greater benefits.

11.8 Conclusion

The most important way to control the lifestyle diseases is to take control over the risk factors associated with it. Moreover, a number of communicable diseases can also be prevented by controlling the behavioral or lifestyle habits associated with those diseases. There are a number of solutions that can be put into operation by the government and other involved groups to reduce the major risk factors. Regular monitoring the tendency of non-communicable diseases and their associated risks is crucial for guiding policies and guidelines.

An overall approach is necessary that involves all affected sectors like health, finance, education, planning and others, to minimize the impact of lifestyle diseases on individuals and society. This approach needs to prompt a collaborative effort to reduce the risks associated with these diseases and at the same time inspire interventions to control and prevent them.

Lifestyle diseases are a hazard to the socioeconomic aspects of nations globally and appropriate actions for their management are the need of the hour. Management of lifestyle diseases is quite a one-way process that involves proper diagnosis, screening and treatment of these diseases in addition to providing palliative care for people who require it. Quality lifestyle disease intervention needs to be delivered through a primary healthcare approach where early detection and proper treatment are prioritized [2].

References

[1]   MM Pandey, S Rastogi, AKS Rawat, Indian traditional ayurvedic system of medicine and nutritional supplementation, Evid Based Complement Alternat Med 2013 (2013) 12.

[2]   SA Tabish, Lifestyle diseases: consequences, characteristics, causes and control, J Cardiol Curr Res 9 (2017) 1–4.

[3]   V Ormazabal, S Nair, O Elfeky, C Aguayo, C Salomon, FA Zuniga, Association between insulin resistance and the development of cardiovascular disease, Cardiovasc Diabetol 17 (2018) https://doi.org/10.1186/s12933-018-0762-4.

[4]   N Choudhary, V Padmalatha, R Nagarathna, A Ram, Prevalence of polycystic ovarian syndome in Indian adoloscents, J Pediatr Adolesc Gynecol 24 (2011) 223–227.

[5]   M Sharma, PK Majumdar, Occupational lifestyle diseases: an emerging issue, Indian J Occup Environ Med 13 (2009) 109–112.

[6]   JS Garrow, Obesity and Related Diseases, Churchill Livingstone, Edinburgh, 1998.

[7]   F Ofei, Obesity- a preventable disease, Ghana Med J 39 (2005) 98–101.

[8]   M De Onis, M Blössner, Prevalence and trends of overweight among preschool children in developing countries, Am J Clin Nutr 72 (2000) 1032–1039.

[9]   MB Howorth, Posture in adolescents and adults, Am J Nurs 56 (1956) 34–36.

[10] T Everett, M Trew, Function of the lower limb, Human Movement, 4th edition, Churchill Livingstone, Edinburgh, 2001, pp. 171–189.

[11] L Brianezi, DC Cajazeiro, LBM Maifrino, Prevalence of postural deviation in school education and professional practice of physical education, J Morphol Sci 28 (2011) 35–36.

[12] DM Fessler, Luteal phase immunosuppression and meat eating, Riv Biol 94 (2001) 403–426.

[13] AM Siega-Riz, LM Bodnar, DA Savitz, What are pregnant women eating? Nutrient and food group differences by race, Am J Obstet Gynecol 186 (2002) 480–486.

[14] C Koebnick, UA Heins, I Hoffmann, Folate status during pregnancy in women is improved by long-term high vegetable intake compared with the average western diet, J Nutr 131 (2001) 733–739.

[15] S Khayat, H Fanaei, A Ghanbarzehi, Minerals in pregnancy and lactation: a review article, J Clin Diagn Res 11 (2017) 1–5.

[16] JW Van der Steeg, P Steures, MJ Eijkemans, Obesity affects spontaneous pregnancy chances in subfertile, ovulatory women, Hum Reprod 23 (2008) 324–328.

[17] AV Stazi, A Mantovani, A risk factor for female fertility and pregnancy: celiac disease, Gynecol Endocrinol 14 (2000) 454–463.

[18] JN Hall, S Moore, SB Harper, JW Lynch, Global variability in fruit and vegetable consumption, Am J Prev Med 36 (2009) 402–409.

[19] M Sharma, PK Majumdar, Occupational lifestyle diseases: An emerging issue, Indian J Occup Environ Med 13 (2009) 109–112.

[20] AB Bhavanani, Role of yoga in prevention and management of lifestyle disorders, Yoga Mimamsa 49 (2017) 42–47.

[21] BKS Iyengar, The Illustrated Light on Yoga, Harpercollins, New Delhi, 1968.

[22] MV Rao, A Text Book of Swasthavritta, Chaukhambha Orientalia, Varanasi, 2017.

[23] H Sharma, Meditation: process and effects, Int J Res Ayurveda 36 (2015) 233–237.

[24] H Yaribeygi, Y Panahi, H Sahraei, TP Johnston, A Sahebkar, The impact of stress on body function: a review, EXCLI J 16 (2017) 1057–1072.

CHAPTER 12

Ethnic foods and concentrates: Its role in health protection

Deepshikha Gupta and Christine Jeyaseelan

Amity Institute of Applied Sciences, Amity University, Noida, India

12.1 Introduction

Food is one of the most important necessities for survival [1]. Being deprived of food for an extended period of time could have detrimental effects on the lifespan of an individual, and thus good nutrition is required for a healthy life [2]. It is not just associated with physical effects but also the emotional toll it could take on one's behavioral response in everyday life. It is something that has been well analyzed by health experts from around the world for many decades now [3,4].

However, food has shifted from merely being a necessity to getting optimized for targeted health benefits to suit different lifestyles [5]. The world has progressed from an era of agriculture-based development marked by the green revolution [6] to industrialization involving use of advanced techniques for food production [7] and back to the ethnic or local produce to make food more accessible and healthier for all individuals [8,9]. The best processes developed during industrialization were used for making locally available food into healthier options [10]. The increasing global trend towards "organic food" has brought people back to ethnic and traditional ways of food production and consumption [11]. This has started what we could call an "era of food ethics" around the world in the past few decades with people realizing the importance of ethnic ways of food production [12] and even acknowledging the potential of Ayurveda in food consumption ethics, the study of which might get one interested in Ayurnutrigenomics [13]. There has been an increasing interest in slow foods that is, food eaten for pleasure rather than survival especially in the developed countries while also keeping in mind the ecological positivity that the idea brings with it [14]. This has increased the demand for what we call "ethnic foods" in the US to almost double in the past decade [15,16].

12.1.1 Ethnic foods

Ethnic foods are those which originate from a culture and heritage, with the use of knowledge of local ingredients of plant and animal sources [17]. Thus, ethnic foods represent a country's or region's traditions in not just the way of consumption but how intricately, keeping in mind the rich heritage of the nation, the foods are prepared with old age cooking methodologies [18]. Thus, Indian food, Korean food, Mexican food, American food are all categorized as ethnic foods in their respective nations and countries outside as a symbol of the country's intangible cultural heritage and traditions [19]. Foods identified with different religions like Buddhism, Hinduism, Christianism, Islam also fall under the category of ethnic foods with regards to the customary beliefs in food consumption, with imposing restrictions on consumption of certain food items, and the traditional preparation methods [17,20]. It is of no surprise that with the diversity in geographical locations, belief systems and preferences, availability of plant and animal sources, the uniqueness range of the ethnic foods would be quite massive.

A study on consumer behavior, examined the trend towards this shift, as people enjoy ethnic foods as a part of diving into other cultures, showing an increasing interest in world cuisine as the world becomes more and more inclusive. There are others who did it for passion towards ethnic cooking and food ingredients [21]. However, apart from their appealing exotic factor these delicacies have been in the limelight for the health benefits that they bring. Not only are these foods delicious, but also suit different goals of the growing health-conscious populations [22].

People are understanding the importance of healthy eating to enhance one's quality of life. The population has understood the importance of nutritional value of food to prevent or delay the onset of deadly diseases enhancing the quality of life with healthy aging [23].

In the industrialization era, cropping of high macronutrient-yielding cultivars gradually with more economic benefit are replacing low yielding traditional food plants. Replacing traditional food resources with foods having high concentrations of refined hyper-processed carbohydrate and fat resulted in a major shift in lifestyle and in dietary patterns of American native communities. Loss of access to whole and traditional fresh foods and drastic changes in food habits resulted in rapid rise of obesity-linked diseases and cardiovascular diseases or CVDs [24,25].

12.2 Classification of ethnic food

Food is anything that can be consumed by living organisms for its survival while cuisine is the special style of preparing the food depending on its country or place of origin. Food classification can be done based on region, style, ethnicity/religion or those having historical significance.

On the basis of the region, foods can be classified into global and regional foods. Global are those which are consumed and adopted around the world while regional may depend on the specific nation, state or local region. The regional difference depends on the ingredients

Table 12.1: Popular dishes of some selected traditional cuisines of world.

Type of cuisine	Popular dishes & drinks
Mexican	Chili con carne (tomato-meat soup), Tacos (crispy corn chips), Quessadilla (wrap), Guacamole (dip made by ripen avocado paste), Enchilladas (Mexican lasagna), Empanada (fried flaky chicken pastry), Margarita, and Tequila (drinks)
Italian	Pizza, Pasta, Caprese, Panna cotta (dessert), Focaccia, Red wine, White wine, Sparkling wine
Chinese	Noodles, Soy puff, tofu, Bird's nest soup, Chop suey, Chinese beer and green tea.
Japanese	Sushi, Sahimi, Unagi (eel-based dish), Tempura, Wagashi (sweets), Tonkatsu (pork cutlets), Soba-Udon (noodles), Yakitori beer, Sake, and Traditional style of Japanese Tea
Korean	Dolsotbap (Rice cooked in stone pot), Ganjang (soy sauce), Kimchi (dish made by fermented vegetable), Kongguksu (noodle dish), Jokbal (pig's feet), Korean liquor-soju
Vietnamese	Goi cuon (spring rolls), Banh mi (sandwiches), Banh Xeo (Pancakes), Pho (noodle soup), Cha ca (butter sautéed fish), Bia Hoi (local beer), Jasmine tea, Lemon tea.
Thai	Tom Yam Goong, Pad Thai, Som Tam, Sang som (local rum)
Greek	Greak Salad, Souvlaki (pork-based dish), Mousaka (egg plant & potato dish), Ouzo, tsipouro (drink)
Indian	Chole bhature, butter chicken, chana masala, Palak paneer, dal makhani, kofta, korma, kebab, naan, pav bhaji, paneer butter masala, rogan josh, Hyderabadi biryani, daal baati, sweets, and delicacies.
French	Croissant, Charcuterie, Bouillabaisse (fish soup), Roquefort (blue cheese), Tartar, Croque-Monsieur (egg-ham sandwich), French Wine

available locally, the trade, climate, cultural differences and cooking traditions. Based on style, foods are classified into Fusion (A combination of foods from different countries, cultures and traditions), Haute (generally found in high level establishments, which have detailed preparation, appealing presentation and high price), Nouvelle (related to French with lighter dishes and greater emphasis on presentation), Vegan (without animal products), Vegetarian (all plant product as well as dairy products and eggs included). Based on ethnicity or religion it is classified depending on the region like Indian, Chinese, Mexican, Thai and religions like Christian, Buddhist, Hinduism. On historical basis some of the common foods are divided into ancient Greek, ancient Roman, Aztec, Maya, Medieval, etc.

A general classification of the different types of cuisine, based on their regions and the main ingredients that they contain are given in Table 12.1. Fig. 12.1 depicts some important traditional and ethnic food around the world such as Peking duck, Beijing (China); Dosa (southern India); Sushi (Japan); Greek Salad (Mediterranean region); Orecchiette with Broccoli Rabe (Italian); Tom yum goong, Thailand.

12.3 Traditional and folkloric ethnic foods of different provinces of world

Traditional food belongs to a particular geographical area, involving different culture, that have been consuming it locally and regionally for long time period. Traditional foods are

Figure 12.1
Ethnic foods from around the world.

prepared by using folkloric methods of a country or a region. A cuisine is a traditional and characteristic style of cooking practices [26], often associated with a specific region, country, or specific culture [27]. A local, regional, or national cuisine must spread around the world to become a global cuisine. Many factors have influenced traditional foods over the period of time. The availability of raw materials is one of these factors. As a result, agricultural practices and location have an impact on traditional food. Lower altitude regions, for example, have different vegetation than higher altitude regions; countries without access to the sea typically have a lower availability of fish and seafood than those with a large coastal area. Dietary patterns are influenced by local food availability as well as the cultural and socioeconomic environment, but there is a trend for new habits to be transferred and assimilate between countries. In the 1960s, Mediterranean populations' diets were characterized by a high consumption of fruits and vegetables, in contrast to Northern European countries' low consumption of these foods. Many traditional foods are at risk of extinction as the food market continues to globalize and internationalize. Documenting traditional foods and dishes is critical for preserving traditional foods, which are an important part of cultural heritage.

Fresh fruit, dark leafy vegetables, olive oil, lentils, grains, high-fiber beans, and omega-3-fatty acids rich fish are used in the Mediterranean diet, which delivers a lot of immune-boosting, anti-inflammatory, and cancer-fighting ingredients that reduce the risks of obesity, diabetes, heart disease, and other diet-related ailments [28]. A traditional Mediterranean-style diet, according to Harvard University research, is associated with a 25% lower risk of death from cancer and heart disease. Another study conducted by a Harvard University research group found that people lose more weight and feel more satisfied when they eat a diet high in healthy fat rather than a low-fat traditional diet. There is mounting evidence that the Mediterranean diet is cardioprotective [29].

Californian style of cooking is based on use of seasonal, locally grown simple preparations with plenty of low-calorie food and nutrient rich green vegetables and fresh fruits which is protective from disease. Important dishes are Italian parsley, Orecchiette with mushrooms, grilled veggie skewers, broccoli rabe with hot pepper, olive oil, and Parmesan cheese. High fat cheese is also used to make vegetables tastier.

Traditional Vietnamese flavors (such as cilantro, mint, Thai basil, star anise, and red chili) have long been used as alternative treatments for a variety of ailments. Cilantro and anise have been shown to reduce inflammation and aid digestion. Fresh herbs, a lot of vegetables and seafood, and cooking techniques that use water or broth instead of oils are some of the techniques that contribute to the high quality of Vietnamese food. They rely on traditional cooking methods, such as less frying and herb-infused coconut milk sauces. One of the healthiest and most delicious Vietnamese dishes is pho (pronounced "fuh,") an aromatic broth-based noodle soup full of antioxidant-rich spices.

Traditional Japanese cuisine consumed on the island of Okinawa is said to promote longevity, as the average life expectancy is more than 100 years. Japanese people cook in the healthiest ways possible, from light steam to stir fry, and they eat a lot of cancer-fighting fruits and vegetables. They believe in eating healthy, but they have an 80 percent full stomach. Japanese staples include cruciferous calcium-rich vegetables like bok choy, as well as antioxidant-rich yams and green tea, which protect against breast and colon cancer and are extremely beneficial to one's health. Iodine-rich seaweed (good for thyroid); omega-3-rich seafood; shiitake mushrooms (high in iron, potassium, zinc, copper, and folate) and whole-soy foods, tofu, edamame, miso and tempeh made of fermented soybeans are part of their regular diet.

The traditional Italian cuisine has star ingredients like olive oil, tomatoes, oregano, parsley, garlic and basil. Vitamins A and C are found in garlic and traditional Italian herbs. They are thought to be beneficial to digestion. The liberal use of lycopene-rich tomatoes in Italian dishes is thought to protect against cancer. Olive oil aids in the reduction of cholesterol, the prevention of heart disease, and the reduction of belly fat. The common use of grated Parmesan cheese in Italian dishes adds a lot of flavor.

The Spanish food plate is called tapas. The Spanish culinary practices make extensive use of fresh seafood, vegetables, and olive oil in the preparation of super healthy dishes such as gazpacho (contains tomatoes) and paella (rich in fresh vegetables, seafood and rice).

Mexican cuisine can be both heart healthy and slimming. A Mexican diet rich in beans, soups, and tomato-based sauces has been shown to reduce the risk of breast cancer in women. The emphasis in Mexican cuisine is on eating foods that are slowly digested, such as beans and fresh ground corn, which may protect against type 2 diabetes.

South America has a diverse culinary repertoire with 12 countries within its borders. The traditional South American diet includes fresh fruits and vegetables (including legumes) as well as high-protein grains such as quinoa. A South American meal of rice and beans provides an excellent source of protein. This seafood medley includes a variety of healthy spices and ingredients, including cilantro, chile peppers, tomatoes, and onions. Fried items such as sausage, yams, and bananas are frequently served in Brazilian or Argentine restaurants [16].

Traditional African diets include a variety of grains, legumes, plantains, tubers, vegetables, spices, oil seeds, lean meat, and game meat – all of which are well-balanced, flavorful, and low in fat and sodium. In ashes, food is mostly boiled, steamed, grilled, roasted, or baked. Ethiopian cuisine, in particular, is gaining popularity in the West. Teff flour, which is high in protein, fiber, iron, and vitamin C, is used to make injera, the traditional Ethiopian flat bread. The food is mostly unprocessed, and the produce is mostly organic from this region.

In Thai cuisine, it is quite important to prepare balanced dishes. Thailand is known to grow 3500 kinds of rice and because of the abundance of rice and fish, they form the basis of Thai diet. Especially Pad Thai, Tom Yung Goong, green curry, som tam, massaman curry, and tom kha kai are among the most well-known dishes. Thai [30] favorite soup called *Tom Yung Goong* is said to prevent cancer. It is made with shrimp, coriander, lemongrass, ginger, and other herbs and spices with properties 100 times more effective than other antioxidants in inhibiting cancerous-tumor growth. Pad rice is the rice noodle dish cooked with eggs, fish sauce, shrimps, garlic and red pepper. Som tum, the green papaya salad, is prepared with raw papayas. Season fruits like durian, rambutan, mangosteen, pineapple, papaya, starfruit, sapodilla and guava have antioxidant and anti-inflammatory effects in human body. [26,27,31] After noticing that the incidence of digestive tract and other cancers was lower in Thailand than in other countries, researchers at Thailand's Kasetsart University and Japan's Kyoto and Kinki Universities became interested in the soup's immune-boosting properties.

Chinese cuisine is reported to have about ten thousand kind of foods due to their climatic and geographical diversity. Chinese preparation of balanced dishes is believed to use naturalness in meals using low heat cooking, boiling, quick frying or baking techniques. They use lot of sauces, abundant in spices, garlic, green onions, sesame oil, vinegar and soy-based ingredients like soy sauce, soy milk, bean, sprouts and tofu cheese. Unique vegetables used in Chinese cuisine include cabbage, broccoli and water chestnut. Chinese use a lot of meat varieties like chicken, pork, cat and dog meat, snakes, insects and sea food. Chinese foods are very low on sugar, refined carbs and high fat. They also balance the food intake by focusing much on liquid foods (soups and watery porridges). All Chinese foods are made on the principle of

Yin (wet and moist foods that cool you down) and Yang (dry and crisp foods that heat you up). Almost all foods in Chinese cuisine are made to have equilibrium between yin and yang ingredients. Protein foods are seen as Yang, while carbohydrates are yin. Yin and Yang approach to food also stabilizes the health of our metabolisms, with chilies that promote digestion, garlic to fight toxins and many others. Chinese green tea, in particular, is very famous for health benefits like fighting heart diseases, digestion and lowering chance of cancer. Chinese foods use almost zero milk products with great focus on rice, noodles and vegetables. Chinese green tea is very famous for health benefits like fighting heart diseases, digestion and lowering chance of cancer [32].

Indian traditional foods and their dietary guidelines are prescribed in Ayurveda [33]. Food is thought to affect both the mind and the body in Ayurveda that is, "we are what we eat." By understanding which foods are best suited to our minds and bodies, we can use nutrition as a source of healing. Traditionally, Indian foods are divided into three groups: Satvika, Tamasika, and Rajsika. Aromatic spices such as turmeric, ginger, red chilies, and garam masala are common in Indian cuisine (a mixture of cumin, cardamom, black pepper, cinnamon, coriander, and other spices). Turmeric and ginger help fight Alzheimer's. Researchers point out that Alzheimer's disease rates in India are four times lower than in the United States, possibly due to the use of turmeric powder in the preparation of curry on a daily basis. Turmeric, a key ingredient in curry, is said to be anti-inflammatory and healing. Other ingredients in Indian cuisine include yoghurt and lentils, which contain high levels of folate and magnesium and may help to stabilize blood sugar levels. Dal is a dish made from lentils and Indian spices that is typically served as a side dish. Use of fried food is also common. Street food like samosas (pastry puffs), kachori, golgappa, bhel, vada pao, and heavy curries made with lots of cream and butter are also very popular.

12.4 Religious ethnic foods

Religious ethnic foods are processed by religious groups having cultural aspects of ritualistic processes. They are also sometimes created and established in order to circumvent food taboos. Every community has a distinct dietary culture that represents its heritage as well as sociocultural aspects of its ethnicity. Food prepared by various ethnic groups of people is unique and distinct due to geographical location, environmental factors, food preferences, and the availability of plant or animal sources. Some of the characteristics that contribute to the description of a culture are customary beliefs, food rules and laws, religions, and social groupings, while ethnicity is the affiliation with a race, people, or cultural group. Some ethnic foods have been mentioned in holy books such as the Bible, Buddhist Scripture, Koran and Bhagavad Geeta, and as a result, most of the ethnic foods have been influenced by religion.

Nonvegetarian foods are not strictly forbidden according to the Buddhist religious dietary code, and if animal flesh is consumed, non-Buddhists should kill the animal. Monks follow a strict dietary regimen. They don't eat anything substantial in the afternoon. Every month, fasting for the entire day is expected on the new and full moons. Buddhists typically eat together at home with their families. Noodles in soup, skiu, or momo (small dumplings of wheat flour with meats), baked potatoes, tsampa (ground roasted barley grains), and other dishes are common among Tibetan Buddhists [34]. The consumption of animal flesh and alcoholic beverages is not prohibited in Buddhism. Tibetans do not eat small animals like chickens, ducks, goats, and pigs because they believe killing many small animals is more sinful than killing a single large animal (yak or cow), which is more practical. Fish consumption is uncommon among Tibetan Buddhists, who worship fish as a symbol of longevity and prosperity. Nepali Buddhism is a synthesis of Tibetan Buddhism and Hinduism, with elements of nature and ancestor worship thrown in for good measure. In South-East Asia, Buddhists eat fish and soybean products.

In Christian food culture, after family prayers, all family members sit together at a table and eat together. The majority of Christians' cultural foods include a variety of ethnic foods such as loaf, cheese, and sausage. Eggs are commonly decorated and featured by Christians throughout North America and Northern Europe as a symbol of Christ's Resurrection [34].

Muslim foods are prepared in accordance with dietary laws, and no alcoholic beverages are served. For Muslims, food consumption is governed by strict dietary laws. Food prohibitions include not eating swine, carrion (dead animals' flesh), blood in any form, food previously offered to gods, and alcohol and any intoxicant. Traditionally Muslims women and children may eat separately after the male members finish their meals. During Ramadan, a month-long fast, family members, friends, and relatives share meals together after sunset. Halal food laws have been enacted to maintain strict guidelines regarding Muslim foods [35].

Sattvika, Raajasika, and Taamasika are three different types of food based on their property, quality, and sanctity, according to Hindu sacred books. Sattvika food refers to food that promotes prosperity, longevity, intelligence, strength, health, and happiness. Fruits, vegetables, legumes, cereals, and sweets are examples of this category. Raajasika food represents activity, passion, and restlessness, and includes foods that are hot, sour, spicy, and salty. Taamasika food is intoxicating and unhealthy, resulting in dullness and inertia. Sattvika foods include cooked vegetables, milk, fresh fruits, and honey, which are intended for the truly wise. Taamasika foods are those that bring out the worst, ugliest aspects of human behavior, such as meat, liquor, garlic, and spicy and sour foods. Raajasika foods are those that provide enough energy to carry out daily tasks [36]. Hindu foods follow the concept of purity and pollution which determines inter-personal and intercaste relationships [37]. Although Hindus are traditionally vegetarians, many nonBrahmins are not. Beef is not eaten by Hindus because the cow is considered sacred. Fish is considered more palatable than other animal flesh foods. Garlic, onion, and intoxicants are forbidden to Brahmin Hindus. On religious occasions, foods are offered to temples for the worship of gods, the possession of spirits, and the feeding of domestic and some wild animals, including birds. In Hindu food culture, it is customary to serve meals first to the family's elder male members. Traditionally, female Hindu family members eat after the men.

12.5 Role of spices in ethnic food

Spices have been used to enhance the flavor of food, provide health benefits, and act as food preservatives since the Mesopotamian civilization, which dates back to 5000 BC. Spices are a type of functional food. Spices are known to have medicinal properties such as purgative, laxative, expectorant, carminative, diuretic, and so on. Spices have been used for a variety of medicinal purposes since ancient times and continue to be so today. Spices like turmeric, fenugreek, mustard, ginger, onion, and garlic have a wide range of bio functions and are likely to protect the human body from a variety of ailments due to their additive or synergistic actions [38]. The list of major spices used in South Asian subcontinent is given in Table 12.2 and few selected spices are pictured in Fig. 12.2.

12.6 Food concentrates

Concentrates consist of dried plant and animal products to which sugar, spices, fats, or salt have been added to help in preparing cooked dishes in a fast manner. They are packs of food containing all the benefits of plant, vegetable and animal matter into small packages. They can be treated as quick meal substituents in today's day and age when people are so busy that they hardly have time to cook food and make healthy dishes for daily consumption. The dried plant material includes vegetables, fruits and cereals. The animal products consist of meat, dairy and fish. These food concentrates are available for any mealtime like breakfast, lunch, supper/dinner. For people who like to take their meals in different courses, the food concentrates are available for first course which may include soups, some starters (hash browns, oats, cornflakes, fried potatoes, nuggets) and even vegetables. Concentrates like noodles, omelets, stews, porridge could be part of the second course followed by gelatins, mousses, dried cakes, and other desserts as the final course.

Food concentrates can be broadly classified into fruit juice concentrates, fiber concentrates, protein, and fat concentrates. Fruit juice concentrates which are thick, syrupy liquids are prepared by removing the water (which is about 90% of the juice) by filtration, evaporation and pasteurization and may or may not include additives. This helps to reduce bacterial growth, thus increasing the storage life, decreasing the cost of storage transportation, and packaging. The different types of juice concentrates are 100% fruit concentrate, concentrated fruit cocktail, beverage or punch, or powdered juice concentrates. Fiber concentrates are prepared from fruits like apple and citrus fruit residues, to use them as sources of fiber in the enrichment of foods. These fiber concentrates were found to have high dietary fiber content. The texture was

Table 12.2: Commonly used spices of Indian sub-continent with their scientific names, part of plant used, and common benefits.

Spices Name	Scientific name	Part of plant used as spice	Uses and benefits	References
Asafoetida	*Ferula asafoetida*	Oleo gum resin extracted from its thick roots and rhizome	Used in tadka and for seasoning, It expels gas and counteracts spasmodic disorders; digestive agent, sedative and nerve stimulant.	[38]
Bay leaf	*Cinnamomum tamala*	leaves	Added in curries in beginning to give flavour. Bay leaf oil possesses antifungal and anti-bacterial action.	[38–40]
Cardamom	*Elettaria cardamomum*	Green pods and seeds	Also known as "queen of spices." It has an aromatic, carminative and stimulant effect. Used as flavouring agent in food preparations. Also helps in vomiting, indigestion, pulmonary disorders with copious phlegm and also as laxative.	[41,42]
Chilli	*Capsicum annum*	fruit	Provides hot flavour to food. The capsaicin antioxidants present in chilli help to cope with cholesterol. It increased secretion of gastric acid and stimulated the nerve endings in the skin.	[43]

(*continued on next page*)

Table 12.2: Commonly used spices of Indian sub-continent with their scientific names, part of plant used, and common benefits—cont'd

Spices Name	Scientific name	Part of plant used as spice	Uses and benefits	References
Cinnamon	*Cinnamomum zeylanicum*	bark	Used as flavouring agent. It supports natural production of insulin and reduces blood cholesterol and act as powerful antioxidant. Cinnamon oil possesses antimicrobial and insecticidal activity	[44,45]
Clove	*Syzygium aromaticum*	bud	Clove oil is beneficial for coping with tooth ache and sore gums. Eugenol present in clove oil imparts strong antimicrobial activity. It promotes salivation and digestive juices used as expectorant, antispasmodic agent.	[33]
Coriander	*Coriandrum sativum*	Seed, leaves	It is used as analgesic, carminative, digestive, anti-rheumatic and antispasmodic agent. It is also good for coping with sore throat, allergies, digestion problems, hay fever etc.	[38]

(continued on next page)

Table 12.2: Commonly used spices of Indian sub-continent with their scientific names, part of plant used, and common benefits—cont'd

Spices Name	Scientific name	Part of plant used as spice	Uses and benefits	References
Cumin	*Cuminum cyminum*	seed	Good source of Iron, chemoprotective, antimicrobial, antioxidant, hypoglycemic, hypolipidemic, and anticarcinogenic agent.	[46]
Curry leaves	*Murraya koenigii*	leaves	These leaves are beneficial for reducing blood sugar. Each part of the plant provides some benefit or the other. The dried leaves are extensively used in herbal medicines.	[45,47]
Fennel	*Foeniculum vulgare*	Seeds	Anethole in Fennel has antispasmatic, hepatoprotective, and antimicrobial activity and also relieves nausea.	[46]
Fenugreek	*Trigonella foenum-graecum*	Seeds and leaves	Seeds are bitter, mucilaginous, carminative, galactogogue, astringent, and emollient agent. Also helpful in heart disease.	[45,47]
Garlic	*Allium sativum*	Bulb	It is useful for coping with cough and cold. It also has antibiotic properties.	[38,48]
Ginger	*Zingiber officinale*	Rhizome	Traditionally used to treat stomach upset, diarrhoea and nausea. It is beneficial for coping with cough and cold.	[38]

(continued on next page)

Table 12.2: Commonly used spices of Indian sub-continent with their scientific names, part of plant used, and common benefits—cont'd

Spices Name	Scientific name	Part of plant used as spice	Uses and benefits	References
Gum Arabic	*Acacia senegal*	Sap or exudate	Used in indigenous Indian medicine for diarrhoea, diabetes, sore throat, and dysentery.	[49]
Mustard	*L. Brassica nigra*	Oily seed	Used as seasoning, extracting vegetable oil and leaves are used as vegetable. Mustard oil is good for body massage and even for getting good hair. It consists of omega-3 fatty acids. It is an excellent source of iron, zinc, manganese, calcium, protein etc.	[50]
Nutmeg	*Myristica fragrans*	Dried kernel of seed (Mace, dried aril)	It is used as flavouring agent. It has carminative, stomachic, astringent, and aphrodisiac properties.	[33,38,39]
Pepper corns	*Piper nigrum*	Fruit also known as peppercorn	It helps treatment of muscular pain due to rheumatism, cold, cough and infections. It stimulates saliva, appetite, tones colon muscles.	[38]
Saffron	*Crocus sativus*	Dried Stigma and style of flower	It helps to cope with skin diseases. It is a good remedy for cough, cold and asthma.	[38]

(continued on next page)

Table 12.2: Commonly used spices of Indian sub-continent with their scientific names, part of plant used, and common benefits—cont'd

Spices Name	Scientific name	Part of plant used as spice	Uses and benefits	References
Star anise	*Illicium verum*	Fruit	Star anise oil is beneficial for rheumatism. Also used as flavouring agent.	[51]
Turmeric	*Curcuma longa*	Rhizome	Turmeric powder can be used for wound healing, anti-inflammatory antioxidant, and anticancer properties. Both turmeric and ginger are believed to have anti-Alzheimer.	[45,38]

Figure 12.2

Important spices used in culinary practices (L to R from row 1–3) clove, mace, cumin, chilli, fenugreek, fennel, peppercorns, cardamom, turmeric, nutmeg, chilli flakes, coriander, pomegranate seeds, cinnamon, star anise. (https://commons.wikimedia.org/wiki/File:Indian_Spices.jpg; By Joe mon bkk (Own work) [CC BY-SA 4.0], via Wikimedia Commons).

dependent on the particle size which could be increased by heat treatment. These concentrates could be used for development of fiber-enriched foods [52]. Rapeseed (*Brassica napus)* protein is an oil industry waste which has been used for food application. The protein was treated with alcohol and steamed to deactivate the myrosinase. This process reduced the oil binding capacity, emulsification capacity and protein solubility. When this protein concentrate was added to sausages, it gave a good taste, improved texture and a characteristic aroma [43]. Fat mimetics are substances which have chemical structures different from fats. They have diverse functional properties, but some physiochemical properties are similar and have the desirable qualities of fats. Whey protein concentrates are known as fat mimetics and they have good gelling properties as well as water binding, emulsification, and adhesion. They emulsify well with cream, soup, sauce, mayonnaise, and processed meat [53].

Processing of concentrates: To prepare concentrates, heat processing is the most commonly used method. This includes boiling of the substance to form a pulp followed by blanching and steaming. The final product is flattened and given a thin form so that they can be easily cooked and consumed. Thicker forms take a longer time to cook. There are some concentrates like groats which do not need any cooking and can be consumed directly or by pouring hot boiling water over them. They can be soaked in hot water for a few minutes (depending on the concentrate) which allows them to acquire normal consistency to be eaten. Some of the food products require to be dehydrated before forming the concentrate. The method generally used is sublimation drying which gives a product that is easily prepared and can be stored over a long period of time. The nutritional value of concentrates can be increased by adding protein products obtained from raw material of plants and animals. Juice concentrates are prepared by washing and scrubbing the fruits thoroughly, followed by crushing and blending them to get a pulp. This is then evaporated to remove water. Due to this process the flavor may get diluted, thus, additives to increase flavor are added. Sweeteners like high-fructose corn syrup may be added to fruit juice and sodium added to vegetable blends along with artificial colors and aromas.

The food concentrates are generally available in the form of *granules* or *briquettes*. The granular ones are packaged and sold in paper bags coated with polyethylene or other polymer film like materials. The briquettes weigh around 50–200 g and are packaged in moisture resistant material like vegetable parchment, cellophane, which is then packed in an outer cover before being dispatched to the market.

The temperature for storage of food concentrate is below 20°C with a maximum humidity of 75%. The shelf life of the food concentrates ranges from 3–12 months.

Food concentrates are healthy, rich in important nutrients and provide high energy on consumption. They are light weight and cost effective being easy to store and transport. They can be used in homes, industries, restaurants and even while travelling long distances (in areas where fresh food is not easily accessible). Fruit and vegetable concentrates help people who are sick in recovering by providing major antioxidants, vitamins, and other supplements required to improve health and immunity [43].

12.7 Advantage and disadvantage of ethnic food

It is well documented in many countries worldwide that raw materials and ethnic ingredients can boost the health, medicinal, and nutraceutical values of the foods. Food is responsible for physical and mental well-being and also provides protection from diseases. Many dreadful diseases like hypertension, stroke, diabetes, cardiovascular disease (CVD), and other noncommunicable diseases are being caused by obesity. Therefore, the role of functional ingredients in the development of healthy traditional foods is more important [54]. Traditional food ingredients rich in nutrients may help with weight management, heart health, bone health, and mental health, in addition to providing enhanced immunity to disease [23]. Using diet rich in traditional food have many advantages like- less calories (controls weight), less saturated fat (cardioprotective), more lean meat and fish, more iron, zinc, calcium, vitamin A and vitamin C containing ingredients (immunity booster) and also strengthens cultural values and well-being.

All over the world the traditional ethnic cuisine is healthier than modern cuisine. Modernization, however, rather than wholesome nutrition, has led to a focus on speed and taste. This translates to high-calorie, high-fat, processed, sodium-loaded foods, complex sugars, and questionable additives. Use of food concentrates largely instead of fresh fruits and vegetables also have deleterious effect. As food concentrates lacks fibers and also contain lot of sugar as preservative may lead to various health hazards. The ethnic cuisines are linked to many foodborne illness outbreaks such as use of fresh vegetables are the main cause of outbreak in the case of Mexican food and use of uncooked, raw fish in Japanese food like sushi. According to a study conducted by Lee et al., [16] the fresh and uncooked ingredients mainly cause food borne illness outbreak. In the current day, the food handlers are known to identify the ingredients of ethnic food. Food safety issues are directly related to human health, hence critical which also increase consumer's risk perceptions toward ethnic cuisine which, in turn, can result in a decrease in ethnic foods sales.

Several Mesoamerican traditional food plants and wild edibles are high in phytochemicals relevant to human health, such as phenolic bio-actives, which have a variety of health benefits. These traditional plant-based foods are high in natural antioxidants and may protect animal cells from oxidative damage caused by chronic disease and obesity. Aside from antioxidant properties, the bioactive compounds of many traditional food plants also have antihyperglycemic, antihypertensive, and antidyslipidemia properties, as well as microbiome-supporting benefits for gut health. Therefore, traditional plant-based foods of Native Americans, with their rich source of natural phenolic antioxidants, can be incorporated in dietary intervention strategies to counter chronic oxidative stress and other metabolic dysfunctions commonly associated with type 2 diabetes, cardiovascular diseases and gestational diabetes mellitus [54]. Despite a predominantly vegetarian diet, India has a high prevalence of obesity-related diseases. Traditional ethnic cuisine from around the world is generally nutritious. French culinary arts and French pastry are well-known and popular. The pastries are high in carbohydrates, butter, and sugar. Italian pastas and German currywurst are high in calories and can cause rapid weight gain if consumed in excess. Spaniards and Italians are big fans of dairy and cheese. Traditional Japanese diets are low in fiber because the vegetables are mostly sea vegetables and there isn't enough cereal in this cuisine. Mercury levels in fish (such as tuna and swordfish) are also a source of concern. Beef and pork are consumed in larger quantities and portions in Chinese cuisine. The levels of oil, sodium, and MSG (monosodium glutamate) in such food are quite high, making it a poor choice for those with heart or blood pressure problems.

The American Western diet is high in sodium, saturated and trans-fat, complex sugars, and refined grains. There's way too much dairy, way too much meat, way too much alcohol, and way too little fiber. This cuisine contains very few raw and fermented foods. Large portions are typical of American restaurants, delis and fa(s)t food joints. The majority of the population is obese, with a high risk of Type 2 diabetes, coronary heart disease, colon cancer, breast cancer, and hemorrhoids.

Indians adore sweets and snacks, and nutrition is frequently sacrificed for taste. The majority of snacks are deep fried and high in sodium. Rather than incorporating fruits and nuts, the sweets rely heavily on ghee (clarified butter), sugars, and syrup. Restaurant cuisine is dominated by meat, cheese, cream, and sugar. Restaurant food can be deficient in vegetables, fruits, salad options, and fiber. Genetic factors, as well as a lack of exercise, contribute to the health crisis. In these stressful times, sugary, starchy, and fried foods are also extremely comforting. This is becoming a major cause of rising disease in South East Asia and the Middle East. A reversal to home style cooking and a more cultural tilt to cooking will foster good health and longevity across the continents.

12.8 Conclusion

This chapter gives an overall perspective into ethnic foods and concentrates. The general introduction gives an idea into what ethnic foods are giving a detailed study into traditional and folkloric ethnic foods. Each country/region has its own specific food suited as per their availability and societal trends being followed over the ages. Some of the regions covered are Japanese, Chinese, Mexican, Thai, etc., each having its own specialty. Keeping the religious facets in view the food habits of Muslims, Christians, Buddhists and Hindus are also covered. A description of various spices which add flavor to the food have also been discussed. The new age trend of moving towards food concentrates, that is, foods which are easily available, and consumable has changed the food habits of humans and helped them find shortcuts in food consumption also. Even though these ethnic foods have a number of advantages and is a healthy option for human beings since it helps to maintain a safe lifestyle, there are some disadvantages which also need to be kept in mind.

Acknowledgement

Authors are thankful to Editors *"Herbal medicines: A Boon for healthy human life"* for providing opportunity to submit this book chapter. Authors are also thankful to Amity University Uttar Pradesh for providing necessary facilities to carry out this work.

References

[1] BJ Brown, ME Hanson, DM Liverman, RW Merideth, Global sustainability: toward definition, Environ Manage 11 (6) (1987) 713–719.

[2] SD Ohlhorst, R Russell, D Bier, DM Klurfeld, Z Li, JR Mein, et al., Nutrition research to affect food and a healthy life span. American Society for Nutrition, Am J Clin Nutr 143 (8) (2013) 620–625.

[3] KC Berridge, Motivation concepts in behavioral neuroscience, Physiol Behav 81 (2) (2004) 179–209.

[4] K Mogg, BP Bradley, H Hyare, S Lee, Selective attention to food-related stimuli in hunger: are attentional biases specific to emotional and psychopathological states, or are they also found in normal drive states? Behav Res Ther 36 (2) (1998) 227–237.

[5] AK Sikalidis, From food for survival to food for personalized optimal health: a historical perspective of how food and nutrition gave rise to nutrigenomics, J Am Coll Nutr 38 (1) (2019) 84–95. https://doi.org/10.1080/07315724.2018.1481797.

[6] PL Pingali, Green revolution: impacts, limits, and the path ahead, Proc Natl Acad Sci U S A 109 (31) (2012) 12302–12308.

[7] E Rehber, Industrialization in the agri-food sector and globalization, Agro Food Ind Hi Tech 15 (3) (2004) 46–50.

[8] S Martinez, M Hand, M da Pra, S Pollack, K Ralston, T Smith, et al., Local food systems: concepts, impacts, and issues, Local Food Syst Backgr Issues (97) (2010) 1–75.

[9] JE Pelletier, MN Laska, D Neumark-Sztainer, M Story, Positive attitudes toward organic, local, and sustainable foods are associated with higher dietary quality among young adults, J Acad Nutr Diet 113 (1) (2013) 127–132. http://dx.doi.org/10.1016/j.jand.2012.08.021.

[10] T Lang, Food industrialisation and food power: Implications for food governance, Development Pol Rev 21 (5–6) (2003) 555–568.

[11] J Guthman, Back to the land: the paradox of organic food standards, Environ Plan A 36 (3) (2004) 511–528.

[12] PB Thompson. The emergence of food ethics. Food Ethics. 2016;1(1):61–74. http://dx.doi.org/10.1007/s41055-016-0005-x

[13] S Banerjee, P Debnath, PK Debnath, Ayurnutrigenomics: Ayurveda-inspired personalized nutrition from inception to evidence, J Tradit Complement Med 5 (4) (2015) 228–233. http://dx.doi.org/10.1016/j.jtcme.2014.12.009.

[14] P Jones, P Shears, D Hillier, D Comfort, J Lowell, Return to traditional values? A case study of Slow Food, Br Food J 105 (2003) 297–304.

[15] B Bell, K Adhikari, IV EC, P Cherdchu, T Suwonsichon, Ethnic food awareness and perceptions of consumers in Thailand and the United States, Nutr Food Sci 41 (4) (2011) 268–277.

[16] JH Lee, J Hwang, A Mustapha, Popular ethnic foods in the United States: a historical and safety perspective, Compr Rev Food Sci Food Saf 13 (1) (2014) 2–17.

[17] DY Kwon, What is ethnic food? J Ethn Foods 2 (1) (2015) 1.

[18] C Cotillon, AC Guyot, D Rossi, M Notarfonso, Traditional food: a better compatibility with industry requirements, J Sci Food Agric 93 (14) (2013) 3426–3432.

[19] M Tomić, K Deronja, MT Kalit, Ž Mesić, Consumers' attitudes towards ethnic food consumption Stavovi potrošača o konzumaciji etničke hrane, J Cent Eur Agric 19 (2) (2018) 349–367.

[20] PA Kroon, LF D'Antuono, Special issue: traditional foods: from culture, ecology and diversity, to human health and potential for exploitation, J Sci Food Agric 93 (14) (2013) 3403–3405.

[21] MG Roseman, Y Hoon Kim, Y Zhang, A study of consumers' intention to purchase ethnic food when eating at restaurants, J Foodserv Bus Res 16 (3) (2013) 298–312.

[22] YS Tey, P Arsil, M Brindal, SY Liew, CT Teoh, R Terano, Personal values underlying ethnic food choice: means-end evidence for Japanese food, J Ethn Foods 5 (1) (2018) 33–39. https://doi.org/10.1016/j.jef.2017.12.003.

[23] V Prakash, M-B Olga, K Larry, B Sian, S Astley, H Braun, HL McMahon, Introduction: the importance of traditional and ethnic food in the context of food safety, harmonization, and regulations, Regulating Safety of Traditional and Ethnic Foods, Elsevier Inc., 2016, pp. 1–6. http://dx.doi.org/10.1016/B978-0-12-800605-4/00001-3.

[24] MI Gómez, CB Barrett, T Raney, P Pinstrup-Andersen, J Meerman, A Croppenstedt, et al., Post-green revolution food systems and the triple burden of malnutrition, Food Policy 42 (2013) 129–138.

[25] PJ Jacques, JR Jacques, Monocropping Cultures into ruin: The loss of food varieties and cultural diversity, Sustainability 4 (11) (2012) 2970–2997.

[26] J Leelarungrayub, Potential health benefits of Thai seasonal fruits; sapodilla and star fruit for elderly people, Am J Biomed Sci Res 5 (1) (2019) 49–53.

[27] W Kraikruan, W Klaipook, R Thanumthat, Benefits of local humid tropical fruit trees in Thailand, Acta Hortic 1186 (2017) 235–240. https://doi.org/10.17660/ActaHortic.2017.1186.37.

[28] MA Melinda Smith, R Lawrence, Cancer Prevention Diet A healthy diet can help you prevent or fight cancer. Here's how to lower your risk with cancer-fighting, World Cancer Res Fund Int, pp. 18. https://www.helpguide.org/articles/diets/cancer-prevention-diet.htm#. (Accessed: April 2020).

[29] AS Dontas, NS Zerefos, DB Panagiotakos, DA Valis, Mediterranean diet and prevention of coronary heart disease in the elderly. Clin Int Age 2 (1) (2007 Mar) 109.

[30] S Balıkçıoğlu Dedeoğlu, Ş Aydın, G Onat, A general overview on the far East cuisine: cuisines of Thailand, Korea and China, J Multidiscip Acad Tour 4 (2) (2019) 109–121.

[31] S Sengupta. 8 Incredible Benefits of Mangosteen, The Queen of Fruits [Internet]. 2017. Available from: https://food.ndtv.com/food-drinks/8-incredible-benefits-of-mangosteen-the-queen-of-fruits-1679593. (Accessed 5th Feb 2021).

[32] Great Health Benefits of Chinese Food [Internet]. 2021. Available from: http://www.chinesefoodhistory.com/chinese-food-facts/health-benefits-of-chinese-food/.

[33] P Sarkar, KDH Lohith, C Dhumal, SS Panigrahi, R Choudhary, Traditional and ayurvedic foods of Indian origin, J Ethn Foods 2 (3) (2015) 97–109. http://dx.doi.org/10.1016/j.jef.2015.08.003.

[34] JP Tamang, Himalayan Fermented Foods: Microbiology, Nutrition, and Ethnic Values, Himalayan Fermented Foods: Microbiology, Nutrition, and Ethnic Values, 1st ed., Boca Raton, CRC Press, 2009. https://doi.org/10.1201/9781420093254.

[35] JM Regenstein, MM Chaudry, CE Regenstein, The kosher and halal food laws, Comp Rev Food Sci Food Safety 2 (3) (2003) 111–127.

[36] KG Dubey, The Indian Cuisine, PHI Learning Pvt. Ltd., Delhi, 2010, p. 2010.

[37] DY Kwon, JP Tamang, Religious ethnic foods, J Ethn Foods 2 (2) (2015) 45–46. http://dx.doi.org/10.1016/j.jef.2015.05.001.

[38] AK Sachan, S Kumar, K Kumari, D Singh, C Anupam Kr Sachan, Medicinal uses of spices used in our traditional culture: worldwide. ∼ 116 ∼, J Med Plants Stud 6 (3) (2018) 116–122.

[39] B Shan, YZ Cai, M Sun, H Corke, Antioxidant capacity of 26 spice extracts and characterization of their phenolic constituents. 2005;

[40] S Batool, RA Khera, MA Hanif, MA Ayub, Bay leaf, Medicinal Plants of South Asia (2020) 63–74.

[41] F Anwar, A Abbas, KM Alkharfy, A ul H Gilani, Cardamom (Elettaria cardamomum Maton) oils, Essential Oils in Food Preservation, Flavor and Safety. Ed. Victor Preedy, 1st ed., Academic Press, 2016, pp. 295–301.

[42] L Wilson, Spices and flavoring crops: fruits and seeds, Encyclopedia of Food and Health. Ed. Benjamin Caballero Paul Finglas Fidel Toldrá, Elsevier Inc., 2015, pp. 73–83.

[43] A Esfahani, JM Wong, J Truan, CR Villa, A Mirrahimi, K KC Srichaikul, Health effects of mixed fruit and vegetable concentrates: a systematic review of the clinical interventions, J Am Coll Nutr 30 (5) (2011) 285–294.

[44] J Wang, B Su, H Jiang, N Cui, Z Yu, Y Yang, et al., Traditional uses, phytochemistry and pharmacological activities of the genus cinnamomum (Lauraceae): a review, Fitoterapia 146 (104675) (2020) 1–24. https://doi.org/10.1016/j.fitote.2020.104675.

[45] DS Rajput, DK Dash, AK Sahu, K Mishra, P Kashyap, SP Mishra, Brief update on Indian herbs and spices used for diabetes in rural area of Chhattisgarh, Int J Pharma Chem Anal 4 (1) (2017) 1–4.

[46] I Hinneburg, DamienHJ Dorman, R Hiltunen, Antioxidant activities of extracts from selected culinary herbs and spices, Food Chem 97 (1) (2006) 122–129. https://doi.org/10.1016/j.foodchem.2005.03.028.

[47] M Modak, P Dixit, J Londhe, S Ghaskadbi, TPA Devasagayam. Recent Advances in Indian Herbal Drug Research. In: TPA Devasagayam, Ed. 2007, pp. 163–73.

[48] G El-Saber Batiha, A Magdy Beshbishy, GL Wasef, YH Elewa, A Al-Sagan, A El-Hack, et al., Chemical constituents and pharmacological activities of garlic (Allium sativum L.): A review, Nutrients 12 (3) (2020) 1–21, 872.

[49] AA Ahmed, Health benefits of gum arabic and medical use, Gum Arabic: Structure, Properties, Application and Economics, Academic Press, 2018, pp. 183–210. https://doi.org/10.1016/B978-0-12-812002-6.00016-6.

[50] R Kaur, AK Sharma, R Rani, I Mawlong, P Rai, Medicinal qualities of mustard oil and its role in human health against chronic diseases: a review, Asian J Dairy Food Res 38 (2019) 98–104.

[51] IP Online, MH Shahrajabian, W Sun, Q Cheng, Asian Journal of Medical and Biological Research Chinese star anise and anise, magic herbs in traditional Chinese medicine and modern pharmaceutical science. 2019;5(3):162–79.

[52] F Figuerola, A Mar, Food chemistry fibre concentrates from apple pomace and citrus peel as potential fibre sources for food enrichment. 2005;91:395–401.

[53] BR Johnson, Whey protein concentrates in low-fat applications, Application Monograph. Published by US Dairy Export Council (2000) 1–8.

[54] D Sarkar, J Walker-Swaney, K Shetty, Food diversity and indigenous food systems to combat diet-linked chronic diseases, Curr Dev Nutr 4 (Supplement_1) (2020) 3–11.

CHAPTER 13

Role of natural products as therapeutic option against nonalcoholic fatty liver disease

Bhat M. Aalim[a,b], Sharma R. Raghu[a,b] and Sheikh A. Tasduq[a,b]

[a]*Academy of Scientific & Innovative Research (AcSIR), Ghaziabad-201002, UP, India* [b]*PK-PD and Toxicology Division, CSIR-Indian Institute of Integrative Medicine, Jammu, Jammu & Kashmir, India*

13.1 Introduction

Nonalcoholic fatty liver disease (NAFLD) is exponentially increasing serious chronic liver disease around world, characterized by abnormal accumulation of triglycerides in liver with no or little alcohol consumption, affecting about one third of population in western countries. In US, NAFLD is the second leading cause of liver transplantation and is becoming the second cause of death in general population [2]. NAFLD is an umbrella term encompassing a range of hepatic disorders, leading from simple hepatic steatosis to NASH, and fibrosis to cirrhosis [3]. Sedentary lifestyle along with high caloric diet intake remains main causes of NAFLD in industrialized countries. Various studies have shown that NAFLD is a multisystem disorder augmenting the risk of many other diseases, including T2D, kidney diseases, and cardiovascular diseases (CVD). Notably among NAFLD patients, CVD remains the main cause of demise [4]. Multifactorial nature of NAFLD is associated with multiplex pathogenesis [1,5]. The physiopathology of disease (NAFLD) was earlier elucidated on the basis of *"two hit hypothesis"* proposing that accumulation of TG's or the steatosis is the first hit and the inflammation triggered by the cytokines and the adipokines leading to NASH or fibrosis is the second hit [6]. The *"multiple hit hypothesis"* proposes numerous insults acting simultaneously on predisposed genetic subjects promoting NAFLD. Multiple factors like nutritional factors, insulin resistance, hormones secreted by adipose tissue, gut micro biota, epigenetic along with genetic factors are included in multiple hit hypothesis as plausible key factors in the NAFLD progression and pathogenesis [7].

NAFLD, as known is caused due to abnormal assimilation of excessive triglycerides in hepatocytes [8]. Several cellular and molecular mechanisms, including excess fat (obesity and

Herbal Medicines: A Boon for Healthy Human Life.
DOI: https://doi.org/10.1016/B978-0-323-90572-5.00029-9

overweight), bacterial growth, mitochondrial dysfunction, genetic predisposition, inflammation, insulin resistance, oxidative and endoplasmic reticulum stress play a key role [5,8]. It has been claimed that NAFLD is related to overabundance of body fat, central to obesity and visceral fat. Furthermore, IR acts independently in the development of NAFLD, as insulin suppresses lipolysis of adipose tissue. Resistance to insulin leads to increase in the efflux of free fatty acids (FFA), promoting hepatic injury a key factor in NASH [8]. Another key player in NAFLD etiology is oxidative stress caused due to altered balance between the production of antioxidants and ROS that is linked with lipotoxicity and inflammation [5,9]. Fructose and saturated fatty acids (SFA) rich in western type of diet has been related to NAFLD independently. Energy garnered from lipid, glucose, and fructose are stored in hepatocytes, as liver has subtle capacity of storage. Excessive supply of energy is redistributed to peripheral organs/tissues for the production of energy or for storage in adipose tissue. Dysregulation of lipid trafficking, due to increase in supply of intrahepatic lipids can have serious consequences involving organ failure [5,10,11]. Excessive overload of lipids in white adipose tissue (storage place) leading to metabolic incompetency followed by immune cell (microphages) infiltration, secreting inflammatory factors such as TNFα and IL-6 [12,13]. Genetic polymorphism regarding metabolism of lipids, IR, fibrogenesis, ER stress, oxidative stress, and adipokines/cytokines all do have a positive correlation with NAFLD besides dietary and the environmental factors contributing toward the NASH [6]. Based on literature, current molecular mechanisms linked with NAFLD is represented in Fig. 13.1.

So far, no effective pharmacological treatments are available for NAFLD [14]. Most the drugs available in market such as vitamin E, alleviates only inflammation and hepatic steatosis, but have a minimal effect on the progression of fibrosis [15,16]. Pharmacological drugs such as insulin sensitizers, inhibitors of de novo lipogenesis, antifibrotic, antiinflammatory, and antiapoptotic agents as well as metabolic regulators and antioxidants have yielded limited desirable outcomes and have raised safety concerns in clinical trials. Modern potential drug candidates have not achieved major desirable end points and have displayed very less therapeutic efficacy. U.S. Food and Drug Administration (FDA) have not approved any drug to treat NASH. Therefore, only bariatric surgery and the lifestyle modifications or physical activity remains effective. Thus, the development of drugs with higher efficacy and least possible side effects is highly desirable [17,18]. Natural products/herbal medicines as alternative means have attracted a considerable attention from past decade as a potent agents for the treatment of NAFLD. Traditional Chinese medicines (TCM) are rich in bio-active substances that can be employed to treat various disorders. Nowadays, most of the studies focus on the natural products or herbal extracts to explore their positive effect against NAFLD [19,20]. The main aim of this review is to summarize the available natural products (including traditional Chinese medicines, crude extracts, and pure natural products) as potential natural bioactive molecules in the treatment of NAFLD.

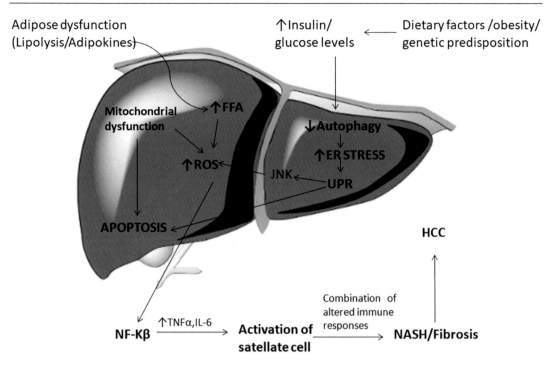

Figure 13.1

Most common mechanisms in NAFLD pathogenesis, integrating dietary, obesity, genetic factors, lipotoxicity (induced by free fatty acids [FFA]), ER stress, apoptosis, and stellate cell activation resulting in hepatic steatosis and fibrosis. In addition, proinflammatory cytokines secreted by adipose tissue and the liver may further augment disease progression. (NF-kβ, nuclear factor kappa beta; ROS, reactive oxygen species; TGF- transforming growth factor alpha; UPR, unfolded protein response).

13.2 Underlying molecular mechanisms in natural product based therapeutic strategy against nonalcoholic fatty liver disease

Currently, no effective pharmacological drugs are available to treat NAFLD. Evidences from animal and cell culture studies has suggested that many natural products have potential to treat NAFLD and to prevents its subsequent progression to steatosis. Traditional medicines treat NAFLD through several ways, including: (1) inhibition of de-novo lipogenesis through attenuation of sterol regulatory element binding protein 1C (SREBP-1C); (2) upregulation of β-fatty acid (FA) oxidation through the upregulation of proliferator activated receptor α (PPARα); (3) inhibition of the inflammatory pathways; (4) enhancing insulin sensitivity and decreasing nuclear factor erythroid 2-related factor (Nrf2). Recently, it has been found that activation of AMPK pathway acts as a common signalling mechanism regulating all these cellular processes; and (5) improving cellular autophagy levels, as the activation of autophagy alleviates NASH and improves several metabolic parameters (including insulin sensitivity, weight loss, blood glucose levels and triglycerides levels) [21]. Autophagy is one among the pathways that are intimately linked with lipid metabolism (lipophagy) [22].

13.3 Natural bioactive compounds against nonalcoholic fatty liver disease

13.3.1 Curcumin

Curcumin is yellow color pigment, which has been used from hundreds of years in Asian countries as spice. Curcumin is a naturally occurring polyphenol, present in *Curcuma longa L,* plant and it's one of the active compounds in turmeric [23]. Curcumin has been reported for it's antioxidant, antitumor, antithrombotic, anti-inflammatory, analgesic, antidepressant, chemoprotective, and chemosensing, cardioprotective, and antiatherosclerotic, pulmonoprotective, antirheumatic, and lipid regulating properties [24–26]. In addition, curcumin plays a very effective role in metabolic syndromes like obesity, hypertension, insulin resistance, and hypertriglyceridemia [24]. All the properties of curcumin are considered to be associated with its antioxidant, insulin sensitizing, antisteatotic, lipid modifying, antifibrotic, and anti-inflammatory activities [27]. Therefore, curcumin has attracted a considerable attention in the treatment of NAFD. Efficacy of the curcumin has been analyzed in several animal studies and in human NAFLD patients. Treatment of curcumin for eight weeks have significantly reduced the serum levels of AST, ALT, cholesterol, TG, HDL-C, and LDL-C as compared to control group [28,29].

Molecular mechanisms underlying curcumin involves inhibition of nuclear factor-kappa – B (NF-κB), which in turn suppresses the activation of NF-κB regulated genes, such as cyclooxygenase 2, cell adhesion molecule and monocyte chemotactic protein [30]. Curcumin reduces the production of ROS and alpha-smooth muscle actin levels in a NASH mice model. Murine preclinical trials have shown, curcumin attenuates the NAFLD development however the clinical trials has shown, curcumin as auspicious, but not the substantiate treatment [31,32]. Curcumin also decreases HFD induced steatosis through the activation of AMPK pathway and by upregulation PPARα signaling [33]. Thus, the attenuation of NAFLD occurs through activation of AMPK and NRF2 pathways, while blocking the activation of NF-κB to regulate inflammation and to suppress the hepatosteatosis [34].

13.3.2 Resveratrol

Resveratrol is a naturally occurring polyphenol found in different plants. Red wine, grapes, berries and peanuts are rich in resverateol [35]. Damage caused by osmotic stress and inflammation to the hepatocytes, brain and kidney can be relieved by the anti-inflammatory and antioxidant properties of resveratrol [36]. As such resveratrol may be useful in blocking the development of various diseases associated with inflammation and oxidative stress involving cancer, diabetes, and various other metabolic disorders [37–41]. Resveratrol improves pathologies linked with NAFLD. Experimental studies have reported that administration of resveratrol for 18 weeks, to the rats maintained under high fat diet, alleviated the hepatocyte ballooning and steatosis [41]. In addition, another study has demonstrated that effect of resveratrol on hepatic steatosis is achieved due to activation of autophagy [42]. Intravenous administration of resveratrol in NASH model developed through methionine-choline dificient (MCD) diet, for four weeks, significantly reduces triglycerides in liver and simultaneously decreasing p62 expression while increasing LC3II levels, indicating the activation of autophagy [43]. Therapeutic efficacy of resveratrol in NAFLD patients were investigated and it was reported that ALT, IL6 and TNFα, NF-κB activity, cytokeratin-18M30 and hepatic steatosis were greatly reduced in resveratrol treated group as compared to control group [44,45]. Resveratrol was shown to stimulate NRF2, providing the defense against oxidative stress [45,46]. Thus, the positive effects of the resveratrol on metabolic disorders can be attributed to it's potential to activate AMPK pathway, stimulating antioxidant response through the upregulation of NRF2, decrease in lipogenesis and increase in FA oxidation [47–50].

13.3.3 Berberine

Berberine is an alkaloid, isolated from the natural plant *Caplis chinensis* and several other Chinese medicines. It has been commonly used in China to treat diarrhea, but various experimental studies have proven a beneficial effect of berberine in NAFLD animal models and NAFLD patients [51,52]. Treatment of berberine to NAFLD patients for 16 weeks significantly decreased BMI, bodyweight, WC, liver fat TG, total serum cholesterol (TC), apolipoprotein B, AST and ALT as compared to control group [53]. Possible molecular mechanism underlying berberine against NAFLD progression involves activation of AMPK pathway, insulin sensitivity, regulating gut environment, improving mitochondrial function, and reducing oxidative stress. Berberine has also been found to be associated with suppression of ER stress and hepatic inflammation in MCD fed mice, ApoE$^{-/-}$ mice, or in db$^-$/db$^-$ (leptin deficient receptor) mice [54–56]. Berberine is also known to suppress palmitate induced hepatoxicity by inducing autophagy and it also attenuates the high fat diet induced IR In-vivo. Hepatic steatosis induced by HFD in rats leading to change in gene expression pattern globally can be reversed by the treatment of berberine [57]. In addition, berberine is also known to induce NRF2 signaling and PPARG and blocking oxidative stress induced hepatic injuries in rodents. Furthermore, as the berberine has a very prominent role in the activation of the peripheral, central and adipose AMPK pathway, berberine can have beneficial role in fatty liver and diabetes treatment [58–61].

13.3.4 Cinnamon

Cinnamon, a spice derived from the inner bark of various plants from genus *Cimamomum*. It has been used in Asian countries from thousands of years as food flavoring agent with various health benefits [62]. Cinnamon is known to have several beneficial effects on health such as insulin sensitizing and antioxidant properties. As such, cinnamon has attracted a considerable attention from scientific world [63,64]. Cinnamon has been reported to increase insulin sensitivity, insulin release, and regulating protein tyrosine phosphates 1B(PTP1P) and insulin receptors kinase, demonstrating insulinotropic activities of cinnamon [64,65]. A number of studies have confirmed the cinnamon's therapeutic role in ameliorating lipid profile and in blood glucose levels [63–67]. To investigate whether cinnamon may have potential effect in treating NAFLD, an experimental study was conducted in which NAFD patients were given cinnamon each day for 12 weeks. A significant decrease in serum cholesterol, TG, AST, ALT, and GGT were observed in cinnamon group as compared to control group [68], various studies have suggested that the ability of cinnamon to increase insulin sensitivity, improving lipid levels and glycemic control is due to the number of mechanism involving activation of insulin receptor through autophosphorylation, activation and up regulation of synthesis of glucose transporter 4(GLUT 4) receptor [69,70].

13.3.5 Garlic

Garlic is a well-known plant that has been used in herbal medicines from ancient times. This plant is the hub of biologically active compounds, such as flavonoids responsible for antioxidant properties and lipid soluble organosulfur compounds, including *allicin*. Garlic provides protection against oxidative stress and neutralizes ROS due to the phytochemicals with known antioxidative properties [70]. Earlier studies on garlic have demonstrated positive effects in the treatment of certain cancers and the prevention of CVD because of its effects on hypercholesterolemia and blood pressure. Additionally, certain In-vitro studies have shown that garlic has anti-thrombotic, anti-atherosclerotic, anticarcinogenic, antidiabetic, and antiobesity properties [71,72]. One of the key factor in NAFLD pathophysiology is oxidative stress in hepatocytes. Therefore, the utility of garlic in NAFLD treatment has developed an immense interest due to the antioxidative properties of garlic [73]. Therapeutic efficacy of garlic in the treatment of NAFLD was analyzed in animal and human studies due to antioxidant properties of garlic and a significant decreases in body fat and hepatic fat was obsreved in treatment group as compared to control group. However, in this particular experimental study, other biochemical parameters were not narrated [74,75].

13.3.6 Silyamarin

Silymarin is the extract of *Silybum marianum*, milk thistle plant, which is used in the treatment of liver diseases [76]. Silyamarin consists of the many flavonoids and flavonolignans. One of the most important biologically active component of silymarin is silybin [77]. Silymarin/silybin is used to treat various liver disorders and has displayed a hepatoprotective role in suppressing NAFLD. The hepatoprotective activity of silymarin/silybin is attributed to its potential in attenuating mitochondrial dysfunction, oxidative stress, liver fat accumulation and insulin resistance. In addition to that, silybin is considered to act via inhibition of intrahepatic gluconeogenesis and glycolysis and regulation of apoptosis, fibrogenesis, and inflammation [76,77]. Due to the low solubility of silymarin in water, intestinal absorption, and bioavailability is very low. Therefore, a complex of phosphatidylcholine with silymarin is used to enhance water solubility [77]. Experimental studies have analyzed efficacy of silymarin in animal as well as in human subjects. It has been reported that treatment of silymarin for six months have significantly reduced WC and hip circumference, TG, total cholesterol, and fatty hepatic index in comparison to normal (control) group. However, there were no significant changes between the two groups in other biochemical parameters, including ALT, AST, GGT, LDL-C, and HDL-C [78–82].

13.3.7 Green tea

Green tea is obtained from the leaves of *Camella sinesis* plant. Green tea originated from china and is most common beverage used throughout Asia. Green tea is rich in flavonoids such as catechins which is the major constituent in green tea [83]. Green tea along with catechins has become subject of immenses interest as several studies has reported it's antioxidant [84], antiarteriosclerotic [83], antihypersensitive [85], antiglycemic [86], anti-inflammatory, and hypolipidemic [87,88], antitumor, and anticancer properties [83,89]. Green tea is considered as good beverage to improve fatty liver index and hepatic function in NAFLD patients. In this regard, effect of green tea on NAFLD was analyzed and it has been reported that consumption of green tea with high density of catechins for 12 weeks significantly reduced BMI and body weight. Biochemical parameters like serum ALT and AST were also reduced significantly as compared to other groups [83]. Therapeutic uses of catechins (green tea) are due to antioxidant and ROS scavenging effects. Catechins has been reported to modify apoptosis in a sequence at several points, involving the change in expression patterns of proapoptotic and antiapoptotic genes. Anti-inflammatory property of green tea is due to the capacity of catechins to modulate isoform of Nitric oxide synthase [90].

13.3.8 Flaxseeds

Flaxseed is a well known plant derived functional food, rich in w-3-α linolenic acid, fibers (soluble and insoluble) and lignans. Health benefits of flaxseeds is due to its antioxidant properties [91]. Also, flaxseeds have a beneficial role in CVD, diabetes, dyslipidemia, cancer, osteoporosis, atherosclerosis, neurological, and autoimmune disorders [91–95]. Impact of flaxseeds on NAFLD was analyzed in a study, in which brown milled flaxseeds were consumed by NAFLD patients for 12 weeks and a significant decrease in BMI, serum AST, ALT, GGT, TNFα, insulin, and glucose levels, fibrosis and steatosis score and IR were observed in treatment group in comparison to control group [96].

13.3.9 Licorice

Licorice is an ancient herb that grows in different parts of Asia and Europe and roots of this plant are used as medicine from times immemorial. Glycyrrhizin is main principal constituent of licorice root extract. Health benefits of this plant include antiulcer, spasmolytic, antidepressive, anti-inflammatory, hepatoprotective, immune-modulating, and antioxidative properties. Higher amounts of glycyrrhizin are associated with side effects such as hypermineral corticoid related side effects. Earlier studies have observed therapeutic effects of licorice on the liver disorders due its anti-inflammation and hepatoprotective properties [97–100]. Therefore, efficacy of licorice root extract in NAFLD patients was conducted and it was observed that serum ALT and AST significantly decreased in licorice treated group as compared to the control group. However, BMI change was insignificant in both the groups [101].

13.3.10 Bayberry

Bayberry is a shrub, native to china and some parts of Asia, also known by the names Myrica, candleberry, wax myrtle, sweet gale, and bay rum. Bayberry is known for its antioxidant capacity due to the presence of various phenolic acids, including salicylic acid and anthocyanin [102,103]. In addition to that several studies have demonstrated that berries with higher amount of anthocyanins may be used to treat NAFLD to alleviate inflammation, dyslipidemia and liver steatosis [104–106]. To investigate, such effects, NAFLD patients were given bayberry juice for 4weeks. No effect was seen in serum ALT, AST, TG, and fasting glucose levels. However oxidative stress, inflammation and apoptosis markers were significantly reduced in bayberry juice group as compared to normal group [107].

13.3.11 Ampelopsin

Ampelopsis is also known as dihydromyricetin (DHM), is a flavanonol, a type of flavonoid present in *Ampelopsis* species. From thousands of years this herb has been utilized in china to treat hepatitis, common cold, and sore throat [108]. It has been considered that DHM has several health beneficial effects which involves hepatoprotective, anti-inflammatory, antioxidant, and glucose regulating activities [109,110]. Studies have demonstrated that DHM has improved insulin sensitivity and serum lipid profile [111,112]. Considering such properties, DHM could be a potential compound in the treatment of NAFLD. In this regard, an experimental study was performed in which NAFLD patients were given capsules of DHM for three months. Results of this experimental study has displayed serum AST, ALT, LDL-C, GGT, cytokeratin-18, and fibroblast growth factor were significantly decreased as compared to the control group. However, no significant change was observed in BMI, blood pressure TG, TNFα, serum ALP, cholesterol, and apolipoprotien A-1 [113].

Table 13.1: List of traditional Chinese medicines for the treatment of nonalcoholic fatty liver disease.

S. no.	Herb name	Biochemical Parameters	Reference
1.	Xuezhikang	Decreases serum TG, total cholesterol (TC), LDL-C, and increases HDL-C	[122,123]
2.	Qianggan	Anti-fibrotic, and decreases serum ALT levels	[124,125]
3.	Danning pian	Reduces body fat and ALT levels	[126]
4.	Yiqi Sanju	Decreases BMI, WCs, IR, TG, total cholesterol, LDL-C, ALT, AST, TNF-α, and increases HDL-C levels	[127]
5.	Qushi Huayu Decoction	Lowers SCD1, FAS, lipid droplets, and inflammatory response	[128]
6.	Yinchenhao Decoction	Decreases PPARγ expression	[119]
7.	Sini San	Decreases ALT and AST levels and prevents progression to steatosis	[129]
8.	Ganzhixiao Decoction	Decreases ALT, TG, and TC levels	[130]
9.	Cigu Xiaozhi Pill	Decreases ALT, AST, TC, and TG levels	[131]
10.	Hugan Qingzhi Tablet	Decreases ALT, AST, TC, TG IL-6, and P65 expression level	[132]
11.	Gegenqinlian Decoction	Decreases LDL-C, HDL-C levels, and PPARγ expression	[133]
12.	LiGan ShiLiuBaWei San	Increase PPARα expression while decreases ALT, AST, TC, TG, and FFA levels	[134]
13.	Huanglianjiedu Decoction	Decreases TC, TG, LDL-C, and HDL-C levels	[135]
14	Lingguizhugan Decoction	Decreases TC, TG, and LDL-C levels	[136]

In addition to the above-described natural products several oils rich in monounsaturated fatty acids (MUFA), have been found to be effective in the treatment of liver diseases. Olive oil, rich in the MUFA and phenolic compounds has been reported in increasing hepatic glucose transporter-2 expression and decreasing the accumulation of hepatic triglycerides. Such effects of olive oil is believed to be mediated through a number of mechanisms involving suppression of NF-κB activation, downregulation of LDL oxidation and suppression of proinflammatory cytokine release (IL6, TNFα), and ameliorating IR [114–117].

Furthermore, traditional Chinese medicines (TCM) with a capability to treat NAFLD have also been reported. These different types of traditional Chinese medicines reported previously in the treatmen of NAFLD are given in Table 13.1. TCM is the hub of biologically active compounds that have been reported to reduce hepatic TG accumulation, enhance adiponectin secretion, increase in PPARγ expression, decreasing the levels of FFA and ameliorating hepatic inflammation and hepatic steatosis [118–136].

13.4 Conclusion

NAFLD is multifactorial, chronic liver disease, mainly associated with over nutrition along with sedentary lifestyle. So far, the effective pharmacological treatment for NAFLD is lacking. Experimental evidences have demonstrated that natural products can prevent liver diseases including NAFLD. Therapeutic activities exhibited by natural products are mainly mediated through multiple pathways. Underlying molecular mechanisms through which herbal products exert their effects are ambiguous. Therefore, the underlying pharmacological mechanisms of natural/herbal products are far from being a clear. Further research is need of the hour, in examining the biologically active components in natural products and exploring their underlying molecular mechanisms of action against NAFLD pathophysiology. Further, there is a need to evaluate these bioactive natural molecules for their potential therapeutic efficacy against NAFLD using modern biological techniques.

Abbreviations

NAFLD	non-alcoholic fatty liver disease;
NASH	non-alcoholic steatohepatitis;
TG	triglycerides;
Apo	apolipoprotein;
TNF-α	tumor necrosis factor alpha;
WC	waist circumference.
ALP	alkaline phosphatase;
ALT	alanine aminotransferase;
AST	aspartate aminotransferase;
BMI	body mass index;
CVD	cardiovascular diseases;
FFA	free fatty acid;
HDL-C	high-density lipoprotein cholesterol;
TC	total cholesterol;
IR	Insulin resistance;
IL-6	interleukin 6;
LDL-C	low-density lipoprotein cholesterol;
NF-κB	nuclear factor κB;
T2D	type 2 diabetes

Acknowledgements

Council of Scientific and Industrial research (CSIR), New Delhi and University Grants Commission (UGC), New Delhi are acknowledged for financial assistance. Authors are also thankful to Sheikh A. Umar and Shahid H. Naikoo for their help in manuscript revision.

Conflict of interest

The authors declare no conflict of interest.

References

[1]   GC Farrell, VW Wong, S Chitturi, NAFLD in Asia–as common and important as in the West, Nat Rev Gastroenterol Hepatol 10 (5) (2013) 307–318.

[2] ZM Younossi, et al., Global epidemiology of nonalcoholic fatty liver disease-Meta-analytic assessment of prevalence, incidence, and outcomes, Hepatology 64 (1) (2016) 73–84.

[3] E Cobbina, F Akhlaghi, Non-alcoholic fatty liver disease (NAFLD) - pathogenesis, classification, and effect on drug metabolizing enzymes and transporters, Drug Metab Rev 49 (2) (2017) 197–211.

[4] CD Byrne, G Targher, NAFLD: a multisystem disease, J Hepatol 62 (1 Suppl) (2015) S47–S64.

[5] DG Tiniakos, MB Vos, EM Brunt, Nonalcoholic Fatty Liver Disease: Pathology and Pathogenesis, Ann Rev Pathol 5 (1) (2010) 145–171.

[6] JK Dowman, JW Tomlinson, PN Newsome, Pathogenesis of non-alcoholic fatty liver disease, QJM 103 (2) (2010) 71–83.

[7] E Buzzetti, M Pinzani, EA Tsochatzis, The multiple-hit pathogenesis of non-alcoholic fatty liver disease (NAFLD), Metabolism 65 (8) (2016) 1038–1048.

[8] S Petta, C Muratore, A Craxì, Non-alcoholic fatty liver disease pathogenesis: the present and the future, Dig Liver Dis 41 (9) (2009) 615–625.

[9] CJ McClain, S Barve, I Deaciuc, Good fat/bad fat, Hepatology 45 (6) (2007) 1343–1346.

[10] AM El-Badry, R Graf, P-A Clavien, Omega 3–Omega 6: what is right for the liver? J Hepatol 47 (5) (2007) 718–725.

[11] S de Ferranti, D Mozaffarian, The perfect storm: obesity, adipocyte dysfunction, and metabolic consequences, Clin Chem 54 (6) (2008) 945–955.

[12] JP Bastard, et al., Recent advances in the relationship between obesity, inflammation, and insulin resistance, Eur Cytokine Netw 17 (1) (2006) 4–12.

[13] SE Shoelson, L Herrero, A Naaz, Obesity, inflammation, and insulin resistance, Gastroenterology 132 (6) (2007) 2169–2180.

[14] M Asrih, FR Jornayvaz, Diets and nonalcoholic fatty liver disease: the good and the bad, Clin Nutr 33 (2) (2014) 186–190.

[15] AM Oseini, AJ Sanyal, Therapies in non-alcoholic steatohepatitis (NASH), Liver Int 37 (Suppl 1) (2017) 97–103.

[16] S Cassidy, BA Syed, Nonalcoholic steatohepatitis (NASH) drugs market, Nat Rev Drug Discov 15 (11) (2016) 745–746.

[17] E Vilar-Gomez, et al., Weight loss through lifestyle modification significantly reduces features of nonalcoholic steatohepatitis, Gastroenterology 149 (2) (2015) 367–378 e5; quiz e14-5.

[18] G Lassailly, et al., Bariatric surgery reduces features of nonalcoholic steatohepatitis in morbidly obese patients, Gastroenterology 149 (2) (2015) 379–388.

[19] X Xu, et al., Research advances in the relationship between nonalcoholic fatty liver disease and atherosclerosis, Lipids Health Dis 14 (2015) 158.

[20] Y Yang, et al., Alpha-lipoic acid improves high-fat diet-induced hepatic steatosis by modulating the transcription factors SREBP-1, FoxO1 and Nrf2 via the SIRT1/LKB1/AMPK pathway, J Nutr Biochem 25 (11) (2014) 1207–1217.

[21] S Lee, et al., Dysregulated expression of proteins associated with ER stress, autophagy and apoptosis in tissues from nonalcoholic fatty liver disease, Oncotarget 8 (38) (2017) 63370–63381.

[22] WJ Kwanten, et al., Role of autophagy in the pathophysiology of nonalcoholic fatty liver disease: a controversial issue, World J Gastroenterol 20 (23) (2014) 7325–7338.

[23] RC Martin, et al., Effect on pro-inflammatory and antioxidant genes and bioavailable distribution of whole turmeric vs curcumin: Similar root but different effects, Food Chem Toxicol 50 (2) (2012) 227–231.

[24] A Sahebkar, Why it is necessary to translate curcumin into clinical practice for the prevention and treatment of metabolic syndrome? Biofactors 39 (2) (2013) 197–208.

[25] Y Panahi, et al., Antioxidant and anti-inflammatory effects of curcuminoid-piperine combination in subjects with metabolic syndrome: a randomized controlled trial and an updated meta-analysis, Clin Nutr 34 (6) (2015) 1101–1108.

[26] D Lelli, et al., Curcumin use in pulmonary diseases: state of the art and future perspectives, Pharmacol Res 115 (2017) 133–148.

[27]   NA Zabihi, et al., Is there a role for curcumin supplementation in the treatment of non-alcoholic fatty liver disease? The data suggest yes, Curr Pharm Des 23 (7) (2017) 969–982.

[28]   Y Panahi, et al., Curcumin lowers serum lipids and uric acid in subjects with nonalcoholic fatty liver disease: a randomized controlled trial, J Cardiovasc Pharmacol 68 (3) (2016) 223–229.

[29]   Y Panahi, et al., Efficacy and safety of phytosomal curcumin in non-alcoholic fatty liver disease: a randomized controlled trial, Drug Res (Stuttg) 67 (4) (2017) 244–251.

[30]   F Vizzutti, et al., Curcumin limits the fibrogenic evolution of experimental steatohepatitis, Lab Investig 90 (1) (2009) 104–115.

[31]   MH Farzaei, et al., Curcumin in liver diseases: a systematic review of the cellular mechanisms of oxidative stress and clinical perspective, Nutrients 10 (7) (2018) 1–28.

[32]   CM White, JY Lee, The impact of turmeric or its curcumin extract on nonalcoholic fatty liver disease: a systematic review of clinical trials, Pharm Pract (Granada) 17 (1) (2019) 1350.

[33]   MY Um, et al., Curcumin attenuates diet-induced hepatic steatosis by activating AMP-activated protein kinase, Basic Clin Pharmacol Toxicol 113 (3) (2013) 152–157.

[34]   H Yin, et al., Curcumin suppresses IL-1β secretion and prevents inflammation through inhibition of the NLRP3 inflammasome, J Immunol 200 (8) (2018) 2835–2846.

[35]   VA Brown, et al., Repeat dose study of the cancer chemopreventive agent resveratrol in healthy volunteers: safety, pharmacokinetics, and effect on the insulin-like growth factor axis, Cancer Res 70 (22) (2010) 9003–9011.

[36]   R Schmatz, et al., Effects of resveratrol on biomarkers of oxidative stress and on the activity of delta aminolevulinic acid dehydratase in liver and kidney of streptozotocin-induced diabetic rats, Biochimie 94 (2) (2012) 374–383.

[37]   DJ Luther, et al., Chemopreventive doses of resveratrol do not produce cardiotoxicity in a rodent model of hepatocellular carcinoma, Invest New Drugs 29 (2) (2011) 380–391.

[38]   JA Baur, DA Sinclair, Therapeutic potential of resveratrol: the in vivo evidence, Nat Rev Drug Discov 5 (6) (2006) 493–506.

[39]   S Shankar, G Singh, RK Srivastava, Chemoprevention by resveratrol: molecular mechanisms and therapeutic potential, Front Biosci 12 (2007) 4839–4854.

[40]   P Saiko, et al., Resveratrol and its analogs: defense against cancer, coronary disease and neurodegenerative maladies or just a fad? Mutat Res 658 (1-2) (2008) 68–94.

[41]   M Shakibaei, KB Harikumar, BB Aggarwal, Resveratrol addiction: to die or not to die, Mol Nutr Food Res 53 (1) (2009) 115–128.

[42]   Z Zhang, et al., Autophagy regulates turnover of lipid droplets via ROS-dependent Rab25 activation in hepatic stellate cell, Redox Biol 11 (2017) 322–334.

[43]   G Ji, et al., Resveratrol ameliorates hepatic steatosis and inflammation in methionine/choline-deficient diet-induced steatohepatitis through regulating autophagy, Lipids Health Dis 14 (2015) 134.

[44]   F Faghihzadeh, et al., Resveratrol supplementation improves inflammatory biomarkers in patients with nonalcoholic fatty liver disease, Nutr Res 34 (10) (2014) 837–843.

[45]   F Faghihzadeh, P Adibi, A Hekmatdoost, The effects of resveratrol supplementation on cardiovascular risk factors in patients with non-alcoholic fatty liver disease: a randomised, double-blind, placebo-controlled study, Br J Nutr 114 (5) (2015) 796–803.

[46]   Z Ungvari, et al., Resveratrol confers endothelial protection via activation of the antioxidant transcription factor Nrf2, Am J Physiol Heart Circ Physiol 299 (1) (2010) H18–H24.

[47]   T Charytoniuk, et al., Alternative treatment methods attenuate the development of NAFLD: A review of resveratrol molecular mechanisms and clinical trials, Nutrition 34 (2017) 108–117.

[48]   J Yoshino, et al., Resveratrol supplementation does not improve metabolic function in nonobese women with normal glucose tolerance, Cell Metab 16 (5) (2012) 658–664.

[49]   MM Poulsen, et al., High-dose resveratrol supplementation in obese men: an investigator-initiated, randomized, placebo-controlled clinical trial of substrate metabolism, insulin sensitivity, and body composition, Diabetes 62 (4) (2013) 1186–1195.

[50]   A Elgebaly, et al., Resveratrol supplementation in patients with non-alcoholic fatty liver disease: systematic review and meta-analysis, J Gastrointestin Liver Dis 26 (1) (2017) 59–67.

[51]   W Kong, et al., Berberine is a novel cholesterol-lowering drug working through a unique mechanism distinct from statins, Nat Med 10 (12) (2004) 1344–1351.

[52]   Y Liu, et al., Update on berberine in nonalcoholic fatty liver disease, Evid Based Complement Alternat Med 2013 (2013) 1–8 308134.

[53]   HM Yan, et al., Efficacy of berberine in patients with non-alcoholic fatty liver disease, PLoS One 10 (8) (2015) e0134172.

[54]   AM Mahmoud, WG Hozayen, SM Ramadan, Berberine ameliorates methotrexate-induced liver injury by activating Nrf2/HO-1 pathway and PPARγ, and suppressing oxidative stress and apoptosis in rats, Biomed Pharmacother 94 (2017) 280–291.

[55]   J Yang, et al., Berberine ameliorates non-alcoholic steatohepatitis in ApoE(-/-) mice, Exp Ther Med 14 (5) (2017) 4134–4140.

[56]   T Guo, et al., *Berberine ameliorates hepatic steatosis and suppresses liver and adipose tissue inflammation in mice with diet-induced* obesity, Sci Rep 6 (2016) 22612.

[57]   X Yuan, et al., Berberine ameliorates nonalcoholic fatty liver disease by a global modulation of hepatic mRNA and lncRNA expression profiles, J Transl Med 13 (2015) 24.

[58]   P Dinesh, M Rasool, Berberine, an isoquinoline alkaloid suppresses TXNIP mediated NLRP3 inflammasome activation in MSU crystal stimulated RAW 264.7 macrophages through the upregulation of Nrf2 transcription factor and alleviates MSU crystal induced inflammation in rats, Int Immunopharmacol 44 (2017) 26–37.

[59]   Z Zhang, et al., Berberine activates thermogenesis in white and brown adipose tissue, Nat Commun 5 (2014) 5493.

[60]   YS Lee, et al., Berberine, a natural plant product, activates AMP-activated protein kinase with beneficial metabolic effects in diabetic and insulin-resistant states, Diabetes 55 (8) (2006) 2256–2264.

[61]   WS Kim, et al., Berberine improves lipid dysregulation in obesity by controlling central and peripheral AMPK activity, Am J Physiol Endocrinol Metab 296 (4) (2009) E812–E819.

[62]   K Singletary, Cinnamon: overview of health benefits, Nutr Today 43 (2008) 263–266.

[63]   A Khan, et al., Cinnamon improves glucose and lipids of people with type 2 diabetes, Diabetes Care 26 (12) (2003) 3215–3218.

[64]   RW Allen, et al., Cinnamon use in type 2 diabetes: an updated systematic review and meta-analysis, Ann Fam Med 11 (5) (2013) 452–459.

[65]   X Sheng, et al., Improved insulin resistance and lipid metabolism by cinnamon extract through activation of peroxisome proliferator-activated receptors, PPAR Res 2008 (2008) 1–9 581348.

[66]   P Crawford, Effectiveness of cinnamon for lowering hemoglobin A1C in patients with type 2 diabetes: a randomized, controlled trial, J Am Board Fam Med 22 (5) (2009) 507–512.

[67]   B Qin, KS Panickar, RA Anderson, Cinnamon: potential role in the prevention of insulin resistance, metabolic syndrome, and type 2 diabetes, J Diabetes Sci Technol 4 (3) (2010) 685–693.

[68]   F Askari, B Rashidkhani, A Hekmatdoost, Cinnamon may have therapeutic benefits on lipid profile, liver enzymes, insulin resistance, and high-sensitivity C-reactive protein in nonalcoholic fatty liver disease patients, Nutr Res 34 (2) (2014) 143–148.

[69]   Y Shen, et al., Cinnamon extract enhances glucose uptake in 3T3-L1 adipocytes and C2C12 myocytes by inducing LKB1-AMP-activated protein kinase signaling, PLoS One 9 (2) (2014) e87894.

[70]   RA Anderson, et al., Cinnamon extract lowers glucose, insulin and cholesterol in people with elevated serum glucose, J Tradit Complement Med 6 (4) (2016) 332–336.

[71]   C-W Tsai, et al., Garlic: health benefits and actions, Biomedicine 2 (1) (2012) 17–29.

[72]   HAR Suleria, et al., Garlic (Allium sativum): diet based therapy of 21st century–a review, Asian Pac J Trop Dis 5 (4) (2015) 271–278.

[73]   C Borek, Antioxidant health effects of aged garlic extract, J Nutr 131 (3s) (2001) 1010s–1015s.

[74] HN Kim, et al., Efficacy and safety of fermented garlic extract on hepatic function in adults with elevated serum gamma-glutamyl transpeptidase levels: a double-blind, randomized, placebo-controlled trial, Eur J Nutr 56 (5) (2017) 1993–2002.

[75] D Soleimani, et al., Effect of garlic powder consumption on body composition in patients with nonalcoholic fatty liver disease: a randomized, double-blind, placebo-controlled trial, Adv Biomed Res 5 (2016) 2.

[76] L Abenavoli, et al., Milk thistle in liver diseases: past, present, future, Phytother Res 24 (10) (2010) 1423–1432.

[77] A Federico, M Dallio, C Loguercio, Silymarin/silybin and chronic liver disease: a marriage of many years, Molecules 22 (2) (2017) 1–16.

[78] L Abenavoli, et al., Effects of Mediterranean diet supplemented with silybin-vitamin E-phospholipid complex in overweight patients with non-alcoholic fatty liver disease, Expert Rev Gastroenterol Hepatol 9 (4) (2015) 519–527.

[79] C Loguercio, et al., Silybin combined with phosphatidylcholine and vitamin E in patients with non-alcoholic fatty liver disease: a randomized controlled trial, Free Radic Biol Med 52 (9) (2012) 1658–1665.

[80] A Federico, et al., A new silybin-vitamin E-phospholipid complex improves insulin resistance and liver damage in patients with non-alcoholic fatty liver disease: preliminary observations, Gut 55 (6) (2006) 901–902.

[81] R Aller, et al., Effect of silymarin plus vitamin E in patients with non-alcoholic fatty liver disease. A randomized clinical pilot study, Eur Rev Med Pharmacol Sci 19 (16) (2015) 3118–3124.

[82] G Sorrentino, et al., Efficacy of lifestyle changes in subjects with non-alcoholic liver steatosis and metabolic syndrome may be improved with an antioxidant nutraceutical: a controlled clinical study, Drugs R D 15 (1) (2015) 21–25.

[83] R Sakata, et al., Green tea with high-density catechins improves liver function and fat infiltration in non-alcoholic fatty liver disease (NAFLD) patients: a double-blind placebo-controlled study, Int J Mol Med 32 (5) (2013) 989–994.

[84] IA Hakim, et al., Effect of increased tea consumption on oxidative DNA damage among smokers: a randomized controlled study, J Nutr 133 (10) (2003) 3303s–3309s.

[85] JA Kim, et al., Epigallocatechin gallate, a green tea polyphenol, mediates NO-dependent vasodilation using signaling pathways in vascular endothelium requiring reactive oxygen species and Fyn, J Biol Chem 282 (18) (2007) 13736–13745.

[86] V Stangl, M Lorenz, K Stangl, The role of tea and tea flavonoids in cardiovascular health, Mol Nutr Food Res 50 (2) (2006) 218–228.

[87] SI Koo, SK Noh, Green tea as inhibitor of the intestinal absorption of lipids: potential mechanism for its lipid-lowering effect, J Nutr Biochem 18 (3) (2007) 179–183.

[88] T Murase, et al., Green tea extract improves running endurance in mice by stimulating lipid utilization during exercise, Am J Physiol Regul Integr Comp Physiol 290 (6) (2006) R1550–R1556.

[89] N Khan, H Mukhtar, Multitargeted therapy of cancer by green tea polyphenols, Cancer Lett 269 (2) (2008) 269–280.

[90] BA Sutherland, RM Rahman, I Appleton, Mechanisms of action of green tea catechins, with a focus on ischemia-induced neurodegeneration, J Nutr Biochem 17 (5) (2006) 291–306.

[91] S Fukumitsu, et al., Flaxseed lignan lowers blood cholesterol and decreases liver disease risk factors in moderately hypercholesterolemic men, Nutr Res 30 (7) (2010) 441–446.

[92] A Goyal, et al., Flax and flaxseed oil: an ancient medicine & modern functional food, J Food Sci Technol 51 (9) (2014) 1633–1653.

[93] A Pan, et al., Meta-analysis of the effects of flaxseed interventions on blood lipids, Am J Clin Nutr 90 (2) (2009) 288–297.

[94] AM Hutchins, et al., Daily flaxseed consumption improves glycemic control in obese men and women with pre-diabetes: a randomized study, Nutr Res 33 (5) (2013) 367–375.

[95]   LH Brant, et al., Impact of flaxseed intake upon metabolic syndrome indicators in female Wistar rats, Acta Cir Bras 27 (8) (2012) 537–543.

[96]   Z Yari, et al., Flaxseed supplementation in non-alcoholic fatty liver disease: a pilot randomized, open labeled, controlled study, Int J Food Sci Nutr 67 (4) (2016) 461–469.

[97]   RA Isbrucker, GA Burdock, Risk and safety assessment on the consumption of Licorice root (Glycyrrhiza sp.), its extract and powder as a food ingredient, with emphasis on the pharmacology and toxicology of glycyrrhizin, Regul Toxicol Pharmacol 46 (3) (2006) 167–192.

[98]   G Dastagir, MA Rizvi, Review - Glycyrrhiza glabra L. (Liquorice), Pak J Pharm Sci 29 (5) (2016) 1727–1733.

[99]   F Aoki, et al., Suppression by licorice flavonoids of abdominal fat accumulation and body weight gain in high-fat diet-induced obese C57BL/6J mice, Biosci Biotechnol Biochem 71 (1) (2007) 206–214.

[100]   X Wu, et al., Prevention of free fatty acid-induced hepatic lipotoxicity by 18beta-glycyrrhetinic acid through lysosomal and mitochondrial pathways, Hepatology 47 (6) (2008) 1905–1915.

[101]   AA Hajiaghamohammadi, A Ziaee, R Samimi, The efficacy of licorice root extract in decreasing transaminase activities in non-alcoholic fatty liver disease: a randomized controlled clinical trial, Phytother Res 26 (9) (2012) 1381–1384.

[102]   J Bao, et al., Anthocyanins, flavonols, and free radical scavenging activity of Chinese bayberry (Myrica rubra) extracts and their color properties and stability, J Agric Food Chem 53 (6) (2005) 2327–2332.

[103]   W Kang, et al., Characterization of aroma compounds in Chinese bayberry (Myrica rubra Sieb. et Zucc.) by gas chromatography mass spectrometry (GC-MS) and olfactometry (GC-O), J Food Sci 77 (10) (2012) C1030–C1035.

[104]   L Wang, X Meng, F Zhang, Raspberry ketone protects rats fed high-fat diets against nonalcoholic steatohepatitis, J Med Food 15 (5) (2012) 495–503.

[105]   PR Ferreira de Araujo, et al., Benefits of blackϰBerry nectar (Rubus spp.) relative to hypercholesterolemia and lipid peroxidation, Nutr Hosp 26 (5) (2011) 984–990.

[106]   AO Ferreira, et al., Grape juice concentrate modulates p16 expression in high fat diet-induced liver steatosis in Wistar rats, Toxicol Mech Methods 22 (3) (2012) 218–224.

[107]   H Guo, et al., Effects of bayberry juice on inflammatory and apoptotic markers in young adults with features of non-alcoholic fatty liver disease, Nutrition 30 (2) (2014) 198–203.

[108]   J Gao, et al., Characterization and antioxidant activity of flavonoid-rich extracts from leaves of Ampelopsis grossedentata, J Food Biochem 33 (6) (2009) 808–820.

[109]   RCV Carneiro, et al., Vine tea (Ampelopsis grossedentata): A review of chemical composition, functional properties, and potential food applications, J Funct Foods 76 (2021) 104317.

[110]   Y Zhou, et al., Ampelopsin induces cell growth inhibition and apoptosis in breast cancer cells through ROS generation and endoplasmic reticulum stress pathway, PLoS One 9 (2) (2014) e89021.

[111]   L Yu-shan, Effects of Enshi-Ampelopsis Grossede on serum lipid and blood rheology and oxygen free radical of hyperlipidemia model rats, J Hu Ins Nat (Medical Edition) 1 (2) (2006) 1–15.

[112]   C Yu-qiong, CQ N.D.-j., H Hai-bo, M Yan, WU Mou-cheng, Study on the hypolipidemic effect of flavones and dihydromyricetin from Tengcha, J Tea Sci 27 (3) (2007) 221–225.

[113]   S Chen, et al., Dihydromyricetin improves glucose and lipid metabolism and exerts anti-inflammatory effects in nonalcoholic fatty liver disease: a randomized controlled trial, Pharmacol Res 99 (2015) 74–81.

[114]   P Priore, et al., Modulation of hepatic lipid metabolism by olive oil and its phenols in nonalcoholic fatty liver disease, IUBMB Life 67 (1) (2015) 9–17.

[115]   N Assy, et al., Olive oil consumption and non-alcoholic fatty liver disease, World J Gastroenterol 15 (15) (2009) 1809–1815.

[116]   M Kruse, et al., Dietary rapeseed/canola-oil supplementation reduces serum lipids and liver enzymes and alters postprandial inflammatory responses in adipose tissue compared to olive-oil supplementation in obese men, Mol Nutr Food Res 59 (3) (2015) 507–519.

[117]   L Lin, et al., Evidence of health benefits of canola oil, Nutr Rev 71 (6) (2013) 370–385.

[118]   TY Lee, et al., Alleviation of hepatic oxidative stress by Chinese herbal medicine Yin-Chen-Hao-Tang in obese mice with steatosis, Int J Mol Med 25 (6) (2010) 837–844.

[119]   HB Zhou, et al., Apoptosis of human pancreatic carcinoma cell-1 cells induced by Yin Chen Hao Decoction, World J Gastroenterol 21 (27) (2015) 8352–8357.

[120]   HS Li, Q Feng, YY Hu, Effect of qushl huayu decoction on high-fat diet induced hepatic lipid deposition in rate, Zhongguo Zhong Xi Yi Jie He Za Zhi 29 (12) (2009) 1092–1095.

[121]   H Zhang et al., Effects of Qushi Huayu Decoction on cathepsin B and tumor necrosis factor-alpha expression in rats with non-alcoholic steatohepatitis. Zhong Xi Yi Jie He Xue Bao 6 (9) (2008) 928–933.

[122]   L Chen, et al., Treating non-alcoholic fatty liver disease accompanied by hyperlipidemia with Xuezhikang capsule, Chin Tradit Patent Med 03 (1992) 1–17.

[123]   XF Fan, et al., Effect of Xuezhikang capsule on serum tumor necrosis factor-alpha and interleukin-6 in patients with nonalcoholic fatty liver disease and hyperlipidemia, Chin J Integr Med 16 (2) (2010) 119–123.

[124]   H Wang, YL Zhao, KC Xu, Clinical and pathological study on effects of Qianggan capsule combined lamivudine on hepatic fibrosis in patients with chronic hepatitis B Zhongguo Zhong Xi Yi Jie He Za Zhi 26 (11) (2006) 978–980.

[125]   L Li, et al., Treatment of non-alcoholic fatty liver disease by Qianggan Capsule, Chin J Integr Med 16 (1) (2010) 23–27.

[126]   JG Fan, Evaluating the efficacy and safety of Danning Pian in the short-term treatment of patients with non-alcoholic fatty liver disease: a multicenter clinical trial, Hepatobiliary Pancreat Dis Int 3 (3) (2004) 375–380.

[127]   SY Lou et al., Effects of Yiqi Sanju Formula on non-alcoholic fatty liver disease: a randomized controlled trial. Zhong Xi Yi Jie He Xue Bao 6(8) (2008) 793–798.

[128]   H Zhang, YY Hu, Q Feng, Inhibitory effects of Qushi Huayu Decoction on fatty deposition and tumor necrosis factor alpha secretion in HepG2 cells induced by free fatty acid. Zhongguo Zhong Xi Yi Jie He Za Zhi 27 (12) (2007) 1105–1109.

[129]   N Wang, et al., Application of proton magnetic resonance spectroscopy and computerized tomography in the diagnosis and treatment of nonalcoholic fatty liver disease, J Huazhong Univ Sci Technolog Med Sci 28 (3) (2008) 295–298.

[130]   Y Jiang, et al., Li-Gan-Shi-Liu-Ba-Wei-San improves non-alcoholic fatty liver disease through enhancing lipid oxidation and alleviating oxidation stress, J Ethnopharmacol 176 (2015) 499–507.

[131]   T Li et al., Huanglian jiedu decoction regulated and controlled differentiation of monocytes, macrophages, and foam cells: an experimental study. Zhongguo Zhong Xi Yi Jie He Za Zhi 34(9) (2014) 1096–1102.

[132]   Y Ma, et al., Cigu Xiaozhi pills's influence on lipid peroxidation and TNF-alpha expression in liver tissues of rats with nonalcoholic steatohepatitis. Zhongguo Zhong Yao Za Zhi 35(10) (2010) 1292–1297.

[133]   YL Wang, et al., Intervening TNF-α via PPARγ with gegenqinlian decoction in experimental nonalcoholic fatty liver disease, Evid Based Complement Alternat Med (2015) 715638.

[134]   W Yuanyuan, et al., Effect of a combination of calorie-restriction therapy and Lingguizhugan decoction on levels of fasting blood lipid and inflammatory cytokines in a high-fat diet induced hyperlipidemia rat model, J Tradit Chin Med 35 (2) (2015) 218–221.

[135]   Q Zhang, et al., Effect of Sinai san decoction on the development of non-alcoholic steatohepatitis in rats, World J Gastroenterol 11 (9) (2005) 1392–1395.

[136]   W Tang, et al., Hugan Qingzhi exerts anti-inflammatory effects in a rat model of nonalcoholic fatty liver disease, Evid Based Complement Alternat Med (2015) 810369.

PART 4

Skin disorders

CHAPTER 14

Herbal medicines and skin disorders

Shoaib Shoaib[a], Gurmanpreet Kaur[b], Khurram Yusuf[c] and Nabiha Yusuf[b,d]

[a]Department of Biochemistry, Faculty of Medicine, Aligarh Muslim University, Aligarh, India [b]School of Public Health, University of Alabama at Birmingham, AL, United States [c]Kendriya Vidyalaya, Jabalpur, India [d]Department of Dermatology, School of Medicine, University of Alabama at Birmingham, AL, United States

14.1 Introduction

Skin disorders are the most frequently befalling diseases affecting the health status of the individual of all ages. Skin is affected by various factors and can bear a wide range of skin disorders. Psoriasis, atopic dermatitis, acne, dermatophytosis are common skin disorders and their prevalence is increasing worldwide. Melanoma and non-melanoma skin cancers are the most fatal skin disorders and their prevalence and mortality are increasing worldwide [1]. The emerging data suggests that skin disorders impose psychological, physical, economic and social burden on thousands of people worldwide affecting the quality of life in patients [2–5]. There are numerous skin disorders and each of them may be diagnosed on the basis of its specific symptoms. The symptoms and severity of each skin disorder varies greatly in the affected individuals. The well identified symptoms in many skin disorders may include pimples, redness, itching, burning, allergy, dryness, rashes, blisters, swollen skin etc. Evidence-based studies have reported that the skin disorders may be permanent or temporary, painful or painless, genetic or non-genetic, minor or major and superficial or serious. A skin disorder may start in response of a genetic cause while most of the skin disorders correlate with non-genetic causes like bacterial, fungal, viral and parasitic infections [6,7]. There are other associated factors of skin problems like persistent stress and inflammation can lead to the development of skin cancer such as basal cell carcinoma and squamous cell carcinoma [8,9]. Various therapeutic strategies have shown symptomatic relief in many skin disorders while others are known to incomplete cure. Globally, commonly used therapies may include surgery, topical and systemic drugs, radiation therapy, immunotherapy, chemotherapy, and sometimes their combinations are also applied in the treatment of several skin disorders. Importantly, many of the therapeutic strategies are associated with the occurrence of many adverse reactions. Recent growing interest in the utilization of herbal medicines is supposed

Herbal Medicines: A Boon for Healthy Human Life.
DOI: https://doi.org/10.1016/B978-0-323-90572-5.00014-7

due to their beneficial aspects and also because of associated adverse reactions appeared in response to the applied treatment.

14.2 Psoriasis

Psoriasis is a well characterized chronic inflammatory dermatosis occurring as a result of the T-lymphocyte immunological response. Population-based studies performed on incidence and prevalence of psoriasis indicates that it depends on age and geographical area. Its prevalence among adults shows significant variation as it is evident from the data of United States (0.91%) and Norway (8.5%) while in children the prevalence of psoriasis ranges from 0% (in Taiwan) to 2.1% (in Italy) [10]. Clinical perspectives suggest that psoriasis is characterized by presence of various clinical subtypes, indicating a wide spectrum of skin manifestations. However, psoriasis has been classified as pustular and non-pustular psoriasis and this clinical classification becomes an important factor in prescribing the therapeutic regimens. Any of the regions of the body can be affected but the lumbosacral region, elbows, knees, genital portion and scalp are most frequently affected regions in psoriasis which are characterized by the presence of round, oval and polycyclic lesions with features of erythema, thickening and squamae [11]. However, the clinical investigations provided the evidences of high frequency of several symptoms in subgroups of psoriasis patients experiencing itching, irritation, stinging, sensitivity, pain and bleeding [12]. Previous clinical studies show that patients of psoriasis often encounter various problems, which include psychological disturbance, imbalanced emotional attitude and limitations in social activities [13].There are many therapies that are approved for psoriasis, including the use of biologics, which have been very effective. Such therapies can provide symptomatic relief but do not cure the disease completely. Currently available systemic therapy for psoriasis includes the use of various drugs such as methotrexate, cyclosporine, retinoid, mycophenolate, tacrolimus, mofetil and sulfasalazine [14]. The debilitating disease is also treated by using broadband and narrowband ultraviolet B (UVB) and the prescribing corticosteroids for such patients remains the most preferred choice, claiming the improved ability to treat and manage the disease effectively and safely [15]. Recently, in a clinical trial the efficacy and safety of the human interleukin-12/23 monoclonal antibody was also evaluated against the psoriasis [16].

14.3 Atopic dermatitis

Atopic dermatitis is another common skin disorder that displays itchy, highly pruritic and chronic inflammatory conditions, developed in response to immunological and allergic mechanisms [17]. The disease etiology suggests that it is a major health concern worldwide, affecting particularly children of early age and can persist or start in adulthood. The recent interest towards epidemiological studies has highlighted the rising prevalence of atopic dermatitis which is ranging from 10% to 20% in children while in adults its prevalence is around 1% to 3% [18]. The diminished recruitment of innate immune cells such as natural killer cells, neutrophils and plasmacytoid dendritic cells to skin, decreased antimicrobial peptides, and disruption in epithelial barrier are considered as susceptible factors for cutaneous pathogenic infections with *Staphylococcus aureus*, vaccinia virus and herpes simplex virus in atopic dermatitis [19]. Interestingly, the defects in innate immune system are the most credible explanation for atopic dermatitis development because they may participate in promotion of inflammation resulting in aggravation of the disease. Diagnosis of atopic dermatitis is based on the clinical features, the cornerstone to follow the treatment protocol. The disorder is symptomatic in its nature and may have a considerable set of symptoms, with a consistent correlation with familial history of atopy, acute and subacute eczematous lesions, pruritic erythematosus, papules, and lichenification [20]. Depending upon the type of symptoms noted in the individual affected, a specific therapeutic regimen, such as phototherapy [21], use of topical and systemic corticosteroids [22], immunomodulatory agents [23], and allergen immunotherapy [24], often begins for atopic dermatitis. Also, due to the side effects of these agents, the patients seek alternative treatments, which include oral agents, e.g., probiotics, vitamins, oils, Chinese herbal treatments and ayurveda [25].

14.4 Acne

Evidence-based studies indicate that acne is a multifactorial skin disorder and presence of bacteria (e.g., *Propionibaterium*), genetic factors and hormones are considered as notable causes of the disease. Both primary lesions (papules, pustules, and comedones) and secondary lesions (post-inflammatory hyperpigmentation, erythema and scarring) occur in acne [26]. Acne vulgaris is presented as the eight most common skin disorder which is characterized by the presence of papules, open and closed comedones (white and blackheads) and pustules [27]. However, the presence of hyperpigmentation, scars, erythema and inflamed regions filled with pus, and sometimes nodules and cyst are observed in acne patients indicating the severity of the disease [28]. Global burden of the disease data suggested high prevalence of acne vulgaris for all age groups that ranges up to 9%. Moreover, the estimates indicate that 100% of the adolescents are affected with the acne problems at some point [29]. The acne problem is associated with several negative impacts on the patients due to clinical symptoms of the disease resulting in the adverse effects on economic and social status which considerably co-morbid with the psychological imbalance in response of depression and anxiety [30,31]. Interestingly, analytical approaches have identified numerous risk factors influencing the onset of the acne problems. Prominently, the most evocable risk factors for acne vulgaris presentation include age, familial history, menstrual history, body mass index (BMI), diet, and smoking [32,33]. Several studies have also focused on the involvement of the demographic factors and indicated the onset of acne is strongly correlates with puberty and teenage [34].

However, the growing interest also suggests its prevalence is high in women than man at younger age owing to onset of the puberty in females [35]. Various promising therapeutic approaches are available for its treatment but choosing a most effective therapy mainly depends up on many factors such as type of acne, patient factors, disease severity and location [36]. Globally, photodynamic therapy [37], surgical therapy [38], topical agents (retinoid, adapalene and tazarotene) [39], topical antibiotics (benzoyl peroxide, clindamycin) [40], systemic antibiotics (doxycycline, minocycline and erythromicin) [41] and hormonal agents (estrogen, spironolactone) [42] are most accessible and preferred therapeutic options for acne patients. The problem of resistance due to antibiotics is continuously elevating worldwide and *Propionibacterium* strains have gained resistance against topical macrolides which limits the efficacy of the drugs, the stumbling block in treatment and management of acne patients [43]. Thus, there is a high demand for the use of natural products and holistic approaches in patients with acne.

14.5 Nonmelanoma skin cancers

Emerging data has suggested that skin cancer is a diverse group of cancers arising of different origin within the skin. Depending on the type of skin, the skin cancers can be classified as basal cell carcinoma (BCC), squamous cell carcinoma (SCC) and melanoma. The white skinned people are more prone for the development of melanoma and keratinocyte skin cancer (KSC). The most considerable risk factor for development of cutaneous melanoma and KSC (primary cutaneous neoplasms) is the exposure of the skin to the ultraviolet radiation [44], attributed to the rising incidence of melanoma and KSC worldwide. However, the incidence rates of the BCC and SCC are reported to be much higher than the melanomas, causing the significant morbidity [45]. The time and level of exposure of ultraviolet (UV) radiation both act as a determining factor for the development of melanoma or non-melanoma skin cancer. For example, intensive ultraviolet exposure during childhood and teenage promotes BCC while chronic, cumulative and long term UV exposure results in the development of SCC [44]. Epidemiological studies indicate that the incidence rates of cutaneous melanoma is rapidly increasing in United States (age-specific incidence) [46], European countries [47], Australia (sex-specific incidence) [48] and New Zealand [49]. However, another study noted the stabilizing and declining incidence rates of cutaneous melanoma in Australia and New Zealand owing to preventive campaigns focused to reduce the UV exposure [50]. Diagnosis of the melanoma or non-melanoma skin begins with visual examinations and lead to advanced testing through biopsy and imaging (MRI, CT-scan and X-ray). Dermoscopy is an effective and valuable diagnostic tool which is widely used for the clinical assessment of non-melanoma skin cancers, remarkably enhancing the diagnostic accuracy [51]. Skin cancers have been documented to involve various genetic and non-genetic risk factors. Previous data indicates the credibility of many risk-factors including sunburns, type of skin, premalignant skin lesions, typical and atypical moles and history of melanoma in cutaneous melanoma [52]. The increased incidence and morbidity of the primary cutaneous neoplasms has gained great attention earlier in order to understand the pathogenesis of the disease, demanding novel non-invasive treatment strategies. Previous studies have provided evidences of UVB and UVA involvement because they induce DNA damage through direct formation of photoproducts and indirect photo-oxidative stress causing production of reactive oxygen species, thus they potentially participate in carcinogenesis of skin cancer [53,54]. Previously, impaired DNA repair, prevention of UV-induced apoptosis, oxidative stress and genetic alterations in p53 gene has been strongly correlated with pathogenesis of non-melanoma skin cancers [55–57]. There are many medicaments used for treating skin cancers, e.g., surgical therapy [58] and radiation therapy [59], topical therapy (fluorouracil and imiquimod) [60] and photodynamic therapy (systemic or topical) [61]. However, each of them is associated with adverse side effects, motivating to search the efficacious and safe therapeutics with improved toxicity profile.

14.6 Melanoma

Melanoma is a most serious type of skin cancer which accounts for its greater impact on white skin and older age people. Numerous epidemiological studies have demonstrated that the prevalence of melanoma is rapidly increasing and in past few decades it has become much more common among the people of Europe, United States, New Zealand and Australia [62–64].However, the incidence and mortality of melanoma was reported to be highest in New Zealand and Australia as the citizens of both countries experience more sun exposure and tanned skin [63]. Particularly, younger, black skin people and women are less susceptible for developing melanoma and the data indicates that currently the lifetime risk in Americans is around 1 in 87 [65]. Melanoma diagnosis may include physical examination, clinical markers, prognostic markers, non-invasive imaging [66]. Despite investing billions of dollars, and numerous efforts put to enhance the prevention and advanced diagnosis methods still the problem is continued to increase and having several bad impacts on the life quality of the people worldwide. Interestingly, melanoma shows variety of risk factors which includes age, gender, ethnicity, indoor tanning, ultraviolet exposure, familial history, socioeconomic status etc. [67]. A study was conducted in Italy to investigate the possible role of presence of dysplastic nevi, light eye and skin color which were having positive correlation with melanoma [68]. Additionally, there are many studies which have identified genetic factors increasing risk of developing melanoma. Notably, mutations in CDKN2A and CDK4 and single nucleotide polymorphisms (SNPs) melanocortin 1 receptor (MC1R), agouti signaling protein gene (ASIP), tyrosinase (oculocutaneous albinism IA) (TYR) and tyrosinase related protein 1 (TYRP1) are most known genetic factors [69]. While mutation in BRAF gene promotes constitutive activation of kinase activity of B-Raf protein (BRAF) and many studies have demonstrated its contribution in most of the melanomas [70]. Molecular basis of melanoma also indicated mutated forms of TP53 and PTEN in majority of melanomas [71,72]. In view of the melanoma treatment, various approaches are applied including surgical resections, chemotherapy, radiation therapy, targeted therapy and immunotherapy [73–75]. There is emerging evidences on combined therapy prescribed for melanoma treatment which may include biochemotherapy or chemoimmunotherapy [76]. However, each of the therapy is associated with different kind of limitations and toxicities during patient treatment.

14.7 Dermatophytosis

Dermatophytosis is a common disorder of skin, developed in response of a fungal infection which is characterized by erythema, itching and rashes on any of the region of the body and skin-to-skin contact is considered as a critical mode for its transmission [77].Generally, dermatophytosis is diagnosed with many clinical features but gradual appearance annular erythematous, pustules, hair loss, inflammation and itching are the most dominating symptoms of the disease [78]. Dermatophytes display a wide range of fungal strains, predominantly surviving on the keratin-rich matter, therefore, they are considered as keratinous fungi. Dermatophytosis may begin in response to the infective activity of the members of *Tricophyton, Epidermophyton,* and *Microsporum* genera [79]. The prevalence of the dermatophytosis is largely affected by several risk factors which may include immunological status, region-type, humidity and temperature. Moreover, the prevalence in terms of geographic distribution of dermatophytosis varies greatly country-to-country and is affected by socioeconomic status, immunosuppressive status, lifestyle, use of antifungal agents and travel records [80]. It is evident from previous studies that systemic and topical antifungal therapeutics may reduce the load of the dermatophytosis and help in the management of the disease. Several investigations have demonstrated that itraconazole, terbinafine and fluconazole are the most common oral drugs used frequently in the treatment of various clinical forms of dermatophytosis [81]. However, ketaconazole and griseofulvin are considered as old oral antifungal drugs, approved for treating tinea pedis, tinea cruris and tinea corporis with limitation of shorter treatment periods [82,83]. Recently, *in vitro* study was conducted in order to treat dermatophytosis by through terbinafine and griseofulvin combinations [84]. More recently, another study presented recalcitrant infections of tinea corporis and cruris as the most difficult to treat due to drug resistance; therefore, they are increasing worldwide [85]. In line with observational studies, it is indicated that the burden of the skin disorders is rapidly increasing due to poor treatment outcomes.

14.8 Herbal medicines in skin disorders

In past few decades, herbal plant has gained tremendous interest towards their application against many skin disorders due to their efficacy, safety and improved toxicity profiles. Herbal medicine is considered to have healing properties among many cultures throughout history. In herbal medicine, plant-based herbal products are used to treat the illness and these products have various forms such as powdered, dried, minced, capsulated, or creams that can either be ingested or applied. Helanalin, $11\alpha,13$-dihydrohelenalin, and chamissonolid are active ingredients of arnica that reduce inflammation by inhibition of transcription nuclear factor which in turn inhibits interleukin (IL)-1, IL-2, IL-6, IL-8, adhesion molecules, and tumor necrosis factor (TNF)α. The German chamomile (*Matricaria recutita*) contains sesquiterpene alcohol, α-bisabolol, chamazulene, and flavonoids, which inhibit cyclooxygenase and lipoxygenase and exhibit anti-inflammatory and antispasmodic properties [86]. The use of nutraceuticals provides a useful alternative, which can be used as supplements. These include vitamins, herbal extracts, phytochemicals, and other dietary supplements [87]. Aloe vera (*Aloe barbadensis miller*) is one of the most important plants with medicinal properties. Skin conditions like drying, flaking, itching and cracking and healing can be healed by use of aloe vera. Neem (*Azadirachta indica*) contains nimbdin that has anti-inflammatory properties.

14.9 Herbal medicines in psoriasis

A growing body of interests has indicated that the many herbal formulations have been used to treat psoriasis, implicating their potential. Previously, clinical and experimental studies were conducted to evaluate the efficacy of two herbal medicines and based on the outcomes it was concluded that capsaicin and *Mahonia aquifolium* are most promising herbal medicines used topically for psoriasis [88]. Fatty acid containing oils were also used for treating psoriasis. Interestingly, eicosapentaenoic acid (EPA) and docosahexaenoic acid (DHA) have shown beneficial effects in the treatment and management of psoriasis. Another study shows that a combined formulation of curcumin and EPA synergistically inhibited inflammatory effects [89]. Similarly, effect of Chinese herbal medicines in combination of acitretin was investigated for its effectiveness against psoriasis and clinical evidences indicate better efficacy of herbal medicines in treating psoriasis which is confirmed by their safety profile and less adverse side effects [90]. Moreover, Chinese herbal medicines were also administered in order to treat psoriasis which includes roots of *Rehmannia glutinosa, Salvia miltiorrhiza* and *Lithospermum erythrorhizon* displaying their anti-inflammatory effects in experimental studies [91,92]. Developing herbal medication also include use of *Glycyrrhizae Radix, Saposhnikoviae Radix* and *Rehmanniae Radix* which are frequently prescribed in clinical treatment of psoriasis [93]. The active chemical constituent of aloe vera is lignin, which allows aloe vera to penetrate into the deeper layers of skin [94]. Aloe vera is available in several forms to treat psoriasis, which includes shampoos, breams, gels that allow for water retention in skin. Aloe vera juice can also be consumed orally and acts as an internal cleanser [94]. An aqueous extract of neem leaves was shown to improve psoriasis area and severity index (PASI) in patients compared to the placebo group [95]. Curcumin (an active component of turmeric) has many medicinal properties. In a clinical trial, turmeric microemulgel was able to improve the PASI score in psoriasis patients and improve their quality of like compared to the untreated group [96]. In another randomized clinical trial, curcumin could effectively decrease the levels of IL-22 in the serum of psoriasis patients [97]. Genistein, the main isoflavone from Soybean (*Barbadensis mill*) showed reduction in skin thickening and erythema in psoriasis patients receiving psoralen-ultraviolet A (PUVA) therapy [98]. Vitamin D3 (cholecalciferol) produced in skin is able to reduce the duration and severity of outbreak of psoriasis in skin. A vitamin D analog (MC903) in cream was able to clear the psoriatic lesions effectively without any adverse drug reaction [99].The ethanolic extract of *Nigella sativa* (NS) was able to reduce epidermal thickness in patients with psoriasis. Black cumin (*Nigella sativa*) has many active components like thymoquinone, thymohydroquinone, dithymoquinone etc., which have anti-psoriatic activity [100].

14.10 Herbal medicines in atopic dermatitis

Earlier, phytotherapeutic potential of several herbal medicines was explored for treating different subtypes of atopic dermatitis. Numerous studies were previously conducted to assess the preventive and therapeutic ability of natural herbal medicines and their formulations against mild-to-serious atopic dermatitis [101]. Several studies revealed that the application of oral herbal medicines in the form of tablets or syrup display promising effects in the treatment and management of atopic dermatitis. Clinical data justifies the safety and efficacy of Chinese herbal medicines (herbal extracts, decoctions) which may be given alone or in combination in the treatment of atopic dermatitis in children and adults [102]. Moreover, in a randomized controlled study, a formulation of nine herbal plants (*Radix glycyrrhizae, Rhizoma dioscoreae, Forsythia suspensa, Radix pseudostellariae, Concha ostreae, Herba lophatheri, Medulla Junci* and *Semen coicis*) was found to induce positive effects in atopic dermatitis patients [103]. In another randomized controlled study 85 children with mild-to-severe atopic dermatitis were prescribed to take three oral tablets twice a day which were composed of traditional Chinese herbal medicines (*Herba menthae, Cortex moutan, Rhizoma atractylodis, Flos lonicerae* and *Cortex phellodendri*) [104]. Previously, several times herbal medicines and western medicines were investigated experimentally for their efficiencies when given in combination which includes clinical trials and randomized controlled trials. The clinical and experimental data presented earlier suggest that combinations of herbal medicines with the western medicines have benefits because of their critical role in the prevention of adverse reactions induced by western medicines and combined treatment of herbal medicines with western medicines also enhanced efficacy [105]. *In vivo* and pharmacodynamic studies were performed to understand the clinical relevance of using traditional Chinese medicines in the treatment of atopic dermatitis. Therefore, it was clarified that using traditional Chinese medicines in atopic dermatitis significantly cause the activation of anti-inflammatory and immunomodulatory mechanisms which helped in the management and treatment of the disease [106]. Bioactive compounds from Chinese herbal medicines were also investigated to report their impacts on the mast cells, eosinophils and cytokines production in atopic dermatitis as one of the clinical study reported that the extract of *Sanguisorbae Radix* efficiently reduce the accumulation of eosinophils in mouse model [107]. Berberine, palmatine, liquiritigenin and liquiritin the bioactive compounds of *Coptidis rhizoma* and *Glycyrrhiza uralensis* were reported to decrease the levels of degranulated mast cells and inhibit the activities of TNF-α, COX-2 (cyclooxygenase-2) and iNOS ((inducible nitric oxide synthase) [108]. In a study, anti-atopic dermatitis activities of *Angelicae dahuricae Radix* were evaluated and found that it exerted its inhibitory effects by decreasing the infiltration of mast cells and inflammatory cells into the lesions of atopic dermatitis which resulted in the suppression of expression of TNF-α, IL-6, IL-4 and IL-10 [109]. Gallic acid from *Cortex Moutan* and Berberine from *Cortex Phellodendri*, both were demonstrated to inhibit the secretion of pro-inflammatory cytokine IL-6 and chemokines CCL7 (Chemokine (C-C motif) ligand 7) and CXCL8 (Chemokine (C-X-C motif) ligand 8) [110]. Another study indicated that oral administration of the extract of *Poria cocus* displayed beneficial effects as it exerted immunosuppressive effects. Moreover, *in vivo* studies performed to the assess the impact of topical application of *Forsynthia suspensa* extract which was reported to prevent the migration and infiltration of leucocytes into the lesions of atopic dermatitis through reducing the production of IL-4, ICAM-1 (intercellular adhesion molecule-1) and VCAM-1 (vascular cell adhesion molecule-1) [111]. On further, the effect of topical application of chlorogenic acid and isochlorogenic acid extracted from *Artemisia capilliaris* was also assessed on RAW264.7 cells which resulted in the suppression of production of nitric oxide and histamine [112]. There is an evidence of decreased mRNA expression of many proinflammatory cytokines in the atopic dermatitis lesions upon administering crude extract of *Angelicae dahuricae* [113]. Morus alba, a plant rich in terpenes, flavonoids, glycosides, alkaloids, tannins and volatile oils, was showed to inhibit nitric oxide (NO) and prostaglandin E2 (PGE2) production in HaCaT cells and mice by using its ethanolic extract [114]. *In vitro* and *in vivo* studies also revealed that *Pseudostellaria heterophylla* and Ginseng are capable of down-regulating the expression of NF-κB which may lead to the inhibition of many cytokines [115]. In an earlier study of traditional Chinese herbal medicines, it was concluded that combinations of many of these herbal medicines improve quality of life effectively and minimize the use of topical corticosteroids in atopic dermatitis children [116].

14.11 Herbal medicines in acne

Recent intensive research highlights the associated problems of existing treatment strategies because none of the above-mentioned medications of acne vulgaris is free of adverse side effects which become one of the important reasons for searching natural regimens with less adverse side reactions and importantly, displaying improved safety and toxicity profiles. Interestingly, herbal medicines have a wide range of bioactive ingredients with broad scope of their use in the treatment of acne vulgaris. Therefore, numerous studies have showed their excellent efficacy profiles in treating the acne vulgaris. The possible explanation for gained attention towards the use of the herbal medicines may include low cost, fewer side effects, better efficacy and tolerance. In this direction, we are more concerned of the herbal medicines used alone and in combination with synthetic drugs for treating the disease. The herbal products have been found to exert anti-inflammatory, antibacterial and antioxidant effects in several studies that help in the treatment of acne vulgaris. More importantly, extraction of many tannin containing plants has been used topically to treat acne which includes *Quercus alba, Juglans regia, Ladum latifolium, Syzygium cuminum, Verbascum thapsus, Rumex crispus, Alchemilla mollis* etc [117]. Previously, the silicic acid from Equisetum species and anthranoids-containing extract of *Aloe ferox* were recommended to apply topically for the treatment of acne [118,119]. *Melissa officinalis* is an important medicinal plant which has phenolics, flavonoids, rosmarinic acid, terpenes and essential oils. Hydro-alcoholic extracts of this plant was used to relieve the inflammation related problems in acne and pimples, improving the status of acne infection, hypersecretion of sebaceous gland and neurological problems [120]. *Melaleuca alternifolia, Aloe capensis, Aloe barbadensis, Aloe ferox* and *Solanum dulcamara* L. were investigated for their beneficial properties which on further confirmed anti-acne medicinal activities [121]. There are numerous herbal plants which possess many bioactive ingredients such as xanthone, phenolics, tannin, alkaloids, flavonoids and essential oils which are associated with the antibacterial and anti-inflammatory activities. In vitro studies showed antibacterial activities of many Indian medicinal plants which includes *Quercus infectoria, Berberis aristata, Hibiscus syriacus, Ammania baccifera, Mucana prurience* etc [122]. Moreover, studies done with alkaloids of *Coscinium fenestratum*, monoterpenes and terpinen-4-ol of *Anthemis aciphylla* revealed their antibacterial activities in acne [123]. Extracts of *Vitex agnuscastus* [124], *Glycyrrhiza glabra* [125] and *Usnea barbata* [126] were investigated to find their antibacterial activities against acne and the results of in vitro studies suggest them as promising anti-acne substances.

14.12 Herbal medicines in melanoma and nonmelanoma skin cancers

The world is enriched with a wide range of herbal plants and since ancient time use of medicinal herbs is believed to be safe, cost effective and harmless than the synthetic drugs. Numerous studies confirm the role of herbal medicines in the treatment and prevention of skin cancer. Herbal medicines are known to exert their positive impacts on skin cancer through their bioactive compounds which include phenolics, coumarins curcuminoids, tannins, lignans, quinones etc. Importantly, various promising anticancer activities of such compounds have been displayed through intensive research incorporating anti-inflammatory, antioxidant, anti-carcinogenic prospective. Plant-derived products have been investigated for their antioxidant activity in the quest for effective topical photoprotective agents, thereby increasing the use of natural antioxidants in commercial skin care products. In this direction, resveratrol enriched *Polygonum cuspidatum, Veratrum grandiflorum* and *Veratrum formosanum*, apigenin-containing *Calendula officinalis*, luteolin, apigenin and their derivative rich *Artemisia inculata, Arnica montana, Cuminum cyminum, Agrimonia eupatoria, Echinacea purpurea* and *Euphrasia officinalis* were identified for their anti-inflammatory properties [127–128]. Gallic acid is found in a variety of herbal plants such as *Punica granatum, Rhus chinensis, Rheum officinale, Cornus officinalis, Barringtonia racemosa, Cassia auriculata, Sanguisorba officinalis* etc. *In vitro* studies indicate that gallic acid exerted its inhibitory effects on human melanoma cells through the induction of apoptosis and reduced migration and invasion [129]. Recently, lupeol, gallic acid, ursolic acid, ellagic acid, quercitrin and epicatechin were reported to be present in several Nigerian herbs. The extracts of these Nigerian herbal plants were demonstrated to inhibit the proliferative ability of melanoma cancer cells, in which the IC_{50} recorded as 5, 35 and 50 µg/mL for different herbal plant extracts [130]. Previously, there is evidence of green tea for its preventive role in non-melanoma skin cancer through its direct involvement in enhancement of DNA repair [131] while isorhamnetin suppressed skin cancer by inhibiting the activity of MEK (mitogen-activated protein kinase kinase) and PI3K (phosphoinositide 3-kinase) [132]. In a clinical study, the extract of *Euphorbia peplus* was shown to treat SCC and BCC which suggested it for further clinical development [133]. A bioactive compound, hypericin extracted from *Hypericum perforatum* has provided evidence of its efficacy against SCC and BCC indicating the plausibility of its use in the treatment of non-melanoma cancers [134]. Many herbal plants such as *Aloe barbadensis, Luffa cylindrical, Emblica officinalis, Crocus sativus* and *Terminalia chebula* are known to provide photo-protection through their antioxidant properties [135]. Salvianolic acid B from *Radix Salviae Miltiorrhizae* was demonstrated to anticancer potential in melanoma cells through suppressing invasion and angiogenic activities [136].

Interestingly, naturally occurring chemicals from several herbal plants were previously investigated for their antioxidant and anti-inflammatory functions in order to identify effective topical photoprotective herbs-derived products, thereby increasing the use of such herbal plants in skin care. *In vitro* and *in vivo* studies conducted on *Scutellaria barbata*, a Chinese herb indicated remarkable antitumorigenic and anti-inflammatory properties in skin cancer [137]. Previously, 6-gingerol containing *Zingiber officinale* was also demonstrated to exert anti-inflammatory activity by reducing COX2 and NF-κB expression via altering p38-MAPK pathway and also its topical application led to the reduced skin papilloma formation [138,139]. Many herbal plants are enriched with caffeic acid which also shows anticancer, anti-inflammatory and antioxidant activities. Another *in vitro* studies show that caffeic acid provided protection in skin cancer by inhibiting migration and invasion through reducing the level of N-cadherin and vimentin with subsequent increase in E-cadherin level [140,141]. *In vivo* study of silymarin (*Silybum marianum*) showed protective potential through targeting the inhibition of TPA-induced tumor development [142]. *Trifolium pratense* was reported to provide photoprotection because of its inhibitory effects on inflammation and immunological response induced by UV radiation [143].*Ginkgo biloba* extract exerted antioxidant activity [144], induced apoptosis in melanoma cancer cells [145], enhanced photoprotective mechanism against UV exposure, prevented skin cancer and acted as a skin improving agent [146,147]. Earlier, the antioxidant, anticancer and anti-inflammatory properties of *Salvia rosmarinus* and *Salvia officinalis* has been documented in many studies. Additionally, therapeutic elucidation of mechanism of action of cornosol from these two plants provided the clarification for its photchemopreventive ability against UVB exposure because the compound was found sufficient to reduce UVB-induced ROS generation, DNA damage and NF-κB activation in skin keratinocytes [148].*In vitro* studies revealed that *Annona muricata* significantly participated in the inhibition of the proliferation, suppressed colony formation and induced pro-apoptotic effects in non-melanoma skin cancer cells [149]. Apigenin is most commonly found in many of the herbal plants such as *Apium graveolens*, *Petroselinum crispum* and *Matricaria chamomilla*. In several studies the therapeutic and chemopreventive potential has been elucidated. Apigenin significantly inhibited UVB-induced mTOR activity in cultured keratinocytes [150] and reactivated Nrf2 signaling in mouse epidermal cells by modulating epigenetic mechanism [151]. Similarly, there is evidence of apigenin induced inhibition of STAT-3 signaling that supports the suppression of migration and invasion through downregulation of matrix metalloproteinase (MMP-2 and -9) in melanoma cells [152]. Apigenin actively participated in the suppression of proliferation and invasion, activated apoptosis and cell cycle arrest associated signaling pathways in melanoma cells [153].

14.13 Herbal medicines in dermatophytosis

Studies on treatment of dermatophyte infections with herbal medicines have gained keen attention because these herbs are known to possess numerous bioactive compounds which are assumed to have low side effects and improved efficacy against dermatophytosis. There are several herbs which have so far tested for their excellent properties against dermatophytes. The estimated minimum inhibitory concentration (MIC) for the chloroform extracts of *Justicia gendarussa*, *Solanum melongena* L. and *Lawsonia inermis* L. ranged between 3.12-12.5 mg/mL against *Trichophyton mentagrophytes*, *Microsporum gypseum*, *Trichophyton rubrum* and *Microsporum fulvum* confirming their anti-dermatophytosis properties [154]. Various Iranian medicinal herbs were tested for their antimycotic potential against five dermatophytic strains for which the MIC ranged between 0.25-4 mg/mL and the lowest MIC was observed for *Artemisia sieberi* [155]. Previously, in a clinical study, anti-dermatophytosis activities of *Lawsonia inermis* were proven to be most effective against majority of the dermatophytes due to 20 to 50 mm inhibition zone [156]. *Senna alata* (Linn)and *Borreria ocymoides* are enriched with flavonoids, saponins and anthraquinones and in a clinical investigation extracts from both plants were compared with conventional drugs (clotrimazole and griseofulvin)indicating better efficacy against dermatophytes isolated from 840 pupils, offering some hope in the treatment and management of the disease [157]. In other study, antidermatophytic activities of extracts of *Pergularia tomentosa* and *Mitracarpus scaber* at the dose of 10 mg/mL were shown effective against two strains of dermatophytes [158]. Polyherbal preparation of *Lawsonia alba, Azadirachta indica* and *Shorea robusta* combined with castor and sesame oil was investigated for its antifungal efficacy and showed positive outcomes as the preparation inhibited the enzymatic activities in fungal strains [159]. With IC_{50}0.5, 0.8 and 0.9 mg/mL of methanolic leaf extracts of *Cassia alata* L., *Cassia fistulata* L. and *Cassia tora* L. exerted inhibitory effects against *Microsporum gypseum, Penicillium marneffei* and *Trichophyton rubrum* because their extract application efficiently inhibited the growth of hyphae and microconidia [160]. In vitro studies performed to confirm the antifungal activities of some of combined extracts of three traditional Chinese medicines and the results indicated additive inhibitory effects on *Microsporum gypseum, Trichophyton mentagrophytes* and *Microsporum canis* [161]. The latex of *Calotropis procera* exerted inhibitory effects on *Trichophyton* spp in a dose dependent manner through alkaloids, saponins, flavonoids, triterpenoids and tannins present in the plant extract [162]. Previously, formulation prepared from the *Origanum vulgare, Thymus serpillum* and *Rosmarinus officinalis* were applied to treat dermatophytosis caused by *Tricophyton mentagrophytes*, offering a clinical and etiological cure [163].

14.14 Conclusion

The recent growing interest indicates the crucial role of several herbal medicines, confirming the improvement in the symptomatic severity of some skin disorders and many *in vivo* and clinical investigations assured their tolerability and routine use. Interestingly, the herbal medicines, herbal formulations, decoction and bioactive ingredients of various plants have been applied widely in the treatment and photoprotection in different kind of skin diseases. Remarkable studies investigated on herbal plants revealed that using herbal medicines, polyherbal formulations and plant-derived products in treatment of a particular skin order is safe, patient compliant and reported to show less systemic toxicity along with increased effectiveness compared to conventional therapies.

References

[1] R Ossio, R Roldan-Marin, H Martinez-Said, DJ Adams, CD Robles-Espinoza, Melanoma: a global perspective, Nat Rev Cancer 17 (7) (2017) 393–394.

[2] MK Basra, M Shahrukh, Burden of skin diseases, Expert Rev Pharmacoeconomics Outcomes Res 9 (3) (2009) 271–283.

[3] FJ Dalgard, U Gieler, L Tomas-Aragones, L Lien, F Poot, GB Jemec, J Kupfer, The psychological burden of skin diseases: a cross-sectional multicenter study among dermatological out-patients in 13 European countries, J Invest Dermatol 135 (4) (2015) 984–991.

[4] AB Kimball, C Jacobson, S Weiss, MG Vreeland, Y Wu, The psychosocial burden of psoriasis, Am J Clin Dermatol 6 (6) (2005) 383–392.

[5]   AM Layton, D Thiboutot, J Tan, Reviewing the global burden of acne: how could we improve care to reduce the burden? Br J Dermatol 184 (2) (2021) 219–225.

[6]   F Gu, TH Chen, RM Pfeiffer, MC Fargnoli, D Calista, P Ghiorzo, MT Landi, Combining common genetic variants and non-genetic risk factors to predict risk of cutaneous melanoma, Hum Mol Genet 27 (23) (2018) 4145–4156.

[7]   KA Wanat, AR Dominguez, Z Carter, P Legua, B Bustamante, RG Micheletti, Bedside diagnostics in dermatology: viral, bacterial, and fungal infections, J Am Acad Dermatol 77 (2) (2017) 197–218.

[8]   GB Maru, K Gandhi, A Ramchandani, G Kumar, The role of inflammation in skin cancer, Adv Exp Med Biol 816 (2014) 437–469.

[9]   CS Sander, F Hamm, P Elsner, JJ Thiele, Oxidative stress in malignant melanoma and non-melanoma skin cancer, Br J Dermatol 148 (5) (2003) 913–922.

[10]   R Parisi, DP Symmons, CE Griffiths, DM Ashcroft, Global epidemiology of psoriasis: a systematic review of incidence and prevalence, J Invest Dermatol 133 (2) (2013) 377–385.

[11]   G Sarac, TT Koca, T Baglan, A brief summary of clinical types of psoriasis, North Clin Istanb 3 (1) (2016) 79.

[12]   F Sampogna, P Gisondi, CF Melchi, P Amerio, G Girolomoni, D Abeni, et al., Prevalence of symptoms experienced by patients with different clinical types of psoriasis, Br J Dermatol 151 (3) (2004) 594–599.

[13]   J De Korte, FM Mombers, JD Bos, MA Sprangers, Quality of life in patients with psoriasis: a systematic literature review, in: Journal of Investigative Dermatology Symposium Proceedings, 9, Elsevier, 2004, pp. 140–147.

[14]   M Lebwohl, S Ali, Treatment of psoriasis. Part 2. Systemic therapies, J Am Acad Dermatol 45 (5) (2001) 649–664.

[15]   M Lebwohl, S Ali, Treatment of psoriasis. Part 1. Topical therapy and phototherapy, J Am Acad Dermatol 45 (4) (2001) 487–502.

[16]   GG Krueger, RG Langley, C Leonardi, N Yeilding, C Guzzo, Y Wang, M Lebwohl, A human interleukin-12/23 monoclonal antibody for the treatment of psoriasis, N Engl J Med 356 (6) (2007) 580–592.

[17]   H Williams (Ed.), Atopic Dermatitis: The Epidemiology, Causes and Prevention of Atopic Eczema, Cambridge University Press, 2000. 10.1017/CBO9780511545771.

[18]   FS Larsen, JM Hanifin, Epidemiology of atopic dermatitis, Immunol Allergy Clin North Am 22 (1) (2002) 1–24.

[19]   A De Benedetto, R Agnihothri, LY McGirt, LG Bankova, LA Beck, Atopic dermatitis: a disease caused by innate immune defects? J Invest Dermatol 129 (1) (2009) 14–30.

[20]   DY Leung, RA Nicklas, JT Li, IL Bernstein, J Blessing-Moore, M Boguniewicz, SA Tilles, Disease management of atopic dermatitis: an updated practice parameter, Ann Allergy Asthma Immunol 93 (3) (2004) S1–S21.

[21]   J Krutmann, Phototherapy for atopic dermatitis, Clin Exp Dermatol 25 (7) (2000) 552–558.

[22]   AM Drucker, K Eyerich, MS de Bruin-Weller, JP Thyssen, PI Spuls, AD Irvine, et al., Use of systemic corticosteroids for atopic dermatitis: International Eczema Council consensus statement, Br J Dermatol 178 (3) (2018) 768–775.

[23]   AM Drucker, AG Ellis, M Bohdanowicz, S Mashayekhi, ZZ Yiu, B Rochwerg, et al., Systemic immunomodulatory treatments for patients with atopic dermatitis: a systematic review and network meta-analysis, JAMA dermatology 156 (6) (2020) 659–667.

[24]   L Cox, MA Calderon, Allergen immunotherapy for atopic dermatitis: is there room for debate? J Allergy Clin Immunol Pract 4 (3) (2016) 435–444.

[25]   J Fenner, NB Silverberg, Oral supplements in atopic dermatitis, Clin Dermatol 36 (5) (2018) 653–658.

[26]   B Adityan, R Kumari, DM Thappa, Scoring systems in acne vulgaris, Indian J Dermatol Venereol Leprol 75 (3) (2009) 323.

[27]   HC Williams, RP Dellavalle, S Garner, Acne vulgaris, Lancet North Am Ed 379 (9813) (2012) 361–372.

[28]   A Mahto, Acne vulgaris, Medicine (Baltimore) 45 (6) (2017) 386–389.

[29] T Vos, AD Flaxman, M Naghavi, R Lozano, C Michaud, M Ezzati+, et al., Years lived with disability (YLDs) for 1160 sequelae of 289 diseases and injuries 1990–2010: a systematic analysis for the Global Burden of Disease Study 2010, Lancet North Am Ed 380 (9859) (2012) 2163–2196.

[30] Y Kaymak, E Taner, Y Taner, Comparison of depression, anxiety and life quality in acne vulgaris patients who were treated with either isotretinoin or topical agents, Int J Dermatol 48 (1) (2009) 41–46.

[31] AHS Heng, FT Chew, Systematic review of the epidemiology of acne vulgaris, Sci Rep 10 (1) (2020) 1–29.

[32] A Di Landro, S Cazzaniga, F Parazzini, V Ingordo, F Cusano, L Atzori, GISED Acne Study Group, Family history, body mass index, selected dietary factors, menstrual history, and risk of moderate to severe acne in adolescents and young adults, J Am Acad Dermatol 67 (6) (2012) 1129–1135.

[33] Al Hussein, S M., Al Hussein, V H., E C., N Todoran, Al Hussein, CA H., MT Dogaru, Diet, smoking and family history as potential risk factors in acne vulgaris–a community-based study, Acta Medica Marisiensis 62 (2) (2016) 173–181.

[34] K Bhate, HC Williams, Epidemiology of acne vulgaris, Br J Dermatol 168 (3) (2013) 474–485.

[35] DD Lynn, T Umari, CA Dunnick, RP Dellavalle, The epidemiology of acne vulgaris in late adolescence, Adolesc Health Med Ther 7 (2016) 13.

[36] N Benner, D Sammons, Overview of the treatment of acne vulgaris, Osteopath Fam Physician 5 (5) (2013) 185–190.

[37] Y Itoh, Y Ninomiya, S Tajima, A Ishibashi, Photodynamic therapy for acne vulgaris with topical 5-aminolevulinic acid, Arch Dermatol 136 (9) (2000) 1093–1095.

[38] H Kurzen, S Schönfelder-Funcke, W Hartschuh, Surgical treatment of acne inversa at the university of Heidelberg, Coloproctology 22 (2) (2000) 76–80.

[39] HE Baldwin, T Lin, Management of severe acne vulgaris with topical therapy, J Drugs Dermatol 16 (11) (2017) 1134–1138.

[40] A Haider, JC Shaw, Treatment of acne vulgaris, JAMA 292 (6) (2004) 726–735.

[41] F Ochsendorf, Systemic antibiotic therapy of acne vulgaris, J Dtsch Dermatol Ges 4 (10) (2006) 828–841.

[42] D Thiboutot, W Chen, Update and future of hormonal therapy in acne, Dermatology 206 (1) (2003) 57–67.

[43] TR Walsh, J Efthimiou, B Dréno, Systematic review of antibiotic resistance in acne: an increasing topical and oral threat, Lancet Infect Dis 16 (3) (2016) e23–e33.

[44] MS Paulo, B Adam, C Akagwu, I Akparibo, RH Al-Rifai, S Bazrafshan, SM John, WHO/ILO work-related burden of disease and injury: Protocol for systematic reviews of occupational exposure to solar ultraviolet radiation and of the effect of occupational exposure to solar ultraviolet radiation on melanoma and non-melanoma skin cancer, Environ Int 126 (2019) 804–815.

[45] U Leiter, U Keim, C Garbe, Epidemiology of skin cancer: update 2019, Sunlight, Vitamin D and Skin Cancer, Springer, Cham, 2020, pp. 123–139.

[46] KG Paulson, D Gupta, TS Kim, JR Veatch, DR Byrd, S Bhatia, JM Gardner, Age-specific incidence of melanoma in the United States, JAMA Dermatol 156 (1) (2020) 57–64.

[47] E de Vries, JWW Coebergh, Melanoma incidence has risen in Europe, Br Med J 331 (7518) (2005) 698.

[48] CM Olsen, JF Thompson, N Pandeya, DC Whiteman, Evaluation of sex-specific incidence of melanoma, JAMA Dermatol 156 (5) (2020) 553–560.

[49] JJ Liang, E Robinson, RC Martin, Cutaneous melanoma in New Zealand: 2000–2004., ANZ J Surg 80 (5) (2010) 312–316.

[50] MR Iannacone, AC Green, Towards skin cancer prevention and early detection: evolution of skin cancer awareness campaigns in Australia, Melanoma Manag 1 (1) (2014) 75–84.

[51] MC Fargnoli, D Kostaki, A Piccioni, T Micantonio, K Peris, Dermoscopy in the diagnosis and management of non-melanoma skin cancers, Eur J Dermatol 22 (4) (2012) 456–463.

[52] L Belbasis, I Stefanaki, AJ Stratigos, E Evangelou, Non-genetic risk factors for cutaneous melanoma and keratinocyte skin cancers: an umbrella review of meta-analyses, J Dermatol Sci 84 (3) (2016) 330–339.

[53] TM Runger, How different wavelengths of the ultraviolet spectrum contribute to skin carcinogenesis: the role of cellular damage responses, J Invest Dermatol 127 (9) (2007) 2103–2105.

[54] AJ Ridley, JR Whiteside, TJ McMillan, SL Allinson, Cellular and sub-cellular responses to UVA in relation to carcinogenesis, Int J Radiat Biol 85 (3) (2009) 177–195.

[55] DR Bickers, M Athar, Oxidative stress in the pathogenesis of skin disease, J Invest Dermatol 126 (12) (2006) 2565–2575.

[56] CL Benjamin, HN Ananthaswamy, p53 and the pathogenesis of skin cancer, Toxicol Appl Pharmacol 224 (3) (2007) 241–248.

[57] V Madan, JT Lear, RM Szeimies, Non-melanoma skin cancer, Lancet North Am Ed 375 (9715) (2010) 673–685.

[58] CR Rogers-Vizena, DH Lalonde, FJ Menick, ML Bentz, Surgical treatment and reconstruction of non-melanoma facial skin cancers, Plast Reconstr Surg 135 (5) (2015) 895e–908e.

[59] R Hulyalkar, T Rakkhit, J Garcia-Zuazaga, The role of radiation therapy in the management of skin cancers, Dermatol Clin 29 (2) (2011) 287–296.

[60] JK Cullen, JL Simmons, PG Parsons, GM Boyle, Topical treatments for skin cancer, Adv Drug Deliv Rev 153 (2020) 54–64.

[61] FS De Rosa, MVL Bentley, Photodynamic therapy of skin cancers: sensitizers, clinical studies and future directives, Pharm Res 17 (12) (2000) 1447–1455.

[62] AM Forsea, V Del Marmol, E De Vries, EE Bailey, AC Geller, Melanoma incidence and mortality in Europe: new estimates, persistent disparities, Br J Dermatol 167 (5) (2012) 1124–1130.

[63] MJ Sneyd, B Cox, A comparison of trends in melanoma mortality in New Zealand and Australia: the two countries with the highest melanoma incidence and mortality in the world, BMC Cancer 13 (1) (2013) 1–9.

[64] AM Glazer, RR Winkelmann, AS Farberg, DS Rigel, Analysis of trends in US melanoma incidence and mortality, JAMA Dermatol 153 (2) (2017) 225–226.

[65] HI Hall, DR Miller, JD Rogers, B Bewerse, Update on the incidence and mortality from melanoma in the United States, J Am Acad Dermatol 40 (1) (1999) 35–42.

[66] LE Davis, SC Shalin, AJ Tackett, Current state of melanoma diagnosis and treatment, Cancer Biol Ther 20 (11) (2019) 1366–1379.

[67] S Carr, C Smith, J Wernberg, Epidemiology and risk factors of melanoma, Surgical Clinics 100 (1) (2020) 1–12.

[68] MT Landi, A Baccarelli, D Calista, A Pesatori, T Fears, MA Tucker, G Landi, Combined risk factors for melanoma in a Mediterranean population, Br J Cancer 85 (9) (2001) 1304–1310.

[69] KD Meyle, P Guldberg, Genetic risk factors for melanoma, Hum Genet 126 (4) (2009) 499–510.

[70] H Davies, GR Bignell, C Cox, P Stephens, S Edkins, S Clegg, et al., Mutations of the BRAF gene in human cancer, Nature 417 (6892) (2002) 949–954.

[71] A Birck, V Ahrenkiel, J Zeuthen, K Hou-Jensen, P Guldberg, Mutation and allelic loss of the PTEN/MMAC1 gene in primary and metastatic melanoma biopsies, J Invest Dermatol 114 (2) (2000) 277–280.

[72] C Dahl, PER Guldberg, The genome and epigenome of malignant melanoma, APMIS 115 (10) (2007) 1161–1176.

[73] C Garbe, TK Eigentler, U Keilholz, A Hauschild, JM Kirkwood, Systematic review of medical treatment in melanoma: current status and future prospects, Oncologist 16 (1) (2011) 5.

[74] M Chowdhary, KR Patel, HH Danish, DH Lawson, MK Khan, BRAF inhibitors and radiotherapy for melanoma brain metastases: potential advantages and disadvantages of combination therapy, Onco Targets Ther 9 (2016) 7149.

[75] S Raigani, S Cohen, GM Boland, The role of surgery for melanoma in an era of effective systemic therapy, Curr Oncol Rep 19 (3) (2017) 17.

[76]   C Wack, A Kirst, JC Becker, WK Lutz, EB Bröcker, WH Fischer, Chemoimmunotherapy for melanoma with dacarbazine and 2, 4-dinitrochlorobenzene elicits a specific T cell-dependent immune response, Cancer Immunol Immunother 51 (8) (2002) 431–439.

[77]   GS de Hoog, K Dukik, M Monod, A Packeu, D Stubbe, M Hendrickx, et al., Toward a novel multilocus phylogenetic taxonomy for the dermatophytes, Mycopathologia 182 (1-2) (2017) 5–31.

[78]   H Degreef, Clinical forms of dermatophytosis (ringworm infection), Mycopathologia 166 (5-6) (2008) 257.

[79]   AA AL-Janabi, Dermatophytosis: causes, clinical features, signs and treatment, J Symptoms Signs 3 (3) (2014) 200–203.

[80]   M Ameen, Epidemiology of superficial fungal infections, Clin Dermatol 28 (2) (2010) 197–201.

[81]   S Rand, Overview: the treatment of dermatophytosis, J Am Acad Dermatol 43 (5) (2000) S104–S112.

[82]   FD Choi, ML Juhasz, NA Mesinkovska, Topical ketoconazole: a systematic review of current dermatological applications and future developments, J Dermatol Treat (2019).

[83]   S Singh, U Chandra, VN Anchan, P Verma, R Tilak, Limited effectiveness of four oral antifungal drugs (fluconazole, griseofulvin, itraconazole and terbinafine) in the current epidemic of altered dermatophytosis in India: results of a randomized pragmatic trial, Br J Dermatol 183 (5) (2020) 840–846.

[84]   B Pippi, AJ Lana, RC Moraes, CM Güez, M Machado, LF de Oliveira, et al. In vitro evaluation of the acquisition of resistance, antifungal activity and synergism of Brazilian red propolis with antifungal drugs on Candida spp. J Appl Microbiol. 118 (4) (2015) 839–850.

[85]   A Khurana, K Sardana, A Chowdhary, Antifungal resistance in dermatophytes: recent trends and therapeutic implications, Fungal Genet Biol 132 (2019) 103255.

[86]   O Singh, Z Khanam, N Misra, MK Srivastava, Chamomile (Matricaria chamomilla L.): an overview, Pharmacogn Rev 5 (2011) 82–95.

[87]   G Raut, S Wairkar, Management of psoriasis with nutraceuticals: an update, Complement Ther Clin Pract 31 (2018) 25–30.

[88]   J Reuter, I Merfort, CM Schempp, Botanicals in dermatology, Am J Clin Dermatol 11 (4) (2010) 247–267.

[89]   BH May, AL Zhang, W Zhou, CJ Lu, S Deng, CC Xue, Oral herbal medicines for psoriasis: a review of clinical studies, Chin J Integr Med 18 (3) (2012) 172–178.

[90]   LX Zhang, YP Bai, PH Song, LP You, DQ Yang, Effect of Chinese herbal medicine combined with acitretin capsule in treating psoriasis of blood-heat syndrome type, Chin J Integr Med 15 (2) (2009) 141–144.

[91]   CS Zhang, JJ Yu, S Parker, AL Zhang, B May, C Lu, et al., Oral Chinese herbal medicine combined with pharmacotherapy for psoriasis vulgaris: a systematic review, Int J Dermatol 53 (11) (2014) 1305–1318.

[92]   YS Cho, JH Baek, A review of case studies with pattern identifications and herbal medicines for psoriasis, J Pediatr Korean Med 31 (1) (2017) 1–11.

[93]   SR Vasani, DG Saple, Aloe vera, short review, Indian J Dermatol Venereol Leprol 53 (2008) 163–166.

[94]   VG Avila, D Segura, B Escalante, Anti-inflammatory activity of extracts from aloe vera gel, J Ethnopharmacol 55 (1996) 69–75.

[95]   SS Pandey, AK Jha, V Kaur, Aqueous extract of neem leaves in treatment of Psoriasis vulgaris, Indian J Dermatol Venereol Leprol 60 (1994) 63–67.

[96]   G Sarafian, M Afshar, P Mansouri, J Asgarpanah, K Raoufinejad, M Rajabi, Topical turmeric microemulgel in the management of plaque psoriasis; a clinical evaluation, Iran J Pharm Res 14 (2015) 865–876.

[97]   E Antiga, V Bonciolini, W Volpi, E Del Bianco, M Caproni, Oral curcumin (Meriva) is effective as an adjuvant treatment and is able to reduce IL-22 serum levels in patients with psoriasis vulgaris, Biomed Res Int 2015 (2015) 283634.

[98]   EQ Shyong, Y Liu, A Lazinsky, RN Saladi, RG Phelps, LM Austin, M Lebwohl, H Wei, Effects of the isoflavone 4′, 5, 7-trihydroxyisoflavone (genistein) on psoralen plus ultraviolet A radiation (PUVA)-induced photodamage, Carcinogenesis 23 (2002) 317–321.

[99]   JR Staberg, T Menne, Efficacy of topical treatment in psoriasis with MC903, a new vitamin D analogue, Acta Derm 69 (1989) 147–150.

[100]   SH Aljabre, OM Alakloby, MA Randhawa, Dermatological effects of Nigella sativa, J Dermatol Surg Oncol 19 (2015) 92–98.

[101]   J Saary, R Qureshi, V Palda, J DeKoven, M Pratt, S Skotnicki-Grant, et al., A systematic review of contact dermatitis treatment and prevention, J Am Acad Dermatol 53 (5) (2005) 845.e1.

[102]   Z Hussain, HE Thu, AN Shuid, P Kesharwani, S Khan, F Hussain, Phytotherapeutic potential of natural herbal medicines for the treatment of mild-to-severe atopic dermatitis: A review of human clinical studies, Biomed Pharmacother 93 (2017) 596–608.

[103]   J Liu, X Mo, D Wu, A Ou, S Xue, C Liu, et al., Efficacy of a Chinese herbal medicine for the treatment of atopic dermatitis: a randomised controlled study, Complement Ther Med 23 (5) (2015) 644–651.

[104]   KLE Hon, TF Leung, PC Ng, MCA Lam, WYC Kam, KY Wong, et al., Efficacy and tolerability of a Chinese herbal medicine concoction for treatment of atopic dermatitis: a randomized, double-blind, placebo-controlled study, Br J Dermatol 157 (2) (2007) 357–363.

[105]   JH Kim, H Kim, Combination treatment with herbal medicines and Western medicines in atopic dermatitis: Benefits and considerations, Chin J Integr Med 22 (5) (2016) 323–327.

[106]   F Yan, F Li, J Liu, S Ye, Y Zhang, J Jia, et al., The formulae and biologically active ingredients of Chinese herbal medicines for the treatment of atopic dermatitis, Biomed Pharmacother 127 (2020) 110142.

[107]   JH Yang, JM Yoo, WK Cho, JY Ma, Ethanol extract of sanguisorbae radix inhibits mast cell degranulation and suppresses 2, 4-dinitrochlorobenzene-induced atopic dermatitis-like skin lesions, Mediators Inflamm 2016 (2016) 2947390.

[108]   HY Cha, SH Ahn, JH Cheon, IS Park, JT Kim, K Kim, Hataedock treatment has preventive therapeutic effects in atopic dermatitis-induced NC/Nga mice under high-fat diet conditions, Evid Based Complement Alternat Med 2016 (2016) 173960.

[109]   JM Ku, SH Hong, HI Kim, HS Seo, YC Shin, SG Ko, Effects of Angelicae dahuricae Radix on 2,4-dinitrochlorobenzene-induced atopic dermatitis-like skin lesions in mice model, BMC Complement Alternat Med 17 (1) (2017) 1–8.

[110]   MS Tsang, D Jiao, BC Chan, KL Hon, PC Leung, C Lau, CK Wong, Anti-inflammatory activities of pentaherbs formula, berberine, gallic acid and chlorogenic acid in atopic dermatitis-like skin inflammation, Molecules 21 (4) (2016) 519.

[111]   YY Sung, T Yoon, S Jang, HK Kim, Forsythia suspensa suppresses house dust mite extract-induced atopic dermatitis in NC/Nga mice, PLoS One 11 (12) (2016) e0167687.

[112]   H Ha, H Lee, CS Seo, HS Lim, JK Lee, MY Lee, H Shin, Artemisia capillaris inhibits atopic dermatitis-like skin lesions in Dermatophagoides farinae-sensitized Nc/Nga mice, BMC Complement Alternat Med 14 (1) (2014) 1–10.

[113]   H Lee, JK Lee, H Ha, MY Lee, CS Seo, HK Shin, Angelicae dahuricae radix inhibits dust mite extract-induced atopic dermatitis-like skin lesions in NC/Nga mice, Evid Based Complement Alternat Med 2012 (2012) 743075.

[114]   HS Lim, H Ha, H Lee, JK Lee, MY Lee, HK Shin, Morus alba L. suppresses the development of atopic dermatitis induced by the house dust mite in NC/Nga mice, BMC Complement Alternat Med 14 (1) (2014) 1–8.

[115]   JH Choi, SW Jin, BH Park, HG Kim, T Khanal, HJ Han, HG Jeong, Cultivated ginseng inhibits 2,4-dinitrochlorobenzene-induced atopic dermatitis-like skin lesions in NC/Nga mice and TNF-α/IFN-γ-induced TARC activation in HaCaT cells, Food Chem Toxicol 56 (2013) 195–203.

[116]   YY Choi, MH Kim, KS Ahn, JY UMm, SG Lee, WM Yang, Immunomodulatory effects of Pseudostellaria heterophylla (Miquel) Pax on regulation of Th1/Th2 levels in mice with atopic dermatitis, Mol Med Rep 15 (2) (2017) 649–656.

[117]   H Nasri, M Bahmani, N Shahinfard, A Moradi Nafchi, S Saberianpour, M Rafieian Kopaei, Medicinal plants for the treatment of acne vulgaris: a review of recent evidences, Jundishapur J Microbiol 8 (11) (2015) e25580.

[118]   C Fernandez, K Adamson, M Dale, WJ Cunliffe, A randomized double-blind study to assess the effects of silicic acid compared to placebo in patients with mild to moderate acne, J Dermatol Treat 16 (5-6) (2005) 287–294.

[119]   RU Hamzah, EC Egwim, AY Kabiru, & Muazu MB, Phytochemical and in vitro antioxidant properties of the methanolic extract of fruits of Blighia sapida, Vitellaria paradoxa and Vitex doniana. Oxid Antioxid Med Sci 2 (3) (2013) 217–223.

[120]   GS Teymouri, MS Teimouri, The comparative effect of hydro alcoholic and hy121, dro distillation extracts of Melissa officinalis on acne and pimple, Int J Pharmacol, Phytochem Ethnomed 12 (2019) 35–43.

[121]   S Chen, Herbology in Three Traditional Medicines for Acne, Xlibris Corporation, 2011.

[122]   GS Kumar, KN Jayaveera, CK Kumar, UP Sanjay, BM Swamy, DV Kumar, Antimicrobial effects of Indian medicinal plants against acne-inducing bacteria, Trop J Pharm Res 6 (2) (2007) 717–723.

[123]   KHC Baser, B Demirci, G Iscan, T Hashimoto, F Demirci, Y Noma, Y Asakawa, The essential oil constituents and antimicrobial activity of Anthemis aciphylla BOISS. var. discoidea BOISS, Chem Pharm Bull 54 (2) (2006) 222–225.

[124]   W Wuttke, H Jarry, V Christoffel, B Spengler, D Seidlova-Wuttke, Chaste tree (Vitex agnus-castus)–pharmacology and clinical indications, Phytomedicine 10 (4) (2003) 348–357.

[125]   C Nam, S Kim, Y Sim, I Chang, Anti-acne effects of oriental herb extracts: a novel screening method to select anti-acne agents, Skin Pharmacol Physiol 16 (2) (2003) 84–90.

[126]   S Weckesser, K Engel, B Simon-Haarhaus, A Wittmer, K Pelz, CÁ Schempp, Screening of plant extracts for antimicrobial activity against bacteria and yeasts with dermatological relevance, Phytomedicine 14 (7--8) (2007) 508–516.

[127]   BC Vastano, Y Chen, N Zhu, CT Ho, Z Zhou, RT Rosen, Isolation and identification of stilbenes in two varieties of Polygonum c uspidatum, J Agric Food Chem 48 (2) (2000) 253–256.

[128]   RR Korać, KM Khambholja, Potential of herbs in skin protection from ultraviolet radiation, Pharmacogn Rev 5 (10) (2011) 164.

[129]   WY Huang, YZ Cai, Y Zhang, Natural phenolic compounds from medicinal herbs and dietary plants: potential use for cancer prevention, Nutr Cancer 62 (1) (2009) 1–20.

[130]   A AlQathama, UF Ezuruike, AL Mazzari, A Yonbawi, E Chieli, JM Prieto, Effects of selected Nigerian medicinal plants on the viability, mobility, and multidrug-resistant mechanisms in liver, colon, and skin cancer cell lines, Front Pharmacol 11 (2020) 1456.

[131]   SK Katiyar, Green tea prevents non-melanoma skin cancer by enhancing DNA repair, Arch Biochem Biophys 508 (2) (2011) 152–158.

[132]   JE Kim, DE Lee, KW Lee, JE Son, SK Seo, J Li, et al., Isorhamnetin suppresses skin cancer through direct inhibition of MEK1 and PI3-K, Cancer Prev Res 4 (4) (2011) 582–591.

[133]   JR Ramsay, A Suhrbier, JH Aylward, S Ogbourne, SJ Cozzi, MG Poulsen, et al., The sap from Euphorbia peplus is effective against human nonmelanoma skin cancers, Br J Dermatol 164 (3) (2011) 633–636.

[134]   M Alecu, C Ursaciuc, F Halalau, G Coman, W Merlevede, E Waelkens, P de Witte, Photodynamic treatment of basal cell carcinoma and squamous cell carcinoma with hypericin, Anticancer Res 18 (6B) (1998) 4651–4654.

[135]   AK Mishra, A Mishra, P Chattopadhyay, Herbal cosmeceuticals for photoprotection from ultraviolet B radiation: a review, Trop J Pharm Res 10 (3) (2011) 351–360.

[136]   LJ Zhang, L Chen, Y Lu, JM Wu, B Xu, ZG Sun, AY Wang, Danshensu has anti-tumor activity in B16F10 melanoma by inhibiting angiogenesis and tumor cell invasion, Eur J Pharmacol 643 (2-3) (2010) 195–201.

[137]   SJ Suh, JW Yoon, TK Lee, UH Jin, SL Kim, MS Kim, et al., Chemoprevention of Scutellaria bardata on human cancer cells and tumorigenesis in skin cancer, Phytother Res 21 (2) (2007) 135–141.

[138]   KK Park, KS Chun, JM Lee, SS Lee, YJ Surh, Inhibitory effects of [6]-gingerol, a major pungent principle of ginger, on phorbol ester-induced inflammation, epidermal ornithine decarboxylase activity and skin tumor promotion in ICR mice, Cancer Lett 129 (2) (1998) 139–144.

[139]   SO Kim, KS Chun, JK Kundu, YJ Surh, Inhibitory effects of [6]-gingerol on PMA-induced COX-2 expression and activation of NF-κB and p38 MAPK in mouse skin, Biofactors 21 (1-4) (2004) 27–31.

[140]   SJ Tsai, CY Chao, MC Yin, Preventive and therapeutic effects of caffeic acid against inflammatory injury in striatum of MPTP-treated mice, Eur J Pharmacol 670 (2-3) (2011) 441–447.

[141]   Y Yang, Y Li, K Wang, Y Wang, W Yin, L Li, P38/NF-κB/snail pathway is involved in caffeic acid-induced inhibition of cancer stem cells-like properties and migratory capacity in malignant human keratinocyte, PLoS One 8 (3) (2013) e58915.

[142]   R Agarwal, SK Katiyar, DW Lundgren, H Mukhtar, Inhibitory effect of silymarin, an anti-hepatotoxic flavonoid, on 12-O-tetradecanoylphorbol-13-acetate-induced epidermal ornithine decarboxylase activity and mRNA in SENCAR mice, Carcinogenesis 15 (6) (1994) 1099–1103.

[143]   S Widyarini, N Spinks, AJ Husband, VE Reeve, Isoflavonoid compounds from red clover (Trifolium pratense) protect from inflammation and immune suppression induced by UV radiation, Photochem Photobiol 74 (3) (2001) 465–470.

[144]   R Eli, JA Fasciano, An adjunctive preventive treatment for cancer: ultraviolet light and ginkgo biloba, together with other antioxidants, are a safe and powerful, but largely ignored, treatment option for the prevention of cancer, Med Hypotheses 66 (6) (2006) 1152–1156.

[145]   Y Wang, J Lv, Y Cheng, J Du, D Chen, C Li, et al., Apoptosis induced by Ginkgo biloba (EGb761) in melanoma cells is Mcl-1-dependent, PLoS One 10 (4) (2015) e0124812.

[146]   S F'guyer, F Afaq, H Mukhtar, Photochemoprevention of skin cancer by botanical agents, Photodermatol Photoimmunol Photomed 19 (2) (2003) 56–72.

[147]   DG Mercurio, TAL Wagemaker, VM Alves, CG Benevenuto, LR Gaspar, PM Campos, In vivo photo-protective effects of cosmetic formulations containing UV filters, vitamins, Ginkgo biloba and red algae extracts, J Photochem Photobiol, B 153 (2015) 121–126.

[148]   L Tong, S Wu, The mechanisms of carnosol in chemoprevention of ultraviolet B-light-induced non-melanoma skin cancer formation, Sci Rep 8 (1) (2018) 1–9.

[149]   JC Chamcheu, I Rady, RCN Chamcheu, AB Siddique, MB Bloch, S Banang Mbeumi, et al., Graviola (Annona muricata) exerts anti-proliferative, anti-clonogenic and pro-apoptotic effects in human non-melanoma skin cancer UW-BCC1 and A431 cells in vitro: involvement of hedgehog signaling, Int J Mol Sci 19 (6) (2018) 1791.

[150]   BB Bridgeman, P Wang, B Ye, JC Pelling, OV Volpert, X Tong, Inhibition of mTOR by apigenin in UVB-irradiated keratinocytes: a new implication of skin cancer prevention, Cell Signal 28 (5) (2016) 460–468.

[151]   X Paredes-Gonzalez, F Fuentes, ZY Su, ANT Kong, Apigenin reactivates Nrf2 anti-oxidative stress signaling in mouse skin epidermal JB6 P+ cells through epigenetics modifications, AAPS J 16 (4) (2014) 727–735.

[152]   HH Cao, JH Chu, HY Kwan, T Su, H Yu, CY Cheng, ZL Yu, Inhibition of the STAT3 signaling pathway contributes to apigenin-mediated anti-metastatic effect in melanoma, Sci Rep 6 (1) (2016) 1–12.

[153]   G Zhao, X Han, W Cheng, J Ni, Y Zhang, J Lin, Z Song, Apigenin inhibits proliferation and invasion, and induces apoptosis and cell cycle arrest in human melanoma cells, Oncol Rep 37 (4) (2017) 2277–2285.

[154]   KK Sharma, R Saikia, J Kotoky, JC Kalita, R Devi, Antifungal activity of Solanum melongena L, Lawsonia inermis L. and Justicia gendarussa B. against Dermatophytes, Int J PharmTech Res 3 (3) (2011) 1635–1640.

[155]   RA Khosravi, H Shokri, Z Farahnejat, R Chalangari, M Katalin, Antimycotic efficacy of Iranian medicinal plants towards dermatophytes obtained from patients with dermatophytosis, Chin J Nat Med 11 (1) (2013) 43–48.

[156]   GS Gozubuyuk, E Aktas, N Yigit, An ancient plant Lawsonia inermis (henna): determination of in vitro antifungal activity against dermatophytes species, J Mycol Méd 24 (4) (2014) 313–318.

[157]   ME Eja, GE Arikpo, KH Enyi-Idoh, SE Etim, HE Etta, Efficacy of local herbal therapy in the management of dermatophytosis among primary school children in Cross River State, South-south Nigeria, Afr J Med Med Sci 38 (2) (2009) 135–141.

[158] SA Shinkafi, Antidermatophytic activities of column chromatographic fractions and toxicity studies of Pergularia tomentosa L. and Mitracarpus scaber Zucc used in the treatment of dermatophytoses, Adv Med Plant Res 2 (1) (2014) 7–15.

[159] VN Simhadri, M Muniappan, I Kannan, S Viswanathan, Phytochemical analysis and docking study of compounds present in a polyherbal preparation used in the treatment of dermatophytosis, Curr Med Mycol 3 (4) (2017) 6.

[160] S Phongpaichit, N Pujenjob, V Rukachaisirikul, M Ongsakul, Antifungal activity from leaf extracts of Cassia alata L., Cassia fistula L. and Cassia tora L, Songklanakarin J Sci Technol 26 (5) (2004) 741–748.

[161] S Dong-xia, Testing the antifungal activity of the combination of three traditional Chinese medicine extracts in vitro against the common dermatophyte, Chin J Vet Med (2012) 04.

[162] RM Aliyu, MB Abubakar, AB Kasarawa, YU Dabai, N Lawal, MB Bello, et al., Efficacy and phytochemical analysis of latex of Calotropis procera against selected dermatophytes, J Intercult Ethnopharmacol 4 (4) (2015) 314.

[163] L Mugnaini, S Nardoni, L Pistelli, M Leonardi, L Giuliotti, MN Benvenuti, F Mancianti, A herbal antifungal formulation of Thymus serpillum, Origanum vulgare and Rosmarinus officinalis for treating ovine der+matophytosis due to Trichophyton mentagrophytes, Mycoses 56 (3) (2013) 333–337.

Herbal medicine and common dermatologic diseases

Begüm Ünlü

Department of Dermatology, Ağrı State Hospital, Ağrı, Turkey

15.1 Introduction

Complementary and alternative medicine (CAM) represents variety of medical and healthcare practices, which are not involved in conventional treatment modalities [1]. Herbal medicine is one of the most used complementary and alternative therapies. There is a growing tendency toward herbal medicine worldwide among patients with dermatologic diseases. Herbal medicine is most frequently chosen for acne, atopic dermatitis (AD), and psoriasis which are all chronic and sometimes recalcitrant conditions. The prevalence of CAM use by dermatology patients at some point in their lifetime is found 35%–69% in the literature [2]. With the increase in use of herbal medicine, clinicians must be able to inform their patients about safety and efficacy of herbal medicine and give answers to their concerns about these modalities. In this chapter, we discuss herbal medicine and common dermatologic diseases one by one.

15.2 Acne and herbal medicine

Acne is a common, chronic inflammatory disease of pilosebaceous unit, which affects 80% of adolescent, 54% of adult women and 40% of adult men [3]. It affects patients most commonly from 12 years of age [3]. It represents with wide range of severity and commonly leads to psychosocial distress [4]. Acne causes spots to occur simultaneously on several areas of the body, including the face, neck, back, and chest and presents with comedones (blackheads), whiteheads, red papules, pustules, or cysts (Fig. 15.1) [4].

According to classical acne classification, acne lesions are divided into four categories: noninflammatory, mild papular, scarring papular, and nodular. Noninflammatory acne presents as comedons and whiteheads. The latter three are inflammatory acne lesions [5–7]. Because of widespread localization of lesions and chronicity of disease, acne can have significant psychosocial consequences [5]. Poor patient adherence and antibiotic resistance are two major

Herbal Medicines: A Boon for Healthy Human Life.
DOI: https://doi.org/10.1016/B978-0-323-90572-5.00019-6

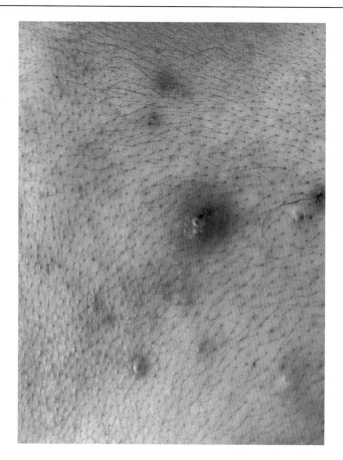

Figure 15.1
Papules and pustules of acne.

reasons of treatment failure [8]. Due to these difficulties that patients with acne face; some of these patients seek for other treatment modalities such as herbal therapies. A summary of herbal medicine use in acne treatment is shown in Table 15.1.

15.2.1 Tea tree oil

Tea tree oil (TTO) is an essential oil which is used for almost 100 years in Australia for its antiseptic and anti-inflammatory features [9]. Now it is available in variable commercial forms worldwide that are marketed for acne treatment. TTO is a well-known, well characterized and standardized herbal treatment [10].

In the literature, there are several clinical studies showed that TTO is effective in acne treatment. In a single blind study of Basset et al. showed that there were significant reductions in inflammatory and comedonal lesions with both 5% TTO and 5% benzoil peroxide (BPO)

Table 15.1: Summary of acne and herbal medicine.

Herbal medicine	Mechanism of action
Tea tree oil	Antiseptic and anti-inflammatory
Witch hazel	Antioxidant
Lactobacillus fermented chamaecyparis obtusa leaf extract	Antibacterial and sebosuppressive
Copaiba oil resin	Anti-inflammatory, wound healing, and antiseptic
Sandalwood album oil	Anti-inflammatory, antimicrobial, and antiproliferative
Rosemary extract	Antioxidant, anti-inflammatory, and anti-antimicrobial
Jeju essential oil	Antibacterial, antioxidant, antielastase, and anti-inflammatory
Korean citrus oil	Antibacterial and anti- inflammatory

treatments [11]. TTO also showed better tolerability than 5% BPO in this study [11]. In a randomized 45-day controlled trial of 5% TTO vs vehicle in 60 subjects with mild to moderate acne, it is found that TTO is effective than vehicle [10]. In addition, contact sensitization to TTO is %0,9 in the North American Contact Dermatitis Group Database so it is well tolerated with a low contact dermatitis risk [12]. Adverse effects, including skin irritation, allergic contact dermatitis, systemic contact dermatitis, linear immunoglobulin A disease, erythema multiforme like reactions, systemic hypersensitivity reactions, and idiopathic male prepubertal gynecomastia have been reported in the literature [13]. It can be potentially toxic if it is ingested at higher doses, but there are no human deaths due to TTO intoxication [13]. As a result, TTO is a well-tolerated, safe, and effective treatment choice for acne patients.

15.2.2 Witch hazel

Witch hazel (Hamamelis virginiana Linnaeus) was used by the Cherokee, Chippewa, Iroquois, and Mohegan nations in treatment of furuncles, bruises, and insect bites, and as an astringent [14]. Despite there is no in vivo studies in the literature about its efficacy in control of acne, in vitro studies showed that it is a potent antioxidant due to its high concentration of tannins [15]. Nowadays, it is used in aftershaves, moisturizers, and in hair sprays [14].

Contact allergy has been reported after exposure to witch hazel in a few numbers of cases [15].

15.2.3 Lactobacillus fermented Chamaecyparis Obtusa leaf (LFCO) extract

Chamaecyparis obtusa is a type of Korean hinoki cypress. This plant is widely used for household and commercial purposes [16]. Fermentation of leaves of this plant by Lactobacillus fermentum yields an extract which has antibacterial effects on Propionibacterium acnes [16]. In an 8-week randomized, controlled split face study that compare LFCO with TTO 5% by Kwon et al. found that acne lesions were reduced by 65.3% with LFCO compared to a 38.2% reduction with TTO [16]. There is also significant faster onset of action and strong sebosuppressive effects that is associated with a reduction in size of sebaceous glands was documented with LFCO [17]. Despite there is limited reports about use this extract LFCO can be safe and effective choice for acne treatment.

15.2.4 Copaiba oil resin

Copaiba oil resin is widely used in traditional medicine of Central and South America due to its anti-inflammatory, wound healing, and antiseptic activities for centuries [18]. It is obtained from the South American tree genus named as Copaifera [18]. The therapeutic actions of Copaiba are primarily attributed to diterpine compounds in the oil which serve the plant as biological defenses against predators and pathogens [18]. In a double-blind study of Copaiba essential oil versus placebo in patients with mild inflammatory acne by da Silva et al., there was a significant improvement of acne lesions in both treated and placebo regions without significant difference between treatment groups [18]. Further reports are needed for understanding efficacy and safety of copaiba plant in acne treatment.

15.2.5 Sandalwood album oil

Sandalwood album oil (SAO) or East Indian sandalwood oil is an essential oil distilled from the *Santalum album* tree and used as a therapeutic agent in many Asian countries [19]. It has an anti-inflammatory, antimicrobial, and antiproliferative properties for this reason, used to treat inflammatory and cutaneous eruptions for long time as traditional medicine in Asia [19]. In the literature there are several clinical trials about SAO treatment of several dermatologic diseases, such as acne, psoriasis, dermatitis, warts, and molluscum contagiosum [16]. SAO has antibacterial, anti-inflammatory, and wound healing effects. In a single center, 8-week open-label study by Moy et al., 42 patients with mild to moderate acne were treated with a combination therapy of 0.5% salicylic acid with sandalwood [19]. Acne lesions improved in 89% of cases when compared with baseline lesions [19]. SAO is usually well tolerated and safe. Only small percent of patients presented mild adverse effects, such as burning, dryness, and stinging [19].

15.2.6 Rosemary extract

Bioactive compounds of rosemary (Rosmarinus officinalis) extract are rosmarinic acid, carnosol, and carnosic acid [20]. Rosemary is a culinary spice which has an antioxidant, anti-inflammatory, anticarcinogenic, and antimicrobial features [20]. This extract modulates cytokine production via several pathways. Tsai and colleagues demonstrated that Rosemary extract reduces P. acnes–induced inflammation in vitro and in vivo [20]. Randomized, controlled studies are needed for understanding the efficacy of rosemary extract in acne treatment.

15.2.7 Jeju essential oil

Jeju essential oil is derived from thymus plants, which are endemic plants of Korea [21]. In a study of Oh et al. from Korea showed that this plant has antibacterial, antioxidant, antielastase, and anti-inflammatory effects with a low toxicity in human cell lines [21]. Further reports are needed for evaluating its efficacy and safety profile.

15.2.8 Korean citrus

Citrus oils obtained from distillation of fruits of Citrus obovoides and Citrus natsudaidai [22]. In an in vitro study by Kim et al., citrus oil has antibacterial and anti-inflammatory activity against *P. acnes* and *S. epidermidis*. This finding suggests potential utility of this oil in acne [22].

15.3 Atopic dermatitis and herbal medicine

AD is a common and chronic type of chronic eczematous skin inflammation that also known as eczema and atopic eczema (Fig. 15.2). It presents as dry inflamed skin, intense pruritus, itching, skin lichenification, sleep disturbance, and emotional distress [23]. AD can occur in any age, but most commonly appears during early childhood and periodically relapses throughout the life of a patient [23]. AD affects up to 20% of children and 10% of adults in high-income countries where this prevalence is almost stabilized [24]. In low-income countries prevalence is continuously increasing [24].

In pathophysiology of AD, there is a complex pathogenic interplay between patient's genetic features, skin barrier abnormalities, immune deregulation and environmental factors [25]. Due to this complex etiology, there is no curative therapy for the treatment of AD. Avoidance of causative allergens, moisturizers, topical anti-inflammatory or immunosuppressant therapies, and antipruritic medications are conventional choices of treatment for AD [25]. Unfortunately, recurrences are not rare in AD prognosis. Because of recalcitrant and chronic nature of AD, treatment may be challenging. For this reason, it is understandable that AD patients frequently seek for herbal medicines [25]. There are numerous reports about use of herbal medicine which applied oral or topical ways in the literature. Herbal medicine use in AD is summarized in Table 15.2.

15.3.1 Chinese herbal medicine

Chinese herbal medicine (CHM) is part of traditional Chinese medicine therapy. Oral and topical application of CHM is used in far east for centuries [26]. In the meta-analysis by Tan et al., there is a significant improvement in disease severity scores by the combination

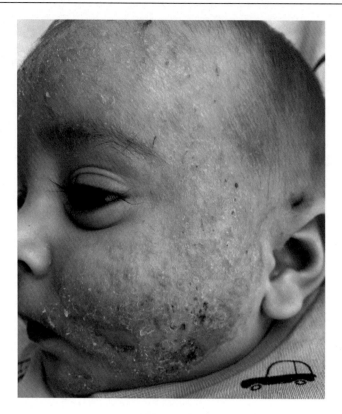

Figure 15.2
Atopic dermatitis lesions of an infant.

Table 15.2: Summary of atopic dermatitis and herbal medicine.

Herbal medicine	Mechanism of action
Chinese herbal medicine	Anti-inflammatory and immunmodulatory
Korean medicine	Anti-inflammatory
Coconut oil	Antimicrobial and anti-inflammatory
Sunflower seed oil	Restoring skin barrier and immune modulation
Honey	Anti-inflammatory, antioxidant, wound healing, antibacterial, antiviral, and antifungal

treatment of CHM and conventional treatment compared with conventional treatment alone [26]. There is also significant improvement in erythema, pruritus, sleep scores, and quality of life and decrease the need of conventional therapy [26].

A multicentered, double-blind, controlled, randomized, prospective, parallel group trial that involved 275 AD patients revealed that scoring of atopic dermatitis (SCORAD) index decreased, and dermatologic quality of life index (DLQI) increased in CHM and conventional therapy (topical 0.1% mometasone Furoate) groups when compared with control group [27]. In this study, well-known traditional Chinese herbal formula, "Pei Tu Qing Xin Tang

(PTQXT)" that was composed of nine different herbs including Radix pseudostellariae, Forsythia suspensa, Ramulus Uncariae cum Uncis, Medulla Junci, Herba lophatheri, Semen coicis, Rhizoma dioscoreae, Concha ostreae, and Radix glycyrrhizae; applied in oral and topical way [27]. Rebound symptoms is also more severe in control group when compared with treatment groups [27]. There is no significant difference in DLQI, SCORAD, and severity of rebound symptoms between CHM and conventional therapy groups [27]. Except that diarrhea was observed and disappeared with discontinuation of treatment, there were no serious side effects of CHM [27].

15.3.2 Korean medicine

Herbal medicine and acupuncture are the main treatment modalities of Korean medicine (KM) [28]. In a clinical trial in three pregnant women by Kim and colleagues, a decoction of KM that was composed of several herbs such as Rehmannia glutinosa, Talcum, Glycyrrhiza glabra, Atractylodes chinensis, Plantago asiatica L., Gentiana scabra Bunge, Akebia quinata Decaisne, Raphanus sativus, Adenophora triphylla, Smilax china L., Scutellaria baicalensis Georgi, and Angelica gigas was administered to patients orally [29]. Herbal wet-wrap dressing impregnated into the decoction of Sophora flavescens, Phellodendron amurense, Schizonepeta tenuifolia Briquet, Liriope platyphylla, and Perilla frutescens var. acuta was also applied to skin lesions [29]. There was improvement in SCORAD index in all patients without drug toxicity [29].

15.3.3 Coconut oil

Coconut oil is very popular in recent years in food and skin products [30]. It is derived from the white lining within the shell, of mature coconuts [30]. Coconut oil contains high levels of lauric acid, which has been shown to have antimicrobial and anti-inflammatory effects [30]. Virgin coconut oil is the coconut oil processed within 24 h of harvest to avoid the formation of fatty acids that may cause skin irritation [30].

In a randomized, controlled trial of 117 mild to moderate AD patients, the effects of topical virgin coconut oil (VCO) vs mineral oil is investigated [31]. Mean SCORAD indices decreased from baseline by 68.23% in the VCO group by 38.13% in the mineral oil group [31]. Decrease in VCO group is significantly lower than mineral oil group [31]. Topical application of VCO was also superior to mineral oil in improvement of skin capacitance (TEWL) [31].

15.3.4 Olive oil

Olive oil is used for skin care from dates to ancient Egypt. Oleic acid is the predominant fatty acid component of olive oil [30]. In a study by Cooke and colleagues, authors randomized 115 neonates to three treatment groups, which are no oil group, four drops of olive oil group, and four drops of sunflower seed oil group [32]. Olive oil treated group had a significantly lower ordering of lipids in the stratum corneum than the no-oil group [32]. In another study with seven adult AD patients, it is demonstrated that TEWL is significantly greater in the olive oil treated patients when compared with control group [33]. It is shown that despite the popularity of olive oil, this herbal oil may not be the treatment of choice for AD. Further wide sample size studies are needed.

15.3.5 Sunflower seed oil

Sunflower seed oil (SSO) is accessible and effective treatment choice for xerosis and AD [25]. SSO is composed of mostly linoleic acid (60% of the oil) [34]. Linoleic acid is important fatty acid to maintain normal barrier function of the epidermis [34]. Linoleic acid converts to arachidonic acid and is a precursor to prostaglandin E2 (PGE2) [34]. PGE2 is a known modulator of cutaneous inflammation [34]. The similarities of SSO lipids and those constituting the epidermal barrier and immune modulation effect of SSO suggest their potential utility in a variety of cutaneous conditions. In infants with linoleic acid deficiency, Friedman and colleagues applied topical linoleic acid to patients and showed that application of topical linoleic acid restored the skin barrier and improved AD [35].

15.3.6 Honey

Honey has been used for centuries for various medicinal purposes [25]. Honey has anti-inflammatory, antioxidant, wound healing, antibacterial, antiviral, and antifungal features [25]. In an open-label, single-blind, randomized controlled trial on 15 adult AD patients, Fingleton et al. demonstrated that Kanuka honey is as effective as aqueous cream in the management of AD [36]. The other popular bee product is beeswax [37]. Topical application of honey, beeswax and olive oil is found to be superior to corticosteroid treatment and well tolerated in AD treatment [37].

15.4 Psoriasis and herbal medicine

Psoriasis is an inflammatory skin disorder with a strong genetic predisposition and influenced by environmental factors [38–39]. Main pathology of psoriasis is uncontrolled keratinocyte proliferation and differentiation mediated by T- helper cells [39]. Prevalence of psoriasis is 2%–3% worldwide. Psoriasis vulgaris is characterized by erythematous, sharply demarcated, pruritic plaques that are covered by scales (Fig. 15.3) [38]. Patients with psoriasis have disturbed psychological and mental health due to social stigmatization that likely results in anxiety, depression, and suicidal ideation [38]. Due to chronic and intractable course of psoriasis, patients usually seek for herbal medicine [40]. Herbal medicine use in psoriasis is summarized in Table 15.3.

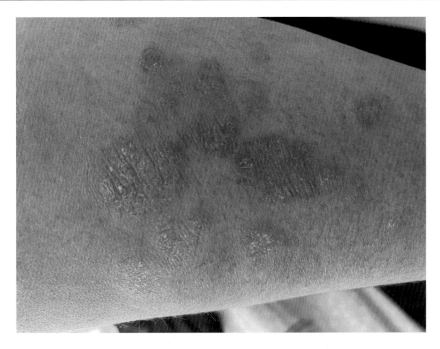

Figure 15.3
Erythematous scaling plaques of psoriasis patient.

Table 15.3: Summary of psoriasis and herbal medicine.

Herbal medicine	Mechanism of action
Aloe vera	Immunomodulating, moisturizing, wound healing, antioxidant, anti-inflammatory, antibacterial, and antifungal
Angelica sinensis	Photosensitizer, antimitotic, and anti-inflammatory
Capsicum annuum	Anti-inflammatory and antiproliferative
Gotu kola	Anti-inflammatory, immunmodulatory, and antioxidant
Day blooming jasmine	İnhibiting abnormal keratinocyte proliferation and inducing keratinocyte differentiation
Turmeric	Anti-inflammatory
American wintergreen	Anti-inflammatory
Oregon grape	Anti-inflammatory
Chamomile	Anti-inflammatory and antibacterial
Psorospermum febrifugum	Immunmodulatory
Milk thistle	Anti-inflammatory and inhibiting abnormal keratinocyte proliferation

15.4.1 *Aloe vera*

Aloe vera Linnaeus or Aloe barbadensis Miller is a tropical succulent plant [25]. The aloe vera gel is obtained from leaves of aloe vera contains 99.5% water and remaining part of gel is composed of carbohydrates, mucopolysaccharides, proteins, enzymes, anthraquinones, chromones, minerals, vitamins, phenolic compounds, and salicylic acid [41]. Aloe vera has immunomodulating, moisturizing, wound healing, antioxidant, anti-inflammatory, antitumor, antibacterial, and antifungal effects [25]. Owing to the salicylic acid component of the gel, it also has a keratolytic effect [25].

In a randomized, double-blind study by Choonhakarn et al., aloe vera is found to be more effective than 0.1% triamcinolone acetonide in treatment of mild to moderate psoriasis vulgaris [42]. In a mouse model study by Dhanabal et al., aloe vera extract demonstrated almost same amount of antipsoriatic activity tazarotene (81.95%–87.94%) [43].

15.4.2 Angelica sinensis

Angelica sinensis, commonly known as Dong quai, is an aromatic herb of the parsley family, native to China and Japan [25]. It contains psoralen, which acts as photosensitizers upon exposure to UV-A. Exposure to UV after psoralen consumption causes decreasing the rate of synthesis of epidermal DNA [25,44]. It has also antimitotic and anti-inflammatory activity results in apoptosis of keratinocytes and lymphocytes [44]. UV treatment with psoralen is shown to be effective than placebo in various studies [25].

15.4.3 Capsicum annuum

Capsicum annuum is known as red pepper, which is a commonly used spice and originated in South America [45]. It consists of capsaicinoids such as capsaicin. Capsaicin reduces inflammation and keratinocyte proliferation *via* inhibiting substance P which is a neurotransmitter [45]. Substance P activates the inflammatory process and produces angiogenesis, vasodilatation, and keratinocyte hyperproliferation in psoratic lesions [46]. Substance P in psoriatic lesions was decreased and efficacy of capsaicin was greater than control group in a study by Wang and coworkers [46]. In an 8-week open trial by Kürkçüoğlu et al. in 10 patients with severe psoriasis vulgaris, seven patients showed marked improvement with capsaicin as regards itching, scaling, and erythema [45]. In also other patients, the itching disappeared [45]. There was no significant change in the placebo-treated lesions [45].

15.4.4 Gotu kola

Gotu kola or Centella asiatica is a medicinal plant that has been used in folk medicine for hundreds of years in oriental medicine [47]. The active compounds of this plant are pentacyclic triterpenes, mainly asiaticoside, madecassoside, asiatic, and madecassic acids [47]. It is effective in improving treatment of small wounds, hypertrophic wounds, keloid scars, burns, psoriasis, and scleroderma [47]. The mechanism of action involves promoting fibroblast proliferation, increasing the synthesis of collagen and intracellular fibronectin content, improvement of the tensile strength of newly formed skin, anti-inflammatory and antioxidant properties [47].

IL-23 and IL-17 are two of the main interleukins which play key roles in psoriasis pathogenesis [48]. In a mouse model study on imiquimod-induced psoriasis-like dermatitis, IL-23 and IL-17 levels were significantly reduced in madecassoside ointment group when compared to control group [48]. Madecassoside ointment was found to be effective in psoriasis treatment through the regulation of IL-23 and IL-17 axis [48]. Several products have marketed extracts of Centella asiatica so it is an affordable choice of treatment.

15.4.5 Day blooming jasmine

Day blooming jasmine or Cestrum diurnum is a plant which is native to the West Indies [49]. It is widely cultivated in gardens throughout India [49]. The leaves of this plant contain free vitamin D3, 25-hydroxycholecalciferol, and 1, 25-dihydroxycholecalciferol [49]. Vitamin D is a local therapeutic agent in psoriasis treatment due to it inhibits keratinocyte proliferation and induces keratinocyte differentiation [50]. It is a corticosteroid-sparing agent that can be choice of treatment, especially in treatment of face and flexural regions [50]. Getting adequate dietary intake of vitamin D or oral vitamin D supplementation in psoriasis treatment is still controversial because of unmet an unmet evidence of its beneficial effects [50]. Today Cestrum diurnum is marketed and used in psoriasis treatment successfully. It is clear that randomized, controlled studies with large patient groups are needed.

15.4.6 Turmeric

Turmeric is named Curcuma longa or Curcuma domestica in Latin. It is used in traditional medicine for ages [25]. Sesquiterpenes, zingiberene, and curcuminoids are main constituents of turmeric plant that have anti-inflammatory properties [25]. Manzoor and colleagues also showed that turmeric also has anticancer and antioxidant features [51].

In a randomized placebo-control clinical trial by Bahrani et al., it is demonstrated that turmeric tonic significantly reduced the erythema, scaling and induration of lesions (PASI score), and improved the patients' quality of life compared to the placebo [52]. The quality of life of the patients is also improved [52]. A novel formulation od of topical turmeric in microemulgel form is studied for treatment for plaque psoriasis [53]. The clinical and therapeutic benefit of this form of turmeric was evaluated on 34 patients with mild to moderate plaque psoriasis in a randomized, prospective intra-individual, right-left comparative, placebo-controlled, double-blind clinical trial [53]. This novel form was found to be effective and improved clinical and quality of life parameters (PASI and DQLI) in treated lesions in comparison with untreated lesions have improved the patients [53].

15.4.7 American wintergreen

American wintergreen is also known as eastern teaberry, boxberry, or checkerberry, which is named as Gaultheria procumbens in binominal name [54]. It is native to northeastern North America from Newfoundland west to southeastern Manitoba, and south to Alabama [25]. Historically, it was used by Native Americans as an analgesic [54]. Leaves of this plant are used to prepare essential oils which contain methyl salicylate, which is responsible for anti-inflammatory activity [54]. Due to salicylic acid component of this plant it can cause systemic effects like tinnitus, vomiting, tachypnea, and acid-base disturbances even it is used topically [54]. Although it has potential anti-inflammatory effects, there are no reports on its efficacy in psoriasis in the literature. Further scientific investigations are need for its safe and effectively use in psoriasis.

15.4.8 Oregon grape

The bionomical name of Oregon grape is Mahonia aquifolium is an evergreen shrub belonging to the family Berberidaceae [54]. It is native to the United States and is useful in treating various inflammatory cutaneous disorders, including psoriasis for ages [54]. Its main constituents are berberine, berbamine, and oxyacanthine, which produce anti-inflammatory effects because of various mechanisms, including inhibition of lipoxygenase, lipid peroxidation, and keratinocyte proliferation. Mahonia aquifolium is also inhibits production of IL-8, IL-1, and TNF-α [54].

In an intraindividual the efficacy of an ointment containing 10% extract of Mahonia aquifolium bark is investigated on moderately severe psoriasis vulgaris patients [55]. It was found that differences could be found for patients' but not for physicians' assessments [55].

15.4.9 Chamomile

Chamomile is a well-known flower that is called Matricaria recutita in binominal name [54]. It is used to treat gastrointestinal ailments for centuries [54]. Main antipsoriatic components of this plant is chamazulene, quercetin, α-bisabolol, and apigenin [54]. Chamazulane produces anti-inflammatory activity by inhibiting lipoxygenase [54,56]. Quercetin is a flavonoid, which has anti-inflammatory, antitumour, antiviral, and antibacterial activities. It inhibits IL, TNF-α, and IFN. Apigenin is also a flavonoid, which has antioxidant and anti-inflammatory activities. It inhibits TNF-α and expression of IL-6 and IL-8 [56]. α-Bisabolol inhibits TNF-α, IL-6, and IL-1β, therefore, it has anti-inflammatory properties [54–56].

Chamomile oil has antimicrobial activity against staphylococcus and candida [54]. There are no studies conducted on effectiveness of chamomile oil in psoriasis in the literature. Further reports are needed.

15.4.10 Psorospermum febrifugum

Psorospermum febrifugum has been used to treat various skin diseases in folk medicine for centuries and native to Africa, Madagascar, and South America [57]. Asogwa and coworkers studied anti psoriatic activity of Psorospermum febrifugum stem bark extract using the rat – dinitrofluorobenzene induced model of psoriasis [57]. They claimed that it has antipsoriatic and immunomodulatory activities [57].

15.4.11 Milk thistle

Silybum marianum is commonly known as milk thistle and used for its liver protective function for ages [54]. It is found mainly in the Mediterranean regions but is now it distributed worldwide. Silybum marianum contains silymarin and taxifolin [58]. It improves endotoxin removal by the liver, inhibits cAMP phosphodiesterase and inhibits leukotriene synthesis [58].

Yuan et al. conducted a mouse model study to determine the antipsoriatic activity of taxifolin that is isolated from the milk thistle plant [59]. It was found that taxifolin may inhibit abnormal keratinocyte proliferation in human keratinocyte [59]. There is no randomized, controlled study about efficacy and safety of milk thistle on psoriasis treatment in the literature.

15.5 Conclusion

Acne, AD, and psoriasis are common and chronic dermatologic diseases. The conventional drugs used to treat it are having side effects commonly. Herbal remedies may be more effective and safer than traditional medicines in some cases. It is not surprising that patients look for alternative medicine for ages. People use the plants that are mentioned in this chapter for cure of their illnesses for centuries. Despite the popularity of herbal medicine, there are no randomized, controlled studies with large patient groups in the literature. Further studies are needed to enlighten clinicians and patients.

References

[1]   AN Kalaaji, DL Wahner-Roedler, A Sood, TY Chon, LL Loehrer, SS Cha, BA Bauer, Use of complementary and alternative medicine by patients seen at the dermatology department of a tertiary care center, Complement Ther Clin Pract 18 (1) (2012) 49–53.
[2]   E Ernst, The usage of complementary therapies by dermatological patients:a systematic review, Br J Dermatol 142 (5) (2000) 857e61.

[3]   M Ramos-e-Silva, SC Carneiro, Acne vulgaris: review and guidelines, Dermatol Nurs 21 (2) (2009) 63–68.

[4]   C Simon, H Everitt, T Kendrick, Oxford Handbook of General Practice, 2nd Edition, Oxford University Press, New York, 2005.

[5]   SF Friedlander, LF Eichenfield, JF Fowler, RG Fried, L ML, GF Webster, Acne epidemiology and pathophysiology, Semin Cutan Med Surg 1 (2010) 2–4.

[6]   GF Webster, Acne vulgaris, BMJ 325 (7362) (2002) 475–479.

[7]   R Fried, A Friedman, Psychosocial sequelae related to acne: looking beyond the physical, J Drugs Dermatol 5 (2010) s50–s52.

[8]   P Lubtikulthum, N Kamanamool, M Udompataikul, A comparative study on the effectiveness of herbal extracts vs 2.5% benzoyl peroxide in the treatment of mild to moderate acne vulgaris, J Cosmetic Dermatol 18 (6) (2019) 1767–1775.

[9]   CF Carson, KA Hammer, TV Riley, Melaleuca alternifolia (tea tree) oil: a review of antimicrobial and other medicinal properties, Clin Microbiol Rev 19 (1) (2006) 50–62.

[10]   S Enshaieh, A Jooya, AH Siadat, F Iraji, The efficacy of 5% topical tea tree oil gel in mild to moderate acne vulgaris: a randomized, double-blind placebo-controlled study, Indian J Dermatol Venereol Leprol 73 (2007) 22–25.

[11]   IB Bassett, P DL, RS Barnetson, A comparative study of tea-tree oil versus benzoylperoxide in the treatment of acne, Med J 153 (1990) 455–458.

[12]   EM Warshaw, KA Zug, DV Belsito, Positive patch-test reactions to essential oils in consecutive patients from North America and Central Europe, Dermatitis 28 (4) (2017) 246–252.

[13]   N Pazyar, R Yaghoobi, N Bagherani, A Kazerouni, A review of applications of tea tree oil in dermatology, Int J Dermatol 52 (7) (2012) 784–790.

[14]   S Colantonio, JK Rivers, Botanicals with dermatologic properties derived from first nations healing: part 2—plants and algae, J Cutan Med Surg 21 (4) (2016) 299–307.

[15]   EA Gurnee, S Kamath, L Kruse, Complementary and alternative therapy for pediatric acne: a review of botanical extracts, dietary interventions, and oral supplements, Pediatr Dermatol 36 (5) (2019) 596–601.

[16]   WJ Winkelman, Aromatherapy, botanicals, and essential oils in acne, Clin Dermatol 36 (3) (2018) 299–305.

[17]   HH Kwon, JY Yoon, SY Park, S Min, DH Suh, Comparison of clinical and histological effects between lactobacillus-fermented Chamaecyparis obtusa and tea tree oil for the treatment of acne: an eight-week double-blind randomized controlled split-face study, Dermatology 229 (2014) 102–109.

[18]   AG da Silva, F Puziol Pde, RN Leitao, TR Gomes, R Scherer, ML Martins, AS Cavalcanti, LC Cavalcanti, Application of the essential oil from copaiba (Copaifera langsdori Desf.) for acne vulgaris: a double-blind, placebo-controlled clinical trial, Altern Med Rev 17 (1) (2012) 69–75.

[19]   RL Moy, C Levenson, JJ So, JA Rock, Single-center, open-label study of a proprietary topical 0.5% salicylic acid-based treatment regimen containing sandalwood oil in adolescents and adults with mild to moderate acne, J Drugs Dermatol 11 (12) (2012) 1403–1408.

[20]   TH Tsai, LT Chuang, TJ Lien, YR Liing, WY Chen, PJ Tsai, Rosmarinus officinalis extract suppresses Propionibacterium acnes-induced inflammatory responses, J Med Food 16 (4) (2013) 324–333.

[21]   TH Oh, SS Kim, WJ Yoon, JY Kim, EJ Yang, L NH, CG Hyun, Chemical composition and biological activities of Jeju Thymus quinquecostatus essential oils against Propionibacterium species inducing acne, J Gen Appl Microbiol 55 (1) (2009) 63–68.

[22]   SS Kim, JS Baik, TH Oh, WJ Yoon, L NH, CG Hyun, Biological activities of Korean Citrus obovoides and Citrus natsudaidai essential oils against acne-inducing bacteria, Biosci Biotechno Biochem 72 (2008) 2507–2513.

[23]   T Bieber, Atopic dermatitis, Ann Dermatol 22 (2) (2010) 125–137.

[24]   SM Langan, I AD, S Weidinger, Atopic dermatitis, Lancet 396 (10247) (2020) 345–360.

[25]   Z Hussain, HE Thu, AN Shuid, P Kesharwani, S Khan, F Hussain, Phytotherapeutic potential of natural herbal medicines for the treatment of mild-to-severe atopic dermatitis: A review of human clinical studies, Biomed Pharmacother 93 (2017) 596–608.

[26]   HY Tan, AL Zhang, D Chen, CC Xue, GB Lenon, Chinese herbal medicine for atopic dermatitis: a systematic review, J Am Acad Dermatol 69 (2) (2013) 295–304.

[27]   J Liu, X Mo, D Wu, A Ou, S Xue, C Liu, H Li, Z Wen, D Chen, Efficacy of a Chinese herbal medicine for the treatment of atopic dermatitis: a randomized controlled study, Complement Ther Med 23 (2015) 644–651.

[28]   JA Lee, J Choi, T-Y Choi, JH Jun, D Lee, S-S Roh, MS Lee, Clinical practice guidelines of Korean medicine on acupuncture and herbal medicine for atopic dermatitis: a GRADE approach, Eur J Integr Med 8 (5) (2016) 854–860.

[29]   M Kim, Y Yun, K seok Kim, I Choi, Three cases of atopic dermatitis in pregnant women successfully treated with Korean medicine, Complement Ther Med 21 (5) (2013) 512–516.

[30]   TK Karagounis, JK Gittler, V Rotemberg, KD Morel, Use of "natural" oils for moisturization: review of olive, coconut, and sunflower seed oil, Pediatr Dermatol 36 (1) (2018) 9–15.

[31]   MT Vangelista, F Abad-Casintahan, L Lopez-Villafuerte, The effect of topical virgin coconut oil on SCORAD index, transepidermal water loss, and skin capacitance in mild to moderate pediatric atopic dermatitis: a randomized, double-blind, clinical trial, Int J Dermatol 53 (1) (2014) 100–108.

[32]   A Cooke, MJ Cork, S Victor, M Campbell, S Danby, J Chittock, T Lavender, Olive oil, sunflower oil or no oil for baby dry skin or massage: a pilot, assessor-blinded, randomized controlled trial (the Oil in Baby SkincaRE [OBSeRvE] Study), Acta Derm Venereol 96 (2016) 323–330.

[33]   SG Danby, T AlEnezi, A Sultan, T Lavender, J Chittock, K Brown, MJ Cork, Effect of olive and sunflower seed oil on the adult skin barrier: implications for neonatal skin care, Pediatr Dermatol 30 (2013) 42–50.

[34]   PM Elias, B BE, VA Ziboh, The permeability barrier in essential fatty acid deficiency: evidence for a direct role for linoleic acid in barrier function, J Invest Dermatol 74 (1980) 230–233.

[35]   Z Friedman, SJ Shochat, MJ Maisels, KH Marks, EL Lamberth, Correction of essential fatty acid deficiency in newborn infants by cutaneous application of sunflower-seed oil, Pediatrics 58 (1976) 650–654.

[36]   J Fingleton, D Sheahan, A Corin, M Weatherall, R Beasley, A randomised controlled trial of topical Kanuka honey for the treatment of psoriasis, JRSM Open 5 (3) (2014) 2042533313518913.

[37]   NS Al-Waili, Topical application of natural honey, beeswax and olive oil mixture for atopic dermatitis or psoriasis: partially controlled, single-blinded study, Complement Ther Med 11 (2003) 226–234.

[38]   N Dabholkar, VK Rapalli, G Singhvi, Potential herbal constituents for psoriasis treatment as protective and effective therapy, Phytother Res 11 (2020) 1–6.

[39]   A Rendon, K Schäkel, Psoriasis pathogenesis and treatment, Int J Mol Sci 20 (6) (2019) 1–28.

[40]   A Cordan Yazıcı, B Ünlü, G İkizoğlu, Complementary and alternative medicine use among patients with psoriasis on different treatment regimens, Arch Dermatol Res 312 (8) (2020) 601–604.

[41]   M Miroddi, M Navarra, F Calapai, F Mancari, SV Giofrè, S Gangemi, G Calapai, Review of clinical pharmacology of Aloe vera L. in the treatment of psoriasis, Phytother Res 29 (5) (2015) 648–655.

[42]   C Choonhakarn, P Busaracome, B Sripanidkulchai, P Sarakarn, A prospective, randomized clinical trial comparing topical Aloe vera with 0.1% triamcinolone acetonide in mild to moderate plaque psoriasis, J Eur Acad Dermatol Venereol 24 (2010) 168–172.

[43]   SP Dhanabal, L Priyanka Dwarampudi, N Muruganantham, R Vadivelan, Evaluation of the antipsoriatic activity of Aloe vera leaf extract using a mouse tail model of psoriasis, Phytother Res 26 (2012) 617–619.

[44]   EG Richard, The science and (lost) art of Psoralen plus UVA phototherapy, Dermatol Clin 38 (2020) 11–23.

[45]   N Kürkçüoğlu, F Alaybeyi, Topical capsaicin for psoriasis, Brit J Dermatol 123 (4) (2006) 549–550.

[46]   W Wang, B Su, S Tan, J Wang, Effect of substance P and epidermal growth factor receptor on the pathogenesis of early psoriasis, Journal of Xi'an Jiaotong University (Medical Sciences) 2 (2004) 130–131.

[47]   W Bylka, P Znajdek-Awiżeń, E Studzińska-Sroka, M Brzezińska, Centella asiatica in cosmetology, Postepy Dermatol Alergol 30 (1) (2013) 46–49.

[48]   Q OuYang, YQ Pan, HQ Luo, CX Xuan, JE Liu, J Liu, MAD ointment ameliorates Imiquimod-induced psoriasiform dermatitis by inhibiting the IL-23/IL-17 axis in mice, Int Immunopharmacol 39 (2016) 369–376.

[49]   TP Prema, N Raghuramulu, Free vitamin D3 metabolites in Cestrum diurnum leaves, Phytochemistry 37 (1994) 677–681.

[50] L Barrea, MC Savanelli, C Di Somma, M Napolitano, M Megna, A Colao, S Savastano, Vitamin D and its role in psoriasis: an overview of the dermatologist and nutritionist, Rev Endocr Metab Disord 18 (2) (2017) 195–205.

[51] A Manzoor, U Qaisar, Z Parveen, S Siddique, AA Sardar, N Ishaq, In vitro anticancer and antioxidant potential of Cestrum species, Pak J Pharm Sci 33 (4) (2020) 1535–1541.

[52] P Bahraini, M Rajabi, P Mansouri, G Sarafian, R Chalangari, Z Azizian, Turmeric tonic as a treatment in scalp psoriasis: a randomized placebo-control clinical trial, J Cosmet Dermatol 17 (3) (2018) 461–466.

[53] G Sarafian, M Afshar, P Mansouri, J Asgarpanah, K Raoufinejad, M Rajabi, Topical turmeric microemulgel in the management of plaque psoriasis; a clinical evaluation, Iran J Pharm Res 14 (3) (2015) 865–876.

[54] KK Singh, S Tripathy, Natural treatment alternative for psoriasis: a review on herbal resources, J Appl Pharm Sci 4 (11) (2014) 114–121.

[55] M Wiesenauer, R Lüdtke, Mahonia aquifolium in patients with Psoriasis vulgaris—an intraindividual study, Phytomedicine 3 (1996) 231–235.

[56] M Bonesi, MR Loizzo, F Menichini, R Tundis, Flavonoids in treating psoriasis, in: S Chatterjee, W Jungraithmayr, D Bagchi (Eds.), Immunity and Inflammation in Health and Disease: Emerging Roles of Nutraceuticals and Functional Foods in Immune Support, Elsevier Science Publishing Co Inc., San Diego, CA, 2017, pp. 281–294.

[57] FC Asogwa, A Ibezim, F Ntie-Kang, CJ Asogwa, COB Okoye, Anti-psoriatic and immunomodulatory evaluation of psorospermum febrifugum spach and its phytochemicals, Sci Afr 7 (2020) e00229.

[58] S Sabir, M Arsshad, S Asif, SK Chaudhari, An insight into medicinal and therapeutic potential of Silybum marianum (L.) Gaertn, Int J Biosci 4 (2014) 104–115.

[59] X Yuan, N Li, M Zhang, C Lu, Z Du, W Zhu, D Wu, Taxifolin attenuates IMQ-induced murine psoriasis-like dermatitis by regulating T helper cell responses via Notch1 and JAK2/STAT3 signal pathways, Biomed Pharmacother 123 (2020) 109747.

CHAPTER 16

Role of herbal products as therapeutic agents against ultraviolet radiation-induced skin disorders

Sajeeda Archoo[a,b], Shahid H. Naikoo[a,b] and Sheikh A. Tasduq[a,b]

[a]*Academy of Scientific & Innovative Research (AcSIR), Ghaziabad-201002, UP, India* [b]*PK-PD and Toxicology Division, CSIR-Indian Institute of Integrative Medicine, Jammu, Jammu and Kashmir, India*

16.1 Introduction

Skin, being the largest and most apparent organ of the human body, represents one's overall health and appearance, which has fueled a centuries-old desire to look young and vibrant. Skin ages inexorably and subtly as a result of the dual advancement of intrinsic and external causes. The epidermis, dermis, and subcutaneous fat are the three layers that make up the skin. It protects against pathogens, maintains fluid and temperature homeostasis by acting as a barrier to the environment [1]. The epidermis contains no nerves and blood vessels, thus depending on the dermal layer for metabolism. The epidermis consists of hardened cells or stratified squamous epithelium which plays an important role in protective function of skin. In the deepest layer of epidermis are melanin producing cells called Melanocytes. Melanin shields the skin from harmful ultraviolet radiations [2]. When the skin is subjected to environmental stress, epidermal stem cells replace injured surface keratinocytes on regular basis. Dermis is the skin's main living tissue which provides nutrients and physical support to the epidermis. The basic substances present in the dermis are elastin, collagen and fibrillin, all three fall off with age [3]. The dermis includes sweat glands, sebaceous glands, nerve endings, blood vessels, and hair follicles scattered amid fibroblast cells. Type I collagen fibers, which are made from procollagen bodies, offer structural integrity, healing, and strength to the skin's dermal layer. Genetics, environmental exposure (mainly UV radiation), metabolic processes (production of reactive chemical compounds such as ROS, sugars, and aldehydes) and hormonal changes are collectively responsible for aging of skin. When these elements are combined, they cause long-term changes in skin function, structure and appearance. The impact of the environment, particularly UV irradiation, on skin aging is most significant.

Herbal Medicines: A Boon for Healthy Human Life.
DOI: https://doi.org/10.1016/B978-0-323-90572-5.00030-5

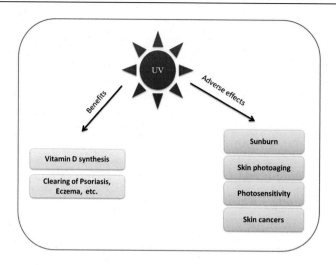

Figure 16.1
Effects of UV on human skin.

The phrase ultraviolet means "beyond violet," and refers to wavelengths greater than X-rays but shorter than visible violet light, as defined by the light spectra. UVR can be helpful when used to treat diseases like psoriasis and creation of pre-vitamin D3 from 7-dehydrocholestrol during an electro cyclic reaction in the epidermal layer of skin but UVR can also be detrimental for human skin (Fig. 16.1). Ultraviolet irradiation leads to skin photoaging, causes or worsens many of the skin associated disorders [4]. The ultraviolet radiation spectrum is categorized into three parts (UVA, UVB, & UVC) with each part having a distinct effect on skin. UVC (100–280) is presently irrelevant to skin diseases because it cannot penetrate the atmospheric ozone layer. All of the UVB radiation is absorbed by the skin epidermis and nearly about 30% of UVA reaches the skin dermis. The UVB radiation (280–315nm) leads to sunburn. UVA (315–400nm) tans and ages the skin (Fig. 16.2). Moreover, different types of human skin (classified into six types by Fitzpatrick), react differently to ultraviolet radiations (Table 16.1). Thus, each type of skin requires different degrees of protection against the sun. UV radiations, particularly UVA and UVB, are known to impair the structural integrity and physiological function of skin, inducing photo-aging or premature skin aging. Photoaging is characterized by uneven pigmentation, decreased suppleness, brown patches, and wrinkle growth [4,5]. Photoaging occurs in addition to chronological aging. Photoaging and chronological skin aging have always been thought of as separate things. Although the representative appearance of chronologically aged and photoaged human skin can be easily distinguished, current evidences indicate that the two share major molecular features, such as altered signal transduction pathways that decrease procollagen synthesis, promote matrix-metalloproteinase (MMP) expression, and connective tissue damage. UV exposure accelerates many major components of the chronological aging process in human skin, based on the agreement of

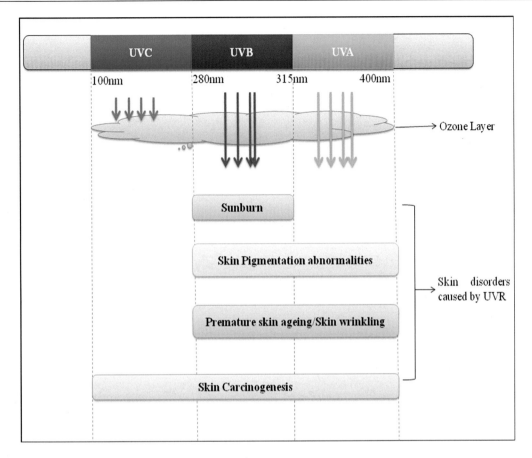

Figure 16.2
Skin disorders and UVR wavelengths that cause them. (Inspired from Clinical Dermatology, *Fourth Edition* © 2008 Richard Weller *et al.* ISBN: 978-1-405-14663-0).

molecular pathways. UV-B destroys biological macromolecules such as deoxyribonucleic acid (DNA), lipids and proteins in a direct way. It also lowers the activity of antioxidant enzymes like superoxide dismutase (SOD) and glutathione peroxidase (GSH-Px), causing the skin to produce reactive oxygen species (ROS). UV protection is mostly achieved by the use of topical sunscreens. Sunscreens work by reflecting, absorbing or dispersing ultraviolet (UV) rays to protect against sunburn and skin cancers. Salicylate-containing sun protection creams were first used in the 1920s, while petroleum was utilized as an active component in the 1930s and 1940s. Sun-protective property of para-amino-benzoic acid (PABA) was identified in the 1940s, but it is rarely used nowadays because of the high risk of allergic contact of dermatitis. In the 1950s, minerals such as titanium dioxide were available. These early sunscreens were among the first to provide UV-B protection.

Table 16.1: Fitzpatrick skin types and their response to ultraviolet radiation (UVR).

Fitzpatrick phototype	Phenotype	Cutaneous response to UVR
I	Unexposed skin is bright white Red/ blond Hair, Pa Blue/light green eyes Fling frequent	Always burns and never tans Peels Frequent Freckling
II	Unexposed skin is white Blue, hazel, or brown eyes Red, blonde, or brown hair	Burns easily Peels Tans poorly
III	Unexposed skin is fair Brown eyes Dark/light brown hair	Burns moderately Average tanning ability
IV	Unexposed skin is light brown Dark eyes Dark hair Mediterranean	Burns minimally Tans easily
V	Unexposed skin is brown Dark eyes Dark hair East India	Rarely burns Tans easily and substantially
VI	Unexposed skin is black deeply pigmente Dark eyes Dark hair	Almost never burns Tans darkly and profusely

16.2 Herbs and phtoprotection

Herbal extracts and plant chemicals have traditionally been used to regenerate the skin, heal skin disorders and protect against UV radiations [6]. Herbal medications that provide protection from the harmful effects of UV radiation are becoming increasingly popular. They may have less adverse effects than chemical sunscreens, which make them appealing to patients who prefer natural products. Medical plant has a variety of chemical ingredients that are both ecologically friendly and chemically complex (e.g., vitamins, proteins, polysaccharides, fatty acids, terpenoids, flavonoids, tannins, glycosides, etc.). Plants produce large amounts of photo-protective phytoconstituents and antioxidative enzymes in response to UVB photons. As a result, a higher concentration of these phytoconstituents would be available per unit of harvested plant tissue, raising the economic status of herbal medications. The most promising natural topical preventions for photoaging are the secondary metabolites biosynthesized as a response to UV radiations in plant cells, such as glycosylated flavonoids, phenyl propanoids and their glycosylated metabolites, leontopodic acids and aglycons. Plants have evolved the biosynthesis of secondary metabolites, which are not involved in their growth, division, and spreading, but are essential for signal transduction, cell functional coordination, environmental defense and adaptation to stresses. This UV protective system in plants is similar to the cutaneous photo-chemical human patterns of photo-screens (Eumelanin, melatonin and aromatic amino acid proteins) and photosensitis agents (pheomelanin, porphyrines, pyrimidines, flavins, and hemoglobin). The barrier's effectiveness depends on the balance of photoprotectors/photo sensitis, which deteriorates progressively in the aging or stressful skin. The plants are now regarded significant sources for the development of cosmeceutical and topical medicinal products for photo-aging and photoprotection.

Secondary metabolites from plants are thought to regulate the interaction of skin and ultraviolet radiations at various critical steps:

(i) Plant secondary metabolites absorb UVA and UVB radiations (screen action)
(ii) Secondary metabolites scavenge UVR induced free radicals and subsequently prevent the extracellular matrix responses towards UVR in skin cells (chain breaking action).
(iii) By providing protection to lipid antioxidants present on skin surface, which include squalene, alpha-tocopherol, and coenzyme Q10, (rescue antioxidant activity).
(iv) By activation of internal antioxidant systems in epidermal keratinocytes and dermal fibroblasts (indirect antioxidant action);
(v) By mitigating the inflammatory response of immune cells present in skin.
(vi) By regulating the UVR mediated proliferative and exorbitant metabolic stress responses.

The essential secondary metabolites such as flavonoids, phenolic acids, terpenes, polyphenols and amino acids act as UV radiation blockers. In addition to their ability to absorb UV radiation, some natural chemicals have been discovered to lower inflammation, oxidative stress, and influence numerous signaling pathways, all of which help to protect the skin from UV-induced damage. As a result of their engagement in several phototoxicity pathways, naturally occurring compounds have gotten a lot of interest for their potential use as effective photo-damage reducers in humans via different pathways. In subsections below we will talk about some of the important herbal natural compounds (Table 16.2) and extracts (Table 16.3) that have scientifically proven photo protective properties.

16.3 Plant bioactive molecules

16.3.1 Silymarin

Silymarin is a standardized flavonolignan extract derived from the seeds of *Silybum marianum*, a milk thistle. Silybin, silidianin, silychristin, and isosilybin are the major components of silymarin. Silybin is thought to be the most physiologically active component of silymarin extract in terms of antioxidant and anti-inflammatory activity [7]. Topical application of silymarin prevents skin from edema and sunburn and prevents the cells from UVR induced apoptosis and has shown a remarkable antitumor effect in mice [8]. Topical application of silymarin was found to provide protection from UVB induced DNA damage as it attenuated the production of cyclobutane-pyrimidine dimers in mice skin. Further, inducible nitric oxide

Table 16.2: Natural plant derived compounds for skin protection against UVR.

Compound	Activity
Silymarin	Protection from UVB induced DNA damage, skin edema, sunburn and UVR induced apoptosis
Tea polyphenoids	Protection against UVB-induced cutaneous edema and erythema, prevent depletion of the epidermal antioxidant-defense enzyme system and lipid peroxidation.
Inhibits the production of CPDs, provides protection against DNA damage.
Protection against human epidermal skin cancers by arresting the transformed cells in G0–G1 phase of cell cycle and promoting their apoptosis.
Genistein	Reduces UVB-induced hydrogen peroxide production, edema and contact hypersensitivity.
Resveratrol	Inhibits tumor initiation, development, and promotion. Inhibits the production of hydrogen peroxide.
Caffeine	Protects phospholipidic membranes from UV-induced peroxidation and prevents human skin from erythema.
Tannin	Protect from UVR mediated free radical damage and reduce the risk of premature aging and skin cancers.
Quercetin	Protects the antioxidant system of skin such as glutathione reductase, glutathione peroxidase catalase, and superoxide dismutase.
Apigenin	Inhibits the cell cycle and cyclin dependent kinases, arrests the cycle cycle in G2/M phase and increases the expression of p53 tumor suppressor protein.
Curcumin	Decrease protein expression of matrix metalloproteinase (MMP) -1 and MMP-3. Inhibits the ROS accumulation.
Kaempferol	Kaempferol inhibits the proliferation of cancerous cells by disrupting the cell cycle at check points.
Vitamin E	Reduce the skin roughness and depth of wrinkles.
Vitamin C	Limits the UVR induced DNA damage, reduces the proinflammatory cytokine release and inhibits apoptosis.
Trigonelline	Prevents the UV-B induced oxidative stress and the resulting ER- stress.
Glycyrrhizic acid	Prevent DNA damage, attenuates the UVB induced apoptosis via inhibiting the translocation of Bax to mitochondria.

synthase-expressing cells and the number of UVB radiation induced hydrogen peroxide-producing cells were also markedly reduced. In cultured human keratinocytes Silymarin inhibited the activation of NF-κB and demonstrated a dose-dependent protective effect against UV radiation-induced skin damage.

16.3.2 Tea polyphenoids

Polyphenols in Tea (Camelia sinensis), one of the world's most popular beverages, is grown in around 30 countries. Epicatechins are the principal antioxidant polyphenols found in black and green tea. (-)-epicatechin (EC), (-)-epicatechin-3-gallate (ECG), (-)-epigallocatechin (EGC), and (-)-epigallocatechin-3-gallate (EGCG) are the primary epicatechins found in green tea. EGCG accounts for 60–70% of the total amount of catechins. These polyphenols have proven to be anti-inflammatory or anticarcinogenic agents [9]. Among all these EGCG is

Table 16.3: Herbal extracts with protective properties against UVR induced damage.

Extract	Effects
Ginkgo biloba extract	Possess antioxidant property, used for treatment of sunburn, stroke, aging, and drug-induced toxicities.
Grape seed extract (GSE)	Possess oxygen free radical scavenging effect, protect lipids from UVB and UVC irradiation induced oxidation, oxidative stress, tissue damage and skin ageing.
Prunus persica flower extract	DNA damage and lipid peroxidation.
Aloe Vera	Augments the collagen production, maintains the membrane integrity.
Walnut	Efficient self-tanning agent, inhibit oxidative damages, inflammation, photo-aging, and tumor growth
Orange	Antioxidant property, prevent UVB induced erythema. Improve physiological antioxidant defenses, decrease the age spots pigmentation
Ginseng	Has antioxidant, anti-inflammatory and antitumor effects, increases the production of procollagen I and suppresses MMP-1 production, inhibits the expression of pro-inflammatory cytokines
Soybean	Antioxidant, anti-inflammatory, depigmentation and antiaging effects, inhibits the formation of CPDS and prevents DNA damage
Neem	Antioxidant, antimicrobial, and anti-inflammatory properties, increases type I procollagen and elastin production, decrease the expression of matrix metalloproteinase

most efficient in protecting against cutaneous inflammatory or carcinogenic responses. Green tea polyphenols (GTP) provide considerable protection against UVB-induced lipid peroxidation, depletion of the epidermal antioxidant-defense enzyme system, cutaneous edema and erythema, and the production of prostaglandin metabolites in hairless mice. Application of GTP to human skin prior to UVB irradiation inhibited the production of CPDs thus provides protection against DNA damage. Polyphenols from green tea have been shown to provide protection against human epidermal skin cancers by arresting the transformed cells in G0–G1 phase of cell cycle and promoting their apoptosis [10].

16.3.3 Genistein

Many herbs contain genistein (5,7,4′-trihydroxyisoflavone). Soybeans, in particular, are a rich source of genitein. Genistein has the ability to regulate cell cycle and prevent cancers [11]. The topical application of genistein and its metabolites namely, equol, isoequol, and dehydroequol significantly reduces UVB-induced hydrogen peroxide production, edema and contact hypersensitivity. Among the metabolites of genistein equol was found to be most effective.

16.3.4 Resveratrol

Resveratrol (trans-3,5,4′-trihydroxystilbene) is a polyphenolic phytoalexin that is present in abundance in the skin and grape seeds [7]. Peanuts and berries are among other plant sources. Resveratrol is a powerful antioxidant with anti-inflammatory and antiproliferative properties. Resveratrol inhibits a range of cellular processes involved in tumor initiation, development, and promotion [12]. In SKH-1 hairless mice, resveratrol reduces the UVB induced skin edema, inhibits the production of hydrogen peroxide and leukocyte infiltration. Moreover, resveratrol inhibits the UVB-mediated activation of NF-κB in human epidermal keratinocytes.

16.3.5 Caffeic acid

Caffeic acid (3,4-dihydroxycinnamic acid) is primarily found conjugated with saccharides in all plant species such as coffee, fruits, cereals and other medicinal plants. Caffeic acid has anti-inflammatory, antioxidant and anticarcinogenic activity. This phenolic acid protects phospholipidic membranes from UV-induced peroxidation and prevents human skin from erythema [13].

16.3.6 Tannin

Tannins are polyphenolic chemicals found across the plant kingdom that are thought to offer plants with a chemical defense against predators and UV radiations. Tannins are classified as condensed tannins (CTs) and Hydrolysable tannins (HTs). Both these groups are potent antioxidants and are thought to protect from UVR mediated free radical damage and reduce the risk of premature aging and skin cancers [14].

16.3.7 Quercetin

One of the most abundant natural flavonoids is quercetin (3,5,7,3′,4′-pentahydroxyflavon). It is found in variety of common fruits and vegetables (grapes, apples, lemons, onions, tomatoes, lettuce, kale, broccoli, cottonseed, and so on) beverages (tea, red wine), herbs (Apocynumvenetum, Gingko biloba Poacynumhendersonii, olive oil, Opuntia ficusindica), and propolis from bee hives. Quercetin possesses a strong antioxidant activity. It is thought to prevent or at least alleviate the detrimental effects of UV radiation on human skin [15]. Quercetin is proven to prevent UVA induced aging in rats by protecting the antioxidant system of skin such as glutathione reductase, glutathione peroxidase catalase, and superoxide dismutase.

16.3.8 Apigenin

Apigenin (5,7,4′-trihydroxyflavone) is a flavonoid, found in fruits (e.g., cherries, apples, grapes), herbs (e.g., endive), vegetables (e.g., beans, broccoli, celery, onions, barley, tomatoes), and beverages (e.g., tea, wine). Apigenin inhibits the cell cycle and cyclin dependent kinases. When applied topically, it reduces tumor incidence and increases tumor free survival by inhibiting UV-mediated production of ornithine decarboxylase activity [16]. Apigenin arrests cell-cycle in G1 phase by decreasing cdk2 kinase activity in UV irradiated fibroblasts. Whereas in mice keratinocytes apigenin arrests the cycle in G2/M phase and increases the expression of p53 tumor suppressor protein.

16.3.9 Curcumin

Curcumin (diferuloylmethane) is a yellow pigment found in the rhizome of *Curcuma longa* (turmeric). Most of the therapeutic benefits of curcumin have been attributed to its anti-inflammatory and antioxidant property [17]. Several studies have reported the photoprotective effects of curcumin against the UVB induced skin carcinogenesis. Curcumin when applied topically prior to UVB exposure has been reported to delay the tumor appearance, multiplicity and size. Furthermore, curcumin protects the human dermal fibroblasts from UVA induced photoaging by inhibiting the ROS accumulation, restoring the activity of antioxidant defense enzymes and attenuating UVA-induced ER stress, and apoptotic signaling. Additionally, curcumin could decrease expression of matrix metalloproteinase (MMP) -1 and MMP-3 thus regulating the collagen metabolism in UVA irradiated Human dermal fibroblasts [18].

16.3.10 Kaempferol

Kaempferol is a member of flavanols, abundantly found in tea, beans, broccoli, apples, and strawberries. It has also been found in medicinal herbs such as Aloe vera (L.), Delile, Euphorbia pekinensis Rupr, Ginkgo biloba L, and Rosmarinus officinalis.Kaempferol has anti-inflammatory and antioxidant activity [19]. It is a potent promoter of apoptosis and regulates a range of signaling pathways inside cells. Kaempferol is relatively less toxic to normal cells than the standard cancer chemotherapy. Moreover, Kaempferol inhibits the proliferation of cancerous cells by disrupting the cell cycle at check points. Kaempferol has a lot of potential as a new chemopreventive drug, and it could be used to treat UVB-related carcinogenesis.

16.3.11 Vitamin E

Vitamin E includes both tocopherols and tocotrienols. It has been reported that both acute and chronic photodamage is reduced by the topical application of these antioxidants. Topical application of natural forms of vitamin E alpha-tocopherol and tocotrienol substantially reduce the skin roughness and depth of wrinkles [20]. Alpha-tocopherol prevents the collagen degradation by reducing the expression of enzyme collagenase. Vitamin E appears to protect dermal fibroblasts and epidermal keratinocytes from UVB damage and it has been used to protect skin from UV damage. Alpha tocotrienol shielded keratinocytes against UVB induced inflammation. UVB induced DNA damage in human epidermis was mitigated by the antioxidant alpha-tocopherol. Many sunscreens and lotions in the market now contain Vitamin E as important antioxidant.

16.3.12 Vitamin C

Vitamin C (L-ascorbic acid) is an essential molecule for skin health because of its antioxidant property and role in synthesis of collagen. Several studies have reported that vitamin C helps in the prevention and treatment of UV induced photodamage [21]. Vitamin C in cultured keratinocytes limits the UV induced DNA damage, reduces the pro-inflammatory cytokine release and inhibits apoptosis. In rodents addition of ascorbic acid to diet have shown to reduce chronic UV induced number and size of dermal neoplasm and skin tumors. Furthermore, in reconstituted human epidermal surface Pre- and post-treatment with ascorbic acid prevents UVB induced cell death, reactive oxygen species (ROS) production, DNA damage, and the inflammatory response by reducing the expression and release of tumor necrosis factor-α (TNF-α) [22].

16.3.13 Trigonelline

Trigonelline a heterocyclic alkaloid particularly found in the seeds of fenugreek & coffee have broad spectrum of pharmalogical activities. Trigonelline has been reported to have photoprotective effect in human dermal fibroblast cells & Balb/C mice irradiated with UV-B radiation [23]. Trigonelline possess antioxidant activity and it prevents the UV-B induced oxidative stress and the resulting ER- stress. Trigonelline was also found to maintain calcium homeostasis in UVB irradiated human dermal fibroblasts.

16.3.14 Glycyrrhizic acid

Glycyrrhizic acid, a triterpenoid saponin glycoside found in the roots and rhizomes of licorice plant (Glycyrrhiza glabra) is one of the most widely known medicines since ancient times. Glycyrrhizic acid has been reported to have antiphotoaging property in Ultraviolet-B irradiated human dermal fibroblasts by scavenging the ROS and blocking the MMP1 activation [24]. Glycyrrhizic acid inhibits the UVB induced production of pro-inflammatory cytokines tumor necrosis factor (TNF)-a, interleukin (IL)-1a, -1b and -6, and prostaglandin E2 (PGE2). Additionally, glycyrrhizic acid attenuates the UVB induced apoptosis via inhibiting the translocation of Bax to mitochondria. Glycyrrhizic acid has also been found to prevent the UVB induced DNA damge (cyclobutane pyrimidine dimers formation and DNA fragmentation) via modulation of autophagic flux in Human dermal fibroblasts [25].

16.4 Plant extracts

16.4.1 Ginkgo biloba extract (EGb 761)

Ginkogo biloba is a native tree to china and the only living member of ancient plant phylum, so sometimes referred as living fossil. Its seeds and leaves have been traditionally used in

Chinese medicine, while modern research is focused on Ginkgo extract which is made from the green leaves of the tree. This extract is a natural blend of flavone glycosides (33%) and terpenes (mainly quercetin and kaempferol derivatives) (6 percent). EGb 761 is a standardized leaf extract of *Ginkogo biloba*. It possesses antioxidant property and has been used in humans for treating pathologic illnesses such as sunburn, stroke, aging, and drug-induced toxicities [26].

16.4.2 Grape seed extract (GSE)

Grape seed extract is made by removing, drying and pulverizing the seeds of grapes. Grapes are high in polyphenols, with grape seeds accounting for 60–70% of total polyphenols. Catechin, epicatechin, and oligomeric proanthocyanidins are flavan-3-ol derivatives found in grape seed polyphenols (GSP).Grape seeds are rich source of antioxidants and have antiinflammatory, antiapoptotic, and anticarcinogenic properties. GSP has been demonstrated to have a much better ROS scavenging effect than vitamin E and C, as well as to protect lipids from UVC and UVB irradiation induced oxidation, oxidative stress, tissue damage and skin aging [27,28].

16.4.3 Prunus persica flower extract

Prunus persica flowers have long been used in East Asia to treat skin problems. Multiflorin B, trifolin, afzelin, and astragalin are four kaempferol glycoside compounds found the P. persica flowers. The flower extract possesses the property to protect keratinocytes and fibroblasts from UVR induced cytotoxicity and reduces the amount of 14C-arachidonic acid/metabolites generated by UVB-irradiated human keratinocytes and inhibits UVC/UVB-induced lipid peroxidation and DNA damage in skin fibroblasts [29]. Topical application of Prunus persica flower extract to guinea pigs reduces the UVB-induced erythema [29].

16.4.4 Aloe vera

Aloe barbadens is a well-known herbal plant that people have been using from decades. Aloe vera is a short stemmed herb that stores water in its leaves. Aloe vera has several medicinal benefits but is eminent in treating skin injuries. Aloe vera has been scientifically demonstrated to help with all types of burns, including radiation, thermal, and sun. It has also been shown that using it before, during, and after these skin-damaging events has a preventive effect. Aloe vera's natural chemical contents can be divided into the following categories: Amino acids, enzymes, anthraquinones, lignin, mono- and polysaccharides, minerals, salicylic acid, sterols, saponins, and vitamins. Aloe vera is known to not only improve the shape of fibroblast cells, but it also augments the collagen production. Aloe vera is a highly powerful skin moisturizer and healing agent [30]. Aloe Vera exhibits a remarkable ability to reduce both invitro and in vivo photodamage. Interestingly, the protection conferred by Aloe Vera is associated with its ability to maintain the membrane integrity of both mimetic membranes and intracellular organelles [31]. Possibly this is why extracts of Aloe Vera safeguard against photodamage at cellular level in both the visible and UV light spectra, leading to its advantageous use as a supplement in protective dermatological formulations. Although isolated plant compounds have a high potential in skin protection, whole herb extracts showed better efficacy due to their complex composition.

16.4.5 Walnut

Walnut (*Juglans regia*) is a reach source of phytochemicals and contains a notable amount of flavonoids, phenols, sterols, tannins, terpenes, phospholipids, essential fatty acids, sphingolipids, hydrocarbons, quinines, and oils. The aqueous extract from the fresh green shells of walnut has been reported to be efficient self-tanning agent. The major chemical component present is Juglone. Juglone reacts with skin keratin proteins and form colored sclerojuglonic compounds, which have UV protection properties. Moreover, the extracts from J. regia L. have been shown to inhibit oxidative damages, inflammation, photo-aging and tumor growth. Additionally, the oil of J. regia nut is an important ingredient of dry skin creams, antiwrinkle and antiaging products [32].

16.4.6 Orange

Red orange (*Citrus snensis*) extracts (ROE) have high levels of anthocyanins, flavanones, ascorbic acid and hydroxycinnamic acids all of these are promising antioxidants. This extract offers excellent photoprotection to skin and is thought to be beneficial in diseases where oxidative stress plays a part. It has been demonstrated that topically applied red orange extract can prevent UVB irradiation induced erythema in healthy humans [33]. Moreover, ROE intake can bolster up the physiological antioxidant defenses of skin, decrease the age spots pigmentation and protect skin from the damaging processes involved in photo-aging, leading to healthy appearance and pigmentation of skin [34].

16.4.7 Ginseng

Panax ginseng is a well-known medicinal plant traditionally used for treatment of many diseases including skin disorders due for its promising healing and restorative properties. Ginsenosides and saponins are the principal constituents responsible for the biological activities of ginseng which include antioxidant, anti-inflammatory, and antitumor effects [35]. Ginseng is usually of two different types, white ginseng and red ginseng based on its processing conditions, each of them is believed to have different bioactive compounds. White ginseng is made by peeling and sun drying of fresh ginseng roots without steaming, increases the production of procollagen I and suppresses MMP-1 production in UVB-irradiated hairless mice and Human Dermal Fibroblasts. Red ginseng which is made by steaming and drying of fresh ginseng improves atopic dermatitis by inhibiting the expression of pro-inflammatory cytokines in mice model [36].

16.4.8 Soybean extract

Soybean (*Glycine max*) is legume species native to East Asia. Soybean extracts are popularly known as nutritional and medicinal ingredients. The functional constituents of soybean include flavonoids, isoflavonoids, tannins, phenolic acids and proanthocyanidins. Soy extracts are beneficial to human skin for their antioxidant, anti-inflammatory, depigmentation and anti-aging effects [37]. Soyabean extract inhibits the formation of CPDS in UVB irradiated human keratinocytes and prevents cells from possible DNA damage. It also reduces the relative ratio of MMP1/ TIMP-1 mRNA, prevents collagen from degradation and protects cells from UV radiation [38].

16.4.9 Neem extract

Azadirachta indica A.Juss. (Neem) is an evergreen tree, used for treating skin disorders in Indian subcontinent from decades. Neem leaf extract possesses various pharmacological activities including antioxidant, antimicrobial, and anti- inflammatory properties. Nimbin, Nimbidine, Quericetin, and B-sitosterol are the vital constituents of neem leaf [39]. Neem leaf extract have been shown to prevent UVB irradiation induced photo-aging in Human dermal fibroblasts and Hairless mice. Neem leaf extract increases type I procollagen and elastin production by regulating the cellular pathways (such as, increasing the transforming growth factor TGF-β1, downregulating MAPK pathway and AP-1 protein expression) that decrease the expression of matrix metalloproteinase thus provides protection against UVB radiation exposure [40].

Acknowledgements

The authors would like to acknowledge Council of Scientific and Industrial Research (CSIR), New Delhi and University Grants Commission (UGC), New Delhi for their financial assistance.

References

[1] CA Brohem, LB Cardeal, M Tiago, MS Soengas, SB Barros, SS Maria-Engler, Artificial skin inperspective: concepts and applications, Pigment Cell Melanoma Res 24 (2011) 35–50.

[2] F Solano, Photoprotection and skin pigmentation: melanin-related molecules and some other new agents obtained from natural sources, Molecules 25 (7) (2020) 1537.

[3] K Scharffetter–Kochanek, P Brenneisen, J Wenk, G Herrmann, W Ma, L Kuhr, et al., Photoaging of the skin from phenotype to mechanisms, Exp Gerontol 35 (3) (2000) 307–316 1.

[4] P Kullavanijaya, HW Lim, Photoprotection, J Am Acad Dermatol. 52 (6) (2005) 937–958.

[5] EM Gil, TH Kim, UV-induced immune suppression and sunscreen, Photodermatol Photoimmunol Photomed 16 (3) (2000) 101–110.

[6] M Cavinato, B Waltenberger, G Baraldo, CV Grade, H Stuppner, P Jansen-Dürr, Plant extracts and natural compounds used against UVB-induced photoaging, Biogerontology 18 (4) (2017) 499–516.

[7] S Saraf, CD Kaur, Phytoconstituents as photoprotective novel cosmetic formulations, Pharmacogn Rev 4 (7) (2010) 1.

[8] SK Katiyar, NJ Korman, H Mukhtar, R Agarwal, Protective effects of silymarin against photocarcinogenesis in a mouse skin model, J Natl Cancer Inst 89 (8) (1997) 556–565.

[9] SK Katiyar, CA Elmets, Green tea polyphenolic antioxidants and skin photoprotection, Int J Oncol 18 (6) (2001) 1307–1313.

[10] JA Nichols, SK Katiyar, Skin photoprotection by natural polyphenols: anti-inflammatory, antioxidant and DNA repair mechanisms, Arch Dermatol Res 302 (2) (2010) 71–83.

[11] HS Tuli, MJ Tuorkey, F Thakral, K Sak, M Kumar, AK Sharma, et al., Molecular mechanisms of action of genistein in cancer: recent advances, Front Pharmacol 10 (2019) 1336 6.

[12] LS Baumann, L Baumann, Cosmetic Dermatology, McGraw-Hill Professional Publishing, New York, New York, United States, 2009.

[13] A Saija, A Tomaino, RL Cascio, D Trombetta, A Proteggente, A De Pasquale, et al., Ferulic and caffeic acids as potential protective agents against photooxidative skin damage, J Sci Food Agric 79 (3) (1999) 476–480.

[14] HU Gali-Muhtasib, SZ Yamout, MM Sidani, Plant tannins as inhibitors of hydroperoxide production and tumor promotion induced by ultraviolet B radiation in mouse skin in vivo, Oncol Rep 6 (4) (1999) 847–900.

[15] B Choquenet, C Couteau, E Paparis, LJ Coiffard, Quercetin and rutin as potential sunscreen agents: determination of efficacy by an in vitro method, J Nat Prod 71 (6) (2008) 1117–1118.

[16] A Svobodová, J Psotová, D Walterová, Natural phenolics in the prevention of UV-induced skin damage. A review, Biomed Pap Med Fac Univ Palacky Olomouc Czech Repub 147 (2) (2003) 137–145.

[17] AM García-Bores, JG Avila, Natural products: molecular mechanisms in the photochemoprevention of skin cancer, Rev Latinoamer Quím 36 (2008) 83–102.

[18] X Liu, R Zhang, H Shi, X Li, Y Li, A Taha, et al., Protective effect of curcumin against ultraviolet A irradiationinduced photoaging in human dermal fibroblasts, Molecular medicine reports 17 (5) (2018) 7227–7237.

[19] MT Mercader-Ros, C Lucas-Abellán, MI Fortea, A Serrano-Martínez, JA Gabaldón, E Núñez-Delicado, Biological activities of kaempferol: effect of cyclodextrins complexation on the properties of kaempferol, Kaempferol: Chemistry, Natural Occurrences and Health Benefits, Nova Science Publishers, Inc., New York, NY, USA, 2013, pp. 1–31.

[20] P Mayer, W Pittermann, S Wallat, The effects of vitamin E on the skin, Cosmetics and toiletries 108 (2) (1993) 99–109.

[21] N Dayan, Skin Aging Handbook: An Integrated Approach to Biochemistry and Product Development (2008).

[22] S Kawashima, T Funakoshi, Y Sato, N Saito, H Ohsawa, K Kurita, et al., Protective effect of pre-and post-vitamin C treatments on UVB-irradiation-induced skin damage, Sci Rep 8 (1) (2018) 1–2.

[23] S Naikoo, SA Tasduq, Trigonelline, a naturally occurring alkaloidal agent protects ultraviolet-B (UV-B) irradiation induced apoptotic cell death in human skin fibroblasts via attenuation of oxidative stress, restoration of cellular calcium homeostasis and prevention of endoplasmic reticulum (ER) stress, J Photochem Photobiol, B 202 (2020) 111720.

[24] Q Afnan, MD Adil, A Nissar-Ul, AR Rafiq, HF Amir, P Kaiser, et al., Glycyrrhizic acid (GA), a triterpenoid saponin glycoside alleviates ultraviolet-B irradiation-induced photoaging in human dermal fibroblasts, Phytomedicine 19 (7) (2012) 658–664.

[25] SA Umar, MA Tanveer, LA Nazir, G Divya, RA Vishwakarma, SA Tasduq, Glycyrrhizic acid prevents oxidative stress mediated DNA damage response through modulation of autophagy in ultraviolet-B-irradiated human primary dermal fibroblasts, Cell Physiol Biochem 53 (1) (2019) 242–257.

[26] SE Dal Belo, LR Gaspar, MC PM, Photoprotective effects of topical formulations containing a combination of Ginkgo biloba and green tea extracts, Phytother Res 25 (12) (2011) 1854–1860.

[27] SD Sharma, SM Meeran, SK Katiyar, Dietary grape seed proanthocyanidins inhibit UVB-induced oxidative stress and activation of mitogen-activated protein kinases and nuclear factor-κB signaling in in vivo SKH-1 hairless mice, Mol Cancer Ther 6 (3) (2007) 995–1005.

[28] JA Vinson, MA Mandarano, DL Shuta, M Bagchi, D Bagchi, Beneficial effects of a novel IH636 grape seed proanthocyanidin extract and a niacin-bound chromium in a hamster atherosclerosis model, Mol Cell Biochem 240 (1) (2002) 99–103.

[29] MY Heo, SH Kim, HE Yang, SH Lee, BK Jo, HP Kim, Protection against ultraviolet B-and C-induced DNA damage and skin carcinogenesis by the flowers of Prunus persica extract, Mutat Res Genet Toxicol Environ Mutagen 496 (1-2) (2001) 47–59.

[30] A Barcroft, A Myskja, Aloe vera: nature's silent healer, Alasdair Aloe Vera; (2003).

[31] D Rodrigues, AC Viotto, R Checchia, A Gomide, D Severino, R Itri, et al., Mechanism of aloe vera extract protection against UVA: shelter of lysosomal membrane avoids photodamage, Photochem Photobiol Sci 15 (3) (2016) 334–350.

[32] LM Tewari, L Rana, SK Arya, G Tewari, N Chopra, NC Pandey, et al., Effect of drying on the essential oil traits and antioxidant potential J. regia L. leaves from Kumaun Himalaya, SN Appl Sci 1 (12) (2019) 1–9.

[33] A Saija, A Tomaino, R Lo Cascio, P Rapisarda, JC Dederen, In vitro antioxidant activity and in vivo photoprotective effect of a red orange extract, Int J Cosmet Sci 20 (6) (1998) 331–342.

[34] C Puglia, A Offerta, A Saija, D Trombetta, C Venera, Protective effect of red orange extract supplementation against UV-induced skin damages: photoaging and solar lentigines, J Cosmet Dermatol 13 (2) (2014) 151–157.

[35] JM Lu, Q Yao, C Chen, Ginseng compounds: an update on their molecular mechanisms and medical applications, Curr Vasc Pharmacol 7 (3) (2009) 293–302.

[36] E Hwang, ZW Sun, TH Lee, HS Shin, SY Park, DG Lee, et al., Enzyme-processed Korean Red Ginseng extracts protects against skin damage induced by UVB irradiation in hairless mice, J Ginseng Res 37 (4) (2013) 425.

[37] MK Waqas, N Akhtar, R Mustafa, M Jamshaid, HM Khan, G Murtaza, Dermatological and cosmeceutical benefits of Glycine max (soybean) and its active components, Acta Pol Pharm 72 (1) (2015) 3–11.

[38] NH Park, JS Park, YG Kang, JH Bae, HK Lee, MH Yeom, et al., Soybean extract showed modulation of retinoic acid-related gene expression of skin and photo-protective effects in keratinocytes, Int J Cosmet Sci 35 (2) (2013) 136–142.

[39] MA Alzohairy, Therapeutics role of Azadirachta indica (Neem) and their active constituents in diseases prevention and treatment, Evid Based Complement Altern Med (2016).

[40] HT Ngo, E Hwang, SA Seo, B Park, ZW Sun, M Zhang, et al., Topical application of neem leaves prevents wrinkles formation in UVB-exposed hairless mice, J Photochem Photobiol, B. 169 (1) (2017) 161–170.

CHAPTER 17

On bioactive compounds and the endophyte community in medicinal plants: Bioprocessing nature's abundance for skin disorder treatment

Siphiwe G. Mahlangu and Siew L. Tai

Centre for Bioprocess Engineering Research, Department of Chemical Engineering, University of Cape Town, Cape Town, South Africa

17.1 Skin overview

Human skin is an important and the largest organ of the human body, with a surface area between $1.5 - 2 \text{ m}^2$. The human skin consists of a sequence of crucial functions officiated by various physical and chemical reactions that occur inside it [1]. Some of these functions include excretory function, acting as a protective barrier by protecting the body from injuries, oxidation and foreign elements, regulation of the human body temperature, and forming an important component of our immune system by protecting us against microbial invasion [2,3].

Skin damage happens with excessive exposure to ultraviolet radiation (UV) rays causing the formation of free radicals such as reactive oxygen species (ROS) and reactive nitrogen species (RNS), which may react with biomolecules of the skin [4]. These free radicals promote sunburn, dermatitis, photo-allergy and photo-toxicity. Skin repairs itself by activating the wound healing process that includes collagen production [5]. The wound healing process is regulated by various cells via the secretion of cytokines, chemokines, molecular factors, and growth factors [6,7].

As humans age, skin repair efficiency drops, and with the continual exposure to physical and chemical ailments, the skin may succumb to various diseases and abnormalities. In rare occasions, where there are biochemical/hormonal imbalance or a genetic defect relating to skin structure and regulation, chronic skin disorder would remain prevalent for a prolonged period affecting the patient at any age or mental/physical health state.

Herbal Medicines: A Boon for Healthy Human Life.
DOI: https://doi.org/10.1016/B978-0-323-90572-5.00001-9

Skin diseases are among the most common health problems in humans. Various types of skin diseases occur in humans, which include infections, auto-inflammatory skin diseases and immune-mediated skin diseases such as psoriasis and skin aging. There are other skin diseases and disorders that are due to environmental factors, chemical imbalance, and genetic defects. Other skin conditions are rare and not yet medically classified. Herein, we only cover some of the most known and common human skin diseases and disorders (Table 17.1).

17.1.1 Synthetic versus natural cosmetics and cosmeceuticals

Consumers are exposed to significant amounts of chemicals found in personal care products/skincare products. In recent years, consumers are moving away from synthetically produced skincare and cosmetics products to natural alternatives. This is primarily due to the negative environmental impacts of synthetic production and the toxic nature of certain chemical ingredients, which subsequently cause harm to human skin [8]. Due to consumer demands for organic skincare products, plant extracts are added to skincare formulations. Plants are known to be associated with pigmentation inhibition, anti-microbial and anti-oxidant properties, which may prevent various skin conditions [9].

The term 'botanicals' is derived from botanical materials such as roots, leaves, stems, herbs and spices. Botanicals are applied in cosmetics and therapeutics and have been used either as extracted plant material, fresh plant material or in dried form [10]. Traditional botanical medicine also known as phytomedicine or herbal medicine are therapeutic alternatives due to their effectiveness in treating skin disorders and their safety for human use compared to conventional therapy [11]. Collectively, these products containing biologically active metabolites and ingredients with medicinal properties are referred to as 'cosmeceuticals.' These products contain essential bioactive molecules, providing the skin with vitamins, proteins, anti-oxidants, essential oils and more [10]. Botanical-based cosmetics have fewer side effects, low toxicity, are biodegradable and environmentally friendly. For this reason, natural skin care formulations are growing in popularity.

17.1.2 Role of medicinal plants in skin health, medicine, biotherapeutics

Different medicinal plants have been used traditionally to treat various skin diseases and disorders. There are two main methods used for medicinal plant preparation, which are decoctions and infusion methods [43]. The decoction method is when the active ingredients are extracted by boiling plant parts either in water or solvents, while infusions involve the immersion of the plant portions in hot water for a certain period and topically apply to the affected area thereafter. Other preparation methods include pastes, juice, grounding of plant parts into powder, poultices, and ointments. Ordinarily, poultices and ointments are applied to the affected area as dressing agents.

Table 17.1: List of skin diseases, their conditions and current treatment.

	Disease type/description	Conditions	Current treatment	References
Infection-related	*Acne vulgaris* Inflammatory and chronic skin disease caused by over secretion of the sebaceous glands. *Propionibacterium* colonization, an immune response to inflammation, increased sebum production, and keratinization of the infrainfundibulum.	Affects all ages but most common in teenagers between 16-20 years age group. Symptoms include rash, stinging and pain. It's non-life threating but can cause psychological and psychiatric challenges	Topical and systematic antibiotic, topical retinol, hormonal agents, isotretinoin and novel therapies. Complementary and alternative medicine (CAM) Salicylic acid, benzoyl peroxide, clay, sulfur soaps, lemons, yoghurt, face powder, rose water, black soap Psychological approaches	[3,12–19]
	Human dermatophytosis Fungal infection mainly associated with *Epidermophyton, Microsporum* and *Trichophyton* species. Tinea pedis (Athlete's foot) and Tinea corpis (ringworm) are the most common human dermatophytic infections.	Can affect skin (any body part), nails and hair. Tinea pedis: itchiness and causes inflammation, redness and cracking of the feet. Tinea corpis: itchy, red, scaly circular rash.	Antifungal creams, powders, or lotions Clotrimazole with glucocorticoids such as betamethasone, miconazole, and ketoconazole. Griseofulvin, itraconazole and fluconazole. Note: increase in anti-fungal resistance on current treatments.	[7,20–27]
	Other pathogen infection-related skin diseases *Malassezia*: common yeast skin pathogen causing *Malassezia* folliculitis.	*Malassezia* folliculitis affects the shoulders and the upper back with symptoms associated with itchiness, follicular rash resembling acne.	Topical (ketoconazole lotion/ shampoo, econazole nitrate cream and clotrimazole cream), oral antifungal agents (ketoconazole, fluconazole and itraconazole).	[7,28,29]
	Pseudomonas aerogenosa Streptococci and *Staphylococci Pasteurella multocida Borrelia burgdorferi Mycobacterium leprae*	Infected open wound and furunculosis Infected open wound and impetigo Cellulitis erythema chronicum migrans Leprosy	Nafcillin, oxacillin, ceftriaxone and clindamycin, vancomycin or penicillin Doxycycline, tetracycline, amoxicillin or cefuroxime axetil Dapsone, rifampin, clofazimine, minocycline or ofloxacin.	[30,31]

(continued on next page)

Table 17.1: List of skin diseases, their conditions and current treatment—cont'd

Disease type/description	Conditions	Current treatment	References
Psoriasis			
Chronic, inflammatory, immune-mediated skin disorder	Environmental, lifestyle, genetic factors, skin lesions and red plaques normally enclosed with silver or white scales.	Conventional therapy, phototherapy, systemic therapies, biologic therapies, target-specific therapies for psoriasis, targeted drug delivery systems and herbal drug formulations.	[2,32,33]
Skin aging			
Intrinsic aging: chronological aging, occurs naturally and may be caused by hormonal shifts, inherited genes, the inability of the body to repair damaged skin and degenerative effects of free radicals. Extrinsic aging: caused by lifestyle and environmental factors such as sun exposure (UV).	Reactive oxygen in the skin causing collagen degradation. UV irradiation degrades skin's elastic fibre leading to loss of moisture retainment. UV irradiation damages skin, causing premature aging.	Topical medical agents (antioxidants and cell regulators), cosmetologic care (correct sun protection and daily skin care), invasive procedures (chemical peelings, visible light devices), systematic agents (antioxidants and hormone replacement therapy), correction of habits and lifestyle (smoking, nutrition, diet, stress)	[34–37]
Autoimmune bullous diseases (AIBD)			
Chronic inflammatory skin disorders characterized by blisters and erosions of the skin or mucous membranes. Divided into two major subtypes: pemphigus and pemphigoid.	Pemphigus: intraepidermal blistering, uncommon but serious, resulting in the formation of intraepithelial blisters. Pemphigoid: subepidermal blistering common, resulting in tense blisters, erosions in the skin or mucous membranes.	Systemic corticosteroid therapy (prednisone or prednisolone), rituximab, azathioprine, mycophenolate, cyclophosphamide. Systemic glucocorticoids, superpotent topical corticosteroids, anti-inflammatory antibiotics (tetracyclines), doxycycline with prednisolone, and anti-IgE monoclonal antibodies.	[38–42]

Immunine-related

Autoinflammatory

The most commonly used plant parts to treat skin disorders are flowers, bulbs, fruits, rhizomes, leaves bark, roots or the whole plant [44]. It has been reported that plants that contain active ingredients such as polysaccharides, flavonoids, polyphenols, oligosaccharides as well as phenolic acids, are suitable for medicinal or cosmetic preparations [45]. Polyphenolic compounds have functional groups that can accept a negative charge of free radicals and bind to proteins, and because of this, polyphenols are regarded as excellent anti-oxidants. Flavonoids are another group of anti-oxidants due to their ability to scavenge free radicals effectively [43].

17.1.2.1 Anti-microbial effects of medicinal plants used to treat skin conditions

Studies have addressed the evaluation of anti-microbial effects of plant species against different pathogens known to cause bacterial and fungal infections (*Bacillus* sp., *Staphylococcus* sp. and *Pseudomonas aerogenosa*), and some of these plant species are listed in Table 17.2. Medicinal plants such as *Allium sativum, Lawson inermis, Citrus limon* and *Acalypha indica* used traditionally for treating fungal infections have been investigated for anti-fungal effects against fungal pathogen species namely, *Curvularia* spp., *Alternaria* spp., *Fusarium* spp., *Geotrichum* spp. and *Trichophyton* spp [46]. An active compound called marrubiin and the essential oil both extracted from *Marrubim vulgare* L. have also been reported to possess anti-fungal properties against pathogenic dermatophyte strains [47]. *Lawson inermis* has anti-microbial properties and has been used to treat various skin infections including tinea. In a study by Gozubuyuk et al. (2014), *Lawson inermis* was proven to have anti-fungal properties against a range of dermatophyte species, specifically those associated with tinea as well as fungal species causing ringworm including *Microsporum canis, Epidermophyton floccosum, Trichophyton rubrum, Trichophyton tonsurans, Trichophyton violaceum, Trichophyton mentagrophytes* and *Microsporum gypseum* [48].

17.1.2.2 Wound healing effects of medicinal plants used to treat skin conditions

The majority of studies have reported on the validation for the wound healing potential of medicinal plants such as *Aloe vera, Albizia lebbeak* and *Gunnera perpensa* (Table 17.2). *Aloe vera* has been used traditionally as a multipurpose skin treatment for many years [49]. The gel from *A. vera* leaves is used to treat wounds, burns, abrasions, surgical wounds, and bruises. *A. vera* contains various bioactive compounds, including saponins, phytol, glycosides, pyrocatechol, and simple as well as complex polysaccharides [50]. In addition, *in vivo* wound healing effects were identified, where notable effects of leaf extracts from medicinal plants *Bulbine frutescens* and *Bulbine notalensis* were observed [51]. Another study reported on the prominent activity of the gel formulation containing organic extracts of *Centella asiatica* which was applied topically on rat models for treatment of open wounds [52].

Table 17.2: Medicinal plants used for antimicrobial, wound healing and anti-inflammatory properties.

	Plant name	Bioactive ingredients/(parts used)	References
Anti-microbial	*Azadirachta indica* L. Neem/nimtree	Polyphenolic flavonoids, nimbin, nimbolinin, salaninin, nimbanene, ascorbic acid, amino acid, nimbiol, n-hexacosanol, 17-hydroxyazadiradione, 7-desacetyl-7-benzoylazadiradione and 7-desacetyl-7-benzoylgedunin. (Leaves)	[60,61]
	Boophane disticha (L.f.) Tumbleweed, windball	Alkaloids, phenolics and flavonoids. (Bulb)	[62–64]
	Chironia baccifera L. Christmas berry	Saponins, phenolics, tannins, chironioside, sweroside, gentiopicroside, swertiamarine and eustomoside. (Whole plant)	[65,66]
	Combretum paniculatum Burning bush	Flavonoids, triterpenoids, glycosides and coumarins. (Leaves)	[67]
	Dichrostachys cinera (l.) Sickle bush	Vitamins, phenolic acids, terpenoid, lignins, tannins, stilbenes, betalains, amines, flavonoids, coumarins, quinones and alkaloids. (Bark)	[68]
	Lepidium sativum Garden cress	Proteins, fatty acids, vitamins, calcium, phosphorus, trace elements, alkaloids, essential oils and polysaccharides. (Seeds)	[69,70]
	Strychnos spinose Lam. Kokiya	Triterpenoids, sterols, alkaloids, essential oils, monoterpenes and secoiridoids. (Leaves)	[71]
	Withania somnifera (L.) Ashwagandha	Glycosides, flavonoids, alkaloids, steroids, saponins, phenolics, phytophenols, amino acids, reducing sugars and acylsteryl glucosides. (Roots)	[72]
Wound[healing]	*Aloe vera* L Barbados aloe	Saponins, flavonoids, methylchromones, aloeride, aloesin, amino acids, anthraquinones, phytol, glycosides, pyrocatechol, polysaccharides, carbohydrates and sterols (Leaves)	[49,50]
	Aloe ferox Miller Bitter aloe	Aloenin, aloeresin A, aloesin, aloin A, aloe-emodin, anthraquinones and chrysophanol. (Leaves)	[43]
	Albizia lebbeak (L.) Lebbeck Tree	Terpenes, saponins, tannins, alkaloids, flavonoids and anthraquinones. (Leaf powder)	[73]
	Allium cepa L. Garden onion	Sulfur compounds, saponins, steroids and flavonoids. (Bulb)	[74,75]
	Curcuma longa L. Turmeric	Curcuminoids, sesquiterpenoids, monoterpenoids, carbohydrates and essential oils. (Rhizome)	[49,76]
	Gunnera perpensa L. River pumpkin or wild rhubarb	Alkaloids, flavonoids, tannins and cardiac glycosides. (Roots)	[77,78]
	Matricaria chamomilla L. Chamomile	Essential oils, terpenoids and flavonoids. (Flowers)	[79]

(continued on next page)

Table 17.2: Medicinal plants used for antimicrobial, wound healing and anti-inflammatory properties—cont'd

	Plant name	Bioactive ingredients/(parts used)	References
Anti-inflammatory	*Baccharis dracunculifolia* Alecrim-do-campo/ rosemary-of-the-field	Terpenoids, flavonoids, phenolic acids and essential oils. (Flowers and leaves)	[56,80,81]
	Casearia decandra Pitumba	Essential oils, flavonoids, phenolic acids, diterpenes, steroids, lignans, amides and cinnamic acids. (Leaves)	[82,83]
	Cyperus rotundus L. Nutgrass	Essential oils, chlorogenic acid, flavonoids, sesquiterpenes, phenylpropanoids, saponins, alkaloids and phenolic acids. (Rhizome)	[56,84]
	Dryopteris filix-mas Male fern	Filixic acid, deaspidin and phloroglucinols. (Ferns)	[85]
	Hyperricum perforatum St John's wort	Hyperforin, naphtodianthrones, phenolic acids, essential oils and flavonoids. (Flowers and leaves)	[86,87]
	Moringa oleifera Horseradish/drumstick tree	Essential oils, saponins, isothiocyanates, tannins, flavonoids, vitamins, and phenolic acids. (Seed, flowers and leaves)	[54,88,89]
	Saussurea medusa Saw-wort	Flavonoids, triterpenoids, sesquiterpenoids, phenolics, lignans, chlorogenic acid, gallic acid, isoquercitrin, syringing and ethyl gallat. (Roots)	[90,91]

17.1.2.3 Anti-inflammatory effects of medicinal plants used to treat skin conditions

Many plants are traditionally used for anti-inflammatory properties and some of them are presented in Table 17.2. Various medicinal plants including *Zantedeschia aethopica*, *Bulbine natalensis*, *Eucomis autumnalis*, *Tetradenia riparia* and *Hypericum aethiopicum* were screened for anti-oxidant, anti-inflammatory and wound healing properties in a study conducted by Ghuman et al. [53]. A study by Pawlowska et al. [54] identified the anti-bacterial and anti-inflammatory activities of *Bistorta officinalis*, a medicinal plant used to treat skin injuries and infections, also popularly used as a topical anti-inflammatory agent in Europe and Asia. A study by Pompermaier et al. [55] identified the anti-inflammatory effects of five medicinal plants, namely, *Acanthus montanus*, *Alchornea cordifolia*, *Annona stenophylla* subsp. *cuneate*, *Chaetocarpus africanus*, and *Cyperus articulates*, where the potential of these plants to inhibit cyclooxygenase (COX)–2 and nitric oxide were observed. Extracts of *Alchornea cordifolia*, *Acanthus montanus* and *Chaetocarpus africanus* were reported as the most active with strong inhibition effects. In addition, the phytochemical analysis of *Cyperus rotundus* L. rhizome extract revealed the presence of chlorogenic acid and flavonoid. These extracts showed anti-inflammatory efficacy on a skin inflammation mice model and may have the potential to treat skin-related disorders including psoriasis [56].

17.1.2.4 Anti-aging effects of medicinal plants used to treat skin conditions

Free radicals such as nitrogen species and reactive oxygen are known to cause several skin problems including hyperpigmentation, inflammation and skin aging. These free radicals are also associated with the activation of matrix metalloproteinases (MMPs) and increased levels of MMPs promote aging. Previous studies have illustrated that anti-oxidants and phenolic acids including epicatechin gallate (ECG), epigallocatechin gallate (EGCG), gallocatechin gallate (GCG), epigallocatechin (EGC) and gallic acid may help to neutralize free radicals and prevent free radicals from causing damages [57]. These polyphenols have also been proven to inhibit anti-aging activity and suppressing melanin production. It is known that green tea (*Camellia sinensis* L Kuntze) detoxifies and is packed with phenolic constituents including gallic acid, EGC, EGCG, and ECG [58]. In a study by Chaikul et al. [34], green tea extract demonstrated anti-skin aging activities, anti-oxidant activity, MMP-2 inhibitory activity, and the inhibition of tyrosinase to suppress melanin production.

In addition, quercetin and caffeic acid have been found to inhibit tyrosinase and reduce oxidative stress. Fruits of *Platycarya strobilacea* have been evaluated in various experimental models and they have been found to exhibit biological activities against oxidative damage and anti-elastase [59]. The crude extracts of *Harungana madagascariensis* and *Psorospermum aurantiacum* resulted in a strong inhibition against anti-tyrosinase, anti-elastase and anti-oxidative activities in the same study.

17.1.3 Plant-endophyte relationship

In nature, plants interact with microorganisms usually bacteria and fungi that inhabit their phyllosphere, endosphere, and rhizosphere [92]. The interaction between microbes and their host plants could be mutualistic or pathogenic [93]. Plants are always exposed to various pathogens and through the assistance of networking pathways, their defence mechanism has evolved against pathogen infections [94]. However, microorganisms like endophytes, are ubiquitous, symbiotic, and mutualistic organisms present (long/short-term residents) in all plants. They occupy the intracellular and intercellular tissues of their host plants and do not cause harm to host plants [95]. Endophytes contribute to their host plant secondary metabolite production and they produce secondary metabolites with the potential to contribute to biotechnological applications with novel and unique valuable therapeutic substances [96].

17.1.3.1 High-value plant secondary metabolites by endophytes as an alternative route

Plants produce a wide range of bioactive compounds (Table 17.2) called secondary metabolites and play a crucial role as defence response during environmental stress factors, particularly response to abiotic and biotic stresses [97]. With over 60,000 plants used for their medicinal properties, an alternate and sustainable source of metabolites is sought after to prevent over-harvesting and over-use. Although plant *in vitro* systems have been established for many years, there has been very limited progress made in this field because the commercial process associated with the production of plant secondary metabolites has several challenges such as high contamination risks, genetic instability, inadequate productivities, scale-up complications as well as the use of costly phytohormones [98]. As a result, the commercial scale of metabolites production from plants has only been successful for few compounds like ginsenosides, diosgenin and taxol. In contrast, camptothecin, vinblastine, vincristine, and podophyllotoxin are still not commercially successful [99].

Endophytes have an attraction because they are able to produce similar secondary metabolites as their host plant. *Taxomyces andreanae* isolated from *Taxus brevifolia* has been reported to be the first endophytic fungus to produce taxol. Research interest in secondary metabolites production by endophytes has increased drastically over the past years [100]. Some of the commercially significant metabolites produced by plants and endophytes include camptothecin, azadirachtin, quinine, and podophyllotoxin [21]. In addition to active compounds, anti-bacterial peptides called munumbicins, a new antibiotic produced by an endophytic isolate *Streptomyces* sp. NRRL 30562 from *Kennedia nigriscans* plant was found to be active against numerous human pathogens [101]. In another study by Gupta et al. [102], a compound asiaticoside which showed potent anti-inflammatory properties was produced by endophyte *Colletotrichum gloeosporioides* from the medicinal plant *Centella asiatica*. Preveena and Bhore. [103] reported on 50 bacterial endophytes associated with *Tridax procumbens* Linn. a medicinal plant traditionally used to treat wounds and injuries. In 2014, Abdalla and a co-worker reported on the status of peptides extracted from endophytes and recently discovered endophytic peptides showing anti-bacterial, anti-fungal, and other novel bioactivities that could be beneficial for skin health [104]. Other documented endophytes exhibiting potential biological activities for the treatment of skin ailments are presented in Table 17.3.

Although numerous studies have reported on the isolation of endophytes and their respective bioactive compounds, the commercial production of compounds from endophytes has not been explored yet [114]. Studies report that endophytes are more economically viable/competitive when cultivated *in vitro* compared to plant cell culture because of the low costs of microbial fermentations such as substrate and nutrient requirements [115]. As opposed to plant cell culture, microbes have a fast growth rate and therefore, much easier to sustain sterility for prolonged cultivation time. Scale-up and optimization are easier with microbes and also adapt easily to exogenous additions such as elicitors and precursors [116]. Thus, the use of endophytes as alternative means of metabolites production could resolve the over-exploitation and extinction of plant sources, and also result in the optimization of the production of metabolites for pharmacological and biotechnological studies [104].

Table 17.3: Endophytes with potential biological activity for treatment of skin ailments.

Source endophyte	Host plant of endophyte	Biological activity	References
Aspergillus austroafricanus CGJ-B3	Zingiber officinale	Antimicrobial and antioxidant activities	[105]
Phomopsis sp., Diaporthe phaseolorum, Diaporthe sp., Sordariomycetes sp. and Cochlibolus sp.	Costus spiralis	Antimicrobial and antioxidant activities	[106]
Streptomyces olivaceous strain BPSAC77, Streptomyces sp. strain BPSAC121 and Streptomyces thermocarboxydus strain BPSAC147	Rhynchotoechum ellipticum	Antimicrobial and antioxidant activities	[107]
Aspergillus niger, Aspergillus flavus, Penicillium sp. and Alternaria alternata	Loranthus micranthus	Antimicrobial and anti-inflammatory activities	[108]
Talaromyces wortmannii	Aloe vera	Antimicrobial and anti-inflammatory activities	[109]
Pestalotiopsis sp. DO14	Dendrobium officinate	Antifungal activity	[110]
Arcopilus aureus, Fusarium equiseti, Fusarium solani and Xylaria psidii	Vitis vinifera	Antifungal and antioxidant activities	[111]
R13, Fusarium solani	Rheum palmatum L.	Antibacterial activity	[112]
Colletotrichum gloeosporioides, Colletotrichum dematium, Curvularia pallescens, Lasiodiplodia theobromae, Nigrospora sphaerica, Pestalotiopsis maculans (URM-6061) and Phomopsis archeri	Indigofera suffruticosa	Antibacterial activity	[113]

17.1.4 Bioprospecting endophytes for identification of useful metabolites

The first step in bioprospecting endophytes is to isolate the endophytes. Surface sterilization is the first step and most commonly used method for the isolation of endophytes from healthy plant tissues or organs. This is to eliminate microbes or contaminants from the plant source, including epiphytes and pathogens. Endophytes may be isolated from different organs such as the leaves, fruits, roots, and stems [117–119]. During isolation of endophytes, the following must be taken into consideration when sterilizing plant organs for isolation of endophytes; (i) all microbes present on the plant surface must be removed, and (ii) the sterilization method must not cause minimum negative effect on the endophytes [120].

Sterilized samples are macerated in sterile nutrient broth or saline and plated on preferable nutrient media. In isolating bacterial endophytes Nutrient Agar (NA) or Luria Bertani Agar (LBA) is used. For fungal endophytes, Potato Dextrose Agar (PDA) or Mycological Agar

(MCA) are used, while S-agar medium is used for endophytic actinomycetes [63,74]. According to previous studies, the preferable incubation temperature for growing bacterial and fungal endophytes is about 28°C. Colonies are isolated from these plates and are further grown in a liquid medium and pure cultures are finally preserved in (20% or 30%) glycerol stocks solution at -80°C or -130°C for long-term storage [121]. Endophytic isolates are revived by inoculating the culture in liquid media or broth and incubated in a shaking incubator until the culture of microbes reaches the mid-exponential stage [122,123].

17.1.4.1 Extraction of metabolites from endophytes

Metabolites are mainly extracted based on the principle of solvent extraction and often solvents such as methanol and ethyl acetate are used. Sharma et al. [122] and Manganyi et al. [123] reported on the extraction of secondary metabolites with ethyl acetate and methanol as organic solvents. With the use of a rotary evaporator, the solvent is evaporated and then redissolved in dimethyl sulphoxide. Synytsya et al. [124] reported the use of acidified ethanol for solvent extraction, following partitioning of the extracts with petroleum ether, ethanol and water. In some cases, in studies conducted by Sebola et al. [125,126], it was reported that sterile XAD-7-HP resin was added to the fermented culture, and then returned to the shaking incubator for 2 hours. A cheesecloth was used to filter the resin which was washed with acetone and the collected acetone extract was concentrated with a rotary evaporator and the remaining water residue was further extracted with ethyl acetate and lastly evaporated to obtain crude extracts of secondary metabolites.

17.1.4.2 Identification and confirmation of metabolites from endophytes

Once metabolites from endophytes have been extracted, this is followed by sample analysis. Currently, metabolomics study is the most commonly utilised approach for the identification or search of novel secondary metabolites from endophytes. Metabolomics is an advanced analytical technique used to quantitatively analyse metabolites in biological samples [110]. Nuclear magnetic resonance (NMR), liquid chromatography (LC) or gas chromatography (GC) coupled to a mass spectrometry (MS) detector are the main platforms used in metabolomics research to identify a large number of metabolites, whether as targeted or untargeted metabolites [110].

Identification of metabolites is achieved by using appropriate software and tools. Online databases such as Feihn metabolomics libraries, NIST and Mainlib database, KEGG, KNApSACK, Pubchem, Dictionary of Natural Products and Chemspider may be used for metabolite annotations. Furthermore, a calculated putative empirical formula derived from MarkerLynx XS™ and spectral fragmentation patterns in comparison to literature may be used to putatively annotate and interpret the secondary metabolites [127–129].

17.1.5 Bioprocess strategies for enhanced metabolite production by endophytes

Various bioprocess optimization strategies have been successful for yield enhancement of metabolites production [130,131]. Several studies have reported many microorganisms produce metabolites in low quantities and high yielding techniques that help to reach demands have been discovered over the past years and have been applied successfully in various processes for primary and secondary metabolite production [132]. The use of optimization strategies such as medium/process condition, culture mode and co-cultivation have been adopted to enhance the yield of high-value secondary metabolites from endophytic isolates [115].

17.1.5.1 Culture/process condition optimization

Fermentation technology is a well-known technique used to provide favourable culture conditions to enhance secondary metabolites production. This is achieved by adjusting fermentation parameters such as physical parameters (i.e. temperature, pressure, culture volume, stirrer speed, etc.), chemical parameters (i.e. dissolved O_2 and CO_2, pH, nutrients, etc.), biochemical parameters (i.e. carbohydrates, total proteins, vitamins, etc.) and biological parameters (i.e. total cell count, biomass concentration, viable cell count, etc.) [133]. The optimization and control of these parameters create an ideal environment for microbes and this may be done either by using One-Factor-at-a-Time (OFAT) method, which involves optimizing each factor separately while ensuring all other factors are kept constant or statistical method, whereby different factors are varied simultaneously and statistical analysis is used to establish interactions [133]. The one-factor method does not allow interaction investigations between variables and can only be used to ascertain performance improvements rather than getting the global maxima [132].

Statistical experimental design, on the other hand, gives more reliable information and can be used to get global maxima, establish the relative significance of various factors, interactions among factors as well as the optimal level of variables [134]. Both approaches have been explored and applied in endophytic isolates. For example, a 10.3-fold increase was observed in the yield of mycoepoxydiene, an anti-fungal compound from endophyte *Phomopsis* sp. Hant25 after single-factor optimization of complex media – between czapek yeast autolysate broth, malt czapek broth and modified M1D medium [135].

In another study, the optimization of carbon and nitrogen sources resulted in an 8-fold enhancement in zofimarin from *Xylaria* sp. Acra L38 after orthogonal array design, Plackett-Burman design and response surface methodology (RSM) [136].

17.1.5.2 Co-cultivation

The co-cultivation approach unlocks the cryptic metabolites in co-cultured cells to enable researchers to identify novel secondary metabolites. This technique has been used to enhance the production of plant cell metabolites and recently has been used for endophytes. Several studies have reported on significant yield enhancement associated with co-cultivation strategy

[137]. The co-cultivation system entails several challenges and therefore the optimization of the following parameters must be taken into consideration to design a successful system; (a) optimum inoculum ratio (b) environmental parameters (c) medium components and (d) reactor design. When these parameters are optimized during co-cultivation, this may lead to enhanced production of the desired secondary metabolites [115]. A study by Soliman and Raizada. [138] reports on a threefold increase in taxol production in co-cultures of endophytic fungus *Paraconiothyrium* with *Taxus* bark fungus *Alternaria* sp. and an eightfold increase in co-cultures of *Paraconiothyrium* with *Alternaria* sp. along with *Taxus* fungus *Phomopsis* sp.

17.1.5.3 Culture mode: Batch, fed-batch, continuous

There are three major cultivation modes, namely; batch, fed-batch and continuous cultivation systems. For predictions of microbial biomass production and final production yield, mathematical models are generally used to regulate these modes [139].

Batch fermentation mode has been regarded as the easiest lab operation system that involves adding medium requirements at the initial phase of the fermentation process to scale-up biomass production and metabolites production. The production of microbial metabolites may occur at either the primary or secondary phase of the cultivation period [140].

Fed-batch fermentation mode is the modified phase of batch fermentation used at the beginning of this mode. Microbes are inoculated and allowed to grow at the batch regime. This is then followed by adding fresh nutrients/medium to the bioreactor at a fixed/variable volume throughout the fermentation period [141]. The feeding strategy can either be done continuously or exponentially at a short or long period in the course of the run, and feedback and operational controls are required to ensure accurate monitoring as well as to prevent catabolism repression and overfeeding the substrate during the process [139].

Continuous fermentation mode is not often utilized on a lab-scale but is common in the bioprocess industry. Fresh medium is added continuously to the bioreactor and used medium and cells are collected concurrently [141]. Following, the removal of used up nutrients and toxic metabolites from the culture, and the culture volume remains constant when this happens. Steady-state can be reached by optimization of the medium exchange rate, and as a result metabolites concentrations remain constant and cultures can last for days even weeks in a steady state. Hence this reduces downtime and makes the process economically effective and efficient [142].

17.1.6 Conclusion and future directions

There are many gaps in bioprospecting endophytes for their metabolites for commercial gain. Although endophytes hold potential for novel bioactives, there remain much to be explored and understood. Furthermore, *in vitro* cultivation of endophytes has been reported to encounter many challenges including difficulties in isolation, identification, scaling-up, risks of contamination, low yields and product attenuation. On the contrary, current endophyte research has demonstrated new possibilities and therefore, exploiting endophytes for bioactive production with the application of bioprocess engineering strategies could help conquer these current limitations. The future potential of endophytes and skin health research looks promising, however, the dynamics of plant-endophyte at metabolic and molecular levels are still necessary as it would provide great insight and understanding when exploiting the potential of endophyte technology at the commercial level.

Acknowledgements

The authors would like to thank the National Research Foundation and the University of Cape Town Postgraduate Funding Office, the University Research Office and the EBE Faculty Research Block Grant. SG Mahlangu received the DST-NRF Innovation Doctoral Scholarship (grant number: MND190510435402) and the University of Cape Town Doctoral Financial Aid.

Conflict of interest

The authors declare no conflict of interest.

References

[1] O Uchechi, JDN Ogbonna, AA Attama, Nanoparticles for dermal and transdermal drug delivery, Appl Nanotechnol Drug Deliv 4 (2014) 193–227.
[2] VK Rapalli, T Waghule, N Hans, A Mahmood, S Gorantla, SK Dubey, et al., Insights of lyotropic liquid crystals in topical drug delivery for targeting various skin disorders, J Mol Liq 315 (2020) 113771.
[3] T Zhang, B Sun, J Guo, M Wang, H Cui, H Mao, et al., Active pharmaceutical ingredient poly(ionic liquid)-based microneedles for the treatment of skin acne infection, Acta Biomater 115 (2020) 136–147.
[4] L Chen, JY Hu, SQ Wang, The role of antioxidants in photoprotection: a critical review, J Am Acad Dermatol 67 (5) (2012) 1013–1024.
[5] X Feng, S Zhou, W Cai, J Guo, The miR-93-3p/ZFP36L1/ZFX axis regulates keratinocyte proliferation and migration during skin wound healing, Mol Ther - Nucleic Acids 23 (158) (2021) 450–463.
[6] A Liubaviciute, T Ivaskiene, G Biziuleviciene, Modulated mesenchymal stromal cells improve skin wound healing, Biologicals 67 (2020) 1–8.
[7] D Wang, B Duncan, X Li, J Shi, The role of NLRP3 inflammasome in infection-related, immune-mediated and autoimmune skin diseases, J Dermatol Sci 98 (3) (2020) 146–151.
[8] A Nepalia, A Singh, N Mathur, S Pareek, Toxicity assessment of popular baby skin care products from Indian market using microbial bioassays and chemical methods, Int J Environ Sci Technol 15 (11) (2018) 2317–2324.
[9] AS Ribeiro, M Estanqueiro, MB Oliveira, JMS Lobo, Main benefits and applicability of plant extracts in skin care products, Cosmetics 2 (2) (2015) 48–65.
[10] J Reuter, I Merfort, CM Schempp, Botanicals in dermatology: an evidence-based review, Am J Clin Dermatol 11 (4) (2010) 247–267.
[11] RA Rosenbloom, J Chaudhary, D Castro-Eschenbach, Traditional botanical medicine: an introduction, Am J Ther 18 (2) (2011) 158–161.

[12] P Duru, Ö Örsal, The effect of acne on quality of life, social appearance anxiety, and use of conventional, complementary, and alternative treatments, Complement Ther Med 56 (2021) 102614.

[13] AL Zaenglein, AL Pathy, BJ Schlosser, A Alikhan, HE Baldwin, DS Berson, et al., Guidelines of care for the management of acne vulgaris, J Am Acad Dermatol 74 (5) (2016) 945–973.

[14] EL Anaba, IR Oaku, Adult female acne: a cross-sectional study of diet, family history, body mass index, and premenstrual flare as risk factors and contributors to severity, Int J Women's Dermatology (2021).

[15] P Magin, Appearance-related bullying and skin disorders, Clin Dermatol 31 (1) (2013) 66–71.

[16] SM Gallitano, DS Berson, How acne bumps cause the blues: the influence of acne vulgaris on self-esteem, Int J Women's Dermatology 4 (1) (2018) 12–17.

[17] N Rumsey, Psychosocial adjustment to skin conditions resulting in visible difference (disfigurement): what do we know? Why don't we know more? How shall we move forward? Int J Women's Dermatology 4 (1) (2018) 2–7.

[18] RB Shah, Impact of collaboration between psychologists and dermatologists: UK hospital system example, Int J Women's Dermatology 4 (1) (2018) 8–11.

[19] AU Tan, BJ Schlosser, AS Paller, A review of diagnosis and treatment of acne in adult female patients, Int J Women's Dermatology 4 (2) (2018) 56–71.

[20] IE Cock, SF Van Vuuren, A review of the traditional use of southern African medicinal plants for the treatment of fungal skin infections, J Ethnopharmacol 251 (2020) 112539.

[21] RK Thapa, JY Choi, SD Han, GH Lee, CS Yong, JH Jun, et al., Therapeutic effects of a novel DA5505 formulation on a guinea pig model of tinea pedis, Dermatologica Sin 35 (2) (2017) 59–65.

[22] IL Ross, GF Weldhagen, SE Kidd, Detection and identification of dermatophyte fungi in clinical samples using a commercial multiplex tandem PCR assay, Pathology 52 (4) (2020) 473–477.

[23] O Sy, K Diongue, O Ba, CB Ahmed, MA Elbechir, MSM Abdallahi, et al., Tinea capitis in school children from Mauritania: A comparative study between urban and rural areas, J Mycol Med 13 (2020) 101048.

[24] X Song, YX Wei, KM Lai, ZD He, HJ Zhang, In vivo antifungal activity of dipyrithione against *Trichophyton rubrum* on guinea pig dermatophytosis models, Biomed Pharmacother 108 (2018) 558–564.

[25] NA El-Zawawy, SS Ali, Pyocyanin as anti-tyrosinase and anti tinea corporis: a novel treatment study, Microb Pathog 100 (2016) 213–220.

[26] AB Zambelli, CA Griffiths, South African report of first case of chromoblastomycosis caused by *Cladosporium* (syn Cladophialophora) *carrionii* infection in a cat with feline immunodeficiency virus and lymphosarcoma, J Feline Med Surg 17 (4) (2015) 375–380.

[27] GD Brown, DW Denning, NAR Gow, SM Levitz, MG Netea, TC White, Hidden killers: human fungal infections, Sci Transl Med 4 (165) (2012) 10–11.

[28] AK Gupta, R Batra, R Bluhm, T Boekhout, TL Dawson, Skin diseases associated with *Malassezia* species, J Am Acad Dermatol 51 (5) (2004) 785–798.

[29] R Hay, Superficial fungal infections, Medicine (Baltimore) 37 (11) (2009) 610–612.

[30] H M. Feder Jr., A M, B M, W-W D, G-K JM, Diagnosis, treatment, and prognosis of *erythema migrans* and *Lyme arthritis*, Clin Dermatol 24 (6) (2006) 509–520.

[31] NA Levin, S Delano, Evaluation and treatment of *Malassezia*-related skin disorders, Cosmet Dermatol 24 (3) (2011) 137–145.

[32] M Pompili, L Bonanni, F Gualtieri, G Trovini, S Persechino, RJ Baldessarini, Suicidal risks with psoriasis and atopic dermatitis: systematic review and meta-analysis, J Psychosom Res 141 (2021) 110347.

[33] SL Jyothi, KL Krishna, VK Ameena Shirin, R Sankar, K Pramod, HV Gangadharappa, Drug delivery systems for the treatment of psoriasis: current status and prospects, J Drug Deliv Sci Technol 62 (2021) 102364.

[34] P Chaikul, T Sripisut, S Chanpirom, N Ditthawutthikul, Anti-skin aging activities of green tea (*Camelliasinensis* (L) Kuntze) in B16F10 melanoma cells and human skin fibroblasts, Eur J Integr Med 40 (2020) 101212.

[35] EA Georgakopoulou, C Valsamidi, D Veroutis, S Havaki, The bright and dark side of skin senescence. Could skin rejuvenation anti-senescence interventions become a "bright" new strategy for the prevention of age-related skin pathologies? Mech Ageing Dev 193 (2021) 111409.

[36]   I Sjerobabski-Masnec, M Šitum, Skin aging, Acta Clin Croat 49 (4) (2010) 515–518.

[37]   CC Zouboulis, R Ganceviciene, AI Liakou, A Theodoridis, R Elewa, E Makrantonaki, Aesthetic aspects of skin aging, prevention, and local treatment, Clin Dermatol 37 (4) (2019) 365–372.

[38]   S Egami, J Yamagami, M Amagai, Autoimmune bullous skin diseases, *pemphigus* and *pemphigoid*, J Allergy Clin Immunol 145 (4) (2020) 1031–1047.

[39]   AS Payne, Y Hanakawa, M Amagai, JR Stanley, Desmosomes and disease: 9d *bullous impetigo*, Curr Opin Cell Biol 16 (5) (2004) 536–543.

[40]   CM Montagnon, JS Lehman, DF Murrell, MJ Camilleri, SN Tolkachjov, Intraepithelial autoimmune bullous dermatoses disease activity assessment and therapy, J Am Acad Dermatol 84 (6) (2021) 1523–1537.

[41]   R Maglie, M Hertl, Pharmacological advances in *pemphigoid*, Curr Opin Pharmacol 46 (2019) 34–43.

[42]   J Yamagami, Recent advances in the understanding and treatment of *pemphigus* and *pemphigoid*, F1000Research. 7 (0) (2018) 1–8.

[43]   NC Dlova, MA Ollengo, Traditional and ethnobotanical dermatology practices in Africa, Clin Dermatol 36 (3) (2018) 353–362.

[44]   MA Aziz, M Adnan, AH Khan, AA Shahat, MS Al-Said, R Ullah, Traditional uses of medicinal plants practiced by the indigenous communities at Mohmand Agency, FATA, Pakistan, J Ethnobiol Ethnomed 14 (1) (2018) 1–16.

[45]   P Palombo, G Fabrizi, V Ruocco, E Ruocco, J Fluhr, R Roberts, et al., Beneficial long-term effects of combined oral/topical antioxidant treatment with the carotenoids lutein and zeaxanthin on human skin: a double-blind, placebo-controlled study, Skin Pharmacol Physiol 20 (4) (2007) 199–210.

[46]   S Abirami, B Edwin Raj, T Soundarya, M Kannan, D Sugapriya, N Al-Dayan, et al., Exploring antifungal activities of acetone extract of selected Indian medicinal plants against human dermal fungal pathogens, Saudi J Biol Sci 28 (4) (2021) 2180–2187.

[47]   M Rezgui, N Majdoub, B Mabrouk, A Baldisserotto, A Bino, LB Ben Kaab, et al., Antioxidant and antifungal activities of marrubiin, extracts and essential oil from *Marrubium vulgare* L. against pathogenic dermatophyte strains, J Mycol Med 30 (1) (2020) 100927.

[48]   GS Gozubuyuk, E Aktas, N Yigit, An ancient plant Lawsonia inermis (henna): determination of *in vitro* antifungal activity against dermatophytes species, J Mycol Med 24 (4) (2014) 313–318.

[49]   TZ Shadi, AZ Talal, A review of four common medicinal plants used to treat eczema, J Med Plants Res 9 (24) (2015) 702–711.

[50]   A Shedoeva, D Leavesley, Z Upton, C Fan, Wound healing and the use of medicinal plants. Evidence-based, Complement Altern Med 2019 (2019) 2684108, doi:10.1155/2019/2684108.

[51]   N Pather, AM Viljoen, B Kramer, A biochemical comparison of the *in vivo* effects of *Bulbine frutescens* and *Bulbine natalensis* on cutaneous wound healing, J Ethnopharmacol 133 (2) (2011) 364–370.

[52]   SA Dahanukar, RA Kulkarni, NN Rege, Pharmacology of medicinal plants and natural products, Indian J Pharmacol 32 (4) (2000) 81–118.

[53]   S Ghuman, B Ncube, JF Finnie, LJ McGaw, E Mfotie Njoya, RM Coopoosamy, et al., Antioxidant, anti-inflammatory and wound healing properties of medicinal plant extracts used to treat wounds and dermatological disorders, South African J Bot 126 (2019) 232–240.

[54]   ABM Cretella, S Soley B da, PL Pawloski, RM Ruziska, DR Scharf, J Ascari, et al., Expanding the anti-inflammatory potential of *Moringa oleifera*: topical effect of seed oil on skin inflammation and hyperpro-liferation, J Ethnopharmacol 254 (2020) 112708, doi:10.1016/j.jep.2020.112708.

[55]   L Pompermaier, S Marzocco, S Adesso, M Monizi, S Schwaiger, C Neinhuis, et al., Medicinal plants of northern Angola and their anti-inflammatory properties, J Ethnopharmacol 216 (2018) 26–36.

[56]   FG Rocha, M Brandenburg M de, PL Pawloski, S Soley B da, SCA Costa, CC Meinerz, et al., Preclinical study of the topical anti-inflammatory activity of *Cyperus rotundus* L. extract (Cyperaceae) in models of skin inflammation, J Ethnopharmacol 254 (2020) 112709.

[57]   S Bhattacharya, AP Sherje, Development of resveratrol and green tea sunscreen formulation for combined photoprotective and antioxidant properties, J Drug Deliv Sci Technol 60 (1) (2020) 102000.

[58] P Meetham, M Kanlayavattanakul, N Lourith, Development and clinical efficacy evaluation of anti-greasy green tea tonner on facial skin, Rev Bras Farmacogn 28 (2) (2018) 214–217.

[59] NJ Manjia, NF Njayou, A Joshi, K Upadhyay, K Shirsath, VR Devkar, et al., The anti-aging potential of medicinal plants in Cameroon - *Harungana madagascariensis* Lam. and *Psorospermum aurantiacum* Engl. prevent in vitro ultraviolet B light-induced skin damage, Eur J Integr Med 29 (2019) 100925.

[60] MA Alzohairy, Therapeutics role of *Azadirachta indica* (Neem) and their active constituents in diseases prevention and treatment, Evid Based Complement Altern Med 2016 (2016).

[61] W Sujarwo, AP Keim, G Caneva, C Toniolo, M Nicoletti, Ethnobotanical uses of neem (*Azadirachta indica* A.Juss.; Meliaceae) leaves in Bali (Indonesia) and the Indian subcontinent in relation with historical background and phytochemical properties, J Ethnopharmacol 189 (2016) 186–193.

[62] HM Heyman, AA Hussein, JJM Meyer, N Lall, Antibacterial activity of South African medicinal plants against methicillin resistant *Staphylococcus aureus*, Pharm Biol 47 (1) (2009) 67–71.

[63] DN Nair, S Padmavathy, Impact of endophytic microorganisms on plants, environment and humans, Sci World J 2014 (2014) 250693, doi:10.1155/2014/250693.

[64] S Tonisi, K Okaiyeto, LV Mabinya, AI Okoh, Evaluation of bioactive compounds, free radical scavenging and anticancer activities of bulb extracts of *Boophone disticha* from Eastern Cape Province, South Africa. Saudi J Biol Sci. 27 (12) (2020) 3559–3569.

[65] TSA Thring, EP Springfield, FM Weitz, Antimicrobial activities of four plant species from the Southern Overberg region of South Africa, African J Biotechnol 6 (15) (2007) 1779–1784.

[66] A Maroyi, Ethnomedicinal uses, phytochemistry and pharmacological properties of Chironia baccifera, J Pharm Sci Res 11 (11) (2019) 3670–3674.

[67] KC Chinsembu, Ethnobotanical study of plants used in the management of HIV/AIDS-related diseases in Livingstone, Southern Province, Zambia, Evid Based Complement Altern Med 2016 (2016) 4238625, doi:10.1155/2016/4238625.

[68] DE Greydanus, R Azmeh, MD Cabral, CA Dickson, DR Patel, Acne in the first three decades of life: an update of a disorder with profound implications for all decades of life, Dis Mon 67 (4) (2021) 101103.

[69] NU Rehman, AU Khan, KM Alkharfy, AH Gilani, Pharmacological basis for the medicinal use of Lepidium sativum in airways disorders. Evidence-based, Complement Altern Med 2012 (2012) 596524, doi:10.1155/2012/596524.

[70] A Ahmad, BL Jan, M Raish, KM Alkharfy, A Ahad, A Khan, et al., Inhibitory effects of Lepidium sativum polysaccharide extracts on TNF-α production in Escherichia coli-stimulated mouse, 3 Biotech 8 (6) (2018) 1–8.

[71] AI Isa, MD Awouafack, JP Dzoyem, M Aliyu, RAS Magaji, JO Ayo, et al., Some *Strychnos spinosa* (Loganiaceae) leaf extracts and fractions have good antimicrobial activities and low cytotoxicities, BMC Complement Altern Med 14 (1) (2014) 1–8.

[72] H Punetha, S Singh, AK Gaur, Antifungal and antibacterial activities of crude withanolides extract from the roots of *Withania somnifera* (L .) Dunal (Ashwagandha), Environ Conserv J 11 (1&2) (2010) 65–69.

[73] M Chaabi, S Benayache, F Benayache, S N'Gom, M Koné, R Anton, et al., Triterpenes and polyphenols from *Anogeissus leiocarpus* (Combretaceae), Biochem Syst Ecol 36 (1) (2008) 59–62.

[74] VC Verma, SK Gond, A Kumar, A Mishra, RN Kharwar, AC Gange, Endophytic actinomycetes from *Azadirachta indica* A. Juss.: Isolation, diversity, and anti-microbial activity, Microb Ecol 57 (4) (2009) 749–756.

[75] CR Thornfeldt, Therapeutic herbs confirmed by evidence-based medicine, Clin Dermatol 36 (3) (2018) 289–298.

[76] A Amalraj, A Pius, S Gopi, S Gopi, Biological activities of curcuminoids, other biomolecules from turmeric and their derivatives – a review, J Tradit Complement Med 7 (2) (2017) 205–233.

[77] M Nkomo, L Kambizi, Antimicrobial activity of *Gunnera perpensa* and *Heteromorpha arborescens* var. abyssinica, J Med Plants Res 3 (12) (2009) 1051–1055.

[78] MBC Simelane, OA Lawal, TG Djarova, AR Opoku, In vitro antioxidant and cytotoxic activity of *Gunnera perpensa* L. (Gunneraceae) from South Africa, J Med Plants Res 4 (21) (2010) 2181–2188.

[79]   JK Srivastava, E Shankar, GS Chamomile, A herbal medicine of the past with a bright future (review), Mol Med Rep 3 (6) (2010) 895–901.

[80]   NSS Guimarães, JC Mello, JS Paiva, PCP Bueno, AA Berretta, RJ Torquato, et al., *Baccharis dracunculifolia*, the main source of green propolis, exhibits potent antioxidant activity and prevents oxidative mitochondrial damage, Food Chem Toxicol 50 (3–4) (2012) 1091–1097.

[81]   LN Cazella, J Glamoclija, M Soković, JE Gonçalves, GA Linde, NB Colauto, et al., Antimicrobial activity of essential oil of *Baccharis dracunculifolia* DC (Asteraceae) aerial parts at flowering period, Front Plant Sci 10 (2019) 1–9.

[82]   GM Vieira Júnior, LA Dutra, RB Torres, N Boralle, S Bolzani V da, DHS Silva, et al., Chemical constituents from *Casearia* spp. (Flacourtiaceae/Salicaceae sensu lato), Rev Bras Farmacogn 27 (6) (2017) 785–787.

[83]   C Camponogara, S Brum E da, BV Belke, TF Brum, S Jesus R da, M Piana, et al., *Casearia decandra* leaves present anti-inflammatory efficacy in a skin inflammation model in mice, J Ethnopharmacol 249 (2020) 112436, doi:10.1016/j.jep.2019.112436.

[84]   GA Mohamed, Iridoids and other constituents from *Cyperus rotundus* L. rhizomes, Bull Fac Pharmacy, Cairo Univ. 53 (1) (2015) 5–9.

[85]   EO Erhirhie, CN Emeghebo, EE Ilodigwe, DL Ajaghaku, BO Umeokoli, PM Eze, et al., *Dryopteris filix-mas* (L.) Schott ethanolic leaf extract and fractions exhibited profound anti-inflammatory activity, Avicenna J Phytomedicine 9 (4) (2019) 396–409.

[86]   AI Oliveira, C Pinho, B Sarmento, ACP Dias, Neuroprotective activity of hypericum perforatum and its major components, Front Plant Sci 7 (2016) 1–15.

[87]   N Kladar, J Mrdanović, G Anačkov, S Šolajić, N Gavarić, B Srdenović, et al., *Hypericum perforatum*: Synthesis of Active Principles during Flowering and Fruitification - Novel Aspects of Biological Potential, Evid Based Complement Altern Med 2017 (2017) 2865610, doi:10.1155/2017/2865610.

[88]   EI Omodanisi, YG Aboua, OO Oguntibeju, RM Lamuela-Raventós, Assessment of the anti-hyperglycaemic, anti-inflammatory and antioxidant activities of the methanol extract of *Moringa oleifera* in diabetes-induced nephrotoxic male wistar rats, Molecules 22 (4) (2017) 1–16.

[89]   M Vergara-Jimenez, MM Almatrafi, ML Fernandez, Bioactive components in *Moringa oleifera* leaves protect against chronic disease, Antioxidants 6 (4) (2017) 1–13.

[90]   JY Fan, HB Chen, L Zhu, HL Chen, ZZ Zhao, T Yi, *Saussurea medusa*, source of the medicinal herb snow lotus: a review of its botany, phytochemistry, pharmacology and toxicology, Phytochem Rev 14 (3) (2015) 353–366.

[91]   T Zhao, SJ Li, ZX Zhang, ML Zhang, QW Shi, YC Gu, et al., Chemical constituents from the genus Saussurea and their biological activities, Heterocycl Commun 23 (5) (2017) 331–358.

[92]   PA Rodriguez, M Rothballer, SP Chowdhury, T Nussbaumer, C Gutjahr, P Falter-Braun, Systems biology of plant-microbiome interactions, Mol Plant 12 (6) (2019) 804–821.

[93]   KN Nataraja, TS Suryanarayanan, RU Shaanker, M Senthil-Kumar, R Oelmüller, Plant–microbe interaction: prospects for crop improvement and management, Plant Physiol Rep 24 (4) (2019) 461–462.

[94]   DV Badri, TL Weir, D van der Lelie, JM Vivanco, Rhizosphere chemical dialogues: plant-microbe interactions, Curr Opin Biotechnol 20 (6) (2009) 642–650.

[95]   DA Bastías, LJ Johnson, SD Card, Symbiotic bacteria of plant-associated fungi: friends or foes? Curr Opin Plant Biol 56 (2020) 1–8.

[96]   JS Singh, Microbial secondary metabolites and plant-microbe communications in the rhizosphere. New and Future Developments in Microbial Biotechnology and Bioengineering: Microbes in Soil, Crop and Environmental Sustainability, Elsevier B.V., India, 2019, pp. 93–111.

[97]   A Khojasteh, MH Mirjalili, MA Alcalde, RM Cusido, R Eibl, J Palazon, Powerful plant antioxidants: a new biosustainable approach to the production of rosmarinic acid, Antioxidants 9 (12) (2020) 1–31.

[98]   H Chandran, M Meena, T Barupal, K Sharma, Plant tissue culture as a perpetual source for production of industrially important bioactive compounds, Biotechnol Rep 26 (2020) e00450.

[99]   G Kai, C Wu, L Gen, L Zhang, L Cui, X Ni, Biosynthesis and biotechnological production of anti-cancer drug Camptothecin, Phytochem Rev 14 (3) (2015) 525–539.

[100]   A Christina, V Christapher, S Bhore, Endophytic bacteria as a source of novel antibiotics: an overview, Pharmacogn Rev 7 (13) (2013) 11–16.

[101]   UF Castillo, GA Strobel, EJ Ford, WM Hess, H Porter, JB Jensen, et al., Munumbicins, wide-spectrum antibiotics produced by *Streptomyces* NRRL 30562, endophytic on *Kennedia nigriscans*, Microbiology 148 (9) (2002) 2675–2685.

[102]   S Gupta, P Bhatt, P Chaturvedi, Determination and quantification of asiaticoside in endophytic fungus from *Centella asiatica* (L.) Urban, World J Microbiol Biotechnol [Internet] 34 (8) (2018) 1–10.

[103]   J Preveena, S Bhore, Identification of bacterial endophytes associated with traditional medicinal plant *Tridax procumbens* Linn, Anc Sci Life 32 (3) (2013) 173.

[104]   MA Abdalla, JC Matasyoh, Endophytes as producers of peptides: an overview about the recently discovered peptides from endophytic microbes, Nat Products Bioprospect 4 (5) (2014) 257–270.

[105]   A Danagoudar, CG Joshi, R Sunil Kumar, J Poyya, T Nivya, MM Hulikere, et al., Molecular profiling and antioxidant as well as anti-bacterial potential of polyphenol producing endophytic fungus-*Aspergillus austroafricanus* CGJ-B3, Mycology 8 (1) (2017) 28–38.

[106]   PG Marson Ascêncio, SD Ascêncio, AA Aguiar, A Fiorini, RS Pimenta, Chemical assessment and antimicrobial and antioxidant activities of endophytic fungi extracts isolated from *Costus spiralis* (Jacq.) Roscoe (Costaceae), Evid Based Complement Altern Med 2014 (2014) 190543, doi:10.1155/2014/190543.

[107]   AK Passari, VK Mishra, G Singh, P Singh, B Kumar, VK Gupta, et al., Insights into the functionality of endophytic actinobacteria with a focus on their biosynthetic potential and secondary metabolites production, Sci Rep 7 (1) (2017) 1–17.

[108]   M Govindappa, R Channabasava, DV Sowmya, J Meenakshi, MR Shreevidya, A Lavanya, et al., Phytochemical screening, antimicrobial and *in vitro* anti-inflammatory activity of endophytic extracts from *Loranthus* sp, Pharmacogn J 3 (25) (2011) 82–90.

[109]   A Pretsch, M Nagl, K Schwendinger, B Kreiseder, M Wiederstein, D Pretsch, et al., Antimicrobial and anti-inflammatory activities of endophytic fungi *Talaromyces wortmannii* extracts against acne-inducing bacteria, PLoS One 9 (6) (2014) e97929, doi:10.1371/journal.pone.0097929.

[110]   C Wu, HK Kim, GP Van Wezel, YH Choi, Metabolomics in the natural products field - a gateway to novel antibiotics, Drug Discov Today Technol 13 (2015) 11–17.

[111]   V Dwibedi, S Saxena, In vitro anti-oxidant, anti-fungal and anti-staphylococcal activity of resveratrol-producing endophytic fungi, Proc Natl Acad Sci India Sect B - Biol Sci 90 (1) (2020) 207–219.

[112]   X You, S Feng, S Luo, D Cong, Z Yu, Z Yang, et al., Studies on a rhein-producing endophytic fungus isolated from *Rheum palmatum* L, Fitoterapia 85 (1) (2013) 161–168.

[113]   IP dos Santos, LCN da Silva, MV da Silva, JM de Araújo, S Cavalcanti M da, M Lima VL de, Antibacterial activity of endophytic fungi from leaves of *Indigofera suffruticosa* Miller (Fabaceae), Front Microbiol 6 (2015) 1–7.

[114]   I Mohinudeen, R Kanumuri, KN Soujanya, RU Shaanker, SK Rayala, S Srivastava, Sustainable production of camptothecin from an Alternaria sp. isolated from *Nothapodytes nimmoniana*, Sci Rep 11 (1) (2021) 1–11.

[115]   A Venugopalan, S Srivastava, Endophytes as in vitro production platforms of high value plant secondary metabolites, Biotechnol Adv 33 (6) (2015) 873–887.

[116]   Zhao J, Zhou L, Wang J, Shan T, Zhong L, Liu X, et al. Endophytic fungi for producing bioactive compounds originally from their host plants. 2010;567–576.

[117]   S Kandel, P Joubert, S Doty, Bacterial Endophyte Colonization and Distribution within Plants, Microorganisms. 5 (4) (2017) 77.

[118]   E Martinez-Klimova, K Rodríguez-Peña, S Sánchez, Endophytes as sources of antibiotics, Biochem Pharmacol 134 (2017) 1–17.

[119]   J Papik, M Folkmanova, M Polivkova-Majorova, J Suman, O Uhlik, The invisible life inside plants: deciphering the riddles of endophytic bacterial diversity, Biotechnol Adv 44 (2020) 107614.

[120]   J Mercado-Blanco, BJJ Lugtenberg, Send orders for reprints to reprints@benthamscience.net biotechno-logical applications of bacterial endophytes, Curr Biotechnol 3 (2014) 60–75.

[121]   RA Tolulope, AI Adeyemi, MA Erute, Abiodun TS. Isolation and screening of endophytic fungi from three plants used in traditional medicine in Nigeria for antimicrobial activity, Int J Green Pharm 9 (1) (2015) 58–62.

[122]   D Sharma, A Pramanik, PK Agrawal, Evaluation of bioactive secondary metabolites from endophytic fungus Pestalotiopsis neglecta BAB-5510 isolated from leaves of Cupressus torulosa D.Don, 3 Biotech 6 (2) (2016) 1–14.

[123]   MC Manganyi, CDK Tchatchouang, T Regnier, CC Bezuidenhout, CN Ateba, Bioactive compound produced by endophytic fungi isolated from *Pelargonium sidoides* against selected bacteria of clinical importance, Mycobiology 47 (3) (2019) 335–339.

[124]   A Synytsya, J Monkai, R Bleha, A Macurkova, T Ruml, J Ahn, et al., Antimicrobial activity of crude extracts prepared from fungal mycelia, Asian Pac J Trop Biomed 7 (3) (2017) 257–261.

[125]   TE Sebola, NC Uche-Okereafor, KI Tapfuma, L Mekuto, E Green, V Mavumengwana, Evaluating an-tibacterial and anticancer activity of crude extracts of bacterial endophytes from *Crinum macowanii* Baker bulbs, Microbiologyopen 8 (12) (2019) 1–10.

[126]   TE Sebola, NC Uche-Okereafor, L Mekuto, MM Makatini, E Green, V Mavumengwana, Antibacterial and anticancer activity and untargeted secondary metabolite profiling of crude bacterial endophyte extracts from *Crinum macowanii* baker leaves, Int J Microbiol (2020) 8839490, doi:10.1155/2020/8839490.

[127]   LW Sumner, A Amberg, D Barrett, MH Beale, R Beger, CA Daykin, et al., Proposed minimum reporting standards for chemical analysis, Metabolomics 3 (3) (2007) 211–221.

[128]   F Matsuda, Y Shinbo, A Oikawa, MY Hirai, O Flehn, S Kanaya, et al., Assessment of metabolome annotation quality: a method for evaluating the false discovery rate of elemental composition searches, PLoS One 4 (10) (2009) 10–15.

[129]   R Jaiswal, H Müller, A Müller, MGE Karar, N Kuhnert, Identification and characterization of chlorogenic acids, chlorogenic acid glycosides and flavonoids from *Lonicera henryi* L. (Caprifoliaceae) leaves by LC-MSn, Phytochemistry 108 (2014) 252–263.

[130]   S Srivastava, AK Srivastava, Hairy root culture for mass-production of high-value secondary metabolites, Crit Rev Biotechnol 27 (1) (2007) 29–43.

[131]   MS Hussain, S Fareed, S Ansari, MA Rahman, IZ Ahmad, M Saeed, Current approaches toward produc-tion of secondary plant metabolites, J Pharm Bioallied Sci 4 (1) (2012) 10–20.

[132]   V Singh, S Haque, R Niwas, A Srivastava, M Pasupuleti, CKM Tripathi, Strategies for fermentation medium optimization: an in-depth review, Front Microbiol 7 (2017).

[133]   M Dinarvand, M Rezaee, M Foroughi, Optimizing culture conditions for production of intra and extra-cellular inulinase and invertase from *Aspergillus niger* ATCC 20611 by response surface methodology (RSM), Brazilian J Microbiol 48 (3) (2017) 427–441.

[134]   XH Chen, WY Lou, MH Zong, TJ Smith, Optimization of culture conditions to produce high yields of active *Acetobacter* sp. CCTCC M209061 cells for anti-Prelog reduction of prochiral ketones, BMC Biotechnol 11 (2011) 1–12.

[135]   N Thammajaruk, N Sriubolmas, D Israngkul, V Meevootisom, S Wiyakrutta, Optimization of culture conditions for mycoepoxydiene production by *Phomopsis* sp. Hant25, J Ind Microbiol Biotechnol 38 (6) (2011) 679–685.

[136]   J Chaichanan, S Wiyakrutta, T Pongtharangkul, D Isarangkul, V Meevootisom, Optimization of zofimarin production by an endophytic fungus, Xylaria sp. Acra L38, Brazilian J Microbiol 45 (1) (2014) 287–293.

[137]   S Pant, D Mishra, S Gupta, P Chaturvedi, Fungal endophytes as a potential source of therapeutically important metabolites. Fungi bio-prospects in sustainable agriculture, environment and nano-technology, INC 3 (1) (2021) 275–314.

[138]   SSM Soliman, MN Raizada, Interactions between co-habitating fungi elicit synthesis of Taxol from an endophytic fungus in host Taxus plants, Front Microbiol 4 (2013) 1–14.

[139]   SH El Moslamy, Application of fed-batch fermentation modes for industrial bioprocess development of microbial behaviour, Ann Biotechnol Bioeng 1 (1) (2019) 1–11.

[140] F Eberhardt, A Aguirre, L Paoletti, G Hails, M Braia, P Ravasi, et al., Pilot-scale process development for low-cost production of a thermostable biodiesel refining enzyme in *Escherichia coli*, Bioprocess Biosyst Eng 41 (4) (2018) 555–564.

[141] T Li, CX Bin, JC Chen, Q Wu, GQ Chen, Open and continuous fermentation: products, conditions and bioprocess economy, Biotechnol J 9 (12) (2014) 1503–1511.

[142] S Sen, PK Roychoudhury, Development of optimal medium for production of commercially important monoclonal antibody 520C9 by hybridoma cell, Cytotechnology 65 (2) (2013) 233–252.

PART 5

Reproductive disorders

CHAPTER 18

Future of herbal medicines in assisted reproduction

Leonard C. D'Souza[a], Jagdish G. Paithankar[a], Hifzur R. Siddique[b] and Anurag Sharma[a]

[a]Nitte (Deemed to be University), Nitte University Centre for Science Education and Research (NUCSER), Division of Environmental Health and Toxicology, Mangaluru, India [b]Molecular Cancer Genetics & Translational Research Lab, Section of Genetics, Department of Zoology, Aligarh Muslim University, Aligarh, India

18.1 Introduction

The origin of human life begins with the fusion of sperm and oocyte, followed by a series of events during embryogenesis and morphogenesis. Reproduction is a phenomenon that orchestrates reticulate behavioral and physiological attributes that decide the structure of the family and the size of the community [1]. The gonads steal the limelight by playing a central role in this process. Testes in males and ovaries in females are the two glands that produce morphologically, and functionally distinct gametes called sperm and oocyte, respectively. Fusion of oocyte and sperm results in fertilization is the most important event when two gametes, sperm from male and oocyte from the female, undergo fusion to form an embryo [2]. The fusion of gametes ensures the dodging of polyspermy by creating a plasma membrane block. A series of biochemical reactions among sperm acrosomal proteins, and zona proteins of the egg activates the egg, leading to zygote formation. Several signaling pathways activate the cell cycle, and the embryo starts its mitotic cleavage. The embryo multiplies exponentially to form a series of stages like blastocyst at five days post fertilization (dpf), implantation at seven dpf in uterus followed by gastrula formation and further development [3–6].

The reproductive system is a composite and sophisticated structure, with several hormones and growth factors playing their due role during reproductive events. This system is prone to damage by several internal and external agents, and even slender damage to any part of the system may affect the success rate of fertilization leading to infertility. According to WHO, around 48 million couples and 168 million individuals have infertility, and these numbers are likely to be increased in the coming years [https://www.who.int/health-topics/infertility]. Sun et al. study focused on the global burden of the disease and observed that female and male infertility rates increased by 0.37% and 0.29 % respectively from 1990 to 2017 [7]. The rate of successful fertilization depends on numerous factors like physiological, socio-psychological, and lifestyle factors. Age is one of the most important factor that affects fertility. Studies reveal that fertility peaks at the early 20s and declines with age in both males and females [8,9].

On the other hand, diet plays a crucial role in maintaining reproductive health. Foods, rich in fibre, carbohydrates, and folate were found to be essential for sperm quality in men. Foods rich in antioxidants can reduce oxidative stress induced by various factors, damaging the structure and function of the sperm [10]. A diet involving chicken or turkey meat was unfavorable for ovulation and may lead to infertility in females [11]. Bodyweight is also known to play a crucial role infertility. Especially in females, obese individuals are prone to irregular menstrual cycles and is evidently observed in PCOD. Obesity accounts for poor spermatogenesis and low sperm levels and in men [12]. Reproductive health is also adversely affected by psychological stress. It was observed that stressful events reduced sperm motility by 48%, sperm count by 36%, and affect overall semen quality [13]. Physical stress like prolonged working hours, psychological stress like anxiety leads to infertility in 30% of the women [14]. According to the health experts, modern-day lifestyle activities like smoking, alcohol abuse, and drug addiction completely dismantle and damages the fertilization events by inducing numerous abnormalities [15–18]. Exposure to environmental toxicants like endocrine disruptors, heavy metals, and pesticides also negatively affects fertility in both men and women. Several treatment strategies, pills, surgery, and assisted reproductive therapy (ART) are being followed to overcome infertility issues. Individuals with irreversible infertility issues undergo various ART methods like artificial insemination, *in vitro* fertilization (IVF), egg donation, gamete intrafallopian transfer (GIFT) etc. [19–21]. However, fertility drugs and ART pose adverse severe effects like impairment of sexual function, changes in libido, and erectile dysfunction [22] and promote carcinogenesis in the ovary, uterus, and breast [23–25]. Thus, an alternate, safe, and more effective therapy is need of the hour.

Plant-based products or extracts have been an integral part of our medication, especially in indigenous medical practices, traditional Chinese medicine, and Ayurveda all over the globe. Due to the rising trend in the infertility rate and severe adverse effects of current medications and practices, phytochemicals emerge as a safe and efficient alternative to treat infertility or reproductive disorders. Several plants and their extracts were evaluated for their potent anti-infertility effects, and several potential therapeutic molecules has emerged in the treatment of infertility. Several plant extracts exert their effect on male and female sexual hormones by acting as antioxidants, anti-inflammatory, targeting endocrine glands, and alleviating toxic effects of environmental substances [26–28]. Moreover, the majority of plant-based compounds studies explain their role in anti-infertility in males. Still, very few plant-based compounds have been evaluated for their effect on female infertility. Meanwhile, most of the studies conducted on male and female infertility using phytochemicals have just reported the effect of phytochemicals in the improvement of sperm count, irregular menstruation, elevated sex hormone levels, etc. [29–33]. This book chapter focuses on the role of phytochemicals in mitigating infertility issues and improvement of reproductive health.

18.2 Men/male associated infertility and the herbal interventions

Due to low sperm count/concentration, poor sperm motility, and defective sperm morphology, males account for almost 45–50% of infertility. Environment and lifestyle, physiology and genetics are critical factors for male-associated sperm dysfunction/infertility [34]. Pre-clinical and clinical evidences have underlined the potential benefits of phytochemicals in treating men's infertility in the recent past (Table 18.1).

Hachimijiogan is a Chinese medicinal herb, and its effect on oligozoospermic (decreased sperm number) men was examined [35]. Hachimijiogan (7.8 mg/day/8-28 weeks) was given to 28 oligozoospermic men, and significantly increased sperm number (78%), motility (56%), and fertility index were recorded. The author concluded that this improved semen parameter/reproductive index could be due to serum estradiol-17β as its increased level was found in those patients after hachimijiogan treatment. In another clinical study, the effect of the Androsten pill (*Tribulus terrestris* extract) has been evaluated on the semen quality of infertile men. Sixty-five men were registered for the study for 12 successive weeks. One pill/eight hours/day was prescribed to each of them, subsequently an increased sperm concentration, motility, and liquefaction time were observed after 84 days [36]. In a pilot study, the spermatogenic activity of *Withania somnifera* (Ashwagandha, a traditional Indian herb) was evaluated in 46 oligospermic males [sperm count < 20 million/mL semen; randomly divided into experimental (*n = 21*) and placebo groups (*n = 25*)] [37]. Ashwagandha was administered orally in a capsule (containing 225 mg of Ashwagandha root extract) three-time/day for 12 weeks, and semen parameters and serum hormone levels were examined at the end of the treatment. Interestingly the study found an increased sperm count (156%), sperm motility (57%), and semen volume (53%) from the baseline. Khani et al. [38] studied the antioxidant intervention of *Sesamum indicum* (sesame (contains lignans)) against semen quality of infertile men. A 0.5 mg/kg sesame was given to 25 infertile men for three months, and a significant increase in sperm count and motility was observed at the end of the treatment. The author further concluded that the antioxidant property of sesame was attributed to the improved sperm parameter. In another clinical study, 20 volunteers (20–40 years) were administered 1.75 gm/day maca (*Lepidium meyenii Walp.*; alimentary supplement) for 12 weeks in a double-blind manner, and increased sperm concentration and motility was observed in comparison to placebo, thus showing a potential herbal medicine for the patients suffering with oligozoospermia [39]. Similarly, randomized, double-blind, placebo-controlled clinical trials showed a significant improvement in the semen parameters (quality, count, and motility) in infertile men after the *Nigella sativa L.* (black cumin) seed oil and *Alpinia officinarum*

Table 18.1: Physiological/molecular effects of phytochemicals/plant formulation-based interventions in animal (male) and cell (derived from males) models with different conditions.

Sr. no.	Model of study	Condition	Phytochemical/plant formulations	Physiological/ molecular effects	Reference
1	Human	Infertility	Androsten pill (*Tribulus terrestris* extract)	sperm number ↑, sperm motility ↑	[36]
2			*Sesamum indicum* (Sesame (lignans))	sperm count ↑, sperm motility ↑	[38]
3			*Nigella sativa* L. (black cumin) seed oil and *Alpinia officinarum*	semen quality ↑, sperm count ↑, sperm motility ↑	[40,41]
4		Idiopathic infertility	Traditional Korean medicine	sperm motility ↑, improved sperm morphology	[42]
5		Oligozoospermia	Maca (*Lepidium meyenii Walp.*)	sperm concentration ↑, sperm motility ↑	[39]
6			*Withania somnifera* (Ashwagandha)	sperm count ↑, sperm motility ↑, semen volume ↑	[37]
7			Hachimijiogan	sperm number ↑, sperm motility ↑, fertility index ↑	[35]
8	Rats	Normal condition	Trans-resveratrol	sperm count ↑	[43]
9		Prediabetic	White tea	sperm concentration ↑, sperm motility ↑, sperm viability ↑, antioxidants ↑	[45,47]
10		Hyperinsulinemic	*Moringa oleifera* (drumstick tree)	testicular function ↑	[48]
11		—	*Bulbine natalensis* (perennial herb)	sexual/reproductive behavior parameters	[50]
12		—	*Apium graveolens* (celery, marshland plant)	sperm count ↑, sertoli cells ↑, primary spermatocytes ↑	[56]
13		—	*Tynnanthus fasciculatus* and *Lepidium meyenii*	sperm count ↑, testicular weight, length of the seminiferous tubules ↑	[57,58]
14		—	Mixture of- *T. terrestris, Curculigo orchioides, Allium tuberosum, Cucurbita pepo, Elephant creeper, Mucuna pruriens,* and *Terminalia catappa*	sperm count ↑, mating behavior ↑, mating performance ↑	[60]

(continued on next page)

Table 18.1: Physiological/molecular effects of phytochemicals/plant formulation-based interventions in animal (male) and cell (derived from males) models with different conditions—cont'd

Sr. no.	Model of study	Condition	Phytochemical/plant formulations	Physiological/molecular effects	Reference
15		Impaired sexual behavior	*Lecaniodiscus cupanioides* (Shrub, a Nigerian folk medicine)	sexual competency ↑	[62]
16		Arsenic induced reproductive toxicity	Ellagic and ferulic acids (food plants and fruits)	sperm quality ↑, testicular impairments ↓, oxidative stress ↓, arsenic accumulation ↓	[28]
17		Cadmium induced reproductive toxicity	*Nigella sativa* L. (thymoquinone) and l-cysteine	testicular modulation, antioxidants ↑	[71]
18		Reproductive toxicity	*Alpinia officinarum* Hance	oxidative stress ↓, testosterone ↑	[73]
19		Aflatoxin B1 induced male reproductive toxicity	Gallic acid (polyphenol in fruits and vegetables)	inflammation ↓, oxidative stress ↓, apoptosis ↓	[75]
20		Busulfan-induced reproductive toxicity	*Anthocleista djalonensis* A. (cabbage tree)	inflammation ↓, intra-testicular interleukin-6 ↓, serum nitrite ↓	[27]
21	Wistar rats	Histopathological changes, oxidative stress	*Launea taraxacifolia* (contains polyphenols)	steroidogenic enzymes ↑, cellular ATP ↑, oxidative stress ↓	[69]
22		Sperm count ↓, sperm motility ↓, abnormal sperm number ↑	Quercetin (a flavonoids)	partial restoration of; sperm count, sperm motility	[70]
23		Testicular damages	*Eruca sativa* (biannual herb)	sperm viability ↑, sperm motility ↑, histological injuries ↓	[72]
24	Albino rats	—	*Fadogia agrestis* (Nigerian shrubbery)	intromission frequency ↑, lengthy ejaculatory latency	[63]
25	Mice	—	*Aplysia dactylomela* (large sea slug)	sexual behavior ↑	[61]
26	Albino mice	Lead induced reproductive toxicity	*Cinnamomum zeylanicum* (cinnamon)	oxidative stress ↓, cell death ↓	[74]
27	Zebrafish	Reproductive impairments	*Putranjiva roxburghii* Wall seeds (Indian herbal medicine)	sperm count ↑, sperm motility ↑	[51]
28	Rabbits	—	*Zingiber officinale* (Ginger) and *Thymus vulgaris* (Thyme)	semen ↑, testosterone ↑, improved testicular structure	[49]
[a]29	Spermatogonial stem cells	Infertility	5H-purin-6-amine (*Sedum sarmentosum*)	self-renewal ↑	[59]

[a] Possible treatment measure in future.

treatment [40,41]. In a case report, sperm motility and normal sperm morphology of an idiopathic infertile male were increased by 13% and 4%, respectively, after three receiving of traditional Korean medicine [42].

Along with clinical studies, a significant rise in animal and *in vitro* based data in examining the reproductive efficacy of phytochemicals/herbal plants has been noticed. Trans-resveratrol is a natural oxidant present in grapes, and peanuts, increase sperm count in healthy rats was observed following its treatment [43]. Spermatogenesis is prone to oxidative stress; thus, the author concluded the increased sperm count could be associated with the anti-oxidative properties of trans-resveratrol. Altered testicular function and increased oxidative stress are attributed to prediabetes conditions [44]. Oliveira et al. [45] found that the administration of white tea to prediabetic rats enhanced the testicular antioxidant levels along with sperm concentration, motility, and viability of the organism. Metabolic disturbance in the prediabetic individual leads to male infertility [46]. White tea administration to prediabetic rats controls the epididymal and testicular metabolism by restoring the testicular lactate content and improved sperm motility and viability [47]. Treatment of *Moringa oleifera* (often called drumstick tree and known for traditional medicinal value) leaf extract to hyperinsulinemic rats (fed high fructose diet) improved the testicular function of the organism [48].

In Asia and Africa region, *Zingiber officinale* (Ginger) and *Thymus vulgaris* (Thyme) are considered as conventional medicine. A study on V-line male rabbits showed that treatment of aqueous extract of ginger and thyme improved the reproductive performance (characteristics of semen, levels of testosterone, and testis structure) of the exposed organism [49]. In another report, the sexual behavior of male rats was examined after the treatment of *Bulbine natalensis* (perennial herb, present in eastern and northern parts of South Africa) [50]. The study revealed that *B. natalensis* extract enhanced sexual/reproductive behavior parameters, i.e., ejaculatory latency, intromission frequency, ejaculation frequency, thus might be used for the treatment of premature ejaculation and erectile dysfunction. Recently, Balkrishna et al. [51] evaluated the effectiveness of *Putranjiva roxburghii* Wall seeds (Indian herbal medicine) on reproductive impairments on N-ethyl-N-nitrosourea (ENU) mutualized zebrafish. They observed an increased sperm count and sperm motility in ENU-mutagenic zebrafish and concluded the efficacy of *P. roxburghii* in reversing the reproductive output in the ENU-mutagenic organism. The freeze-thaw during the *in vitro* storage of sperm, the polyunsaturated fatty acids in the sperm plasma membrane makes the sperm susceptible to oxidative stress, lipid peroxidation, and cell death [52]. In this line, the use of phytochemicals for the long-term preservation of sperm for commercial purposes has been recognized. Supplementation of *Rosmarinus officinalis* (a source of polyphenols) oil before sperm storage showed a beneficial impact on sperm motility of Hubbard commercial broiler [53]. Likewise, Sobeh et al. [54] found anti-oxidative protection of bull semen via *Albizia harveyi* (a rich source of polyphenols) addition during semen storage. Another study characterized and identified the beneficial impact of addition of *Schisandra chinensis* (dispersed in China, Korea, and Russia, traditional medicinal herb) on *in vitro* storage of bovine sperm, which might be attributed to *S. chinensis's* anti-oxidative nature [55]. The effect of hydroalcoholic extract of *Apium graveolens* (celery, marshland plant) was evaluated on male rat's testis and sperm by Kooti et al. [56]. The number of sperm, sertoli cells, and primary spermatocytes was found significantly increased, which indicated a positive effect of *A. graveolens* on organism's spermatogenesis. Moreover, *Tynnanthus fasciculatus* and *Lepidium meyenii* administration to rats has increased spermatogenesis (increased daily sperm count, testicular weight, and total length of the seminiferous tubules) [57,58]. Self-renewal and differentiation are vital characteristics of any stem cell including spermatogonial stem cells. Jung et al. [59] found that 5H-purin-6-amine (isolated from *Sedum sarmentosum*) could have the capacity to maintain self-renewal of spermatogonial stem cells, which might be a new therapeutic strategy for male infertility in the future.

The substance which enhances/better the sexual desire is termed as aphrodisiac. Sahoo et al. [60] tested the aphrodisiac activity of herbal mixture. A formulation of *T. terrestris, Curculigo orchioides, Allium tuberosum, Cucurbita pepo, Elephant creeper, Mucuna pruriens,* and *Terminalia catappa* was administrated to rats, which resulted in increased sperm count, mating behavior, and mating performance in the organism. This might be a synergistic action of herbal mixture formulation. Another study by Hashim et al. [61] found an increased sexual behavior, when the male mice was treated with the lipid extract of *Aplysia dactylomela* (large sea slug). Paroxetine is an antidepressant drug, and its exposure to male rat led to impaired sexual behavior, i.e., mount, intromission and ejaculatory frequency [62]. Further, the intervention of aqueous root extract of *Lecaniodiscus cupanioides* (Shrub, a Nigerian folk medicine) restored the paroxetine-associated sexual competency, possibly through controlling the testosterone levels. Similarly, increased blood testosterone concentration was observed in male albino rats after *Fadogia agrestis* (Nigerian shrubbery) administration, which was concluded a possible reason behind the increased amount and intromission frequency and lengthy ejaculatory latency in the exposed organism [63].

Due to the enormous development in the industrial and agricultural sector, a vast number of chemicals/pollutants (pesticides, heavy metals, dyes) have been detected in different environmental compartments, i.e., soil, water, and air. Several clinical and pre-clinical studies have underlined that acute or chronic exposure to these harmful chemicals negatively impacts organism's health, including humans [64]. In this line, a considerable number of evidences conclude that these chemicals could negatively affect the reproductive output at several levels in the exposed organism [65–68]. The phytochemical intervention to mitigate the chemical-induced reproductive adversities have also been investigated in the different experimental system. Guvvala et al. [28] reported the anti-oxidative property of ellagic and ferulic acids (present in food plants and fruits) might be helpful to mitigate arsenic-induced male reproductive toxicity. The study found that arsenic exposure to male rats promoted structural and functional abnormalities in the exposed organism accompanied with variation in reproduction and oxidative stress associated genes, i.e., nuclear factor erythroid-derived 2-like 2 (Nfe2l2) and steroidogenic acute regulatory protein (StAR), and peroxisome proliferator-activated receptor gamma coactivator 1-alpha (Ppargc1a). Further, the study demonstrated that co-administration of ellagic and ferulic acids with arsenic reduced the testicular impairments and increased the sperm quality via lowering the oxidative stress and arsenic accumulation in the tissue and regulating the Nfe2l2 and StAR, and Ppargc1a expression. Another study showed that exposure of surulere polluted river water (contaminated water due to solid waste dumping) to male Wistar rats caused oxidative stress and histopathological changes, which were diminished by the administration of methanolic extract of *Launea taraxacifolia* (contains polyphenols) [69]. This was possible due to the raise in steroidogenic enzymes and cellular ATP by *L. taraxacifolia*. Likewise, atrazine (a herbicide) exposure to male Wistar rats resulted in lower sperm count, sperm motility, higher abnormal sperm number, along with decreased 3β-hydroxysteroid dehydrogenase and 17β hydroxysteroid dehydrogenase activities, which was partially controlled by quercetin (a flavonoids) administration [70]. In addition, Sayed et al. [71] showed thymoquinone (active constituent of *Nigella sativa L.*), and l-cysteine supplementation leads to alleviating the harmful effects of cadmium on rat's reproductive system through testicular modulation, antioxidants and endocrine system. *Eruca sativa* (biannual herb, also known as rocket/arugula) is being used in salad. A study on Wistar rats showed that aqueous extract of *E. sativa* lowered the Bisphenol A (BPA- used in plastic industries)-induced testicular damages (sperm viability-motility and histological injuries) [72]. In another study, lower testosterone level and increased oxidative stress were recorded in nonylphenol (endocrine disruptor)-exposed male rats, further the co-administration of alcoholic extract of *Alpinia officinarum* Hance mitigated reproductive toxicity by reducing oxidative stress and restoring testosterone level [73]. Lead acetate exposure to male albino mice increased oxidative stress, which resulted in reduced sperm concentration and abnormalities [74]. Subsequently, *Cinnamomum zeylanicum* (cinnamon) intervention showed a protective effect on the lead acetate-induced reproductive parameter, possibly by reducing oxidative stress and cell death. Recently, Owumi et al. [75] examined the efficacy of gallic acid, a polyphenol (present in fruits and vegetables), towards aflatoxin B1 induced male reproductive toxicity. The study found improved reproductive function upon gallic acid co-administration by controlling inflammation, oxidative stress, and apoptotic processes in the organism. The protective effect of *Anthocleista djalonensis A.* (commonly called cabbage tree and used in traditional medicine to treat various illnesses) against lipopolysaccharide and busulfan-induced reproductive toxicity in adult rats. The study found that lipopolysaccharide treatment significantly augmented the inflammation, as evident by intratesticular interleukin-6 and serum nitrite levels, which was significantly inhibited in the ethanolic and methanolic root extracts of *A. djalonensis* [27]. Moreover, protection was also observed in busulfan-induced impaired spermatogenesis in *A. djalonensis* administrated group.

18.3 Women/female-associated infertility and herbal interventions

Lifestyle, physiology, genetics, sexually transmitted disease, and social-economical structure such as delayed marriages can alter the female reproductive behavior, which contributes to female infertility. Polycystic Ovary Syndrome (PCOS), endometriosis, interstitial cystitis, uterine leiomyoma; are the major gynecological disorders. Pre-clinical and clinical studies have suggested that phytochemicals could be an alternative therapy for female fertility (Table 18.2). PCOS is associated with an endocrine and metabolic disbalance in the child-bearing age of women, with the prevalence of ~6–20% [76]. Menstrual dysfunction, infertility, hirsutism, acne, obesity, and metabolic syndrome are typical PCOS manifestations however, the etiology of PCOS is mainly unknown. Insulin resistance and excess inflammation have been observed in PCOS patients [77]. Guizhi Fuling is a Chinese formulation of several herbs commonly used to treat uterine fibroids. The therapeutic effect of Guizhi Fuling was studied on PCOS-insulin resistance rats (induced by letrozole + high-fat diet) [78]. The study found that Guizhi Fuling improves insulin resistance via regulating intestinal flora (abundance of *Alloprevotella, Ruminococcaceae UCG-003*, and *Lachnospiraceae UCG-008* in Guizhi Fuling group) and the inflammation in the organism. Like inflammation, increased oxidative stress is a key characteristic of PCOS patients. Rezvanfar et al. [79] demonstrated that Carvedilol and ANGIPARS™ protect the ovarian function in the PCOS-murine model (induced by letrozole), possibly through suppression of TNF-α overactivation, reduction of oxidative stress, and preserving ovarian steroidogenesis. In another study, letrozole-induced PCOS in female rats displayed a cyst in their ovary accompanied by increased oxidative stress (increased lipid peroxidation and decreased activity of antioxidant enzymes) and inflammation [80]. Further, this study showed that treating with an immunomodulatory drug (mixture of *Rosa canina, Urtica dioica*, and *Tanacetum vulgare*, selenium based) significantly reduced the free radical generation and lessened the inflammation, which might be helpful in the prevention/progression of the PCOS.

Chinese herbal medicine, Liuwei Dihuang Pills (formulation of six commonly used herbs containing *Rehmannia glutinosa, Cornus officinalis Sieb., Common Yam Rhizome, Alisma orientalis, Tree Peony Bark*, and *Poria cocos*), is widely used to treat diabetes [81]. Phosphoinositide 3-kinases/ Protein kinase B (PI3K/Akt) signaling plays a vital role in follicular differentiation, growth, and survival, and its disruption has been observed in PCOS patients. Qiu et al. [82] found that treatment of Liuwei Dihuang Pills to PCOS-female Sprague-Dawley rats (induced by letrozole + high-fat diet) recovered the follicle development and lessened the insulin resistance through regulating Cytochrome P450 family 19 subfamily a member 1 (Cyp19a1) and PI3K/Akt signaling pathway, respectively. Another study found a similar PI3K/Akt signaling mediation of another Chinese herb named Heqi San (used to treat metabolic disease) to treat dehydroepiandrosterone induced

Table 18.2: Physiological/molecular effects of phytochemicals/plant formulation-based interventions in animal (female) and cell (derived from females) models with different conditions.

Sr. no.	Model of study	Condition	Phytochemical/plant formulations	Physiological/ molecular effects	Reference
1	Human	PCOS	Shouwu Jiangqi Decoction (Chinese herbal medicine)	improved insulin resistance	[86]
2			Linum usitatissiumum (Flaxseed)	body weight ↓, benefited insulin metabolism	[88]
3			Heyan Kuntai capsule (Chinese medicine mixture of Radix Rehmanniae, Rhizoma Coptidis, Radix Paeoniae Alba, Radix Scutellariae Baicalensis, Colla Corii Asini, Poria)	improved glucose & lipid metabolism, improved insulin resistance, type 2 diabetes mellitus ↓	[89]
4			Dingkun Pill (Chinese traditional medicine)	improved insulin resistance	[90]
5			Mixture of Mentha spicata, Zingiber officinale, Cinnamomum zeylanicum, and Citrus sinensis	Antioxidants ↑, glycemic control, and pregnancy rate ↑	[91]
6		Endometriosis	Origanum majorana (marjoram herb)	improved hormonal profile	[92]
7			Resveratrol (a polyphenol)	endometriosis ↓	[96]
8			Luteolin (a flavonoid)	arrested cell cycle, apoptosis ↑, cell proliferation	[101]
9		Endometriosis-associated infertility	Chinese medicine	improved live birth rate, follicular development, endometrial receptivity	[104]
10	Uterine leiomyoma cells	Uterine leiomyoma	Deoxyelephantopin extract-Elephantopus scaber (flowering plant)	apoptosis ↑, cell cycle arrest ↑, cell proliferation ↓	[107]
11			Quercetin (indole-3-carbinol treatment)	anti-fibrotic, migration ↓, proliferation ↓	[108]
12			Rhus verniciflua stoke (Fisetin)	cytotoxicity ↑, cell death ↑, cell cycle arrest ↑	[111]
[a]13	Leiomyoma and myometrial cells	—	Mixture of Alba (strawberry cultivar), Romina (strawberry cultivar), fraction anthocyanin extracts	apoptosis ↑, oxidative stress ↑	[109]

(continued on next page)

Table 18.2: Physiological/molecular effects of phytochemicals/plant formulation-based interventions in animal (female) and cell (derived from females) models with different conditions—cont'd

Sr. no.	Model of study	Condition	Phytochemical/plant formulations	Physiological/molecular effects	Reference
14	Rat	PCOS, insulin resistance	Guizhi Fuling	regulate intestinal flora & inflammation	[78]
15		PCOS, Cyst in ovary, oxidative stress ↑, lipid peroxidation ↑, antioxidant enzymes ↓	Mixture of *Rosa canina*, *Urtica dioica*, and *Tanacetum vulgare*, selenium based	inflammation ↓, oxidative stress ↓, PCOS ↓	[80]
16		PCOS	Hochu-ekki-to (Japanese herbal medicine)	reproductive function ↑, immune function ↑	[85]
17		Benzo[a]pyrene and N-methyl-N-nitrosourea induced cytotoxicity	*Calliandra portoricensis* (a shrub)	inflammation ↓, apoptosis ↑, oxidative stress ↓	[29]
18		BPA-induced uterine abnormalities	*Ficus deltoidei* (mistletoe fig)	restore uterine weight, lipid peroxidation ↓, stromal cell space ↑,	[112]
19		Cadmium-induced ovarian toxicity	Tualang honey (Malaysian honey)	maintenance of homeostasis, gonadotropin hormones, ovarian structures	[113]
20		Endometriosis	*Melilotus officinalis* (L.)	endometrial foci ↓, inflammation →	[99]
21			*Anthemis austriaca* Jacq.	endometrial foci ↓, cytokines →	[100]
22			β-caryophyllene	cell death ↑, endometriosis ↓	[103]
23	Murine model	PCOS	Carvedilol and ANGIPARS™	TNF-a ↓, oxidative stress ↓, ovarian steroidogenesis ↑	[79]
24	Sprague-Dawley rats	PCOS, follicular differentiation ↓, growth ↓, survival ↓	Mixture of *Rehmannia glutinosa*, *Cornus officinalis* Sieb., Common Yam Rhizome, *Alisma orientalis*, Tree Peony Bark, and *Poria cocos* (Liuwei Dihuang Pills)	follicle development ↑, insulin resistance ↓	[81,82]
25		PCOS	Heqi San (Chinese herb)	–	[83]
26			Gui Zhu Yi Kun (traditional Chinese medicine)	autophagy ↓	[84]
27	Wistar rats	Cisplatin-induced reproductive toxicity	Resveratrol (a polyphenol)	lipid peroxidation ↓, antioxidants ↑, inflammation ↓, cell death ↓	[114]
28	BALB/c mice	Endometriosis	Curcumin	endometriosis ↓	[97]
29	Mice		*Phaleria macrocarpa*		[98]
30	VK2/E6E7 and End1/E6E7	–	Chrysin (a natural flavone)	apoptosis ↑, proliferation ↓	[102]

a Possible treatment measure in future

PCOS-female Sprague-Dawley rats. Moreover, the study also identified some potential miRNAs, which might have an essential role in Heqi San therapeutics [83]. The phosphorylation/activation of 5' AMP-activated protein kinase (AMPK) by p53 may trigger autophagy, and the activation of autophagy promotes PCOS. The underline mechanism of Gui Zhu Yi Kun (traditional Chinese medicine) induced PCOS protection was studied by Xing et al. [84]. The study revealed that increased autophagy rate in PCOS-granulosa cells in rats was inhibited by Gui Zhu Yi Kun mediated p53/AMPK signaling. Park et al. [85] studied a protective action of Japanese herbal medicine, Hochu-ekki-to (prescribed for viral and bacterial infection), against the stress-induced rat PCOS model. Organisms exposed to adrenocorticotropic hormone (ACTH) or cold temperatures found abnormal follicle development and elevated levels of Hsp90 (stress protein), steroid hormone receptors in ovaries, and reproductive hormone in blood was reduced with the Hochu-ekki-to treatment.

Shouwu Jiangqi Decoction (SWJQD) is a traditional Chinese herbal medicine. In a randomized controlled trial on 81 PCOS patients, three months of SWJQD administration with acupuncture to patients showed improved insulin resistance in these PCOS patients [86]. Flaxseed (*Linum usitatissiumum*) is an ample source of α-linolenic acid and soluble and insoluble dietary fibres [87]. A randomized open-labeled controlled clinical study on 41 PCOS patients was conducted by Haidari et al. [88]. The patients were divided into two groups 1. patients who received Flaxseed along with lifestyle modification and 2. Only lifestyle modification for 12 weeks. The study found that compared to the lifestyle modification group, the Flaxseed supplemented group with lifestyle modification showed a reduced body weight, and benefited insulin metabolism. Heyan Kuntai capsule (HYKT) is a Chinese patented medicine mixture of *Radix Rehmanniae, Rhizoma Coptidis, Radix Paeoniae Alba, Radix Scutellariae Baicalensis, Colla Corii Asini,* and *Poria*. In a randomized double-blinded, placebo-controlled clinical trial, six months of HYKT administration to PCOS patients significantly controlled the glucose and lipid metabolism ailment; improved insulin resistance in the PCOS patients. A nationwide cohort study reported that Chinese herbal medicines (five herbal formulas and two herbs) are effectively lowered the occurrence of type 2 diabetes mellitus in PCOS patients [89]. *Paeonia lactiflora* (Chinese Peony, the herbaceous perennial flowering plant) was common in all five herbal formulas. Dingkun Pill (DKP) is another Chinese traditional medicine, which is commonly prescribed for gynaecological issues. In a randomized controlled study, 117 PCOS patients were divided into three groups, i.e., group 1: administrated 7 gm DKP; group 2: one Diane-35 tablet (medicine for PCOS); group 3: 7 gm DKP plus one Diane-35 tablet for three successive months (seven drug-free days after 21 consecutive days administration). DKP supplementation was found to improve insulin resistance in patients suffering from PCOS [90]. In a single-blinded randomized study, 60 PCOS patients have received herbal mixture (*Mentha spicata, Zingiber officinale, Cinnamomum zeylanicum,* and *Citrus sinensis*) alone or with clomiphene citrate (medicine for women fertility) and showed a positive impact on the antioxidant level, glycemic control, and pregnancy rate in PCOS patients [91]. A randomized controlled pilot study involving a *Origanum majorana* (marjoram herb, controls menstrual cycle) to examine marjoram's efficacy in PCOS was completed [92]. PCOS patients who received marjoram twice a day for one month showed a positive effect on the hormonal profile of PCOS patients.

Endometriosis occurs in the reproductive age of women and is defined by the presence of a similar uterus-like tissue that grows other than the uterine cavity. According to various sources, the prevalence of endometriosis is ~6-15% worldwide [93,94]. The common risk factors and theories associated with endometriosis include family history, uterus or fallopian tube defects, immune disorder, and retrograde menstrual flow [95]. Chronic pelvic pain, reduced fertility, lumbar pain, and dysmenorrhea are the common symptoms associated with endometriosis [94]. According to retrograde menstrual flow theory, growth and angiogenesis are important factors in the progression of endometriosis. Vascular endothelial growth factor (VEGF), transforming growth factor-β (TGF-β), and matrix metalloproteinase-9 (MMP9) are essential for cell adhesion, growth, and angiogenesis. Arablou et al. [96] found that resveratrol (a polyphenol) intervention regulated the expression of VEGF, TGF-β, and MMP9 in endometrial stromal cells, which might lessen the endometriosis progression. Curcumin treatment inhibited Nuclear factor kappa B (NFκB) translocation, matrix metalloproteases (MMP) expression, and inducing mitochondrial-mediated apoptosis in endometriotic developed female BALB/c mice, resulting in endometriosis regression [97].

Similarly, a mice study demonstrated that *Phaleria macrocarpa* (Indonesia native, anti-apoptotic function) treatment reduced growth of endometriosis lesions, possibly via regulating the proliferation and apoptosis status of endometriotic mice [98]. Another recent study has suggested that exposure to methanolic extract of *Melilotus officinalis* (L.) to endometriotic rat lessened the endometrial foci and inflammation, possibly through the glycosylated flavonoids present in the extract [99]. Similarly, methanolic extract of *Anthemis austriaca* Jacq. flowers contain flavonoids and sterols, and its exposure to the endometriosis rat model leads to suppression of endometrial foci, cytokines, and adhesion in the organism [100]. Luteolin is a flavonoid exist in fruits and vegetables. Park et al. [101] revealed luteolin treatment to human VK2/E6E7 and End1/E6E7 cells (an endometriosis model) arrested the cell cycle and promoted apoptosis, which lastly inhibited the cell proliferation. Another experimental study showed a pro-apoptotic and anti-proliferative effect of chrysin (a natural flavone) in VK2/E6E7 and End1/E6E7 via regulating cytosolic calcium level and generation of high cellular reactive oxygen species (ROS) in the cells [102]. β-caryophyllene has a strong anti-inflammatory activity, and its effect on endometriosis, fertility and reproduction was investigated in endometrial implanted female rats [103]. β-caryophyllene promoted significant cell death in the luminal epithelium of the cyst, accompanied by a 52.5% reduction in endometriotic implant growth.

In a randomized, double-blind placebo parallel controlled clinical trial, Chinese medicine improved the following parameters: pregnancy outcome and live birth rate, improve follicular development, and endometrial receptivity in endometriosis patients [104]. Zhao et al. [105] examined the efficacy of Chinese medicine to control the recurrence of pelvic endometriosis and found that a prevented effect of Chinese medicine and better the conception rate.

Uterine leiomyoma is a benign tumor of the female genital tract, with a high prevalence in >40 aged females [106]. Pelvic pain, abnormal vaginal bleeding, and infertility are the common outputs of uterine leiomyoma. Deoxyelephantopin (DOE) is extracted from *Elephantopus scaber* (flowering plant) and possesses an anti-tumor activity. Recently an anti-tumor activity of DOE was reported by Pandey et al. [107]. The study revealed that DOE treatment induces ROS-dependent mitochondrial apoptosis and cell-cycle arrest in the primary cell culture of uterine leiomyoma. Further, the study also observed downregulation of oncogenic lncRNAs (H19, HOTAIR, BANCR and ROR), which might inhibit uterine leiomyoma cell proliferation. In another study, the status of extracellular protein, cell migration, and proliferation rate of myometrial and uterine leiomyoma cells were examined after quercetin or indole-3-carbinol treatment [108]. This study showed that quercetin or indole-3-carbinol modulated the expression of collagen 1A1, fibronectin and showed an anti-fibrotic, anti-migratory, and anti-proliferative effect in the exposed cells. In order to examine the impact of strawberry on uterine leiomyomas, Giampieri et al. [109] exposed the leiomyoma and myometrial cells to methanolic extract of Alba (strawberry cultivar), Romina (strawberry cultivar), and its fraction anthocyanin. They observed more apoptosis and ROS generation in the leiomyoma cells in the Romina anthocyanin-treated group than other tested cultivars. They concluded with the possibility of their use for future therapeutics for uterine leiomyoma. However, in the same study, the strawberry cultivars positively impacted myometrial cells, i.e., increased cell viability and reduced ROS generation. *Rhus verniciflua* stoke is an Asian tree with anti-oxidant, anti-inflammatory, and anti-cancer importance [110]. Fisetin is ane of the bioactive compound of *R. verniciflua* stoke, and its impact on uterine leiomyomas was studied by Lee et al. [111]. The study observed that fisetin treatment exerted cytotoxicity, cell death, and cell cycle arrest to leiomyoma cells. Further, the study demonstrated that the intrinsic apoptotic pathway and autophagy was involved fisetin-induced death of leiomyoma cells.

Like the male, environmental toxicants negatively impact the female reproductive system/function, and several studies have underlined the phytochemical intervention against chemical-induced female reproductive toxicity. An increased uterus and ovary weight was observed after co-exposure of benzo[a]pyrene (BaP) and N-methyl-N-nitrosourea (NMU) to rat [29]. Simultaneously, elevated lipid peroxidation, decreased glutathione S-transferase

(GST), glutathione peroxidase (GPx), and catalase (CAT) activity were found in the organism's ovarian and/or uterine tissues during co-exposure of BaP and NMU. However, simultaneous supplementation with *Calliandra portoricensis* (a large shrub) protected against BaP- and NMU-induced cytotoxicity and weight changes via its anti-inflammatory, anti-apoptotic, and anti-oxidative properties. *Ficus deltoidei* (mistletoe fig) is a large shrub known for its pharmacological properties. BPA is an endocrine disruptor and a potent reproductive toxicant. Zaid et al. [112] found that co-exposure of *F. deltoidei* with BPA effectively reduced uterine abnormalities such as restored the uterine weight, lessen lipid peroxidation, enlargement of stromal cells, expanding the interstitial stromal cells space. Tualang honey is Malaysian honey extracted from the comb of *Apis dorsata* bees. Tualang honey has therapeutic importance as it contains phenolic acids and flavonoids. A study by Ruslee et al. [113] showed that daily Tualang honey supplementation could protect the female reproductive system from cadmium-induced ovarian toxicity by stabilizing the redox homeostasis, restoring the gonadotropin hormones levels, and reducing the ovarian morphological anomalies. Resveratrol (polyphenol, present in edible plants such as grapes and wine) was co-administrated with cisplatin (chemotherapeutic drug, causes reproductive toxicity) to female Wistar rats [114]. Resveratrol was found to abolish cisplatin-induced reproductive toxicity in the exposed organism. Decreased lipid peroxidation and increased glutathione content and GPx, superoxide dismutase, and CAT activity in the ovary and uterus were recorded in the resveratrol+cisplatin group. Along with antioxidant action, down-regulation of NF-κB and Cyclooxygenase-2 (Cox-2), Cleaved Caspase-3 and Bcl-2-associated X protein (Bax), and increased B-cell lymphoma 2 (Bcl-2) protein concluded that resveratrol was countering the cisplatin-associated reproductive toxicity through maintaining redox homeostasis, lowering inflammation and cell death in the exposed organism.

18.4 Conclusion

Like other biological processes/diseases/disorders, the use of phytochemicals/herbal medicine in reproductive biology has been well-recognized. They could significantly positively impact male and female reproduction behaviors in normal and stressful situations (Fig. 18.1). The positive effects on male's reproduction include increased sperm count, quality, motility, shape, hormone, and infection control, whereas female's benefits include control of the PCOS symptoms and uterine leiomyoma, endometriosis regression. Although the different extracts or different parts of plats are used to improve the human reproductive performance, the knowledge of effective bioactive compounds is still largely unknown, which will be a future challenge for the researchers. Moreover, to take herbal medicine to the next level of drug development, their mode of action should be deciphered. In addition, global policy should be developed to regulate the designs and production of herbal medicine/phytochemicals.

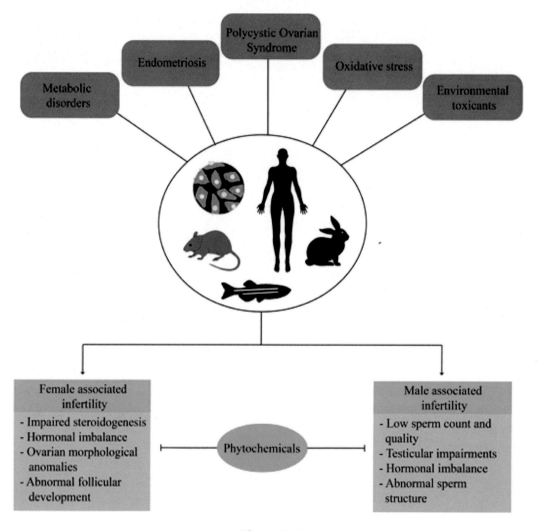

Figure 18.1
Effect of phytochemicals on male and female associated reproductive abnormalities.

Acknowledgement

The authors thank the Director Nitte University Centre for Science Education and Research, Mangalore. The financial support from the Science and Engineering Research Board (ECR/2016/001863) and Nitte Research Grant (NU/DR/NUFR1/NUCSER/2019-20/01) to AS is gratefully acknowledged. Financial support to LCD and JGP as Directorate of Minorities, Karnataka (DOM/FELLOWSHIP/CR-33/2018-19) and CSIR-RA (09/1257(0001)/2019-EMR-I) respectively is thankfully acknowledged

Abbreviations

ACTH	Adrenocorticotropic hormone
AMPK	5′ AMP-activated protein kinase
BaP	Benzo[a]pyrene
Bax	Bcl-2-associated X protein
Bcl-2- B	cell lymphoma 2
BPA	Bisphenol A
CAT	Catalase
Cox-2	Cyclooxygenase-2
Cyp19a1	Cytochrome P450 family 19 subfamily a member 1
DKP	Dingkun Pill
DOE	Deoxyelephantopin
End1/E6E7	Endocervical cell line
ENU	N-ethyl-N-nitrosoure
GPx	Glutathione peroxidase
GST	Glutathione S-transferase
HYKT	Heyan kuntai capsule
MMP	Matrix metalloproteases
MMP9	Matrix metalloproteinase-9
Nfe2l2	Nuclear factor erythroid-derived 2-like 2
NFκB	Nuclear factor kappa B
NMU	N-methyl-N-nitrosourea
PCOS	Polycystic ovary syndrome
PI3K/Akt	Phosphoinositide 3-kinases/ Protein kinase B
Ppargc1a	Peroxisome proliferator-activated receptor gamma coactivator 1-alpha
ROS	Reactive oxygen species
StAR	Steroidogenic acute regulatory protein
SWJQD	Shouwu jiangqi decoction
TGF-β	Transforming growth factor-β
VEGF	Vascular endothelial growth factor
VK2/E6E7	Cell line derived from normal vaginal mucosal tissue

References

[1] SJ Segal, The physiology of human reproduction, Sci Am 231 (3) (1974) 52–62.
[2] B Soygur, L Sati, The role of syncytins in human reproduction and reproductive organ cancers, Reproduction 152 (5) (2016) R167–R178.
[3] AD Burkart, B Xiong, B Baibakov, M Jimenez-Movilla, J Dean, Ovastacin, a cortical granule protease, cleaves ZP2 in the zona pellucida to prevent polyspermy, J Cell Biol 197 (1) (2012) 37–44.
[4] D Clift, M Schuh, Restarting life: fertilization and the transition from meiosis to mitosis, Nat Rev Mol Cell Biol 14 (9) (2013) 549–562.
[5] B Gellersen, JJ Brosens, Cyclic decidualization of the human endometrium in reproductive health and failure, Endocr Rev 35 (6) (2014) 851–905.
[6] SE Wamaitha, KK Niakan, Human pre-gastrulation Development, Curr Top Dev Biol 128 (2018) 295–338.
[7] H Sun, TT Gong, YT Jiang, S Zhang, YH Zhao, QJ Wu, Global, regional, and national prevalence and disability-adjusted life-years for infertility in 195 countries and territories, 1990-2017: results from a global burden of disease study, 2017, Aging (Albany NY) 11 (23) (2019) 10952–10991.
[8] RE Jones, KH Lopez, Infertility, Human Reprod Biol 1 (2014) 283–299.

[9]   R Sharma, KR Biedenharn, JM Fedor, A Agarwal, Lifestyle factors and reproductive health: taking control of your fertility, Reprod Biol Endocrinol 11 (2013) 66.

[10] M Cocuzza, SC Sikka, KS Athayde, A Agarwal, Clinical relevance of oxidative stress and sperm chromatin damage in male infertility: an evidence based analysis, Int Braz J Urol 33 (5) (2007) 603–621.

[11] JE Chavarro, JW Rich-Edwards, BA Rosner, WC Willett, Protein intake and ovulatory infertility, Am J Obstet Gynecol 198 (2) (2008) e1–e7 210.

[12] R Pasquali, L Patton, A Gambineri, Obesity and infertility, Curr Opin Endocrinol Diabetes Obes 14 (6) (2007) 482–487.

[13] G Ragni, A Caccamo, Negative effect of stress of in vitro fertilization program on quality of semen, Acta Eur Fertil 23 (1) (1992) 21–23.

[14] BD Peterson, CR Newton, T Feingold, Anxiety and sexual stress in men and women undergoing infertility treatment, Fertil Steril 88 (4) (2007) 911–914.

[15] N Battista, N Pasquariello, M Di Tommaso, M Maccarrone, Interplay between endocannabinoids, steroids and cytokines in the control of human reproduction, J Neuroendocrinol 20 (Suppl 1) (2008) 82–89.

[16] A Calogero, R Polosa, A Perdichizzi, F Guarino, S La Vignera, A Scarfia, et al., Cigarette smoke extract immobilizes human spermatozoa and induces sperm apoptosis, Reprod Biomed Online 19 (4) (2009) 564–571.

[17] OR Koch, G Pani, S Borrello, R Colavitti, A Cravero, S Farre, et al., Oxidative stress and antioxidant defenses in ethanol-induced cell injury, Mol Aspects Med 25 (1-2) (2004) 191–198.

[18] FI Sharara, SN Beatse, MR Leonardi, D Navot, RT Scott Jr., Cigarette smoking accelerates the development of diminished ovarian reserve as evidenced by the clomiphene citrate challenge test, Fertil Steril 62 (2) (1994) 257–262.

[19] RS Sharma, R Saxena, R Singh, Infertility & assisted reproduction: a historical & modern scientific perspective, Indian J Med Res 148 (Suppl) (2018) S10–S14.

[20] J Stanhiser, AZ Steiner, Psychosocial aspects of fertility and assisted reproductive technology, Obstet Gynecol Clin North Am 45 (3) (2018) 563–574.

[21] M Szamatowicz, Assisted reproductive technology in reproductive medicine - possibilities and limitations, Ginekol Pol 87 (12) (2016) 820–823.

[22] MK Samplaski, AK Nangia, Adverse effects of common medications on male fertility, Nat Rev Urol 12 (7) (2015) 401–413.

[23] N Kanakas, T Mantzavinos, Fertility drugs and gynecologic cancer, Ann N Y Acad Sci 1092 (2006) 265–278.

[24] G Lo Russo, F Tomao, GP Spinelli, AA Prete, V Stati, PB Panici, et al., Fertility drugs and breast cancer risk, Eur J Gynaecol Oncol 36 (2) (2015) 107–113.

[25] F Tomao, G Lo Russo, GP Spinelli, V Stati, AA Prete, N Prinzi, et al., Fertility drugs, reproductive strategies and ovarian cancer risk, J Ovarian Res 7 (2014) 51.

[26] SO Abarikwu, CL Onuah, SK Singh, Plants in the management of male infertility, Andrologia 52 (3) (2020) e13509.

[27] CY Ezirim, SO Abarikwu, AA Uwakwe, CJ Mgbudom-Okah, Protective effects of Anthocleista djalonensis A. Chev root extracts against induced testicular inflammation and impaired spermatogenesis in adult rats, Mol Biol Rep 46 (6) (2019) 5983–5994.

[28] PR Guvvala, JP Ravindra, S Selvaraju, A Arangasamy, KM Venkata, Ellagic and ferulic acids protect arsenic-induced male reproductive toxicity via regulating Nfe2l2, Ppargc1a and StAR expressions in testis, Toxicology 413 (2019) 1–12.

[29] AO Adefisan, JC Madu, SE Owumi, OA Adaramoye, Calliandra portoricensis ameliorates ovarian and uterine oxido-inflammatory responses in N-methyl-N-nitrosourea and benzo[a]pyrene-treated rats, Exp Biol Med (Maywood) 245 (16) (2020) 1490–1503.

[30] N Jaradat, AN Zaid, Herbal remedies used for the treatment of infertility in males and females by traditional healers in the rural areas of the West Bank/Palestine, BMC Complement Altern Med 19 (1) (2019) 194.

[31]   Z Li, Y Yu, Y Li, F Ma, Y Fang, C Ni, et al., Taxifolin attenuates the developmental testicular toxicity induced by di-n-butyl phthalate in fetal male rats, Food Chem Toxicol 142 (2020) 111482.

[32]   M Majid, F Ijaz, MW Baig, B Nasir, MR Khan, IU Haq, Scientific validation of ethnomedicinal use of Ipomoea batatas L. Lam. as aphrodisiac and gonadoprotective agent against bisphenol A induced testicular toxicity in male Sprague Dawley rats, Biomed Res Int 2019 (2019) 8939854.

[33]   GT Mbemya, LA Vieira, FG Canafistula, ODL Pessoa, APR Rodrigues, Reports on in vivo and in vitro contribution of medicinal plants to improve the female reproductive function, Reprodução & Climatério 32 (2) (2017) 109–119.

[34]   A Agarwal, S Baskaran, N Parekh, CL Cho, R Henkel, S Vij, et al., Male infertility, Lancet 397 (10271) (2021) 319–333.

[35]   S Usuki, Hachimijiogan changes serum hormonal circumstance and improves spermatogenesis in oligozoospermic men, Am J Chin Med 14 (1-2) (1986) 37–45.

[36]   RM Salgado, MH Marques-Silva, E Goncalves, AC Mathias, JG Aguiar, P Wolff, Effect of oral administration of Tribulus terrestris extract on semen quality and body fat index of infertile men, Andrologia 49 (5) (2017).

[37]   VR Ambiye, D Langade, S Dongre, P Aptikar, M Kulkarni, A Dongre, Clinical evaluation of the spermatogenic activity of the root extract of ashwagandha (Withania somnifera) in oligospermic males: a pilot study, Evid Based Complement Alternat Med 2013 (2013) 571420.

[38]   B Khani, SR Bidgoli, F Moattar, H Hassani, Effect of sesame on sperm quality of infertile men, J Res Med Sci 18 (3) (2013) 184–187.

[39]   I Melnikovova, T Fait, M Kolarova, EC Fernandez, L Milella, Effect of Lepidium meyenii Walp. on semen parameters and serum hormone levels in healthy adult men: a double-blind, randomized, placebo-controlled pilot study, Evid Based Complement Alternat Med 2015 (2015) 324369.

[40]   M Kolahdooz, S Nasri, SZ Modarres, S Kianbakht, HF Huseini, Effects of Nigella sativa L. seed oil on abnormal semen quality in infertile men: a randomized, double-blind, placebo-controlled clinical trial, Phytomedicine 21 (6) (2014) 901–905.

[41]   F Kolangi, H Shafi, Z Memariani, M Kamalinejad, S Bioos, SGA Jorsaraei, et al., Effect of Alpinia officinarum Hance rhizome extract on spermatogram factors in men with idiopathic infertility: a prospective double-blinded randomised clinical trial, Andrologia 51 (1) (2019) e13172.

[42]   J Jo, SH Lee, JM Lee, UM Jerng, Semen quality improvement in a man with idiopathic infertility treated with traditional Korean medicine: a case report, Explore (NY) 11 (4) (2015) 320–323.

[43]   ME Juan, E Gonzalez-Pons, T Munuera, J Ballester, JE Rodriguez-Gil, JM Planas, Trans-resveratrol, a natural antioxidant from grapes, increases sperm output in healthy rats, J Nutr 135 (4) (2005) 757–760.

[44]   L Rato, AI Duarte, GD Tomas, MS Santos, PI Moreira, S Socorro, et al., Pre-diabetes alters testicular PGC1-alpha/SIRT3 axis modulating mitochondrial bioenergetics and oxidative stress, Biochim Biophys Acta 1837 (3) (2014) 335–344.

[45]   PF Oliveira, GD Tomas, TR Dias, AD Martins, L Rato, MG Alves, et al., White tea consumption restores sperm quality in prediabetic rats preventing testicular oxidative damage, Reprod Biomed Online 31 (4) (2015) 544–556.

[46]   L Rato, MG Alves, AI Duarte, MS Santos, PI Moreira, JE Cavaco, et al., Testosterone deficiency induced by progressive stages of diabetes mellitus impairs glucose metabolism and favors glycogenesis in mature rat Sertoli cells, Int J Biochem Cell Biol 66 (2015) 1–10.

[47]   TR Dias, MG Alves, L Rato, S Casal, BM Silva, PF Oliveira, White tea intake prevents prediabetes-induced metabolic dysfunctions in testis and epididymis preserving sperm quality, J Nutr Biochem 37 (2016) 83–93.

[48]   MA Mohamed, MA Ahmed, RA El Sayed, Molecular effects of Moringa leaf extract on insulin resistance and reproductive function in hyperinsulinemic male rats, J Diabetes Metab Disord 18 (2) (2019) 487–494.

[49]   MA Kandeil, AEH Mohamed, M Abdel Gabbar, RR Ahmed, SM Ali, Ameliorative effects of oral ginger and/or thyme aqueous extracts on productive and reproductive performance of V-line male rabbits, J Anim Physiol Anim Nutr (Berl) 103 (5) (2019) 1437–1446.

[50]   MT Yakubu, AJ Afolayan, Effect of aqueous extract of Bulbine natalensis (Baker) stem on the sexual behaviour of male rats, Int J Androl 32 (6) (2009) 629–636.

[51]   A Balkrishna, P Nain, M Joshi, L Khandrika, A Varshney, Supercritical fluid extract of Putranjiva roxburghii Wall. Seeds mitigates fertility impairment in a zebrafish model, Molecules 26 (4) (2021).

[52]   M Meseguer, N Garrido, C Simon, A Pellicer, J Remohi, Concentration of glutathione and expression of glutathione peroxidases 1 and 4 in fresh sperm provide a forecast of the outcome of cryopreservation of human spermatozoa, J Androl 25 (5) (2004) 773–780.

[53]   L Touazi, B Aberkane, Y Bellik, N Moula, M Iguer-Ouada, Effect of the essential oil of Rosmarinus officinalis (L.) on rooster sperm motility during 4 degrees C short-term storage, Vet World 11 (5) (2018) 590–597.

[54]   M Sobeh, SA Hassan, MA El Raey, WA Khalil, MAE Hassan, M Wink, Polyphenolics from Albizia harveyi exhibit antioxidant activities and counteract oxidative damage and ultra-structural changes of cryopreserved bull semen, Molecules 22 (11) (2017).

[55]   E Tvrda, J Michalko, J Arvay, NL Vukovic, E Ivanisova, M Duracka, et al., Characterization of the omija (Schisandra chinensis) extract and its effects on the bovine sperm vitality and oxidative profile during in vitro storage, Evid Based Complement Alternat Med. 2020 (2020) 7123780.

[56]   W Kooti, E Mansouri, M Ghasemiboroon, M Harizi, D Ashtary-Larky, R Afrisham, The effects of hydroalcoholic extract of Apium graveolens leaf on the number of sexual cells and testicular structure in rat, Jundishapur J Nat Pharm Prod 9 (4) (2014) e17532.

[57]   GF Gonzales, J Nieto, J Rubio, M Gasco, Effect of Black maca (Lepidium meyenii) on one spermatogenic cycle in rats, Andrologia 38 (5) (2006) 166–172.

[58]   FC Melo, SL Matta, TA Paula, ML Gomes, LC Oliveira, The effects of Tynnanthus fasciculatus (Bignoniaceae) infusion on testicular parenchyma of adult Wistar rats, Biol Res 43 (4) (2010) 445–450.

[59]   SE Jung, YH Kim, S Cho, BJ Kim, HS Lee, S Hwang, et al., A phytochemical approach to promotion of self-renewal in murine spermatogonial stem cell by using sedum sarmentosum extract, Sci Rep 7 (1) (2017) 11441.

[60]   HB Sahoo, S Nandy, AK Senapati, SP Sarangi, SK Sahoo, Aphrodisiac activity of polyherbal formulation in experimental models on male rats, Pharmacognosy Res 6 (2) (2014) 120–126.

[61]   R Hashim, NA Roslan, FH Zulkipli, JM Daud, Screening of aphrodisiac property in sea slug, Aplysia dactylomela, Asian Pac J Trop Med 7 (7S1) (2014) S150–S154.

[62]   A TO, QO Nurudeen, MT Yakubu, Aphrodisiac effect of aqueous root extract of Lecaniodiscus cupanioides in sexually impaired rats, J Basic Clin Physiol Pharmacol 25 (2) (2014) 241–248.

[63]   MT Yakubu, MA Akanji, AT Oladiji, Aphrodisiac potentials of the aqueous extract of Fadogia agrestis (Schweinf. Ex Hiern) stem in male albino rats, Asian J Androl 7 (4) (2005) 399–404.

[64]   JG Paithankar, S Saini, S Dwivedi, A Sharma, DK Chowdhuri, Heavy metal associated health hazards: an interplay of oxidative stress and signal transduction, Chemosphere 262 (2021) 128350.

[65]   AE Calogero, M Fiore, F Giacone, M Altomare, P Asero, C Ledda, et al., Exposure to multiple metals/metalloids and human semen quality: a cross-sectional study, Ecotoxicol Environ Saf 215 (2021) 112165.

[66]   Y Gao, J Hong, Y Guo, M Chen, AK Chang, L Xie, et al., Assessment spermatogenic cell apoptosis and the transcript levels of metallothionein and p53 in Meretrix meretrix induced by cadmium, Ecotoxicol Environ Saf 217 (2021) 112230.

[67]   S Misra, A Singh, HR C, V Sharma, MK Reddy Mudiam, KR Ram, Identification of drosophila-based endpoints for the assessment and understanding of xenobiotic-mediated male reproductive adversities, Toxicol Sci 141 (1) (2014) 278–291.

[68]   HR Siddique, K Mitra, VK Bajpai, K Ravi Ram, DK Saxena, DK Chowdhuri, Hazardous effect of tannery solid waste leachates on development and reproduction in Drosophila melanogaster: 70kDa heat shock protein as a marker of cellular damage, Ecotoxicol Environ Saf 72 (6) (2009) 1652–1662.

[69]   JK Akintunde, MO Aina, AA Boligon, Launaea taraxacifolia (Willd.) Amin ex C. Jeffrey inhibits oxidative damage and econucleotidase followed by increased cellular ATP in testicular cells of rats exposed to metropolitan polluted river water, J Basic Clin Physiol Pharmacol 29 (2) (2018) 141–153.

[70]   SO Abarikwu, EO Farombi, Quercetin ameliorates atrazine-induced changes in the testicular function of rats, Toxicol Ind Health 32 (7) (2016) 1278–1285.

[71]   MM Sayed, KMA Hassanein, W Senosy, Protective effects of thymoquinone and l-cysteine on cadmium-induced reproductive toxicity in rats, Toxicol Rep 1 (2014) 612–620.

[72]   D Grami, K Rtibi, I Hammami, S Selmi, L De Toni, C Foresta, et al., Protective action of eruca sativa leaves aqueous extracts against bisphenol a-caused in vivo testicular damages, J Med Food 23 (6) (2020) 600–610.

[73]   M Pirzadeh, M Barary, SM Hosseini, S Kazemi, AA Moghadamnia, Ameliorative effect of Alpinia officinarum Hance extract on nonylphenol-induced reproductive toxicity in male rats, Andrologia 53 (6) (2021) e14063.

[74]   RAR Elgawish, HMA Abdelrazek, Effects of lead acetate on testicular function and caspase-3 expression with respect to the protective effect of cinnamon in albino rats, Toxicol Rep 1 (2014) 795–801.

[75]   SE Owumi, IA Adedara, AP Akomolafe, EO Farombi, AK Oyelere, Gallic acid enhances reproductive function by modulating oxido-inflammatory and apoptosis mediators in rats exposed to aflatoxin-B1, Exp Biol Med (Maywood) 245 (12) (2020) 1016–1028.

[76]   M Szczuko, P Sankowska, M Zapalowska-Chwyc, P Wysokinski, Studies on the quality nutrition in women with polycystic ovary syndrome (PCOS), Rocz Panstw Zakl Hig 68 (1) (2017) 61–67.

[77]   J Rojas, M Chavez, L Olivar, M Rojas, J Morillo, J Mejias, et al., Polycystic ovary syndrome, insulin resistance, and obesity: navigating the pathophysiologic labyrinth, Int J Reprod Med 2014 (2014) 719050.

[78]   Y Zhu, Y Li, M Liu, X Hu, H Zhu, Guizhi Fuling Wan, Chinese herbal medicine, ameliorates insulin sensitivity in PCOS model rats with insulin resistance via remodeling intestinal homeostasis, Front Endocrinol (Lausanne) 11 (2020) 575.

[79]   MA Rezvanfar, S Saeedi, P Mansoori, S Saadat, M Goosheh, HA Shojaei Saadi, et al., Dual targeting of TNF-alpha and free radical toxic stress as a promising strategy to manage experimental polycystic ovary, Pharm Biol 54 (1) (2016) 80–90.

[80]   MA Rezvanfar, MA Rezvanfar, A Ahmadi, HA Shojaei-Saadi, M Baeeri, M Abdollahi, Molecular mechanisms of a novel selenium-based complementary medicine which confers protection against hyperandrogenism-induced polycystic ovary, Theriogenology 78 (3) (2012) 620–631.

[81]   L Lin, Q Wang, Y Yi, S Wang, Z Qiu, Liuwei Dihuang pills enhance the effect of western medicine in treating diabetic nephropathy: a meta-analysis of randomized controlled trials, Evid Based Complement Alternat Med 2016 (2016) 1509063.

[82]   Z Qiu, J Dong, C Xue, X Li, K Liu, B Liu, et al., Liuwei Dihuang Pills alleviate the polycystic ovary syndrome with improved insulin sensitivity through PI3K/Akt signaling pathway, J Ethnopharmacol 250 (2020) 111965.

[83]   H Zhao, D Zhou, Y Chen, D Liu, S Chu, S Zhang, Beneficial effects of Heqi san on rat model of polycystic ovary syndrome through the PI3K/AKT pathway, Daru 25 (1) (2017) 21.

[84]   Y Xing, YX Liu, X Liu, SL Wang, P Li, XH Lin, et al., Effects of Gui Zhu Yi Kun formula on the P53/AMPK pathway of autophagy in granulosa cells of rats with polycystic ovary syndrome, Exp Ther Med 13 (6) (2017) 3567–3573.

[85]   E Park, CW Choi, SJ Kim, YI Kim, S Sin, JP Chu, et al., Hochu-ekki-to treatment improves reproductive and immune modulation in the stress-induced rat model of polycystic ovarian syndrome, Molecules 22 (6) (2017).

[86]   W Xu, M Tang, J Wang, L Wang, Clinical effects of Shou-Wu Jiang-Qi decoction combined acupuncture on the treatment of polycystic ovarian syndrome with kidney deficiency, phlegm and blood stasisness: study protocol clinical trial (SPIRIT Compliant), Medicine (Baltimore) 99 (12) (2020) e19045.

[87]   P Kajla, A Sharma, DR Sood, Flaxseed-a potential functional food source, J Food Sci Technol 52 (4) (2015) 1857–1871.

[88]   F Haidari, N Banaei-Jahromi, M Zakerkish, K Ahmadi, The effects of flaxseed supplementation on metabolic status in women with polycystic ovary syndrome: a randomized open-labeled controlled clinical trial, Nutr J 19 (1) (2020) 8.

[89]   WT Liao, CC Su, MT Lee, CJ Li, CL Lin, JH Chiang, et al., Integrative Chinese herbal medicine therapy reduced the risk of type 2 diabetes mellitus in patients with polycystic ovary syndrome: a nationwide matched cohort study, J Ethnopharmacol 243 (2019) 112091.

[90]   Y Deng, W Xue, YF Wang, XH Liu, SY Zhu, X Ma, et al., Insulin resistance in polycystic ovary syndrome improved by Chinese medicine dingkun pill: a randomized controlled clinical trial, Chin J Integr Med 25 (4) (2019) 246–251.

[91]   N Ainehchi, A Khaki, A Farshbaf-Khalili, M Hammadeh, E Ouladsahebmadarek, The effectiveness of herbal mixture supplements with and without clomiphene citrate in comparison to clomiphene citrate on serum antioxidants and glycemic biomarkers in women with polycystic ovary syndrome willing to be pregnant: a randomized clinical trial, Biomolecules 9 (6) (2019).

[92]   I Haj-Husein, S Tukan, F Alkazaleh, The effect of marjoram (Origanum majorana) tea on the hormonal profile of women with polycystic ovary syndrome: a randomised controlled pilot study, J Hum Nutr Diet 29 (1) (2016) 105–111.

[93]   CV Anastasiu, MA Moga, A Elena Neculau, A Balan, I Scarneciu, RM Dragomir, et al., Biomarkers for the noninvasive diagnosis of endometriosis: state of the art and future perspectives, Int J Mol Sci 21 (5) (2020).

[94]   MA Moga, A Balan, OG Dimienescu, V Burtea, RM Dragomir, CV Anastasiu, Circulating miRNAs as biomarkers for endometriosis and endometriosis-related ovarian cancer-an overview, J Clin Med 8 (5) (2019).

[95]   P Parasar, P Ozcan, KL Terry, Endometriosis: epidemiology, diagnosis and clinical management, Curr Obstet Gynecol Rep 6 (1) (2017) 34–41.

[96]   T Arablou, N Aryaeian, S Khodaverdi, R Kolahdouz-Mohammadi, Z Moradi, N Rashidi, et al., The effects of resveratrol on the expression of VEGF, TGF-beta, and MMP-9 in endometrial stromal cells of women with endometriosis, Sci Rep 11 (1) (2021) 6054.

[97]   S Jana, S Paul, S Swarnakar, Curcumin as anti-endometriotic agent: implication of MMP-3 and intrinsic apoptotic pathway, Biochem Pharmacol 83 (6) (2012) 797–804.

[98]   M Maharani, L Lajuna, C Yuniwati, O Sabrida, S Sutrisno, Phytochemical characteristics from Phaleria macrocarpa and its inhibitory activity on the peritoneal damage of endometriosis, J Ayurveda Integr Med 12 (2) (2021) 229–233.

[99]   M Ilhan, Z Ali, IA Khan, H Tastan, E Kupeli Akkol, The regression of endometriosis with glycosylated flavonoids isolated from Melilotus officinalis (L.) Pall. in an endometriosis rat model, Taiwan J Obstet Gynecol 59 (2) (2020) 211–219.

[100] M Ilhan, Z Ali, IA Khan, H Tastan, E Kupeli Akkol, Promising activity of Anthemis austriaca Jacq. on the endometriosis rat model and isolation of its active constituents, Saudi Pharm J 27 (6) (2019) 889–899.

[101] S Park, W Lim, S You, G Song, Ameliorative effects of luteolin against endometriosis progression in vitro and in vivo, J Nutr Biochem 67 (2019) 161–172.

[102] S Ryu, FW Bazer, W Lim, G Song, Chrysin leads to cell death in endometriosis by regulation of endoplasmic reticulum stress and cytosolic calcium level, J Cell Physiol 234 (3) (2019) 2480–2490.

[103] MA Abbas, MO Taha, MA Zihlif, Disi AM. beta-Caryophyllene causes regression of endometrial implants in a rat model of endometriosis without affecting fertility, Eur J Pharmacol 702 (1-3) (2013) 12–19.

[104] RH Zhao, Y Liu, D Lu, Y Wu, XY Wang, WL Li, et al., Chinese medicine sequential therapy improves pregnancy outcomes after surgery for endometriosis-associated infertility: a multicenter randomized double-blind placebo parallel controlled clinical trial, Chin J Integr Med 26 (2) (2020) 92–99.

[105] RH Zhao, ZP Hao, Y Zhang, FM Lian, WW Sun, Y Liu, et al., Controlling the recurrence of pelvic endometriosis after a conservative operation: comparison between Chinese herbal medicine and western medicine, Chin J Integr Med 19 (11) (2013) 820–825.

[106] H Cook, M Ezzati, JH Segars, K McCarthy, The impact of uterine leiomyomas on reproductive outcomes, Minerva Ginecol 62 (3) (2010) 225–236.

[107] V Pandey, A Tripathi, A Rani, PK Dubey, Deoxyelephantopin, a novel naturally occurring phytochemical impairs growth, induces G2/M arrest, ROS-mediated apoptosis and modulates lncRNA expression against uterine leiomyoma, Biomed Pharmacother 131 (2020) 110751.

[108]   S Greco, MS Islam, A Zannotti, G Delli Carpini, SR Giannubilo, A Ciavattini, et al., Quercetin and indole-3-carbinol inhibit extracellular matrix expression in human primary uterine leiomyoma cells, Reprod Biomed Online 40 (4) (2020) 593–602.

[109]   F Giampieri, MS Islam, S Greco, M Gasparrini, TY Forbes Hernandez, G Delli Carpini, et al., Romina: a powerful strawberry with in vitro efficacy against uterine leiomyoma cells, J Cell Physiol 234 (5) (2019) 7622–7633.

[110]   JH Kim, YC Shin, SG Ko, Integrating traditional medicine into modern inflammatory diseases care: multitargeting by Rhus verniciflua Stokes, Mediators Inflamm 2014 (2014) 154561.

[111]   JW Lee, HJ Choi, EJ Kim, WY Hwang, MH Jung, KS Kim, Fisetin induces apoptosis in uterine leiomyomas through multiple pathways, Sci Rep 10 (1) (2020) 7993.

[112]   SSM Zaid, S Othman, NM Kassim, Protective role of Ficus deltoidea against BPA-induced impairments of the follicular development, estrous cycle, gonadotropin and sex steroid hormones level of prepubertal rats, J Ovarian Res 11 (1) (2018) 99.

[113]   SS Ruslee, SSM Zaid, IH Bakrin, YM Goh, NM Mustapha, Protective effect of Tualang honey against cadmium-induced morphological abnormalities and oxidative stress in the ovary of rats, BMC Complement Med Ther 20 (1) (2020) 160.

[114]   MA Ibrahim, IA Albahlol, FA Wani, A Abd-Eltawab Tammam, MT Kelleni, MU Sayeed, et al., Resveratrol protects against cisplatin-induced ovarian and uterine toxicity in female rats by attenuating oxidative stress, inflammation and apoptosis, Chem Biol Interact 338 (2021) 109402.

CHAPTER 19

Herbal medicine to cure male reproductive dysfunction

Homa Fatma and Hifzur R. Siddique

Molecular Cancer Genetics & Translational Research Lab, Section of Genetics, Department of Zoology, Aligarh Muslim University, Aligarh, Uttar Pradesh, India

19.1 Introduction

Plant-based products have long been a fruitful source of nutrition, decoration, climate control, beverages, and medicines. The use of plant parts as the source of various diseases has been a common trend since the ancient eras. The advancement in science, particularly of the medical field, might have swayed the interest and fancy of the world towards the use of modern-day pharmaceutical drugs. However, herbal medicines as complementary and alternative therapies are still thriving [1]. The fabric of the chemical sphere exploited from the plant kingdom increases the probability of discovering the correct mechanistic targets for various diseases and their underlying mechanisms [2].

Secondary metabolites produced by plants have shown excellent property of interspecies chemical interaction and have drug-like properties that can selectively target various human proteins or interfere with the biological processes of pathogens and parasites [2]. Bioactive compounds of plants show various medicinal effects, such as anti-allergic, anticarcinogenic, anti-inflammatory, anti-microbial, hepato-protective, cardioprotective, antiviral, antibacterial vasodilatory, etc. Some plants are a rich source of antioxidants and are highly marketed as the main chemical ingredient of various herbal drugs [3]. Modern herbal medicine practice deviates slightly from traditional herbalism. Traditional practitioners or paraherbalists allow minimal adulteration to the plant parts and preserves the whole plant's original composition. In contrast, herbalism includes the safe use of plant-derived compounds or active compounds that are effective and not the entire source [4].

The survival of a species depends upon the reproductive robustness of the surviving generation. Reproductive potency ensures the preservation of the genetic characteristics by passing them to the next generation and so on. However, recently, an alarming trend has been observed in female and reproductive health [5]. Deteriorating sexual function is a diligent medical concern that sabotages a biological relationship and mars a person's social and mental

Herbal Medicines: A Boon for Healthy Human Life.
DOI: https://doi.org/10.1016/B978-0-323-90572-5.00023-8

behavior [5]. Several diseases have been reported affecting different male reproductive organs that have been the cause of the overall increase in male reproductive dysfunction worldwide. Hypospadias, cryptorchidism, prostatitis, testicular germ cell cancer, poor semen quality, etc., are few common diseases that are common among males [6]. Few studies indicate the importance of a father's lifestyle in the survival of future generations. It has been suggested that the next generation's health is influenced by the quality of the father's sperms, which may be epigenetically altered by the diet/ lifestyle [7].

The progressive increase in the deteriorating state of male reproductive health has given rise to the need to develop efficient therapies. Herbal medicines have risen as the popular alternative therapy for better results and fewer side effects. In Indian traditional medicine (ITM), a separate group of herbal extracts known as *vajikarna* is classified that have the property to aid, nourish, and stimulate sexual tissues [5]. Other herbal medicines also show a promising effect to improve male reproductive health. *Hedyotis diffusa* wild extract has shown a promising effect against the autoimmune prostatitis mouse model [8]. A decrease in micturition rate, inflammation, and increased pain threshold have been observed in the treated group [8]. In rats suffering from cryptorchidism, the effects of multi-herbal medicine *Ojayeonjonghwan (KH-204)* were tested. A marked decrease in HSP70 level and apoptosis has been observed along with increased testicular fertility and restored sperm counts [9]. In this chapter, we summarize some of the complications of the male reproductive system and herbal medicine's effect on them.

19.2 Herbal medicines

'Even before human civilization started recording history, herbs and plant-based extracts had served as the first-line defense and basic healthiness. In the fossilized bones of stone age man found in Iraq, marshmallow roots, hyacinth, and yarrow were found, which are still used as a demulcent, diuretic, and household remedy for the common cold [10]. However, with the development of medical science and modern techniques, the dependence on plant extracts and plant-based compounds had decreased significantly [10]. World Health Organization (WHO) agrees that herbal medicines can serve as a mainstream therapeutic alternative or complement the synthetic drug in their functions [11]. Modern-day drug development is based on finding well defined-targets and active molecules that can strongly bind to these targets, thus regulating their functions. However, these healthcare systems must deal with the prominent side effects of the active compounds in the drugs. Personalized medicines wrapped up in complex mixtures are more often doomed to be unsuccessful due to the involvement of more than one target for the same disease [12]. Recently, a swift shift has been observed in using herbal medicines in the areas where modern medicine is still only accessible. Plant-based compounds tend to have fewer side-effects and higher efficacy than the processed modern medicines [10]. Plant-based compounds are recognized for their therapeutic effect and are a source of remedial treatment in various forms. They can either be used to produce bioactive compounds of common synthetic drugs and serve as the main compound for the semisynthetic drug with less toxicity, for example, metformin based on morphine or used as a pharmacological tool. Plant as a whole is also used for remedial purposes, such as garlic, or active compounds are isolated to use as a medicine such as morphine [13].

According to a survey based on the U.S. population, the consumption of herbal has increased by 380%, primarily for allergies, insomnia, digestive disorder, respiratory problems, and reproductive dysfunction [14]. Several plant-based compounds are found to have different protective behavior. The bioactive components of species *L. barbarum* berries show hepato-protective activity by reducing reactive oxygen species (ROS), oxidative stress, inflammation, and increased antioxidants activity [15]. Herbal medicines have also shown marked antiviral capability against novel coronavirus (COVID-19), a virus responsible for the pandemic during 2020 and onwards. Before any vaccine was developed against it, herbal or complementary therapy was prevalent in preventing lethal COVID-19 infection and building immunity against it [16]. Herbal drugs can stimulate interferon production in the body in response to viral infection. They can also strengthen immunity in a non-specific manner. Herbal medicines are also observed to be effective against various bacterial infections and pathogens [17]. Yinchareon et al. [18]. have found that ethanol extract of polyherbal such as *Kheaw-Hom, Learning-Pid-Sa-Mud*, and *Ummaluk-kawatee* has anti-infective properties against bacteria *Pseudomonas aeruginosa, Acinetobacter baumanii,* methicillin-resistant *Staphylococcus aureus,* and *Staphylococcus epidermis.* These polyherbal formulations can inhibit almost 90% of bacterial growth at the concentration of 31.3–125 μg/mL.

Active compounds isolated from plants have also found their function as anticarcinogenic drugs [19–23]. It has been estimated that during the period between 1940 to 2002, around 40 % of all the anticarcinogens discovered are either natural products or derived from a natural source [13]. Moreover, some of the drugs are mimics of natural products [13]. Interestingly, many flowers have a therapeutic effect against various diseases and disorders. Flowers of *Stereospermum suaveolens* are effective against respiratory diseases such as bronchitis. They are also found to be useful against malarial pathogens. *Syzygium aromaticum* has antibiotic and antiseptic properties. Its flower bud is the source of clove oil which is helpful in soothing toothaches [24]. A bioactive compound, Acutissimalignan B extracted for *Daphne kiusiana* var *atrocaulis* (Rehd.), has anti-inflammatory properties. Acutissimalignan B inhibits the NF-κB signaling pathway in neurons and inhibits inflammation in both *in vitro* and *in vivo* conditions [25].

Herbal medicines are also effective against various reproductive dysfunction. Various sexual dysfunctions are characterized by a person's inability to maintain sexual vigor regularly and consistently. Both males and females are affected by various sexual dysfunctions. In the male, while some illnesses, such as cryptorchidism, are inherited and affect a person from birth

[26], many others are caused by infections of the reproductive organs, such as epididymitis, prostatitis, and [27–29]. Herbal medicines help cure the root cause of the reproductive disease and relieve various symptoms [30–31]. Common female reproductive disorders are polycystic ovary syndrome, premature ovarian failure, uterine dysfunction, endometriosis, uterine fibroids, etc. [32]. Moreover, pubertal changes, including the onset of menstruation, menopause, post-menopausal syndrome, are some common gynecological disorders. Herbal medicines rich in various bioactive compounds effectively maintain hormonal balance and treat gynecological problems [33].

19.3 Male reproductive disorder

All humans at their embryonic stage are destined to be female by default. However, Y-chromosomes in the genome can reroute embryo differentiation to male phenotypes due to the sex-determining region (SRY) gene. SRY gene is responsible for developing testes that produce Mullerian inhibiting substance (MIS) in Sertoli cells to repress the formation of the female reproductive tract. Moreover, Leydig cells of the testes produce testosterone responsible for developing the male reproductive system [34]. The male reproductive system comprises testicles, penis, and accessory sex organs, including vas deferens, prostate glands, seminal vesicles, and bulbourethral glands [Fig. 19.1; 35]. The primary function of accessory sex organs is to secrete fluid that provides nutrition to the male germ cells and facilitates their movement in the female genital tract [35]. The male reproductive system is susceptible to several infections and diseases; these diseases sometimes cause irreversible fertility. The practice of using herbal medicines for treating male reproductive dysfunction has always been widespread among traditional medicine practitioners, but nowadays, modern clinicians are also advising for its use.

19.3.1 Cryptorchidism

Testes are the primary reproductive organ that makes single-celled gametes in males. The testes are found deep within the belly of the embryo, but they descend from the abdomen to the scrotum, a sac-like structure, right before birth. Testes' descent to the scrotum is a crucial developmental step as the temperature of the scrotal environment is paramount for spermatogenesis [26]. Failure to descend one or both testes to the environment of the scrotum leads to congenital abnormalities among males, referred to as cryptorchidism. The timing of the onset of disease may vary in different males depending upon the descend of the testis in the scrotum [26]. Since birth, the male may have cryptorchidism, or a normal testis may ascend a few days after delivery, referred to as acensus testis. In some recurrent cases, the initially-cryptorchid testis may descend and re-ascend back to the abdomen [26]. Mother health, parity, age, lifestyle is found to be conclusive about cryptorchidism. The primary treatment is orchidopexy, surgical repositioning of the testis, but the risk factor associated with cryptorchidism

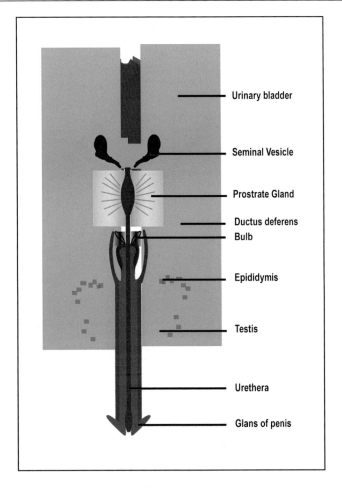

Figure 19.1
Diagrammatic representation of male Reproductive system. The male reproductive tract consists of a pair of testes, epididymis, vas deferens, ejaculatory duct, and accessory sex glands (seminal vesicles, prostate, and bulbourethral glands).

include infertility, and sperm functioning is still not proper. Recently, various medicinal and herbal treatments are also preferred to treat pathophysiologies [Table. 19.1].

In an experiment conducted by Afolabi et al. [36], it was reported that the antioxidative properties of the ginger rhizome could effectively increase testosterone levels and scavenge free radicals. In addition, the extract is also capable of reversing cryptorchidism-induced weight loss in the testis and enhance the spermatogenesis process. Another medicinal plant, *Rubus apetalus,* has potent antioxidant activity and effectively increases the fertility of cryptorchid testis. The aqueous and methanolic extract of *R. apetalus* can release oxidative stress and significantly increased spermatozoa motility, normality, viability, etc. [37]. Similarly, Curcumin

Table 19.1: Essential herbal medicines and their implications in testis.

S.No.	Disease	Herbal Plants	Parts Used	References
1.	Cryptorchidism	*Zingiber officinale*	Rhizome	[36]
		Laurus nobilis, Rubus apetalus	Extract	[37,41]
		Curcuma longa	Curcumin	[38]
		Angelica gigas	Plant parts	[39]
2.	Orchitis	*Callophyllum inophyllum, Justicia betonica, Mimosa pudica, Conocarpus*	Bark & Leaf	[46–48,52]
		Momordica charantia, Litchi chinensis, Lygodium microphyllum, Crotalaria pallida, Cardiospermum halicacabum, Nigella	Plant parts	[30,47,49,52–54]
		Madhua longifolia, Coriandrum sativum	Phytochemicals	[51,55]
3.	Hydrocele	*Cannabis sativa, Holoptelea integrifolia, Butea monosperma*	Bark	[94,97]
		Decaschistia,, Caesalpinia bonduc, Mimosa pudica, Melothria indica, Tabernaemontana divaricate, Nigella	Plant parts	[99,100–102]
		Butea monosperma, Cardiospermum	Roots	[94,99]
		Zingiber officinale, Bambusoidea	Shoots	[99]
		Curcuma longa, Datura metal	Leaves	[98,99]
		Calotropis procera	Oil Extract	[102]
4.	Varicocele	*Cuscuta, Citrus aurantia, Pericarpium citri, Cynanchum otophylum, Angelica sinensis, Cyperus, Cantella*	Plant parts	[104,105]
		Litchi chinensis	Seeds	[105]
		Morinda officinalis	Polysaccharides	[107]
5.	Testicular Torsion	*Matricaria chamomile, Plantago major, Lycium barbarum, Fumaria parviflora, Biebersteinia multifida*	Extract	[110,112,113,116,117]
		Boswellia serrata, Berberis vulgaris	Plant parts	[114,118]
		Capsicum annum	Phytochemical	[115]

has been shown to be a significant and influential medical line for safeguarding against the adverse effects of cryptorchidism. The antioxidant properties of Curcumin may improve post-surgery fertility and prevent various risk factors [38].

In addition, a bioactive compound, Decursin, found in *Angelica gigas,* was found to have antioxidant and ROS scavenging properties. Decursin reduces oxidative stress by upregulating heme oxygenase-1 via nuclear factor erythroid 2-related factor 2 protein in cryptorchid testis and might improve fertility and sperm motility [39]. It has been discovered that Resveratrol can help to reverse the negative effects of cryptorchidism on spermatogenesis and spermatogonia differentiation at the primary spermatocyte stage by the regulation of various sex hormones such as FSH, LH, and Estradiol [40]. Mansour et al. [41] have reported that, in leaves extract of sweet bay leaf, *Laurus nobilis*, have antioxidant properties. The polyphenols present in sweet bay leaf extract can improve the SOD and CAT level in cryptorchid testis and, subsequently, various sperm parameters.

19.3.2 Orchitis

Orchitis is a bacterial or viral infection that causes the collection of inflammatory conditions of the testicles [Fig. 19.2A]. Both testicles may be affected simultaneously, or only one testicle may be affected at a time. Orchitis is generally accompanied by epididymitis, i.e., inflammation of the epididymis [42]. Orchitis is generally caused by various infectious agents such as bacteria and viruses and has several forms: intratubular orchitis, necrotizing orchitis, and granulomatous orchitis [43]. Mumps orchitis is primary orchitis in post-pubertal males, and symptoms are often observed 4 to 8 days after the parotitis onset [44]. The conventional treatment of orchitis depends upon the causative pathogen and are generally bed rest, ice packs, and anti-inflammatory agents [44]. Moreover, a number herbal medicines are implicated in the treatment of orchitis [Table. 19.1]. It has been reported that annual climber *Cardiospermum halicacabum,* found in India, has antioxidant activities that can be implicated in various diseases. The methanol extract of its leave and tender shoots are a potent demulcent in orchitis [45]. Indigenous Indians are said to treat orchitis with a decoction produced from the bark and leaves of *Mimosa pudica*, also known as Lajjabati climbers [46].

The decoction made from the bark of *Calophyllum inophyllum,* a plant found in islands of the Indian ocean, is used to treat orchitis [47]. Whole plant parts of *Momordica charantia,* another plant found on the Indian Ocean islands, appear to treat orchitis effectively [47]. Jereto et al. [48] have reviewed that leaves and flower ash of *Justicia betonica* are used for orchitis by Kenyan people. The common name of the preparations within the locals is *Kipkesio* and are typical for treating cough, diarrhea, and orchitis. In a survey performed in mangrove and beach forests in Thailand, the whole plant or decoction of *Lygodium microphyllum,* commonly known as *li phao yung,* is found to have medicinal properties against orchitis [49]. The aerial parts of popular fruit found in northern India, *Litchi chinensis,* have anti-inflammatory effects. The petroleum extracts of lychee plants comprise Lupeol, Stigmasterol, Betunilic acid, etc. These compounds present in *Litchi chinensis* might exert an anti-inflammatory response against orchitis complications [50]. Various phytochemicals found in *Madhuca longifolia,* such as Saponins, 21-Hydroxy-3-oleanyl myricitate, β-Amyrin acetate, and Urosolic acids, were found to be effective against orchitis [51].

Conocarpus erectus is a decorative plant, but the plant's barks and leaves have found their purpose against the treatment of various diseases, including orchitis [52]. Yaradua et al. [53] studied the ethnobotanical importance of plants falling in genus Crotalaria. The authors reviewed that macerated whole plant of *Crotalaria pallida* and *Crotalaria pallida* var *obovate* can be orally given to orchitis patients. *Nigella sativa,* also known as black cumin, is a common spice in Indian households and an effective folk medicine with a wide range of

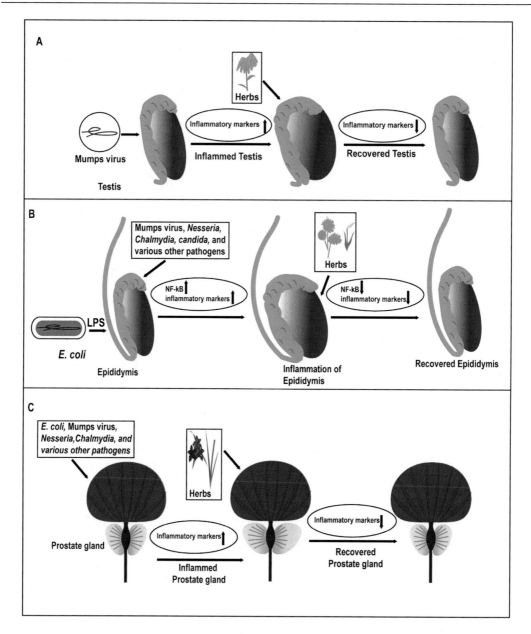

Figure 19.2

Pictorial representation of (A) orchitis: orchitis is the inflammation of the testes due to various pathogens. Treatment of orchitis with herbal medicine can suppress the inflammatory markers causing the inflammation and reverse the condition. (B) Epididymitis: is the inflammation of epididymis. The epididymis is caused by various bacteria, viruses, and protozoans. The application of various medicinal herbs has the potential to reduce epididymitis pathogenesis. (C) Prostatitis: is the inflammation of the prostate gland, which is caused by various pathogens, and herbal treatment can reduce the inflammatory markers causing prostatitis. *E.Coli, Escherichia Coli*; LPS, lipopolysaccharide; NF-κB, nuclear factor kappa light chain enhancer of activated B cells.

pharmacological effects. It is applied topically as an anti-inflammatory agent for the treatment of orchitis [30]. *Cardiospermum halicacabum* was utilized as a demulcent to minimize inflammation and oxidative stress in orchitis patients. In Lipopolysaccharide (LPS)-stimulated macrophages, *Cardiospermum* can regulate NOS and COX-2 expression. In addition, it also reduces TNF-α, cyclophosphamide-induced toxicity and oxidative stress. These pharmacological actions might be the mechanism to bring down inflammatory responses associated with orchitis [54]. The fruit and fresh leaves of *Coriandrum sativum*, a popular Indian household spice and herb, comprised of almost 39 different essential oil such as Dodecanal, Undecanal, Tetradecenal, etc. and polyphenols, including Caffeic acid and Glycitin useful in treating orchitis [55].

19.3.3 Epididymitis

The epididymis is a highly convoluted duct attached to the testis and divided into an initial segment, caput, corpus, and tail. By removing the bulk of the fluid leaving the testis, the epididymis is responsible for increased sperm concentration. In addition, the epididymis is involved in sperm storage during the quiescent period, as well as protection, sperm maturation, motility acquisition, and transport [56]. The epididymis is immunologically privileged and enjoys a tolerogenic environment due to the epididymis-blood barrier. This diligently controlled immune environment protects spermatozoa not only from autoimmunity but also from pathogenic damage [57]. Epididymitis is the swelling of the epididymis, which is accompanied by pain and discomfort [Fig 19.2B]. It may affect men of any age group. Epididymitis can be acute or chronic, depending on the severity of the symptoms and the length of time it lasts [27]. Attack on the epididymis by bacteria such as uropathogenic *Escherichia coli* via the urethra is a common cause of epididymitis, and the bacterial outer-membrane component, LPS that triggers a robust immune response, resulting in epididymitis. LPS caused epididymitis by upregulating NF-κB and inducing proinflammatory cytokines as a result [58]. Other pathogens commonly associated with epididymitis are *Chlamydia trachomatis*, *Candida albicans*, *Histoplasma capsulatum*, *Neisseria gonorrhoeae*, *Pseudomonas* spp., Mumps' virus, Adenovirus, etc. [59].

While conservative anti-microbial therapy is usually sufficient to eliminate the pathogen in most patients, research has shown that complementary and alternative medicine has been shown to be relevant and successful in treating epididymitis [Table. 19.2]. Recently, Jing et al. [60] revealed that *Ningmitai* capsules along with antibiotics are effective against both acute epididymitis and chronic epididymitis. *Ningmitai* capsules are composed of a series of herbal components that contribute to various pharmacological outcomes such as antibacterial, analgesics, and anti-inflammatory. *Martynia annua* Linn (Martyniaceae) is a vital folklore medicinal plant that is used in ITM to treat various diseases. One of the critical uses of *Martynia annua* Linn is its anti-infertility and anti-inflammatory properties that result in a significant

Table 19.2: Essential herbal medicine and their implications in epididymis and prostate gland dysfunction.

S.No	Disease	Herbal medicines	Parts used	References
1.	Epididymitis	*Martynia annua, Panax ginseng, Hoya kerri Vitex negundo*	Plant parts Leaf	[61,63–64] [31]
2.	Prostatitis	*Plantago depressa, Radix astragali, Curcubita pepo, Macrothelypteris oligophlebia, Alisma plantgo-aquativa, Rheum palmatum, Astragalus memberanaceus, Chrysanthemi Indici, Allium sativum, Aurantii immarturus, Paeonae alba, Herba epimedii, Herba leonuri, Cortex phellodendri, Angelica sinensis, Fritillariae thunbergia, Sophorae flavescentis*	Plant parts	[67,69–71,74,76]
		Glycyrrhiza glabra, Bupleuri, Astragalus mongholicus, Achyrantis bidentatae	Roots	[71,74]
		Hedyotis diffusa, Clematis terniflora	Extract	[8,73]
		Curcuma longa, Nigella sativa	Phytochemicals	[67]

decrease in epididymitis [61]. In gonorrheal epididymitis, the use of *Vitex negundo* leaves is reported to be very effective. Leaves of *Vitex negundo* have anti-inflammatory properties that help in the treatment of gonorrheal epididymitis [31].

19.3.4 Epididymo-orchitis

As the epididymis and testis lie close to each other, more often than not, inflammation in both reproductive organs coincides. Epididymo-orchits (EO) is a urological problem marked by epididymis and ipsilateral testis inflammation. It is a consequence of the contiguous spread of infection from the epididymis to the testis [62]. Uropathogenic *E. coli,* chlamydial and gonorrheal infection, and mumps virus is the common cause of EO. In some cases, patients with tuberculosis and brucellosis also cause EO [62]. Abscess development, testicular infarction, testicular atrophy, chronic epididymitis, and infertility are all possible outcomes of EO.

The standard therapy used against bacterial EO is antibiotics, but it cannot fully reverse fertility or impaired spermatogenesis. Use of herbal medicines have shown therapeutic potential in treating EO [Table. 19.2]. Co-exposure of bacterial EO with Korean red ginseng and antibiotics for seven days have protective effects. Korean red ginseng has anti-inflammatory and immune enhancer properties, which help in improve EO-related symptoms and consequences [63]. *Hoya kerri,* commonly called lucky heart or valentine hoya, is an edible herb with antioxidant activities. *Hoya kerri* is a Thai herbal plant used to treat inflammatory-related diseases, including EO [64].

19.3.5 Prostatitis

Prostatitis is the third most common urinary tract disease characterized by the inflammation of prostate glands [Fig. 19.2C]. No generally accepted definition of prostatitis or a fixed set of symptoms to characterize prostatitis. Prostatitis is classified into four major categories depending upon the spectrum of symptoms: acute bacterial prostatitis, chronic bacterial prostatitis, chronic pelvic pain syndrome, and asymptomatic [28]. The most common agent for prostatic infections is gram-negative bacteria *E. Coli, Pseudomonas, Klebsila spps,* and gram-positive bacteria such as *Enterococcus.* Pathogens responsible for sexually transmitted diseases such as gonorrhea, chlamydia, etc., are also involved in prostatitis [65]. In addition to the standard therapy, a number of alternative and complementary medicines are found to be useful for curing prostatitis [Table. 19.2]. Moreover, Paulis, [65] have reviewed the use of herbals and their parts as powerful antioxidants against chronic prostatitis. The author reported that plants like *Epilobium spps. Serenoa repens* and phytochemicals such as Quercetin, Curcumin, Resveratrol have high antioxidant and anti-inflammatory effects of eradicating inflammatory response generated by infectious agents. Furthermore, curcumin can reduce the expressions of proinflammatory cytokines IL-8 and TNF-α in blood and tissues of chronic non-bacterial prostatitis cases [66].

A number of Chinese herbs are implicated as an anti-inflammatory agent against chronic prostatitis and chronic pelvic pain syndrome in traditional Chinese medicine (TCM). *Plantago depressa, Radix astragali, Cucurbita pepo, Curcuma longa, Macrothelypteris oligophlebia, Alisma plantago-aquatica* are a few of the TCM that have anti-chronic prostatitis effects [67] The main constituent of *Nigella sativa* is Thymoquinone which has anti-inflammatory and antioxidant properties in tissues. Thymoquinone has potential therapeutic effects against *E. coli* induced prostatitis. Thymoquinone improves the general histology of the prostate gland and the detoxification process of biochemical enzymes [68]. In TCM, various plant formulated capsules are being used to treat chronic prostatitis or chronic pelvic pain syndrome. *Nan mi qing* capsules formulated by the mixture of *Rheum palmatum* and *Rx. Astragalus membranaceus* ad *Ye Ju Hua Shuan,* made from *Flos Chrysanthemi Indici,* are few formulated capsules that affect inflammation markers [69]. Sohn et al. [70] have reported that garlic exerts an anti-inflammatory and anti-microbial effect on chronic prostatitis caused by bacteria. Moreover, garlic treatment and standard antibiotics, ciprofloxacin generates a synergistic effect and significantly decreases bacterial growth in prostatic fluid.

In another study, it was reported that Chinese herbal formula *Sini San* (SNS) comprising *Aurantii immaturus fructus* (*Zhi-Shi*), *Paeoniae alba* radix (*Bai-Shao*), *Licorice* root (*Gan-Cao*), and *Bupleuri* radix (*Chai-Hu*) with a dose proportion of 1:1:1:1 is effective against chronic prostatitis with high cure rate [71]. Similarly, another standard Chinese herbal formula that has a therapeutic effect on prostatitis is *Bazhengsan.* The *Bazhengsan* decoction was provided intragastrically once a day to the *E. coli* infected rat model, and it was observed that *Bazhengsan* could successfully decrease prostate index, inflammatory markers, and free radicals in prostate tissues [72]. Another study reports that *Clematis terniflora* has anti-inflammatory properties. The ethanol extract of its aerial parts has therapeutic significance against prostatitis with little to no side effects [73]. In carrageenan-induced chronic prostatitis, the remedial impact of *Qian-Yu* decoction has been observed. *Qian-Yu* is a Chinese medicine prepared by mixing *Herba epimedii, Cortex phellodendri, Radix astragali, Bunge var. mongholicus, Herba leonuri,* and *Radix achyranthis bidentatae.* Polysaccharides, flavonoids, and Saponins present in *Qian-Yu* exert a combined effect on inflammatory markers [74]. Multiherbal activating blood circulation formula is considered effective therapy against chronic non-bacterial prostatitis. The formula is prepared by mixing *Angelica sinensis*, peach kernel, cortex moutan, *Radix linderae, Rhizoma corydalis,* and *Rhizoma sparganii* in a boiler followed by filtering. The concentration of the dose of herbs can vary according to the severity of the disease [75]. Another multiherbal formulation *Danggui Beimu Kushen Wan,* used by the Chinese since 1800, has found its efficacy against chronic prostatitis. The formula comprises *Angelicae sinensis radix, Fritillariae thunbergii bulbus, Sophorae flavescentis radix,* and Talcum. The formula downregulates inflammatory markers, and microbial growth increases antioxidant action and regulates sex hormones [76]. Similarly, high dose decoction prepared from the *Jiedu Huoxue* decreased NF-κB in type III prostatitis. Moreover, a significant reduction in glandular hyperplasia and inflammatory markers and marked an increase in Iκ-Bα was observed [77]. Wazir et al. [8] found that the aqueous extract of *Hedyotis diffusa* has anti-inflammatory and immunomodulatory properties. *H. diffusa* reduces TNF-α level, inflammatory lesions, and increased the pain threshold in prostatitis patients.

19.3.6 Erectile dysfunction

Erectile dysfunction (ED), also known as male impotence, is the incompetence to sustain a penile erection due to reduced nitric oxide (NO) production characterized by endothelial and neuronal dysfunction [Fig. 19.3A]. In every traditional medicine discipline, plant based therapy was found to be effective in treating ED and ED-related damages [Table. 19.3]. Aphrodisiac herbs are effective in the treatment of ED. Aphrodisiacs herbs are the plants or plant parts that can increase sexual pleasure, libido, and potency. Some of the aphrodisiac therapies used in ED are *Pausinystalia yohimbe, Lepidium meyeneii, Rubus coreanus, Folium Gingko biloba, Artemisia capillaris, Schisandra chinensis,* Saffron, etc. [78]. Herbal dietary supplement ginseng is effective against ED [79]. In addition, dried roots of herbs like *Angelica sinensis* and *Morinda officalinalis,* dried stems of *Herba cistanche,* dried seeds of *Cuscuta Chinese* Lam, extracts from *Ligusticum chuanxiong* Hort, lotus seeds, *Stephania tetrandra,* and *Panax notoginseng,* common cnidium fruits, *Tribulus terrestris* can enhance NO activity and downstream signaling. Also, these herbs can increase testosterone activity, reduce oxidative stress and calcium ion concentration [80]. Moreover, some Chinese herbal

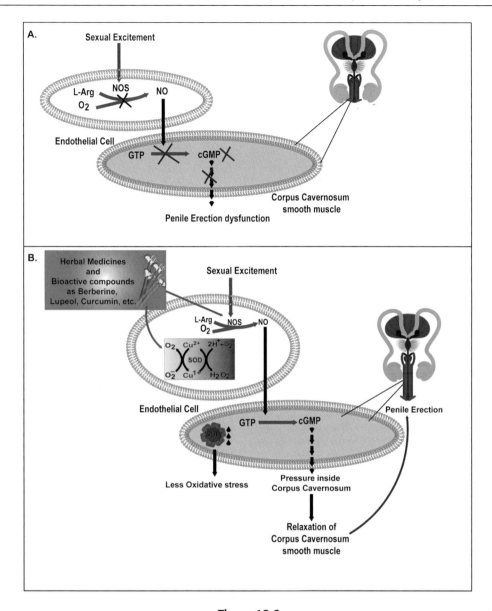

Figure 19.3

Diagram representing the cause of erectile dysfunction. (A) Nitrogen oxide is an essential compound for the proper penile erection. Deficiency of nitrogen oxide in the penile endothelial cells can lead to erectile dysfunction. (B) Application of herbal medicines helps in the proper functioning of NOS which lead to the synthesis of NO. NO production in the endothelial cells increase the level of GTP which subsequently increases intracavernous pressure. Intracavernous pressure further causes the relaxation of cavernous smooth muscles and helps in penile erection. Herbal medicines also increase the level of antioxidants inside the endothelial cells and prevent secondary damage and restore fertility. cGMP, cyclic guanine monophosphate; Cu, copper element; GTP, guanosine triphosphate; H2, hydrogen gas; H_2O_2, hydrogen peroxide; L-Arg, L-arginine; NO, nitrogen oxide; NOS, nitrogen synthase; O_2, oxygen gas; SOD, superoxide dismutase.

Table 19.3: Essential herbal medicine and their implication in penile dysfunction.

S.No	Diseases	Herbal medicine	Parts used	Reference
1.	Erectile dysfunction	*Pausinystalia johimbe, Lepidium meyeneii, Rubus coreanus, Panax quinquefolius, Schisandra chinensis, Artemisia capillaris, Folium Gingko, Stephania tetrandra, Panax notoginseng, Panax notoginseng, Chione venosa, Corchorus depressus, Montana tomentosa, Microdesmis keayana, Monsonia augustifolia, Lecaniodiscus cupanioides, Anethum graveolens, Withania somnifera, Asphaltum, Tribulus terrestris, Mucuna pruriens, Asparagus, Sida cordifolia, Ginseng, Dioscorea tenuipes, Lycium chinense, Curcuma, Cornus, Acorus, Trachyspermum, Cocos, Commiphora, Phaseolus vulgaris, Abrus, Nigella, Zingiber, Achillea, Cyperus, Pistacia terebinthus, Syzygium aromaticum, Smilax, Phoenix, Boswellia, Allium cepa, Allium sativa, Brassica nigra, Rhoicissus tomentosa, Lepdium sativum, Zataria multiflora, Calendula officinalis, Urtica dioica, Phyllanthus emblica, Brassica nigra*	Plant parts	[78–80,83,85–88]
		Angelica sinensis, Morinda officinalis, Herba cistanche	Roots & stems	[80]
		Cusutua Chinensis	Seeds	[80]
		Nelumbo nucifera, Ocimum gratissimum, Cinnamon,	Leaves & barks	[80,83,89]
		Ligusticum striatum, Anacyclus pyrethrum, Garcinia kola, Asparagus, Turnera diffusa, Alpinia calcarata,	Extracts	[80,83]
2.	Balanitis and balanoposthitis	*Azadirachta indica*	Bark & leaves	[122]
		Tinospora cordifolia, Terminalia chebula, Phyllanthus emblica, Smilax, Emblica, Piper longum, Basella alba,	Fruits	[120–121]
		Curcuma longa	Phytochemicals	[123]

formulations, such as *Shuganyiyang* capsules, *Yougui* pill, *Xiaoyao* pill, *Yidiyan* decoction, and so on, are also common in clinical practice [80].

Furthermore, the combination of various TCM such as compound *Xuanju, congrong yishen, Shisanwei ziyinzhuangyang* capsule, and *No.1 Hongjingtian* decoction with standard ED drug tadalafil can improve the erectile function and sexual spontaneity [81]. *KH-204* is a Korean herbal formulation that shows synergistic effects, with extracorporeal shockwave therapy and enhanced antioxidant activity in ED patients. Aside from that, *KH-204*/shockwave therapy can also reduce apoptosis and increase nuclear factor erythroid 2-related factor 2/heme oxygenase-1 and SOD activity [82]. Ethanol extract of *Anacyclus pyrethrum, Garcinia kola,* ethanol extract of roots of *Asparagus adscendens,* stem, bark & roots of *Chione venosa,* the bark of *Cinnamomum cassia, Corchorus depressus,* aqueous extract of *Turnera diffusa, Montanoa tomentosa, Microdesmis kenya, Monsonia angustifolia, Lecaniodiscus cupanioides,* boiled extract of *Alpinia calcarata, Anethum graveolens* exhibit potential pharmacological activity against ED [83]. *Asteracantha longifolia* comprises several isoflavones such as Lupeol, Stigmasterol, and fatty acids. Oral administration of alcoholic extract of *A. longifolia* seeds increases penile erection and mount frequency. Another alkaloid compound, Berberine, found in *Berberis vulgaris, Phellondendrone amurense,* and *Berberis aristana,* is used medicinally to treat various ailments, including ED. Administration of Berberine in cavernous muscles induced pressure inside the muscles, leading to the corpus cavernosum relaxation through NO production in the endothelial cells. Moreover, the relaxation effect is enhanced by the interference of charybdotoxin and 4-AP, the potassium (K+) channel blockers in the corpus cavernosum [Fig. 19.3B] [84].

Penile erection index, ejaculation latency, sperm count, testosterone level, and intromission frequency are reported to be improved by the use of compound formulations of *Withania somnifera* commonly known as Ashwagandha, *Asphaltum (Shilajit), Tribulus terrestris, Mucuna pruriens (klwanch), Asparagus racemosus,* and *Sida cordifolia (Bala* Extract) [85]. In patients with ED, preliminary clinical study on herbal formulation *KBMSI-2* prepared from *Ginseng radix Rubra, Dioscorea tenuipes, Cornus officinalis* Sieb, *Lycium chinense,* and *Curcuma longa* shows significant changes in erectile function and intercourse satisfaction [86]. Another study reported that the herbal concoction *imvusankunzi* prepared by mixing *Elephantorrhiza elephantina, Eriosema kraussianum, Hypoxis hemerocallidea, Drimia elata, Rhoicissus tomentosa,* and *Ash* could be used as a therapeutic option for ED by decreasing the oxidative stress [87]. *Calendula officinalis, Brassica nigra, Phyllanthus emblica, Urtica dioica, Matricaria chamomilla, Allium cepa, Allium sativum, Lepidium sativum, Portulaca oleracea, Zataria multiflora, Semecarpus anacardium, Boswellia sacra, Phoenix dactylifera, Smilax china, Juglans regia, Syzygium aromaticum, Pistacia terebinthus, Cyperus esculentus, Achillea millefolium, Zingiber officinale, Nigella sativa, Abrus precatorius, Phaseolus vulgaris, Commiphora mukul, Cocos nucifera, Trachyspermum ammi,* and *Acorus calamus* are the noted medicinal plants that act as the sexual stimulant and increase NO production which might help in the treatment of ED [88]. In another study, it was found that leaves of *Ocimum gratissimum,* a tropical herbal plant, can reduce enzymes that are associated with ED. A significant inhibitory response against ED-related enzymes such as phosphodiesterase-5 (PDE-5), angiotensin I –converting enzyme (ACE), acetylcholinesterase (AChE), and arginase had been recorded [89].

19.3.7 Hydrocele

A hydrocele accumulates clear fluid in a sac-like structure formed in the scrotum, including the testicle [90]. In various tribal areas hydrocele is still treated with the help of herbs and their parts in the forms of decoction, powder, or paste [Table. 19.1]. In the Eastern Ghats of India, tribal people use medicinal plants as remedies for treating different ailments. *Caesalpinia bonduc* and *Decaschistia crotonifolia* are the plants that are effective against hydrocele [91]. *Oryeong-san* decoction is used to treat spermatic cord hydrocele, a variant of hydrocele found in children where fluid is collected along the spermatic cord [92]. The herbal paste made from the root of *Boerhaavia diffusa* L. (Nyctaginaceae), *Mookkirattai*, is applied topically to treat hydrocele [93]. The bark of *Butea monosperma* is bitter, acrid, oily, and anti-inflammatory. It is used in the reduction of inflammatory response due to hydrocele. Another plant from the family Ulmaceae, *Holoptelea integrifolia, also implicated hydrocele treatment* [94]. Similarly, the whole plant of *Mimosa pudica* and *Melothria indica, the* bark of *Holoptelea integrifolia,* and leaf paste of *Datura metal* are also used in hydrocele treatment in some Indian regions [95–98] Turmeric has a wide variety of pharmacological benefits, including antirheumatic, antifibrotic, antiviral, and anti-inflammatory properties. Decoction prepared with turmeric leaves in combination with cumin, bamboo shoots, dried ginger, and roots of *Cardiospermum* is effective against hydrocele [99].

In addition, the herbal plant *Cannabis sativa*, commonly known as *ganja* or *bhang*, has anti-inflammatory properties. The bark of *Cannabis* is used to suppress the inflammatory response in hydrocele [100]. Paste of whole plant part of *Tabernaemontana divaricata* (L.) R.Br. ex Roem. and Schult is prepared and applied topically in the case of hydrocele. In Ayurveda, *Tabernaemontana divaricata* (L.) R.Br. ex Roem. and Schult are known to have anti-inflammatory and pain-relieving qualities [101]. Similarly, topical application of oil extract from *Calotropis procera,* or castor oil, over the inflamed testis causes relief in inflammation and treats hydrocele [102]. In Bangladesh, tribal people use, *Croctus specius* for the treatment of hydrocele. Leaves are boiled, macerated, and applied topically to bring down inflammation and relieve the [103].

19.3.8 Varicocele

Similar to a varicose vein in the legs, varicocele is hypogonadism that originated in the testis, which leads to abnormal enlargement or dilation of the scrotal vein, a common abnormality leading to male infertility which can be cured with herbal medicines [Table. 19.1]. Asiatic acid, found in the plant, can be used to treat varicocele. Moreover, dried and powdered *Cantella asiatica* helps in curing varicocele [104]. According to TCM, the liver is a primary target organ for treating varicocele. Decoction of liver-regulating herb compound composed of *Cyperus rotundus*, seeds of *Litchi chinensis, Angelica sinensis, Cynanchum otophyllum, Pericarpium citri reticulatae, Citrus aurantium,* and of *Cuscuta australis* and water is used against varicocele for treatment [105]. The bark of *Ailanthus excelsa,* a common ingredient in the ayurvedic ointment, comprises Ailantic acid, Sitosterol, and Quassinoids and is implicated in the treatment of varicocele [106]. *Morinda officinalis,* a herb used in TCM, regulates oxidative stress and has therapeutic potential against varicocele. Polysaccharides of *Morinda officinalis* reduce the varicocele-associated damage in the epithelium and decrease cytokines. It also regulates the production of sex hormones in varicocele-damaged testis [107].

In TCM, varicocele-induced infertility is thought to be reversed by using robust antioxidants that can regulate the function of testosterone. *Wu-Zi-Yan-Zong-Wan*, a herbal formulation, can relieve oxidative stress and mitochondrial DNA damage, thus reversing infertility. Another medicine, Palm-leaf raspberry fruit or *Fu-Pen-Zi,* can regulate testosterone and Estradiol level, *Radix morindae officinalis* regulate Testosterone and oxidative stress, and *Semen cuscutae* regulate sperm motility. These herbal medicines may help in relieving varicocele-induced infertility [108]. In traditional Korean medicine (TKM), one of the popular methods for the treatment of varicocele is herbal therapy with acupuncture. The elevated scrotal temperature and oxidative stress are the pathophysiological responses associated with varicocele-related damage that the user can treat of pharmacopuncture therapies. TKM herbs effective against varicoceles and related risk factors are *Cornus officinalis* Sieb. Et Zucc, *Schizandra chinensis* Baillon, *Rubus coreanus* Miquel, *Cuscuta chinensis* Lam., and *Lycium chinense* Mill. [109].

19.3.9 Testicular torsion

Testicular torsion, described as an irregular twisting of the spermatic cord caused by a testes' rotation, is a dangerous pathogenesis condition that causes extreme scrotal and testes distress and is considered an emergency [110]. One of the significant pathophysiologies of testicular torsion is ischemia-reperfusion (IR) damage that leads to oxidative stress. Many herbal medicines and decoctions are said to be anti-inflammatory which can help in reversing the damage and relieving the IR-related pain [Table. 19.1]. *Psoralea coryfolia* is a potent antioxidant that protects testicular injury by reducing reactive oxygen species [111]. It was reported that hydroalcoholic extract of perennial plants *Matricaria chamomile* could lessen the damage done to the testes due to torsion. *Matricaria chamomile* flower extract can increase testosterone and antioxidants levels in serum while reducing lipid peroxidation [110].

The leaf extract of *Plantago major* can reduce lipid peroxidation and peroxidase activity in the damaged tissue and increase CAT levels. The high level of scavenging capability of *Plantago major* leaf extract can minimize the IR injury [112]. A study on *Lycium barbarum*, commonly known as goji berry or wolfberry extract, found that the lower dose of goji berry can increase total antioxidant capacity in testicular cells and reduce the ischemic injury caused by testis torsion [113]. The main active constituent of *Boswellia serrata* resin, acetyl-11-keto-boswellic acid (AKBA), has a significant inhibitory effect on the 5-lipoxygenase enzyme through 5-LOX/LTB4 and p38-MAPK/JNK/Bax pathways, which transforms arachidonic acid into inflammatory mediators. The inhibition of AKBA may help lessen the ailment's severity [114].

In addition, testicular injury due to torsion involves increased mammalian target of rapamycin (mTOR) phosphorylation and apoptosis which can be minimized or lessened using capsaicin present in red peppers [115]. Methanolic extract of *Bieberesteinia multifida* is a potent antioxidant that can scavenge free radicals from the testis damaged by torsion and act as a potential therapeutic effect on testicular injury [116]. Hydroalcoholic extracts of another medicinal plant, *Fumaria parviflora* was found to effectively scavenge free radicals, regulating serum testosterone level and apoptosis index in injury associated with torsion/detorsion of the testis [117]. A natural alkaloid present in the root, rhizome, and stem barks of *Berberis vulgaris,* has dose-dependent pharmacological benefits against ischemic injury [118]. Apigenin, a dietary flavone found in a variety of medicinal plants and fruits, has also been reported to improve the antioxidant profile of IR injury. Apigenin inhibits the inflammatory cascade by decreasing proinflammatory cytokines TNF-α and increasing the endogenous production of the anti-inflammatory cytokine IL-10 [119].

19.3.10 Balanitis and balanoposthitis

Balanitis and balanoposthitis is an inflammatory disease of the male genitalia that is widespread in the uncircumcised population and is characterized by inflammation of the glans penis and the prepuce internal portion of the skin genitalia [29]. A list of polyherbal formulations has been reported in Ayurveda medicines with anti-microbial properties and can improve various attributes of balanoposthitis [Table. 19.3]. *Nimabdi churan (formulated powder)* composed of Neem, Giloyi, Haritaki, Amla, and *Chopchinyadi* Churan containing Chopchini, Triphala, Pipali, etc., Gandhak Rasayan, Guggulu is a few of the traditional formulations that have healing properties without any comorbidities [120]. An anti-inflammatory green leafy vegetable, *Basella alba,* commonly known as Niwithi has various bioactive compounds such as Betacyanin, Lupeol, Carotenoids have therapeutic potential against balanitis [121]. *Basella alba* is another ethnoveterinary plant that comprises several phytoconstituents necessary to treat male reproductive diseases such as balanitis and gonorrhoea [121]. In Indian tribal areas, *Basella alba* is used for the treatment of balanitis. In some regions, whole plants are used for their anti-microbial properties, while in some regions, juice made from leaves is utilized to reduce infection and inflammation [122]. *Candida albicans* is responsible for various genital and non-genital infections. Balanitis and balanoposthitis are common genital diseases caused by *Candida* infection. A bioactive compound, Curcumin shows anti-*Candida* properties by modulating TUP-1, an essential negative regulator of hyphae induction [123].

19.4 Conclusion

Several factors influence normal functioning of the male reproductive organ. Various pathophysiologic mechanisms are implicated in developing reproductive diseases, including oxidative stress, hormonal disorders, pathogen infection, and an unhygienic lifestyle. Owing the side effects of modern medicine prompted clinicians and researchers to divulge the benefit of medicinal plants, herbs, and their bioactive compounds. Human civilizations have used several medicinal plants for a long time. Even today, people living in tribal areas rely on the therapeutic use of medicinal plants. At the same time, there has been a slow but steady shift toward herbal medicine, many questions that need to be addressed before herbal medicines can be considered the safest option. These include questions about the long-term benefits and drawbacks of herbal therapy, the prevention of complications, the standardization of extracts, and the use of herbal therapy in conjunction with "mainstream" drugs, and elucidation of molecular mechanisms. In addition, more preclinical and randomized clinical trials with a large number of cases need to be studied.

Conflict of interest

None

Acknowledgment

The authors are thankful to the Department of Zoology, AMU, Aligarh for providing the necessary facilities. HF expresses her sincere gratitude to MANF (UGC), India, for Fellowship. HRS is thankful to the UGC [Grant no. F.30-377/2017(BSR)] and DST-SERB (Grant no. EMR/2017/001758), New Delhi, for providing financial help.

Abbreviations

COVID-19	Corona Virus Disease 2019
CAT	Catalase
COX-2	Cyclooxygenase
FSH	Follicle stimulating hormone
ED	Erectile Dysfunction
EO	Epididymo-Orchitis
HSP70	Heat Shock Protein 70
IK-β	Inhibitor of κB
IL	Interleukin
IR	Ischemic-Repurfusion
LH	Luteinizing Hormone
LPS	Lipopolysaccharide
ITM	Indian Traditional medicine
NF-κB	Nuclear Factor kappa light chain enhancer of activated B cells
NO	Nitrogen Oxide
NOS	Nitrogen Oxide Synthase
SOD	Superoxide Dismutase
SRY	Sex-determining Region Y
TCM	Traditional Chinese Medicine
TKM	Traditional Korean Medicine
TNF-α	Tumor Necrosis Factor alpha

References

[1]   AN Welz, A Emberger-Klein, K Menrad, Why people use herbal medicine: insights from a focus-group study in Germany, BMC ComplemenT Altern Med 18 (1) (2018) 92. https://doi.org/10.1186/s12906-018-2160-6.

[2]   FS Li, JK Weng, Demystifying traditional herbal medicine with modern approach, Nat Plants 3 (2017) 17109. https://doi.org/10.1038/nplants.2017.109.

[3]   M Sellami, O Slimeni, A Pokrywka, G Kuvačić, L D Hayes, M Milic, et al., J Int Soc Sports Nutr 15 (15) (2018) 14. https://doi.org/10.1186/s12970-018-0218-y.

[4]   CC Falzon, A Balabanova, Phytotherapy: an introduction to herbal medicine, Prim Care 44 (2) (2017) 217–227. https://doi.org/10.1016/j.pop.2017.02.001.

[5]   S Dutta, P Sengupta, Medicinal herbs in the management of male infertility, J Pregnancy Reprod 2 (2018) 1–6.

[6]   JS Xing, ZM Bai, Is testicular dysgenesis syndrome a genetic, endocrine, or environmental disease, or an unexplained reproductive disorder? Life Sci 194 (2018) 120–129. https://doi.org/10.1016/j.lfs.2017.11.039.

[7]   C Barratt, CJ De Jonge, RM Sharpe, Man up': the importance and strategy for placing male reproductive health centre stage in the political and research agenda, Hum Reprod 33 (4) (2018) 541–545. https://doi.org/10.1093/humrep/dey020.

[8]   J Wazir, R Ullah, P Khongorzul, MA Hossain, Khan, N Aktar, X Cui, et al., The effectiveness of hedyotis diffusa wild extract in a mouse model of experimental autoimmune prostatitis, Andrologia 53 (2021) e13913. https://doi.org/10.1111/and.13913.

[9]   WJ Bae, US Ha, KS Kim, SJ Kim, HJ Cho, SH Hong, et al., Effects of KH-204 on the expression of heat shock protein 70 and germ cell apoptosis in infertility rat models, BMC Complement Altern Med 14 (2014) 367. https://doi.org/10.1186/1472-6882-14-367.

[10]   H Nasri, Herbal drugs and new concepts on its use, J Prev Epidemiol 2016 (1) (2016) e01.

[11]   B Yang, Y Xie, M Guo, MH Rosner, H Yang, C Ronco, Nephrotoxicity and Chinese Herbal Medicine, Clin J Am Soc Nephrol 13 (10) (2018) 1605–1611. https://doi.org/10.2215/CJN.11571017.

[12]   R Verpoorte, Medicinal plants: a renewable resource for novel leads and drugs, in: K. Ramawat (Ed.), Herbal Drugs: Ethnomedicine to Modern Medicine, Springer, Berlin, Heidelberg, 2009. https://doi.org/10.1007/978-3-540-79116-4_1.

[13]   A Nahata, Anticancer agents: a review of relevant information on important herbal drugs, IJCPT 6 (2) (2017) 250–255 2017.

[14]   E Ernst, The efficacy of herbal medicine –an overview, Fundam Clin Pharmacol 19 (2005) 405–409. https://doi.org/10.1111/j.1472-8206.2005.00335.x.

[15]   M Ali, T Khan, K Fatima, Q Ali, M Ovais, AT Khalil, et al., Selected hepatoprotective herbal medicines: Evidence from ethnomedicinal applications, animal models, and possible mechanism of actions, Phytother Res 32 (2) (2018) 199–215. https://doi.org/10.1002/ptr.5957.

[16]   S Panyod, CT Ho, LY Sheen, Dietary therapy and herbal medicine for COVID-19 prevention: a review and perspective, J Tradit Complement Med 10 (4) (2020) 420–427. https://doi.org/10.1016/j.jtcme.2020.05.004.

[17]   Zhu F (2020). A review on the application of herbal medicines in the disease control of aquatic animals. 526:735422. https://doi.org/10.1016/j.aquaculture.2020.735422.

[18]   Yinchareon K, Adekoya AE, Chokpaisarn J, Kunworarath N, Jaisamut P, Limsuwan S, et al. (2021). Anti-infective effects of traditional household remedies described in the national list of essential medicines, Thailand, on important human pathogens. 26:100401. https://doi.org/10.1016/j.hermed.2020.100401.

[19]   HR Siddique, M Saleem, Beneficial health effects of lupeol triterpene: a review of preclinical studies, Life Sci 88 (7-8) (2011) 285–293. https://doi.org/10.1016/j.lfs.2010.11.020.

[20]   HR Siddique, S Nanda, A Parray, M Saleem, Androgen receptor in human health: a potential therapeutic target, Curr Drug Targets 13 (14) (2012) 1907–1916. https://doi.org/10.2174/138945012804545579.

[21] D Singh, MA Khan, HR Siddique, Apigenin, a plant flavone playing noble roles in cancer prevention via modulation of key cell signaling networks, Recent Pat Anticancer Drug Discov 14 (4) (2019) 298–311. https://doi.org/10.2174/1574892814666191026095728.

[22] HR Siddique, H Fatma, MA Khan, Medicinal properties of saffron with special reference to cancer-a review of preclinical studies, Saffron, First Ed., Elsevier, Amsterdam, 2020, pp. 233–244.

[23] SK Maurya, G Shadab, HR Siddique, Chemosensitization of therapy resistant tumors: targeting multiple cell signaling pathways by lupeol, a pentacyclic triterpene, Curr Pharm Des 26 (4) (2020) 455–465. https://doi.org/10.2174/1381612826666200122122804.

[24] S Gunawardana, W Jayasuriya, Medicinally important herbal flowers in Sri Lanka, Evid Based Complement Altern Med 2019 (2019) 2321961. https://doi.org/10.1155/2019/2321961.

[25] Z Bai, J Liu, Y Mi, D Zhou, G Chen, D Liang, et al., Acutissimalignan B from traditional herbal medicine Daphne kiusiana var. atrocaulis (Rehd.) F. Maekawa inhibits neuroinflammation via NF-κB Signaling pathway, Phytomedicine 84 (2021) 153508. https://doi.org/10.1016/j.phymed.2021.153508.

[26] JK Gurney, KA McGlynn, J Stanley, T Merriman, V Signal, C Shaw, et al., Risk factors for cryptorchidism, Nat Rev Urol 14 (9) (2017) 534–548. https://doi.org/10.1038/nrurol.2017.90.

[27] M Çek, L Sturdza, A Pilatz, Acute and chronic epididymitis, Eur Urol Suppl 16 (4) (2017) 124–131. https://doi.org/10.1016/j.eursup.2017.01.003.

[28] V Magri, M Boltri, T Cai, R Colombo, S Cuzzocrea, P De Visschere, et al., Multidisciplinary approach to prostatitis, Arch Ital Urol 90 (4) (2019) 227–248. https://doi.org/10.4081/aiua.2018.4.227.

[29] MAB Fahmy, Posthitis and balanoposthitis, Normal & Abnormal Perpuce, First ed., Springer International Publishing, Switzerland AG, 2020, pp. 195–203. https://doi.org/10.1007/978-3-030-37621-5_20.

[30] EM Yimer, KB Tuem, A Karim, N Ur-Rehman, F Anwar, Nigella sativa L. (black cumin): a promising natural remedy for wide range of illnesses, Evid Based Complement Altern Med 2019 (2019) 1528635. https://doi.org/10.1155/2019/1528635.

[31] KK Smrity, S Sharmin, MA Hassan, ML Hossain, Medicinal activity of Vitex negundo L.(Family: Lamiaceae) leaves extract: assessment of phytochemical and pharmacological properties, J Pharmacogn Phytochem 8 (3) (2019) 3571–3575.

[32] DA Crain, SJ Janssen, TM Edwards, J Heindel, SM Ho, P Hunt, et al., Female reproductive disorders: the roles of endocrine-disrupting compounds and developmental timing, Fertil Steril 90 (4) (2008) 911–940. https://doi.org/10.1016/j.fertnstert.2008.08.067.

[33] S Lodh, MK Swamy, Phytochemical aspects of medicinal plants of Northeast India to improve the gynaecological disorders: an update, in: M. Swamy, M. Akhtar (Eds.), Natural Bio-active Compounds, Springer, Singapore, 2019. https://doi.org/10.1007/978-981-13-7205-6_15.

[34] PE Gurung, E Yetiskul, I Jalal, Physiology, male reproductive system, StatPearls. Internet (2020).

[35] M Mawhinney, A Mariotti, Physiology, pathology and pharmacology of the male reproductive system, Periodontol. 2000 61 (1) (2013) 232–251. https://doi.org/10.1111/j.1600-0757.2011.00408.x.

[36] AO Afolabi, IA Alagbonsi, TA Oyebanji, Beneficial effects of ethanol extract of Zingiber officinale (Ginger) rhizome on epididymalsperm and plasma oxidative stress parametersin experimentally cryptorchid rats, Annu Res Rev Biol 4 (9) (2014) 1448–1460.

[37] DA Munyali, ACT Momo, GRB Fozin, PBD Defo, YP Tchatat, B Lieunang, et al., Rubus apetalus (Rosaceae) improves spermatozoa characteristics, antioxidant enzymes and fertility potential in unilateral cryptorchid rats, Basic Clin Androl 30 (8) (2020). https://doi.org/10.1186/s12610-020-00107-3.

[38] MA Abd-El-Hafez, MD El-Shafee, SH Omar, AA Aburahma, SS Kamar, The ameliorative effect of curcumin on cryptorchid and non-cryptorchid testes in induced unilateral cryptorchidism in albino rat: histological evaluation, Folia Morphologica 80 (3) (2021) 596–604. https://doi.org/10.5603/FM.a2020.0084.

[39] WJ Bae, US Ha, JB Choi, KS Kim, SJ Kim, HJ Cho, et al., Protective effect of decursin extracted from Angelica gigas in male infertility via Nrf2/HO-1 signaling pathway, Oxidative medicine and cellular longevity 2016 (2016) 5901098. https://doi.org/10.1155/2016/5901098.

[40] E Li, Y Guo, G Wang, F Chen, Q Li, Effect of resveratrol on restoring spermatogenesis in experimental cryptorchid mice and analysis of related differentially expressed proteins, Cell Biol Int 39 (6) (2015) 733–740. https://doi.org/10.1002/cbin.10441.

[41] O Mansour, M Darwish, G Ismail, ZA Douba, A Ismaeel, KS Eldair, Review study on the physiological properties and chemical composition of the Laurus nobilis, The Pharmaceutical and Chemical Journal 5 (1) (2018) 225–231.

[42] NE MacDonald, WR Bowie, Epididymitis, Orchitis, and Prostatitis, Principles and Practice of Pediatric Infectious Disease, Fifth Ed., Elsevier, Amsterdam, 2018, pp. E1371–E1373. https://doi.org/10.1016/B978-0-323-40181-4.00053-0.

[43] RA Foster, Male Reproductive System, Pathologic Basis of Veterinary Disease, Sixth Ed., Elsevier, Amsterdam, 2017, pp. E11194–E11222. https://doi.org/10.1016/B978-0-323-35775-3.00019-9.

[44] YA Maldonado, AK Shetty, Mumps virus, Principles and Practice of Pediatric Infectious Disease, Fifth Ed, Elsevier, Amsterdam, 2018, pp. E21157–E21162.

[45] AJ Kumaran, J Karunakaran, Antioxidant activities of the methanol extract of Cardiospermum halicacabum, Pharmacetical Biology 44 (2) (2006) 146–151. https://doi.org/10.1080/13880200600596302.

[46] G Shukla, S Chakravarty, Ethnobotanical plant use of Chilapatta reserved forest in West Bengal, Indian Forester 138 (12) (2012) 1116.

[47] SK Jain, S Srivastava, Traditional uses of some Indian plants among islanders of the Indian Ocean, Indian Journal of Traditional knowledge 4 (4) (2005) 345–357. https://nopr.niscair.res.in/handle/123456789/8532.

[48] P Jeruto, C Lukhoba, G Ouma, D Otieno, C Mutai, An ethnobotanical study of medicinal plants used by the Nandi people in Kenya, J Ethnopharmacol 116 (2) (2008) 370–376. https://doi.org/10.1016/j.jep.2007.11.041.

[49] O Neamsuvan, P Singdam, K Yingcharoen, Sengnon, N. A survey of medicinal plants in mangrove and beach forests from sating Phra Peninsula, Songkhla Province, Thailand, Journal of Medicinal Plants Research 6 (12) (2012) 2421–2437. https://doi.org/10.5897/JMPR11.1395.

[50] I Malik, VU Ahmad, S Anjum, FZ Basha, A pentacyclic triterpene from Litchi chinensis, Natural product communications 5 (4) (2010) 529–530.

[51] NH Sofia, HVM Kumari, Anti-diabetic polyherbal siddha formulation Atthippattaiyathi Kasayam: a review, International Journal of Pharmaceutical Sciences Reviews and Research 28 (2014) 169–174.

[52] R Khalil, Q Ali, MM Hafeez, A Malik, Phenolic acid profiling by Rp-HPLC: evaluation of antibacterial and anticancer activities of Conocarpus Erectus plant extracts, Biological & Clinical Sciences Research Journal 2021 (2021) e010.

[53] SS Yaradua, M Shah, Ethnobotanical studies of the genus Crotalaria L.(Crotalarieae, Fabaceae) in Katsina State, Nigeria. Pure Applied Biology. 7 (2) (2018) 882–889. http://dx.doi.org/10.19045/bspab.2018.700107.

[54] R Gaziano, E Campione, F Iacovelli, D Marino, F Pica, P Di Francesco, et al., Antifungal activity of Cardiospermum halicacabum L. (Sapindaceae) against Trichophyton rubrum occurs through molecular interaction with fungal Hsp90, Drug Des Dev Ther 12 (2018) 2185–2193. https://doi.org/10.2147/DDDT.S155610.

[55] G Fatima, A Siddiqui, A Jamal, S Chaudhary, Coriandrum sativum Linn. Traditional, biochemical and biological activities: an overview, Journal of Drug Delivery and Therapeutics 9 (3) (2019) 513–516. http://dx.doi.org/10.22270/jddt.v9i3.2.

[56] ER James, DT Carrell, KI Aston, TG Jenkins, M Yeste, A Salas-Huetos, The role of the epididymis and the contribution of epididymosomes to mammalian reproduction, Int J Mol Sci 21 (15) (2020) 5377. https://doi.org/10.3390/ijms21155377.

[57] H Zhao, C Yu, C He, C Mei, A Liao, D Huang, The immune characteristics of the epididymis and the immune pathway of the epididymitis caused by different pathogens, Front Immunol 11 (2020) 2115. https://doi.org/10.3389/fimmu.2020.02115.

[58] X Song, NH Lin, YL Wang, B Chen, HX Wang, K Hu, Comprehensive transcriptome analysis based on RNA sequencing identifies critical genes for lipopolysaccharide-induced epididymitis in a rat model, Asian J Androl 21 (6) (2019) 605–611. https://doi.org/10.4103/aja.aja_21_19.

[59]   V Michel, A Pilatz, MP Hedger, A Meinhardt, Epididymitis: revelations at the convergence of clinical and basic sciences, Asian J Androl 17 (5) (2015) 756–763. https://doi.org/10.4103/1008-682X.155770.

[60]   Z Jing, G Liying, W Zhenqing, Z Hui, L Shuai, S Dingqi, et al., Efficacy and safety of Ningmitai capsules in patients with chronic epididymitis: a prospective, parallel randomized controlled clinical trial, Evidence-based complementary and alternative medicine: eCAM 2021 (2021) 9752592. https://doi.org/10.1155/2021/9752592.

[61]   RK Gupta, M Deogade., RK Gupta, M Deogade, A critical review on ethnobotanical, phytochemical and pharmacological investigations of Martynia Annua Linn, International Journal of Ayurvedic Medicine 9 (3) (2018) 136–143.

[62]   O Banyra, O Nikitin, I Ventskivska, Acute epididymo-orchitis: relevance of local classification and partner's follow-up, Central European journal of urology 72 (3) (2019) 324–329. https://doi.org/10.5173/ceju.2019.1973.

[63]   M Eskandari, S Jani, M Kazemi, H Zeighami, A Yazdinezhad, S Mazloomi, S Shokri, Ameliorating effect of ginseng on epididymo-orchitis inducing alterations in sperm quality and spermatogenic cells apoptosis following infection by uropathogenic Escherichia coli in rats, Cell journal 18 (3) (2016) 446–457. https://doi.org/10.22074/cellj.2016.4573.

[64]   P Sittisart, B Dunkhunthod, C Chuea-nongthon., Antioxidant and Anti-inflammatory Activities of Ethanolic Extract from Hoya kerrii Craib, Chiang Mai Journal of Science 47 (5) (2020) 912–925. http://epg.science.cmu.ac.th/ejournal.

[65]   G Paulis, Inflammatory mechanisms and oxidative stress in prostatitis: the possible role of antioxidant therapy, Res Rep urol 10 (2018) 75–87. https://doi.org/10.2147/RRU.S170400.

[66]   QY Zhang, ZN Mo, XD Liu, Reducing effect of curcumin on expressions of TNF-alpha, IL-6 and IL-8 in rats with chronic nonbacterial prostatitis, National Journal of Andrology 16 (1) (2010) 84–88.

[67]   O Dashdondov, J Wazir, G Sukhbaatar, R Mikrani, B Dorjsuren, N Aktar, et al., Herbal nutraceutical treatment of chronic prostatitis-chronic pelvic pain syndrome: a literature review, Int Urol Nephrol 53 (8) (2021) 1515–1528. https://doi.org/10.1007/s11255-021-02868-w.

[68]   M Inci, M Davarci, M Inci, S Motor, FR Yalcinkaya, E Nacar, et al., Anti-inflammatory and antioxidant activity of thymoquinone in a rat model of acute bacterial prostatitis, Hum Exp Toxicol 32 (4) (2013) 354–361. https://doi.org/10.1177/0960327112455068.

[69]   JL Capodice, DL Bemis, R Buttyan, SA Kaplan, AE Katz, Complementary and alternative medicine for chronic prostatitis/chronic pelvic pain syndrome., Evidence-based complementary and alternative medicine: eCAM 2 (4) (2005) 495–501. https://doi.org/10.1093/ecam/neh128.

[70]   DW Sohn, CH Han, YS Jung, SI Kim, SW Kim, YH Cho, Anti-inflammatory and anti-microbial effects of garlic and synergistic effect between garlic and ciprofloxacin in a chronic bacterial prostatitis rat model, International journal of anti-microbial agents 34 (3) (2009) 215–219. https://doi.org/10.1016/j.ijantimicag.2009.02.012.

[71]   GM Zou, J Wu, D He, The effect of combined treatment of Sini San and Chinese herbs enema on NIH-CPSI of chronic prostatitis and the clinical observation, Liaoning Journal of Traditional Chinese Medicine 8 (2010) 1517–1519.

[72]   Y Xiong, X Qiu, W Shi, H Yu, X Zhang, Anti-inflammatory and antioxidant effect of modified Bazhengsan in a rat model of chronic bacterial prostatitis, J Ethnopharmacol 198 (2017) 73–80. https://doi.org/10.1016/j.jep.2016.12.039.

[73]   RZ Chen, L Cui, YJ Guo, YM Rong, XH Lu, MY Sun, et al., In vivo study of four preparative extracts of Clematis terniflora DC. for antinociceptive activity and anti-inflammatory activity in rat model of carrageenan-induced chronic non-bacterial prostatitis, J Ethnopharmacol 134 (3) (2011) 1018–1023. https://doi.org/10.1016/j.jep.2011.01.004.

[74]   K Zhang, X Zeng, Y Chen, R Zhao, H Wang, J Wu, Therapeutic effects of Qian-Yu decoction and its three extracts on carrageenan-induced chronic prostatitis/chronic pelvic pain syndrome in rats, BMC complementary and alternative medicine 17 (1) (2017) 75. https://doi.org/10.1186/s12906-016-1553-7.

[75] YJ Liu, GH Song, GT Liu, Investigation of the effect of traditional Chinese medicine on pain and inflammation in chronic non-bacterial prostatitis in rats, Andrologia 48 (6) (2016) 714–722. https://doi.org/10.1111/and.12544.

[76] H Li, A Hung, A Yang, A classic herbal formula Danggui Beimu Kushen Wan for chronic prostatitis: from traditional knowledge to scientific exploration., Evidence-based complementary and alternative medicine: eCAM 2018 (2018) 1612948. https://doi.org/10.1155/2018/1612948.

[77] Z Yan, C Huang, G Huang, Y Wu, J Wang, J Yi, et al., Jiedu Huoxue decoction improves inflammation in rat type III prostatitis: The importance of the NF-κB signalling pathway, Andrologia 51 (5) (2019) e13245. https://doi.org/10.1111/and.13245.

[78] B Goel, NK Maurya, Aphrodisiac herbal therapy for Erectile Dysfunction, Archives of Pharmacy practice 11 (1) (2020) 1–6.

[79] F Borrelli, C Colalto, DV Delfino, M Iriti, AA Izzo, Herbal dietary supplements for erectile dysfunction: a systematic review and meta-analysis, Drugs 78 (6) (2018) 643–673. https://doi.org/10.1007/s40265-018-0897-3.

[80] H Li, H Jiang, J Liu, Traditional Chinese medical therapy for erectile dysfunction, Translational andrology and urology 6 (2) (2017) 192–198. https://doi.org/10.21037/tau.2017.03.02.

[81] YL Wang, LG Geng, CB He, SY Yuan, Chinese herbal medicine combined with tadalafil for erectile dysfunction: a systematic review and meta-analysis, Andrology 8 (2) (2020) 268–276. https://doi.org/10.1111/andr.12696.

[82] SH Jeon, WJ Bae, GQ Zhu, W Tian, EB Kwon, GE Kim, et al., Combined treatment with extracorporeal shockwaves therapy and an herbal formulation for activation of penile progenitor cells and antioxidant activity in diabetic erectile dysfunction, Translational andrology and urology 9 (2) (2020) 416–427. https://doi.org/10.21037/tau.2020.01.23.

[83] L Chen, GR Shi, DD Huang, Y Li, CC Ma, M Shi, et al., Male sexual dysfunction: a review of literature on its pathological mechanisms, potential risk factors, and herbal drug intervention, Biomed Pharmacother 112 (2019) 108585. https://doi.org/10.1016/j.biopha.2019.01.046.

[84] NP Masuku, JO Unuofin, SL Lebelo, Promising role of medicinal plants in the regulation and management of male erectile dysfunction, Biomed Pharmacother 130 (2020) 110555. https://doi.org/10.1016/j.biopha.2020.110555.

[85] R Tiwari, S Choudhary, G Gaba, Pharmacological and toxicological evaluation of herbal extracts in rat model of erectile dysfunction, International Journal of Herbal Medicine 6 (4) (2018) 09–16.

[86] NC Park, SW Kim, SY Hwang, HJ Park, Efficacy and safety of an herbal formula (KBMSI-2) in the treatment of erectile dysfunction: A preliminary clinical study, Investigative and clinical urology 60 (4) (2019) 275–284. https://doi.org/10.4111/icu.2019.60.4.275.

[87] RM Maboa, TL Rasakanya, T Mdaka, Evaluation of the antioxidant effects of Imvusankunzi (a herbal remedy used in the treatment of erectile dysfunction) using neutrophils, in: In Proceedings for Annual Meeting of The Japanese Pharmacological Society WCP2018 (The 18th World Congress of Basic and Clinical Pharmacology), Japanese Pharmacological Society, 2018 PO2-14.

[88] M Nimrouzi, AM Jaladat, MM Zarshenas, A panoramic view of medicinal plants traditionally applied for impotence and erectile dysfunction in Persian medicine, Journal of traditional and complementary medicine 10 (1) (2018) 7–12. https://doi.org/10.1016/j.jtcme.2017.08.008.

[89] OA Ojo, AB Ojo, BE Oyinloye, BO Ajiboye, OO Anifowose, A Akawa, et al., Ocimum gratissimum Linn. Leaves reduce the key enzymes activities relevant to erectile dysfunction in isolated penile and testicular tissues of rats, BMC Complement Altern Med 19 (1) (2019 Mar 19) 71. https://doi.org/10.1186/s12906-019-2481-0.

[90] J Rioja, FM Sánchez-Margallo, J Usón, LA Rioja, Adult hydrocele and spermatocele, BJU Int 107 (11) (2011) 1852–1864. https://doi.org/10.1111/j.1464-410X.2011.10353.x.

[91] AM Reddy, MVS Babu, RR Rao, Ethnobotanical study of traditional herbal plants used by local people of Seshachalam Biosphere Reserve in Eastern Ghats, India. Herba Polonica 65 (1) (2019) 40–54.

[92] EJ Seok, SY Jeon, WB Ki, DJ Kim, SI Lee, An analysis of clinical studies on Oryeong-San, Herbal formula science 26 (4) (2018) 341–362. https://doi.org/10.14374/HFS.2018.26.4.341.

[93]   S Shanmugam, CP Muthupandi, VM Eswaran, K Rajendran, Medicinal plants from Vettangudi water bird sanctuary in Sivagangai district of Tamil Nadu, Southern India. 2021, J Drug Delivery and Therapeutics 11 (2) (2021) 135–140. http://dx.doi.org/10.22270/jddt.v11i2.4608.

[94]   Gitte TA, Kare MA, Deshmukh AM (2012). Ethno- medicinal studies on barks of some medicinal plants in Marathwada (M. S.), India –I. 4(10): 08-10

[95]   K Singh, S Gupta, PK Mathur, Investigation on ethnomedicinal plants of district Firozabad, Journal of Advanced Laboratory Research in Biology 1 (1) (2010) 64–66.

[96]   A Singh, MK Singh, R Singh, Traditional medicinal flora of the district Buxar (Bihar, India), Journal of Pharmacognosy and Phytochemistry 2 (2) (2013) 41–49.

[97]   D Bose, JG Roy, SD Mahapatra, T Datta, SD Mahapatra, H Biswas, Medicinal plants used by tribals in Jalpaiguri district, West Bengal, India, Journal of Medicinal Plants studies 3 (2015) 15–21.

[98]   K Chandrashekara, HM Somashekarappa, (210)Po and (210)Pb in medicinal plants in the region of Karnataka, Southern India, J Environ Radioact 160 (2016) 87–92. https://doi.org/10.1016/j.jenvrad. 2016.04.036.

[99]   KC Velayudhan, N Dikshit, & MA Nizar, (2012). Ethnobotany of turmeric (Curcuma longa L.). 11(4):607-614.

[100] B Sharma, AN Dey, A Kumari, NA Pala, Documentation of medico–religious plants of Dooars region of West Bengal, International Journal of Forest Usufructs Management 15 (2) (2014) 3–15.

[101] BB Shiddamallayya Nagayya, A Janardhanan, G Anku, T Borah, AK Tripathi, C Rath, et al (2020). Ethnomedical importance of traditional medicinal plants among the indigenous people of Kanchanpur subdivision of North Tripura district. 5(2):88–97.

[102] A Singh, GS Singh, PK Singh, Medico-ethnobotanical inventory of Renukoot forest division of district Sonbhadra, Uttar Pradesh, India, Indian Journal of Natural Products and Resources 3 (3) (2012) 448–457.

[103] S Hossain, S Rahman, MT Morshed, M Haque, S Jahan, R Jahan, et al., Tribal cross-talk as an effective way for ethnobotanical knowledge transfer-inference from Costus specious as a case study, American-Eurasian Journal of Sustainable Agriculture 7 (5) (2013) 373–390.

[104] R Jahan, S Hossain, S Seraj, D Nasrin, Z Khatun, PR Das, et al., Centella asiatica (L.) Urb.: Ethnomedicinal uses and their scientific validations, American-Eurasian Journal of Sustainable Agriculture 6 (4) (2012) 261–270.

[105] X Lu, J Liu, H Yin, C Ding, Y Wang, F Zhang, et al., Effects of liver-regulating herb compounds on testicular morphological and ultrastructural changes in varicocele rats through SCF/C-KIT pathway, Andrologia 52 (9) (2020) e13658. https://doi.org/10.1111/and.13658.

[106] D Kumar, ZA Bhat, P Singh, MY Shah, SS Bhujbal, Ailanthus excelsa Roxb. is really a plant of heaven, International Journal of Pharmacology 6 (5) (2010) 535–550.

[107] L Zhang, X Zhao, F Wang, Q Lin, W Wang, Effects of Morinda officinalis polysaccharide on experimental varicocele rats, Evidence-based complementary and alternative medicine: eCAM 2016 (2016) 5365291. https://doi.org/10.1155/2016/5365291.

[108] RL Dun, M Yao, L Yang, XJ Cui, JM Mao, Y Peng, et al., Traditional Chinese herb combined with surgery versus surgery for varicocele infertility: a systematic review and meta-analysis, Evidence-based complementary and alternative medicine: eCAM 2015 (2015) 689056. https://doi.org/10.1155/2015/689056.

[109] J Jo, H Kim, UM Jerng, Improvements in scrotal thermoregulation in patients with varicoceles treated by using traditional Korean medicine: two case reports, Journal of acupuncture and meridian studies 9 (3) (2016) 156–160. https://doi.org/10.1016/j.jams.2015.12.001.

[110] M Soltani, M Moghimian, SH Abtahi-Eivari, H Shoorei, A Khaki, M Shokoohi, Protective effects of Matricaria chamomilla extract on torsion/detorsion-induced tissue damage and oxidative stress in adult rat testis, International journal of fertility & sterility 12 (3) (2018) 242–248. https://doi.org/10.22074/ijfs.2018.5324.

[111] SM Wei, ZZ Yan, J Zhou, Psoralea corylifolia protects against testicular torsion/detorsion-induced ischemia/reperfusion injury, J Ethnopharmacol 137 (1) (2011) 568–574. https://doi.org/10.1016/j.jep.2011.06.010.

[112] M Moradi-Ozarlou, S Javanmardi, H Tayefi-Nasrabadi, Antioxidant property of Plantago major leaf extracts reduces testicular torsion/detorsion-induced ischemia/reperfusion injury in rats, Vet Res Forum 11 (1) (2020) 27–33. https://doi.org/10.30466/vrf.2019.102182.2432.

[113] R Dursun, Y Zengin, E Gündüz, M İçer, HM Durgun, M Dağgulli, et al., The protective effect of goji berry extract in ischemic reperfusion in testis torsion, Int J Clin Exp Med 8 (2) (2015) 2727–2733.

[114] M Ahmed, A Ahmed, El Morsy, M E., Acetyl-11-keto-β-boswellic acid prevents testicular torsion/detorsion injury in rats by modulating 5-LOX/LTB4 and p38-MAPK/JNK/Bax/Caspase-3 pathways, Life Sci 260 (2020) 118472. https://doi.org/10.1016/j.lfs.2020.118472.

[115] N Javdan, SA Ayatollahi, MI Choudhary, S Al-Hasani, F Kobarfard, A Athar, et al., Capsaicin protects against testicular torsion injury through mTOR-dependent mechanism, Theriogenology 113 (2018) 247–252. https://doi.org/10.1016/j.theriogenology.2018.03.012.

[116] Y Kamali, S Gholami, AR Jahromi, F Namazi, A Khorsandi, Biebersteinia multifida DC a potential savior against testicular torsion-induced reperfusion injury, Compar Clin Pathol 25 (5) (2016) 1001–1005.

[117] M Shokoohi, H Shoorei, M Soltani, SH Abtahi-Eivari, R Salimnejad, M Moghimian, Protective effects of the hydroalcoholic extract of Fumaria parviflora on testicular injury induced by torsion/detorsion in adult rats, Andrologia 50 (7) (2018) e13047. https://doi.org/10.1111/and.13047.

[118] M Moradi-Ozarlou, M Ashrafizadeh, S Javanmardi, The ameliorative impacts of berberine on testicularischemic/reperfusion injury in rats: an experimental study, Iran J Vet Surg 16 (1) (2021) 19–23. https://doi.org/10.30500/IVSA.2021.263095.1239.

[119] I Skondras, M Lambropoulou, A Tsaroucha, S Gardikis, G Tripsianis, C Simopoulos, et al., The role of Apigenin in testicular damage in experimental ischemia-reperfusion injury in rats, Hippokratia 19 (3) (2015) 225–230.

[120] C Sharma, A Sharma, An ayurvedic approach for balanoposthitis-a case study, International Journal of Ayurveda and Pharma Research 8 (1) (2020) 67–71 2020. https://doi.org/10.47070/ijapr.v8iSupply1.1650 .

[121] Ganegoda C, & Senevirathne W (2020). Medicinal properties of selected green leaves used in Sri Lankan cuisine. 1:15. https://doi.org/10.1186/s12906-016-1455-8.

[122] SA Deshmukh, DK Gaikwad, A review of the taxonomy, ethnobotany, phytochemistry, and pharmacology of Basella alba (Basellaceae), Journal of Applied Pharmaceutical Science 4 (01) (2014) 153–165. https://doi.org/10.7324/JAPS.2014.40125.

[123] AD Thakre, SV Mulange, SS Kodgire, GB Zore, SM Karuppayil, Effects of cinnamaldehyde, ocimene, camphene, curcumin, and farnesene on Candida albicans, Advances in Microbiology 6 (09) (2016) 627.

CHAPTER 20

Polycystic ovarian syndrome: Causes and therapies by herbal medicine

Payal Poojari[a], Additiya Paramanya[a], Dipty Singh[b] and Ahmad Ali[a]

[a]University Department of Life Sciences, University of Mumbai, Mumbai, India [b]Neuroendocrinology, ICMR- National Institute for Research in Reproductive Health, Mumbai, India

20.1 Introduction

Polycystic ovarian syndrome (PCOS) is one of the prevailing endocrine disorders causing reproductive as well as metabolic dysfunctions in women. First described by Stein and Leventhal in 1935, as a condition where approximately 10 small cysts ($\sim$2 and 9 mm diameter) on one or both ovaries and/or the ovarian volume exceeds 10 ml of at least one ovary [1]. According to WHO and Rotterdam diagnostic criteria, the global prevalence of the disorder in adolescents ranges from a minimum of 2.2% to as high as 26% of the population [2]. However, the childhood prevalence of PCOS still remains unknown [3]. Therefore, this condition demands more attention towards accurate and early diagnosis and management [4].

PCOS is characterized as an endocrine disorder, which itself suggests hormonal abnormality is associated with the disorder. The human reproductive system is completely under the control of endocrine hormones secreted by the hypothalamus and pituitary. The ovarian follicular growth is completely dependent on the reproductive hormones such as LH and FSH. The majority of PCOS cases witness higher levels of androgen synthesis [5]. This disrupted hormone secretion is believed to be due to disrupted GnRH secretion. Hence, causing an imbalance in the ovary thereby affecting its morphology [6]. Women's reproductive health is also directly associated with their psychological health or vice versa.

There are overall multiple factors contributing to the manifestation of the disorder which is many times also associated with various comorbidities (Fig. 20.1). Although the main cause of the disorder is not well understood yet, over the year's attempts are made to understand the pathophysiology of the disorder [5]. Many studies suggest perturbed steroidogenesis being the main cause of the disorder whereas few studies characterized it as a genetic disorder, or a lifestyle associated disorder [6]. Albeit all the limitations in understanding pathophysiology,

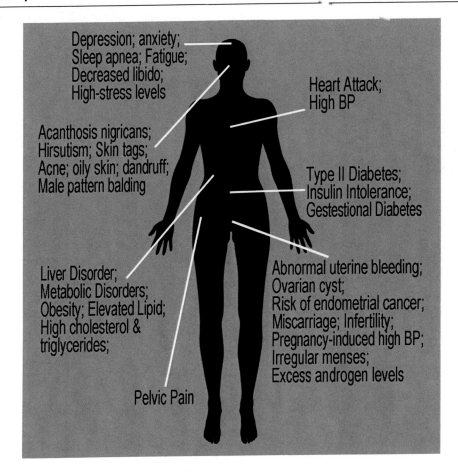

Figure 20.1
Causes and consequences of polycystic ovarian syndrome.

many researchers have made successful attempts for the management of this disorder [4]. In many cases, medicinal side effects and resistance towards the treatment have been reported making it challenging to control. This highlights the unmet need to naturally control and treat the disorder. Many researchers have attempted to showcase the usefulness of traditional herbal medicines as a therapeutic aid [7]. Traditional herbs have shown very encouraging results in the management of PCOS [7]. Therefore, understanding the natural ways of treating the disorder may prove to be a boon. This chapter includes available knowledge about the disorder, pathophysiology, its various symptoms, and natural therapeutic aids suggested for ovarian health.

20.2 Pathophysiology

To understand the pathophysiology of the disorder, it is essential to decipher the root cause and different factors that are involved in PCOS. The most updated knowledge of pathophysiology describes PCOS as a multifactorial disorder that includes aberrant ovarian steroidogenesis, severe increase in the level of oxidants, faulty insulin signaling, excessive oxidative stress, and genetic and/or environmental conditions [8]. As evident in various studies, around 60% of women are observed with excess androgen, often mentioned as a salient characteristic of the disorder [9]. Hyperandrogenism is commonly understood by free elevated testosterone, most commonly observed and has a major contribution to the pathophysiology of PCOS. Excessive androgen is also produced by the ovary and adrenal gland. In a wide number of cases, high levels of ovarian androgen production are most commonly reported and excessive adrenal androgen production is also present in a few cases. Elevation of androgen suppresses sex hormone-binding globulin (SHBG) concentration facilitating higher free testosterone concentrations [10].

A typical human folliculogenesis involves various factors involving follicular growth and development, wherein a single primordial follicle is designated to grow and develop into a Graafian follicle fated to either ovulate its egg during the mid-cycle or to die by the process of atresia. However, transformational changes are required for a primordial follicle to enter into an ovulatory stage which requires a span of 365 days [11]. The process of folliculogenesis involves cell proliferation for follicular growth and follicular fluid formation. Cytodifferentiation leads to the development of the cells and tissue of the follicle. Follicles that fail to survive the process of cytodifferentiation die due to apoptosis. The phases involved in follicular growth are influenced by various hormones and growth factors [12].

A subtle balance between dormant and growing follicles are maintained by androgens, gonadotropins and glycoprotein hormones. Secreted androgen, follicle stimulating hormone (FSH) along with antimullerian hormone (AMH) are crucially important for initial follicular growth. Hence, a disrupted balance between 3 factors causes an early follicular arrest [13]. Excess LH secreted stimulates theca cells to produce excess androgen. A disrupted GnRH pathway secretes excess LH and the level of FSH substantially drops. Low concentration of FSH and insufficient conversion of androgens to estradiol results in failed selection of dominant follicle developing a chronic type of anovulation [14]. AMH secreted by granulosa cells plays a determining role in moderating this balance as it inhibits the transition from primordial to primary follicles. Increased growth of small follicles and subsequent growth arrest develops a polycystic morphology in the ovary. A difference in the follicles of the PCOS ovary and the normal ovary has also been observed suggesting an inherent difference in the ovarian follicles and an intrinsic ovarian abnormality [15].

To understand the ovarian abnormality, it is important to understand the ovarian biochemistry, ovary functionality and the external influence associated with it. It is also observed that the concentration of testosterone is higher than healthy ovaries. Ovary being the ultimate site for follicular maturation is also responsible for the excess of androgen due to the anomaly of steroidogenesis. The ovary and adrenal cortex are the two main organs responsible for circulating androgens. Theca Interna layer of the ovarian follicle is responsible for the androgen production in the ovary, on the other hand, the adrenal cortex is responsible for adrenal androgens. Both androgen producing glands use cholesterol as an initial substrate for the production of androstenedione. These glands are completely under the endocrine control of LH (ovary) and adrenocorticotropic hormone (ACTH) of the adrenal glands [16].

20.2.1 Biosynthesis of steroid hormone

The synthesis of steroid hormones is majorly dependent on cholesterol. The initial step of biosynthesis involves the conversion of cholesterol to pregnenolone. Pregnenolone is known as an endogenous steroid hormone and a metabolic intermediate that is responsible for the biosynthesis of different steroid hormones such as estrogen, progesterone, androgens, gluco- and minerelo-corticoids. This conversion takes place in two stages which involves cholesterol side-chain cleavage enzyme and acute steroidogenic acute regulatory protein (StAR). Cholesterol scc (side cleavage chain) is a mitochondrial enzyme also known as P450scc, catalyzing the conversion of cholesterol to pregnenolone with three monooxygenase reactions. Cholesterol sidechain undergoes hydroxylation twice, which initially generates 22R-hydroxycholesterol and later generates 20α,22R-dihydroxycholesterol. The final step involves bond cleavage between C20 and C22 generating pregnenolone. The three monooxygenase reactions take place through electron transport. Each step demands 2 electrons while NADPH being the initial source of electrons [17]. Electron transport is initiated from NADPH to P450scc involving two electron transport proteins, namely, adrenodoxin and adrenodoxin reductase [18,19]. Altogether the three proteins constitute the cholesterol side-chain cleavage complex. The mitochondrial P450scc is always active, the activity of which can be regulated with a limited supply of cholesterol in the inner membrane. The supply of cholesterol from the outer mitochondrial membrane to the inner mitochondrial membrane is considered as a rate-limiting step in steroidogenesis. Primarily, this step is mediated by StAR, encoded by the STAR gene in humans. When a cell receives stimulation for steroid production, the availability of StAR to transfer cholesterol to the inner membrane limits the pace of the reaction [20].

The synthesis of steroid hormone involves two different pathways: Δ5- steroid pathway and Δ4- steroid pathway. The formation of pregnenolone leads to its immediate conversion to dehydroepiandrosterone (DHEA) by a two-step process along with the Δ5- steroid pathway. The conversion is catalyzed by cytochrome P450c17α. The enzyme P450c17α is regulated by the P450c17 gene in humans, which also has a minor role in regulating the activity of 17,20-lyase in the Δ4- steroid pathway. On the other hand, progesterone undergoes parallel transformation forming androstenedione as the final product following the Δ4- steroid pathway. The formation of progesterone is initially converted 17-hydroxyprogesterone, catalyzed by the enzyme 17α-hydroxylase, is either converted to cortisol or 17-ketosteroid, depending on whether it undergoes 21-hydroxylation or 17,20 lysis, respectively. The 17-ketosteroid formed is then converted to testosterone, dihydrotestosterone, and estradiol by the action of 17-B-hydroxy dehydrogenase [9].

The formation of intraovarian androgen is essential for normal follicular growth as well as for the synthesis of estradiol. However, impairment in the synthesis of androgen according to the needs of the developing follicle triggers poor maturation of the follicle and increased follicular atresia is observed. In a healthy ovary, the LH secreted acts on the theca-interstitial-stromal cells whereas FSH acts on the granulosa cells. In estrogen biosynthesis, the thecal compartment secretes androgens in response to LH. However, the androstenedione formed is converted to estrogen by the action of aromatase in the granulosa cells. The enzyme aromatase is however under the influence of FSH. When a dominant follicle is liberated, the level of estrogens dominates the level of androgen. Moreover, the regulation of intraovarian androgen synthesis by LH plays a major regulatory role. As the LH stimulation increases a simultaneous desensitization sets in. However, overstimulation and dose-dependent release causes downregulation of LH receptors, reducing cholesterol scc activity, 17,20 lyase and finally reducing the activity of 12-hydroxylase increasing the 17-hydroxyprogesterone to androgen activity [21].

20.2.2 Insulin sensitivity

The steroidogenic response to the trophic hormones is modulated by different classes of small peptides, such as insulin and insulin-like growth factors (IGFs). Women suffering from PCOS gain insulin resistance (IR) [9]. Although it is much more prevalent in obese women than lean women with PCOS. Now insulin plays an important role in maintaining the overall glucose homeostasis. Obtaining insulin resistance on the other hand causes hyperinsulinemia. However, hyperinsulinemia is accounted as one of the most important factors in maintaining hyperandrogenemia [22]. Hyperandrogenemia directly induces excessive production of androgen through stimulating thecal cells. Elevation of Insulin causes metabolic disorder by affecting the steroidogenesis and endocrine secretion.

The molecular mechanism responsible for IR in PCOS involves defective post-receptor insulin activity, elevated levels of free fatty acids, increased cytokine secretion, and increased androgens. Increased release of free fatty acids and cytokine secretion, e.g., TNF-α, IL-6, leptin, and resistin are majorly observed in intra-abdominal adipocytes [9]. The leakage of free fatty acids to the liver subsequently affects the secretion, metabolism and the peripheral action of insulin. Therefore, accumulation of fats is highly associated with PCOS rather than obesity or increased BMI. The functioning of insulin in the body is maintained by insulin transduction signals. Disturbed secretion of insulin leads to hypoglycemia, hyperglycemia and T2DM. The insulin transduction pathway (ITP) works on the coordination of different enzymes, receptors and hormones. These factors ensure a proper functioning of insulin in the body. ITP also generates a cascade of signals in the cells in response to its environment. Although, an impairment in the ITP facilitates cellular inflammatory pathways, being the prime cause of insulin resistance [23]. Besides, a decreased strength of IRS/PI-3K signaling pathway is also considered as a main cause for insulin resistance, however, its authenticity is yet under scrutiny. Decreased function of IRS/PI-3K takes place due to increased expression of p85α and serine phosphorylation of insulin receptor substrate (IRS)-1 [23]. Women with PCOS have a higher degree of serine phosphorylation of the insulin receptor and IRS-1 resulting in impaired insulin signal transduction and intrinsic IR independent of total or fat-free body mass [23].

PI-3K also helps in maintenance of insulin sensitivity in the liver. The two subunits, namely, the regulatory subunit (p85) is tightly bound to a catalytic subunit (p110) forming a complex heterodimer (p85-p110). The regulatory subunit (p85) is responsible for regulating the activity of PI-3K by binding to the active site of IRS-1. Over or under expression of p85α competes with the p85-p110 heterodimer on the binding site of IRS-1 leading to either increase or decrease in PI-3K activity [23].

Insulin is also known to indirectly potentiate the steroidogenic response to gonadotropins, by acting on the pituitary. Hence, increasing gonadotrope sensitivity to GnRH. In PCOS, an increased level of androgen is associated with a decreased secretion of adiponectin by adipocytes, resulting in decreased insulin sensitivity and subsequently increasing insulin. Insulin is a precursor for adipose androgen by stimulating the release of aldo-keto reductase 1C3 (AKR1C3) activity in subcutaneous adipose tissue in females [24]. Obesity in PCOS is generally associated with insulin resistance and hyperinsulinemia [25]. Obesity associated with IR is linked with predisposal of developing prediabetes and Type 2 Diabetes during puberty [26]. Studies also suggest association of visceral obesity, proinflammatory markers, elevated levels of fasting and glucose-stimulated insulin and IR in women suffering with PCOS. The mechanism behind obesity associated with IR is not quite clear. Although ectopic accumulation of fatty acids in organs and tissues which are not known to store large amounts of fats appear to play a minor role. However, ectopic accumulation of fats can also appear in absence of obesity in women [27].

Association of IR with lipids shows irregularities in insulin signaling and decreased insulin-stimulated glucose transport resulting due to lipid-induced IR in skeletal muscle. Activation of protein kinase C (isoform PKCθ) is known to be associated with muscle lipid accumulation altering insulin intracellular signals [28,29]. Lipid induced IR is however mediated via diacylglycerol (DAG) activating PKCθ promoting insulin signal impairment. Intracellular ceramides lead to the activation of Akt, an important mediator of insulin sensitivity, limiting the translocation of GLUT4-containing storage vesicles to the plasma membrane decreasing the glucose uptake therefore decreasing insulin mediated glycogen synthesis. Some data has put further, strong evidence between disrupted insulin signaling in the central nervous system with the development of obesity and impaired ovarian follicular maturation, suggesting a solid link between IR, hyperinsulinemia, obesity and PCOS [30].

20.2.3 Endocrine alteration

Apart from steroidogenesis that takes place in the ovary and the adrenal glands, which are under the influence of the trophic hormones. Gonadotropins such as LH and FSH, also initiate the stimulation of steroidogenesis. The LH and FSH are under the influence of the pituitary gland which is in turn under the influence of hypothalamus [31]. The hypothalamus stimulates the pituitary through gonadotropin releasing hormone (GnRH). A disrupted secretion of GnRH leads to overproduction of LH triggering over production of ovarian androgen and disturbing the LH/FSH ratio, hence, inhibiting the cycle and causing altered morphology. However, an increased number of GnRH pulse secretion favors the secretion of LH over FSH by the pituitary. Although genetic mutation in GnRH pulses can lead to perturbation on key modulators of GnRH neurosecretion, which includes secretions of insulin and androgen, whose levels are altered in PCOS [32].

As mentioned earlier, hyperandrogenism being a primary marker of PCOS, researchers have demanded considerable attention for investing in a potential mechanism for deregulated androgen secretion leading to the neuroendocrine alteration of the syndrome. Evidence have made compelling suggestion, that elevated androgen potentially destroys the capacity of sex steroids to regulate GnRH/LH secretion following a classical feedback loops [33]. This further results in a diminished negative feedback action on the ovarian steroids such as estrogens and progesterone which perpetuate the hypersecretion of LH giving rise to another characteristic feature in PCOS. Clinical data has also linked diminished ovarian negative feedback to excess of androgen, playing a major role in elevation of LH pulses in PCOS. However, a decreased sensitivity to progesterone-negative feedback, during the primary onset of hyperandrogenism, increases LH secretion in women with PCOS, however, the probability of which is 50% [34]. Notably, GnRH neurons appear to be devoid of major sex steroid receptors generating a negative feedback, on the other hand, the feedback role of estrogen receptor-β (ERβ) in controlling the GnRH neurons remains under scrutiny [35].

One of the afferent neurons of GnRH, Kiss1 neurons, responsible for producing kisspeptins (encoded by KISS1 gene), which emerged as a master regulator of GnRH neurosecretion and ovulation [35]. They are known as the most potent activators of GnRH neurons. A highly conserved group of KISS1 neurons found on the arcuate nucleus (ARC) of the mediobasal hypothalamus or on the infundibular region in humans. This ARC kiss1 are known to operate the mediating center of the negative feedback effects of sex steroids, as they constantly suppress the expression of Kiss 1 at the particular site. Conversely, a population of rostral hypothalamic groups of Kiss1 neurons may initiate a positive feedback, as the presence of estrogen enhances the expression of Kiss1 at this site. A fraction of the Kiss1 neurons in the

ARC region co-expresses other neurotransmitters such as, neurokinin B (NKB) and dynorphin, which plays an important role in the regulation of GnRH/gonadotropin secretion. The receptor (NKB3R) of NKB has also been expressed in Kiss1 neurons. The group of neurons that co-express kisspeptins, NKB, and dynorphin is classified as KNDy neurons. The action of NKB and dynorphins predominantly stimulate and inhibit the secretion of LH respectively. However, the interconnection of KNDy neurons with the ARC, shows that NKB and dynorphin participate in harmony to regulate the kisspeptin output to GnRH neurons, therefore generating GnRH pulses [36]. As KNDy neurons are sensitive towards sex steroids and helps in modulating the generation of GnRH pulses highlight that deregulated function of Kiss1 group might contribute to the neuroendocrine alteration of PCOS. However, there is not enough evidence to support this hypothesis. Some studies have reported alteration in Kiss1 expression in PCOS animal models have shown to generate excessive androgen during the developmental stage. The upregulation or downregulation of the Kiss1 group is somewhat dependent on the developmental stage and the exposure of androgen. Various KNDy neuropeptides might also be responsible for the pathophysiology of the neuroendocrine alteration of PCOS. NKB levels are also known to impact the GnRH and LH profiles. However, studies have shown that NKB alone can not alter circulating gonadotropin levels. Although their association with PCOS is yet to be clarified [37].

Other than kisspeptin and KNDy signaling, alteration of gamma-aminobutyric acid (GABA) signaling is linked with neuroendocrine alteration in PCOS, which was first identified in mice. GABA can depolarize GnRH neurons directly via GABA-A receptors [38]. The neuronal pathway of GABA originates from the ARC transmit feedback action of sex steroids. An increase in GABA is due to suppressed expression of progesterone receptors in ARC GABA neurons towards the GnRH neurons, thus giving out restraint GABA signaling to GnRH neurons hence consequently leading to elevated levels of GnRH neurosecretion. However, the role of GABA in the case of PCOS remains unclear [39].

20.3 Genetic and epigenetic basis of polycystic ovarian syndrome

Being a multifactorial syndromic disorder, the genetic and epigenetic basis of PCOS has also been investigated [40]. Several genes affecting fertility either directly or indirectly are reported to be associated with PCOS. The complexity and heterogeneity of this disorder makes it very difficult to be linked with a single gene or related genes in a single family. Thus, familial studies conducted on PCOS patients failed to identify a fully penetrant variant(s) [41,42]. Patients from the same family differ in the degree of genetic susceptibility of different genes. In spite of the multiple familial genetic studies, not a true penetrance of the single gene mutation has been reported so far. Furthermore, in order to understand the reason for high androgen levels, several genes associated with this pathology have been reported, such as: CYP11a, CYP21, CYP17, CYP19, Androgen Receptor Gene, Sex Hormone-Binding Globulin

Gene, Lutein Hormone (LH) and Its Receptor Gene, *AMH, FSHR, INSR*, Insulin Receptor Substrate Proteins, Calpain10 Gene, Fat Mass Obesity gene, PCOS1, *SRD5A2 and SRD5A1* [40].

Epigenetic changes are the trans generationally and mitotically heritable changes in gene expression that are not due to DNA sequence variability. Studies investigating the epigenetic contributions to the PCOS pathophysiology are very few in comparison to genetic studies, but this field is evolving fast. An altered DNA methylation pattern has been reported in PCOS patients as compared to healthy fertile control groups [43]. Another study reported the differential methylation of CPG island in *PPARG1, NCOR1* genes of granulosa cells that causes hyperandrogenism-induced epigenetic alteration, leading to the development of ovarian dysfunction [44]. In a recent study, Sagvekar et al. has performed the DNA methylome profiling of granulosa cells which revealed altered methylation in genes regulating vital ovarian functions in PCOS. Their findings highlighted that the epigenetic dysregulation of genes associated with follicular development may contribute to ovarian defects in women with PCOS [45]. The involvement of the epigenetic mechanisms in PCOS brings out the complexity of the disease.

20.4 Therapeutic approaches

The severity of the disorder, along with its potential to give rise to various metabolic, physiological, psychological and endocrine symptoms has fueled the need to develop a stable and reliable treatment or management for PCOS. As mentioned earlier, the symptoms of PCOS include an array of disorders that eventually result in minor to major health related diseases. Altered steroidogenesis is often remarked as the key culprit in the disorder, making it the major focus in control and treating the disorder. Elevated levels of free testosterone in the blood are detected through androgen profiling to diagnose hyperandrogenemia in women suffering from PCOS. Although, androgen profile alone cannot be related to the degree of hirsutism [46]. The therapeutic approaches towards hyperandrogenism and hirsutism majorly focus on reducing endogenous production of androgen. Rapid production of androgen rates and its subsequent tissue availability portrays the main pathophysiology of hirsutism. Depending on the degree of hirsutism, its treatment includes cosmetic procedures, often suggested for mild symptoms and localized hirsutism. Estro-progestins (EPs) and other combinations of hormones along with antiandrogen and Ethinyl estradiol. Antiandrogen can also be combined with insulin sensitizers such as Metformin. Treatment of hirsutism can also be brought about through epilatory methods, providing a variety of options namely electrolysis, Thermolysis and laser treatments. EPs are majorly used for a reasonable improvement in hyperandrogenism [47]. EPs have a unique ability to suppress LH and thus suppress the production of ovarian androgen through progestins, estrogen (ethinylestradiol) increases SHBG and reduces free androgen. EPs also reduce adrenal steroid through adrenal steroid synthesis.

Progestin also has some antagonizing effect on the androgen receptor and to the inhibition of 5α-reductase activity. The efficacy of EPs is quite promising in terms of reducing hirsutism. Drospirenone/ethinyl estradiol pills seem to be more efficacious in reducing hirsutism. Antiandrogen such as spironolactone, flutamide and finasteride are androgen receptor blocker and 5α-reductase inhibitors respectively. Although use of flutamide is prescribed under attention as it has shown a negative effect on liver function in adolescent and adult PCOS women. Therefore, the use of flutamide is restricted due its potential risk factor. However, usage of antiandrogen in pregnant women is forbidden due to high risk of feminization of male fetuses. Insulin sensitizers such as metformin or thiazolidinediones, when included individually or combined with EPs showed significant effect on insulin and produced a modest effect on hyperandrogenism or hirsutism in particular. Side effects of the above-mentioned drug ranges from mild to modest, with manageable or reversible effects. Although long term use of the drug might create higher risk. Adolescent women suffering from PCOS may develop abnormal symptoms. The treatment of PCOS is still controversial, however the use of metformin is more recommended in adolescent women than adult women suffering from PCOS. Modern approaches include use of aromatase inhibitors; it directly acts on HPOA. Letrozole is an aromatase inhibitor that is showing significant results in infertility treatment. Apart from the available drugs, lifestyle modification, weight management have also shown to be effective in PCOS management.

20.5 Alternative medicines

Higher side effects resulting from the intake of PCOS medications have prompted researchers to look for natural alternatives. Herbal medicine can be a treatment option for women with PCOS as an adjunct or alternative treatment to pharmaceuticals with greater acceptability by PCOS women. Similar to pharmaceutical approaches, herbal medicines aim at the relief of PCOS symptoms and improvement of sexual well-being among patients.

20.5.1 Chinese Herbal Medicines (CHM)

There is no classification of PCOS in Traditional Chinese Medicine. However, its symptoms that is, amenorrhea (failure to menstruate) and infertility are treated using CHMs. They are prescribed either as a comprehensive mixture or as a single herb. An observational study by Mei-Juin and colleagues (2019) identifies 20 such mixtures and 10 single herbs. Xiang-Fu (*Cyperus rotundus* L) was found to be the most prescribed single herb followed by Da-Huang (*Rheum officinale*) containing emodin [48,49]. Emodin at concentrations as low as 3μM can inhibit enzymes leading to diet-induced obesity such as the 11β-hydroxysteroid dehydrogenase type 1 enzyme and cholesterol 7α-hydroxylase enzyme [50]. Jia-Wei-Xiao-Yao-San, a medicinal formula was analyzed for its protein-protein interaction, and it was proved that the herbal formula could effectively lower LH, regulate ovarian gene expression, lower body weight in addition to other functions [51].

20.5.2 Aloe vera gel

Desai *et al.* (2012) reported 1ml/day of aloe vera gel orally for 30 days could lower the ovarian weight, reduce body weight, lower liver cholesterol level along with reduction of activities of HMG-CoA reductase in virgin female Charles Foster rats [52]. This was analogous to the findings of Maharjan *et al.* (2010) wherein the rats were given doses of 1mg/day for 45 days [53]. The non-polar fraction of Aloe vera gel containing β-sitosterol along with lupeol and stigmasterol was found to decrease the expression of stAR proteins in rodent models.

20.5.3 Ginger

Improvement in FSH, LH, estrogen, and progesterone levels in serum of female rats was observed with a dosage of 30mg/kg/day for 89 days. Gingerols and prostaglandins in ginger supposedly inhibit prostaglandin production by inhibiting arachidonic acid production [54].

20.5.4 Chamomile

Intraperitoneal dosage of 50 mg/kg/day of chamomile in virgin adult Wistar rats showed a decrease in size of ovarian cysts with better endometrial tissue arrangements and increase in dominant follicles [55]. A randomized control trial in 80 PCOS patients showed chamomile capsules 3/day could reduce testosterone levels while no significant changes were noticed in LH/FSH ratio and lipid parameters [56].

20.5.5 Cinnamon

High oxidant capability of cinnamon was found to reduce oxidative stress in PCOS patients. The findings of Borzoei *et al.* (2017) suggest that supplementation of cinnamon (1.5g) for 2 weeks is beneficial for improving antioxidant status and lipid profile in PCOS patients [57]. In a similar randomized double blind controlled trial among overweight and obese PCOS women, it was observed that consumption of cinnamon could significantly reduce total cholesterol, weight, insulin level, serum fasting blood glucose in comparison with placebo [58].

20.5.6 Fenugreek

Fenugreek seed extracts (or furocyst) has been known to contain various saponins and flavonoids which helps in regulation of androgen levels in the body, Bashtian *et al.* (2013)

conducted a double-blind placebo-controlled study and found that fenugreek alone had no effect on the insulin resistance of PCOS patients [59]. However, ultrasound scans after 8 weeks of fenugreek supplementation with metformin could decrease the cyst size in ovaries. In another study, 12% of the patients got pregnant with supplementation of novel *Trigonella foenum-graecum* seed extract. Almost 71% of treated individuals had regular menstrual cycles by the end of the study [60]. Another study found that the hydroethanolic extract of the Fenugreek seeds showed considerable decline in depression among PCOS patients without any adverse effect in the patient. A considerable decline in vasomotor symptoms such as vaginal dryness, hot flashes and flushes were also observed [61].

20.5.7 Ayurvedic formulations

Siriwardene *et al.* (2010) used several herbal formulations for the treatment of symptoms of PCOS. *Triphala*, a mixture of Vibhitaki (*Terminalia bellirica*), Amla *(Emblica officinalis)* and Haritaki (*Terminalia chebula*) could effectively reduce weight in PCOS patients [62]. Seeds of *Pelargonium graveolens* and oil of *Nigella sativa* were found to be effective in the reduction of ovarian cysts size and enhancing follicular maturity. Another study found that *Ikshvaku* seeds induced therapeutic vomiting followed by consumption of *Shatapushpadi Ghanavati* for 45days could considerably provide relief to PCOS patients. 80-85% of patients showed regularity in their menstrual cycle and improvement in BMI [63].

20.5.8 Herbal mixtures

Ainehchi *et al.* (2020) tested the effect of the herbal mixture on the treatment of PCOS [64]. The mixture contained spearmint (*Mentha spicata*), ginger (*Zingiber officinale*), cinnamon (*Cinnamomum zeylanicum*) and sweet orange (*Citrus sinensis*). This mixture in combination with clomiphene citrate could significantly reduce lipid profile and an increase in pregnancy rate in PCOS patients was observed without any significant side effects.

20.6 Concluding remarks

Conclusively, PCOS is a multifactorial disorder that requires more holistic treatment approaches. The current treatment modalities using pharmaceuticals predisposes patients to a range of side effects. Hence, there is a prevailing need for alternative medicines or approaches towards the management of PCOS. Various preclinical and clinical studies have provided evidence that herbal medicines can be helpful for women with oligo/amenorrhea, hyperandrogenism and PCOS. However, the amount of data on alternative medicines is limited, and has variable clinical evidence. Further investigations deciphering the mechanisms of action of herbal medicines are required to establish its endocrinological and reproductive effects in women with PCOS.

References

[1] A Balen, M Rajkowha, Polycystic ovary syndrome–a systemic disorder? Best Pract Res Clin Obstet Gynaecol 17 (2) (2003) 263–274. https://doi.org/10.1016/s1521-6934(02)00119-0. PMID: 12758099.

[2] R Vidya Bharathi, S Swetha, J Neerajaa, J Varsha Madhavica, D Janani, S Rekha, et al., An epidemiological survey: effect of predisposing factors for PCOS in Indian urban and rural population, Middle East Fertil Soc J 22 (4) (2017) 313–316. https://doi.org/10.1016/j.mefs.2017.05.007.

[3] F Kamangar, JP Okhovat, T Schmidt, A Beshay, L Pasch, MI Cedars, et al., Polycystic ovary syndrome: special diagnostic and therapeutic considerations for children, Pediatr Dermatol 32 (5) (2015) 571–578. https://doi.org/10.1111/pde.12566.

[4] MT Sheehan, Polycystic ovarian syndrome: diagnosis and management, Clin Med Res 2 (1) (2004) 13–27. https://doi.org/10.3121/cmr.2.1.13.

[5] RL Rosenfield, DA Ehrmann, The pathogenesis of polycystic ovary syndrome (PCOS): the hypothesis of PCOS as functional ovarian hyperandrogenism revisited, Endocr Rev 37 (5) (2016) 467–520. https://doi.org/10.1210/er.2015-1104.

[6] R Norman, M Davies, J Lord, L Moran, The role of lifestyle modification in polycystic ovary syndrome, Trends Endocrinol Metab 13 (6) (2002) 251–257. https://doi.org/10.1016/s1043-2760(02)00612-4.

[7] S Arentz, JA Abbott, CA Smith, A Bensoussan, Herbal medicine for the management of polycystic ovary syndrome (PCOS) and associated oligo/amenorrhoea and hyperandrogenism; a review of the laboratory evidence for effects with corroborative clinical findings, BMC Complement Altern Med 14 (2014) 511. https://doi.org/10.1186/1472-6882-14-511.

[8] S El Hayek, L Bitar, LH Hamdar, FG Mirza, G Daoud, Poly cystic ovarian syndrome: an updated overview, Front Physiol 7 (2016) 124. https://doi.org/10.3389/fphys.2016.00124.

[9] A Balen, The pathophysiology of polycystic ovary syndrome: trying to understand PCOS and its endocrinology, Best Pract Res Clin Obstet Gynaecol 18 (5) (2004) 685–706. https://doi.org/10.1016/j.bpobgyn.2004.05.004.

[10] A Chang, S Abdullah, T Jain, H Stanek, S Das, D McGuire, et al., Associations among androgens, estrogens, and natriuretic peptides in young women, J Am Coll Cardiol 49 (1) (2007) 109–116. https://doi.org/10.1016/j.jacc.2006.10.040.

[11] A Hsueh, K Kawamura, Y Cheng, B Fauser, Intraovarian control of early folliculogenesis, Endocr Rev 36 (1) (2015) 1–24. https://doi.org/10.1210/er.2014-1020.

[12] Erickson G (2020). *Follicle growth and development | glowm*. Available from: https://www.glowm.com/section_view/item/288. [accessed 27 November 2020].

[13] S Franks, J Stark, K Hardy, Follicle dynamics and anovulation in polycystic ovary syndrome, Hum Reprod Update 14 (4) (2008) 367–378. https://doi.org/10.1093/humupd/dmn015.

[14] M Lebbe, T Woodruff, Involvement of androgens in ovarian health and disease, Mol Hum Reprod 19 (12) (2013) 828–837. https://doi.org/10.1093/molehr/gat065.

[15] L Webber, S Stubbs, J Stark, G Trew, R Margara, K Hardy, et al., Formation and early development of follicles in the polycystic ovary, Lancet 362 (9389) (2003) 1017–1021. https://doi.org/10.1016/s0140-6736(03)14410-8.

[16] M Kirschner, C Wayne Bardin, Androgen production and metabolism in normal and virilized women, Metabolism 21 (7) (1972) 667–688. https://doi.org/10.1016/0026-0495(72)90090-x.

[17] I Hanukoglu, R Rapoport, Routes and regulation of NADPH production in steroidogenic mitochondria, Endocr Res 21 (1-2) (1995) 231–241. https://doi.org/10.3109/07435809509030439.

[18] I Hanukoglu, T Gutfinger, M Haniu, J Shively, Isolation of a cDNA for adrenodoxin reductase (ferredoxin -NADP+ reductase). Implications for mitochondrial cytochrome P-450 systems, Eur J Biochem 169 (3) (1987) 449–455. https://doi.org/10.1111/j.1432-1033.1987.tb13632.x.

[19] I Hanukoglu, V Spitsberg, J Bumpus, K Dus, C Jefcoate, Adrenal mitochondrial cytochrome P-450scc. Cholesterol and adrenodoxin interactions at equilibrium and during turnover, J Biol Chem 256 (9) (1981) 4321–4328. https://doi.org/10.1016/s0021-9258(19)69436-6.

[20]   A Arukwe, Steroidogenic acute regulatory (StAR) protein and cholesterol side-chain cleavage (P450scc)-regulated steroidogenesis as an organ-specific molecular and cellular target for endocrine disrupting chemicals in fish, Cell Biol Toxicol 24 (6) (2008) 527–540. https://doi.org/10.1007/s10565-008-9069-7.

[21]   D White, A Leigh, C Wilson, A Donaldson, S Franks, Gonadotropin and gonadal steroid response to a single dose of a long-acting agonist of gonadotropin-releasing hormone in ovulatory and anovulatory women with polycystic ovary syndrome, Clin Endocrinol 42 (5) (1995) 475–481. https://doi.org/10.1111/j.1365-2265.1995.tb02665.x.

[22]   C Burt Solorzano, K Knudsen, A Anderson, E Hutchens, J Collins, J Patrie, et al., Insulin resistance, hyperinsulinemia, and LH: relative roles in peripubertal and increased expression of p85: the two sides of a coin, Diabetes 55 (8) (2018) 2392–2397. https://doi.org/10.2337/db06-0391.

[23]   B Draznin, Molecular mechanisms of insulin resistance: serine phosphorylation of insulin receptor substrate-1 obesity-associated hyperandrogenemia, J Clin Endocrinol Metab 103 (7) (2006) 2571–2582. https://doi.org/10.1210/jc.2018-00131.

[24]   M O'Reilly, L Gathercole, F Capper, W Arlt, J Tomlinson, Effect of insulin on AKR1C3 expression in female adipose tissue: in-vivo and in-vitro study of adipose androgen generation in polycystic ovary syndrome, Lancet 385 (2015) S16. https://doi.org/10.1016/S0140-6736(15)60331-2.

[25]   S Lim, M Davies, R Norman, L Moran, Overweight, obesity and central obesity in women with polycystic ovary syndrome: a systematic review and meta-analysis, Hum Reprod Update 18 (6) (2012) 618–637. https://doi.org/10.1093/humupd/dms030.

[26]   A Paramanya, Y Jain, A Ali, Obesity: its complications and available medications, Kocaeli Üniversitesi Sağlık Bilimleri Dergisi 6 (1) (2020) 68–76. https://www.researchgate.net/publication/339068350_OBESITY_ITS_COMPLICATIONS_AND_AVAILABLE_MEDICATIONS.

[27]   E Fabbrini, F Magkos, B Mohammed, T Pietka, N Abumrad, B Patterson, et al., Intrahepatic fat, not visceral fat, is linked with metabolic complications of obesity, Proc Natl Acad Sci 106 (36) (2009) 15430–15435. https://doi.org/10.1073/pnas.0904944106.

[28]   V Samuel, G Shulman, Mechanisms for insulin resistance: common threads and missing links, Cell 148 (5) (2012) 852–871. https://doi.org/10.1016/j.cell.2012.02.017.

[29]   V Samuel, K Petersen, G Shulman, Lipid-induced insulin resistance: unravelling the mechanism, Lancet 375 (9733) (2010) 2267–2277. https://doi.org/10.1016/s0140-6736(10)60408-4.

[30]   WT Garvey, L Maianu, JH Zhu, G Brechtel-Hook, P Wallace, AD Baron, Evidence for defects in the trafficking and translocation of GLUT4 glucose transporters in skeletal muscle as a cause of human insulin resistance, J Clin Investig 101 (11) (1998) 2377–2386. https://doi.org/10.1172/JCI1557.

[31]   AD Rawindraraj, H Basit, I Jialal, Physiology, Anterior Pituitary, StatPearls [Internet]. Treasure Island (FL), StatPearls Publishing, 2021 2021 Jan-. Available from. https://www.ncbi.nlm.nih.gov/books/NBK499898/.

[32]   A Moore, R Campbell, The neuroendocrine genesis of polycystic ovary syndrome: a role for arcuate nucleus GABA neurons, J Steroid Biochem Mol Biol 160 (2016) 106–117. https://doi.org/10.1016/j.jsbmb.2015.10.002.

[33]   I Thompson, U Kaiser, GnRH pulse frequency-dependent differential regulation of LH and FSH gene expression, Mol Cell Endocrinol 385 (1-2) (2014) 28–35. http://doi.org/10.1016/j.mce.2013.09.012.

[34]   C Eagleson, M Gingrich, C Pastor, T Arora, C Burt, W Evans, et al., Polycystic ovarian syndrome: evidence that flutamide restores sensitivity of the gonadotropin-releasing hormone pulse generator to inhibition by estradiol and progesterone1, J Clin Endocrinol Metab 85 (11) (2000) 4047–4052. http://doi.org/10.1210/jcem.85.11.6992.

[35]   L Pinilla, E Aguilar, C Dieguez, RP Millar, M Tena-Sempere, Kisspeptins and reproduction: physiological roles and regulatory mechanisms, Physiol Rev 92 (3) (2012) 1235–1316. https://doi.org/10.1152/physrev.00037.2010.

[36]   V Navarro, M Tena-Sempere, Neuroendocrine control by kisspeptins: role in metabolic regulation of fertility, Nat Rev Endocrinol 8 (1) (2011) 40–53. http://doi.org/10.1038/nrendo.2011.147.

[37]   S Narayanaswamy, J Prague, C Jayasena, D Papadopoulou, M Mizamtsidi, A Shah, et al., Investigating the KNDy hypothesis in humans by coadministration of kisspeptin, neurokinin B, and naltrexone in men, J Clin Endocrinol Metab 101 (9) (2016) 3429–3436. http://doi.org/10.1210/jc.2016-1911.

[38]   A Herbison, S Moenter, Depolarising and hyperpolarising actions of GABAA receptor activation on gonadotrophin-releasing hormone neurones: towards an emerging consensus, J Neuroendocrinol 23 (7) (2011) 557–569. http://doi.org/10.1111/j.1365-2826.2011.02145.x.

[39]   A Moore, M Prescott, C Marshall, S Yip, R Campbell, Enhancement of a robust arcuate GABAergic input to gonadotropin-releasing hormone neurons in a model of polycystic ovarian syndrome, Proc Natl Acad Sci 112 (2) (2014) 596–601. http://doi.org/10.1073/pnas.1415038112.

[40]   MJ Khan, A Ullah, S Basit, Genetic basis of polycystic ovary syndrome (PCOS): current perspectives, Appl Clin Genet 12 (2019) 249–260. http://doi.org/10.2147/TACG.S200341. PMID: 31920361; PMCID: PMC6935309.

[41]   M Urbanek, RS Legro, DA Driscoll, R Azziz, DA Ehrmann, RJ Norman, et al., Thirty-seven candidate genes for polycystic ovary syndrome: strongest evidence for linkage is with follistatin, Proc Natl Acad Sci U S A 96 (15) (1999) 8573–8578. http://doi.org/10.1073/pnas.96.15.8573. PMID: 10411917; PMCID: PMC17558.

[42]   S Franks, C Gilling-Smith, N Gharani, M McCarthy, Pathogenesis of polycystic ovary syndrome: evidence for a genetically determined disorder of ovarian androgen production, Hum Fertil (Camb) 3 (2) (2000) 77–79. http://doi.org/10.1080/1464727002000198731. PMID: 11844358.

[43]   N Xu, R Azziz, MO Goodarzi, Epigenetics in polycystic ovary syndrome: a pilot study of global DNA methylation, Fertil Steril 94 (2010) 781–783 e1. http://doi.org/10.1016/j.fertnstert.2009.10.020 .

[44]   F Qu, FF Wang, R Yin, et al., A molecular mechanism underlying ovarian dysfunction of polycystic ovary syndrome: hyperandrogenism induces epigenetic alterations in the granulosa cells, J Mol Med 90 (2012) 911–923. http://doi.org/10.1007/s00109-012-0881-4.

[45]   P Sagvekar, P Kumar, V Mangoli, S Desai, S Mukherjee, DNA methylome profiling of granulosa cells reveals altered methylation in genes regulating vital ovarian functions in polycystic ovary syndrome, Clin Epigenetics 11 (1) (2019) 61. http://doi.org/10.1186/s13148-019-0657-6. PMID: 30975191; PMCID: PMC6458760.

[46]   R Pasquali, Contemporary approaches to the management of polycystic ovary syndrome, Ther Adv Endocrinol Metab 9 (4) (2018) 123–134. https://doi.org/10.1177/2042018818756790.

[47]   J Vrbíková, D Cibula, Combined oral contraceptives in the treatment of polycystic ovary syndrome, Hum Reprod Update 11 (3) (2005) 277–291. http://doi.org/10.1093/humupd/dmi005.

[48]   MJ Lin, HW Chen, PH Liu, WJ Cheng, SL Kuo, MC Kao, The prescription patterns of traditional Chinese medicine for women with polycystic ovary syndrome in Taiwan: a nationwide population-based study, Medicine (Baltimore) 98 (24) (2019) e15890. https://doi.org/10.1097/MD.0000000000015890.

[49]   YJ Wang, SL Huang, Y Feng, MM Ning, Y Leng, Emodin, an 11β-hydroxysteroid dehydrogenase type 1 inhibitor, regulates adipocyte function in vitro and exerts anti-diabetic effect in ob/ob mice, Acta Pharmacol Sin 33 (9) (2012) 1195–1203. https://doi.org/10.1038/aps.2012.87.

[50]   J Wang, J Ji, Z Song, W Zhang, X He, F Li, et al., Hypocholesterolemic effect of emodin by simultaneous determination of in vitro and in vivo bile salts binding, Fitoterapia 110 (2016) 116–122 2016 Apr. http://doi.org/10.1016/j.fitote.2016.03.007 .

[51]   B Leclerc, YC Wu, P Wu, A network pharmacological approach to evaluate Jia-Wei-Xiao-Yao-san formula's mechanistic pathways and its implication in the symptomatology of polycystic ovarian syndrome, Int J Complement Alt Med 13 (5) (2020) 200–213. http://doi.org/10.15406/ijcam.2020.13.00517.

[52]   BN Desai, RH Maharjan, LP Nampoothiri, Aloe barbadensis Mill. formulation restores lipid profile to normal in a letrozole-induced polycystic ovarian syndrome rat model, Pharmacognosy Res 4 (2) (2012) 109–115. http://doi.org/10.4103/0974-8490.94736.

[53]   R Maharjan, PS Nagar, L Nampoothiri, Effect of Aloe barbadensis Mill. formulation on Letrozole induced polycystic ovarian syndrome rat model, J Ayurveda Integr Med 1 (4) (2010) 273–279. http://doi.org/10.4103/0975-9476.74090.

[54] S Atashpour, H Kargar Jahromi, Z Kargar Jahromi, M Maleknasab, Comparison of the effects of ginger extract with clomiphene citrate on sex hormones in rats with polycystic ovarian syndrome, Int J Reprod Biomed 15 (9) (2017) 561–568.

[55] ZZ Farideh, M Bagher, A Ashraf, A Akram, M Kazem, Effects of chamomile extract on biochemical and clinical parameters in a rat model of polycystic ovary syndrome, J Reprod Infertil 11 (3) (2010) 169–174 PMCID: PMC3719301.

[56] M Heidary, Z Yazdanpanahi, MH Dabbaghmanesh, ME Parsanezhad, M Emamghoreishi, M Akbarzadeh, Effect of chamomile capsule on lipid- and hormonal-related parameters among women of reproductive age with polycystic ovary syndrome, J Res Med Sci 23 (2018) 33. https://doi.org/10.4103/jrms.JRMS_90_17.

[57] A Borzoei, M Rafraf, S Niromanesh, L Farzadi, F Narimani, F Doostan, Effects of cinnamon supplementation on antioxidant status and serum lipids in women with polycystic ovary syndrome, J Tradit Complement Med 8 (1) (2017) 128–133. https://doi.org/10.1016/j.jtcme.2017.04.008.

[58] A Borzoei, M Rafraf, M Asghari-Jafarabadi, Cinnamon improves metabolic factors without detectable effects on adiponectin in women with polycystic ovary syndrome, Asia Pac J Clin Nutr 27 (3) (2018) 556–563. https://doi.org/10.6133/apjcn.062017.13.

[59] M Hassanzadeh Bashtian, SA Emami, N Mousavifar, HA Esmaily, M Mahmoudi, AH Mohammad Poor, Evaluation of fenugreek (Trigonella foenum-graceum L.), effects seeds extract on insulin resistance in women with polycystic ovarian syndrome, Iran J Pharm Res 12 (2) (2013) 475–481 PMCID: PMC3813238.

[60] A Swaroop, AS Jaipuriar, SK Gupta, M Bagchi, P Kumar, HG Preuss, et al., Efficacy of a novel fenugreek seed extract (Trigonella foenum-graecum, Furocyst) in polycystic ovary syndrome (PCOS), Int J Med Sci 12 (10) (2015) 825–831. https://doi.org/10.7150/ijms.13024.

[61] A Khanna, F John, S Das, J Thomas, J Rao, B Maliakel, et al., Efficacy of a novel extract of fenugreek seeds in alleviating vasomotor symptoms and depression in perimenopausal women: A randomized, double-blinded, placebo-controlled study, J Food Biochem 44 (12) (2020) e13507. https://doi.org/10.1111/jfbc.13507.

[62] SA Dayani Siriwardene, LP Karunathilaka, ND Kodituwakku, YA Karunarathne, Clinical efficacy of Ayurveda treatment regimen on Subfertility with Poly Cystic Ovarian Syndrome (PCOS), Ayu 31 (1) (2010) 24–27. https://doi.org/10.4103/0974-8520.68203.

[63] KB Bhingardive, DD Sarvade, S Bhatted, Clinical efficacy of *Vamana Karma* with *Ikshwaaku Beeja Yoga* followed by *Shatapushpadi Ghanavati* in the management of *Artava Kshaya* w. s. r to polycystic ovarian syndrome, Ayu 38 (3-4) (2017) 127–132. https://doi.org/10.4103/ayu.AYU_192_16.

[64] N Ainehchi, A Khaki, E Ouladsahebmadarek, M Hammadeh, L Farzadi, A Farshbaf-Khalili, et al., The effect of clomiphene citrate, herbal mixture, and herbal mixture along with clomiphene citrate on clinical and para-clinical parameters in infertile women with polycystic ovary syndrome: a randomized controlled clinical trial. *Archives of medical science*, AMS 16 (6) (2020) 1304–1318. https://doi.org/10.5114/aoms.2020.93271.

PART 6

Infectious diseases

CHAPTER 21

Emerging natural product based alternative therapeutics for tuberculosis

Vipul K. Singh, Abhishek Mishra, Chinnaswamy Jagannath and Arshad Khan

Department of Pathology and Genomic Medicine, Houston Methodist Research Institute, Houston, TX, United States

21.1 Introduction: Recent directions in the therapeutics for tuberculosis and underlying challenges

Tuberculosis (TB) is a major public health issue worldwide and is a leading cause of death from an infectious disease, accountable for more than a million deaths annually [1]. Its global incidence is increasing by approximately 1% per year, with an estimated 10 million new cases being identified in 2020 (WHO tuberculosis report 2020). More frighteningly, one third of the world's population is estimated to have latent TB infection, and diabetic or human immunodeficiency virus (HIV) positive patients, or those undergoing an immunotherapy have several time increased risks of converting from latent TB cases to active infection [2–4]. As TB is primarily a disease of the poor and the TB burden disproportionately affects the lower income countries, the pharmaceutical industry has shown relatively little interest in developing new products for the treatment of TB in the last few decades. TB prevention through vaccine has also been proved to be inefficient largely due to the poor efficacy of only available BCG vaccine [5]. Apart from new vaccine development hurdles, the increased number of drug resistant (MDR/XDR) strains, HIV co-infection and the failure of the conventional regimens against these strains are the other major challenges of the coming decades [6]. Non-communicable diseases such as diabetes mellitus are also gaining attention as potential risk factors for TB in both poor and wealthy settings, thus raising new issues on prevention and early diagnosis even in developed countries now [7]. These emerging and challenging problems have contributed in recent years to a resurgence of interest in developing improved therapeutic and prevention strategies for TB. As a result, there are now a larger number of TB drugs and vaccines under development than at any previous point in history. Concerted efforts of stakeholders, advocates, and researchers are advancing to develop shorter course, more effective, safer, and

Herbal Medicines: A Boon for Healthy Human Life.
DOI: https://doi.org/10.1016/B978-0-323-90572-5.00017-2

better tolerated treatment regimens along with more effective vaccines and adjuvants for TB. The developing pipeline and landscape of new and repurposed TB drugs, treatment regimens, vaccines and other alternative host-directed therapies (HDTs) for drug-sensitive and drug-resistant TB as well, have increased the belief that perhaps elimination of TB will be a reality in near future. 14 candidate drugs for drug-susceptible, drug-resistant, and latent TB are in clinical stages of drug development; nine are now in phase 1 and 2 trials, and three new drugs are in advanced stages of development for MDR TB [8]. Apart from that, many candidate vaccines have also shown improved efficacy over conventional BCG vaccine in recent clinical trials [9]. However, this euphoria of eliminating TB could be short lived if we do not address the underlying challenges that still exist and hinder the development of novel treatment and vaccine strategy for TB. At present, even the vaccines in the pipeline have not shown a desired efficacy that can prevent the TB infection across the board in all age groups as well as in different geographical regions with variable susceptibility to TB. Identification of drugs that will shorten treatment and thereby improve adherence is also still far from reality. Ideally, finding such drugs and vaccines would be based on knowledge of the underlying mechanisms of mycobacterial persistence, enabling identification of crucial targets. At present, both a clear understanding of persistence mechanisms and fully validated animal models that reliably predict human infection outcomes are lacking, and thus so is an efficient path to developing novel vaccines and drugs for effectively preventing infection or shortening treatment. In the absence of fundamental biological understanding of the mechanism and immune correlate of protection, a preventive or therapeutic option that can drastically change the TB control measure is thus likely to remain a distant goal if we overlook the other alternative options that can help find us boost our current TB management.

21.2 Recent development in natural product based alternative tuberculosis therapeutic options

The impact of natural products (NPs) on the well-being of mankind has been massive, and their study continues to influence research in the multidisciplinary fields of chemistry, biology, and medicine. Historically, many of our medicines originate from NPs and their synthetic derivatives and have taught us valuable lessons about the drug discovery and biology. While advances in synthetic and combinatorial chemistry have given rise to notable successes in the development of new drugs, the perceived value of NPs has not waned when it comes to treating various deadly diseases. NPs have been extensively used at the origin of several established drugs for TB as well throughout the long history of this disease. Although the current clinical TB pipeline rarely features any candidate drugs derived from new NP scaffolds, numerous novel NPs or their analogs have been discovered in the recent past with promising activity against *Mycobacterium tuberculosis* [10–12]. This includes newly discovered structures as well as known NP classes that had not been previously recognized to be active against *M. tuberculosis*. These compounds could help to replenish the dry clinical TB pipeline and, thus, contribute to improvements in the treatment of a devastating disease. In this section we discuss recent advances in the development of TB drug leads from NPs, with a special focus on anti-mycobacterial compounds in late-stage preclinical and clinical development. The section also highlights the importance of collaboration between phytochemistry, medicinal chemistry, and physical chemistry, which is very important for the development of these natural compounds for alternative TB therapy. The ideal natural product must display high potency, particularly against drug-resistant strains, and possess an adequate safety profile. In addition, drugs should be active against latent and replicating forms of *M. tuberculosis* and have limited drug/drug interactions, particularly with anti-retroviral agents. Many NPs have multiple activities against multiple targets, which may help in combating drug-resistance issue of TB since it is likely that multiple genetic mutations would be required for the pathogen to become resistant against such drugs. Nonetheless, promiscuous compounds with multiple mode of action, whether natural or synthetic, also likely to have some toxicity issues. Identification of bioactive molecules from natural sources involves a defined series of steps to characterize/synthesize the products of interest. In addition, NPs are often bioactive molecules that may display high degrees of bioavailability, thus increasing their capacity to access their site of action within target cells. We have kept the definition of NPs relatively broad in order to include the major TB drug candidates in development. Due to the scope of this chapter, we have only included 10 most promising natural products/their derivatives that demonstrate strong evidences for potent antimycobacterial activity (Table 21.1).

21.3 Candidate natural products and their derivatives for alternative tuberculosis therapy

21.3.1 Griselimycin

The cyclic peptide griselimycin (GM) was first discovered by Noufflard-Guy-Loe and Berteaux in 1965 and currently being actively pursued for treatment of TB [13]. Cyclic peptide GM was isolated in the 1960s from *Streptomyces* and its antimycobacterial activity was evaluated in the early 1970s [14]. Due to unfavorable pharmacokinetic properties and especially after rifampicin was approved at that time, the interest in development of GM as an anti-TB drug was diluted and hence it was not actively pursued further. However, interest in the GM was renewed with some very promising results of this compound against *M. tuberculosis* in a more recent study [15]. By increasing the metabolic stability of GM through alkylation of the proline residue in position 8, a cyclohexyl griselimycin (CGM) derivative, obtained via a newly developed total synthesis, was found to be metabolically more stable, and exhibited enhanced penetration of the thick mycobacterial cell wall. The minimum inhibitory concentration (MIC) values of this modified GM were 0.06 and 0.2 μg/ml for the

Table 21.1: Potential alternative drugs for tuberculosis treatment.

S.No.	Name	Structure	Mode of action	Antimycobacterial activity
1	**Griselimycin**		DNA polymerase inhibitor	Axenic MIC = 0.06μg/ml In Macrophage = 0.2 μg/ml In chronic model = 100mg/Kg
2	**Ecumicin**		ClpC1 protein inhibitor	Susceptible MIC = 0.16 to 0.62 μM MDR 0.12–0.31 μg/ml XDR 0.31–0.62 μg/ml In cell line IC50 values of <32–<63 μM In vivo 20 or 32 mg/kg (12 doses)

(continued on next page)

3	Clofazimine		DNA replication inhibitor	Mouse model = 20 mg/kg
4	Piperine		efflux pump inhibitor	Susceptible and MDR strains 50 to >100 μg/mL
5	Rifapentine		RNA polymerase inhibitor	Mouse model = 20 mg/kg

(continued on next page)

Table 21.1: Potential alternative drugs for tuberculosis treatment—cont'd

S.No.	Name	Structure	Mode of action	Antimycobacterial activity
6	Teixobactin		bacterial cell wall synthesis inhibitor	MICs for most bacterial pathogens is well below 1 µg/ml
7	Sansanmycin		Lipid biosynthesis inhibitor	MIC 8.0 to 20 µg/ml

(continued on next page)

| 8 | Cyclomarin A | | binding to the N-terminal domain of ClpC1 | MIC Replicating and non-replicating 0.3 and 2.5 μm, respectively |
| 9 | Lariatins | | Inhibitory peptides | For *M. smegmatis* MIC = 3.13 and 6.25 μg/ml
For Mtb
MIC = 0.39 μg/ml |

(continued on next page)

Table 21.1: Potential alternative drugs for tuberculosis treatment—cont'd

S.No.	Name	Structure	Mode of action	Antimycobacterial activity
10	**Trichoderins**		ATP synthase inhibitor	*M. smegmatis* MIC = 0.1 μg/ml, *M. bovis* BCG MIC = 0.02 μg/ml, susceptible Mtb MIC = 0.12 μg/ml

drug susceptible *M. tuberculosis* strain H37Rv in broth culture and within macrophage-like (RAW264.7) cells, respectively. CGM also exhibited time dependent bactericidal activity in vitro. CGM was found to be highly active against *M. tuberculosis*, both in vitro and in vivo, via inhibiting the DNA polymerase sliding clamp DnaN. At the treatment dose of 50 mg/kg in mouse based acute infection model, CGM was found to prevent bacterial growth and gross lung lesions. In a chronic model, which is used to test for antimicrobial activity against a stable bacterial population in vivo, the CGM was found to be as effective as rifampicin as the treatment with CGM at 100 mg/kg resulted in a decrease in lung CFU counts with a kinetics similar to RIF at 10 mg/kg. In both cases, no adverse or toxic effects of CGM administration were observed in treated mice indicating its safety in preclinical studies. Nonetheless, a more robust pharmacokinetics/pharmacodynamics studies is warranted to assure its safety and efficacy in clinical studies.

21.3.2 Ecumicin

Ecumicin is a cyclic tridecapeptide that consists of mainly N-methylated amino acids, produced by a genetically distinct *Nonomuraea sp.*, which is a more recently discovered species of actinomycetes [16]. Ecumicin has been shown to possess potent anti-TB activity against MDR and XDR *M. tuberculosis* as well as drug-susceptible *M. tuberculosis* in vitro. The target molecule of these antibiotics is ClpC1, a protein that is essential for the growth of M. tuberculosis [17]. With MIC values ranging from 0.16 to 0.62 µM, Ecumicin could effectively kill *M. tuberculosis* at a minimal bactericidal concentration of 1.5 µM at a much faster rate than standard drugs, hence indicating even the possibility of shortening the duration of treatment [18]. Ecumicin exhibited MIC values ranging from 0.16–0.58 µg/ml against drug sensitive *M. tuberculosis* strains whereas the MIC values against MDR and extensively drug-resistant (XDR) strains of *M. tuberculosis* were slightly higher in the range of 0.12–0.31 µg/ml and 0.31–0.62 µg/ml. Other actinomycetes produced compounds such as cyclomarin A and lassomycin also exhibit antimycobacterial activity via inhibiting ClpC1, though their mode of action is slightly different with engagement of different adaptor molecules of the host [17]. Since the target molecule of Ecumicin, the ClpC1 protein of *M. tuberculosis*, is distinct from those of existing anti-TB drugs, it could be an effective therapy for treatment of MDR TB. Ecumicin is essentially nontoxic to mammalian cells (IC50 values of <63 µM and <32 µM against Vero and J774 cells, respectively). In vivo confirmation of antitubercular activity of Ecumicin with a complete inhibition of *M. tuberculosis* growth in the lungs of mice was obtained following 12 doses at 20 or 32 mg/kg after subcutaneous administration [18]. Altogether, Ecumicin targeting ClpC1 supports the *M. tuberculosis* ClpC1 protein as a potentially new therapeutic target for the development of anti-TB agents. The development of Ecumicin and other NPs targeting the *M. tuberculosis* ClpC1 protein could thus be a promising strategy for the treatment of MDR and XDR TB resistant to existing anti-TB drugs.

21.3.3 Clofazimine and its derivatives

Clofazimine is a fat-soluble riminophenazine dye discovered in 1954 as a structural derivative of diploicin which is extracted from *Buellia canescens* [19]. Although clofazimine was originally developed as an anti-TB drug, it was not found to be very effective against TB in pre-clinical studies and clinical trials done later [12]. Indeed, clofazimine was found to be more effective for leprosy than TB and was an approved medication used together with rifampicin and dapsone to treat leprosy [20]. Clofazimine is known to bind to the guanine bases of bacterial DNA, thereby blocking the template function of the DNA, resulting in inhibition of bacterial proliferation [21]. However, due to cytotoxicity and side effects of skin pigmentation at the prescribed dose, the drug was orphaned later. Immunomodulatory and ion channel blocking effect of clofazimine were reported in later studies and it was reported to be effective in treating some of the autoimmune diseases [22]. In more recent studies, it has shown potentiality in vitro and in vivo activity as a sterilizing drug to treat MDR-TB when used in combination with other drugs [23]. In another study when clofazimine was used as a TB drug in a combination with gatifloxacin, ethambutol, pyrazinamide, prothionamide, kanamycin, and high-dose isoniazid for 9 months, and clofazimine was able to treat 88% of MDR-TB patients studied [24]. A combination of clofazimine with other drugs also resulted in shortened treatment period for drug-susceptible TB [25]. The side effect of Clofazimine that includes accumulation within the cells and tissues in high concentration were found to be due to its very long half-life [21,22]. Thus, to minimize the side effects, TBI-166, clofazimine analogues have been invented. One of the clofazimine derivative TBI-166, has been shown to have a potent in vitro activity against drug susceptible *M. tuberculosis* in axenic culture as well as in intracellular conditions with low cytotoxicity [26]. It was also found to be effective against drug-resistant clinical isolates of *M. tuberculosis* though not against non-replicating *M. tuberculosis* in vitro. In a mouse based experimental model of TB, a dose of 20 mg/kg of TBI-166 via oral route in *M. tuberculosis* H37Rv-infected mice, resulted in greater than one log10 CFU/mL reduction when compared with infected mice treated with clofazimine control. Pharmacokinetic studies in rats and beagle dogs also showed the improved safety profile of TBI-166 over clofazimine where TBI-166 was eliminated from the body much more rapidly than clofazimine [27]. More analogs of clofazimine that can further reduce its side effect could thus be very useful options to treat drug resistant TB.

21.3.4 Piperine and derivatives

Piperine (PIP) is a bioactive compound of *Piper nigrum* (Black pepper) and *Piper longum* (Long pepper) and has been considered as possible alternative for the treatment of TB [28]. Past studies spanning over several decades have suggested that PIP is an efflux pump inhibitor (EPI) and may thus increase the efficacy of some antimicrobials via blocking their efflux. Apart from that PIP has also been reported to contain anti-inflammatory, antimicrobial, antifungal, analgesic, antipyretic, antioxidant, and anticarcinogenic activity [29–33]. It is also known to increase the bioavailability of some of the drugs while also increasing the activation of detoxification enzymes [34,35]. The investigation of activity of PIP against mycobacteria has shown some promising results in recent past and has also provided more details about its possible mechanism of action against mycobacteria [36,37]. Various studies have tested the activity of PIP obtained from different extracts and reported MICs ranging from 50 to >100 μg/mL against the *M. tuberculosis* reference strain H37Rv as well as MDR strains of the pathogen. Considering the high MIC value and relatively low antimycobacterial activity of PIP, many studies have evaluated its activity combined with other antimicrobial drugs. A combination of PIP with antitubercular drug rifampin has been found to have a synergistic effect on the antimycobacterial activity of both the compounds in many in vitro and in vivo studies [38,39]. Overall, most studies have shown that PIP works through inhibition of efflux pumps to increase the potency of standard antitubercular drugs and at the same time it can also modulate the immune system to strengthen our natural ability to fight TB. The mechanisms of action of PIP seems more pleotropic and should be further explored to develop possible alternative treatments for TB.

Piperidine (PIPD) which is a derivative of PIP has also been reported to have potent anti-TB activity. The MIC of PIPD has been found to be around 0.07 μM against *M. tuberculosis* in vitro which is much lower than PIP [40]. In vivo studies with experimental TB in animals have shown a decrease in the bacterial load in lung and spleen tissues of about 1.30 and 3.73-log10 respectively. BTZ043 is another derivative of PIPD and discovered via screening of sulfur-containing heterocycles against *M. smegmatis, M. aurum*, and *M. fortuitum* in vitro [41]. BTZ403 has been found to be highly active against a total of 240 sensitive and MDR clinical isolates of *M. tuberculosis*. BTZ043 was found to inhibit DprE1 (Rv3790), a key enzyme in the arabinogalactan and arabinomannan synthesis pathway. BTZ043 has been also found to work synergistically with recently added second line antitubercular drug bedaquiline, pretomanid, moxifloxacin, meropenem, and SQ-109 in vitro to inhibit the growth of *M. tuberculosis* H37Rv.

21.3.5 Rifapentine

Rifapentine is a derivative of first line antitubercular drug rifamycin (Rifampin), which is a fermentation product of *Streptomyces mediterranei sp* [42,43]. Rifapentine was first developed by Sanofi-Aventis and was found to contain similar activity like rifampin. Rifapentine persists in the blood at therapeutic levels for much longer (72 h post-antibiotic effect) then rifampin [44]. It also exhibited promising result for the treatment of infections caused by *Mycobacterium avium* complex (MAC). In a beige mouse model, rifapentine (20 mg/kg each) showed promising protection against few MAC isolates [45]. The activity of rifapentine against intracellular *M. tuberculosis* was found to be much more effective as compared to rifampin. The prolonged and improved efficacy of rifapentine was suggested to be due to better accumulation of this drug in macrophages which were 4-5-fold higher than those found for rifampin [46]. High-dose rifapentine regimen in combination with other drugs is currently undergoing Phase III clinical trials for the treatment of drug-susceptible as well as drug resistant TB [47].

21.3.6 Teixobactin

Teixobactin is recently discovered natural compound that is known to have antimicrobial activity against a wide range of MDR gram positive bacterial pathogens [48]. It is an antimicrobial peptide (AMP), discovered from a screen of previously unculturable gram negative bacteria provisionally named Eleftheria terrae with the use of iChip culturing technology [49]. Once synthesized inside the bacterial cells, teixobactin is exported across their outer membrane where it exerts antimicrobial action against other microorganisms. Teixobactin does not exhibit any cytotoxicity to mammalian cells even at a very high concentration of 100µg/ml [49]. Teixobactin has potent broad-spectrum activity, against Gram-positive bacteria with MICs for most bacterial pathogens well below 1 µg/ml. Teixobactin interferes with bacterial cell wall synthesis by binding to a highly conserved motif of lipid II (precursor of peptidoglycan) and lipid III (precursor of cell wall teichoic acid) [50]. Apart from its antibacterial activity against methicillin resistant staphylococcus aureus, Enterococcus sp., it has been found to inhibit the growth of *M. tuberculosis* as well at a concentration below 1ug/ml. Teixobactin is also active against *Clostridium difficile* and *Bacillus anthracis* with the MIC of 5 and 20 ng/ml respectively [51]. Remarkably, no drug resistance emerged in either S. aureus or *M. tuberculosis* culture when grown in the presence of the compound over a period of 27 days which indicates that resistance through horizontal gene transfer from the producing organism is less likely. Texiobactin can now also be synthesized with synthetic natural product cyclic depsipeptide by solid-phase peptide synthesis [52]. The synthetic natural product also displayed the potent antibacterial activity against *M. tuberculosis* and methicillin-resistant *Staphylococcus aureus* (MRSA). While the preliminary resistance studies of teixobactin against *M. tuberculosis* and other pathogens has been found to be very promising so far, there is a need for rigorous resistance screening against different strains of *M. tuberculosis* resistant to multiple antibiotics. Certainly, given the unique recipe of its activity against *M. tuberculosis* coupled with its inability to elicit resistance make teixobactin a very attractive molecule for new alternative therapeutic development for TB.

21.3.7 Sansanmycins

Sansanmycins is a nucleosidyl-peptide antibiotic isolated from an unidentified *Streptomyces* sp [53]. The structure of sansanmycin was revealed through the analysis of its alkaline hydrolysate and spectroscopic investigations. A series of sansanmycin analogues have now been synthesized that exhibit potent and selective activity against the virulent H37Rv as well as its MDR strains [54]. MIC values of sansanmycins against *M. tuberculosis* H37Ra and *Pseudomonas aeruginosa* were found to be 10 and 12.5 µg/ml, respectively [55]. Sansanmycins was also found to be effective against intracellular *M. tuberculosis* and *Psuedomonas aeruginosa*. Sansanmycins inhibits *M. tuberculosis* growth through disrupting the activity of *M. tuberculosis* phospho-MurNAc-pentapeptide translocase (MurX/MarY/ translocase I), the integral membrane enzyme responsible for the biosynthesis of lipid I, a key component in mycobacterial peptidoglycan synthesis [54]. Among various derivatives of this compound, Sansanmycin A exhbitis anti-mycobacterial activity against drug susceptible and MDR strains of *M. tuberculosis* [55]. Sansanmycin B was active against drug-susceptible and MDR *M. tuberculosis*, with a much lower MIC values ranging from 8.0 to 20 µg/ml. In addition, sansanmycin MX-2, and MX-4 displays significantly improved stability than sansanmycin A. Various approaches have been used in recent past to expand the diversity of the N-terminus of the peptide, with potential to yield more novel compounds with improved activity and/or other properties. These improved properties of sansanmycins derivatives may promote the novel anti-TB drug investigation targeting a clinically unexploited target MraY [56].

21.3.8 Cyclomarin A and derivatives

Cyclomarin A (CymA) is a cyclic peptide isolated from marine *Streptomyces* CN3-982 [57]. CymA was identified as a potent antitubercular compound from a whole cell-based screening assays of various NPs [58]. It exhibited a significant bactericidal activity against axenic culture as well as intracellular culture of *M. tuberculosis*. Importantly, it was found to be potent against growing as well as dormant, non-replicating mycobacteria. The bactericidal concentration was determined as 0.3 and 2.5 µM, respectively for actively replicating and non-replicating organisms. It was identified through a chemical-proteomic approach that the caseinolytic protein C1 (ClpC1) of *M. tuberculosis* was the target of CymA [58–60]. Antimycobacterial activity of CymA is due to its binding to the N-terminal domain of ClpC1. CymA mimics adapter binding and enables autonomous protein degradation by ClpC1. Thus, CymA causes gain of uncontrolled proteolytic activity in *M. tuberculosis* resulting in the cell death of organisms. Cyclomarin C along with other CymA derivatives that have been synthesized in recent past have also shown promising activity against drugs susceptible as well as drug resistant strains of *M. tuberculosis* [61]. Due to a remarkable simplification via replacement of replacing non-canonical amino acids, some of these new CymA derivatives are much easier to synthesized and offer highly attractive natural-product-derived lead structure for combating *M. tuberculosis*.

21.3.9 Lariatins/lassomycins

Lariatins, also called as Lasso peptides, are ribosomally synthesized and post-translationally modified NPs of microbial origin [62]. Lasso peptides, lariatins A and B, were first identified through the separation by HP-20 and ODS column chromatographies and purified by HPLC from the culture broth of bacterial species *Rhodococcus jostii* K01-B0171, which was isolated from soil aggregates collected in Yunnan, China. Lariatins A and B peptides are made of 18 - 20 amino acid residues with a macrolactone ring through the linkage between Gly1 and Glu8.

Lariatins A and B showed growth inhibition against *M. smegmatis* with MIC values of 3.13 and 6.25 µg/ml respectively. Lariatin A inhibited the growth of *M. tuberculosis* with a MIC of 0.39 µg/ml and is the most promising lead currently in early-stage development as potential drug leads for TB. The subsequent study on structure-activity relationships confirmed that amino acid residues Tyr6, Gly11 and Asn14 are responsible for the anti-mycobacterial activity and the residues at positions 15, 16 and 18 in lariatin A are critical for enhancing its activity [63]. Due to the exclusive threaded structure and the unusual bactericidal mechanism toward *M. tuberculosis*, these peptides have drawn significant interest in synthetic methods that can provide higher yield and simplified production of these peptides and their derivatives [64]. Recent discovery of newer and novel biosynthetic pathways for lasso peptides has laid the groundwork for combinatorial biosynthesis of their analogs, which provides new perspectives for the production of novel anti-TB compounds of natural product original.

21.3.10 Trichoderins

Trichoderins are a new class of aminolipopeptides that were initially isolated from a culture of marine sponge-derived fungus of *Trichoderma sp* [65]. This family of aminolipopeptide antibiotics is consisted of three unique compounds named as, trichoderins A, A1 and B and their chemical structures were determined on the basis of spectroscopic study. Importantly, they exhibited antimycobacterial activity against active as well as dormant bacilli. Trichoderins A, A1, and B showed excellent anti-mycobacterial activity against a wide range of mycobacterial species including *M. smegmatis, M. bovis* BCG, and *M. tuberculosis* H37Rv under aerobic and dormancy inducing hypoxic, with MIC values in the range of 0.02–2.0 µg/mL. Among them, trichoderin A showed the most promising anti-microbial activity against the non-pathogenic strain of *M. smegmatis* (MIC 0.1 µg/ml), *M. bovis* BCG (MIC 0.02 µg/ml), and the drug-susceptible laboratory strain of *M. tuberculosis* H37Rv (MIC0.12 µg/ml) under both replicating and non-replicating conditions. Interestingly, the potency of Trichoderin remains unchanged under hypoxic conditions in which the organism is known to be dormant. Trichoderins are known to inhibit the growth and multiplication of *M. tuberculosis* via blocking its ATP synthesis machinery [66]. Given the broad-spectrum activity of these compounds against different mycobacterial species and their ability to inhibit even non replicating mycobacterial pathogens, trichoderins are being actively pursued for their development as potential drug leads for TB therapy [67].

References

[1]   A MacNeil, P Glaziou, C Sismanidis, A Date, S Maloney, K Floyd, Global epidemiology of tuberculosis and progress toward meeting global targets — worldwide, 2018, MMWR Morb Mortal Wkly Rep 69 (2020) 281–285. http://doi.org/10.15585/mmwr.mm6911a2.
[2]   RMGJ Houben, PJ Dodd, The global burden of latent tuberculosis infection: a re-estimation using mathematical modelling, PLOS Med 13 (2016) e1002152. http://doi.org/10.1371/journal.pmed.1002152.

[3] CK Kwan, JD Ernst, HIV and tuberculosis: a deadly human syndemic, Clin Microbiol Rev 24 (2011) 351–376. http://doi.org/10.1128/CMR.00042-10.

[4] J Zaemes, C Kim, Immune checkpoint inhibitor use and tuberculosis: a systematic review of the literature, Eur J Cancer 132 (2020) 168–175. http://doi.org/10.1016/j.ejca.2020.03.015.

[5] P Andersen, TM Doherty, The success and failure of BCG — implications for a novel tuberculosis vaccine, Nat Rev Microbiol 3 (2005) 656–662. http://doi.org/10.1038/nrmicro1211.

[6] A Singh, R Prasad, V Balasubramanian, N Gupta, Drug-resistant tuberculosis and HIV infection: current perspectives, HIV/AIDS - Res Palliat Care 12 (2020) 9–31. http://doi.org/10.2147/HIV.S193059.

[7] KE Dooley, RE Chaisson, Tuberculosis and diabetes mellitus: convergence of two epidemics, Lancet Infect Dis 9 (2009) 737–746. http://doi.org/10.1016/S1473-3099(09)70282-8.

[8] MD J Libardo, HI Boshoff, CE Barry, The present state of the tuberculosis drug development pipeline, Curr Opin Pharmacol 42 (2018) 81–94. http://doi.org/10.1016/j.coph.2018.08.001.

[9] G Voss, D Casimiro, O Neyrolles, A Williams, SHE Kaufmann, H McShane, et al., Progress and challenges in TB vaccine development, F1000Research 7 (2018) 199. http://doi.org/10.12688/f1000research.13588.1.

[10] M Maiolini, S Gause, J Taylor, T Steakin, G Shipp, P Lamichhane, et al., The war against tuberculosis: a review of natural compounds and their derivatives, Molecules 25 (2020) 3011. http://doi.org/10.3390/molecules25133011.

[11] M Dong, B Pfeiffer, K-H Altmann, Recent developments in natural product-based drug discovery for tuberculosis, Drug Discov Today 22 (2017) 585–591. http://doi.org/10.1016/j.drudis.2016.11.015.

[12] D Quan, G Nagalingam, R Payne, JA Triccas, New tuberculosis drug leads from naturally occurring compounds, Int J Infect Dis 56 (2017) 212–220. http://doi.org/10.1016/j.ijid.2016.12.024.

[13] U Holzgrabe, New griselimycins for treatment of tuberculosis, Chem Biol 22 (2015) 981–982. http://doi.org/10.1016/j.chembiol.2015.08.002.

[14] J Herrmann, J Rybniker, R Müller, Novel and revisited approaches in antituberculosis drug discovery, Curr Opin Biotechnol 48 (2017) 94–101. http://doi.org/10.1016/j.copbio.2017.03.023.

[15] A Kling, P Lukat, DV Almeida, A Bauer, E Fontaine, S Sordello, et al., Targeting DnaN for tuberculosis therapy using novel griselimycins, Science 348 (*80*) (2015) 1106–1112. http://doi.org/10.1126/science.aaa4690.

[16] W Gao, J-Y Kim, S-N Chen, S-H Cho, J Choi, BU Jaki, et al., Discovery and characterization of the tuberculosis drug lead ecumicin, Org Lett 16 (2014) 6044–6047. http://doi.org/10.1021/ol5026603.

[17] H Lee, J-W Suh, Anti-tuberculosis lead molecules from natural products targeting Mycobacterium tuberculosis ClpC1, J Ind Microbiol Biotechnol 43 (2016) 205–212. http://doi.org/10.1007/s10295-015-1709-3.

[18] W Gao, J-Y Kim, JR Anderson, T Akopian, S Hong, Y-Y Jin, et al., The cyclic peptide ecumicin targeting ClpC1 is active against mycobacterium tuberculosis in vivo, Antimicrob Agents Chemother 59 (2015) 880–889. http://doi.org/10.1128/AAC.04054-14.

[19] R O'connor, JF O'sullivan, R O'kennedy, The pharmacology, metabolism, and chemistry of clofazimine, Drug Metab Rev 27 (1995) 591–614. http://doi.org/10.3109/03602539508994208.

[20] Garrelts JC Clofazimine, A review of its use in leprosy and mycobacterium avium complex infection, DICP 25 (1991) 525–531. http://doi.org/10.1177/106002809102500513.

[21] JL Arbiser, SL Moschella, Clofazimine: a review of its medical uses and mechanisms of action, J Am Acad Dermatol 32 (1995) 241–247. http://doi.org/10.1016/0190-9622(95)90134-5.

[22] MC Cholo, HC Steel, PB Fourie, WA Germishuizen, R Anderson, Clofazimine: current status and future prospects, J Antimicrob Chemother 67 (2012) 290–298. http://doi.org/10.1093/jac/dkr444.

[23] S Tang, L Yao, X Hao, Y Liu, L Zeng, G Liu, et al., Clofazimine for the treatment of multidrug-resistant tuberculosis: prospective, multicenter, randomized controlled study in China, Clin Infect Dis 60 (9) (2015) 1361–1367. http://doi.org/10.1093/cid/civ027.

[24] C Kuaban, J Noeske, HL Rieder, N Aït-Khaled, JL Abena Foe, A Trébucq, High effectiveness of a 12-month regimen for MDR-TB patients in Cameroon, Int J Tuberc Lung Dis 19 (2015) 517–524. http://doi.org/10.5588/ijtld.14.0535.

[25] A Van Deun, AKJ Maug, MAH Salim, PK Das, MR Sarker, P Daru, et al., Short, highly effective, and inexpensive standardized treatment of multidrug-resistant tuberculosis, Am J Respir Crit Care Med 182 (2010) 684–692. http://doi.org/10.1164/rccm.201001-0077OC.

[26] Y Lu, M Zheng, B Wang, L Fu, W Zhao, P Li, et al., Clofazimine analogs with efficacy against experimental tuberculosis and reduced potential for accumulation, Antimicrob Agents Chemother 55 (2011) 5185–5193. http://doi.org/10.1128/AAC.00699-11.

[27] D Li, L Sheng, X Liu, S Yang, Z Liu, Y Li, Determination of TBI-166, a novel antituberculotic in Rat Plasma by Liquid Chromatography–Tandem Mass Spectrometry, Chromatographia 77 (2014) 1697–1703. http://doi.org/10.1007/s10337-014-2771-0.

[28] LA Hegeto, KR Caleffi-Ferracioli, Perez de Souza J, Almeida AL de, SS Nakamura de Vasconcelos, ILE Barros, et al., Promising antituberculosis activity of piperine combined with antimicrobials: a systematic review, Microb Drug Resist 25 (2019) 120–126. http://doi.org/10.1089/mdr.2018.0107.

[29] W Zhai, Z Zhang, N Xu, Y Guo, C Qiu, C Li, G Deng, M Guo, Piperine plays an anti-inflammatory role in staphylococcus aureus endometritis by inhibiting activation of NF-κ B and MAPK pathways in mice, Evid Based Complement Altern Med 2016 (2016) 1–10. http://doi.org/10.1155/2016/8597208.

[30] DM Hikal, Antibacterial activity of piperine and black pepper oil, Biosci Biotechnol Res Asia 15 (2018) 877–880. http://doi.org/10.13005/bbra/2697.

[31] Y-S Moon, W-S Choi, E-S Park, I Bae, S-D Choi, O Paek, et al., Antifungal and antiaflatoxigenic methylenedioxy-containing compounds and piperine-like synthetic compounds, Toxins (Basel) 8 (2016) 240. http://doi.org/10.3390/toxins8080240.

[32] A Khajuria, N Thusu, U Zutshi, KL Bedi, Piperine modulation of carcinogen induced oxidative stress in intestinal mucosa, Mol Cell Biochem 189 (1998) 113–118. http://doi.org/10.1023/a:1006877614411.

[33] AF Majdalawieh, RI Carr, In vitro investigation of the potential immunomodulatory and anti-cancer activities of black pepper (Piper nigrum) and cardamom (Elettaria cardamomum), J Med Food 13 (2010) 371–381. http://doi.org/10.1089/jmf.2009.1131.

[34] VM Patil, S Das, K Balasubramanian, Quantum chemical and docking insights into bioavailability enhancement of curcumin by piperine in pepper, J Phys Chem A 120 (2016) 3643–3653. http://doi.org/10.1021/acs.jpca.6b01434.

[35] K Selvendiran, J Prince Vijeya Singh, D Sakthisekaran, In vivo effect of piperine on serum and tissue glycoprotein levels in benzo(a)pyrene induced lung carcinogenesis in Swiss albino mice, Pulm Pharmacol Ther 19 (2006) 107–111. http://doi.org/10.1016/j.pupt.2005.04.002.

[36] K Dhama, R Tiwari, S Chakraborty, M Saminathan, A Kumar, K Karthik, et al., Evidence based antibacterial potentials of medicinal plants and herbs countering bacterial pathogens especially in the era of emerging drug resistance: an integrated update, Int J Pharmacol 10 (2013) 1–43. http://doi.org/10.3923/ijp.2014.1.43.

[37] J Jin, J Zhang, N Guo, H Feng, L Li, J Liang, et al., The plant alkaloid piperine as a potential inhibitor of ethidium bromide efflux in Mycobacterium smegmatis, J Med Microbiol 60 (2011) 223–229. http://doi.org/10.1099/jmm.0.025734-0.

[38] LS Murase, JV Perez de Souza, JE Meneguello, FAV Seixas, LA Hegeto, L Scodro RB de, et al., Possible binding of piperine in Mycobacterium tuberculosis RNA polymerase and rifampin synergism, Antimicrob Agents Chemother 63 (11) (2019) e02520–18. http://doi.org/10.1128/AAC.02520-18.

[39] S Sharma, M Kumar, S Sharma, A Nargotra, S Koul, IA Khan, Piperine as an inhibitor of Rv1258c, a putative multidrug efflux pump of Mycobacterium tuberculosis, J Antimicrob Chemother 65 (2010) 1694–1701. http://doi.org/10.1093/jac/dkq186.

[40] RR Kumar, S Perumal, P Senthilkumar, P Yogeeswari, D Sriram, Discovery of antimycobacterial spiro-piperidin-4-ones: an atom economic, stereoselective synthesis, and biological intervention, J Med Chem 51 (2008) 5731–5735. http://doi.org/10.1021/jm800545k.

[41] MR Pasca, G Degiacomi, JL Ribeiro AL de, F Zara, P De Mori, B Heym, et al., Clinical isolates of mycobacterium tuberculosis in four european hospitals are uniformly susceptible to benzothiazinones, Antimicrob Agents Chemother 54 (2010) 1616–1618. http://doi.org/10.1128/AAC.01676-09.

[42] P Margalith, G Beretta, Rifomycin. XI. taxonomic study on streptomyces mediterranei nov. sp, Mycopathol Mycol Appl 13 (1960) 321–330. http://doi.org/10.1007/BF02089930.

[43] Rothstein DM Rifamycins, Alone and in combination, Cold Spring Harb Perspect Med 6 (2016) a027011. http://doi.org/10.1101/cshperspect.a027011.

[44] M Weiner, W Burman, A Vernon, D Benator, CA Peloquin, A Khan, et al., Low isoniazid concentrations and outcome of tuberculosis treatment with once-weekly isoniazid and rifapentine, Am J Respir Crit Care Med 167 (2003) 1341–1347. http://doi.org/10.1164/rccm.200208-951OC.

[45] SP Klemens, MA Grossi, MH Cynamon, Comparative in vivo activities of rifabutin and rifapentine against Mycobacterium avium complex, Antimicrob Agents Chemother 38 (1994) 234–237. http://doi.org/10.1128/AAC.38.2.234.

[46] N Mor, B Simon, N Mezo, L Heifets, Comparison of activities of rifapentine and rifampin against Mycobacterium tuberculosis residing in human macrophages, Antimicrob Agents Chemother 39 (1995) 2073–2077. http://doi.org/10.1128/AAC.39.9.2073.

[47] SE Dorman, P Nahid, EV Kurbatova, PPJ Phillips, K Bryant, KE Dooley, et al., Four-month rifapentine regimens with or without moxifloxacin for tuberculosis, N Engl J Med 384 (2021) 1705–1718. http://doi.org/10.1056/NEJMoa2033400.

[48] WD Fiers, M Craighead, I Singh, Teixobactin and Its analogues: a new hope in antibiotic discovery, ACS Infect Dis 3 (2017) 688–690. http://doi.org/10.1021/acsinfecdis.7b00108.

[49] LL Ling, T Schneider, AJ Peoples, AL Spoering, I Engels, BP Conlon, et al., A new antibiotic kills pathogens without detectable resistance, Nature 517 (2015) 455–459. http://doi.org/10.1038/nature14098.

[50] T Homma, A Nuxoll, AB Gandt, P Ebner, I Engels, T Schneider, et al., Dual targeting of cell wall precursors by teixobactin leads to cell lysis, Antimicrob Agents Chemother 60 (2016) 6510–6517. http://doi.org/10.1128/AAC.01050-16.

[51] C Guo, D Mandalapu, X Ji, J Gao, Q Zhang, Chemistry and biology of teixobactin, Chem - A Eur J 24 (2018) 5406–5422. http://doi.org/10.1002/chem.201704167.

[52] AM Giltrap, LJ Dowman, G Nagalingam, JL Ochoa, RG Linington, WJ Britton, Total synthesis of teixobactin, Org Lett 18 (2016) 2788–2791. http://doi.org/10.1021/acs.orglett.6b01324.

[53] Y Xie, R Chen, S Si, C Sun, H Xu, A new nucleosidyl-peptide antibiotic, sansanmycin, J Antibiot (Tokyo) 60 (2007) 158–161. http://doi.org/10.1038/ja.2007.16.

[54] AT Tran, EE Watson, V Pujari, T Conroy, LJ Dowman, AM Giltrap, et al., Sansanmycin natural product analogues as potent and selective anti-mycobacterials that inhibit lipid I biosynthesis, Nat Commun 8 (2017) 14414. http://doi.org/10.1038/ncomms14414.

[55] Y-B Li, Y-Y Xie, N-N Du, Y Lu, H-Z Xu, B Wang, et al., Synthesis and in vitro antitubercular evaluation of novel sansanmycin derivatives, Bioorg Med Chem Lett 21 (2011) 6804–6807. http://doi.org/10.1016/j.bmcl.2011.09.031.

[56] Y Shi, Z Jiang, X Lei, N Zhang, Q Cai, Q Li, et al., Improving the N-terminal diversity of sansanmycin through mutasynthesis, Microb Cell Fact 15 (2016) 77. http://doi.org/10.1186/s12934-016-0471-1.

[57] MK Renner, Y-C Shen, X-C Cheng, PR Jensen, W Frankmoelle, CA Kauffman, et al., New antiinflammatory cyclic peptides produced by a marine bacterium (Streptomyces sp.), J Am Chem Soc 121 (1999) 11273–11276. http://doi.org/10.1021/ja992482o.

[58] EK Schmitt, M Riwanto, V Sambandamurthy, S Roggo, C Miault, C Zwingelstein, et al., The natural product cyclomarin kills mycobacterium tuberculosis by targeting the ClpC1 subunit of the caseinolytic protease, Angew Chemie Int Ed 50 (2011) 5889–5891. http://doi.org/10.1002/anie.201101740.

[59] D Vasudevan, SPS Rao, CG Noble, Structural basis of mycobacterial inhibition by cyclomarin A, J Biol Chem 288 (2013) 30883–30891. http://doi.org/10.1074/jbc.M113.493767.

[60] N Bürstner, S Roggo, N Ostermann, J Blank, C Delmas, F Freuler, et al., Gift from nature: cyclomarin a kills mycobacteria and malaria parasites by distinct modes of action, ChemBioChem 16 (2015) 2433–2436. http://doi.org/10.1002/cbic.201500472.

[61] A Kiefer, CD Bader, J Held, A Esser, J Rybniker, M Empting, et al., Synthesis of new cyclomarin deriva-
      tives and their biological evaluation towards Mycobacterium Tuberculosis and Plasmodium Falciparum,
      Chem – A Eur J 25 (37) (2019) 8894–8902. http://doi.org/10.1002/chem.201901640.

[62] M Iwatsuki, R Uchida, Y Takakusagi, A Matsumoto, C-L Jiang, Y Takahashi, et al., Lariatins, novel anti-
      mycobacterial peptides with a lasso structure, produced by Rhodococcus jostii K01-B0171, J Antibiot
      (Tokyo) 60 (2007) 357–363. http://doi.org/10.1038/ja.2007.48.

[63] J Inokoshi, N Koyama, M Miyake, Y Shimizu, H Tomoda, Structure-activity analysis of gram-positive
      bacterium-producing lasso peptides with anti-mycobacterial activity, Sci Rep 6 (2016) 30375. http://doi.org/
      10.1038/srep30375.

[64] S Zhu, Y Su, S Shams, Y Feng, Y Tong, G Zheng, Lassomycin and lariatin lasso peptides as suitable antibi-
      otics for combating mycobacterial infections: current state of biosynthesis and perspectives for production,
      Appl Microbiol Biotechnol 103 (2019) 3931–3940. http://doi.org/10.1007/s00253-019-09771-6.

[65] P Pruksakorn, M Arai, N Kotoku, C Vilchèze, AD Baughn, P Moodley, et al., Trichoderins, novel amino-
      lipopeptides from a marine sponge-derived Trichoderma sp., are active against dormant mycobacteria,
      Bioorg Med Chem Lett 20 (2010) 3658–3663. http://doi.org/10.1016/j.bmcl.2010.04.100.

[66] P Pruksakorn, M Arai, L Liu, P Moodley, WR Jacobs Jr, Kobayashi M, Action-mechanism of Trichoderin
      A, an anti-dormant mycobacterial aminolipopeptide from marine sponge-derived Trichoderma sp, Biol
      Pharm Bull 34 (2011) 1287–1290. http://doi.org/10.1248/bpb.34.1287.

[67] I Kavianinia, L Kunalingam, PWR Harris, GM Cook, MA Brimble, Total synthesis and stereochemical
      revision of the anti-tuberculosis Peptaibol Trichoderin A, Org Lett 18 (2016) 3878–3881. http://doi.org/
      10.1021/acs.orglett.6b01886.

CHAPTER 22

Unraveling of inhibitory potential of phytochemicals against SARS-CoV-2 using in-silico approach

Deepak Ganjewala, Hina Bansal, Ruchika Mittal and Gauri Srivastava

Amity Institute of Biotechnology, Amity University Uttar Pradesh, Noida, Uttar Pradesh, India

22.1 Introduction

Currently, the entire world is facing the devastating threat from COVID-19 pandemic. The COVID-19 is affecting 219 countries and territories. At the time of writing, over 2.6 million fresh cases were reported as of February 28, 2021, a 7% increase compared to the previous week (www.who.int). According to https://www.worldometers.info, the total number of confirmed COVID-19 cases reported on March 28, 2021, globally is 127.8 MM of which 103 MM recovered while 28 MM have died. On March 11, 2020, World Health Organization (WHO) had declared the COVID-19 as a pandemic [1]. Wuhan in China was the first epicenter of COVID-19 from where the first case was reported in 2019 end. Soon, the disease has spread in almost every country of the globe. A novel coronavirus named as Severe Acute Respiratory Syndrome Coronavirus-2 (SARS-CoV-2) recognized as the causative agent of COVID-19. Major characteristics of COVID-19 disease are respiratory disorders with flu-like symptoms such as a sore throat, fever, cold, cough, and severe pneumonia in more critical cases [2]. The genome of this virus has been found similar to the previously known coronavirus SARS-CoV or Middle East Respiratory Syndrome-coronavirus (MERS-CoV) [3–5]. The primary route of its transmission is respiratory droplets, which are transmitted by contact [6,7].

Since the outbreak of COVID-19, WHO has been fully active kept updating recent development of the diagnosis and issuing guidelines for patient monitoring, sample collection and treatment [8]. In the initial days of outbreak, when very little or no information was available about the virus and its treatment, many countries have imposed total lockdown as per the advices and recommendations of experts as the only means to control the spread. All the malls, markets, schools, colleges, public and private organizations, means of transports, meetings, social and religious gathering declared closed to curb the spread of SARS-CoV2 among the people by restricting human-to-human contact. The lockdown prolonged for several weeks to months in 2020, according to the spread in different countries [6,7]. In addition, other restrictions such as the use of personal protective equipments (masks, gloves, antibacterial gels, hand sanitizers), self-isolation or quarantine, and even borders closing were imposed to effectively bring down the spread of infection. Because of the prolonged lockdown, economy in many countries has been collapsed. After the first wave of the COVID-19, when the numbers of active cases going down, several countries started unlocking process in phases to resume all businesses as usual to bring back economy to the normalcy. Meanwhile, a fresh wave of COVID-19 has been emerging in several countries because of new mutated variants of SARS-CoV-2. This trend made us difficult to predict how long this virus will continue with us. In view of this, we should take very seriously the new threats posed by the new lethal variants of SARS-CoV-2. Currently, the most effective strategy to protect ourselves from COVID-19 is to break transmission chain of SARS-CoV-2 particles. For this, people must adhere with the WHO guidelines emphasizing on hygiene, sanitization (frequent hand wash), social distancing, and wearing mask in public places.

The dramatic changes of events with the unprecedented COVID-19 have prompted an exponential increase of scientific interest in SARS-CoV-2 globally. Now, several vaccines have been developed with which it may be possible to control COVID-19. However, all vaccines have several limitations. They may have possible side effects and low efficacy so they are still under the scanner. It is also not clear whether vaccines will provide protection from new variant of the virus. The SARS-CoV-2 may develop resistance against recent COVID vaccines because of its mutating character. The ability of any virus to mutate rapidly has always been a major challenge for researchers who are involved in developing a single broad-spectrum vaccine. Because of the limitation of COVID vaccines, many researchers have advocated plant-derived natural products as promising bioactive compounds for the development of novel drugs for COVID-19 treatment [9–11]. With the onset of the pandemic, researchers have undertaken *in-silico* studies to evaluate potential of many phytochemicals with promises to the development of natural inhibitors of major protein targets of SARS-CoV-2 to hinder its multiplication [9–14]. The major targets of the virus are host cell receptor, that is, angiotensin-converting enzyme (ACE2) and several viral proteins, for example, spike glycoprotein (S), chymotrypsin-like cysteine protease (3CLpro), papain-like cysteine protease (PLpro), and RNA-dependent RNA polymerase (RdRp). Recently, few review articles have provided insights into drug targets of SARS-CoV and MERS-CoV for phytochemicals, which can be developed as safe and cost-effective drugs for COVID-19 [13,15–17]. Mani et al. [14] have documented information on plant extracts and compound(s) having antiviral potential to inhibit the growth of coronaviruses. Similarly, Lin et al. [18] presented information on naturally occurring antivirals acting against general coronaviruses. Several other articles available have covered many phytochemicals displaying inhibitory potential against major target proteins with their mechanisms of action, suggesting their usefulness to treat COVID-19 [9–13,19,20]. *In-silico* studies conducted have been underpinning breakthroughs for the development of drugs and therapeutics against many diseases and ailments including the current corona pandemic for a lomg time. In this context, *in-silico* studies have revealed plentiful anti-SARS-CoV-2 phytochemicals, however validation of these compounds by *in-vivo* and *in-vitro* experimental procedures in the laboratory is highly recommended. After their validation only they can be translated into novel drugs and therapeutics for COVID-19. Phytochemicals-based drugs and therapeutics are currently under development will have potential to control COVID-19 effectively, but it has to be ensured that they do not pose any serious side effects in the patients. They must have highest efficacy, and long life-span and cost-effective as well. In this chapter, we have reviewed *in-silico* based reports published recently and presented their immense role in the discovery of anti-SARS-CoV-2 phytochemicals. In the light of the current COVID-19, the information summarized here would be of utmost importance for the researchers working on phytochemicals-based drug discovery targeting COVID-19.

22.2 Overview of SARS-CoV-2 structure and infection

SARS-CoV-2 virus belongs to the family Coronaviridae distributed in humans, and the subfamily Orthocoronaviridae [21]. They are enveloped, positive-sense single-stranded ribonucleic acid (RNA) viruses that can infect both animals and humans. Coronaviruses have large genome size of 26–32 kb [22]. The genome size of SARS-CoV-2 is approximately 30 kb [3,23–25], which is almost 70% identical to SARS-CoV [26], hence it has been named as SARS-CoV-2. There are four genera of coronaviruses namely, β-coronavirus, α-coronavirus, gamma-coronavirus, and delta-coronavirus [27]. Studies have revealed that SARS-CoV-2 is the fifth strain of β-coronaviruses, which include SARS-CoV-1 and MERS-CoV [4]. The 30 kb genome of SARS-CoV-2 codes 16 nonstructural and 4 structural proteins, among which some are necessary for entry into the host cell and replication [28]. Comparative genomics studies revealed that the SARS-CoV-2 has 96% identity with the related bat coronavirus [5] and 91% to the Pangolin-CoV, raising the possibility that the latter acted as an intermediate zoonotic host between bats and humans [29].

In the life cycle and infection of SARS-CoV-2, different molecular targets, such as host cell receptor-angiotensin-converting enzyme (ACE2) and viral S protein, PLpro, 3CLpro, helicases, and RdRp play crucial roles. The SARS-CoV-2 showed stronger binding affinity for ACE2 membrane receptor of human host than SARS-CoV [30]. Coronavirus infects host in three stages. First, the virus infects host by attaching the transmembrane spike glycoprotein through ACE2 in the host so that they form a complex between S-glycoprotein and ACE2 with the help of TMPRSS2 produced by host cells. The second is the replication of viral RNA using RdRp, and the third stage is the maturation of the virus replication in the host cell using proteases, such as 3CLpro and PLpro [31]. All the major proteins/enzymes of the SARS-CoV-2 and the host cell essential for the multiplication and survival of the virus are presented in Table 22.1. For further detail of the virology, clinical and molecular epidemiology, diagnosis, pathogenesis, and potential therapeutics for the treatment of COVID-19 readers may refer to an article by Jin et al. [24].

22.3 COVID-19 vaccines v/s anti-SARS-Cov-2 phytochemicals

Since the outbreak of the corona pandemic, we have been witnessing diverse approaches involving combinations of different drugs and therapeutics being used for the effective control of SARS-CoV-2 infection. Among them, developing the vaccine by blocking ACE2 receptor and searching for phytochemicals inhibiting the major proteins/enzymes of the virus are the most important strategies [32]. Several pharmaceutical giants, namely, Pfizer, The Oxford AstraZeneca, Moderna, Sinovac, Bharat Biotech from different countries have undertaken projects for the development of vaccine against COVID-19. By now, several effective vaccines have been developed and approved by FDA, and concerned authorities. For the first time in the history of the vaccine development, COVID-19 vaccines have been developed in such a short duration of 2 years. Vaccines modulate the immune system of a person to make antibodies against specific diseases. Simply, vaccine makes the immune system enable to recognize the disease as the foreign body had already encountered the disease. An effective vaccine modulates the immune system without kicking it into overdrive. However, developing an effective vaccine with no side effects has always been a major challenge. Several of the vaccines developed have got approval from the Health authority of respective countries for vaccination. In the United States, two vaccines Pfizer-BioNTech COVID-19 and the Moderna COVID-19 have been approved by FDA. In India, two vaccines, namely, Covaxin and Covishield, developed by Bharat Biotech have been approved by health authority. In India, currently, these vaccines are being injected to the people of ages above 18 yeras under the world's largest vaccination drive conducted by the government of India and private hospitals. In China, Coronavac developed by Sinovac has been approved for use while in Russia Sputnik V vaccine has been approved by the health authoruty. In the UK, the Oxford AstraZeneca vaccine is approved for use. Currently, the Novavax is under the phase 3 trials, as is Janssen's COVID-19 vaccine. Companies based in the United States developed both these vaccines. At present, in many countries, vaccination is on in full swing, while some coutries have vaccinatied entire population and some will achieve it soon. After vaccination, the person develops immunity and can fight off the infection if get exposed to the SARS-CoV-2 infection.

Vaccines are the best tools against COVID-19. However, we need to deal with types of challenges like its production, storage, transportation, necessary healthcare infrastructures, and trained healthcare professionals to inject vaccine. In view of these challenges, it probably will take longer time to vaccinate the whole world's population and perhaps by then many people

Table 22.1: Major structural proteins and enzymes of the SARS-CoV-2 and their roles.

SARS-CoV-2 proteins/enzymes	PDB ID	Targets	Functions	References
Spike glycoprotein	6XM0	ACE2	• Helps the virus to enter the cells by interacting directly with ACE2	[5,6,53,90]
Transmembrane protease, serine 2 (TMPRSS2)	1Z8G	Proteins S	• Hydrolysis of protein S results in membrane fusion • Membrane fusion • Helps the virus to enter host cells	[9]
3-chymotrypsin-like cysteine protease (3CLpro) sometimes called main protease (Mpro)	6LU7	Replicase polyprotein	• Cleaves replicase polyproteins to release 16 nonstructural proteins (nsps) including the RdRp or nsp12	[91–93]
Papain like protease (PLpro)	4OW0	Replicase polyprotein	• Cleaves N-terminus of the replicase polyprotein to release several nsps, including the nsp3 • Important role in the viral replication and suppressing the host innate immune response • Essential in the virus replication correction • Posttranslational modification activities	[45–47]
RNA-dependent RNA polymerase (RdRp),	6NUR		• It plays an essential role in directing the replication and transcription of SARS-CoV-2 genome.	[45,87,94]
Non-structural proteins (nsps)		Host cell surface	• Helps the virus to enter into the host cells	[56]

would succmbed to death. Other important points of concerns may be the side effects and capability of vaccine raising immune response against new mutant variants of SARS-CoV-2. Recent observation of COVID-19 patients revealed that this virus has tendency to undergo mutation rapidly. The mutation caused changes in the structure and virulence properties of SARS-CoV-2, thus make it resistant against the body's immune system even after the immune system has been intensified by the COVID-19 vaccine. The immune-defence system rendered by the specific vaccine may not be as effective as it has to be against the new variant of the virus. Covid-19 vaccines also pose several side effects such as flu-like symptoms, pain and swelling at the injection site, fatigue, headache and muscle pain, and common fever. The side effects probably get more intensified after the second dose of vaccine because the body's immune response will be intensified. At present, the specific long-term effects of the covid vaccine are also not known. However, several anonymous reports have indicated that covid-19 vaccines may be effective only up to a very limited time, thereafter the person may need fresh dose of the vaccine to maintain the antibody titer to safeguard him against COVID-19. Also, the efficacy of the vaccines lies in the range of 65%–95%, so even after the vaccination a significant proportion of the population may not be able to develop the immunity. In view of these limitations and concerns regarding to the COVID-19 vaccines, intensified studies are required to evaluate the threats of vaccines, their efficacy, life-span, and side-effects toward the goal of producing safest broad-spectrum vaccine to provide protection against the current SARS-CoV-2 and the future mutant variants.

Any vaccine must be safe for everyone; however, all available covid-19 vaccines have some limitations. Therefore, we see tremendous scope for alternative treatment strategies mainly the plants and natural products with aniviral properties offering powerful lead molecules for the development of novel drugs and therapeutics for COVID-19. The phytochemical-based drugs and therapeutics may be very useful for a person who has allergy to COVID vaccine and chances of developing serious side effects after vaccination. Already, by this time, a significant research work has been carried out involving identification of phytoconstituents and developing new therapeutics and drugs, and still continuing [33]. In the early phase of COVID-19 outbreak, when scientific and medical fraternity had no strategy or medicine to control or treat the disease, people in China, gave several herbal traditional medicines to COVID-19 patients, which has been found highly effective with the recovery rate of 90% [35,36]. Some of these traditional medicines prevented SARS-CoV-2 infection of healthy persons and improved the condition of covid-19 patients with mild to severe symptoms [34,35]. Since the natural products are safe and easily available, they offer new scopes for the control and prevention of various viral infections including the SARS-CoV-2 [36–39].

22.4 Molecular docking and simulations

Molecular docking is a technique used to predict the binding affinity and the type of interactions that how a protein (or enzyme) interacts with small molecules (ligands) or how

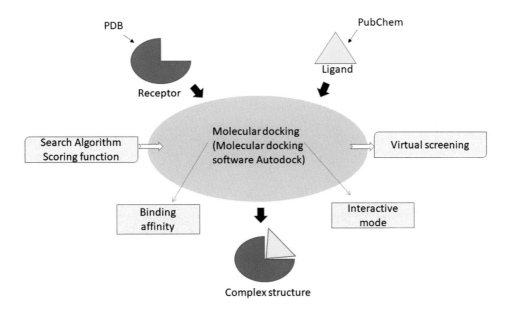

Figure 22.1
Schematic diagram showing various steps of molecular docking.

two or more molecular structures fit together. It is one of the most applied virtual screening methods, especially when the 3D structure of the target protein is available. It is an important component of the drug discovery toolbox, and its relatively low-cost implications and perceived simplicity of use have stimulated an ever-increasing popularity within academic communities. The ability of a protein and nucleic acid to interact with small molecules to form a supramolecular complex plays a major role in the protein's dynamics, which may enhance or inhibit its biological function. Molecular docking can describe the behavior of small molecules in the binding pockets of target proteins. The method aims to identify correct poses of ligands in the binding pocket of a protein and to predict the affinity between the ligand and the protein (Fig. 22.1). Based on the ligand, we can classify docking as protein–small molecule (ligand) docking, protein–nucleic acid docking, and protein–protein docking. The docking process involves two basic steps: prediction of the ligand conformation and its position and orientation within these sites and assessment of the binding affinity. In this section, we have discussed utility of molecular docking and simulation dynamics in characterizing anti-SARS-CoV-2 phytochemicals.

Initially, only remdesivir a broad-spectrum antiviral drug was approved to treat severe COVID-19 patients. Also, some reports indicated that hydroxyclorquinine a known antimalarial drug was found to be effective in COVID-19. The present day antivirals act as fusion inhibitors, protease inhibitors, neuraminidase inhibitors, and M2 ion channel protein blockers [40]. In the beginning, when we lack knowledge about the novel SARS-CoV-2 virus and its molecular mechanism of the infection, and spread, the development of vaccines and drugs for COVID-19 could not be initiated. Hence, unraveling of the viral mechanism of infection and spread has been crucial in the development of vaccines and drugs. Soon, researchers have sequenced the genome of the virus and gained the knowledge of COVID-19 and underlying molecular mechanisms of infection as well as the target proteins crucial for its entry in the host. Having this information, researchers have started exploring all the resources mainly plants and traditional medicines to find out specific, effective bioactive molecules with potential to supress the growth of SARS-CoV2. Plants are a paramount source of many phytochemicals, which may be useful as lead molecules for drug development for COVID-19 [11,13,14,41,42]. Plant phytochemicals have wide advantages for drugs and therapeutics development because they display potential for synergism, a crucial factor in the drug development. Among phytochemicals, polyphenols found in various plant species inhibit SARS-CoV-2 enzyme, which is essential for replication and infection [42]. Currently, *in-silico* approaches such as molecular docking, quantum mechanical simulation by modern tools ONIOM and others have been promoting the discovery of potential anti SARS-CoV-2 phytochemicals [10,32,43,44]. The computer-based *in-silico* methods are becoming more popular owing to their rapidity and cost effectiveness as compared to experimental methods based on trial and errors. Researchers are using *in-silico* search engines to explain protein–drug interaction to identify the energetically most suitable and optimized drug-target protein complex structures. Molecular docking analysis provides newer information on binding affinity and the type of interactions between the prospective phytochemicals and target proteins of SARS-CoV-2 [14,44]. Major protein targets crucial for survival of SARS-CoV-2 have already been identified; they are host cell receptor-angiotensin-converting enzyme ACE2, and viral proteins, such as S protein (containing S1 and S2 domains) [32], various cysteine proteases, papain-like cysteine protease (PLpro) [45–47], and chymotrypsin like protease (3CLpro) [48], helicases and RdRp [15–17]. Tables 22.1 and 22.2 present information on SARS-Co-V2 target proteins. Till date researchers have identified a myriad of phytochemicals using *in-silico* methods, which can arrest the growth of SARS-CoV-2 and elucidated the mechanisms of action by molecular docking tools. In all such studies, efficacy of the phytochemicals to inhibit the SARS-CoV-2 target protein were determined by comparing with the efficacy of the remdesivir [4,49]. Remdesivir is a nucleotide analog prodrug of (GS-441524), which get converted into an active triphosphate form inside the cell. Beside remdesivir, some studies have used other antivirals, such as lopinavir [50] and arbidol [51] equally effective as remdesivir. For the study of anti-SARS-CoV2 phytochemicals, several popular molecular docking tool, namely, Autodock vina and Swissdock, an open-source docking software has been used frequently, which are designed by the Molecular Graphics Laboratory at The Scripps Research Institute, la Jolla. Another most widely used web service is the SwissDock, which predict the molecular interactions occurring between a target protein and a small molecule. Molecular dynamics (MDs) simulation (100 ns) studies provide the data for the stability, flexibility, and binding affinity of the phytoconstituents toward main proteins and enzymes

Table 22.2: Summary of *in-silico* studies unraveling anti-SARS-CoV-2 phytochemicals and their major targets.

Source of phytochemicals	Bioactive phytochemicals	Class of Phytochemicals	SARS-CoV-2 Targets proteins/ enzymes	Computational tools	References
Phytochemicals derived from traditional Saudi medicinal plants	Luteolin 7-rutinoside Chrysophanol 8-(6-galloylglucoside) Kaempferol 7-(600-galloylglucoside) Chrysophanol 8-(6-galloylglucoside) 3,4,5-tri-O-galloylquinic acid Mulberrofuran G Withanolide A Isocodonocarpine Calonysterone	Flavonoids, Steroids	RdRp Mpro/3CLpro and PLpro	Autodock vina 1.1.2 program (Trott and Olson, 2010): for MD Autodock tools 1.5.6 program for MDy GROMACS 2018 software OPLS-AA/L force field: for atom molecular dynamic simulation of ligand–protein complexes	[10]
Indian and Chinese medicinal plants	Hesperidin Emodin Chrysin	Flavonoid and Anthraquinone	Spike protein ACE2 protein	SWISSDOCK: for MD GenBank database https://www.ncbi.nlm. nih.gov/prote in/QHR63 250.2	[32]
Moroccan Medicinal plants	Crocin Digitoxigenin β-Eudesmol	Glycosides	3CLpro/Mpro	Autodock 1.5.4: for MD protein databank (http://www.rcsb.org)	[54]
Cryptolepis sanguinolenta	Cryptomisrine, Cryptospirolepine, Cryptoquindoline Biscryptolepine	Alkaloid	3CLpro/Mpro and RdRp	Molecular docking	[55]
Asparagus racemosus (Willd.)	Asparaside-C Asparaside-D Asparaside -F	Glycoside	NSP15 Endoribonuclease and spike receptor-binding domain	Schrodinger Glide SP module For docking simulation AMBER18 software package for MD simulations	[56]

(continued on next page)

Table 22.2: Summary of *in-silico* studies unraveling anti-SARS-CoV-2 phytochemicals and their major targets—cont'd

Source of phytochemicals	Bioactive phytochemicals	Class of Phytochemicals	SARS-CoV-2 Targets proteins/ enzymes	Computational tools	References
Tinospora cordifolia	Berberine	Alkaloid	Mpro/3CLpro	SWISS-ADME software (https://www.swissadme.ch): For identification of drug molecule GROMACS 5.1 Package: for determination of thermodynamics stability of the ligand: protein complex CHARMM36 all atom force field: for MD simulations	[43]
Calendula officinals	Rutin, Isorhamnetin-3-O-β-D Calendoflaside, Narcissin Calendulaglycoside B Calenduloside Calendoflavoside	Flavonoids	Mpro/3CLpro	GROMACS. 2019.4 package: for MD simulation	[44]
Acacia seyal, Cymbopogon commutatus, and Indigofera caerulea	β-Sitosterol Quercetin Catechin Lupeol Rutin Kaempferol Gallic acid, Piperitone Limonene	Phenolics and monoterpenes	Mpro/3CLpro, and human furin protease	LigPrep: for ligand preparation. Schrodinger suite-Maestro (v 11.1). Discovery Studio (v 4.1) software wissADME, (http://www.swissadme.ch/): for biochemical properties	[59]

(continued on next page)

		SARS CoV-2 Mpro		
Justicia adhatoda	Vasicoline Vasicolinone Vasicinone Vasicine Adhatodine Anisotine	Alkaloids	AutoDock Vina: for MD GROMACS 2019 with GROMOS9653a6 force field and SPC water model: for molecular simulations PubChem database server https://pubchem.ncbi. nlm.nih.gov SwissADME: for prediction of different pharmacokinetic properties	[60]
Natural herbs and fruits	Forty phytochemicals	Flavonoids, Alkaloids, Glucosides, and Polyphenolics	Mpro/3CLpro	AutoDock Vina and GOLD suite: for MD YASARA dynamics software: for MD simulation Gaussian 09 software package: For Optimization of the phytochemicals and calculation of vibrational frequency	[38]
Clerodendrum species	Taraxerol Friedelin Stigmasterol	Terpenoid	Spike protein, Mpro and RdRp	Open Babel software: to generate 3D structures of phytochemicals AutoDock Vina software: for MD	[61]

(continued on next page)

Table 22.2: Summary of *in-silico* studies unraveling anti-SARS-CoV-2 phytochemicals and their major targets—cont'd

Source of phytochemicals	Bioactive phytochemicals	Class of Phytochemicals	SARS-CoV-2 Targets proteins/ enzymes	Computational tools	References
Kabasura Kudineer Chooranam	Chrysoeriol Luteolin	Flavonoid	Spike protein	Cresset Flare software: for MD	[95]
Thyme, cinnamon, Clove, Star anise, Basil, Holy basil, Eucalyptus, Geranium, Oregano and Ajwain	Anethole Cinnamaldehyde Carvacrol Geraniol Cinnamyl acetate L-4-terpineol Thymol Pulegone	Monoterpenes, Polyphenols	Spike protein	https://pubchem.ncbi.nlm. nih.gov/ for the 3D structure of the ligands http://www.scfbioiitd.res. in/dock/ ActiveSite.jsp: For active site prediction Autodock vina in PyRx: For MD	[65]
Chemical database	Sarsasapogenin Ursonic acid Curcumin Ajmalicine Novobiocin Silymarin Aranotin Piperine Gingerol, Rosmarinic acid α-terpinyl acetate	Flavonoid	Nsp15 protein	Autodock software: For MD SwissDock server: to perform the *in-silico* docking UCSF chimera: for visualization and analysis of docking results	[25]

(continued on next page)

Caesalpinia sappan L.) Citrus sp.	Asiatic acid Andrographolide Apigenin Brazilein Brazilin Catechin Curcumin Gingerol Hesperidin Hesperetin Kaemferol Luteolin Myricetin Naringenin Quercetin	Flavonoids	ACE2, TMPRSS2, RdRp, 3CLpro and PLpro	AutoDock Tools 1.5.6: For MD http://www.rcsb.org: for Target protein structures	[31]
Phytochemical and Drug Target Database (PDTDB)	Ergosterol peroxide Punicalin Oleanolic acid Naringin Diosmin	Flavonoids	Mpro/3CLpro	AutoDock Vina: for MD Open Babel software: for minimizing the structure of phytochemicals Biovia Discovery Studio software: to visualize the result of docking and protein–ligand interaction	[64]
Tea, and green tea	Epigallocatechin-3-gallate Theaflavins	Polyphenols	Spike protein RBD binding domain	AutoDock Vina: For MD	[42]
NPACT database and MPD3 database	One thousand plant bioactive terpenes compounds	Terpenes		MD: Autodock Vina MDS: GROMACS 5.1.4	[65]

(continued on next page)

Table 22.2: Summary of *in-silico* studies unraveling anti-SARS-CoV-2 phytochemicals and their major targets—cont'd

Source of phytochemicals	Bioactive phytochemicals	Class of Phytochemicals	SARS-CoV-2 Targets proteins/ enzymes	Computational tools	References
Alangium salviifolium	Salviifoside-A	Glycoside	Spike protein	Autodock (vina) in PyRx-phython prescription 0.8: MD ARGUSLAB software ArgusLab-ww.arguslab com/ for Ligand drug potential	[66]
PubChem Database	Glycyrrhizin, Bicylogermecrene, Tryptanthrine, β-sitosterol Indirubin Indican Indigo Hesperetin Crysophanic acid Rhein Berberine β-caryophyllene	Alkloids, Terpenoids, Flavonoids	Mpro/3CLpro	Auto Dock Vina 1.0: For MD Pymol 1.8.6.0 and Discovery Studio Visualizer 4.0: for visualization of the docking result	[67]
PubChem database	Fisetin Quercetin Kamferol	Flavonoids	Spike protein	Auto Dock Vina 1.0: For MD GROMACS-2018.1 package: For MDS MM-PBSA (Molecular Mechanics-Poisson-Boltzmann Surface Area): for binding free energies for ligands with S protein.	[68]

(continued on next page)

Source	Compounds	Target	Methods/Software	Reference
Cinnamon Indian Medicinal Plants, Phytochemistry, and also Therapeutics data source	Tenufolin Pavetannin C1		Autodock vina and PyRx software: For MD admetSAR, as well as DruLiTo servers: for drug-likeness prophecy	[69]
Azadiracta indica, Plectranthus amboinicus, Calligonum polygonoides, Crateva adansonii, Scutellaria baicalensis Fragaria ananassa Isatis indigotica Torreya nucifera Cinnamomum verum Litchi chinensis), Ganoderma lucidum Tinospora cardifolia	Tetratriacontane Baicalein Hesperetin Naringenin Myricetin 3,4-dichloroisocoumarin 4-Amidinophenylmethanesulfonyl fluoride hydrochloride Aesculitannin B Proanthocynidine A2 Ganoderiol D Ganodermanontriol Columbin Ecdysterone Ciscapsaicin Dihydrocapsaicin	Flavonoids TMPRSS2	Swiss-model server: for homology modelling and validation of TMPRSS2 Protein Data Bank (PDB) UCSF Chimera Procheck ProSA, ProQ Protter Polyview AutoDock HADDOCK ClusPro server	[70]

(continued on next page)

Table 22.2: Summary of *in-silico* studies unraveling anti-SARS-CoV-2 phytochemicals and their major targets—cont'd

Source of phytochemicals	Bioactive phytochemicals	Class of Phytochemicals	SARS-CoV-2 Targets proteins/enzymes	Computational tools	References
Tinospora crispa	Imidazolidin-4-ne 2-imino-1-(4-methoxy-6-dimethylamino-1,3,5-triazin-2-yl) Spirodec-6-en-1-ol 2,6,10,10-tetramethyl, 3.betahydroxy-5-cholen-24-oic acid Androstan-17-one 3-ethyl-3-hydroxy-(5.α) camphenol, (-)-Globulol, Yangambin Nordazepam TMS derivative Benzeneethanamine	Flavonoids	Mpro/3CLpro	Auto Dock Vina 1.0: for MD	[58]
PubChem	Albireodelphin Apigenin 7-(600-malonylglucoside) Cyanidin-3-(p-coumaroyl)-rutinoside-5-glucoside Delphinidin 3-O-β-D-glucoside 5-O-(6-coumaroyl-β-D-glucoside) (-)-Maackiain-3-O-glucosyl-600-O-malonate	Flavonoids	Spike protein Mpro/3CLpro and RdRp	AutoDock 4.1 software and CoDockPP: for MD	[71]

MD: Molecular docking; MDS: Molecular docking simulation; MDy: Molecular docking dynamics; 3CLpro: 3-chymotrypsin-like cysteine protease; Mpro: main protease; nps: Non-structural proteins; RdRp: RNA-dependent RNA polymerase; TMPRSS2: Transmembrane protease, serine 2.

targets of the SARS-CoV2. We can predict and evaluate ligands through ADMET descriptors to know whether it can be used for medicinal purposes or not and it is toxic or non-toxic. Based on the ADMET profiling, it is convenient to evaluate the drug-like properties and promising therapeutic potential of the docked compounds. Protein–ligand docking and MD simulation are powerful computational tools for rapid identification of the bright molecules. These tools help to determine and analyze binding energies of the protein–ligand complexes for the assessment of the stability of protein targets-phytochemical complex. Certainly, *in-silico* studies have been playing vital role in the areas of drug discovery and development, but first the outcomes of these studies needed to be verified by *in-vitro* and *in-vivo* experiments. Afterward, the validated phytochemicals may be exploited for drug development programs, which goes through a long process of clinical trials before its approval. Finally, follow-up studies are required to evaluate the effectiveness of phytochemicals/extracts and monitor side effects or other clinical complications in patients.

22.5 Screening of potential anti-SARS-CoV-2 phytochemicals using in-silico *approach*

In recent years, importance and the scope of *in-silico* studies have been expeditiously expanding in phytochemical-based drug discovery. Numerous *in-silico* studies have provided valuable information about the potential phytochemicals which can be useful for controlling the currently spreading COVID-19 and other devastating viral diseases which humans have been encountering for a long time. With the help of *in-silico* methods several novel anti-SARS-CoV-2 phytochemicals have been identified from Chinese and Indian traditional medicinal systems [52,53] and many more are being identified. Alamri et al. [10] have performed *in-silico* screening of ~1000 phytochemicals from 60 Saudi medicinal plants, which already known to possess antiviral activities leading to identification of only 9 compounds showing potential inhibitory effects against three major targets, RdRp, 3CLpro, and PLpro. Out of the nine, three compounds found based on the binding affinity scores ranged 9.9 to 6.5 kcal/mol have been tagged highly effective. They are luteolin 7-rutinoside, chrysophanol 8-(6-galloylglucoside) and kaempferol 7-(600-galloylglucoside), which bind efficiently to RdRp, while chrysophanol 8-(6- galloylglucoside), 3,4,5-tri-O-galloylquinic acid and mulberrofuran G strongly interacted with 3CLpro, and withanolide A, isocodonocarpine and calonysterone bound tightly to PLpro. Molecular docking study revealed antiviral activity of flavonoid and anthraquinone (hesperidin, emodin, and chrysin) on a complex of spike protein of SARS-CoV-2 and human ACE2. Thus, these compounds can be used as competent natural products to treat COVID-19. Among them, hesperidin binds with ACE2 protein and a complex of ACE2 and spike protein non-competitively. The ACE2 has separate binding sites for spike protein and hesperidin. Binding of the ligand (spike protein) induces a conformational change in the three-dimensional structure of ACE2 as revealed by molecular docking and MDs simulation studies. Hesperidin modulates the binding energy of the ACE2 and spike protein complex, making the complex unstable. The antiviral activity of the five natural compounds have also been confirmed by QSAR study. Moroccan medicinal plants also showed inhibitory activity against main protease of SARS-CoV-2 [54]. Three compounds crocin, digitoxigenin, and b-eudesmol of the 67 tested, possessed inhibitory effects against the coronavirus by blocking the main protease. Borquaye et al. [55] have reported 13 alkaloids from *Cryptolepis sanguinolenta*, which owned substantial inhibitory action against the main proteins, PLpro and the RdRp. Molecular docking analysis revealed that among the alkaloids tested, cryptomisrine, cryptospirolepine, cryptoquindoline, and biscryptolepine found to be the strongest inhibitor of PLpro and RdRp. All the 13 alkaloids had strong binding affinity to PLpro and RdRp based on the binding energy values ranged -6.7 to -10.6 kcal/mol and formation of a stable protein–ligand complex. So, the alkaloids from *C sanguinolenta* may be exploited for the development of new drugs or therapeutics for COVID-19. A well-known medicinal plant *Asparagus racemosus* which has several bioactivities like antiviral, immunomodulatory, anti-inflammatory, and antioxidant also exhibited anti-SARS-CoV-2 property as revealed by *in-silico* studies. The plant contains phytochemicals, namely, asparoside-C, asparoside-D, and asparoside-F which has been corelated with antiviral properties against SARS-CoV-2 [56]. Three compounds asparoside-C, asparoside-D, and asparoside inhibited NSP15 endoribonuclease € and spike receptor-binding domain of SARS-CoV-2. MDs simulations and MM-GBSA binding free energy calculations indicated that the asparoside-D and asparoside-C favourably binds with spike receptor-binding domain, whereas the asparoside-D and asparoside-F with NSP15 endoribonucelase. Based on the binding affinity, selected phytochemicals were found on par with remdesivir. Secondary metabolites such as polyphenols also reported to possess strong anti-SARS-CoV-2 activity [57]. A very popular Indian medicinal herb giloy (*Tinospora cordifolia*) is bestowed with alkaloids, steroids, glycosides, aliphatics all of which are highly effective against SARS-CoV-2. However, the berberine has been identified as a strongest inhibitor of the 3CLpro and RdRp [43]. The berberine, b-sitosterol, octacosanol, tetrahydropalmatine, choline showed the best binding docking poses with 3CLpro targets I, II of the main protease enzymes suggesting that they can be promising molecules as raw drug material. Molecular docking and MD analysis revealed that berberine has lesser binding energy and higher nonbonded interaction capability as compared to other molecules with better binding mode of interactions. The berberine docked with 3CLpro to form a stable complex, thus act as better inhibitor toward the SARS-CoV2 proteins compared to other inhibitors [43]. On the similar line, some phytoconstituents from *T. crispa* reported to be useful in the management of COVID-19 [58]. Methanol extract of *T. crispa* comprises of a mixture of compounds, such as imidazolidin-4-ne, 2-imino-1-(4-methoxy-6-dimethylamino-1,3,5-triazin-2-yl), spiro dec-6-en-1-ol, 2,6,10,10-tetramethyl, β-hydroxy-5-cholen-24-oic acid, androstan-17-one, 3-ethyl-3-hydroxy-(5.α), camphenol, Globulol, yangambin, nordazepam, TMS derivative, benzeneethanamine all of which displayed superior binding affinity to Mpro than the nelfinavir and lopinavir. They all can alter the activity of the Mpro enzyme, therefore can be implicated in the development of new antiviral agents for COVID-19. Das et al. [44] have identified flavonoids from *Calendula officinalis* using *in-silico* approach, which also have potential to inhibiting Mpro. Seven compounds, namely, rutin, isorhamnetin-3-O-b-D, calendoflaside, narcissin, calendulaglycoside B, calenduloside, calendoflavoside from *C. officinals* displayed strong binding affinity toward MPro as compared to co-crystal ligand inhibitor N3 using virtual molecular docking. The protease-ligand docked complexes formed was stable as compared to Mpro-native ligand (inhibitor N3) complex. Evaluation of the apo-form of apo-Mpro revealed no significant difference in dynamic properties between apo-Mpro and its docked complexes, that is, Mpro-rutin, Mpro-isorhamnetin-3-O-b-D, and Mpro-calendoflaside. The study concluded that two main compounds namely, calendoflaside and rutin, have capabilities to inhibit the main protease of SARS-CoV2. The rutin is already being used as a drug and the others may be translated into new drugs against SARS-CoV2. Djiboutian medicinal plants, such as *Acacia seyal, Cymbopogon commutatus*, and *Indigofera caerulea* produce bioactive compounds possessing medicinal properties against SARS-CoV-2 [59]. An *in-silico* study conducted on these plants identified nine biomolecules, β-sitosterol, quercetin, catechin, lupeol, rutin, kaempferol, gallic acid, piperitone and limonene, which displayed stronger affinity toward the Mpro, SARS-CoV2 receptor-binding domain (RBD) and human furin protease with respect to the hydroxychloroquine and remdesivir. Docking scores ascertained rutin as a potent competitor to hydroxychloroquine and remdesivir with comparatively better binding energy score against three protein targets. Ghosh et al. [60] have reported alkaloids from leaf extracts of *Justicia adhatoda* as the powerful inhibitor of the Mpro using *in-silico* approach. Molecular docking and MD simulations (100 ns) showed that only alkaloid anisotine displayed strong binding affinity and stability of Mpro-anisotine complex than the Mpro-darunavir/lopinavir complex, formed with two anti-HIV drugs lopinavir and darunavir. Islama et al. [38] have identified five potent antiviral phytochemicals hypericin, cyanidin 3-glucoside, baicalin, glabridin, and a-ketoamide-11r dis inhibiting the Mpro using molecular docking and MD simulations 100 ns tools. These phytochemicals displayed highest binding affinity and strong interactions with the Mpro through Cys145 and His41. Similarly, natural compounds from *Clerodendrum* spp. have shown inhibitory effects against SARS-CoV-2 [61]. However, taraxerol was profoundly useful molecule against SARS-CoV-2 spike, Mpro and RdRp proteins exhibiting signifiant binding energy scores with all three targets proteins than the reference drugs. In addition, friedelin and stigmasterol also found extremely valuable to supress SARS-CoV-2 proteins. Thus, *Clerodendrum* spp. natural products hold bright prospects for the therapeutic development. Kulkarni et al. [62] have illustrated potential inhibitory effects of essential oils and their major constituents against SARS-CoV-2 spike protein. Several monoterpenes constituents such as anethole, cinnamaldehyde, carvacrol, geraniol, cinnamyl acetate, L-4-terpineol, thymol, and pulegone of essential oils of many plants demonstrated strong activity against SARS-CoV2 spike protein [63]. Kumar et al. [63] described at least 50 phytochemicals capable of blocking the SARS-CoV 2 replication by binding with Nsp15 protein. Among them, sarsasapogenin, ursonic acid, curcumin, ajmalicine, novobiocin, silymarin and aranotin, piperine, gingerol, rosmarinic acid, and α-terpinyl acetate showed strong binding affinity to Nsp15. Similarly, phytoconstituents from Secang (*Caesalpinia sappan* L.) wood and *Citrus* spp. can be used as preventive agent and supportive therapy for SARS-CoV-2 [31]. Two constituents, brazilein and brazilin from secang woods, can inhibit ACE2 and Mpro and hesperidin from *Citrus* spp. showed impressive anti-SARS-CoV-2 activity [31]. Brazilein and brazilin have greater binding affinity to ACE2 with very low binding energy score as compared to chloroquine, arbidol, remdesivir, ribavirin, and lopinavir, while hesperidin showed excellent affinity to TMPRSS2. Thakur et al. [64] have evaluated 191 phytochemicals using molecular docking for their ability to combat Mpro. Several of them namely, ergosterol peroxide, punicalin, oleanolic acid, naringin, and diosmin showed strongest affinity to Mpro. Among them, ergosterol peroxide extracted from fungi offer promises to treat COVID-19, provided that they have been validated by *in vitro* and *in vivo* studies [64]. Mhatre et al. have discussed antiviral activity of green tea and black tea polyphenols in prophylaxis and the treatment of COVID-19 [42]. Tea polyphenols exhibited antiviral activities against various viruses, especially positive-sense single-stranded RNA viruses [42]. Two polyphenols, epigallocatechin-3-gallate from green tea and theaflavins from black tea, have immense potential in the management of COVID-19. Like flavonoids, terpenes also act as inhibitors of spike RBD, that help SARS-CoV-2 to bind with human ACE2 receptor thus can block the entry of coronaviruses into the host cells [65]. Phyto-derivatives from *Alangium salvifolium* used as a fusion inhibitor targeting *HR1* domain in spike protein inducing conformational changes restricting the entry of SARS-CoV-2 [66]. Several natural products such as glycyrrhizin, bicylogermecrene, tryptanthrine, β-sitosterol, indirubin, indican, indigo, hesperetin, crysophanic acid, rhein, berberine, and β-caryophyllene evaluated by *in-silico* approaches have been established as the powerful antivirals against SARS-CoV-2 [67]. They all hinder the activity of SARS-CoV-2 Mpro. However, more intensified studies are needed to investigate the plants for the presence of these natural products follwed by their *in-vivo* studies. Many *in-silico* studies revealed the phytochemicals targeting spike protein [68]. Naturally occurring fisetin, quercetin, and kamferol tend to bind with the hACE2-S complex and collapse the binding of hACE2-S complex [68]. They showed all the properties ascertained by Lipinski's rule, hence holds bright future in the drug development for COVID-19. Phytoconstituents, tenufolin, and pavetannin C1 from *Cinnamon* have shown superb binding affinity with main protease and spike proteins [69]. Traditionally, cinnamon is used as a remedy for many diseases and disorders of humans because of its excellent bioactivities. An *in-silico* study conceded several phytochemicals like nafamostat, meloxicam, ganodermanontriol, columbin, myricetin, proanthocyanidin A2, jatrorrhizine, and baicalein having strong binding affinity toward TMPRSS2 enzyme [70]. Another *in-silico* study provided several flavonoid compounds such as albireodelphin, apigenin 7-(600-malonylglucoside), cyanidin-3-(p-cou maroyl)-rutinoside-5-glucoside, delphinidin 3-O-β-Dglucoside 5-O-(6-coumaroyl-β-D-glucoside), and (-)-Maackiain-3-O-glucosyl-600-O-malonate, which demonstrated highest binding energy values against the Mpro, RdRp, and S-proteins [71]. Istiflia et al. [72] reported 23 different flavonoids, which can bind with the RBD of the spike glyco-protein of SARS-CoV-2 and cellular protease TMPRSS2, cathepsin B and L (CatB/L). How-ever, the (-)-epicatechin gallate interacted strongly with all the proteins and fulfilled Lipinski's rule of five. The (-)-epicatechin can be used as a lead molecule for drug development against Covid-19.

In recent times, combination of network pharmacology and molecular docking have emerged as potential methods for evaluating medicinal and biological prospects of medicinal plants mitigating COVID-19 [73]. A large-scale screening of the medicinal plants involving 16,500 constituents revealed their ability to inhibit 3Clpro, PLpro, and RdRp. These proteins are involved in RNA synthesis and replication and modulating host's immunity through pro-duction of virulence factors. The study identified some phytochemicals like rocymosin B, verbascoside, rutin, caftaric acid, luteolin 7-rutinoside, fenugreekine, and cyanidin 3-(600-malonylglucoside) as the promising anti-SARS-CoV2 molecules. Several medicinal plants, *Glycyrrhiza glabra, Hibiscus sabdariffa, Cichorium intybus, Chrysanthemum coronarium, Nigella sativa, Anastatica hierochuntica, Euphorbia* species, *Psidium guajava,* and *Epilobium hirsutum* accumulating quercetin, ursolic acid, kaempferol, isorhamnetin, luteolin, glycerrhizin, and apigenin attack multitargets proteins prostaglandin-endoperoxide synthase 2 (PTGS2), interleukin-2 (IL2), interleukin 1 beta also known as leukocytic pyrogen (IL1b), vascular cell adhesion molecule 1 (VCAM1), and tumor necrosis factor and pathways includ-ing cytokine-cytokine receptor interaction, tumor necrosis factor signaling, the nucleotide-binding oligomerization domain-like receptor signaling, toll-like receptor signaling, nu-clear factor kappa B signaling, and Janus kinases (JAKs), signal transducer, and activator of transcription proteins (JAK-STAT3) signaling pathways, all of them are closely related to inflammatory, innate, and adaptive immune responses [74]. Thus, *G. glabra* (liquorice) could be a valuable plant in the management of the COVID-19. The plant has long been used in the treatment of several viral infections. Other plants, like *H. sabdariffa* and *C. intybus* having ample caffeic acid content also exhibited antiviral properties. In addition, combina-tions of plants, such as *G. glabra, H. sabdariffa,* and *C. intybus* which exhibit synergism have been found more beneficial in recovering from Covid-19. Two bioactive ingredients, withanoside V and somniferine, from *W. somnifera,* one, tinocordiside from *T. cordifolia* and three, vicenin, isorientin 40-O-glucoside 200-O-p-hydroxybenzoagte and ursolic acid from *O. sanctum* hit SARS-CoV-2 Mpro protein as revealed by the molecular docking studies [75]. This reflected that molecular docking studies are paving new ways for the development of novel drugs/therapeutics for the treatment of COVID-19, which has become a global conta-gion. Curcumin, a traditional spice component possess several pharmacological properties selectively, anti-inflammatory has come out as one of the promising antiviral agent against SARS-CoV-2. It modulate the processes involved in cellular entry, replication, and molecular cascade manifesting pathophysiological consequences of SARS-CoV-2 [78]. Given that curcumin has promising therapeutic potential in controlling COVID-19, researchers have been insisting on its clinical trials [76,77]. Swargiarya et al. [78] have screened 32 phyto-compounds using *in-silico* tools, out of which amentoflavone and gallocatechin gallate displayed outstanding binding affinity to 3CLpro and PLpro. In addition, savinin, theaflavin-3,3-digallate, and kazinol-A showed strong affinity to these protein targets. Thus, amentoflavone and gallocatechin gallate may be useful as effective drug candidates for COVID-19. The limonoids and triterpenoids have been evaluated for their binding affinity with five major protein targets 3CLpro, PLpro, SGp-RBD (spike glycoprotein-receptor binding domain), RdRp, and ACE2 using *in-silico* methods [79]. Four of them, namely, 7- deacetyl-7-benzoylgedunin, glycyrrhizic acid, limonin, and obacunone have been found to bound with main protease 3CL-pro, while four others, obacunone, glycyrrhizic acid, ursolic acid, and 7-deacetylgedunin, with PLpro. Further, six constituents, glycyrrhizic acid, limonin, 7-deacetyl-7-benzoylgedunin, limonin glucoside, 7-deacetylgedunin, and obacunone, bonded with RdRp. Phytochemicals like maslinic acid, glycyrrhizic acid, corosolic acid, 2-hydroxyseneganolide, oleanane, and gedunin bonded firmly through multiple noncovalent interactions. Seven phytochemicals, glycyrrhizinic acid, maslinic acid, obacunone, epoxyazadiradione, azadiradionolide, urso-lic acid, and gedunin binds, at the catalytic site and/or the RBD site of Mpro. Based on the docking score and given medicinal properties, they proposed the combination of seven phy-toconstituents 7-deacetyl-7-benzoylgedunin, glycyrrhizic acid, limonin, obacunone, ursolic acid, corosolic acid, and masilinic acid most likely be advantageous against SARS-CoV-2. The glycyrrhizic acid comes out to be a promising molecule because it can bind with all the five protein targets of SARS-CoV-2. Similarly, a number of flavonoids and a sub-group indole-chalcones included in the diets and traditional medicines have been found efficacious in recovery of many diseases [80]. Vijaykuamran et al. [80] have studied 23 naturally occur-ring flavonoids and 25 synthetic indole-chalcones using molecular docking and dynamics. In total, 30 of the total 48 had an ability to suppress SARS-CoV-2 by inhibiting functionality of RdRp, Mpro, and Spike (S) protein. The cyanidin has been identified to block the RdRp function and quercetin chocked interaction sites on the viral spike protein. The quercetin is a well-known antiviral agent for dengue and influenza virus. It strongly binds with Mpro of SARS-CoV-2 through interactions with Glu290 and Asp289, and receptor binding do-main of the viral spike protein as revealed by computational analysis. Like the curcumin, quercetin may have tremendous scope and future in the management of COVID-19. Concur-rently, limnoids, terpenoids, flavonoids, and sub-group indole chalcones may be promoted for the development of therapeutics to control SARS-CoV-2. Apart, some phytochemicals inhibit human proteases TMPRSS2 and cathepsin L, which play a key role in host cell en-try of SARS-CoV-2 [81]. TMPRSS2 is a transmembrane serine protease that cleaves both ACE2 and the S protein. Thus, such phytochemicals can be used to suppress SARS-CoV-2 infection and it may be another effective strategy to control SARS-CoV-2 infection. In the light of these facts, Vivek-Ananth et al. [81] have performed large-scale virtual screening of 14,011 phytochemicals from Indian medicinal plants to identify phytochemicals, which inhibit TMPRSS2 and cathepsin L. From 14,011, 96 phytoconstituents identified to inhibit

TMPRSS2 and 9 to cathepsin L. Three phytochemicals qingdainone, edgeworoside C and adlumidine were identified as the potent inhibitor of TMPRSS2, and three ararobinol, (+)-oxoturkiyenine and 3,17-cinchophylline of cathepsin L based on MD simulations analysis. Most of the phytochemicals identified owned antiviral and anti-inflammatory properties and used in traditional medicines. These findings will speed up the ongoing researches on the drug development targeting the TMPRSS2 and cathepsin L for COVID-19. Xu et al. [82] have studied anti-SARS-CoV-2 bioactives from food sources, red wine, Chinese hawthorn, and blackberry, which can block the activity of 3CLpro and ACE2. They have attributed these activities to polyphenols, quercetin, luteolin, and isorhamnetin present in red wine, Chinese hawthorn, and blackberry. We can recommend these foods as preventive supplements, because they contain anti-SARS-CoV polyphenols. Yepes-Perez et al. [83] evaluated *Uncaria tomentosa* (Cat's Claw) phytoconstituents for inhibitory effects against the main protease 3CLpro of SARS-CoV-2. They predicted three bioactive compounds namely, speciophylline, cadambine, and proanthocyanidin B2 with a strong affinity toward 3CLpro using the computational tools. Thus, Cat's claw may be a promising herb to control COVID-19. Similarly, phytochemicals from Saudi medicinal plants are reported to inhibit RdRp, 3CLpro, and PLpro [84,85]. Many more natural bioactive compounds inhibiting RdRp, 3CLpro, and PLpro have been identified using *in-silico* approaches [54,86,87]. Theaflavin documented in Chinese traditional medicines, displayed strong antiviral activity against SARS-CoV-2 by inhibiting the RdRp [88]. Zhang et al. [88] screened 115 compounds commonly used in viral respiratory infection in China for their activities against SARS-CoV-2. A group of researchers have also investigated marine natural products by virtual screening and molecular modeling resulting seventeen molecules with potent inhibitory action against 3CLpro [89].

22.6 Conclusion

Plants are the reservoir of uncountable secondary metabolites with immense bioactive properties against viral and bacterial infections, diseases, and ailments. After the onset of the COVID-19, a great deal of *in-silico* research work has been carried out investigating antiviral activity of many plants to combat SARS-CoV-2. *In-silico* studies in combination of molecular docking and dynamic simulations are facilitating prediction of the potential anti-SARS-CoV2 bioactive phytochemicals. In the past two years, *in-silico* studies have delivered many promising anti-SARS-CoV-2 phytochemicals, however, all of them needed to be validated by *in-vivo* studies for further use in drug development. Without the *in-vivo* experimental data, it would not be possible to use the bioactive compounds of *in-silico* studies for drug or therapeutic develoment for COVID-19. However, a plenty of herbal medicines and decoctions mentioned in the traditional medicinal systems have been recommended as the first line of protection against SARS-CoV-2 infection. At present, several traditional herbal mixtures and ayurvedic preparations available in the market claim to be immune modulatory to combat SARS-CoV-2. They showed beneficial effects in COVID-19 patients by improving their health status when given according to the stages of the disease. So, many experts have recommended them to be include in the list of medinces for the treatment and prevention of COVID-19. Several medicines, like the advanced antivirals, lopinavir, remdesivir, and arbidol have been recommended for COVID-19 patients, but in the advanced stages of the disease. In addition, hydroxychloroquine (an antimalarial drug), choloroquine phosphate were also found effective; however, no substantial evidences are available to support their effectiveness in the recovery from COVID-19. Now, several effective COVID-19 vaccines are available. Most of the vaccines are based on the target S protein, papain-like cysteine protease (PLpro) or Chymotrypsin like nprotease (3CLpro or Mpro), and RNA-dependent RNA polymerase (RdRp) of SARS-CoV-2 as they play essential roles in survival and infection. However, all COVID-19 vaccines have pros and cons mainly, side-effects, efficacy, and acceptability by all age people and those with some allergies and other health related complications. In view of these, plant derived phytochemicals are still highly desirable for the development of novel drug for the coronavirus. We need to go hand-in hand with plant-natural product-based therapies for coronavirus as they tend to rapidly undergo mutation producing new variants, which may not get recognized by an immune response induced by current COVID-19 vaccines. Already, new mutant variants of SARS-CoV-2 have been reported and the effectiveness of the current vaccines is yet to be tested against them. Thus, it is worthwhile to continue drug discovery from plant natural products. In this chapter, we have highlighted importance of *in-silico* tools in searching for phytochemicals with anti-SARS-CoV2 properties. However, researchers will need to conduct extensive *in-vitro* and *in-vivo* experimental studies to validate all anti-SARS-CoV2 phytochemicals. Nevertheless, the drug discovery process and development of effective drug and therapeutic agents for SARS-CoV-2 will continue as long as the virus conitnue mutatition in the genome and we find the best treatment for them. This chapter provided valuable information to researchers engaged in the drug discovery program for COVID-19.

Contribution

Deepak Ganjewala: Formulated the chapter, the main conceptual ideas and proof outline. Wrote the manuscript.

Hina Bansal: Carried out all computer-based analyses and numerical calculations.

Ruchika Mittal and Gauri Srivastava are research scholars who helped collecting and organizing reports.

Acknowledgment

Authors would like to thank Dr. Ashok K. Chauhan, Founder and Dr. Atul Chauhan, Chancellor, Amity University Uttar Pradesh, Noida, for providing facilities.

References

[1] LT Ho, KK Chan, VC Chung, TH Leung, Highlights of traditional Chinese medicine frontline expert advice in the China national guideline for COVID-19, Eur J Integr Med 36 (2020) 101116.

[2] N Chen, M Zhou, X Dong, J Qu, F Gong, Y Han, et al., Epidemiological and clinical characteristics of 99 cases of 2019 novel coronavirus pneumonia in Wuhan, China: a descriptive study, Lancet North Am Ed 395 (10223) (2020) 507–513.

[3] Y Chen, Q Liu, D Guo, Emerging coronaviruses: genome structure, replication, and pathogenesis, J Med Virol 92 (4) (2020) 418–423.

[4] M Wang, R Cao, L Zhang, X Yang, J Liu, M Xu, et al., Remdesivir and chloroquine effectively inhibit the recently emerged novel coronavirus (2019-nCoV) in vitro, Cell Res 30 (3) (2020) 269–271.

[5] P Zhou, XL Yang, XG Wang, B Hu, L Zhang, W Zhang, et al., A pneumonia outbreak associated with a new coronavirus of probable bat origin, Nature 579 (7798) (2020) 270–273.

[6] N Zhu, D Zhang, W Wang, X Li, B Yang, J Song, et al., A novel coronavirus from patients with pneumonia in China, 2019, N Engl J Med 382 (8) (2020) 723–733.

[7] Y Zhu, J Xie, F Huang, L Cao, Association between short-term exposure to air pollution and COVID-19 infection: evidence from China, Sci Total Environ 727 (2020) 138704.

[8] VJ Munster, M Koopmans, N van Doremalen, D van Riel, E de Wit, A novel coronavirus emerging in China—key questions for impact assessment, N Engl J Med 382 (8) (2020) 692–694.

[9] AJ Akindele, FO Agunbiade, MO Sofidiya, O Awodele, A Sowemimo, O Ade-Ademilua, et al., COVID-19 pandemic: a case for phytomedicines, Natural Product Communications 15 (8) (2020) 1–9 1934578X20945086.

[10] MA Alamri, A Altharawi, AB Alabbas, MA Alossaimi, SM Alqahtani, Structure-based virtual screening and molecular dynamics of phytochemicals derived from Saudi medicinal plants to identify potential COVID-19 therapeutics, Arabian J Chem 13 (9) (2020) 7224–7234.

[11] B Benarba, A Pandiella, Medicinal plants as sources of active molecules against COVID-19, Frontiers in Pharmacology 11 (2020) 1189.

[12] MN Boukhatem, WN Setzer, Aromatic herbs, medicinal plant-derived essential oils, and phytochemical extracts as potential therapies for coronaviruses: future perspectives, Plants 9 (6) (2020) 800.

[13] P Khare, U Sahu, SC Pandey, M Samant, Current approaches for target-specific drug discovery using natural compounds against SARS-CoV-2 infection, Virus Res 290 (2020) 198169.

[14] JS Mani, JB Johnson, JC Steel, DA Broszczak, PM Neilsen, KB Walsh, et al., Natural product-derived phytochemicals as potential agents against coronaviruses: a review, Virus Res 284 (2020) 197989.

[15] YM Baez-Santos, SE John, AD Mesecar, The SARS-coronavirus papain-like protease: structure, function and inhibition by designed antiviral compounds, Antiviral Res 115 (2015) 21–38.

[16] R Ramajayam, KP Tan, PH Liang, Recent development of 3C and 3CL protease inhibitors for anti-coronavirus and anti-picornavirus drug discovery, Biochem Soc Trans 39 (5) (2011) 1371–1375.

[17] Z Ren, L Yan, N Zhang, Y Guo, C Yang, Z Lou, et al., The newly emerged SARS-like coronavirus HCoV-EMC also has an "Achilles' heel": current effective inhibitor targeting a 3Clike protease, Protein Cell 4 (4) (2013) 248.

[18] LT Lin, WC Hsu, CC Lin, Antiviral natural products and herbal medicines, J Tradit Complement Med 4 (1) (2014) 24–35.

[19] J Pang, MX Wang, IYH Ang, SHX Tan, RF Lewis, JI Chen, et al., Potential rapid diagnostics, vaccine and therapeutics for 2019 novel Coronavirus (2019-ncoV): a systematic review, J Clin Med 9 (3) (2020) 623.

[20] R Lu, X Zhao, J Li, P Niu, B Yang, H Wu, et al., Genomic characterization and epidemiology of 2019 novel coronavirus: implications for virus origins and receptor binding, Lancet 395 (10224) (2020) 565–574.

[21] SD Richman, K Southward, P Chambers, D Cross, J Barrett, G Hemmings, et al., HER2 overexpression and amplification as a potential therapeutic target in colorectal cancer: analysis of 3256 patients enrolled in the QUASAR, FOCUS and PICCOLO colorectal cancer trials, J Pathol 238 (4) (2020) 562–570.

[22] F Wu, S Zhao, B Yu, YM Chen, W Wang, ZG Song, et al., A new coronavirus associated with human respiratory disease in China, Nature 579 (7798) (2020) 265–269.

[23] UJ Kim, EJ Won, SJ Kee, SI Jung, HC Jang, Combination therapy with lopinavir/ritonavir, ribavirin and interferon-alpha for Middle East respiratory syndrome, Antiviral therapy 21 (2016) 455–459.

[24] Y Jin, H Yang, W Ji, W Wu, S Chen, W Zhang, et al., Virology, epidemiology, pathogenesis, and control of COVID-19, Viruses 12 (4) (2020) 372. https://doi.org/10.3390/v12040372.

[25] S Kumar, R Nyodu, V K Maurya, SK Saxena, Morphology, genome organization, replication, and pathogenesis of severe acute respiratory syndrome coronavirus 2 (SARS-CoV-2). Coronavirus Disease 2019 (COVID-19), Epidemiology, Pathogenesis, Diagnosis, and Therapeutics (2020) 23–31 https://doi.org/ 10.1007/978-981-15-4814-7_3. [Accessed: April 30, 2020].

[26] KP Hui, MC Cheung, RA Perera, KC Ng, CH Bui, et al., Tropism, replication competence, and innate immune responses of the coronavirus SARS-CoV-2 in human respiratory tract and conjunctiva: an analysis in ex-vivo and in-vitro cultures, The Lancet Respiratory Medicine 8 (7) (2020) 687–695.

[27] DA Schwartz, AL Graham, Potential maternal and infant outcomes from (Wuhan) coronavirus 2019-nCoV infecting pregnant women: lessons from SARS, MERS, and other human coronavirus infections, Viruses 12 (2) (2020) 194.

[28] A Jimenez-Alberto, RM Ribas-Aparicio, G Aparicio-Ozores, JA Castelan-Vega, Virtual screening of approved drugs as potential SARS-CoV-2 main protease inhibitors, Comput Biol Chem 88 (2020) 107325.

[29] DH Zhang, KL Wu, X Zhang, SQ Deng, B Peng, In silico screening of Chinese herbal medicines with the potential to directly inhibit 2019 novel coronavirus, J Integr Med 18 (2) (2020) 152–158.

[30] W Liu, QI Zhang, J Chen, R Xiang, H Song, S Shu, et al., Detection of COVID-19 in children in early January 2020 in Wuhan, China, N Engl J Med 382 (14) (2020) 1370–1371.

[31] NP Laksmiani, LP Larasanty, AA Santika, PA Prayoga, AA Dewi, NP Dewi, Active compounds fctivity from the medicinal plants Against SARS-CoV-2 using in Silico assay, Biomed Pharmacol J 13 (2) (2020) 873–881.

[32] A Basu, A Sarkar, U Maulik, Molecular docking study of potential phytochemicals and their effects on the complex of SARS-CoV2 spike protein and human ACE2, Sci Rep 10 (1) (2020) 1–5.

[33] YF Tu, CS Chien, AA Yarmishyn, YY Lin, YH Luo, YT Lin, et al., A review of SARS-CoV-2 and the ongoing clinical trials, Int J Mol Sci 21 (7) (2020) 2657.

[34] YR Guo, QD Cao, ZS Hong, YY Tan, SD Chen, HJ Jin, et al., The origin, transmission and clinical therapies on coronavirus disease 2019 (COVID-19) outbreak—an update on the status, Military Medical Research 7 (2020) 11.

[35] J Xu, L Gao, H Liang, SD Chen, In silico screening of potential anti-COVID-19 bioactive natural constituents from food sources by molecular docking, Nutrition 82 (2021) 111049.

[36] H Chen, Q Du, Potential natural compounds for preventing SARS-CoV-2 (2019-nCoV) infection, Preprints (2020) 2020010358, doi:10.20944/preprints202001.0358.v3.

[37] RK Ganjhu, PP Mudgal, H Maity, D Dowarha, S Devadiga, S Nag, et al., Herbal plants and plant preparations as remedial approach for viral diseases, Virus Disease 26 (4) (2015) 225–236.

[38] R Islam, MR Parves, AS Paul, N Uddin, MS Rahman, AA Mamun, et al., A molecular modeling approach to identify effective antiviral phytochemicals against the main protease of SARS-CoV-2, J Biomol Struct Dyn 7 (2020) 1–2.

[39] S Jo, S Kim, DH Shin, MS Kim, Inhibition of SARS-CoV 3CL protease by flavonoids, J Enzyme Inhib Med Chem 35 (1) (2020) 145–151.

[40] A Frediansyah, R Tiwari, K Sharun, K Dhama, H Harapan, Antivirals for COVID-19: a critical review, Clinical Epidemiology and Global Health 9 (2020) 90–98.

[41] A Al-Shawi, M Hameed, Perspective study of exploring some medicinal plants to manage the pandemic COVID-19, European Journal of Medical and Health Science 2 (4) (2020) 1–5.

[42] S Mhatre, T Srivastava, S Naik, V Patravale, Antiviral activity of green tea and black tea polyphenols in prophylaxis and treatment of COVID-19: a review, Phytomedicine 17 (2020) 153286.

[43] P Chowdhury, *In silico* investigation of phytoconstituents from Indian medicinal herb 'Tinospora cordifolia (giloy)' against SARS-CoV-2 (COVID-19) by molecular dynamics approach, J Biomol Struct Dyn 6 (2020) 1–8.

[44] P Das, R Majumder, M Mandal, P Basak, *In-Silico* approach for identification of effective and stable inhibitors for COVID-19 main protease (Mpro) from flavonoid based phytochemical constituents of Calendula officinalis, J Biomol Struct Dyn 23 (2020) 1–6.

[45] Y W Chen, C-P B Yiu, K-Y Wong, Prediction of the SARSCoV-2 (2019-nCoV) 3C-like protease (3CL-pro) structure: virtual screening reveals velpatasvir, ledipasvir, and other drug repurposing candidates, F1000Research 9 (2020) 129.

[46] J Lei, Y Kusov, R Hilgenfeld, Nsp3 of coronaviruses: structures and functions of a large multi-domain protein, Antiviral Res 149 (2018) 58–74.

[47] L Yuan, Z Chen, S Song, S Wang, C Tian, G Xing, et al., p53 degradation by a coronavirus papain-like protease suppresses type I interferon signaling, J Biol Chem 290 (5) (2015) 3172–3182.

[48] MT Ul Qamar, F Shahid, S Aslam, UA Ashfaq, S Aslam, I Fatima, et al., Reverse vaccinology assisted designing of multiepitope-based subunit vaccine against SARS-CoV-2, Infectious Diseases of Poverty 9 (1) (2020) 1–4.

[49] ML Holshue, C DeBolt, S Lindquist, KH Lofy, J Wiesman, et al., First case of 2019 novel coronavirus in the United States, N Engl J Med 382 (2019) 926–936.

[50] X Yao, F Ye, M Zhang, C Cui, B Huang, P Niu, et al., In vitro antiviral activity and projection of optimized dosing design of hydroxychloroquine for the treatment of severe acute respiratory syndrome coronavirus 2 (SARS-CoV-2), Clin Infect Dis 28 (15) (2020) 732–739 71.

[51] RA Khamitov, S Loginova, VN Shchukina, et al., Antiviral activity of arbidol and its derivatives against the pathogen of severe acute respiratory syndrome in the cell cultures, Vopr Virusology 53 (2008) 9–13.

[52] W Li, C Zhang, J Sui, JH Kuhn, MJ Moore, S Luo, et al., Receptor and viral determinants of SARS-coronavirus adaptation to human ACE2, EMBO J 24 (8) (2005) 1634–1643.

[53] ZM Wang, XX Zhu, XL Cui, AH Liang, GH Du, JX Ruan, Screening of traditional Chinese remedies for SARS treatment, Zhongguo Zhong Yao ZaZhi 28 (6) (2003) 484–487.

[54] I Aanouz, A Belhassan, K El-Khatabi, T Lakhlifi, M El-Ldrissi, M Bouachrine, Moroccan medicinal plants as inhibitors against SARS-CoV-2 main protease: Computational investigations, J Biomol Struct Dyn 5 (2020) 1–9.

[55] LS Borquaye, EN Gasu, GB Ampomah, LK Kyei, MA Amarh, CN Mensah, et al., Alkaloids from *Cryptolepis sanguinolenta* as potential inhibitors of SARS-CoV-2 viral proteins: an in silico study, Biomed Res Int 2020 (2020) 5324560.

[56] RV Chikhale, SK Sinha, RB Patil, SK Prasad, A Shakya, et al., *In-silico* investigation of phytochemicals from *Asparagus racemosus* as plausible antiviral agent in COVID-19, J Biomol Struct Dyn 23 (2020) 1–5.

[57] K Chojnacka, A Witek-Krowiak, D Skrzypczak, K Mikula, P Młynarz, Phytochemicals containing biologically active polyphenols as an effective agent against COVID-19-inducing coronavirus, J Funct Foods 30 (2020) 104146.

[58] A Rakib, A Paul, M Chy, N Uddin, SA Sami, SK Baral, et al., Biochemical and computational approach of selected phytocompounds from *Tinospora crispa* in the management of COVID-19, Molecules 25 (17) (2020) 3936.

[59] A Elmi, S Al Jawad Sayem, M Ahmed, F Mohamed, Natural compounds from Djiboutian medicinal plants as inbitors of COVID-19 by in *Silico* Investigations, Chem Rxiv (2020) https://doi.org/10.26434/chemrxiv.12325844.v1. [Accessed: July 04, 2020].

[60] R Ghosh, A Chakraborty, A Biswas, S Chowdhuri, Identification of alkaloids from *Justicia adhatoda* as potent SARS CoV-2 main protease inhibitors: an in silico perspective, J Mol Struct 1229 (2021) 129489.

[61] P Kar, NR Sharma, B Singh, A Sen, A Roy, Natural compounds from *Clerodendrum* spp. as possible therapeutic candidates against SARS-CoV-2: an in silico investigation, J Biomol Struct Dyn 18 (2020) 1–2.

[62] S A Kulkarni, SK Nagarajan, V Ramesh, V Palaniyandi, SP Selvam, T Madhavan, Computational evaluation of major components from plant essential oils as potent inhibitors of SARS-CoV-2 spike protein, J Mol Struct 1221 (2020) 128823.

[63] S Kumar, P Kashyap, S Chowdhury, S Kumar, A Panwar, A Kumar, Identification of phytochemicals as potential therapeutic agents that binds to Nsp15 protein target of coronavirus (SARS-CoV-2) that are capable of inhibiting virus replication, Phytomedicine 85 (2021) 153317.

[64] S Thakur, B Sarkar, AJ Ansari, A Khandelwal, A Arya, R Poduri, et al., Exploring the magic bullets to identify Achilles' heel in SARS-CoV-2: delving deeper into the sea of possible therapeutic options in COVID-19 disease: an update, Food Chemistry Toxicology 147 (2020) (2020) 111887.

[65] ZT Muhseen, AR Hameed, HM Al-Hasani, MT Ul Qamar, G Li, Promising terpenes as SARS-CoV-2 spike receptor-binding domain (RBD) attachment inhibitors to the human ACE2 receptor: integrated computational approach, J Mol Liq 320 (2020) 114493.

[66] MK Nadeem, *In-silico* study to elucidate corona virus by plant Phytoderivatives that hits as a fusion inhibitor targeting *HR1* domain in spike protein which conformational changes efficiently inhibit entry COVID-19, Translational Biomedicine 11 (2020). http://doi.org/10.36648/2172-0479.11.3.1.

[67] RR Narkhede, AV Pise, RS Cheke, SD Shinde, Recognition of natural products as potential inhibitors of COVID-19 main protease (Mpro): in-silico evidences, Natural Products and Bioprospecting 10 (5) (2020) 297–306.

[68] P Pandey, JS Rane, A Chatterjee, A Kumar, R Khan, A Prakash, et al., Targeting SARS-CoV-2 spike protein of COVID-19 with naturally occurring phytochemicals: an in silico study for drug development, J Biomol Struct Dyn 22 (2020) 1.

[69] DS Prasanth, M Murahari, V Chandramohan, SP Panda, LR Atmakuri, C Guntupalli, In silico identification of potential inhibitors from *Cinnamon* against main protease and spike glycoprotein of SARS CoV-2, J Biomol Struct Dyn 19 (2020) 1–5.

[70] M Pooja, GJ Reddy, K Hema, S Dodoala, B Koganti, Unravelling high-affinity binding compounds towards transmembrane protease serine 2 enzyme in treating SARS-CoV-2 infection using molecular modelling and docking studies, Eur J Pharmacol 890 (2021) 173688.

[71] RM Rmeshkumar, P Indu, N Arunagirinathan, B Venkatadri, HA El-Serehy, A Ahmad, Computational selection of flavonoid compounds as inhibitors against SARS-CoV-2 main protease, RNA-dependent RNA polymerase and spike proteins: a molecular docking study, Saudi Journal of Bio Sciences 28 (1) (2021) 448–458.

[72] ES Istifli, PA Netz, A Sihoglu, Tepe, MT Husunet, C Sarikurkcu, B Tepe, *In silico* analysis of the interactions of certain flavonoids with the receptor-binding domain of 2019 novel coronavirus and cellular proteases and their pharmacokinetic properties, J Biomol Struct Dyn 26 (2020) 1–5.

[73] E Shawky, AA Nada, RS Ibrahim, Potential role of medicinal plants and their constituents in the mitigation of SARS-CoV-2: identifying related therapeutic targets using network pharmacology and molecular docking analyses, RSC Adv 10 (47) (2020) 27961–27983.

[74] P Shree, P Mishra, C Selvaraj, SK Singh, R Chaube, N Garg, et al., Targeting COVID-19 (SARS-CoV-2) main protease through active phytochemicals of ayurvedic medicinal plants: *Withaniasomnifera* (Ashwagandha), *Tinosporacordifolia* (Giloy) and *Ocimum sanctum* (Tulsi): a molecular docking study, J Biomol Struct Dyn 26 (2020) 1–4.

[75] VK Soni, A Mehta, YK Ratre, AK Tiwari, A Amit, RP Singh, et al., Curcumin, a traditional spice component, can hold the promise against COVID-19, Eur J Pharmacol 12 (2020) 173551.

[76] PL Hooper, COVID-19 and heme oxygenase: novel insight into the disease and potential therapies, Cell Stress and Chaperons (2020) 1–4. http://doi.org/10.1007/s12192-020-01126-9. Advance online publication.

[77] RI Horowitz, PR Freeman, J Bruzzese, Efficacy of glutathione therapy in relieving dyspnea associated with COVID-19 pneumonia: a report of 2 cases, Respiratory Medicine Case Reports 30 (2020) 101063.

[78] A Swargiary, S Mahmud, MA Saleh, Screening of phytochemicals as potent inhibitor of 3-chymotrypsin and papain-like proteases of SARS-CoV2: an in silico approach to combat COVID-19, J Biomol Struct Dyn 1 (2020) 1–5.

[79]   S Vardhan, SK Sahoo, In silico ADMET and molecular docking study on searching potential inhibitors from limonoids and triterpenoids for COVID-19, Comput Biol Med 124 (2020) 103936.

[80]   BG Vijayakumar, D Ramesh, A Joji, T Kannan, In silico pharmacokinetic and molecular docking studies of natural flavonoids and synthetic indole chalcones against essential proteins of SARS-CoV-2, Eur J Pharmacol 886 (2020) 173448.

[81]   RP Vivek-Ananth, A Rana, N Rajan, HS Biswal, A Samal, In silico identification of potential natural product inhibitors of human proteases key to SARS-CoV-2 infection, Molecules 25 (17) (2020) 3822.

[82]   J Xu, L Gao, H Liang, SD Chen, In silico screening of potential anti-COVID-19 bioactive natural constituents from food sources by molecular docking, Nutrition 82 (2021) 111049.

[83]   AF Yepes-Perez, O Herrera-Calderon, JE Sanchez-Aparicio, L Tiessler-Sala, JD Marechal, W Cardona-G, Investigating potential inhibitory effect of *Uncariatomentosa* (Cat's Claw) against the main protease 3CL-pro of SARS-CoV-2 by molecular modeling, Evidence-Based Complementary and Alternative Medicine 2020 (2020) ID4932572.

[84]   H Aati, A El-Gamal, H Shaheen, O Kayser, Traditional use of ethnomedicinal native plants in the Kingdom of Saudi Arabia, J Ethnobiol Ethnomed 15 (1) (2019) 1–9.

[85]   AH Arbab, MK Parvez, MS AlDosari, AJ AlRehaily, In vitro evaluation of novel antiviral activities of 60 medicinal plants extracts against hepatitis B virus, Experimental and Therapeutic Medicine 14 (1) (2017) 626–634.

[86]   W Zhang, Y Huai, Z Miao, A Qian, Y Wang, Systems pharmacology for investigation of the mechanisms of action of traditional Chinese medicine in drug discovery, Frontier Pharmacology 10 (2019) 743.

[87]   J Lung, YS Lin, YH Yang, YL Chou, LH Shu, YC Cheng, et al., The potential chemical structure of anti-SARS-CoV-2 RNA-dependent RNA polymerase, J Med Virol 92 (10) (2020) 2248. http://doi.org/10.1002/jmv.26176.

[88]   DH Zhang, KL Wu, X Zhang, SQ Deng, B Peng, In silico screening of Chinese herbal medicines with the potential to directly inhibit 2019 novel coronavirus, Journal of Integrative Medicine 18 (2) (2020) 152–158.

[89]   D Gentile, V Patamia, A Scala, MT Sciortino, A Piperno, A Rescifina, Putative Inhibitors of SARS-CoV-2 main protease from a library of marine natural products: a virtual screening and molecular modeling study, Marine Drugs *18* (2020) 225. https://doi.org/10.3390/md18040225.

[90]   L Du, Y He, Y Zhou, S Liu, BJ Zheng, S Jiang, The spike protein of SARS-CoV-a target for vaccine and therapeutic development, Nat Rev Microbiol 7 (3) (2009) 226–236.

[91]   V Thiel, KA Ivanov, A Putics, T Hertzig, B Schelle, S Bayer, et al., Mechanisms and enzymes involved in SARS coronavirus genome expression, J Gen Virol 84 (9) (2020) 2305–2315.

[92]   J Ziebuhr, Molecular biology of severe acute respiratory syndrome coronavirus, Curr Opin Microbiol 7 (4) (2004) 412–419.

[93]   D Needle, GT Lountos, DS Waugh, Structures of the Middle East respiratory syndrome coronavirus 3C-like protease reveal insights into substrate specificity, Acta Crystallography D Biology Crystallography 71 (2015) 1102–1111.

[94]   A Zumla, J FW Chan, EI Azhar, DSC Hui, KY Yuen, Coronaviruses: drug discovery and therapeutic options, Nat Rev Drug Discovery 15 (5) (2016) 327–347.

[95]   G Kiran, L Karthik, MS Shree Devi, P Sathiyarajeswaran, K Kanakavalli, KM Kumar, et al., In silico computational screening of Kabasura Kudineer-official Siddha formulation and JACOM-Novel Herbal Coded Formulation Against SARS-CoV-2 Spike protein, J. Ayurveda Integr. Med (2020) I, doi:10.1016/j.jaim.2020.05.009.

CHAPTER 23

Ethnopharmacological reports on herbs used in the management of tuberculosis

Pragya Pandey[a], Rajendra Awasthi[a], Neerupma Dhiman[a], Bhupesh Sharma[a] and Giriraj T. Kulkarni[b]

[a]Amity Institute of Pharmacy, Amity University Uttar Pradesh, Noida, Uttar Pradesh, India
[b]Gokaraju Rangaraju College of Pharmacy, Hyderabad, Telangana, India

23.1 Introduction

Tuberculosis (TB) is caused by *Mycobacterium tuberculosis* (MTB), which has infected about 1.7 billion of the world population. India is one of the leading countries in the case of active TB. World Health Organization (WHO) estimated about 10 million of TB cases worldwide in a year. About 64%, new cases were accounted in China, Indonesia, India, Pakistan, The Philippines, Nigeria, and South Africa [1]. US$ 9.2 billion was estimated funding to be invested in TB epidemic [2].

TB can be easily diagnosed by Mantoux test or Mendel–Mantoux test, sputum microscopy, Genexpert test, etc. TB can be treated using combination therapy for at least 6 months in two phases (1) 2 months intensive first phase—treatment with isoniazid, rifampicin, ethambutol, and pyrazinamide followed by (2) a 4 months continuation phase second—treatment with rifampicin and isoniazid [3].

Drug resistance in TB can be classified as multidrug resistant TB (MDR TB), and extensively drug-resistant TB (XDR TB). MDR TB indicates resistant to both rifampicin and isoniazid. MDR TB can occur when the patient is noncompliant toward the regimen or using the wrong prescription. To overcome this problem, WHO recommended prescribed by shorter regimen. XDR TB is the condition when patient resists to rifampicin, isonizid, fluoroquinolones, and also at least one of the injectable anti-TB drugs like kanamycin, amikacin, etc [3].

The TB patients have various health issues and weak immunity due to which the body has a high oxidative stress. *Tuberculosis mycobacterium* produces a large number of reactive oxygen species due to activated macrophage which worsen the case of TB patients. So, antioxidants are important in maintaining a defense mechanism of a host. Adaptogens prevent

Table 23.1: Side effects of synthetic drugs used in the treatment of tuberculosis.

Drugs	Side effects
Isoniazid	Nausea, epigastric pain, arthralgia, psychosis, hematological alterations, hepatitis, etc.
Rifampicin	Hepatotoxicity, nausea, anorexia, abdominal pain, orange colored tears, sweat and urine, skin reactions, exanthema, and immunological reactions
Pyrazinamide	Hepatotoxicity, gastrointestinal symptoms, exanthema, dermatitis, etc.
Ethambutol	Retrobulbar neuritis, gastrointestinal symptoms, neurological symptoms, cardiovascular symptoms, hematological symptoms, hypersensitivity, etc.

the damage to the body occurred due to the stress by cope up with stress. All the physical, chemical, and biological damages can be combated by using adaptogen. Also, the number of white blood cells increased by adaptogens hence can be used as an immunomodulator. There are two important hormones that are responsible for stress, that is, corticotropin-releasing hormone and arginine vassopressin. corticotropin releasing hormone releases corticotropin which further releases corticosteroids and cortisone, which is responsible for producing stress response. Vasopressin leads to the reuptake of water by the kidney, thereby increased blood pressure due to the vasoconstriction. The system together activates hypothalamus–pituitary–adrenal axis.

23.2 Rationale behind herbal treatment over regular treatment

The normal drug regimen includes first line and second line drugs. But these drugs are associated with some serious side effects which make their use for the patients unbearable. Some of the major side effects associated with the drugs are listed in Table 23.1.

Synthetic drugs mentioned in Table 23.1 have their respective resistance in the human body by the bacteria. Fewer than three drugs in a regimen can be a reason for drug resistance in a patient, specifically if the patient is noncomplaint toward the regimen. Another reason for the resistance can be the treatment with a single drug. However, the resistance can be overcome by using second line drugs like fluoroquinolones, but this also involves some risk factors. To overcome these side effects and the drug resistance problems, herbal drugs can be opted, which have shown promising results in combating TB. Herbal drugs can act as both anti-TB drugs which can kill the organism, MTB, and can also act as an adjuvant therapy to reduce the side effects caused by the anti-TB drugs.

The properties and traditional uses of some of the herbs used in the treatment of TB are described below.

23.2.1 Sutherlandia frutescens

Sutherlandia frutescens is a Southern African legume that has been used as an indigenous medicine for a variety of ailments. It is a shrub with bitter, aromatic leaves. Red-orange flowers appear in spring to midsummer. It has been reported to have four key compounds, which contribute to the efficacy of this plant, that is, the nonprotein amino acid L-canavanine, pinitol, GABA (gamma-aminobutyric acid), and asparagines [4]. Triterpenoid glycoside has been also isolated and characterized [4].

Traditionally, it has been used as a tonic for enhancing well-being, enhances immunity, promotes longevity, fights stress, depression, and anxiety. The other reported therapeutic benefits of this plant extract are in the treatment of viral hepatitis, asthma, rheumatoid arthritis, wasting from cancer, TB, acquired immune deficiency syndrome (AIDS), etc. It has been reported that the alpha linolenic acid purified from DCM: MeOH (1:1) extract obtained from *S. frustescens,* inhibits shikimate kinase, which targets the *M. tuberculosis*. The IC_{50} value of extract and alpha-linolenic acid (purified inhibitor) was 0.1 µg/mL of MtbSK (shikimate kinase enzyme) and 3.7 µg/mL, respectively [5]. The study suggested that *S. frustescens* extract inhibits shikimate kinase (a good drug target for *M. tuberculosis*).

23.2.2 *Allium sativum*

Allium sativum is a bulbous plant growing up to 1.2 m (4 ft) in height. Its hardiness is United States Department of Agriculture Zone 8. It produces hermaphrodite flowers. It is pollinated by bees and other insects. It belongs to the family Amaryllidacea. Alliin, allicin, allixin, (E)-ajoene, diallyldisulfide, (Z)-ajoene, γ-glutamyl-S-2-propenyl cysteine, methyl allyl disulfide, S-allyl-cysteine, 1,2-vinyldiithin are the reported active constituents of this plant [6].

Traditionally, it was used as antioxidant, immunomodulator, and anti-inflammatory agent. It also increases peristalsis, reduce pain in the hip, in psoas, iliac muscles, etc. The extract of this plant is effective in the treatment of cancer, AIDS, candidiasis, Herpes, *Cryptococcus* infection. It is also effective against gram-positive and gram-negative bacteria [7,8].

Antitubercular activity of cefoperazone (CEF)-allicin extracted from *A. sativum* has been reported against drug susceptible and resistant clinical isolates of *M. tuberculosis*. The study reported 25 µg/mL minimum inhibitory concentration (MIC) of CEF-allicin for *M. tuberculosis* and isoniazid-resistant clinical isolate TRC-C 1193 [9]. Dini and coworkers observed 80–160 µg/mL and 100–200 µg/mL MIC value for the susceptible strain and resistant strain, respectively, using an ethanolic extract of *A. sativum*. The water extract also inhibited protein synthesis by suppressing 14 C glycine incorporated into the whole cells [10]. Gupta et al. reported a synergistic effect of *A. sativum* in combination with antitubercular therapy [11]. Hannan and coworkers evaluated the effectiveness of garlic against non-MDR and MDR isolates of *M. tuberculosis*. They observed MIC values between 1 and 3 mg/mL indicating inhibitory effects of *A. sativum* against both non-MDR and MDR *M. tuberculosis* isolates [12].

23.2.3 Citrullus colocynthis

It is a fruit-bearing plant of known as bitter cucumber, bitter apple, or desert gourd. It is a desert viny plant of Cucurbits family present in Asia, Turkey, etc. The seeds and fruits of this plant at low doses can reduce blood glucose level, whereas higher dose is associated with side effects like colonic inflammation and rectal bleeding [13]. Major phytochemical constituents of this plant extract are carbohydrate, protein, separated amino acid, tannins, saponins, phenolic, flavanoids, terpenoids, alkaloids, anthranol, steroids, cucurbitacin A, B, C, D, E (A-Elaterin), J, L, caffeic acid, and cardic glycoloids [14]. Seeds of *Citrullus colocynthis* contain proteins, crude fibers, moisture, A-tocopherol, Δ-tocopherol, and fixed oil with high amount of unsaturated fatty acids (linoleic acid, oleic acid) and very low N-3 poly-unsaturated fatty acid level. The seed fat contains palmitic acid, stearic acid, arachidic acid, oleic acid, linoleic acid, and linolenic acid. It also contains minerals like Ca, Mg, K, Na, and P [14].

Traditionally, the plant has shown anticancer, antidiabetic, antibacterial, and antiasthmatic effects [14]. *C. colocynthis* extract is used as anti-inflammatory drug for the treatment of breast infection, joint pain, and uterine pain [15]. It has been also reported to use to treat constipation and as abortifacient [16]. Blackening of hairs is the nontherapeutic application of *C. colocynthis* plant [17]. It is also used to treat boils and pimples. The root paste is applied to treat enlarged abdomen of children [18]. The ripe deseeded fruits possess anti-TB activity. The methanolic extract has shown MIC ≤ 62.5 µg/mL. The bioactive has shown MIC 31.2 µg/mL against Mycobacterium H37Rv [19]. In an assay (tube dilution assay, radiometric BACTEC 460 TB system), the ethanolic extract of fruit has shown MIC/IC_{50} 31.2 µg/mL [20].

23.2.4 Artemisia capillaries

It is also known as wormwood. It is a perennial plant found in China, Korea, Japan, and The Philippines. It belongs to the family Asteraceae. Major chemical constituents are α-Pinene, β-Pinene, Limonene, 1, 8-Cineole, Piperitone, β-Caryophyllene, and Capillin [21]. Traditionally, it is used as an antioxidant, anti-inflammatory agent, antipyretic agent, aromatic agent, and diuretic herb that act as a tonic for the liver and gall bladder. It also acts as antitumor, antisteatotic, choleretic agent, antiviral, and antifibrotic agent. Other therapeutic benefits of this plant extract are in the treatment of cirrhosis, hepatitis, and hepatocellular carcinoma [22]. Jyoti et al. (2016) reported antitubercular activity of this plant extract. Ursolic acid and hydroquinone components of methanolic extract are active against susceptible and resistant strains of *M. tuberculosis*. The MIC values of ursolic acid and hydroquinone components were found to be 12.5 µg/mL against susceptible strains and 12.5-25 µg/mL against MDR/XDR MTB [23].

23.2.5 *Kaempfera galanga*

It is also known as aromatic ginger, resurrection lily, and belongs to family Zingiberaceae. It is mainly found in India, China, Cambodia, and Taiwan [24]. It contains volatile oil, alkaloids, starch, protein, amino acids, minerals, and fatty matter. Essential oil of *Kaempfera galangal* contains ethyl–trans p–methoxy cinnamate, pentadecane, 1,8-cineole, g-carene, and borneole. It also contains camphene, kaempferol, kaempferide, cinnamaldehyde, pmethoxycinnamic acid, and ethyl cinnamate [25]. Traditional uses of *K. galangal* are stimulant, immunobooster, carminative, anti-inflammatory, diuretic, depurative, dyspepsia, leprosies, febrifuge, rheumatism, asthma, helminthiasis, ulcer, hemorrhoids, etc [26]. It has been demonstrated that sequential extraction of rhizomes shown the active compound Ethyl p-methoxycinnamate is present in the rhizome extract, which is responsible for inhibiting MTB. Ethyl p-methoxycinnamate by the resazurin microtitre assay inhibited MTB strain H37Ra, H37Rv, drug susceptibility, and MDR with a MIC value of 0.242–0.485 mM [27].

23.2.6 *Acalypha indica*

It is a common herb growing up to 75 cm (30 in) tall with ovate leaves. Flowers are green, unisexual found in catkin inflorescence. It belongs to family Euphorbiaceae [28]. Major chemical constituents obtained from *Acalypha indica* are alkaloids, catechols, flavonoids, phenolic compounds, saponins, and steroids [29]. It has been used as analgesic, anti-inflammatory, antihelmintic, antibacterial, antitubercular, neuroprotective, antivenom, postcoital infertility activity, and traditionally acts as an antioxidant [29]. The aqueous extract of leaves of *A. indica* has shown antitubercular activity in the L–J medium. The extract showed 95% inhibition for MDR isolate DKU-156 and 68% for MDR isolate JAL-1236 and 68% for H37Rv [30]. The n-hexane extract (25 μg/mL) and dichloromethane extract (50 μg/mL) of *A. indica* showed antitubercular activity against H37Rv strain using Micro Plate Alamer Blue Assay [31].

23.2.7 *Adhatoda vesica*

It has a lance-shaped leaves 10–15 cm in length by four wide, found in Asia and belongs to family Acanthaceae [32]. Vasicin, L-vasicinone, deoxyvasicine, maiontone, vasicinolone, and vasicinol are the major chemical constituents isolated from this plant [33]. Traditionally it is used as antiasthmatic, antiulcer, cholagogue agent, antiallergic, antitubercular, abortifacient agent, and insecticidal agent [33]. Water extract of *Adhatoda vesica* leaves showed 55% and 70% inhibition against H37Rv (susceptible strain of *M. tuberculosis*) at 2 and 4% v/v plant extract, respectively. In case of (MDR isolate) DKU-156, 25% and 32% inhibition has been reported at 2% and 4% v/v plant extract, respectively. In the case of MDR isolates JAL-1236, 86% inhibition was observed at both 2% and 4% v/v of plant extract [30]. It has been reported that the vasicine produces bromhexine and ambroxol, which are used as mucolytics. It shows pH dependent inhibitory effect on MTB. It increases lysozyme and rifampicin level in bronchial secretion and sputum [33].

23.2.8 Allium cepa

It is also known as the bulb onion or common onion. It is most widely cultivated species and belongs to family Amaryllidaceae [34]. It was reported to contain organosulfur compounds, quercetin, fructose, quercetin-3-glucoside, isorhamnetin-4-glucoside, allylsulfides, xylose, galactose, glucose, mannose, flavonoids, flavenols, S-alk(en)yl cysteine sulfoxides, cycloalliin, selenium, thiosulfinates, and sulfur and seleno compounds [35]. It is used as anticancer, antidiabetic, **antimutagenic, osteoclastic, and antihypercholesterolemic effects.** It is traditionally used as an antioxidant and to promote hair growth and increase appetite [36]. The aqueous extract of *Allium cepa* bulb showed 39% and 35% inhibition against H37Rv at 2% v/v and 4% v/v, respectively. This effect was 47% and 37% at 2% v/v and 4% v/v, respectively, for against DKU-156. The inhibition effect against JAL-1236 was 76% and 79% at 2% and 4% v/v, respectively [30]. It has been reported that the constituent like sulphur, quercetin, fructose, galactose, glucose, mannose, allylsulfide, flavonoids, and seleno compounds present in the onion possess antibiotic activity against both gram positive and negative bacteria and can be used to prevent tuberculosis [36]. The aqueous and ethanolic extract of *Allium cepa* tissue showed MIC of 100 μg/mL against H37Ra. It has been proved that *Allium cepa* can be used as an adjuvant therapy for TB [37].

23.2.9 Aloe vera

Aloe vera is a species of the genus Aloe and belongs to family Asphodelaceae. It grows in tropical climates throughout the world. It is used to treat skin disorders [38]. It contains vitamin A, C, and E, which exhibits antioxidant property. It also contains vitamin B12, folic acid, and choline), enzymes (aliiase, alkaline phosphatase, amylase, bradykinase, carboxypeptidase, catalase, cellulase, lipase, and peroxidase), minerals (calcium, chromium, copper, selenium, magnesium, manganese, potassium, sodium, and zinc), sugars, anthraquinones (aloin and amodin), fatty acids (cholesterol, campesterol, B-sisosterol, and lupeol), and hormones (auxins and gibberellins) [39]. Traditionally, it is used as antioxidant, immunomodulator, and moisturizer in skincare products. Its wound healing, stress-relieving, anti-inflammatory, analgesic, antimicrobial, anticancer, antidiabetic, and antiseptic properties are well documented [39,40].

Leaf gel of *A. vera* inhibited H37Rv to 10% and 41% at 2% and 4% v/v, respectively. The percentage inhibition against DKU-156 was 25% and 32% at 2% v/v and 4% v/v of

A. vera extract, respectively. The percentage inhibition against JAL-1236 was 79% and 85% at 2% and 4% v/v, respectively [30]. It is used for the treatment of TB due to its antioxidant property, which reduces the adverse effects of antitubercular drugs [40]. It has successfully decreased the tumor necrosis factor-alpha and Th 17 cells, therefore; it can also use as an antitubercular drug [41].

23.2.10 Eclipta alba

Eclipta alba (bhringaraj, Family: Asteraceae) is an annual, erect, branched, and creeping herb with rooting at nodes. It has white colored flowers. It is a native of India and Southwest America and grows in moist places throughout the world. It occurs as a weed [42]. Major phytoconstituents of *E. alba* are wedelolactone, desmethylwedelolactone, desmethylwedelolactone-7-glucoside, stigmasterol in leaves; hentriacontanol, heptacosanol, and stigmasterol, ecliptal, eclalbatin in roots; β-amyrin and luteolin-7-0-glucoside, apigenin, cinnaroside, sulphur compounds, eclalbasaponins I-VI in aerial parts. Wedelolactone in stems, sterols, ecliptalbine (alkaloid) in seeds; resin, ecliptine, reducing sugar, nicotine, stigmaster in whole plant [43]. Traditionally, it is used as an immunomodulator, antioxidant, hair growth promoter, for rejuvenation, dehydration, and skin disorders. The reported uses include anticancer, hepatoprotective, antiulcer, antihyperlipedemic, anthelmintic, anti-inflammatory, analgesic, and antidiabetic activity [42,43]. It has been reported that by giving bhringaraj along with DOTs (directly observed therapy) to the patients, resulted in better, safer, and faster relief. Also, all the patients showed sputum conversion within stipulated time. The relief was more than 75% of the symptoms like a burning sensation in the palms and soles, pyrexia, anorexia, dyspnoea, cough, hemoptysis, etc. It has been also reported to stimulate the reticuloendothelial system [44].

23.2.11 Withania somnifera

Withania somnifera belongs to the family Solanaceae. Other common names of *W. somnifera* are ashwagandha, Indian ginseng, poison gooseberryor, and winter cherry. Sitoindosides VII-X, withaferin-A, 5-dehydroxy withanolide-R, and Withasomniferin-A are the major components isolated from *W. somnifera* [45]. It is traditionally used as neuroprotective, astringent, immunomodulator, and stress reliever. It is also used in insomnia and to improve cognitive function. Other reported applications of this plant are aphrodisiac, diuretic, anthelmintic, thermogenic, antirheumatic, antiulcerogenic, and anti-inflammatory agent. It is also used to treat diseases like Parkinson's, dementia, memory loss, stress-induced diseases, malignoma, etc [46]. The use of Ashwagandha increases body weight and IgM of TB patients [47]. It has been reported to decrease erythrocyte sedimentation rate and IgA. The bioavailability of isoniazid and pyrazinamide was found to increase along with percentage eosinophila. Ashwagandha was found to increase hemoglobin, WBC count, and decreased bacterial load with time [47].

23.2.12 Carum carvi

Caraway (*Carum carvi*, Family: Apiaceae) also known as Meridian fennel and Persian cumin. It is a biennial plant, native to Western Asia, Europe, and North Africa. The active chemical constituents of *C. carvi* are α-pinene, α-terpineol, α-farnesene, β-caryophyllene, β-myrcene, β-ocimene, β-pinene, γ-terpinene, carvone, carvone, camphene, citronellol, cuminaldhyde, eugenol, germacrene-D, limonene, linalyl acetate, nerol, p-cymene, terpinene-4-ol, and thymol. It is used as an antimicrobial, anticancer, hypolipidemic, antidiabetic, bronchodilator, diuretic, hepatoprotective, analgesic, antioxidant, carminative, and antistress agent [48]. *C. carvi* acts as a bioenhancer for rifampicin, isoniazid, pyrazinamide in fixed-dose combination [49].

23.2.13 Curcuma longa

Curcuma longa belongs to family Zingiberaceae. Curcumin, curcuminoids, ar-turmerone, α-turmerone, β-turmerone, and (z) β-ocimene, α-phellantrene, terpinolene, 1,8-cinceole, undecanol, and p-cymene are the major active constituents isolated from *C. longa* [50]. Curcumin is a bright yellow compound responsible for antitubercular effect. *C. longa* is used as an antiplatelet aggregation agent, antiproliferative, and antioxidant. It also exhibits lipid lowering, gastroprotective, hepatoprotective, antiviral, antiprotozoal, antibacterial, anti-inflammatory, antioxidant actions, and anti-inflammatory properties [51]. Curcumin is an inducer of caspase-3-dependent apoptosis and autophagy. It inhibits nuclear factor-kappa B activation [52]. It accelerates pathogen clearance, attains sterile immunity, and restores host protective memory and hence, prevents reinfection and reactivation. It protects against INH induced hepatotoxicity [53]. It decreases the sputum smear positive, and increases hemoglobin. It inhibits the production of tumor necrosis factor-alpha, interleukin (IL) 1-b, and the activation of NF-κB in human monocytic derived cells [54].

23.2.14 Phyllanthus emblica

Phyllanthus emblica fruit (Amla, Family: Phyllanthaceae) is spherical in shape, light greenish yellow in color, smooth, and hard on appearance, with six vertical stripes. It is planted throughout the deciduous of tropical India and on the hill slopes up to 2000 m. It is commercially cultivated in Uttar Pradesh, Tamil Nadu, Rajasthan, and Madhya Pradesh. It contains emblicanin A and B21, gallic acid, ellagic acid, glucose, phyllemblin, corilagin, furosin, geraniin, quercetin, phyllantine, phyllantidine, amino acids, carbohydrates, and vitamin C. *Phyllanthus emblica* fruits are used as antioxidant, immunity booster, promotes vigor, hair tonic, memory enhancer, hepatoprotective, antidiabetic, ulcer protective, antimicrobial, anticancer, anti-inflammatory, and cardioprotective agent [55]. Jawarish amla is a sweet, semisolid, and granular Unani formulation. Its chief ingredient is Amla, which is processed in cow milk and sugar.

It has been reported that patients who were receiving Jawarish amla had significantly reduced symptoms of nausea, vomiting, abdominal pain, itching, and burning sensation in feet. Also, jaundice and skin rashes were less. It is used to mask the bitter taste of the drugs used in DOT [56]. It also decreases the incidences of diarrhea. It decreases acidity and also has antispasmodic activity. It possesses antiemetic, blood purifying, and antihistaminic properties. The values of ALT (alanine transaminase), AST (aspartate transaminase), and alkaline phosphatase decreases. It also has a strengthening effect on brain [56]. The antioxidant effect of 50% hydroalcoholic extract of Amla has shown hepatoprotective activity [57].

The methanolic extract of the leaves at 200,400 mg/kg p.o. given to Sprague dawley rats for 60 days demonstrated scavenged DPPH radical having an IC_{50} of 39.73 μg/mL and nitric oxide (NO) (IC_{50} of 39.14 μg/mL). Also there were reduced levels of antioxidant enzyme activities, namely, superoxide dismutase (SOD), catalase (CAT), glutathione peroxidase (GSH-Px), and reduced glutathione (GSH) whereas enhanced levels of total extractable proteins, lipid peroxides (TBARS), nitrite [58]. It has been demonstrated that gallic acid, ellagic acid, pyrogallol, corilagin, geraniin, elaeocarpusin, and prodelphinidins B1 and B2 possess antineoplastic effects. It also possesses radiomodulatory, chemomodulatory, chemopreventive effects, free radical scavenging, antioxidant, anti-inflammatory, antimutagenic, etc [59]. It suggests that amla protects against liver damage and dysfunction. It is a good remedy during indigestion, flatulence and constipation. It supports the skin texture and protects it against harmful environment. It is also known as a cooling adaptogen for warm weather [60].

23.2.15 Cuminum cyminum

Cuminum cyminum (Cumin, Family: Apiaceae) is a flowering plant, native of East Mediterranean to South Asia. The seeds of *C. cyminum* are used in the cuisines of many different cultures, in both whole and ground forms. α-Pinene, Limonene, 1,8-Cineole, Linalool, Linalyl acetate, and α-Terpineole are the major components of the essential oil [61]. It is used to treat gastrointestinal, gynecological, respiratory disorders, toothache, diarrhea, epilepsy, and also used as anticancer and antibacterial. Traditionally it is used as condiment, stimulant, carminative, antioxidant, and astringent [62]. The aqueous seeds extract is reported to increase the bioavailability of rifampicin in rats. A flavonoid glycoside, 3′,5-dihydroxyflavone 7-O-beta-D-galacturonide 4′-O-beta-D-glucopyranoside (CC-I) is responsible for the enhancement of bioavailability of Rifampicin [63]. An essential oil from *C. cyminum* showed the MIC between 6.25 and 12.5 μg/mL against the anti-TB strain [64].

23.2.16 Camellia sinensis

Green tea is obtained from the leaves of *Camellia sinensis* (Family: Theaceae) that have not been subjected to withering and oxidation process which is used to prepare Oolong and black tea. Green tea was originated in China, but its production has been spread to many countries in Asia. The important chemical constituents of green tea are Potassium, Catechins, Epigallocatechin gallate, Epicatechin 3-gallate (EGC), Epigallocatechin (ECG), Epicatechin (EC), and Caffeine [65]. Green tea has antiproliferative, antitumor, hypoglycemic, cardioprotective, neuroprotective, and antioxidant properties. It is used to reduce the body weight and protect from alcohol intoxication [66]. Administration of green tea significantly decreased the delay in sputum smear conversion time, but there was no effect on weight gain. The hazard ratio of the relative risk of delay in sputum smear conversion was 3.7 [67].

Green tea can significantly decrease the microbial load associated with TB as compared to other teas. Epigallocatechin gallate present in green tea inhibits *Tubercle bacillus* by inhibiting Inh A (a target of the antituberculous drug isoniazid,), enoyl acyl carrier protein reductase and also it weakens the transcription of tryptophan-aspartate containing coat (TACO) genes in human macrophage by inhibiting SP1 transcription factor [68].

A crude catechin extract from green tea can significantly lower the oxidative stress in pulmonary TB. There is a decrease in LPO (lipid peroxidation) levels, catalase, GPx (glutathione peroxidase) level, and sulfhydryl group (SH) level, whereas a significant increase in the level of NO, SOD, glutathione (GSH) [69]. Polyphenols is the main component responsible for antioxidant properties of *C. sinensis*. It acts by scavenging ROS or by chelating transition metals. It can also act as a pro-oxidant by producing hydrogen peroxides which help in preventing cancer [70].

23.2.17 Centella asiatica

Centella asiatica is herbaceous, frost-tender perennial flowering plant of the family Apiaceae. It is native to Wetlands in Asia. Centellin, asiaticin, and centellicin are the major compounds isolated from *C. asiatica*. In the traditional system of medicine this plant extract has been used as an antioxidant, wound healing, increase cognition reduces anxiety, strengthens the weakened veins, stimulatory-nervine tonic, rejuvenant, and sedative properties. The other reported uses are radioprotective, antinociceptive, anti-inflammatory, treat gastric ulcer, and antiepileptic properties [71]. Disturbance on haematological parameters, increased oxidative stress and adverse effect on kidney and liver functioning are the reported side effects of isoniazid. Ethanolic extract of *C. asiatica* leaves (100 mg/kg b.w.) helps to overcome the side effects caused by isoniazid [72].

23.2.18 Zingiber officinale

Zingiber officinale (Family: Zingiberaceae) is an herbaceous perennial which grows annually and has about a meter tall stems bearing narrow green leaves and yellow flowers. Rhizomes (ginger) of *Z. officinale* are widely used in folk medicine as strong antioxidant [73]. It is widely used to treat metabolic and age-related degenerative disorders. Major identified components are gingerol and shogaol, zingerone, and paradol. Other reported traditional uses are antiemetic, anti-inflammatory, antitumor, antimicrobial, antidiabetic, neuroprotective, protect against osteoarthritis, gastroprotective, hepatoprotective, and protective against migraine [74]. It has been demonstrated that the anemic condition in pulmonary TB can be controlled by giving ginger supplements to the patient along with DOTs. Significant increase in CRP, ferritin, and serum iron levels have been observed with an increase in total iron-binding capacity. Based on these observations, it is recommended to use ginger with antitubercular therapy to obtain synergistic effect [75]. Ginger extract has demonstrated anti-inflammatory and antioxidant effect in TB patients. Lipid peroxidation can cause tissue damage due to the release of cytokinins. Ginger extract successfully reduced the levels of TNF alpha, Malondialdehyde (MDA), and ferritin in the patients administered with 3 g of ginger extract for 1 month. The extract also demonstrated free radical scavenging property [76].

23.2.19 Panax ginseng

Panax ginseng belongs to family Araliaceae [77]. Ginsenosides, panaxosides, and gintonin are the major constituents isolated from *P. ginseng*. Traditionally, it is used as anti-inflammatory, antidiabetic, antioxidant, adaptogen, memory enhancer, athletic enhancement, vigor, improvement in mood, and psychophysical performance. It also has antiproliferative activity [78]. Red ginseng along with antitubercular therapy has significantly increased the level of antioxidant enzymes and decreased DNA damage and lipid peroxidation within 8 weeks. A higher level of antioxidant enzymes has been reported in the postmenopausal women receiving 3 g/day of red ginseng for 12 weeks when compared to the placebo group. The study also reported a decreased level of MDA [79].

P. ginseng extract acts as an adaptogen or an actoprotector, which increases the physical ability and mental capability. It also possesses antifatigue activity. It improves endurance time to exhaustion, lowers mean blood pressure. It possesses ergogenic property, improves psychomotor activity, improved oxygen uptake and work output, increased aerobic capacity, and improved breathing, auditory and visual reaction time. It improves vitality, alertness, rigidity, and concentration. Improvement of cognitive deficit in Alzheimer's patient has been also reported [80]. The adaptogenic property of ginseng is due to its effects on hypothalamic—pituitary—adrenal axis, resulting in elevated levels of corticotropin and corticosteroids levels in plasma [81].

23.2.20 Piper sarmentosum Roxb

Piper sarmentosum (family: Piperaceae) is a perennial tree, grows in tropical climate. It grows in India and South Asia. Main reported chemical constituents are sarmentamide (A, B, and

C), guineensine, brachystamide B, brachyamide B, sesamin, asaricin [82], benzyl benzoate, benzyl alcohol, 2-hydroxybenzoic acid phenylmethyl ester, 2-butenylbenzene, myristicine, transcaryophyllene, and 4H-pyran-4-one,2,3-dihydro-3,5-dihydroxy-6-methyl- (DDMP).

Traditionally, it is used as an antioxidant and antibacterial agent. The reported uses involved hypoglycemic effect, anti-inflammatory, antiosteoporosis, antiatherosclerosis, neuromascular blocking activity, antiplasmodial, and anti-TB activity [83]. It has been observed that the ethanolic extract exhibited better antioxidant activity [84]. The ethanolic extract of the leaves was investigated for a dose-response relationship and its EC_{50} was found to be 38 µg/mL. The methanolic and ethyl acetate extract of *P. sarmentosum* leaves exhibited anti-TB activity with MIC 3.12 µg/mL while MIC/MBC of isoniazid (INH) was found to be 0.5 µg/mL [84].

23.2.21 Tulbaghia violacea

Tulbaghia violacea (family: Alliaceae) is a flowering plant indigenous to Southern Africa and naturalized in Tanzania and Mexico [85]. It is an aphrodisiac medicine. It is used to treat the cancer of oesophagus and as a snake repellent. Reported chemical constituent of *T. violacea* are acetamide, 2-cyano, chlorodifluoro acetamide, σ- xylene, (E)-2-heptenoic acid, ρ-xylol, ρ-xylene, thiodiglycol, 2,4-dithiapentane, chloromethylmethyl sulfide, acetamide, phthalic acid 2-ethylhexyl isobutyl ester, phthalic acid, heptyl2-methylallyl ester, nonadecane, hepta-cosane, and tetracosane [86]. Traditionally, it is used as an antioxidant and for the treatment of diabetes and hypolipidemic condition. Moodley et al. reported a decrease in fasting blood glucose and increased plasma insulin and glycogen levels was observed in diabetic rates treated with 60 and 120 mg/kg b.w. of *T. violacea* extract for 42 days. The plant extract reduced TBARS levels and increased SOD and GPx levels. It shows antioxidant, hypoglycemic, hepatoprotective, and hypolipidemic activities [87].

23.2.22 Allophylus edulis

It belongs to the family Sapindaceae. It is used to make poles, cabinet, etc. The wood is used for fuel and to make charcoal. The fruits are also used to make fermented drink. Reported chemical constituent are 6,7-epoxicaryophyllene, spathulenol, sitosterone and sitosterol, lupeol, clerodane diterpene [88]. Traditionally, it is used as antihypertensive, insecticidal, refresher, digestant, nourishing tonic, antidiabetic, and anti-inflammatory agent. It is also used to treat intestinal disorder and hepatitis [89]. Viridoflorol, the component isolated from *Allophylus edulis*, possess anti-TB and antioxidant activities [90].

23.2.23 Euadenia eminens

Euadenia eminens (family: Capparaceae) has irregular flower, two of the petals having advanced to huge proportions over the rest. *E. eminens* plant contains glycosides, coumarins, flavonoids, carotenoids, and alkaloids. Traditionally, it is used as an antioxidant, aphrodisiac, antianemic, and antidote. It is used to treat HIV/AIDS, TB, and inflammation [91]. Ethanolic extract (70%) of *E. eminens* inhibited inflammation by 74.18% at 30 mg/kg. The root extract showed antibacterial activity against *Bacillus subtilis* and *Bacillus thurigiensis*. The amount of phenolic compounds which act as a free radical terminator has been expressed in terms of tannic acid, that is, 7.25 mg/kg dry wt. Scavenging of free radical resulted in 1.175 mg/mL IC_{50} value. Also, the roots of a plant possess a strong scavenging activity by reducing the oxidative stress in infectious or inflammatory ailments [92].

23.2.24 Spirulina

Spirulina is a photosynthetic cyanobacterium. The cell wall of spirulina lacks cellulose but has mycosaccharides. It grows mainly in Central Asia. It has a high nutritive value due to the presence of protein and other compounds in it. It belongs to family Phormidiaceae. It contains essential amino acids, vitamins, β-carotene, minerals, polysaccharides, fatty acids, and sulfolipids [93]. Traditionally, it is used to detoxify heavy metals, boost energy, and alleviates sinus related problems. It is also used to treat AIDS, hypertension, and cancer. Spirulina decreases the intoxication syndrome regression time and reduces the adverse reaction caused due to antitubercular therapy [94]. Spirulina prevents accumulation of cadmium and prevents the toxicity caused by cadmium [95].

23.2.25 Ocimum sanctum

In Ayurveda, *Ocimum sanctum* (Tulsi, family: Lamiaceae) is known as "elixir of life" [96]. Main reported chemical constituents are cirsilineol, circimaritin, isothymusin, apigenin, rosameric acid, eugenol, carvacrol, caryophyllene, orientin, and andvicenin. Traditionally used to increase longevity, as a condiment, treat common cold, headache, skin disease, insomnia, night blindness, sharpens memory, and immunobooster. Reportedly, it is used as anticancer, antifertility, antidiabetic, antilipidemic, and antibacterial [97]. It has demonstrated adaptogenic, antiseptic, germicidal, and antiasthamatic activity. It possesses anti-TB substance. Dried powder of tulsi plant can prevent stress induced hyperacidity and peptic ulcer. It also arrests milk induces leucocytosis [98]. It reduces the stress level and promotes wellbeing. It protects and detoxifies the body against chemicals and also induces apoptosis. It also reduces physical, metabolic, and mental stress [99].

23.2.26 Astragalus

Astragalus belongs to Legume family Fabaceae. Astragalosides I-IV, trigonosides I-III, formononetin, ononin, calycosin, saponins, triterpenoids, and saponins are the major components isolated from Astragalus plant. Traditionally, it is used as antioxidant, cellular protective, immune-booster, growth hormone enhancer and reportedly used as cardioprotective, antineo-plastic, antidiabetic, anti-inflammatory, and antimicrobial agent [100]. It has been suggested that Astragalus is a potent immune booster. It increases T cell and NK cell activity and also promotes interferon production. It is also used for chronic diarrhea and ulcers. It can also be used in treating skin disorders [60]. It has been demonstrated that it acts as an antiageing remedy, balances hormones, inhibit tumor growth, and also provide resistance to stress. It also reduces the sugar level and stimulates WBC [101]. It contains polysaccharides and saponins, which improve heart function. It is also used to treat asthma and restore immunity and adrenal function [102].

23.2.27 Glycyrrhiza glabra

Liquorice or licorice is the root of *Glycyrrhiza glabra* (family: Legumes). *G. glabra* is herba-ceous perennial legume native to Southern Europe and parts of Asia, such as India. Reported active chemical constituents extracted from liquorice are glycyrrhizin, isoliquiritigenin, isoliquiritin, liquiritigenin, and liquiritin. Traditionally, this plant extract is used as a healing agent, soothing agent, regulate hormones, and immunity booster reportedly it is used to treat acid reflux, ulcer and Addison's disease [103]. It helps in weight loss and stimulates gastric and bile secretions [104]. It maintains cortisol levels in the body, thus maintain the adrenals [105].

23.2.28 Rhodiola rosea

Rhodiola rosea is a flowering plant of the family Crassulaceae. It grows naturally in wild Arctic regions of Europe, including Britain, Asia and North America and can be propagated as a groundcover. Main chemical constituents are salidroside, kaempferol-7-O-alpha-L-rhamnopyranoside, herbacetin-7-O-alpha-L-rhamnopyr-anoside, herbace-tin-7-0-(3″-O-beta-D-glucopyran-oside)-alpha-L-rhamnopyranoside; 5, 7, 3′, 5′-tetrahydroxy-flavanone, rosavin, rosin, rosarin, rosavin, rosin, rosarin, rodiolin, rodionin, rodiosin. Traditionally, it is used to increase physical endurance, work productivity, stimulant, antioxidant, antistress, antihypoxic, longevity, resistance to high altitude sickness, improves brain function and treat fatigue, depression, anemia, impotence, gastrointestinal ailments, infections, and nervous system disorders. Reportedly used as anticancer and anti-TB, cardioprotective, treat hernia, leucorrhoea, hysteria, and anti-inflammatory [106].

It vitalizes the nervous system and decreases depression. These effects are due to the pres-ence of salidroside, rosavins, and p-tyrosol. It acts as a performance enhancer. It has been suggested that rosiridin inhibit monoamine oxidase A and B and shows positive effects toward depression and senile dementia. Tyrosol increases the phosphorylation of eNOS, FOXO3a. It also protects brain neurons from serious injuries. It improves attention and cognitive function. It is a good remedy for asthenia, neurosis, and schizophrenia. It helps in reducing the side effects caused by psychotropic therapy in schizophrenia [107].

23.2.29 Boerhaavia diffusa

Boerhavia diffusa (family: Nyctaginaceae) is a species of flowering plant in the 4 o'clock family, which is commonly known as punarnava, red spiderling, spreading hogweed, or tarvine [108]. Reported chemical constituents present in *B. diffusa* extract are arachidic acids, boeravinone A-F, boerhavic acid boerhavin, β-ecdysone, β-sitosterol, C-methylflavone 5,7-dihydroxy-3',4'-dimethoxy-6,8-dimethylflavone, hypoxanthine 9-L-arabinofuranoside, hentriacontane, punarnavine, punarnavoside, palmitic acid, stearic acid, ursolic acid, etc. It has immunomodulatory, antioxidant, protects from gamma radiation, antifibrinolytic, adaptogenic, antibacterial activities. Reported uses of *B. diffusa* extract include anticonvulsant, antitubercular, nonteratogenic, analgesic, anticancer, hepatoprotective, and antidiabetic property [109].

It shows immunomodulation effect due to the presence of quercetin, punernavine, and syringaresinol mono-β-D glucoside [110]. Hydroethanolic extract (80%) of *Boerhaavia diffusa* normalizes the triglyceride levels and affecting the other parameters like Serum Glutamic Oxaloacetic Transaminase (SGOT) and Serum Glutamic Pyruvic Transaminase (SGPT) which ultimately cause antistress effect [111]. Ethanolic extract of *B. diffusa* roots showed antistress and immunomodulatory activities by increasing the carbon clearance, which leads to stimulate Reticulo Endothelial System. It also has a stimulatory effect on lymphocytes and accessory cell types [112]. Rifampicin induced hepatotoxicity and increased levels of alanine aminotransferase (ALT), aspartate aminotransferase (AST), alkaline phosphatase (ALP), gamaglutamyl transpeptidase (GGT), lactate dehydrogenase (LDH), and bilirubin has been significantly reduced with *B. diffusa* therapy [113].

23.2.30 Vitis vinifera (Grape seed)

Grapes are juicy and pulpy berries of woody wines. Reported chemical components of *Vitis vinifera* are catechin, tannins like procyanidins and leucoanthocyanins, epicatehins and flavonoids. Traditionally, it is used as a potent antioxidant. It also balances the vata and pitta of the body and has a purgative action. It shows antibacterial property and can kill gram positive bacteria completely and gram negative bacteria to a lesser extent. It also has anticancer activity [114]. The aqueous extract of grape seeds (100–300 mg/kg) can act as a stress relieving, radical scavenging, and nootropic agent. Antioxidants are also isolated from *V. vinifera* [115]. Hepatoprotective activity has been reported by ethanolic extracts of seeds of *V. vinifera* against diethylnitrosamine in TB [116].

23.2.31 Rhodiola heterodonta

Rhodiola heterodonta (family: Crassulaceae) is a perennial herb and the color of its flower is purple. The main ingredients isolated from this plant are rhodiocyanoside, Epigallocatechin gallate (dimer), rosavin, rosarin, salidroside, tyrosol, heterodontoside, viridoside, monghroside. Adaptogenic properties of this plant are due to the presence of catechins and proantocyanidins [118–119].

23.2.32 *Andrographis paniculata*

It is also known as green chiretta belongs to family Acanthaceae. It is an annual herbaceous plant. The main reported chemical constituents are angrapholide, 14-deoxy-11,12-didehydroandrographolide, 14-deoxyandrographolide, 3,14-dideoxyandrographolide, 14-deoxy-11-oxoandrographolide, neoandrographolide. Traditionally, it is used as a bitter tonic, stomachic, antidote for snake's poison, antimalarial, and immunomodulatory. Other reported uses are in jaundice, colic dysentery, dyspepsia, anthelminitic, anti-HIV, anticancer, and anti-asthmatic. It is effective against hypnosis caused by pentobarbital. It inhibits the GABA-A site triggered by benzodiazepine, which is responsible for stress induction [120]. Organic solvent extract like ethanolic and acetone extract of *Andrographis paniculata* has shown 80%–95% inhibition of MTB [121]. An aqueous extract of the plant has shown 100% inhibition at 5 mg/mL against H37Rv, a MTB strain [122]. It also reduces the sputum conversion time from 12 weeks to 4–6 weeks [123,124].

23.3 Conclusion

TB is a highly infectious disease declared a global health emergency by the WHO. TB treatment requires an intensive phase and a continuation phase treatment (~20 months) using current synthetic therapeutics, such as fluoroquinolones, kanamycin, capreomycin, and amikacin, which are toxic and less efficient. The potential of phytoconstituents for the treatment of TB has been studied extensively for improved clinical benefits. However, this needs to further explore with special emphasis on the mechanism of their action.

References

[1]   CD Tweed, Toxicity related to the treatment of pulmonary tuberculosis. 2019. Doctoral Thesis, University College London. Available from: https://discovery.ucl.ac.uk/id/eprint/10074856/, Accessed on January 05, 2020.
[2]   CL Daley, The global fight against tuberculosis, Thorac Surg Clin 29 (1) (2019) 19–25.
[3]   G Sotgiu, R Centis, L D'ambrosio, GB Migliori, Tuberculosis treatment and drug regimens, Cold Spring Harb Perspect Med 5 (5) (2015) a017822.
[4]   Campbell J The impact of storage time and seasonal harvesting on biomarker levels of lessertia frutescens. Doctoral dissertation, University of Western Cape. 2012. http://hdl.handle.net/11394/3304, Accessed on January 15, 2020.
[5]   P Masoko, IH Mabusa, RL Howard, Isolation of alpha-linolenic acid from *Sutherlandia frutescens* and its inhibition of Mycobacterium tuberculosis' shikimate kinase enzyme, BMC Complementary and Alternative Medicine 16 (1) (2016) 366.

[6] N Martins, S Petropoulos, IC Ferreira, Chemical composition and bioactive compounds of garlic (*Allium sativum* L.) as affected by pre-and post-harvest conditions: a review, Food Chem 211 (2016) 41–50.

[7] G Gebreyohannes, M Gebreyohannes, Medicinal values of garlic: a review, International Journal of Medicine and Medical Sciences 5 (9) (2013) 401–408.

[8] L Bayan, PH Koulivand, A Gorji, Garlic: a review of potential therapeutic effects, Avicenna Journal of Phytomedicine 4 (1) (2014) 1–14.

[9] PS Murthy, P Ratnakar, DV Gadre, V Talwar, HC Gupta, RL Gupta, Trifluoperazine and CEF-allicin from garlic (Allium sativum) as potential new antitubercular drugs active against drug resistant *Mycobacterium tuberculosis*, Indian J Clin Biochem 12 (1) (1997) 72–75.

[10] C Dini, A Fabbri, A Geraci, The potential role of garlic (*Allium sativum*) against the multi-drug resistant tuberculosis pandemic: a review, Annali dell'Istituto superiore di sanità 47 (2011) 465–473.

[11] RL Gupta, S Jain, V Talwar, HC Gupta, PS Murthy, Antitubercular activity of garlic (*Allium sativum*) extract on combination with conventional antitubercular drugs in tubercular lymphadenitis, Indian J Clin Biochem 14 (1) (1999) 12.

[12] Hannan A, Ullah MI, Usman M, Hussain S, Absar M, Javed K Anti-mycobacterial activity of garlic (Allium sativum) against multi-drug resistant and non-multi-drug resistant Mycobacterium tuberculosis. Pak J Pharm Sci 24(1):81-85.

[13] BB Aggarwal, S Prasad, S Reuter, R Kannappan, VR Yadav, B Park, et al., Identification of novel anti-inflammatory agents from Ayurvedic medicine for prevention of chronic diseases: "reverse pharmacology" and "bedside to bench" approach, Curr Drug Targets 12 (11) (2011) 1595–1653.

[14] AE Al-Snafi, Chemical constituents and pharmacological effects of Cynodon dactylon—a review, IOSR Journal of Pharmacy 6 (7) (2016) 17–31.

[15] R Chandrasekar, S Chandrasekar, Natural herbal treatment for rheumatoid arthritis—a review, International Journal of Pharmaceutical Sciences and Research 8 (2) (2017) 368–384.

[16] A Delazar, S Gibbons, AR Kosari, H Nazemiyeh, M Modarresi, L Nahar, SD Sarker, Flavone C-glycosides and cucurbitacin glycosides from *Citrullus colocynthis*, DARU Journal of Pharmaceutical Sciences 14 (3) (2006) 109–114.

[17] MC Meena, RK Meena, V Patni, Ethnobotanical studies of Citrullus colocynthis (Linn.) Schrad. An important threatened medicinal herb, Journal of Medicinal Plants 2 (2) (2014) 15–22.

[18] MC Meena, R Meena, V Patni, High frequency plant regeneration from shoot tip explants of Citrullus colocynthis (Linn.) Schrad.—an important medicinal herb, African Journal of Biotechnology 9 (31) (2010) 5037–5041.

[19] A Mehta, G Srivastva, S Kachhwaha, M Sharma, SL Kothari, Antimycobacterial activity of *Citrullus colocynthis* (L.) Schrad. against drug sensitive and drug resistant *Mycobacterium tuberculosis* and MOTT clinical isolates, J Ethnopharmacol 149 (1) (2013) 195–200.

[20] JP Thakur, PP Gothwal, Edible plants as a source of antitubercular agents, Journal of Pharmacognosy and Phytochemistry 4 (1) (2015) 228–234.

[21] C Yang, DH Hu, Y Feng, Antibacterial activity and mode of action of the Artemisia capillaris essential oil and its constituents against respiratory tract infection-causing pathogens, Molecular Medicine Reports 11 (4) (2015) 2852–2860.

[22] E Jang, BJ Kim, KT Lee, KS Inn, JH Lee , A survey of therapeutic effects of Artemisia capillaris in liver diseases, Evidence-Based Complementary and Alternative Medicine 2015 (2015). Article ID 728137, doi:10.1155/2015/728137.

[23] MA Jyoti, KW Nam, WS Jang, YH Kim, SK Kim, BE Lee, HY Song, Antimycobacterial activity of methanolic plant extract of Artemisia capillaris containing ursolic acid and hydroquinone against *Mycobacterium tuberculosis*, J Infect Chemother 22 (4) (2016) 200–208.

[24] M Wu, P Guo, SW Tsui, H Chen, Z Zhao, An ethnobotanical survey of medicinal spices used in Chinese hotpot, Food Res Int 48 (1) (2012) 226–232.

[25] AP Raina, Z Abraham, Chemical profiling of essential oil of *Kaempferia galanga* L. germplasm from India, J Essent Oil Res 28 (1) (2016) 29–34.

[26]   TS Preetha, AS Hemanthakumar, PN Krishnan, A comprehensive review of *Kaempferia galanga* L.(Zingiberaceae): a high sought medicinal plant in Tropical Asia, Journal of Medicinal Plants Studies 4 (3) (2016) 270–276.

[27]   D Lakshmanan, J Werngren, L Jose, KP Suja, MS Nair, RL Varma, et al., Ethyl p-methoxycinnamate isolated from a traditional anti-tuberculosis medicinal herb inhibits drug resistant strains of *Mycobacterium tuberculosis in vitro*, Fitoterapia 82 (5) (2011) 757–761.

[28]   P Saravanan, G Chandramohan, J Mariajancyrani, P Shanmugasundaram, Extraction and application of eco–friendly natural dye obtained from leaves of *Acalypha indica* Linn on cotton fabric, International Research Journal of Environment Science 2 (12) (2013) 1–5.

[29]   S Thenmozhi, S Rajan, Screening of Antibacterial and Phytochemical activity of *Acalypha indica* Linn against isolated respiratory pathogens, Research in Plant Biology 2 (1) (2012) 1–6.

[30]   R Gupta, B Thakur, P Singh, HB Singh, VD Sharma, VM Katoch, et al., Anti-tuberculosis activity of selected medicinal plants against multi-drug resistant *Mycobacterium tuberculosis* isolates, Indian J Med Res 131 (6) (2010) 809–813.

[31]   N Satyanarayan, W Abaadani, SP Shekhar, S Harishkumar, Anti-tubercular activity of various solvent extracts of *Acalypha indica* l. against drug susceptible h37rv strain, World J Pharm Pharm Sci 5 (8) (2016) 957–965.

[32]   MT Hossain, MO Hoq, Therapeutic use of *Adhatoda vasica*, Asian Journal of Medical and Biological Research 2 (2) (2016) 156–163.

[33]   JM Grange, NJ Snell, Activity of bromhexine and ambroxol, semi-synthetic derivatives of vasicine from the Indian shrub *Adhatoda vasica*, against *Mycobacterium tuberculosis in vitro*, J Ethnopharmacol 50 (1) (1996) 49–53.

[34]   RK Upadhyay, Nutraceutical, pharmaceutical and therapeutic uses of *Allium cepa*: a review, International Journal of Green Pharmacy 10 (1) (2016) S46–S64.

[35]   HA Suleria, MS Butt, FM Anjum, F Saeed, N Khalid, Onion: nature protection against physiological threats, Crit Rev Food Sci Nutr 55 (1) (2015) 50–66.

[36]   V Kuete, Alliumcepa, Medicinal Spices and Vegetables from Africa, Academic Press, London EC2Y 5AS, United Kingdom, 2017, pp. 353–361.

[37]   A Sivakumar, G Jayaraman, Anti-tuberculosis activity of commonly used medicinal plants of south India, Journal of Medicinal Plants Research 5 (31) (2011) 6881–6884.

[38]   T Tesfahun, N Tsehaye, Review on therapeutic and medicinal use of *Aloe vera*, Cancer Biology 7 (4) (2017) 29–38.

[39]   A Surjushe, R Vasani, DG Saple, *Aloe vera*: a short review, Indian J Dermatol 53 (4) (2008) 163–166.

[40]   PK Sahu, DD Giri, R Singh, P Pandey, S Gupta, AK Shrivastava, et al., Therapeutic and medicinal uses of *Aloe vera*: a review, Pharmacology & Pharmacy 4 (08) (2013) 599–610.

[41]   H Mawarti, M Rajin, Z Asumta, The effects of *Aloe vera* on TNF-a levels, the percentage of nk cells and th 17 cells in rat that received izoniazid and rifampycin, Medical Archives 71 (5) (2017) 308–311.

[42]   NP Minh, Dried herbal tea production from *Eclipta prostrata*, Journal of Pharmaceutical Sciences and Research 11 (3) (2019) 684–687.

[43]   C Wiart, Medicinal Plants of Bangladesh and West Bengal: Botany, Natural Products, & Ethnopharmacology, CRC Press, Boca Raton, Florida, USA, 2019.

[44]   SN Dornala, SS Dornala, Clinical efficacy of Bhringarajasava as Naimittika Rasayana in Rajayakshma with special reference to pulmonary tuberculosis, AYU 33 (4) (2012) 523–529.

[45]   V Kumar, A Dey, MB Hadimani, T Marcović, M Emerald, Chemistry and pharmacology of *Withania somnifera*: an update, TANG 5 (1) (2015) 1–3.

[46]   N Singh, M Bhalla, P de Jager, M Gilca, An overview on ashwagandha: a Rasayana (Rejuvenator) of Ayurveda, African Journal of Traditional, Complementary and Alternative Medicines 8 (5S) (2011) 208–213.

[47] PK Debnath, J Chattopadhyay, A Mitra, A Adhikari, MS Alam, SK Bandopadhyay, J Hazra, Adjunct therapy of Ayurvedic medicine with anti tubercular drugs on the therapeutic management of pulmonary tuberculosis, Journal of Ayurveda and Integrative Medicine 3 (3) (2012) 141–149.

[48] AE Al-Snafi, The chemical constituents and pharmacological effects of Carum carvi—a review, Indian Journal of Pharmaceutical Science and Research 5 (2) (2015) 72–82.

[49] N Choudhary, V Khajuria, ZH Gillani, VR Tandon, E Arora, Effect of Carum carvi, a herbal bioenhancer on pharmacokinetics of antitubercular drugs: a study in healthy human volunteers, Perspectives in Clinical Research 5 (2) (2014) 80–84.

[50] PK Awasthi, SC Dixit, Chemical composition of *Curcuma longa* leaves and rhizome oil from the plains of Northern India, Journal of Young Pharmacists 1 (4) (2009) 312–326.

[51] LK Omosa, JO Midiwo, V Kuete, Curcuma longa, Medicinal Spices and Vegetables from Africa, Academic Press, London EC2Y 5AS, United Kingdom, 2017, pp. 425–435.

[52] X Bai, RE Oberley-Deegan, A Bai, AR Ovrutsky, WH Kinney, M Weaver, et al., Curcumin enhances human macrophage control of Mycobacterium tuberculosis infection, Respirology 21 (5) (2016) 951–957.

[53] S Tousif, DK Singh, S Mukherjee, S Ahmad, R Arya, R Nanda, et al., nanoparticle-formulated curcumin prevents Posttherapeutic disease reactivation and reinfection with Mycobacterium tuberculosis following isoniazid therapy, Front Immunol 8 (2017) 739.

[54] MR Adhvaryu, NM Reddy, BC Vakharia, Prevention of hepatotoxicity due to anti tuberculosis treatment: a novel integrative approach, World J Gastroenterol 14 (30) (2008) 4753–4762.

[55] S Dasaroju, KM Gottumukkala, Current trends in the research of *Emblica officinalis* (Amla): a pharmacological perspective, Int J Pharm Sci Rev Res 24 (2) (2014) 150–159.

[56] AM Sherwani, R Mohammad Zulkifle, A pilot trial of Jawarish Amla as adjuvant to anti-tubercular treatment drugs for control of adverse reactions in DOTS regime in pulmonary TB, J IMA 44 (1) (2012) 9.

[57] SA Tasduq, P Kaisar, DK Gupta, BK Kapahi, S Jyotsna, HS Maheshwari, et al., Protective effect of a 50% hydroalcoholic fruit extract of *Emblica officinalis* against anti-tuberculosis drugs induced liver toxicity, Phytother Res 19 (3) (2005) 193–197.

[58] I Tahir, MR Khan, NA Shah, M Aftab, Evaluation of phytochemicals, antioxidant activity and amelioration of pulmonary fibrosis with *Phyllanthus emblica* leaves, BMC Complementary and Alternative Medicine 16 (1) (2016) 406.

[59] MS Baliga, JJ Dsouza, Amla (*Emblica officinalis* Gaertn), a wonder berry in the treatment and prevention of cancer, Eur J Cancer Prev 20 (3) (2011) 225–239.

[60] S Mirunalini, M Krishnaveni, Therapeutic potential of *Phyllanthus emblica* (amla): the ayurvedic wonder, J Basic Clin Physiol Pharmacol 21 (1) (2010) 93–105.

[61] H Mohammadpour, E Moghimipour, I Rasooli, MH Fakoor, SA Astaneh, SS Moosaie, et al., Chemical composition and antifungal activity of *cuminum cyminum* L. essential oil from alborz mountain against Aspergillus species, Jundishapur Journal of Natural Pharmaceutical Products 7 (2) (2012) 50–55.

[62] AR Gohari, S Saeidnia, A review on phytochemistry of *Cuminum cyminum* seeds and its standards from field to market, Pharmacognosy Journal 3 (25) (2011) 1–5.

[63] BS Sachin, SC Sharma, S Sethi, SA Tasduq, MK Tikoo, AK Tikoo, et al., Herbal modulation of drug bioavailability: enhancement of rifampicin levels in plasma by herbal products and a flavonoid glycoside derived from *Cuminum cyminum*, Phytother Res 21 (2) (2007) 157–163.

[64] AO Sergio, CV Fabiola, NM Guadalupe, RC Blanca, HO León, Evaluation of antimycobacterium activity of the essential oils of cumin (*Cuminum cyminum*), clove (*Eugenia caryophyllata*), cinnamon (*Cinnamomum verum*), laurel (*Laurus nobilis*) and anis (*Pimpinella anisum*) against Mycobacterium tuberculosis, Advances in Biological Chemistry 3 (05) (2013) 480–484.

[65] M Reto, ME Figueira, HM Filipe, CM Almeida, Chemical composition of green tea (*Camellia sinensis*) infusions commercialized in Portugal, Plant Foods Hum Nutr 62 (4) (2007) 139.

[66] SM Chacko, PT Thambi, R Kuttan, I Nishigaki, Beneficial effects of green tea: a literature review, Chinese Medicine 5 (1) (2010) 1–9.

[67]   MR Honarvar, S Eghtesadi, P Gill, S Jazayeri, MA Vakili, MR Shamsardekani, et al., The effect of green tea extract supplementation on sputum smear conversion and weight changes in pulmonary TB patients: a randomized controlled trial, Medical Journal of the Islamic Republic of Iran 30 (2016) 381.

[68]   M Chen, J Deng, W Li, D Lin, C Su, M Wang, et al., Impact of tea drinking upon tuberculosis: a neglected issue, BMC Public Health 15 (1) (2015) 515.

[69]   A Agarwal, R Prasad, A Jain, Effect of green tea extract (catechins) in reducing oxidative stress seen in patients of pulmonary tuberculosis on DOTS Cat I regimen. Phytomedicine 17 (1) 2010 23–7.

[70]   SC Forester, JD Lambert, The role of antioxidant versus pro-oxidant effects of green tea polyphenols in cancer prevention, Mol Nutr Food Res 55 (6) (2011) 844–854.

[71]   KJ Gohil, JA Patel, AK Gajjar, Pharmacological review on *Centella asiatica*: a potential herbal cure-all, Indian Journal of Pharmaceutical Sciences 72 (5) (2010) 546–556.

[72]   K Ghosh, N Indra, G Jagadeesan, The ameliorating effect of *Centella asiatica* ethanolic extract on albino rats treated with isoniazid, J Basic Clin Physiol Pharmacol 28 (1) (2017) 67–77.

[73]   K Srinivasan, Ginger rhizomes (*Zingiber officinale*): a spice with multiple health beneficial potentials, Pharma Nutrition 5 (1) (2017) 18–28.

[74]   AH Rahmani, Active ingredients of ginger as potential candidates in the prevention and treatment of diseases via modulation of biological activities, International Journal of Physiology, Pathophysiology and Pharmacology 6 (2) (2014) 125–136.

[75]   S Kumar, UN Singh, K Saxena, R Saxena, Supplementation of ginger with anti-tuberculosis treatment (ATT): a better approach to treat anemic pulmonary tuberculosis patients, Int J Herb Med 1 (2013) 17–20.

[76]   RA Kulkarni, AR Deshpande, Anti-inflammatory and antioxidant effect of ginger in tuberculosis, Journal of Complementary and Integrative Medicine 13 (2) (2016) 201–206.

[77]   BK Shin, SW Kwon, JH Park, Chemical diversity of ginseng saponins from *Panax ginseng*, Journal of Ginseng Research 39 (4) (2015) 287–298.

[78]   T Lakshmi, A Roy, RV Geetha, *Panax ginseng*—a universal panacea in the herbal medicine with diverse pharmacological spectrum–a review, Asian Journal of Pharmaceutical and Clinical Research 4 (1) (2011) 14–18.

[79]   YM Lee, H Yoon, HM Park, BC Song, KJ Yeum, Implications of red *Panax ginseng* in oxidative stress associated chronic diseases, Journal of Ginseng Research 41 (2) (2017) 113–119.

[80]   S Oliynyk, S Oh, Actoprotective effect of ginseng: improving mental and physical performance, Journal of Ginseng Research 37 (2) (2013) 144–166.

[81]   E Nocerino, M Amato, AA Izzo, The aphrodisiac and adaptogenic properties of ginseng, Fitoterapia 71 (2000) S1–S5.

[82]   SF Rahman, K Sijam, D Omar, *Piper sarmentosum* Roxb.: A mini review of ethnobotany, phytochemistry and pharmacology, Journal of Analytical & Pharmaceutical Research 2 (5) (2016) 1–3.

[83]   EWE Chan, SK Wong, Phytochemistry and pharmacology of three Piper species: an update, J Phcog 1 (9) (2014) 534–544.

[84]   K Hussain, Z Ismail, A Sadikun, P Ibrahim, Antioxidant, anti-TB activities, phenolic and amide contents of standardised extracts of Piper sarmentosum Roxb, Nat Prod Res 23 (3) (2009) 238–249.

[85]   S Takaidza, M Pillay, FM Mtunzi, Biological activities of species in the genus Tulbaghia: a review, African Journal of Biotechnology 14 (45) (2015) 3037–3043.

[86]   I Raji, K Obikeze, P Mugabo, Potential beneficial effects of *Tulbaghia violacea* William Henry Harvey (Alliaceae) on cardiovascular system—a review, Tropical Journal of Pharmaceutical Research 14 (6) (2015) 1111–1117.

[87]   K Moodley, K Joseph, Y Naidoo, S Islam, I Mackraj, Antioxidant, antidiabetic and hypolipidemic effects of *Tulbaghia violacea* Harv. (wild garlic) rhizome methanolic extract in a diabetic rat model, BMC Complementary and Alternative Medicine 15 (1) (2015) 408.

[88]   M Díaz, L Castillo, CE Díaz, R Guillermo'Alvarez, A González-Coloma, C Rossini, Differential deterrent activity of natural products isolated from *Allophylus edulis* (Sapindaceae), Advances in Biological Chemistry 4 (2) (2014) 168–179.

[89] RB Chavan, DK Gaikwad, The ethnobotany, phytochemistry and biological properties of Allophylus species used in traditional medicine: a review, World Journal of Pharmacy and Pharmaceutical Sciences 5 (11) (2016) 664–682.

[90] LN Trevizan, KF do Nascimento, JA Santos, CA Kassuya, CA Cardoso, Carmo do, et al., Anti-inflammatory, antioxidant and anti-*Mycobacterium tuberculosis* activity of viridiflorol: the major constituent of *Allophylus edulis* (A. St.-Hil., A. Juss. & Cambess.) Radlk, J Ethnopharmacol 192 (2016) 510–515.

[91] RHMJ Lemmens, *Euadenia eminens* Hook.f, in: G.H. Schmelzer, A. Gurib-Fakim (Eds.), Prota 11(2): Medicinal plants/Plantes médicinales 2, PROTA, Wageningen, Netherlands, 2013. http://uses.plantnet-project.org/en/Euadenia_eminens_(PROTA) Accessed April 8, 2020.

[92] RA Dickson, TC Fleischer, E Ekuadzi, G Komlaga, Anti-inflammatory, antioxidant, and selective antibacterial effects of *Euadenia eminens* root bark, African Journal of Traditional, Complementary and Alternative Medicines 9 (2) (2012) 271–276.

[93] AS Babadzhanov, N Abdusamatova, FM Yusupova, N Faizullaeva, LG Mezhlumyan, MK Malikova, Chemical composition of *Spirulina platensis* cultivated in Uzbekistan, Chem Nat Compd 40 (3) (2004) 276–279.

[94] VP Kostromina, OV Derkach, NV Symonenkova, OO Riechkina, AO Otroshchenko, Evaluation of the efficacy of a plant adaptogen (spirulina) in the pathognic therapy of primary tuberculosis in children, Likars' ka Sprava (5-6) (2003) 102–105.

[95] K Bharavi, AG Reddy, GS Rao, PR Kumar, DS Kumar, PP Prasadini, Prevention of cadmium bioaccumulation by herbal adaptogens, Indian J Pharmacol 43 (1) (2011) 45.

[96] B Joseph, VM Nair, Ethanopharmacological and phytochemical aspects of *Ocimum sanctum* Linn-the elixir of life, British Journal of Pharmaceutical Research 3 (2) (2013) 273–292.

[97] S Verma, Chemical constituents and pharmacological action of *Ocimum sanctum* (Indian holy basil-Tulsi), Int J Phytopharm 5 (5) (2016) 205–207.

[98] MK Khosla, Sacred tulsi (*Ocimum sanctum* L.) in traditional medicine and pharmacology, Ancient Science of Life 15 (1) (1995) 53–61.

[99] MM Cohen, Tulsi—*Ocimum sanctum*: A herb for all reasons, J Ayurveda Integr Med 5 (4) (2014) 251–259.

[100] DJ McKenna, K Hughes, K Jones, Astragalus, Altern Ther Health Med 8 (6) (2002) 34.

[101] S Sinclair, Chinese herbs: a clinical review of Astragalus, Ligusticum, and Schizandrae, Altern Med Rev 3 (1998) 338–344.

[102] Teachey S Herbal wisdom: 8 benefits of adaptogens. https://nectarapothecary.com/tag/ginseng/, Accessed December 11, 2019.

[103] L Dalton, Licorice, Chem Eng News 80 (32) (2002) 37.

[104] M Koithan, K Niemeyer, Using herbal remedies to maintain optimal weight, The Journal for Nurse Practitioner 6 (2) (2010) 153–154.

[105] Morris S 15 Supplements for Leaky Gut (Including 3 Key Essential Oils). https://www.zyto.com/15-supplements-for-leaky-gut, Accessed December 11, 2019).

[106] RP Brown, PL Gerbarg, Z Ramazanov, *Rhodiola rosea*. A phytomedicinal overview, HerbalGram 56 (2002) 40–52.

[107] F Khanum, AS Bawa, B Singh, *Rhodiola rosea*: a versatile adaptogen, Comprehensive Reviews in Food Science and Food Safety 4 (3) (2005) 55–62.

[108] A Panossian, G Wikman, Effects of adaptogens on the central nervous system and the molecular mechanisms associated with their stress—protective activity, Pharmaceuticals 3 (1) (2010) 188–224.

[109] G Chaudhary, PK Dantu, Morphological, phytochemical and pharmacological, studies on *Boerhaavia diffusa* L, Journal of Medicinal Plants Research 5 (11) (2011) 2125–2130.

[110] P Nayak, M Thirunavoukkarasu, A review of the plant *Boerhaavia diffusa*: its chemistry, pharmacology and therapeutical potential, J. Phytopharmacol 5 (2) (2016) 83–92.

[111]   S Mishra, V Aeri, PK Gaur, SM Jachak, Phytochemical, therapeutic, and ethnopharmacological overview for a traditionally important herb: *Boerhavia diffusa* Linn, Biomed Res Int 2014 (2014). Article ID 808302, doi:10.1155/2014/808302.

[112]   SK Desai, SM Desai, S Navdeep, P Arya, T Pooja, Antistress activity of *Boerhaavia diffusa* root extract and a polyherbal formulation containing *Boerhaavia diffusa* using cold restraint stress model, Int J Pharm Pharm Sci 3 (1) (2011) 130–132.

[113]   NM Patel, NJ Dhimmar, V Lambole, Pharmacological activities of *Boerhaavia diffusa*: a review, Research Journal of Pharmacy and Technology 8 (4) (2015) 496–502.

[114]   M Muthulingam, Antihepatotoxic role of *Boerhaavia diffusa* (Linn.) against antituberculosis drug rifampicin induced hepatotoxicity in male Albino wistar rats, Journal of Pharmacy Research 8 (9) (2014) 1226–1232.

[115]   GK Jayaprakasha, T Selvi, KK Sakariah, Antibacterial and antioxidant activities of grape (*Vitis vinifera*) seed extracts, Food Res Int 36 (2) (2003) 117–122.

[116]   S Sreemantula, S Nammi, R Kolanukonda, S Koppula, KM Boini, Adaptogenic and nootropic activities of aqueous extract of *Vitis vinifera* (grape seed): an experimental study in rat model, BMC Complementary and Alternative Medicine 5 (1) (2005) 1–8.

[117]   MA Jiménez-Arellanes, GA Gutiérrez-Rebolledo, M Meckes-Fischer, R León-Díaz, Medical plant extracts and natural compounds with a hepatoprotective effect against damage caused by antitubercular drugs: a review, Asian Pacific Journal of Tropical Medicine 9 (12) (2016) 1141–1149.

[118]   Y Akhmad, I Shakhista, Phytochemical characterization of *Rhodiola heterodonta* dry extract, European Science Review (7-8) (2015) 45–46.

[119]   MH Grace, GG Yousef, AG Kurmukov, I Raskin, MA Lila, Phytochemical characterization of an adaptogenic preparation from *Rhodiola heterodonta*, Natural Product Communications 4 (8) (2009) 1053–1058.

[120]   P Kulyal, UK Tiwari, A Shukla, AK Gaur, Chemical constituents isolated from *Andrographis paniculata*, Indian J Chem 49B (2010) 356–359.

[121]   AK Thakur, SS Chatterjee, V Kumar, Adaptogenic potential of andrographolide: an active principle of the king of bitters (*Andrographis paniculata*), Journal of traditional and Complementary Medicine 5 (1) (2015) 42–50.

[122]   P Bhatter, P Gupta, P Daswani, P Tetali, T Birdi, Antimycobacterial efficacy of Andrographis paniculata leaf extracts under intracellular and hypoxic conditions, Journal of Evidence-based Integrative Medicine 20 (1) (2015) 3–8.

[123]   M Radji, M Kurniati, A Kiranasari, Comparative antimycobacterial activity of some Indonesian medicinal plants against multi-drug resistant *Mycobacterium tuberculosis*, Journal of Applied Pharmaceutical Science 5 (1) (2015) 19–22.

[124]   R Widhawati, E Hanani, J Zaini, *Andrographis Paniculata* (Burm. F.) Nees induces clinical and sputum conversion in pulmonary tuberculosis patients, Medicine Science International Medical Journal 4 (1) (2015) 1869–1875.

PART 7

Cardiovascular disorders

CHAPTER 24

Cardiovascular disorders and herbal medicines in India

Anurag Mishra[a], Sivakumar Vijayaraghavalu[b] and Munish Kumar[a]

[a]Department of Biochemistry, University of Allahabad, Prayagraj, Uttar Pradesh, India [b]Narayana Translational Research Centre, Narayana Medical College and Hospital, Nellore, Andhra Pradesh, India

24.1 Introduction

Cardiovascular disease (CVD) is one of the groups of disorders that include the disease of the heart and blood vessels. CVD is one of the major leading causes of death (approximately 31% of deaths) worldwide. According to the World Health Organisation, billions of dollars are expended in the healthcare cost by various nations for CVD management in the world and it is expected to exceed shortly in the near future [1]. In the last 1 year due to the COVID-19 pandemic, the risk of CVD in the elderly population has significantly increased around the world [2]. CVD is a noncommunicable disease affecting both genders, different age groups, psychosocial values, and socioeconomic status; with the death toll of more than a million each year globally. In the Indian subcontinent, approximately 63% of deaths are caused by various noncommunicable diseases annually out of which about 27% attributed to CVD that affects almost 45% of people in the age group of 40–69 years. Though recent trend shows that the young generation in the country is strongly affected by the circumference of this disease. The socioeconomic burdens due to sudden and premature deaths reduce the quality of life in the affected population tremendously in ongoing developing countries like India [3]. Conventional cardiovascular disorder is strongly attributed to the changes in lifestyle and altered metabolic activity [4].

Several associated disorders of CVD includes coronary heart disease (CHD), peripheral arterial disease (PAD), cerebrovascular disease, rheumatic heart disease (RHD), deep vein thrombosis (DVT) and pulmonary embolism (PE), congenital heart disease (CHD), atherosclerosis, arrhythmias, cardiac arrest, etc. [5]. In India, the common symptoms of CVD prominently appear in the urban population as compared to the rural population because of high cholesterol and household air pollution causing important risk factors for various disorders of CVD. The young population is now affected by the CVD that create a huge economic burden, social and moral imbalance in the developing country like India [6].

Herbal Medicines: A Boon for Healthy Human Life.
DOI: https://doi.org/10.1016/B978-0-323-90572-5.00005-6

525

A list of major risk factors has been discovered and identified which induce, stimulate, activate, and accelerate the rate of CVD progression in any individual. Some major risk factors for CVD have been discussed that include alcohol abuse, smoking, diabetes, physical inactivity, stress, high blood pressure (BP), unhealthy diet, use of certain medicine, family history, age, and ethnicity or race. The above factors attribute and play a central role in severe risk for cardiovascular disorders [2,7].

For the prevention and management of cardiovascular disorders in the country, a variety of modern allopathic medicines are being used widely for the last many decades but still, further research is required to look out for other alternatives. The high cost and major side effects of modern allopathic medicines are directed to the use of an alternative therapeutic option that is strongly related to our traditional medicine system [8]. Recently, the popularity and value of interest in the application of alternative medicines and natural plant products increased beyond the expectancy in the country. A variety of herbs and herbal products (medicinal plants or extracts) called natural remedies have been used in the patients as remedies for various abnormalities and adverse conditions of CVD [9]. Nowadays the values of herbal medicines and natural products have been reviewed and revised the interest of traditional remedies that had been used in immemorable time to provide better treatment and easier recovery for CVD [10]. Although our traditional therapeutic options are highly effective in improving the variety of harmful symptoms, decelerating adverse cardiac remodeling, and reducing morbidity and mortality rate [11]. Herbal medicines and natural plant products or extracts express unique features like create no or minimal cytotoxicity, genotoxicity, and side effects in the CVD-affected populations. Its availability, affordable cost, and easier to use; all these commercial properties are leading for the huge cultivation, production, and demand in the global market. Many traditional and regional herbs and plant products are significantly used as a CVD remedy by the local population without expending any effective cost or effort. The above-discussed values are increasing the interest in conservation and cultivation of such medicinally important natural remedies in the massive region across the world [12]. Many species of medicinal plants (herbs) play a crucial role in reducing the cause of CVD and provide immense protections to fight against CVD abnormalities; such medicinal plants are as Garlic (*Allium sativum*), Guggul (*Commiphora wightii*), Hawthorn (*Crataegus oxyacantha*), Arjuna (*Terminalia arjuna*), Amla (*Emblica officinalis*), Onion (*Allium sepa*), Ginger (*Zingiber officinale*) Triphala, etc. these can decelerate the risk of CVD effects and assist in improving health conditions [13]. Natural remedies contain a list of plant compounds like polyphenols, flavonoids, indirubin, resveratrol, icariin, ascorbic acid bergamot, etc. these are obtained or extracted from various medicinally important plants and have good efficacy to act as anti-low-density lipoprotein (LDL), anti-TAG, antihypertensive, and in improving high-density lipoprotein (HDL) level in CVD affected patients. Many substantial pharmacological studies are providing great evidence that shows the acceptability, reliability, effectivity, and major connectivity of herbal medicines on CVD management strategy [14].

Although with the high global prevalence and prominent use of herbal medicines in the prevention and management of CVD as well as in other aspects, the information regarding the therapeutic use or safety of herbal remedies is usually obtained from ancient books, traditional and nonreliable sources rather than relying on existing parameters of scientific research [15].

24.1.1 Cardiovascular disease study and herbal medicines

The alarm of increasing CVD number in the world population and its associated economic burden poses a major concern for governments across the world. In particular, for the developing and underdeveloped countries, efforts/measures taken so far were not able to control the incidence or reduce the morbidity and mortality rate significantly; this issue urges to review, revise and highlight all the effective parameters either in conventional or alternative therapy for CVD management strategy at the ground zero level [1,2]. In this chapter, the use of herbal medicines that were routinely administered or medicinal plants useful in managing CVD is being reviewed.

As per the 2018 WHO report, 91% of people from South East Asia region including the Indian subcontinent were dependent on traditional medicines and conventional medicines (T & CM); similarly, 87% from the African continent; 80% from the Americas; 90% from Easter Mediterranean countries and 89% from European countries are relying on T & CM for their ailments [16].

Our ongoing traditional medicine system is focusing on the use of various herbs, plants, plant products, or extracts which are usually obtained from a variety of natural as well artificial habitats. In addition, the use of minerals and metals extracted from the land and marine is too effective in reducing morbidity and mortality caused due to CVD [13,60]. Medicinally important herbs or plants products mainly targeting cardiovascular disorders by acting as an antioxidant, anti-inflammatory, antihypertensive, antihypercholesterolemic, and anti-TAG; these potentially provide an excellent defensive system as well as reducing and remodeling the adverse effects of heart and blood vessels caused by CVD and help to overcome CVD symptoms with no or minimal side effects. Due to cos-effectiveness, increased efficacy, local availability are some of the important features that popularize the usage of medicinal plants among the people and it has drawn the attention of WHO as well as the governments of many other countries across the globe [15,61].

24.2 Cardiovascular disease and disorders

A group of associated disorders like CHD, peripheral artery disease (PAD), cerebrovascular disease, RHD, DVT and PE, CHD, arrhythmias, and atherosclerosis is all are entirely integrated with CVD. One of these associated disorders has the efficacy to trigger and raise the level of CVD in any affected individuals [17]. CVD disorders are discussed below:

24.2.1 Coronary heart disease

In the Indian subcontinent, CHD or coronary artery disease (CAD) is the condition of narrowing or blockage of coronary arteries that carries oxygen to the heart. This is one of the major constituents of CVD which is attributed to approximately two-thirds of the total burdens of CVD in the country [18]. Within India, CAD numbers in rural and urban populations make a huge difference. CAD prevalence in the rural population is just half as compared with the urban population [19]. A productive age group of the urban population is strongly affected by CHD. Under the age of 50 years, the risk of CHD is increased by having a positive family history of CHD. Family history (includes numbers of relatives, age, and gender) is a significant constituent that shares common features like family behavior, and genetic heritage which are effectively used for prior prediction of CHD in any affected person [20]. Smoking, use of tobacco, high BP, high cholesterol level in blood, diabetes, lack of exercise, and obesity are the major risk factors that trigger and frequently raised the rate of CHD in older age [21].

In many recent studies, three main types of CHD are significantly observed. These include:

- Obstructive CAD.
- Nonobstructive coronary disease.
- Coronary microvascular disease.

In both obstructive and nonobstructive CAD, the normal function of large arteries is affected on the surface of the heart by the action of CHD whereas coronary microvascular disease affects the tiny arteries in the heart muscles [22]. CHD is usually caused by atherosclerosis (a process of build-up of fatty materials and plaque inside the coronary arteries) but the actual cause depends on the type of CAD [21]. The common symptoms of CHD are chest pain, shortness of breathing during exercise, and heart attack. Some other symptoms also appear during CHD; these are included as angina, cold sweats, dizziness, light-headedness, neck pain, and sleep disturbances [23].

24.2.2 Peripheral artery disease

PAD is generally referred to as the disease of the blood vessels (that includes both arteries and veins), which are located outside the heart and brain. It is one of the common circulatory problems in which narrowed blood vessels (arteries) reduced blood flow to the limbs [24]. When PAD develops in someone's health, legs or arms usually don't receive sufficient blood flow to keep up with its demand that may cause leg pain during walking (claudication). A PAD is also known as peripheral vascular disease. The common sign or symptoms of PAD included pain in one or both hips, thighs, or calf muscles after a certain activity, the appearance of shiny skin on legs, getting pain during the use of arms such as manual tasks, coldness in lower leg or foot as compare with other side and appearance of sores on toes, feet or legs which would not heal easily [25].

A PAD is mostly caused by the build-up of fatty deposits in the walls of arteries and reduces blood flow to the limbs or other body parts. Many Risk factors trigger the normal circulatory function of peripheral arteries throughout the body and commonly lead to reduced blood flow. Factors that increase the risk of developing PAD include high BP, high cholesterol level, age, atherosclerosis, smoking, diabetes, obesity, a family history of PAD, and heart diseases. If the PAD is caused by a build-up of plaque in the blood vessels which can cause a greater risk of developing following adverse conditions [26].

- Critical limb ischemia.
- Stroke and heart attack.

24.2.3 Cerebrovascular disease

In common terms Cerebrovascular is a combination of two words—"Cerebro" which refers to the larger part of the brain and "vascular" which means arteries and veins, imply the blood flow in the region of the brain [27]. Cerebrovascular disease includes all those adverse conditions in which a particular area of the brain is temporarily or permanently affected by bleeding, clotting, or ischemia and pathologically brain changes that correlate to cognitive functions like Alzheimer's disease and vascular Dementia. Cerebrovascular disease is also included like carotid stenosis, vertebral stenosis, intracranial stenosis, stroke, vascular malformations, aneurysms, etc. [28]. The risk factors for cerebrovascular disease are generally two types, for example, some risk factors cannot be controlled by any individual, and these are included as age, gender, and family history. However other risk factors are integrated into lifestyle changes that can be efficiently controlled at the individual level. A CVD-affected person can reduce the risk of cerebrovascular disorders by altering the common lifestyle activity. Lifestyle-related factors that can accelerate the developing risk of cerebrovascular disease include high blood cholesterol level, heavy drinking of alcohol, high BP, smoking, diabetes, fat and salt-rich diet but low in fiber, fruit, and vegetables, lack of regular exercise, and obesity [29].

Various signs and symptoms appear in a patient who is suffering from cerebrovascular disease and some of these are used to access through physical examinations. These signs may include some or all of the symptoms that are usually sudden like dizziness, nausea or vomiting, severe headache, confusion and disorientation, memory loss, loss of vision or difficulty seeing, loss of coordination or the ability to walk and move, numbness, weakness in an arm, leg or the face (especially on one side), abnormal, or slurred speech and difficulty with comprehension [30].

24.2.4 Rheumatic heart disease

RHD is an adverse condition that effects on heart valve which has been permanently damaged by rheumatic fever and rheumatic carditis. If the streptococcal infection on strep throat or scarlet fever is untreated or under-treated, heart valve damage may start shortly [31]. An inflammatory condition in the body caused by an autoimmune response can lead to ongoing heart valve damage and it also affects many connective tissues in the joints, skin, or brain. RHD includes embolic stroke, heart failure, arterial fibrillation, and endocarditic [32]. On an inflammatory response, heart valves become scarred over a period resulting in narrowing or leaking of valves that make it harder for normal heart functions [33].

The risk factors for RHD include age, gender, and the environmental effect these act as potent factors for RHD. But some other associated conditions such as untreated or under-treated streptococcal infections can also raise the risk for RHD. Children who frequently get streptococcal throat infections are having more chances for rheumatic fever and RHD [31]. The symptoms of RHD are depending on the degree of valve damage. The most common symptoms are fever, weakness, swollen, red tender, extremely painful joints, shortness of breath, chest discomfort, uncontrolled movements of arms, legs, or facial muscles, lack of coordination or balance, and nodules (lumps under the skin) [34].

24.2.5 Deep vein thrombosis and pulmonary embolism

In India as well as in several other countries, deep venous thrombosis (DVT), and PE are the two important manifestations of venous thromboembolism (VTE), which are the most common life-threatening CVD. DVT refers to the formation of blood clots (single blood clot is called a "thrombus" while multiple clots are called "thrombi") in the body's large veins, mostly in the lower limbs (e.g., lower leg or calf). Blood clots can cause a partial or complete blockage of blood circulation in the vein. DVT leads to generate pain, swelling, tenderness, discoloration, or redness in the affected region [35–36].

PE is a serious life-threatening complication that arises from DVT which occurs in every one-third of DVT patients. PE commonly occurs when a portion of the blood clot breaks loose and started to move in the bloodstream, initially it reaches the heart and after that to the lungs, where it can partially or fully block the pulmonary artery or its branches [37]. PE frequently causes sudden death, when single or multiple vessels that supply blood to the lungs are completely blocked by the effect of the clot. But some people show survival conditions because the body's natural mechanisms tend to reabsorb or lyses of blood clot [35,38].

Many risk factors for DVT are involved in the increasing rate of blood clot formation in the vessels which can partially or completely block the circulation of blood throughout the body. Risk factors for DVT include **age, sitting for a long time, prolonged bed rest, injury**

or surgery, pregnancy, birth control pills or hormone replacement therapy, obesity, smoking, cancer, heart failure, inflammatory bowel disease, a personal or family history of DVT or PE and genetics. A person that develops or suffering from DVT and PE shows common signs and symptoms that may include shortness of breath, rapid heartbeat, sweating, sharp chest pain (especially during deep breathing), cough up blood, and low BP [39].

24.2.6 Congenital heart disease

CHD is one of the most common congenital malformations diagnosed in newborn baby and the problem can be associated mainly with one of the following adverse heart conditions.

- The heart walls.
- The heart valves.
- The blood vessels.

Despite the above three abnormalities, other numerous types of congenital heart defects have also been detected in the newborn baby, a defects may be simple that does not cause symptoms or complex that may cause severe life-threatening symptoms. In the past several decades the survival numbers and improving conditions of patients suffering from CHD is increased. Now it becomes possible by early diagnosis and advances in cardiac surgery [40]. The combination of anatomic abnormalities, clinical intervention, and increased risk factors of CVD among CHD survivors can have more chances to increase the risk to develop CVD throughout life. A recent study suggests that the common risk factors of CVD such as hypertension and obesity are more prevalent in people affected with CHD than those without CHD [41].

Most of the congenital heart defects problems linked with the early growth and development of a child's heart, the main cause of which is unknown. However certain environmental and genetic factors trigger CHD. The risk factors for CHD included as use of certain medications, drinking alcohol during pregnancy, smoking, heredity, rubella (German measles), and diabetes [42]. In some cases, the common symptoms of congenital heart defects may not be appearing until shortly after birth. Newborns with congenital heart defects may experiences such symptoms included as breathlessness or trouble in breathing, feeding difficulties, low birth weight, chest pain, delayed growth, bluish lips, skins, etc. In other cases, the symptoms of congenital heart defects may not appear until many years of birth. Whenever symptoms begin to develop, they may cause abnormal heart rhythms, swelling, fatigue, dizziness, trouble breathing, and fainting [43].

24.2.7 Arrhythmias

An arrhythmia is an abnormality that relates to the rate or rhythm of the heartbeat. In arrhythmic conditions, the heart may beat too fast, too slowly, or be found with an irregular rhythm.

When a heartbeat is very fast, the condition is called Tachycardia. When a heartbeat is too slow, the condition is generally called Bradycardia. Arrhythmia or heartbeats problem occurs when the electrical impulses that coordinate with heartbeats are not working properly and causing the heart to beat too fast, too slow, or in an irregular way. In a healthy human, being heart beats approximately 60–100 beats per minute but in arrhythmic disorder, the heart is beating beyond the given expected range and causes cardiac arrest, stroke, etc. Certain conditions and many risk factors trigger normal heart's function that stimulates and increase the developing risk of arrhythmia (abnormal and irregular heartbeats), these factors include CAD, previous heart surgery, high BP, CHD, thyroid problems, diabetes, electrolyte imbalance, drugs and supplements, drinking too much alcohol, caffeine, nicotine, and illegal drug use [44].

Arrhythmia may not cause any serious signs or symptoms but few noticeable symptoms may be included as a fluttering in the chest, a racing heartbeat (tachycardia), a slow heartbeat (bradycardia), chest pain, shortness in breathing, anxiety, fatigue, light-headedness or dizziness, sweating, etc. [45].

24.2.8 Atherosclerosis

Atherosclerosis is a lifelong progression of thickening and hardening of the walls of large and medium-sized arteries as the result of fat deposits. The abnormal deposition (build-up) of fats, cholesterol, cellular waste products, calcium, and fibrin on the inner lining of arteries is called plaque. The plaque generally causes the blood vessel to thickens by which arteries become narrower and block general blood circulation (blood flow) throughout the body. It can also burst and leading to the formation of a blood clot. A Plaque may partially or completely block blood flow through large or medium-sized arteries in the heart, brain, kidneys, pelvis, legs, or arms. Although the term atherosclerosis is often considered a cardiac problem and it may also affect the overall arteries of the entire circulation [46].

Some persisting conditions developed by plaque as CHD, carotid artery disease, PAD, angina, chronic kidney disease, and Aneurysms. The basic cause of atherosclerosis is not completely known whereas major possible expected damage included elevated cholesterol and triglycerides (TG) in the blood, high BP, cigarette smoking, type 1 diabetes, obesity, physical inactivity, high saturated fat diet, high levels of C-reactive protein (a marker of inflammation), etc. Smoking plays a significant role in the continuous deposition of fats (accelerate plaque) and leads to the progression of atherosclerosis in the aorta, coronary arteries as well arteries in the legs. The common symptoms of atherosclerosis include chest pain or angina, pain in the leg, arm, etc., shortness of breath, fatigue, confusion (which occurs if the blockage affects circulation to the brain), and muscle weakness (due to lack of circulation) [47].

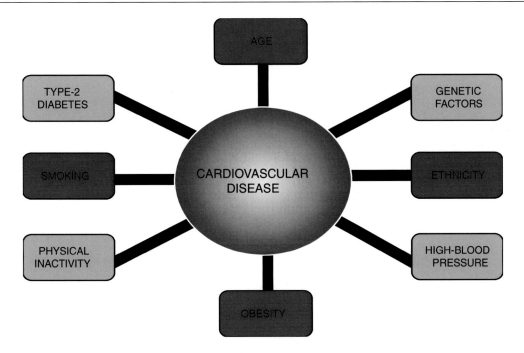

Figure 24.1
Major risk factors for cardiovascular disease.

24.3 Major risk factors for cardiovascular disease

CVD is a setup of abnormal conditions or disorders that strongly affect the normal functions of heart and blood vessels that are prominently included with high BPs (hypertension), cardiac arrest, stroke, atherosclerosis (accumulation and deposition of fat in the artery), etc., and can be found in any individuals at any age group throughout the life, including children [48]. Several risk factors are interrelated, responsible to stimulate and activate the rate of progression of CVD. Stimulating or inducing risk factors triggered with CVD or one of its associated disorders such as CHD, cerebrovascular disease, PAD, arrhythmias, heart attack, etc. [49]. The complex interaction between genes and the genetic variant with the environment can also accelerate the cause of CVD [50].

It has been seen that major risk factors are directly associated with the cause of the severe risk of CVD. In the absence of risk factors, mortality and morbidity remain unaffected but its involvement increasing the incidence of CVD [51]. Frequently observed risk factors (Fig. 24.1) that causing CVD are diabetes mellitus, smoking, use of alcohol, high BP, obesity, genetic disorders, physical inactivity, age, ethnicity, etc. Many recent studies suggested that both triacylglycerol (TAG) and cholesterol play a dominant role in causing various adverse health conditions and stabilize powerful association with impaired functions of heart and blood vessels [52].

Over the past few decades, the field of medicine and surgery has made a big boon to mankind in the area of diagnosis, prevention, and treatment of CVD. The two major studies of Framingham Heart Study and the Seven Countries' Study made an important contribution in predicting and identifying major risk factors causing CVD. According to Framingham risk score is a widely accepted and recognized tool used to calculate 10-year cardiovascular disorders in any individual and classify them as the risk of myocardial infarction or coronary death. For a prior indication of major CHD across ethnic groups and races, the Framingham risk score has been used effectively [53]. All the risk factors for CVD have been classified into two broad categories.

24.3.1 Existing or traditional (conventional) risk factors for cardiovascular disease

Today more than 300 existing risk factors for CVD have been discovered across the world. The study of existing risk factors is based on three such criteria [54].

a.   High prevalence in different populations.
b.   Significant independent impact on the risk of CHD and stroke.
c.   Reduced risk of cardiovascular disorder with treatment and control.

Existing risk factors of any individual or a group of affected population mostly fall into two categories:

24.3.1.1 Modifiable risk factors

As shown in Table 24.1, modifiable risk factors are those which can be treated or controlled and reduce the major risk for CVD in affected ones. In many developed countries, several modified risk factors are found that include high BP, high blood cholesterol level, obesity, tobacco consumption, and diabetes mellitus; these contributed to approximately every one-third of entire CVD cases. While in other developing countries, in addition to all these conventional risk factors some other factors also integrating like low vegetable and fruit intake (unhealthy diet), regular alcohol use, low socioeconomic status, etc. All major conventional risk factors (modifiable and nonmodifiable) for CVD in both developed and underdeveloped countries are listed (Table 24.1) [54,56].

24.3.1.2 Nonmodifiable risk factors

Nonmodifiable risk factors (Table 24.1) could not be modified to reduce the effect of the CVD burden that includes specific age groups, gender, family history, etc. In a certain study, approximately 30%–50% of CHD patients have been found asymptomatic with no appearance

Table 24.1: List of major existing or traditional risk factors for cardiovascular disease.

Modifiable risk factors	Nonmodifiable risk factors
h Dyslipidemia	h Specific age group
h Tobacco (smoking/ chewing)	h Gender
h Lack of physical activity	h Family history
h Regular alcohol intake	h Ethnicity or race
h Hypertension	
h Diabetes Mellitus	
h Abdominal obesity	
h Psychosocial factors	
h Low social economic status	
h Mental ill-health conditions	
h Left ventricular hypertrophy	
h Certain medications	
h Unhealthy diet	

of any conventional risk factors and it has also been reported that even the modest elevation in BP, glucose levels, and cholesterol would predispose an individual to a CVD risk [55].

Another way of explaining major risk factors as established risk factors and emerging risk factors and their meaningful role in CVD is shown in Fig. 24.2.

24.3.2 Emerging risk factors for cardiovascular disease

Substantial body evidence supported that the reduction of risk of existing or traditional risk factors (modifiable or nonmodifiable) directing the search for new emerging risk factors that also play a principal role in CVD management. Recently, over 100 emerging risk factors have been discovered globally, most of these are the existing risk factors that have their independent risk predictive potential. Any new risk factor or prospective biomarker will be referred to as an emerging risk factor if it is capable to address the following three questions [54,57,58].

1. Is a prospective biomarker is readily measured or not?
2. Does a prospective biomarker add value to existing tests and improve its risk predictive efficiency?
3. Will a prospective biomarker enhance clinician's decision-making power and improve patient management?

Thus above processed prospective biomarkers will be considered as emerging risk factors since they are readily associated with increased risk of CVD. While their quantitative, causative, and independent contributions to CVD are not fully understood concerning the

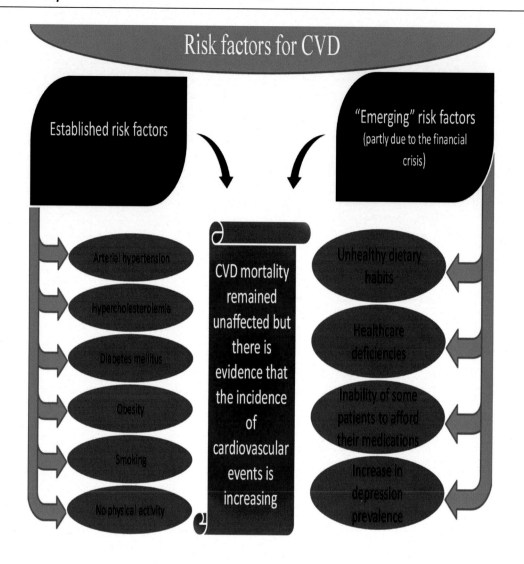

Figure 24.2
Established and Emerging risk factors for CVD. *CVD*, Cardiovascular disease.

existing risk factors. The emerging risk factors contribute to reclassify as intermediate patients' risk for major CHD, which required more aggressive risk reduction [58]. Emerging risk factors are broadly categorized as either early markers of risk due to existing disease or markers of genetic polymorphisms. Name of all emerging risk factors which are associated either with the marker of an existing disease or with the marker of genetic polymorphism listed in Table 24.2 [59].

Table 24.2: List of emerging risk factors for cardiovascular disease.

Markers of existing disease and genetic polymorphism
h C-reactive protein
h Lipoproteins
h Apolipoproteins A and B
h Fasting glucose level
h Leucocytes count (TLC)
h Coronary artery calcium level
h Vitamin-D
h Homocysteine
h Periodontal disease
h Fibrinogen
h Lipoprotein phospholipase A2
h Anchal bronchial index
h Myeloperoxidase
h F2-isoprostanes
h Brain (B-type) natriuretic peptide
h Carotid intima-media thickness
h Serum homocysteine
h Positive family history

24.4 Herbal treatment

The present use of herbs and herbal medicines in the treatments of various diseases is not a modern system but it has conserved an interesting ancient history behind its successful application and Ayurveda is expressing great evidence that supports the therapeutic use of herbal medicines in India. Since the beginning of human civilization herbs have been an integral part of our society and mostly used in both aspects as culinary and as medicines [60]. The use of several natural compounds and plant products in the treatment of many diseases was started from the ancient time around 3900 BC when Indians, Egyptians, Greek, Chinese, Oakes, and Gahlin employed herbal remedies and started writing their properties as well as applied their use. About 60,000 years ago, the disclosures of possible therapeutic advantage of natural remedies traced and knowledge about natural products spread from person to person gradually. Today's traditional natural remedies include several plants handled by different populations worldwide. For instance, a plant genus Salvia was burned by Indian tribes to obtain hot ashes to cover up the body of the expectant mother during childbirth in the regions of Northern Mexico and the United States. While the root of Salvia miltiorrhiza was used by the Chinese as a circulatory stimulant or sedative and as to dilate coronary arteries or as an an-tiangina drug [61]. A long ago, our Ayurvedic therapies suggested that the root of Rauwolfla serpentine was used as a remedy for hypertension (high BP) while the root of Panax ginseng had been employed in the patients suffering from CADs and angina [62].

Today the importance of alternate therapeutic options revised the interest of traditional reme-dies that had been used ago for the treatment of various disorders of CVD [63]. In India, over

100 medicinal plants have been used in the production of herbal medicines that are excellently recommended for patients as remedies suffering from various sets of CVD. The popularity of herbal medicines and natural plant products increased subsequently across the country due to high efficacy rate, low cost, availability, and suitability in use. Although the natural remedies (traditional therapeutic options) is highly effective in improving a variety of sign and symptoms, decelerating adverse cardiac remodeling, inhibiting circulatory abnormalities, reducing early morbidity and mortality rate. Thus the herbal medicines and natural plant products or extracts generate no or less cytotoxicity, genotoxicity, and side effects in the vital systems, and its affordable cost, easy access, availability, storage, long-lasting all are enabling to build a big roadmap for huge production and demand in the market [64].

The medicinal use of different natural compounds and plants product such as ascorbic acid, polyphenols, indirubin, resveratrol, icariin, bergamot, etc., seem to play a particularly intriguing role in the effective treatment of various forms of cardiovascular disorders. In addition, many remedies such as flavonoids, alkaloids oleuropein, ω-3 fatty acids, and lycopene are natural products found in several vegetables and plants that are used in the prevention and CVD management strategy. Most of the natural compounds mainly trigger cardiovascular disorders as an antioxidant, anti-inflammatory, antihypertensive, antihypercholesterolemia, anticoagulant, antiobesity, antidiabetes, antidyslipidemia, etc., and counter the risk conditions systematically. Notably, the herbal medicines exhibit excellent efficacy to modulate the ratio of HDL/ LDL in the blood vascular systems and also protecting the endothelial functions and hence natural compounds potentially counteracting the major risk factors and early pathological CVD conditions and contributing to better cardiac recovery [61,65,66].

24.5 Indian medicinal plants (used for cardiovascular disease management)

Although India is a rich subcontinent where many thousands of medicinal plant species are conserved, cultivated, and grown tremendously. Many plant species contain a variety of compounds like flavonoids, alkaloids, oleuropein, lycopene, ascorbic acid, and polyphenols that have an efficient role in the prevention and treatment of CVD. Some frequently used and easily available herbs employed in CVD treatment include Garlic (A. sativum), Guggul (C. wightii), Hawthorn (C. oxyacantha), Arjuna (T. arjuna), Amla (E. officinalis), Onion (A. sepa), Ginger (Z. officinale) Triphala and so on. Similarly, these are also effective in the treatment of ischemic heart disease, congestive heart failure, arrhythmias, hypertension, etc. [60,67].

The presence of several above discussed naturally occurring compounds in many plant species provide a beneficial effect over the different symptoms of cardiovascular disorders and potentially contribute to the betterment of an individuals' health as well as helpful in saving thousands of lives across the world. Some Indian medicinal plants and their significant role in the improvement of cardiovascular disorders are listed below (Table 24.3) [68,71].

24.6 Medicinal plants and cardiovascular disease

India has its traditional importance in the field of conservation, cultivation, production, processing, and application of various medicinal plants in its different parts. Ayurveda is one of the great Indian popular mythological books that certify the history of Indian medicinal plants and their use is not modern but it has a long historic journey. We are rich in different natural resources, over many hundreds of medicinal plants are found in India and some of effectively used as natural remedies for various cardiovascular disorders. Many thousands of herbal medicine is produced from medicinally important plants in different forms such as extracts, leaves, roots, seeds, fruits, stems, barks, flowers, etc. [60]. The role of certain medicinal plants (herbs) and their pharmacological properties and therapeutic evaluation for CVD management are discussed below:

24.6.1 Garlic

Botanical name: *A. sativum*; Family: *Liliaceae*

Pharmacological properties: Hypertensive, hypocholesterolemic, hypolipidaemic, anticoagulant, fibrinolytic, hepatoprotective, antioxidant, cardioprotective, and cardiovascular depressant.

Therapeutic evaluation: *A. sativum* is a classic example of medicinal plants effectively used in CVD management. It is known for its multifaceted properties against various CVD-associated conditions like hypertension, inflammation, atherosclerosis, hyperlipidemia, oxidative stress, etc. [69]. The beneficial effects and protective mechanism of garlic in CVD are achieved by suppressing LDL oxidation, increasing HDL, decreasing the content of lipid, for example, total cholesterol (TC) and TG in arterial cells. It also has been seen that garlic supplementation reduces BP significantly in the patient suffering from hypertension but it doesn't appreciably affect the patient's normal BP [70]. Garlic has an active role in systolic and diastolic arterial tension in the hypertensive and atherosclerotic patient. It also decreases blood cholesterol levels. The beneficial effects of garlic on cardiovascular disorders are includes [71].

- Reduction in atherosclerosis.
- Antihyperlipidemic effect.
- Inhibition of platelet aggregation.
- Regulation of blood glucose in diabetic conditions.
- Regulation of BP.
- Significant antiarrhythmic effect (both ventricular and supraventricular).
- Prevents oxidative stress.

Table 24.3: List of important Indian medicinal plants and their respective role in the prevention and management of cardiovascular disease [60,68,71].

S.No.	Name of medicinal plants	Major use on cardiovascular disorders
1	*Allium sativum* (Garlic)	Atherosclerosis, hyperlipidemia, thrombosis, hypertension, and diabetes
2	*Terminalia arjuna* (Arjuna)	Anginal pain, hypertension, congestive heart failure, and dyslipidemia
3	*Allium cepa* (Onion)	Hyperlipidemia, hypertension, hyperglycemia, and platelet aggregation
4	*Emblica officinalis* (Amla)	Cardiotoxicity, heart failure, hyperlipidemia, hypertension, myocardial dysfunction, etc.
5	*Nardostachys jatamansi*	Cardioprotective activity, antioxidant and hypolipidemic activity
6	*Evolvulus alsinoides* (Shankhpushpi)	Diabetes, artery blood flow, blood pressure, heart disease, atherosclerosis, hypolipidaemia, etc.
7	*Boerhavia diffusa* (Punarnava)	Cardiotonic, anti-hypertensive potential, and heart failure
8	*Zingiber officinale* (Ginger)	Anti-inflammatory, antioxidant, anti-platelet, hypertensive, and hypolipidemic effect
9	*Nigella sativa*	Reduced blood pressure, hypertension, hypertrophy, cardioprotective, etc.
10	*Commiphora wightii* (Guggul)	Inflammation, gout, rheumatism, obesity, and disorders of lipids metabolism, etc.
11	*Crataegus oxyacantha* (Hawthorn)	Angina attacks, lower blood pressure, serum cholesterol, coronary vessel dilation, etc.
12	*Rauwolfia serpentina* (Sarpagandha)	High blood pressure (Hypertension) and vascular diseases
13	*Fumaria indica* (Parpata)	Rheumatoid arthritis, asthma, inflammatory bowel disease, etc.
14	*Daucus carota* (Carrot seeds)	Congestive heart failure, hypertension, and anti-inflammatory
15	*Cassia absus* (Chaksu)	Diabetes, hemorrhoids, wound healing, anti-inflammatory, etc.
16	*Acorus calamus* (Vacha)	Cardioprotective action and heart disease
17	*Tribulus terrestris* (Gokshura)	Dilating coronary artery, improving coronary circulation, improving myocardial ischemia
18	*Phyllanthus niruri* (Bhumi amalaki)	Regulation of heart function and blood circulation, heart disease, blood flow, etc.
19	*Tinospora cordifolia* (Guduchi)	Cardioprotective, and anti-inflammatory response
20	*Taraxacum officinale* (Dugdha)	Anti-diabetic, cardiac glycolysis, anti-oxidant, etc.
21	*Saussurea lappa* (Kushtha)	Angina, heart disease, blood pressure, cardiac arrest, etc.
22	*Sida cordifolia* (Bala)	High blood pressure, heart attacks, muscle disorders, strokes, irregular heartbeat, etc.
23	*Digitalis purpurea* (Hatapatri)	Congestive heart failure, heart rhythm problems, increase blood flow, etc.
24	*Withania somnifera* (Ashwagandha)	Cardiac function, cardiac protection, cardiac arrest, etc.
25	*Bacopa monniera* (Brahmi)	Cardiac function, coronary blood flow, heart ischemia, and myocardial infarction

Figure 24.3
Terminalia arjuna bark and dried fruits.

24.6.2 Arjuna

Botanical name: *T. arjuna;* Family: *Combretaceae*

Pharmacological properties: Cardioprotective, spasmogenic, hepatoprotective, and antianginal.

Therapeutic evaluation: Arjuna stem bark and dried fruits (Fig. 24.3) is highly utilized for the treatment of cardiovascular sickness conditions including heart illnesses, chest problems, hypertension, and elevated blood cholesterol level. Various clinical studies indicate that it is interminable helpful in CVD patients by providing stable curable actions on angina, endothelial brokenness, heart disappointment, and even ischemic mitral spewing forth [72]. In India, its bark decoction is being widely used for anginal agony, hypertension, congestive heart disappointment, and dyslipidemia [73]. It sustains and reinforces the heart muscle and advances cardiac working by directing the pulse and maintaining cholesterol [74] *T. arjuna* essentially diminishes TC, LDL, TG levels, and also expands HDL and decreases atherosclerosis [75]. *T. arjuna* stem bark is effective as a cardioprotective and intense cancer-preventing agent [76].

24.6.3 Onion

Botanical name: *Allium cepa (A. cepa)*; Family: *Amaryllidaceous*

Pharmacological properties: Anticancer, preventive cardiovascular abnormalities, heart diseases, anti-inflammation, antiobesity, antidiabetes, antioxidants, antimicrobial activity, neuroprotective, etc.

Figure 24.4
Amla (*Emblica officinalis*) green fruits.

Therapeutic evaluation: The supplementation of onion strip extract is beneficial in diminishing the likelihood of creating key hazard factors for cardiovascular ailment by modifying the lipid profiles in most young females [77]. Garlic oil and onion oil have hostile to stoutness properties that can neutralize the impacts of persistent depressive disorder on various abnormalities like body weight, fat tissue weight, and serum lipid profiles [78]. The onion strip contains quercetin that is a real flavonoid and has promising effects against cardiovascular sicknesses [79]. Onion has advantageous effects on hyperglycemia that has been recognized as a noteworthy hazard factor for cardiovascular [80].

24.6.4 Amla

Botanical name: *E. officinalis*; Family: Phyllanthaceae

Pharmacological properties: Antihypertensive, anti-inflammatory, antioxidant, antiatherogenic, anticoagulant, antidyslipidemic, antiplatelet, vasodilatory, and lipid deposition inhibitory effects.

Therapeutic evaluation: The *E. officinalis* (Fig. 24.4) has traditionally been considered as a cardioactive medication and demonstrated remarkable cardiovascular effects as well as has potential effects for prevention and therapy of CVD. It delivered a critical hypolipidemic impact alongside a decrease in circulatory strain. Amla acts as an accessible hypolipidemic treatment against atherosclerosis and coronary conduit infection, with a decrease in the portion and antagonistic impacts on various hypolipidemic operators [81]. It has useful effects

Figure 24.5
Nardostachys jatamansi rooted plants.

on dyslipidemia and cardiac autonomic capacities treated with a high-fat eating regimen [82]. It has also been demonstrated that the impacts of fruits on diverse cardiovascular conditions and other abnormalities are cardioprotective, antiatherogenic, hypoglycemic, hypolipidemic, hepatoprotective, antioxidant, antipyretic, analgesic, antimicrobial, diuretic, and laxative. Amla's removal appearing potential in decreasing TC and TG levels just as lipid proportions in the circulation [83-84].

24.6.5 Nardostachys jatamansi

Botanical name: *Nardostachys jatamansi (N. jatamansi);* Family: *Caprifoliaceae*

Pharmacological properties: Antioxidant, hepatoprotective, cardioprotective, neuroprotective, antimicrobial, etc.

Therapeutic evaluation: *N. jatamansi* (Fig. 24.5) possesses a significant antistress activity which might be because of its cancer prevention action and it is considered by estimating the free radical searching movement. Accompanied with the way of life alterations and psychotherapy is sheltered, it provides an effective solution against hypertension in all age bunches. Major manifestations of *N. jatamansi* include migraine, BP, sleep deprivation, and energy indicated stamped enhancement. It has the potential to prevent and protect the cardiovascular system, regularize the proper circulation and inhibit oxidation of important molecules by capturing free radicals [85]

Figure 24.6
Ashwagandha (*Withania somnifera*) green plant with leaves.

24.6.6 Ashwagandha

Botanical name: *Withania somnifera (W. somnifera)*; Family: *Solanaceae*

Pharmacological properties: Antiplatelet, antioxidant, antihypertensive, hypoglycaemic, hypolipidemic, and cardioprotective.

Therapeutic evaluation: Ashwagandha (Fig. 24.6) is also known as "Indian winter cherry" and "Indian ginseng." It is one of the most important herbs, traditionally used as medicine in India. It provides a variety of medicinal effects attributed to its use [86]. Ashwagandha contains various alkaloids, lactones steroidal, and saponins, and has been used as antistress, adaptogenic, antitumor, anxiolytic, anti-inflammatory, and antiarthritic conditions. Its root extricate upgrades the cardiorespiratory perseverance [87]. It additionally seems to apply for stimulating a positive effect on the endocrine, cardiopulmonary, and focal sensory systems [88].

24.6.7 Ginger

Botanical name: *Z. officinale*; Family: *Zingiberaceae*

Pharmacological properties: Anti-inflammatory, antioxidant, antiplatelet, hypotensive, hypolipidemic, cell reinforcement, pain-relieving, mitigating, and antipyretic.

Therapeutic evaluation: Ginger is one of the most important natural gifts to humankind. It is now exciting considerable interest for its potential to treat many aspects of cardiovascular disorder. The excellent pharmacological exercises of ginger have been mostly ascribed to the presence of its dynamic Phyto compounds 6-gingerol, 6-shogaol, zingerone adjacent to different phenolics and flavonoids. 6-Gingerol has accounted for as the most plentiful bioactive compound in ginger with a variety of pharmacological impacts including cell reinforcement, pain-relieving, antioxidants, anti-inflammatory, antipyretic, and mitigating properties. Similarly, different observations demonstrated that 6-shogaol with the most reduced fixation in ginger speaks to all the more natural actives contrasted with the above-mentioned 6-gingerol. Ginger offers a characteristic elective dietary supplementation to the traditional enemy of hypertensive operators. Thus increasing human preliminaries of ginger on hypertensive patients utilizing diverse measurements of concentrate which are required [89]. The significant impact of ginger on serum cholesterol might be because of the inhibitory impact of the cholesterol biosynthetic pathway and thus the conversion of cholesterol into bile acids by raising the movement of hepatic cholesterol 7 alpha-hydroxylase. Moreover, bringing down serum triglyceride level, expanding leeway of VLDL, improving hepatic take-up of LDL, and hindering cholesterol, the combination is potentially done by the action of ginger niacin supplementing [90].

24.6.8 Guggul

Botanical name: *C. wightii*; Family: *Burseraceae*

According to Ayurveda "Guggul" is a medicine that has enormous use. It is a gum resin obtained from two different plant species Commiphora and Boswellia. It is produced by drying the white sap of a 15–20 years old tree for a year [91].

Pharmacological properties: Anti-inflammatory, antioxidant, cardioprotective hepatoprotective, neuroprotective, hypolipidemia, hypertension, ischemia, anti-diabetic, etc.

Therapeutic evaluation: Guggul (Fig. 24.7) exhibits a profound cardioprotective effect. It also affects diverse chronic defects such as Alzheimer's disease, arthritis, cancer, diabetes, infectious diseases, respiratory diseases, etc. In addition to these, it also exerts anti-inflammatory, antioxidant, hepatoprotective, neuroprotective, hypolipidemia, and thyroid stimulatory effects by targeting multiple signaling pathways [92]. It decreases the level of lipid peroxide, creatine phosphokinase, phospholipase, xanthine oxidase, and TC level in the serum. It may increase the concentration of superoxide dismutase, myocardial antioxidants, glutathione peroxidase, catalase (CAT), and reduces glutathione (GSH), creatine-phosphokinase, and lactate dehydrogenase as well as reverse the cardiac damage induced by isoproterenol [93]. The hypolipidemic effect of guggul has also been well studied in different animals.

Figure 24.7
Guggul (*Commiphora wightii*) resins.

24.6.9 Gokshura

Botanical name: *Tribulis terrestris* (*T. terrestris*); Family: *Zygophyllaceae*

Pharmacological properties: Cardioprotective, hypotensive, cardio-tonics, hepatoprotective, diuretic, and muscle relaxant.

Therapeutic evaluation: Saponin of Gokshura (Fig. 24.8) has been found as an action of dilating coronary artery and improving coronary blood circulation. *T. terrestris* improves cardiac function and attenuates myocardial infarction in the animal. The possible underlying mechanism of the cardioprotective effect of *T. terrestris* could be due to restoration of endogenous myocardial antioxidant status, and scavenging activity of free radicals along with the correction of altered hemodynamic parameters, and preservation of histo-architectural and ultrastructural alteration [94]. It possesses antihypertensive activity by controlling the artery circulation. The useful biological properties of *T. terrestris* extracts include diuretic property, increased release of nitric oxide from the endothelium, and nerve endings; it effectively relaxes smooth muscles and increases angiotensin-converting enzyme inhibition and hence it reduces the hypertensive conditions [95].

24.6.10 Triphala (a combination of three herbs)

Triphala is a perceived, polyherbal Ayurvedic medicine in the equivalent extent of 1:1:1. It is a combination product of these three plant species *E. officinalis* (Amalaki), *Terminalia bellerica* (Bibhitaki), and *Terminalia chebula* (Haritaki), (Fig. 24.9) [96].

Figure 24.8
Gokshura (*Tribulis terrestris*) plants with leaves and flowers.

Figure 24.9
Terminalia chebula (Haritaki) dried fruits.

Pharmacological properties: Antiaging, anti-inflammatory, antioxidant, antineoplastic activity, radioprotective, antimicrobial potential, hypoglycemic, chemoprotective, anticardiovascular disorders, etc.

Therapeutic evaluation: In CVD hypercholesterolemia is an important risk factor. Many recent studies on animals have proven that Triphala reduced the elevated cholesterol levels, low-density lipoprotein, very low-density lipoprotein, and free fatty acid content [97]. Triphala use can counter the impacts of high dietary admission of fats and have the potential for use as antiobesity operators with alluring lipid-profile balancing properties [98]. In another study, Haritaki is one of the combination herbs in Triphala, induced hypolipidemic effects in the herb-treated group, a reduction in TC, TG, total protein, and elevation of high-density lipoprotein, has been found in the herb-treated group compared with the control group [99]. *T. bellerica* is a constituent of Triphala that has a good impact on maintaining cholesterol level, for example, increment the dimension of HDL and abatement LDL, and valuable in the treatment of coronary supply route malady [100]. Triphala is a powerful combination drug to address imbalance conditions in the gastrointestinal tract and cardiovascular system and should be more widely studied in the context of these common disorders. Triphala tablets help enhances the course and is a successful equation for hypercholesterolemia improvement [101].

24.7 Conclusions

A cardiovascular disorder is a great global meaningful challenge and in the upcoming few years, it will be raised beyond the predicted criteria across the world. Today, CVD has become one of the most highlighting and preferring fields globally. Dominantly, we are facing crises in saving millions of lives and billions of dollars affected by CVD worldwide. Current efforts will provide a good result and a better future for surviving population as well the upcoming generation. All national and international healthcare bodies have to realize and evaluate the current circumferences taken against the risk of CVD and to neutralize its complications and it is also necessary to announce effective guidelines for the betterment of human health and wealth. It is too necessary to take effective measures at the individual level to protect and prevent the seriousness of CVD. According to recent global data, a large population is suffering from the severe risk of CVD or one of its associated disorders like CHD, PAD, cardiac arrest, cerebrovascular disease, stroke, arrhythmias, and atherosclerosis. Whenever these major risk conditions are developing in any individual it directly hits its health as physically, morally, psychologically, and wealth together. Indeed governments should have to revealed and identify such strongly affected populations or poorly developed areas and start to set up a healthcare unit at the local level that will assist in early diagnosis and helpful in providing better treatment to the patient. Such small initiatives will help to save both healthcare costs and many lives at a time. It has been seen that even one member of each family is suffering from CVD and a big part of its saving is expended on healthcare cost that almost affects all other fundamental needs of the human being. Additionally, it is also being observed that a big part of the current population like the labor of unorganized sectors or daily workers lost their lives without proper treatment or due to lack of financial support. Most CVDs can be efficiently prevented by addressing some regular behavioral factors like tobacco consumption, imbalance and unhealthy diet, drinking of alcohol, physical inactivity, etc. The risk of CVD can be asses at the individual level by observing certain symptoms to own health and it may be helpful for early prediction and prevention. The effective treatment of CVD is much expensive and all affected candidates cannot be effort lifelong medication only a few of them get proper treatment. It is due to the high cost of modern allopathic medicines and surgery that directed to open a new door to review and revise the use of herbal medicines and various plant products or extracts as natural remedies for CVD management. At present over thousands of herbal medicines are producing by a variety of plant species and successfully applied in preventing and providing better treatment to the affected individuals.

The upcoming challenges will concern a deeper understanding of the underlying mechanisms by which herbal medicines can modulate, prevent, and cure cardiovascular disorders. More research is needed in the field of herbal medicines and their major connectivity with various sets of cardiovascular disorders and addressing the relevant problems concern with human health. Additionally, government funding allocations and support are needed for further and ongoing studies to validate its therapeutic uses in human clinical trials and to define the biological mechanisms relevant to this plant-based medicine. More widespread education of the general public and medical providers on clinical Ayurvedic medicine and complementary therapies is required to increase awareness of herbal treatments for both clinical and healthy populations.

References

[1] WHO cardiovascular diseases (CVDs). http://www.who.int/en/news-room/fact-sheets/detail/cardiovascular-diseases (accessed June 1, 2019).

[2] D Mozaffarian, EJ Benjamin, AS Go, DK Arnett, MJ Blaha, M Cushman, et al., Executive summary: heart disease and stroke statistics—2015 update. A report from the American Heart Association, Circulation 131 (2015) 434–441.

[3] World Health Organization, Noncommunicable diseases country profiles 2018. https://apps.who.int/iris/handle/10665/274512. (Accessed on Sep 09, 2018).

[4] D Mozaffarian, P Wilson, W Kannel, Beyond established and novel risk factors: lifestyle risk factors for cardiovascular disease, Circulation 117 (2008) 3031–3038. http://doi.org/10.1161/CIRCULATIONAHA.107.738732.

[5] Writing Group Members, Heart disease and stroke statistics—2006 update: a report from the American Heart Association Statistics Committee and Stroke Statistics Subcommittee, Circulation 113 (6) (2006) e85–e151.

[6] T Nag, A Ghosh, Cardiovascular disease risk factors in Asian Indian population: a systematic review, J Cardiovasc Dis Res 4 (4) (2013) 222–228. http://doi.org/10.1016/j.jcdr.2014.01.004.

[7] Institute of Medicine (US) Committee on preventing the global epidemic of cardiovascular disease: meeting the challenges in developing countries, in: V. Fuster, B.B. Kelly (Eds.)Promoting Cardiovascular Health in the Developing World: A Critical Challenge to Achieve Global Health, National Academies Press, Washington, DC, 2010.

[8] NH Mashour, GI Lin, WH Frishman, Herbal medicine for the treatment of cardiovascular disease: clinical considerations, Arch Intern Med 158 (20) (1998) 2225–2234. http://doi.org/10.1001/archinte.158.20.2225.

[9]   A Tachjian, V Maria, A Jahangir, Use of herbal products and potential interactions in patients with cardiovascular diseases, J Am Coll Cardiol 55 (6) (Feb 9, 2010) 515–525, doi:10.1016/j.jacc.2009.07.074.

[10]   M Ekor, The growing use of herbal medicines: issues relating to adverse reactions and challenges in monitoring safety, Front Pharmacol 4 (2014) 177. http://doi.org/10.3389/fphar.2013.00177.

[11]   HS Rahman, HH Othman, NI Hammadi, SK Yeap, KM Amin, N Abdul Samad, NB Alitheen, Novel Drug Delivery Systems for Loading of Natural Plant Extracts and Their Biomedical Applications, Int J Nanomedicine 15 (Apr 15, 2020) 2439–2483, doi:10.2147/IJN.S227805.

[12]   M Greenwell, PK Rahman, Medicinal plants: their use in anticancer treatment, Int J Pharm Sci Res 6 (10) (2015) 4103–4112. http://doi.org/10.13040/IJPSR.0975-8232.6(10).4103-12.

[13]   R Walden, B Tomlinson, Cardiovascular Disease, in: IFF Benzie, S Wachtel-Galor (Eds.), Herbal Medicine: Biomolecular and Clinical Aspects, CRC Press/Taylor & Francis, Boca Raton, FL, 2011.

[14]   Natural sources of antioxidants, J Funct Foods 18 Part B 2015, 1756–4646, https://doi.org/10.1016/j.jff.2015.03.005.

[15]   S Wachtel-Galor, IFF Benzie, Chapter 1: Herbal medicine: an introduction to its history, usage, regulation, current trends, and research needs, in: IFF Benzie, S Wachtel-Galor (Eds.), Herbal Medicine: Biomolecular and Clinical Aspects, CRC Press/Taylor & Francis, Boca Raton, FL, 2011. https://www.ncbi.nlm.nih.gov/books/NBK92773.

[16]   Making a case for quantitative assessment of cardiovascular risk, Toth PPJ ClinLipidol 1 (4) (2007) 234–241.

[17]   DR Ardeshna, T Bob-Manuel, A Nanda, A Sharma, WP Skelton 4th, M Skelton, RN. Khouzam, Asian-Indians: a review of coronary artery disease in this understudied cohort in the United States, Ann Transl Med 6 (1) (Jan 2018) 12, doi:10.21037/atm.2017.10.18.

[18]   R Gupta, H Prakash, S Majumdar, et al., Prevalence of coronary heart disease and coronary risk factors in an urban population of Rajasthan, Indian Heart J 47 (1995) 331–338.

[19]   M Chacko, PS Sarma, S Harikrishnan, G Zachariah, P. Jeemon, Family history of cardiovascular disease and risk of premature coronary heart disease: A matched case-control study, Wellcome Open Res 5 (Jun 12, 2020) 70, doi:10.12688/wellcomeopenres.15829.2.

[20]   JC Brown, TE Gerhardt, E Kwon, Risk factors for coronary artery disease, StatPearls [Internet], StatPearls Publishing, Treasure Island, FL, 2021. https://www.ncbi.nlm.nih.gov/books/NBK554410 accessed June 6, 2020.

[21]   EA Ashley, J Niebauer, Chapter 5: Coronary artery disease, Cardiology Explained, Remedica, London, 2004. https://www.ncbi.nlm.nih.gov/books/NBK2216.

[22]   S Bösner, A Becker, M Abu Hani, H Keller, AC Sönnichsen, J Haasenritter, K Karatolios, JR Schaefer, E Baum, N Donner-Banzhoff, Accuracy of symptoms and signs for coronary heart disease assessed in primary care, Br J Gen Pract 60 (575) (Jun 2010) e246–e257, doi:10.3399/bjgp10X502137.

[23]   R Bauersachs, U Zeymer, JB Brière, C Marre, K Bowrin, M. Huelsebeck, Burden of Coronary Artery Disease and Peripheral Artery Disease: A Literature Review, Cardiovasc Ther (Nov 26, 2019) 8295054, doi:10.1155/2019/8295054.

[24]   MR Zemaitis, JM Boll, MA Dreyer, Peripheral arterial disease, StatPearls [Internet], StatPearls Publishing, Treasure Island, FL, 2021. https://www.ncbi.nlm.nih.gov/books/NBK430745 accessed March 17, 2021.

[25]   F Simon, A Oberhuber, N Floros, P Düppers, H Schelzig, M Duran, Pathophysiology of chronic limb ischemia, Gefasschirurgie 23 (Suppl 1) (2018) 13–18.

[26]   AS Khaku, P Tadi, Cerebrovascular disease, StatPearls [Internet], StatPearls Publishing, Treasure Island, FL, 2021. https://www.ncbi.nlm.nih.gov/books/NBK430927 accessed January 31, 2021.

[27]   L Lahousse, H Tiemeier, MA Ikram, GG Brusselle, Chronic obstructive pulmonary disease and cerebrovascular disease: a comprehensive review, Respirat Med 109 (11) (2015) 1371–1380.

[28]   TT van Sloten, CDA Stehouwer, Carotid stiffness: a novel cerebrovascular disease risk factor, Pulse 4 (1) (2016) 24–27.

[29] DC Steffens, KR Krishnan, C Crump, GL. Burke, Cerebrovascular disease and evolution of depressive symptoms in the cardiovascular health study, Stroke 33 (6) (Jun 2002) 1636–1644, doi:10.1161/01.str. 0000018405.59799.d5.

[30] Sika-Paotonu D, Beaton A, Raghu A, Steer A, Carapetis J. Acute Rheumatic Fever and Rheumatic Heart Disease. Apr 3, 2017.

[31] S Ahmed, P Padhan, R Misra, D Danda, Update on post-streptococcal reactive arthritis: narrative review of a forgotten disease, CurrRheumatol Rep 23 (3) (2021) 19. http://doi.org/10.1007/s11926-021-00982-3. PMID: 33569668.

[32] JG Lawrence, JR Carapetis, K Griffiths, K Edwards, JR Condon, Acute rheumatic fever and rheumatic heart disease: incidence and progression in the Northern Territory of Australia, 1997 to 2010, Circulation 128 (2013) 492–501.

[33] MH Gewitz, RS Baltimore, LY Tani, CA Sable, ST Shulman, J Carapetis, B Remenyi, KA Taubert, AF Bolger, L Beerman, BM Mayosi, A Beaton, NG Pandian, EL Kaplan, American Heart Association Committee on Rheumatic Fever, Endocarditis, and Kawasaki Disease of the Council on Cardiovascular Disease in the Young. Revision of the Jones Criteria for the diagnosis of acute rheumatic fever in the era of Doppler echocardiography: a scientific statement from the American Heart Association, Circulation 131 (20) (May 19, 2015) 1806–1818, doi:10.1161/CIR.0000000000000205.

[34] MB Streiff, G Agnelli, JM Connors, M Crowther, S Eichinger, R Lopes, RD McBane, S Moll, J Ansell, Guidance for the treatment of deep vein thrombosis and pulmonary embolism, J Thromb Thrombolysis 41 (1) (Jan 2016) 32–67, doi:10.1007/s11239-015-1317-0.

[35] WM Lijfering, FR Rosendaal, SC Cannegieter, Risk factors for venous thrombosis—current understanding from an epidemiological point of view, Br J Haematol 149 (6) (2010) 824–833. http://doi.org/10.1111/j.1365-2141.2010.08206.x.

[36] PD Stein, RD Hull, KC Patel, RE Olson, WA Ghali, R Brant, RK Biel, V Bharadia, NK. Kalra, D-dimer for the exclusion of acute venous thrombosis and pulmonary embolism: a systematic review, Ann Intern Med 140 (8) (Apr 20, 2004) 589–602, doi:10.7326/0003-4819-140-8-200404200-00005.

[37] KK Narani, Deep vein thrombosis and pulmonary embolism—prevention, management, and anaesthetic considerations, Indian J Anaesth 54 (1) (2010) 8–17. http://doi.org/10.4103/0019-5049. 60490.

[38] W Chang, B Wang, Q Li, Y Zhang, W Xie, Study on the risk factors of preoperative deep vein thrombosis (DVT) in patients with lower extremity fracture, ClinApplThrombHemost 27 (2021) 10760296211002900. http://doi.org/10.1177/10760296211002900.

[39] MR Carazo, MS Kolodziej, ES DeWitt, NA Kasparian, JW Newburger, VE Duarte, MN Singh, AR. Opotowsky, Prevalence and Prognostic Association of a Clinical Diagnosis of Depression in Adult Congenital Heart Disease: Results of the Boston Adult Congenital Heart Disease Biobank, J Am Heart Assoc 9 (9) (May 5, 2020) e014820, doi:10.1161/JAHA.119.014820.

[40] DL Hare, SR Toukhsati, P Johansson, T Jaarsma, Depression and cardiovascular disease: a clinical review, Eur Heart J 35 (2014) 1365–1372.

[41] YY Chen, P Xu, Y Wang, TJ Song, N Luo, LJ Zhao, Prevalence of and risk factors for anxiety after coronary heart disease: Systematic review and meta-analysis, Medicine (Baltimore) 98 (38) (2019) e16973. http://doi.org/10.1097/MD.0000000000016973.

[42] J Xiao, Y Borné, X Bao, M Persson, A Gottsäter, S Acosta, G. Engström, Comparisons of Risk Factors for Abdominal Aortic Aneurysm and Coronary Heart Disease: A Prospective Cohort Study, Angiology 72 (1) (Jan 2021) 24–31, doi:10.1177/0003319720946976.

[43] C Antzelevitch, A Burashnikov, Overview of basic mechanisms of cardiac arrhythmia, Card Electrophysiol Clin 3 (1) (2011) 23–45. http://doi.org/10.1016/j.ccep.2010.10.012.

[44] A Alessandra, B Mohamed, S AromolaranAdemuyiwa, Cardiolipotoxicity, inflammation, and arrhythmias: role for interleukin-6 molecular mechanisms, Front Physiol 9 (2019). http://doi.org/10.3389/fphys.2018.01866.

[45] MCS Wong, HHX Wang, Rapid emergence of atherosclerosis in Asia: a systematic review of coronary atherosclerotic heart disease epidemiology and implications for prevention and control strategies, Curr Opinion Lipidol 26 (4) (2015) 257–269.

[46] MG Santos, M Pegoraro, F Sandrini, EC. Macuco, Risk factors for the development of atherosclerosis in childhood and adolescence, Arq Bras Cardiol 90 (4) (Apr 2008) 276–283, doi:10.1590/s0066-782x2008000400012.

[47] T Nag, A Ghosh, Cardiovascular disease risk factors in Asian Indian population: a systematic review, J Cardiovas Dis Res 4 (4) (2013) 222–228.

[48] P Malambo, AP Kengne, A De Villiers, EV Lambert, T. Puoane, Built Environment, Selected Risk Factors and Major Cardiovascular Disease Outcomes: A Systematic Review, PLoS One 11 (11) (2016) e0166846 Nov 23, doi:10.1371/journal.pone.0166846.

[49] AJ Marian, B Asatryan, XHT Wehrens, Genetic basis and molecular biology of cardiac arrhythmias in cardiomyopathies, Cardiovas Res 116 (9) (2020) 1600–1619. https://doi.org/10.1093/cvr/cvaa116.

[50] RM Carney, KE Freedland, GE Miller, AS. Jaffe, Depression as a risk factor for cardiac mortality and morbidity: a review of potential mechanisms, J Psychosom Res 53 (4) (Oct 2002) 897–902, doi:10.1016/s0022-3999(02)00311-2.

[51] H Ueshima, A Sekikawa, K Miura, TC Turin, N Takashima, Y Kita, M Watanabe, A Kadota, N Okuda, T Kadowaki, Y Nakamura, T. Okamura, Cardiovascular disease and risk factors in Asia: a selected review, Circulation 118 (25) (2008) 2702–2709 Dec 16, doi:10.1161/CIRCULATIONAHA.108.790048.

[52] NJ Bosomworth, Practical use of the Framingham risk score in primary prevention: Canadian perspective, Canad Family Physician 57 (4) (2011) 417–423.

[53] S Gupta, K Gaurav, R Gudapati, M Bhise, Emerging risk factors for cardiovascular diseases: Indian context, Ind J Endocrinol Metabol 17 (5) (2013) 806.

[54] Masoud Mohammadnezhad, Tamara Mangum, William May, Joshua Jeffrey Lucas, Stanley Ailson, published by, World Journal of Cardiovascular Surgery 6 (2016) 11, doi:10.4236/wjcs.2016.611022.

[55] SC Smith, P Greenland, SM Grundy, AHA Conference Proceedings: Prevention Conference V: beyond secondary prevention: identifying the high-risk patient for primary prevention: executive summary: American Heart Association, Circulation 101 (2000) 111–116.

[56] M Helfand, DI Buckley, M Freeman, R Fu, K Rogers, C Fleming, et al., Emerging risk factors for coronary heart disease: a summary of systematic reviews conducted for the U.S. Preventive Services Task Force, Ann Intern Med 151 (2009) 496–497.

[57] DA Morrow, JA de Lemos, Benchmarks for the assessment of novel cardiovascular biomarkers, Circulation 115 (2007) 949–952.

[58] G Thanassoulis, RS Vasan, Genetic cardiovascular risk prediction: Will we get there? Circulation 122 (2010) 2323–2334.

[59] S Rastogi, MM Pandey, AK Rawat, Traditional herbs: a remedy for cardiovascular disorders, Phytomedicine 23 (11) (Oct 15, 2016) 1082–1089, doi:10.1016/j.phymed.2015.10.012.

[60] C Carresi, M Scicchitano, F Scarano, R Macrì, F Bosco, S Nucera, et al., The potential properties of natural compounds in cardiac stem cell activation: their role in myocardial regeneration, Nutrients 13 (2021) 275. https://doi.org/10.3390/nu13010275.

[61] S Rastogi, MM Pandey, AK Rawat, Traditional herbs: a remedy for cardiovascular disorders, Phytomedicine 23 (2016) 1082–1089.

[62] R Liperoti, DL Vetrano, R Bernabei, G Onder, Herbal Medications in cardiovascular medicine, J Amer College Cardiol 69 (9) (2017).

[63] A Shaito, DTB Thuan, HT Phu, THD Nguyen, H Hasan, S Halabi, S Abdelhady, GK Nasrallah, AH Eid, G. Pintus, Herbal Medicine for Cardiovascular Diseases: Efficacy, Mechanisms, and Safety, Front Pharmacol 11 (2020) 422 Apr 7, doi:10.3389/fphar.2020.00422.

[64] J Blanco-Salas, FM Vazquez, MP Hortigón-Vinagre, T Ruiz-Tellez, Bioactive phytochemicals from Mercurialis spp. used in traditional Spanish medicine, Plants 8 (2019) 193.

[65] X Chang, T Zhang, W Zhang, Z Zhao, J Sun, Natural Drugs as a Treatment Strategy for Cardiovascular Disease through the Regulation of Oxidative Stress, Oxid Med Cell Longev (Sep 27, 2020) 5430407, doi:10.1155/2020/5430407.

[66] HR Vasanthi, RP Parameswari, Indian spices for healthy heart - an overview, Curr Cardiol Rev 6 (4) (2010) 274–279. http://doi.org/10.2174/157340310793566172.

[67] L Li, X Zhou, N Li, M Sun, J Lv, Z. Xu, Herbal drugs against cardiovascular disease: traditional medicine and modern development, Drug Discov Today 20 (9) (Sep 2015) 1074–1086, doi:10.1016/j.drudis.2015.04.009.

[68] E Shabani, S Korosh, M Mohammadtaghi, The effect of garlic on lipid profile and glucose parameters in diabetic patients: a systematic review and meta-analysis, Primary Care Diab 13 (1) (2019) 28–42.

[69] A Shang, SY Cao, XY Xu, RY Gan, GY Tang, H Corke, V Mavumengwana, HB. Li, Bioactive Compounds and Biological Functions of Garlic (Allium sativum L.), Foods 8 (7) (2019) 246 Jul 5, doi:10.3390/foods8070246.

[70] Dr Subrat Bhutia MD (Ayu)*, Prof (Dr) Kamdev Das MD, PhD (Ayu), Dr Baby Ganeriwala MD (Ayu). ROLE OF CERTAIN MEDICINAL PLANTS ON CARDIOVASCULAR DISORDERS, Indian Journal of Medical Research and Pharmaceutical Sciences. 3 (2) (February 2016).

[71] N Kaur, N Shafiq, H Negi, A Pandey, S Reddy, H Kaur, N Chadha, S. Malhotra, Terminalia arjuna in Chronic Stable Angina: Systematic Review and Meta-Analysis, Cardiol Res Pract (2014) 281483, doi:10.1155/2014/281483.

[72] SA Shengule, S Mishra, K Joshi, K Apte, D Patil, P Kale, T Shah, M Deshpande, A Puranik, Anti-hyperglycemic and anti-hyperlipidaemic effect of Arjunarishta in high-fat fed animals, J Ayurveda Integr Med 9 (1) (2018) 45–52, doi:10.1016/j.jaim.2017.07.004.

[73] S Subramaniam, R Subramaniam, S Rajapandian, S Uthrapathi, VR Gnanamanickam, GP. Dubey, Anti-Atherogenic Activity of Ethanolic Fraction of Terminalia arjuna Bark on Hypercholesterolemic Rabbits, Evid Based Complement Alternat Med 487916 (2011), doi:10.1093/ecam/neq003.

[74] D Gaikwad, N Jadhav, A review on biogenic properties of stem bark of Terminalia Arjuna: an update, Asian J Pharm Clin Res 11 (8) (2018) 35–39.

[75] J Kim, YJ Cha, KH Lee, E. Park, Effect of onion peel extract supplementation on the lipid profile and antioxidative status of healthy young women: a randomized, placebo-controlled, double-blind, crossover trial, Nutr Res Pract 7 (5) (Oct 2013) 373–379, doi:10.4162/nrp.2013.7.5.373.

[76] C Yang, L Li, L Yang, H Lu, S Wang, G. Sun, Anti-obesity and Hypolipidemic effects of garlic oil and onion oil in rats fed a high-fat diet, Nutr Metab (Lond) 15 (Jun 20, 2018) 43, doi:10.1186/s12986-018-0275-x.

[77] JY Ro, JH Ryu, HJ Park, HJ. Cho, Onion (Allium cepa L.) peel extract has anti-platelet effects in rat platelets, Springerplus 4 (2015) 17 Jan 13, doi:10.1186/s40064-015-0786-0.

[78] YR Kang, HY Choi, JY Lee, SI Jang, H Kang, JB Oh, HD Jang, YI. Kwon, Calorie Restriction Effect of Heat-Processed Onion Extract (ONI) Using In Vitro and In Vivo Animal Models, Int J Mol Sci 19 (3) (2018) 874 Mar 15, doi:10.3390/ijms19030874.

[79] N Aslani, MH Entezari, G Askari, Z Maghsoudi, MR. Maracy, Effect of Garlic and Lemon Juice Mixture on Lipid Profile and Some Cardiovascular Risk Factors in People 30-60 Years Old with Moderate Hyperlipidaemia: A Randomized Clinical Trial, Int J Prev Med 7 (Jul 29, 2016) 95, doi:10.4103/2008-7802.187248.

[80] PS Kanthe, BS Patil, SC Bagali, RC Reddy, MR Aithala, KK. Das, Protective effects of Ethanolic Extract of Emblica officinalis (amla) on Cardiovascular Pathophysiology of Rats, Fed with High Fat Diet, J Clin Diagn Res 11 (9) (Sep 2017) CC05–CC09, doi:10.7860/JCDR/2017/28474.

[81] S Khanna, A Das, J Spieldenner, C Rink, S. Roy, Supplementation of a standardized extract from Phyllanthus emblica improves cardiovascular risk factors and platelet aggregation in overweight/class-1 obese adults, J Med Food 18 (4) (Apr 2015) 415–420, doi:10.1089/jmf.2014.0178.

[82] S Mirunalini, V Vaithiyanathan, M Krishnaveni, Amla: a novel Ayurvedic herb as a functional food for health benefits—a mini review, Int J Pharm Pharm Sci 5 (Suppl1) (2013) 1–4.

[83] R Jain, R Pandey, RN Mahant, DS Rathore, A review on medicinal importance of Emblicaofficinalis, Int J PaharmSci Res 6 (1) (2015) 72–84. http://doi.org/10.13040/IJPSR.0975-8232.6. (1).72-84.

[84] R Nandha, H Singh, P Moudgill, G. Kular, A pilot study to clinically evaluate the role of herbomineral compound "Rakatchap Har" in the management of essential hypertension, Ayu 32 (3) (Jul 2011) 329–332, doi:10.4103/0974-8520.93908.

[85] J Singh, NS Mishra, S Banerjee, YC Sharma, Comparative studies of physical characteristics of raw and modified sawdust for their use as adsorbents for removal of acid dye, Bioresources 6 (3) (2011) 2732–2743.

[86] A Deshpande, N Irani, R Balakrishnan, Study protocol and rationale for a prospective, randomized, double-blind, placebo-controlled study to evaluate the effects of Ashwagandha (Withaniasomnifera) extract on nonrestorative sleep, Medicine (Baltimore) 97 (26) (2018) e11299.

[87] N Singh, M Bhalla, P de Jager, M. Gilca, An overview on ashwagandha: a Rasayana (rejuvenator) of Ayurveda, Afr J Tradit Complement Altern Med 8 (5 Suppl) (2011) 208–213, doi:10.4314/ajtcam.v8i5S.9.

[88] M Torabi, F Naeemzadeh, V Ebrahimi, N Taleschian-Tabrizi, F Pashazadeh, H. Nazemie, 133: THE EFFECT OF ZINGIBER OFFICINALE (GINGER) ON HYPERTENSION; A SYSTEMATIC REVIEW OF RANDOMISED CONTROLLED TRIALS, BMJ Open 7 (1) (2017), doi:10.1136/bmjopen-2016-015415.133.

[89] J Zhu, H Chen, Z Song, X Wang, Z. Sun, Effects of Ginger (Zingiber officinale Roscoe) on Type 2 Diabetes Mellitus and Components of the Metabolic Syndrome: A Systematic Review and Meta-Analysis of Randomized Controlled Trials, Evid Based Complement Alternat Med (Jan 9, 2018) 5692962, doi:10.1155/2018/5692962.

[90] LO Hanuš, T Řezanka, VM Dembitsky, A Moussaieff, Myrrh-commiphora chemistry, Biomed Papers 149 (1) (2005) 3–28.

[91] AB Kunnumakkara, K Banik, D Bordoloi, C Harsha, BL Sailo, G Padmavathi, NK Roy, SC Gupta, BB. Aggarwal, Googling the Guggul (Commiphora and Boswellia) for Prevention of Chronic Diseases, Front Pharmacol 9 (Aug 6, 2018) 686, doi:10.3389/fphar.2018.00686.

[92] MA Ahmad, M Mujeeb, M Akhtar, M Khushtar, M Arif, MR. Haque, Guggulipid: A Promising Multi-Purpose Herbal Medicinal Agent, Drug Res (Stuttg) 70 (4) (Apr 2020) 123–130, doi:10.1055/a-1115-4669.

[93] SK Ojha, M Nandave, S Arora, R Narang, AK Dinda and DS Arya, Chronic administration of Tribulusterrestris Linn. extract improves cardiac function and attenuates myocardial infarction in rats, Int J Pharmacol, 4: 1-10.

[94] L Fatima MS (U), A Sultana MD (U), S Ahmed, MD(U) and S Sultana MD(U), "Pharmacological activities of tribulus terrestris linn: a systemic review, World J Pharm Pharmaceut Sci", 4, 02, 136-150, ISSN 2278 –4357.

[95] CT Peterson, K Denniston, D Chopra, Therapeutic uses of triphala in ayurvedic medicine, J Altern Complement Med 23 (8) (2017) 607–614.

[96] S Saravanan, R Srikumar, S Manikandan, N Jeya Parthasarathy, R Sheela Devi, Hypolipidemic effect of triphala in experimentally induced hypercholesteremic rats, Yakugaku Zasshi 127 (2) (Feb 2007) 385–388, doi:10.1248/yakushi.127.385.

[97] S Gurjar, A Pal, S Kapur, Triphala and its constituents ameliorate visceral adiposity from a high-fat diet in mice with diet-induced obesity, AlternTher Health Med 18 (6) (2012) 38–45.

[98] V Maruthappan, KS Shree, Hypolipidemic activity of haritaki (*Terminalia chebula*) in atherogenic diet induced hyperlipidemic rats, J Adv Pharm Technol Res 1 (2010) 229–235.

[99] R Walden, B Tomlinson, Cardiovascular disease, in: IFF Benzie, S Wachtel-Galor (Eds.), Herbal Medicine: Biomolecular and Clinical Aspects, CRC Press/Taylor & Francis, Boca Raton, FL, 2011.

[100] A Vinaya Kumar, Ayurvedic Clinical Medicine, 1st ed, Sri Satguru publication, Delhi, 1997.

[101] CT Peterson, K Denniston, D Chopra, Therapeutic uses of Triphala in Ayurvedic medicine, J Altern Complement Med 23 (8) (2017) 607–614, doi:10.1089/acm.2017.0083.

PART 8

Cancer

CHAPTER 25

Targeting metabolism with herbal therapy: A preventative approach toward cancer

Deepti Singh and Hifzur R. Siddique

Molecular Cancer Genetics & Translational Research Lab, Section of Genetics, Department of Zoology, Aligarh Muslim University, Aligarh, India

25.1 Introduction

Cancer development involves alterations or mutations within a cell, enabling the cell to bypass checkpoints required for normal cell death, thereby making the cell cancerous. Mutations or dysregulations in the expression of oncogenes and tumor suppressor genes alter the metabolism of cells, thereby driving carcinogenic pathogenicity. Cancer metabolism is a concept introduced by German physiologist Otto Warburg [1]. The Warburg effect is considered a major hallmark of dysregulated metabolism in cancer [2]. Cancer cells undergo reprogramming of their metabolism to fulfill their bioenergetic and biosynthetic requirements. Cancer cell metabolism is a common feature of cancer cells that rely mainly on aerobic glycolysis, fatty acid synthesis, and glutaminolysis for their proliferation [3]. In 1956, Warburg noticed an abnormally high rate of glycolysis in cancer cells. However, only a tiny fraction of this glucose was broken down by oxidative phosphorylation, suggesting that cancer cells preferably undergo glycolytic breakdown of glucose for energy rather than mitochondrial oxidative phosphorylation.

Cancer cells undergo increased synthesis of fatty acids and glutamine metabolism, providing the lipids required for membrane biogenesis to the rapidly dividing cancer cells, helping them in growth and survival [4]. Cancer cells are also sensitive to a deprived glutamine environment and cannot proliferate without a glutamine-rich environment. The addition of glutamine is responsible for enhanced by-products such as amino-acid precursors required for rapid proliferation [5]. Thus, cancer initiation and progression involve the reprogramming of cellular metabolism, particularly for supporting the increase in demand for ATP production, macromolecular biosynthesis, and the regulation of redox balance. Cancer cells have undergone several alterations in the metabolic pathways and the metabolic interactions with the microenvironment for the modulation of nutrient uptake [6]. These differences in the metabolism of cancer cells suggest that the targeting of cancer cell metabolism could be a practical approach in the treatment of cancer patients. Several metabolic targets are being explored for their therapeutic potential, and several metabolism-targeted drugs are being tested both in preclinical cancer models and clinical studies [3,5]. Identifying metabolic reprogramming as the major hallmark of carcinogenesis has made to the development of novel metabolism-targeted cancer therapeutics.

The discovery of new drugs from natural products has gained interest in the past years due to their diverse structures and multiple targets [7,8]. Several herbal compounds are being studied for their anticancerous effects in cancer models [8–12,14]. The use of herbal compounds has several promising advantages, such as being used alone and in combination with standard therapies, thereby promoting synergistic effects and reducing the undesired side effects. Plant-derived products have been reported to be involved in modulating several carcinogenic pathways targeting multiple molecules and increasing the chances of patient survival. A prominent, interesting feature is an interconnection between the cancer-specific signaling pathways and metabolic adaptations [12–14]. Therefore, significant research is being conducted to explore how herbal compounds may modulate cancer cell metabolism and exert their anticancer effects. This chapter provides a background of the metabolic proteins and enzymes that are differentially expressed in cancer cells. Further, we explore the effects of herbal compounds on the metabolic pathways of cancer cells (Fig. 25.1). Our chapter contributes to expanding the scientific community's knowledge about the efficiency of herbal compounds in modulating the metabolic targets and using these compounds as anticancer therapeutics.

25.2 Major targets of metabolic reprogramming in malignant cells

Cancer cells undergo metabolic reprogramming, which drives growth, survival, and proliferation, thereby supporting carcinogenesis. Some of the most relevant cancer cell bioenergetics and metabolism changes include increased glycolysis, increased glutaminolysis flux, upregulated lipid and amino acid metabolism, and enhanced mitochondrial biogenesis. Preclinical and clinical evidence suggests that drugs targeting metabolism could inhibit cancer progression and make this approach a promising area of research for the development of effective anticancer therapeutic strategies [15]. This has resulted in the exploration and identification of novel therapeutic targets within the cancer cell metabolism.

25.2.1 Glucose metabolism

25.2.1.1 Glucose transporters

The primary substrate for cellular metabolism is glucose which is abundantly present in the blood. The increase in glucose uptake is a characteristic feature of a cancer cell. Transport of glucose across mammalian cell membrane involves glucose transporters (GLUTs).

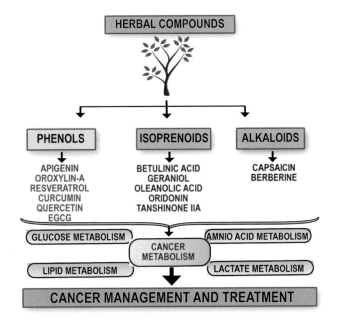

Figure 25.1

A flow chart summarizing the various plant-derived herbal compounds, namely, phenolic compounds, alkaloids, and isoprenoids, which can be used to target cancer metabolic reprogramming.

GLUTs are the major facilitator superfamily, which have four main isoforms viz GLUT1-4. The GLUT isoforms differ in substrate recognition and tissue distribution [16]. The most widely distributed GLUT is GLUT1 which is associated with poor prognosis and is overexpressed in cancer cells [17]. The upregulation of GLUT1 transporters has been correlated with PI3K/AKT, HIF, KRAS, and BRAF in most cancers. Therefore, GLUT1 is a potential anticancer target for which several anticancerous compounds are currently under investigation [18].

25.2.1.2 Glycolytic enzymes

The ten metabolic reactions in glycolysis are catalyzed by several glycolytic enzymes such as hexokinases, phospho-fructokinases (PFKs), pyruvate kinases (PKs), etc. The various glycolytic enzymes that are dysregulated in tumor cells include the following.

25.2.1.2.1 Hexokinases

Cancer cells undergo increased glycolytic flux and production of lactate which is advantageous in terms of increased demand of energy production, accumulation of biomass, redox homeostasis, and tumor invasion [3]. Numerous isoforms of rate-limiting glycolytic enzymes have been reported to be overexpressed in cancer cells. These include hexokinases (HK) which catalyze the first reaction of glycolysis involving the conversion and phosphorylation of glucose to glucose-6-phosphate (G6P). There are five isoforms of hexokinases viz HK1-4 and HKDC1 [19]. Among them, the most ubiquitous is HK1, whereas HK2-4 has tissue-specific distribution. HK2 is found to be upregulated in many cancers. HK2 is responsible for driving the coupling of glycolysis and oxidative phosphorylation via enhancing the glucose flux into several other metabolic pathways [19]. Both HK1 and HK2 are under the allosteric regulation of G6P. Further, HK1 is also regulated via the PI3K pathway. The outer mitochondrial membrane-bound HK2 has been reported to elevate the glycolytic rate, and this rate is affected when HK2 is removed. Therefore, the targeting of HK2 could serve as an effective anticancer therapy. Targeting HK2 is effective in killing cancer cells, including hepatocellular carcinoma (HCC). The ablation of HK2 has also been reported to inhibit mouse tumor growth, thus acting as a novel strategy for cancer treatment [20].

25.2.1.2.2 Phospho-fructokinases (PFKs)

PFKs are the following rate-limiting enzyme in glycolysis involved in the catalyzation of the third reaction of glycolysis, i.e., the conversion of fructose-6-phosphate (F6P) to fructose-1,6-bisphosphate (F1,6BP) via transferring a phosphate group from ATP. PFK is under the allosteric regulation of ATP. High ATP in cells inhibits the activity of PFK by binding to its regulatory site. Oncogenes or HIF-1 α can switch on PFK1, and its activity has been reported to be enhanced in cancer cells [21]. The activated PFK ultimately results in increasing the glycolytic flux. Fructose 2,6-bisphosphate (F2,6BP) is also responsible for the activation of PFK and is the most crucial allosteric activator of PFK [22]. The level of F2,6BP is under the control of bifunctional enzymes having phosphofructo-2-kinase and fructose 2,6-bisphosphate (PFKFBs) activities. PFKFB3 is modulated and phosphorylated by kinases such as PTEN and HIF-1α [21,23] and plays a crucial role in regulating the cell cycle and cell proliferation [22]. PFKFB3 is reported to be upregulated in many cancers [23]. Thus, PFK can be used as a metabolic target in cancer therapy.

25.2.1.2.3 Pyruvate kinases (PKs)

The final step of glycolysis, that is, PK catalyzes phosphoenolpyruvate (PEP) conversion to pyruvate with resultant production of ATP. PK has four mammalian isoenzymes having tissue-specific expression. These include L (present in liver), R (red blood cells), M1 (in heart, brain, and muscle), and M2 (in some of the differentiated tissues such as lungs and adipose tissue and the cells of the intestinal epithelium). Both PKM1 and PKM2 result from alternative splicing of the same gene, and their expression decides whether the cell undergoes glycolysis or oxidative phosphorylation [24]. PKM1 is expressed in normal cells, whereas PKM2 is expressed in rapidly proliferating tumor cells and nonmalignant proliferating cells [25]. Cancer cell growth is reported to be promoted by both metabolic and non-metabolic functions of PKM2. PKM2 is a low activity enzyme, creating a build-up of glycolytic intermediates transported in other metabolic pathways required for cell proliferation [26]. Further, PKM2 is activated by both F1,6BP and serine [26] and is reported mainly in colon cancer [26]. PKM2 is also a coactivator of HIF-1α where PKM2 hydroxylation activates HIF-1 α generating a positive feedback loop that results in its production [27]. The involvement of PKM2 in cancer anticipates the idea that a loss of PKM2 inhibits cancer development; however, removing PKM2 did not lead to inhibition of tumor metabolism [28]. Instead, it was observed that PKM2 plays a pro-cancerous role by converting into the dominant isoform PKM1, which is involved in controlling ATP production more actively [29]. The outcome is the production of NADPH via shuttling the intermediates of glycolysis into the pentose phosphate pathways (PPP).

25.2.2 Lactate metabolism

Tumor cells have been reported to show enhanced lactate production from pyruvate; this conversion is catalyzed by lactate dehydrogenases (LDH), resulting in the production of NAD^+ to maintain the glycolytic flux. In human tissues, there are five active isoenzymatic forms of LDH which are a combination of two different subunits, namely, H and M encoded by LDHA and LDHB genes, respectively [30]. In tumor cells, the LDHA isoform has been reported to show overexpression in several tumors. Further, the inhibition of LDHA activity has been reported to show inhibition in tumor cells' metastatic and invasive potential [30]. Thus, LDHA has gained focus as a predictive biomarker and metabolic target for cancer treatment. A high level of LDH in the serum has also been reported to correlate with an increase in the risk of death in the case of prostate, colorectal, and several other cancers [31]. If lactate builds up, it can be poisonous to the cell and contributes to tumor acidity and the production of CO_2 [32]. As the cancer cells take up more glucose, secretion of lactate is enhanced [33]. Excess lactate is secreted by monocarboxylate transporters 1-4 (MCT1-4). These MCTs have been reported to be upregulated in cancer; specifically, MCT4 correlated with poor prognosis in several cancer types [34]. Due to their crucial role in maintaining the metabolic phenotype of a cancer cell, MCTs serve as potential molecular targets for anticancer therapy [35].

25.2.3 Pentose phosphate pathway or hexose monophosphate shunt

Increased uptake of glucose and glycolytic flux along with the removal of the last step of glycolysis fuels the anabolic pathways required for the growth and proliferation of tumor cells. PPP is responsible for diversifying G6P from glycolysis for the biosynthesis of NADPH, ribose-5-phosphate (R5P), and other glycolytic intermediates. Cancer cells express elevated glucose-6-phosphate dehydrogenase (G6PD) levels involved in the oxidative phase of PPP. Further, some tumor suppressor genes regulate PPP via G6PD activity [36]. Further, another enzyme, 6-Phosphogluconate dehydrogenase (6PGD), shows upregulation in human cancer types, and the inhibition of this enzyme has been reported to show a reduction in the levels of lipogenesis and biosynthesis of RNA along with an increase in the generation of reactive oxygen species [37]. This phenomenon occurs via ribulose-5-phosphate by inhibiting the AMPK signaling pathway. Transketolase is another enzyme associated with the nonoxidative phases of PPP and is involved in the *de novo* generation of ribonucleotides for DNA and RNA synthesis [38] and has been reported in the development of cancer [39]. Trans-aldolase (TALDO1) also functions in the nonoxidative branches of PPP and has been reported to be increased in several cancers such as brain, bladder, breast, and head and neck squamous cell carcinomas [39], thereby acting as a potential target for anticancer therapeutics.

25.2.4 Tricarboxylic acid cycle

Pyruvate produced from glycolysis which is not used in the production of lactate is transported to the mitochondrial matrix and undergoes oxidation resulting in the production of CO_2. Instead of directly entering the Kreb cycle, pyruvate is first oxidatively decarboxylated by PDC into acetyl CoA, necessary for the tricarboxylic acid cycle (TCA) cycle and in the *de novo* pathway of lipogenesis. PDC is under negative regulation of pyruvate dehydrogenase kinase (PDKs), which phosphorylates PDCα subunit and inactivates it. PDKs have been reported to be overexpressed in malignant cells, and this overexpression is also correlated with HIF expression [40]. The inhibition of PDKs with compounds such as dichloroacetate induces the pyruvate to acetyl CoA flux and promotes the TCA cycle [41]. Various other compounds are also being used to specifically target PDKs, such as DCA derivatives, namely, N-(3-iodophenyl)-2,2-dichloroacetamide and Mito-DCA both having the property of inhibiting PDK-1 at micromolar level [42]. Other enzymes of the TCA cycle such as aconitase (AH), isocitrate dehydrogenase (IDH), succinate dehydrogenase, and fumarate dehydrogenase (FH) which are involved in a series of reactions producing isocitrate, α-ketoglutarate (α-KG), NADH, and FADH2 are also dysregulated in cancerous tissues [5]. Mutations in the TCA cycle enzymes are responsible for accelerating the cycle reactions and growth of cells. Therefore, the targeting of these enzymes is under investigation [5]. Further, Electron transport chain components are also mutated in cancer, affecting the glutamine-mediated synthesis of oxaloacetate [29]. Therefore, the TCA cycle and electron transport chain complex enzymes have been explored as potential anticancer therapeutic targets [3].

25.2.5 Amino acid metabolism

Serine serves as a building block for proteins, and the serine, glycine one-carbon network has been implicated in nucleotide synthesis, lipid, and protein synthesis, methylation metabolism, and polyamine metabolism [43]. The *de novo* metabolism of serine is elevated in malignant cells and adds to the one-carbon metabolism, producing NADPH and glutathione [44]. It has been reported that in the case of breast and colorectal cancer, cellular proliferation and survival depend on serine, and the enhanced activity of the serine, glycine one-carbon network induces cancer advancement mainly due to increased synthesis of nucleotides [44]. The first reaction in the serine biosynthesis is the conversion of 3-phosphoglycerate (from glycolysis) to 3-phosphohydroxypyruvate (precursor of serine), which is catalyzed by 3-phosphoglycerate dehydrogenase (PHGDH) [45]. PHGDH enzyme can show upregulation or overexpression in many cancers [46]. Cancer cells utilize the PHGDH activity for its proliferative functions as the products of PHGDH catalyzed reaction are also used in replenishing the α-KG supply of the TCA cycle [45]. Inhibitors such as CBR-5884, NCT-502, and NCT-503 could help in reducing serine levels in the cells having PHGDH overexpression, and inhibition of PHGDH has also been reported to reduce tumor growth *in vivo* tumor xenograft studies [47]. Hence, the inhibition of PHGDH has been an area of investigation and could serve as an effective approach in targeting the tumor cells, having enhanced *de novo* synthesis of serine [43]. Glutamine is another nonessential amino acid that is crucial for cancer cell growth and metabolism. Glutamine supplies nitrogen for other amino acids and nucleic acid synthesis, thereby replenishing the reduced intermediates and contributing to energy generation [48]. Glutamine also acts as a carbon donor for lipid synthesis via the reductive carboxylation to citrate and is involved in the synthesis of glutathione, an antioxidant present in the cells [48]. Hence, targeting the serine and glutamine metabolic pathways could be used as a therapeutic approach.

25.2.6 Lipid metabolism

Fatty acids provide the additional substrates required for fulfilling cancer cell requirements and play a crucial role in the synthesis of phospholipid bilayers, the production and storage of energy, and the synthesis of signaling molecules. Therefore, the enzymes involved in synthesizing fatty acids serve as potential targets for inhibiting cancer progression [49]. One such enzyme is ATP citrate lyase (ACL), the first enzyme involved in converting citrate into the lipogenic precursor acetyl-CoA, linking glucose and glutamine metabolisms to fatty acid synthesis. ACL has been reported to be overexpressed in several human cancers and is under the PI3K/AKT pathway [50]. The following enzyme is acetyl-CoA carboxylase (ACC) which brings about the conversion of acetyl-CoA into malonyl-CoA. ACC1 is overexpressed in several tumors and correlate with poor prognosis and tumor progression [51]. AMPK has been reported to phosphorylate and inactivate ACC1 [52]. Another critical enzyme in lipogenesis is Fatty acid synthase (FAS), a complex multifunctional enzyme catalyzing the terminal steps of *de novo* synthesis of fatty acids, the condensation reaction converting malonyl-CoA and acetyl-CoA substrates into palmitate [53]. FAS has been observed to be overexpressed in many cancer cells, leading to FAS inhibitor development [49]. Other enzymes which are upregulated in cancer include acyl-CoA synthetases and stearoyl-CoA desaturases (SCDs) [51,54]. Further, 3-hydroxy-3-methylglutaryl-CoA reductase (HMGCR) and choline kinase, which are involved in cholesterol and choline synthesis, respectively, have been insinuated in cancer development, confirming the need for their inhibitors to be evaluated in preclinical trials [3]. Therefore, an effective approach to cancer treatment should target the above metabolic enzymes and proteins.

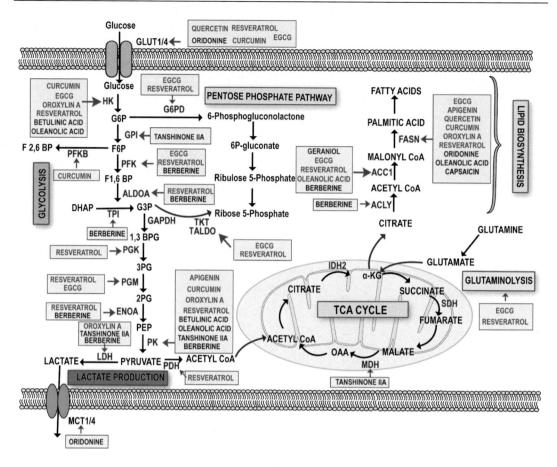

Figure 25.2

Diagrammatic illustration of the dysregulations in the cancer metabolic pathways and the key cancer metabolic targets of herbal compounds for efficient cancer management and treatment.

25.3 Herbal compounds in modulating cancer metabolic reprogramming

According to WHO, approximately 80% of people worldwide use traditional treatments [55]. Several plant-derived natural compounds are also being studied extensively for their anticancerous effects in cancer models [8–10,12,14]. Herbal compounds are promising due to their diverse structures and multitargeting activity. They also have demonstrated profound anticancer effects via targeting dysregulated cancer metabolism (Fig. 25.2 and Table 25.1).

25.3.1 Phenolic compounds in targeting cancer metabolism

Phenols are a class of plant secondary metabolites having hydroxylated aromatic rings comprising phenolic acids, cinnamic acids, coumarins, flavonoids, xanthone, and stilbenes [56]. Phenolic compounds have multiple effects such as antioxidant, anti-inflammatory,

Table 25.1: Effects of herbal compounds in the modulation of metabolic pathways in various cancers.

S. No.	Herbal compound	Cancer type	Cell line	Effect on metabolism	References
1	Apigenin	Lung	H1299, H460	Lowers GLUT1, glucose uptake, lactate production and generation of ATP and NADPH	[57]
		Laryngeal	Hep-2	Lowers GLUT1 expression	[58]
		Adenoid carcinoma	ACC-2	Lowers GLUT1 expression	[59]
		Colon	HCT116, HT29, DLD-1	Lowers PKM2 proteins, glucose uptake, lactate production and ATP generation	[60]
2	Curcumin	Lung	A549	Lowers GLUT1 expression	[66]
		Colon	HCT116, HT-29	Lowers HK2 protein activity and to some extent PFK, PGM, LDH	[67]
		Oesophageal	Ec109	Lowers the expression of GLUT4, HK2, PKM2 and PFKFB3	[68]
		Breast	SK-BR-3	Downregulation of FAS activity	[69]
3	Quercetin	Breast	MCF-7	Downregulation of glucose uptake and lactate production, elevation of GLUT1 expression levels	[71]
		Colon	RKO, HCT15	Downregulation of glucose uptake and lactate production	[72]
		HCC	HepG2, Huh7, Hep3B	Upregulation of GLUT1 membrane expression	[70]
		Breast	MDA-MB-157, MDA-MB-231	Lowering of FAS protein levels	[73]
		HCC	HepG2	Lowering the levels of intracellular Fatty acids and FAS activity	[74]
4	Oroxylin-A	HCC	HepG2	Decreased glucose uptake and lactate production	[78]
		Breast	MDA-MB-231, MCF-7	HK2 dissociation from mitochondria	[75]
		Colon	HCT116	Downregulation of ADRP, Srebp1, and FAS expression levels	[79]
5	Resveratrol	Breast	MCF-7	Upregulation of glucose oxidation and downregulation of lactate production	[83]
		Breast	SKBR-3	Decrease the level of FAS	[84]
		Leukaemia	K-562	Increase cellular levels of ceramide and decrease levels of sphingomyelin, S1P	[85]
		Ovarian	PA-1, OVCAR-3, MDAH2774, SK-OV-3	Decreased uptake of glucose and inhibition of GLUT1 plasma membrane localisation	[81]
6	Geraniol	HCC	HepG2	Decreased HMGR expression, fatty acid metabolism, and mevalonate pathway	[90]
7	Oridonin	Uveal melanoma	OCM-1, MUM2B	Decreased levels of FAS	[92]

(continued on next page)

Table 25.1: Effects of herbal compounds in the modulation of metabolic pathways in various cancers—cont'd

S. No.	Herbal compound	Cancer type	Cell line	Effect on metabolism	References
8	Tanshinone IIA	Gastric	AGS	Downregulation of GPI, LDHB, MDH1 and upregulation of PEPCK 2	[93]
		Oesophageal	Ec109	Downregulation of PKM2 levels	[94]
9	Betulinic acid	Pancreatic	MIA PaCa-2	Downregulation of PKM2 levels	[95]
		Breast	SK-BR-3	Downregulation of PKM2 levels and lactate production	[96]
		Cervical	HeLa	Downregulation of SCD-1 activity and increased incorporation of saturated fatty acids in Cardiolipin	[97]
10	Oleonilic acid	Breast; Prostate	MCF-7; PC-3	Decreased glucose uptake and lactate production; decreased levels of PKM2	[98]
		Breast	MDA-MB-231	Decreased levels of HK, PKM2, LDHA, PK, glucose uptake and lactate production	[100]
11	Berberine	Breast	MCF-7	Decreased levels of TP1, Aldolase A, and ENOA	[104]
		Lung	A549	Increase p-ACC and ATP generation	[106]
		Colon	HCT116	Decrease PKM2 activity	[105]
12	Capsaicin	HCC	HepG2	Decreased levels of FAS and de novo fatty acid synthesis	[107]

ACC, acetyl-CoA carboxylase; FAS, Fatty acid synthase; HK, hexokinases; LDH, lactate dehydrogenases; PFK, phospho-fructokinases, SCD, stearoyl-CoA desaturases.

antibacterial, antiviral, and anticancer. Phenolic compounds have gained focus due to their anticancer effects. Some of the plant-derived phenolic compounds having metabolic targeting potential are described below.

Apigenin is a plant flavone found in fruits, vegetables, and has many properties such as antioxidant, anti-inflammatory, and anticancer properties [reviewed by 12]. This compound inhibits cancer cell growth and functions via the downregulation of glucose uptake. In lung carcinoma cells, Apigenin resulted in downregulation of GLUT1 expression along with a reduction in the glycolytic flux and ATP production [57]. A similar study has also reported that the overexpression of GLUT1 resulted in conferring resistance to Apigenin-mediated apoptosis, which is in confirmation with the studies where Apigenin-induced downregulation of GLUT1 resulted in inhibition of cell growth, apoptosis, and increased the sensitivity to chemotherapeutic drugs [58,59]. In a recent study, Apigenin was found to inhibit the proliferation of colon cancer cells via acting as an allosteric inhibitor of PKM2 and regulating PKM2 expression along with a decrease in glucose consumption and production of lactate and generation of ATP [60]. Epigallocatechin gallate (EGCG) is another bioactive polyphenol obtained from green tea, affecting cancer cell metabolism in several ways. EGCG has been reported to decrease glucose uptake and increase glutamine synthesis while reducing the enrichment of glutamate [61]. In tongue carcinoma cells, treatment with EGCG was shown to inhibit glycolysis via the downregulation of HK2 through the AKT signaling pathway [62]. EGCG treated HCC cells also demonstrated a reduction in HK2 and PFK enzyme expression levels. These expression levels of proteins and the activity of enzymes confirm the EGCG induced glycolytic inhibition of HCC cells [63]. In another study, EGCG has been identified as a potent inhibitor of phosphoglycerate mutase 1 (PGAM1), which is an enzyme involved in the conversion of 3-phosphoglycerate (3-PG) to 2-phosphoglycerate (2-PG), thereby playing a key role in PPP flux [64]. Furthermore, in pancreatic adenocarcinoma cells, ECGC treatment reduces acetyl-CoA, leading to a reduction in palmitate (component of cellular membranes) synthesis [65].

Curcumin is another polyphenol obtained from turmeric (*Curcuma longa*) and affects glucose uptake and transport. Curcumin treatment decreases the GLUT1 expression in breast and lung carcinoma cells and prevents cancer invasiveness [66]. Curcumin has been reported to reduce glucose uptake, lactate production, and ATP generation in colorectal cancer, all associated with HK2 activity [67]. Further, another study also demonstrated a more significant impact of curcumin treatment on glycolytic enzymes, decreasing the expression of GLUT4, HK2, PFKFB3, and PKM2 via AMPK-mediated regulation [68]. Further, Curcumin has also been reported to play a role in the downregulation of FAS lipogenic enzyme in HCC and breast cancer cells [69]. Quercetin is a flavonol widely studied for its role in inhibiting glucose uptake and lactate efflux [70]. In liver cancer, Quercetin has been reported in recruiting GLUT1 from the cytoplasm to the plasma membrane and decrease its function via competitive inhibition [70]. Moreira et al. [71] also reported the competitive inhibition of GLUT1 by Quercetin in breast cancer cells. Further, the Quercetin mediated inhibition of lactate transportation has been reported to increase the cytotoxic potential of chemotherapeutic drug 5-fluorouracil, highlighting the synergistic effect of using natural anticancer compounds with standard anticancer drugs [72]. Quercetin leads to a decrease in the FAS protein levels in triple-negative breast cancer along with a decrease in β-catenin, leading to apoptosis of cancer cells [73]. In HepG2 cells, Quercetin also decreased the expression of FAS and intracellular fatty acids [74].

Another flavonoid Oroxylin-A obtained from the roots of *Scutellaria* has been reported to affect glucose metabolism. For instance, in lung adenocarcinoma cells, Oroxylin-A decreases HK2 protein expression leading to downregulation of glycolysis. It also leads to subsequent induction of pathways involved in detachment-induced apoptosis or anoikis via inhibiting the binding of HK2 to voltage-dependent anion channel [75]. Under hypoxic conditions, Oroxylin-A treated HCC cells showed reduced HK2, LDHA, PDK, and PKM2 [76]. This effect was dependent on Oroxylin-A-mediated degradation of HIF-1α via increased expression of the prolyl-hydroxylase domain (PHD2) and von-Hippel Lindau tumor suppressors [76].

The tumor suppressor gene, p53, plays a significant role in mediating cellular responses such as cell growth, DNA damage response, cell cycle control, and cell death [77]. Interestingly, p53 has been reported to act as a modulator of Oroxylin-A-mediated glycolytic inhibition in cancer cells [78]. In HepG2 cells, Oroxylin-A has been reported to increase the expression of p53 regulated key metabolic effectors such as TP53-induced glycolysis and apoptosis regulator (TIGAR) and SCO2 cytochrome c oxidase assembly [78]. Oroxylin-A has also been reported to affect lipid metabolism. For instance, in colon cancer, this compound results in metabolic reprogramming via inactivating HIF-1α, downregulating lipid uptake and synthesis, and upregulating fatty acid oxidation, decreasing the intracellular fatty acid levels [79]. This reduction leads to the inactivation of canonical WNT signaling as well as cell cycle arrest and growth inhibition. Resveratrol is another polyphenol obtained from grapes, berries, and other dietary compounds profoundly affecting cancer metabolism. In HCC, Resveratrol has been reported to decrease glucose uptake and production of lactate [80]. Further, in ovarian cancer cells, Resveratrol leads to decrease glucose uptake via impairing GLUT1 trafficking through AKT-mediated inhibition of GLUT1 membrane localization [81]. In nonsmall cell lung cancer, Resveratrol decreases the expression of HK2 [82]. Resveratrol treatment leads to glycolytic remodeling in colon cancer cells, decreasing pyruvate dehydrogenase enzyme activity, and this remodeling is regulated by PDP1 gene expression [83]. Resveratrol is further implicated in the regulation of lipid metabolism. This is supported by the study where Resveratrol treatment suppresses FAS expression, suggesting its anticancer effect [84]. The effect of Resveratrol has also been implicated in increasing ceramide, an important component of sphingolipids that act as secondary messengers during cell proliferation and apoptosis [85]. In human erythroleukemia cell lines, Resveratrol treatment leads to decreased sphingomyelin and sphingosine-1-phosphate, suggesting the upregulation of the sphingomyelin degradation pathway on treatment [85].

25.3.2 Isoprenoids in targeting cancer metabolism

Isoprenoids are plant-derived natural products synthesized from five-carbon precursors, namely, isopentenyl pyrophosphate and its isomer dimethylallyl diphosphate [86]. Based on the number of five-carbon building blocks, these include monoterpenoids (C10), sesquiterpenoids (C15), diterpenoids (C20), triterpenoids (C30), and tetraterpenoids (C40) [87]. Isoprenoids are gaining wide attention due to their anticancer potential. For instance, Paclitaxel (C20; Taxol) obtained from *Taxus brevifolia* is widely used in breast, ovarian, and lung cancer therapeutics [88]. Geraniol (C10) present in essential oils of aromatic plants affects cancer metabolism [89]. In HCC cells, it has been demonstrated that Geraniol decreases the fatty acid metabolism and impairs the phosphatidylcholine (PC) synthesis due to the inhibition of CTP-PC cytidylyl transferase. Further, Geraniol-induced inhibition of cell proliferation and induction of apoptosis has been implicated via inhibition of the mevalonate pathway and HMGCR [90]. Oridonin is another diterpenoid isolated from *Rabdosia rubescens*. It has been reported to induce autophagy in colorectal cancer cells via inhibition of glucose metabolism [91]. In another study, Oridonin reduces FAS protein levels, leading to apoptosis of uveal melanoma cell lines [92]. Another isoprenoid, namely Tanshinone IIA, is a major lipophilic component obtained from the roots of *Salvia miltiorrhiza* Bunge, which alters the glucose metabolism leading to suppression of cancer progression [93]. In human stomach adenocarcinoma, this compound has been implicated in the downregulation of glucose-6-phosphate isomerase (GPI), resulting in reduced glucose consumption and pyruvate production. The inhibitory effect of Tanshinone IIA on glycolysis has been attributed to a reduction in intracellular ATP and AKT expression levels along with an elevation in the expression of tumor suppressor p53. Further, Tanshinone IIA has also been reported in the dysregulation of gluconeogenesis via suppressing LDHB and malate dehydrogenase 1 (MDH1) as well as upregulation of PCK2 expression. In the oesophageal cancer cell line, this isoprenoid has also been implicated in the downregulation of PKM2 expression [94].

Another pentacyclic triterpenoid, namely, Betulinic acid is present in the outer barks of birch trees of Betula spp. It is studied widely for its anticancer activity and also in the modulation of cancer cell metabolism. A combination of Betulinic acid (20 μM) with chemotherapeutic drug Gemcitabine on cancer metabolism has been studied. The observed results showed that Betulinic acid could increase the activity of glycolytic enzyme PKM2. Further, the combination showed synergistic effects and enhanced the cytotoxic effects of the chemotherapeutic drug Gemcitabine in pancreatic cancer cells [95]. Betulinic acid has been reported to show a decreased abundance of PKM2 and lactate levels in breast cancer cells [96]. Further, In MCF-7 cells, this triterpenoid showed a reduction in the protein levels of glycolytic pathway modulators, HK and PKM2. The effect of Betulinic acid on fatty acid metabolism has also been reported where it inhibited the activity of the SCD-1, enzyme which plays a role in the *de novo* synthesis of fatty acids [97]. This leads to an enhanced accumulation of saturated fatty acids in the cardiolipin, resulting in structural changes in mitochondria and ultimate cell death. Lupeol is another pentacyclic triterpene obtained from fruits, vegetables, and medicinal plants such as *Tamarindus indica, Celastrus paniculatus, Zanthoxylum riedelianum, Allanblackia monticola, Himatanthus sucuuba, Leptadenia* hastata, Crataeva nurvala, *Bombax ceiba, Sebastiania adenophora, Aegle marmelos, and Emblica officinalis.* Lupeol has been reported to function as a hepatoprotective agent by significantly alleviating the altered liver function via restoring normal lipid metabolizing enzymes [9,14]. Lupeol has also been reported in the management of metabolic syndromes, thus indicating its potential to target cancer cell metabolism [9]. Further, another pentacyclic triterpenoid, Oleanolic acid derived from *Olea europaea* (Olive trees) of the Oleaceae family, shows anticancerous effects via modulating the cancer metabolism. This compound reduces glucose uptake and lactate generation [98,99]. This was followed by a decrease in PKM2 proteins along with an increase in the PKM1 expression in a dose-dependent manner [98].

Additionally, Oleanolic acid has also been reported to prevent fatty acid synthesis via activation of AMPK. This has been attributed to phosphorylation of ACC1, HMGCR, and reduction in FAS protein level [99]. Recently, it has been studied in breast cancer cells that Oleanolic acid can reverse the Warburg effect induced by high-salt mediated osmotic stress [100]. The hypertonic conditions could enhance aerobic glycolysis increasing glucose uptake and production of lactate. This effect was reversed by treating with Oleanolic acid. Oleanolic acid was also implicated in reducing the protein levels of glycolytic enzymes such as HK, PKM2, LDK, and PK [100].

25.3.3 Alkaloids in targeting cancer metabolism

Alkaloids are naturally occurring molecules having a basic nitrogen atom in their structure at any location except in amide or peptide bonds [101]. Alkaloids are obtained from bacteria, fungi, plants, and animals. Alkaloids have been reported to show anticancer effects via regulating GLUT, LDH, PK, PFK, and FAS expression. Further, alkaloids interfere with p53 and MAPK/ caspase-3, thereby inhibiting cancer growth. These compounds also target BECLIN-1, ATG5, and ATG7, which are the regulators of autophagy [102].

Berberine is an isoquinoline quaternary alkaloid obtained from plants that inhibits cancer cell proliferation and induces apoptosis [103]. A proteomic study has demonstrated the anticancer effect of Berberine in breast cancer cells where it downregulates the glycolytic enzymes, including triosephosphate isomerase, fructose-bisphosphate aldolase A, and enolase [104]. These dysregulations lead to a flux of carbohydrates from glycolysis to pathways which generate reducing power in the form of NADPH to cope up with the reactive oxygen species induced oxidative stress [104]. Another study reported Berberine-induced alteration in the glycolytic enzymes, such as reduction in LDH and PFK as well as an increase in PKM2. In colorectal and cervical cancer cells, Berberine has been reported to decrease PKM2 activity [105]. Berberine also affects lipid metabolism by enhancing the levels of p-ACC via the AMPK pathway activation in lung cancer [106]. Capsaicin is another alkaloid derived from pepper plants and has broad anticancer effects. In HCC, treatment with capsaicin resulted in downregulation of FAS protein level [107]. Further, *de novo* fatty acid synthesis was also inhibited by capsaicin, as indicated by a reduction in the levels of intracellular long-chain fatty acids and triglycerides [107].

25.4 Clinical studies on the use of herbal compounds in cancer management

The major cancer therapies such as chemotherapy, radiotherapy, and surgery have shown side effects and resistance in clinical studies. Therefore, the investigation of herbal compounds in cancer management is critically required. However, most of the studies are in a preclinical stage where herbal compounds are tested on early stage cancers, but some are also evaluated in advanced-stage cancers. Recent reports of the ongoing/completed clinical trials have highlighted the promising role of herbal compounds in cancer management. For instance, in a clinical study, coadministration of Curcumin and Quercetin for 6 months resulted in a decrease in the number and size of polyps in familial adenomatous polyposis without any side effects [108]. Further, in clinical trials, Curcumin is being tested for the prevention and treatment of colon (NCT00973869, NCT00027495, NCT00295035), lung (NCT02321293), breast (NCT01740323), pancreas (NCT00192842, NCT00094445, NCT00486460), prostate (NCT03211104, NCT02724618, NCT02138955) and osteosarcoma (NCT00689195) [109,110]. Further, EGCG is already being used in the prevention of colorectal cancer (NCT02891538) [71]. Resveratrol has been reported to show promising anticancer effects in colon cancer (NCT00256334), gastrointestinal cancer (NCT01476592) with fewer side effects [111,112]. In another random open-label clinical trial, Berberine showed anticancer effects (NCT03486496) via the regulation of glycolysis and mitochondrial phosphorylation [113]. Betulinic acid has also been reported to show promising anticancer efficacy in clinical trials via induction of mitochondrial apoptotic pathways and downregulation of BCL-2 as well as upregulation of pro-apoptotic cytochrome-c and caspase [114]. Thus, the above studies support herbal compounds as promising agents in targeting abnormal cancer cell metabolism to prevent cancer cell growth.

25.5 Conclusions

Cancer metabolism is an emerging hallmark of cancer. Dysregulations in metabolic enzymes and processes have been related to cancer progression and metastasis. Targeting dysregulated metabolic enzymes and pathways could therefore serve as an effective anticancer therapeutic approach. The main strategy involves regulating or inhibiting pathways involved in supplying nutrition for the production of energy and the inhibition of molecular biosynthesis required for cell growth, proliferation, and metastasis. Despite targeted therapies, many cancers could not be treated with single targeted therapies, thereby generating the need to identify multitargeted therapeutic agents. Herbal compounds have shown the efficacy to act as multitargeted agents targeting the cancer metabolic pathways with fewer side effects, thus making them promising compounds in cancer management. Despite their effectiveness, these compounds also suffer from instability, poor bioavailability, low solubility, low absorption, and low selectivity. Further, the degradation and rapid metabolism of these compounds result in their low plasma concentration. These pharmacokinetic limitations can be overcome by nanotechnology, which overcomes these limitations and enhances the bioavailability, cellular uptake, specificity, and efficacy of these compounds [115]. Researchers should focus on identifying more cancer metabolism pathways and developing novel formulations and drug delivery molecules to evaluate herbal compounds with reduced cost, toxicity, and chemoresistance, thereby leading to the effective management of cancer.

Abbreviations

2-DG	2-deoxy-D-Glucose
2-PG	2-phosphoglycerate
3-PGDH	3-phosphoglycerate dehydrogenase
ACC	Acetyl-CoA carboxylase
AMPK	AMP-activated Protein Kinase
EGCG	Epigallocatechin Gallate
FAS	Fatty Acid Synthase
G6PD	Glucose-6-phosphate Dehydrogenase
HMGCR	3-Hydroxy-3- Methylglutaryl-CoA Reductase
NADPH	Nicotinamide Adenine Dinucleotide Phosphate
PDC	Pyruvate Dehydrogenase Complex
PK	Pyruvate Kinase
TALDO1	Trans-Aldolase
TIGAR	TP53-Induced Glycolysis and Apoptosis Regulator

Acknowledgment

The authors are thankful to the Department of Zoology, Aligarh Muslim University, for providing the necessary facilities. D.S. also expresses her gratitude to UGC [524/(CSIR-UGC NET JUNE 2019)] India for fellowship. H.R.S. is thankful to UGC [Grant no. F.30-377/2017(BSR)] and DST-SERB (Grant no. EMR/2017/001758), New Delhi, for providing financial help.

Conflict of interest

The authors declare no conflict of interest.

References

[1]   O Warburg, On the origin of cancer cells, Science 123 (1956) 309–314.
[2]   AM Otto, Warburg effect(s)—A biographical sketch of Otto Warburg and his impacts on tumor metabolism, Cancer Metab 4 (2016) 5.
[3]   UE Martinez-Outschoorn, M Peiris-Pagés, RG Pestell, F Sotgia, MP Lisanti, Cancer metabolism: a therapeutic perspective, Nat Rev Clin Oncol 14 (2017) 11–31.
[4]   MT Snaebjornsson, S Janaki-Raman, A Schulze, Greasing the wheels of the cancer machine: the role of lipid metabolism in cancer, Cell Metab 31 (2020) 62–76.
[5]   A Luengo, DY Gui, MG Vander Heiden, Targeting metabolism for cancer therapy, Cell Chem Biol 24 (2017) 1161–1180.
[6]   NN Pavlova, C,B Thompson, The emerging hallmarks of cancer metabolism, Cell Metab 23 (2016) 27–47.
[7]   DJ Newman, GM Cragg, Natural Products As Sources Of New Drugs from 1981 to 2014, J Nat Prod 79 (2016) 629–661.
[8]   HR Siddique, SK Mishra, RJ Karnes, M Saleem, Lupeol, a novel androgen receptor inhibitor: implications in prostate cancer therapy, Clin Cancer Res 17 (2011) 5379–5391.
[9]   HR Siddique, M Saleem, Beneficial health effects of lupeol triterpene: a review of preclinical studies, Life Sci 8 (2018) 285–293.

[10] AML Seca, DCGA Pinto, Plant secondary metabolites as anticancer agents: successes in clinical trials and therapeutic application, Int J Mol Sci 19 (2018) 263.

[11] MA Khan, D Singh, HR Siddiqui, Animal models in cancer chemoprevention, Int J Zoo Animal Bio 2 (2019) 2639-216X.

[12] D Singh, MA Khan, HR Siddique, Apigenin, a plant flavone playing noble roles in cancer prevention via modulation of key cell signaling networks, Recent Pat Anticancer Drug Discov 14 (2019) 298–311.

[13] P Danhier, P Bański, VL Payen, D Grasso, L Ippolito, P Sonveaux, et al., Cancer metabolism in space and time: beyond the Warburg effect, Biochim Biophys Acta Bioenerg 858 (2017) 556–572.

[14] SK Maurya, GGHA Shadab, HR Siddique, Chemosensitization of therapy resistant tumors: targeting multiple cell signaling pathways by lupeol, a pentacyclic triterpene, Curr Pharm Des 26 (2020) 455–465.

[15] A Khan, S Siddiqui, SA Husain, S Mazurek, MA Iqbal, Phytocompounds targeting metabolic reprogramming in cancer: an assessment of role, mechanisms, pathways, and therapeutic relevance, J Agric Food Chem 69 (25) (2021) 6897–6928.

[16] K Zeng, G Ju, H Wang, J Huang, GLUT1/3/4 as novel biomarkers for the prognosis of human breast cancer, Transl Cancer Res 9 (2020) 2363–2377.

[17] C Yin, B Gao, J Yang, J Wu, Glucose transporter-1 (GLUT-1) expression is associated with tumor size and poor prognosis in locally advanced gastric cancer, Med Sci Monit Basic Res 26 (2020) e920778.

[18] CC Barron, PJ Bilan, T Tsakiridis, E Tsiani, Facilitative glucose transporters: implications for cancer detection, prognosis and treatment, Metabolism 65 (2016) 124–139.

[19] N Hay, Reprogramming glucose metabolism in cancer: can it be exploited for cancer therapy? Nat Rev Cancer 16 (2016) 635–649.

[20] KC Patra, Q Wang, PT Bhaskar, L Miller, Z Wang, W Wheaton, et al., Hexokinase 2 is required for tumor initiation and maintenance and its systemic deletion is therapeutic in mouse models of cancer, Cancer Cell 24 (2013) 213–228.

[21] A Yalcin, BF Clem, Y Imbert-Fernandez, SC Ozcan, S Peker, J O'Neal, et al., 6-Phosphofructo-2-kinase (PFKFB3) promotes cell cycle progression and suppresses apoptosis via Cdk1-mediated phosphorylation of p27, Cell Death Dis 5 (2014) e1337.

[22] H Lincet, P Icard, How do glycolytic enzymes favour cancer cell proliferation by non-metabolic functions? Oncogene 34 (2015) 3751–3759.

[23] X Ge, P Lyu, Y Gu, L Li, J Li, Y Wang, et al., Sonic hedgehog stimulates glycolysis and proliferation of breast cancer cells: modulation of PFKFB3 activation, Biochem Biophys Res Commun 464 (2015) 862–868.

[24] X He, S Du, T Lei, X Li, Y Liu, H Wang, et al., PKM2 in carcinogenesis and oncotherapy, Oncotarget 8 (2017) 110656–110670.

[25] TL Dayton, T Jacks, MG Vander Heiden, PKM2, cancer metabolism, and the road ahead, EMBO Rep 17 (2016) 1721–1730.

[26] WJ Israelsen, MG Vander Heiden, Pyruvate kinase: function, regulation and role in cancer, Semin Cell Dev Biol 43 (2015) 43–51.

[27] SJ Bensinger, HR Christofk, New aspects of the Warburg effect in cancer cell biology, Semin Cell Dev Biol 23 (2012) 352–361.

[28] M Cortés-Cros, C Hemmerlin, S Ferretti, J Zhang, JS Gounarides, H Yin, et al., M2 isoform of pyruvate kinase is dispensable for tumor maintenance and growth, Proc Natl Acad Sci U S A 110 (2013) 489–494.

[29] MD Hirschey, RJ DeBerardinis, AME Diehl, JE Drew, C Frezza, MF Green, et al., Dysregulated metabolism contributes to oncogenesis, Semin Cancer Biol 35 (2015) S129–S150.

[30] P Miao, S Sheng, X Sun, J Liu, G Huang, Lactate dehydrogenase A in cancer: a promising target for diagnosis and therapy, IUBMB Life 65 (2013) 904–910.

[31] W Wulaningsih, L Holmberg, H Garmo, H Malmstrom, M Lambe, N Hammar, et al., Serum lactate dehydrogenase and survival following cancer diagnosis, Br J Cancer 113 (2015) 1389–1396.

[32] I Marchiq, J Pouysségur, Hypoxia, cancer metabolism and the therapeutic benefit of targeting lactate/H(+) symporters, J Mol Med (Berl) 94 (2016) 155–171.

[33] YJ Chen, X Huang, NG Mahieu, K Cho, J Schaefer, GJ Patti, Differential incorporation of glucose into biomass during Warburg metabolism, Biochemistry 53 (2014) 4755–4757.

[34] VL Payen, E Mina, VF Van Hée, PE Porporato, P Sonveaux, Monocarboxylate transporters in cancer, Mol Metab 33 (2020) 48–66.

[35] F Baltazar, C Pinheiro, F Morais-Santos, J Azevedo-Silva, O Queirós, A Preto, et al., Monocarboxylate transporters as targets and mediators in cancer therapy response, Histol Histopathol 29 (2014) 1511–1524.

[36] ES Cho, YH Cha, HS Kim, NH Kim, JI Yook, The pentose phosphate pathway as a potential target for cancer therapy, Biomol Ther (Seoul) 26 (2018) 29–38.

[37] R Lin, S Elf, C Shan, HB Kang, Q Ji, L Zhou, et al., 6-Phosphogluconate dehydrogenase links oxidative PPP, lipogenesis, and tumour growth by inhibiting LKB1-AMPK signalling, Nat Cell Biol 17 (2015) 1484–1496.

[38] KC Patra, N Hay, The pentose phosphate pathway and cancer, Trends Biochem Sci 39 (2014) 347–354.

[39] MA Kowalik, A Columbano, A Perra, Emerging role of the pentose phosphate pathway in hepatocellular carcinoma, Front Oncol 7 (2017) 87.

[40] AD Zimmer, G Walbrecq, I Kozar, I Behrmann, C Haan, Phosphorylation of the pyruvate dehydrogenase complex precedes HIF-1-mediated effects and pyruvate dehydrogenase kinase 1 upregulation during the first hours of hypoxic treatment in hepatocellular carcinoma cells, Hypoxia (Auckl) 4 (2016) 135–145.

[41] E Saunier, C Benelli, S Bortoli, The pyruvate dehydrogenase complex in cancer: an old metabolic gatekeeper regulated by new pathways and pharmacological agents, Int J Cancer 138 (2016) 809–817.

[42] W Zhang, SL Zhang, X Hu, KY Tam, Targeting tumor metabolism for cancer treatment: is pyruvate dehydrogenase kinases (PDKs) a viable anti-cancer target? Int J Biol Sci 11 (2015) 1390–1400.

[43] JW Locasale, Serine, glycine and one-carbon units: cancer metabolism in full circle, Nat Rev Cancer 13 (2013) 572–583.

[44] M Mehrmohamadi, JW Locasale, Context dependent utilization of serine in cancer, Mol Cell Oncol 2 (2015) e996418.

[45] I Amelio, F Cutruzzolá, A Antonov, M Agostini, G Melino, Serine and glycine metabolism in cancer, Trends Biochem Sci 39 (2014) 191–198.

[46] SL Nowotarski, PM Woster, RA Jr. Casero, Polyamines and cancer: implications for chemotherapy and chemoprevention, Expert Rev Mol Med 15 (2013) e3.

[47] ME Pacold, KR Brimacombe, SH Chan, JM Rohde, CA Lewis, LJ Swier, et al., A PHGDH inhibitor reveals coordination of serine synthesis and one-carbon unit fate, Nat Chem Biol 12 (2016) 452–458.

[48] KP Michalak, A Maćkowska-Kędziora, B Sobolewski, P Woźniak, Key roles of glutamine pathways in reprogramming the cancer metabolism, Oxid Med Cell Longev 2015 (2015) 964321.

[49] Q Liu, Q Luo, A Halim, G Song, Targeting lipid metabolism of cancer cells: a promising therapeutic strategy for cancer, Cancer Lett 401 (2017) 39–45.

[50] M Chypre, N Zaidi, K Smans, ATP-citrate lyase: a mini-review, Biochem Biophys Res Commun 422 (2012) 1–4.

[51] E Currie, A Schulze, R Zechner, TC Walther, RV Farese Jr., Cellular fatty acid metabolism and cancer, Cell Metab 18 (2013) 153–161.

[52] Z Li, H Zhang, Reprogramming of glucose, fatty acid and amino acid metabolism for cancer progression. *Cell Mol*, Life Sci 73 (2016) 377–392.

[53] R Flavin, S Peluso, PL Nguyen, M Loda, Fatty acid synthase as a potential therapeutic target in cancer, Future Oncol 6 (2010) 551–562.

[54] RA Igal, Stearoyl CoA desaturase-1: new insights into a central regulator of cancer metabolism, Biochim Biophys Acta 1861 (2016) 1865–1880.

[55] CZ Wang, T Calway, CS Yuan, Herbal medicines as adjuvants for cancer therapeutics, Am J Chin Med 40 (2012) 657–669.

[56] W Vermerris, R Nicholson, Families of phenolic compounds and means of classification, Phenolic Compound Biochemistry, 2008, *Springer*, Dordrecht, 2008, p. 1–34.

[57] YM Lee, G Lee, TI Oh, BM Kim, DW Shim, KH Lee, et al., Inhibition of glutamine utilization sensitizes lung cancer cells to apigenin-induced apoptosis resulting from metabolic and oxidative stress, Int J Oncol 48 (2016) 399–408.

[58] YY Xu, TT Wu, SH Zhou, YY Bao, QY Wang, J Fan, et al., Apigenin suppresses GLUT-1 and p-AKT expression to enhance the chemosensitivity to cisplatin of laryngeal carcinoma Hep-2 cells: an in vitro study, Int J Clin Exp Pathol 7 (2014) 3938–3947.

[59] J Fang, YY Bao, SH Zhou, J Fan, Apigenin inhibits the proliferation of adenoid cystic carcinoma via suppression of glucose transporter-1, Mol Med Rep 12 (2015) 6461–6466.

[60] S Shan, J Shi, P Yang, B Jia, H Wu, X Zhang, et al., Apigenin restrains colon cancer cell proliferation via targeted blocking of pyruvate kinase M2-dependent glycolysis, J Agric Food Chem 65 (2017) 8136–8144.

[61] S Sánchez-Tena, G Alcarraz-Vizán, S Marín, JL Torres, M Cascante, Epicatechin gallate impairs colon cancer cell metabolic productivity, J Agric Food Chem 61 (2013) 4310–4317.

[62] F Gao, M Li, WB Liu, ZS Zhou, R Zhang, JL Li, et al., Epigallocatechin gallate inhibits human tongue carcinoma cells via HK2mediated glycolysis, Oncol Rep 33 (2015) 1533–1539.

[63] S Li, L Wu, J Feng, J Li, T Liu, R Zhang, et al., In vitro and in vivo study of epigallocatechin-3-gallate-induced apoptosis in aerobic glycolytic hepatocellular carcinoma cells involving inhibition of phosphofructokinase activity, Sci Rep 6 (2016) 28479.

[64] X Li, S Tang, QQ Wang, EL Leung, H Jin, Y Huang, et al., Identification of Epigallocatechin-3- Gallate as an Inhibitor of Phosphoglycerate Mutase 1, Front Pharmacol 8 (2017) 325.

[65] QY Lu, L Zhang, JK Yee, VW Go, WN Lee, Metabolic Consequences of LDHA inhibition by Epigallocatechin Gallate and Oxamate in MIA PaCa-2 pancreatic cancer cells, Metabolomics 11 (2015) 71–80.

[66] H Liao, Z Wang, Z Deng, H Ren, X Li, Curcumin inhibits lung cancer invasion and metastasis by attenuating GLUT1/MT1-MMP/MMP2 pathway, Int J Clin Exp Med 8 (2015) 8948–8957.

[67] K Wang, H Fan, Q Chen, G Ma, M Zhu, X Zhang, et al., Curcumin inhibits aerobic glycolysis and induces mitochondrial-mediated apoptosis through hexokinase II in human colorectal cancer cells in vitro, Anticancer Drugs 26 (2015) 15–24.

[68] FJ Zhang, HS Zhang, Y Liu, YH Huang, Curcumin inhibits Ec109 cell growth via an AMPK-mediated metabolic switch, Life Sci 134 (2015) 49–55.

[69] O Younesian, F Kazerouni, N Dehghan-Nayeri, D Omrani, A Rahimipour, M Shanaki, et al., Effect of curcumin on fatty acid synthase expression and enzyme activity in breast cancer cell line SKBR3, Int J Cancer Management 10 (2017) e8173.

[70] AF Brito, M Ribeiro, AM Abrantes, AC Mamede, M Laranjo, JE Casalta-Lopes, et al., New approach for treatment of primary liver tumors: the role of Quercetin, Nutr Cancer 68 (2016) 250–266.

[71] L Moreira, I Araújo, T Costa, A Correia-Branco, A Faria, F Martel, et al., Quercetin and epigallocatechin gallate inhibit glucose uptake and metabolism by breast cancer cells by an estrogen receptor-independent mechanism, Exp Cell Res 319 (2013) 1784–1795.

[72] R Amorim, C Pinheiro, V Miranda-Gonçalves, H Pereira, MP Moyer, A Preto, et al., Monocarboxylate transport inhibition potentiates the cytotoxic effect of 5-fluorouracil in colorectal cancer cells, Cancer Lett 365 (2015) 68–78.

[73] AS Sultan, MIM Khalil, BM Sami, AF Alkhuriji, O Sadek, Quercetin induces apoptosis in triple-negative breast cancer cells via inhibiting fatty acid synthase and beta-catenin, Int J Clin Exp Pathol 10 (2017) 156–172.

[74] P Zhao, JM Mao, SY Zhang, ZQ Zhou, Y Tan, Y Zhang, Quercetin induces HepG2 cell apoptosis by inhibiting fatty acid biosynthesis, Oncol Lett 8 (2014) 765–769.

[75] L Wei, Q Dai, Y Zhou, M Zou, Z Li, N Lu, et al., Oroxylin A sensitizes non-small cell lung cancer cells to anoikis via glucose-deprivation-like mechanisms: c-Src and hexokinase II, Biochim Biophys Acta 1830 (2013) 3835–3845.

[76]   Q Dai, Q Yin, L Wei, Y Zhou, C Qiao, Y Guo, et al., Oroxylin A regulates glucose metabolism in response to hypoxic stress with the involvement of Hypoxia-inducible factor-1 in human hepatoma HepG2 cells, Mol Carcinog 55 (2016) 1275–1289.

[77]   D Singh, MA Khan, HR Siddique, Role of p53-miRNAs circuitry in immune surveillance and cancer development: a potential avenue for therapeutic intervention, Semin Cell Dev Biol (2021) S1084-9521–9528, doi:10.1016/j.semcdb.2021.04.003.

[78]   Q Dai, Y Yin, W Liu, L Wei, Y Zhou, Z Li, et al., Two p53-related metabolic regulators, TIGAR and SCO2, contribute to oroxylin A-mediated glucose metabolism in human hepatoma HepG2 cells, Int J Biochem Cell Biol 45 (2013) 1468–1478.

[79]   T Ni, Z He, Y Dai, J Yao, Q Guo, L Wei, Oroxylin A suppresses the development and growth of colorectal cancer through reprogram of HIF1α-modulated fatty acid metabolism, Cell Death Dis 8 (2017) e2865.

[80]   W Dai, F Wang, J Lu, Y Xia, L He, K Chen, et al., By reducing hexokinase 2, resveratrol induces apoptosis in HCC cells addicted to aerobic glycolysis and inhibits tumor growth in mice, Oncotarget 6 (2015) 13703–13717.

[81]   H Gwak, G Haegeman, BK Tsang, YS Song, Cancer-specific interruption of glucose metabolism by resveratrol is mediated through inhibition of Akt/GLUT1 axis in ovarian cancer cells, Mol Carcinog 54 (2015) 1529–1540.

[82]   W Li, X Ma, N Li, H Liu, Q Dong, J Zhang, et al., Resveratrol inhibits Hexokinases II mediated glycolysis in non-small cell lung cancer via targeting Akt signaling pathway, Exp Cell Res 349 (2016) 320–327.

[83]   E Saunier, S Antonio, A Regazzetti, N Auzeil, O Laprévote, JW Shay, et al., Resveratrol reverses the Warburg effect by targeting the pyruvate dehydrogenase complex in colon cancer cells, Sci Rep 7 (2017) 6945.

[84]   A Khan, AN Aljarbou, YH Aldebasi, SM Faisal, MA Khan, Resveratrol suppresses the proliferation of breast cancer cells by inhibiting fatty acid synthase signaling pathway, Cancer Epidemiol 38 (2014) 765–772.

[85]   N Mizutani, Y Omori, Y Kawamoto, S Sobue, M Ichihara, M Suzuki, et al., Resveratrol-induced transcriptional up-regulation of ASMase (SMPD1) of human leukemia and cancer cells, Biochem Biophys Res Commun 470 (2016) 851–856.

[86]   ST Withers, JD Keasling, Biosynthesis and engineering of isoprenoid small molecules, Appl Microbiol Biotechnol 73 (2007) 980–990.

[87]   RMA Domingues, AR Guerra, M Duarte, CSR Freire, CP Neto, CMS Silva, et al., Bioactive triterpenic acids: from agroforestry biomass residues to promising therapeutic tools, Mini-Rev Org Chem 11 (2014) 382–399.

[88]   E Bernabeu, M Cagel, E Lagomarsino, M Moretton, DA Chiappetta, Paclitaxel: what has been done and the challenges remain ahead, Int J Pharm 526 (2017) 474–495.

[89]   M Cho, I So, JN Chun, JH Jeon, The antitumor effects of Geraniol: modulation of cancer hallmark pathways (Review), Int J Oncol 48 (2016) 1772–1782.

[90]   R Crespo, S Montero Villegas, MC Abba, MG de Bravo, MP Polo, Transcriptional and posttranscriptional inhibition of HMGCR and PC biosynthesis by Geraniol in 2 Hep-G2 cell proliferation linked pathways, Biochem Cell Biol 91 (2013) 131–139.

[91]   Z Yao, F Xie, M Li, Z Liang, W Xu, J Yang, et al., Oridonin induces autophagy via inhibition of glucose metabolism in p53-mutated colorectal cancer cells, Cell Death Dis 8 (2017) e2633.

[92]   Z Gu, X Wang, R Qi, L Wei, Y Huo, Y Ma, et al., Oridonin induces apoptosis in uveal melanoma cells by upregulation of Bim and downregulation of Fatty Acid Synthase, Biochem Biophys Res Commun 457 (2015) 187–193.

[93]   LL Lin, CR Hsia, CL Hsu, HC Huang, HF Juan, Integrating transcriptomics and proteomics to show that tanshinone IIA suppresses cell growth by blocking glucose metabolism in gastric cancer cells, BMC Genomics 16 (2015) 41.

[94]   HS Zhang, FJ Zhang, H Li, Y Liu, GY Du, YH Huang, Tanshinone aA inhibits human esophageal cancer cell growth through miR-122-mediated PKM2 down-regulation, Arch Biochem Biophys 598 (2016) 50–56.

[95]   A Pandita, B Kumar, S Manvati, S Vaishnavi, SK Singh, RN Bamezai, Synergistic combination of gemcitabine and dietary molecule induces apoptosis in pancreatic cancer cells and down regulates PKM2 expression, PLoS One 9 (2014) e107154.

[96]   A Lewinska, J Adamczyk-Grochala, E Kwasniewicz, A Deregowska, M Wnuk, Ursolic acid-mediated changes in glycolytic pathway promote cytotoxic autophagy and apoptosis in phenotypically different breast cancer cells, Apoptosis 22 (2017) 800–815.

[97]   L Potze, S Di Franco, C Grandela, ML Pras-Raves, DI Picavet, HA van Veen, et al., Betulinic acid induces a novel cell death pathway that depends on cardiolipin modification, Oncogene 35 (2016) 427–437.

[98]   J Liu, N Wu, L Ma, M Liu, G Liu, Y Zhang, et al., Oleanolic acid suppresses aerobic glycolysis in cancer cells by switching pyruvate kinase type M isoforms, PLoS One 9 (2014) e91606.

[99]   J Liu, L Zheng, N Wu, L Ma, J Zhong, G Liu, et al., Oleanolic acid induces metabolic adaptation in cancer cells by activating the AMP-activated protein kinase pathway, J Agric Food Chem 62 (2014) 5528–5537.

[100] S Amara, M Zheng, V Tiriveedhi, Oleanolic acid inhibits high salt-induced exaggeration of Warburg-like metabolism in breast cancer cells, Cell Biochem Biophys 74 (2016) 427–434.

[101] S E O'Connor, 1.25 - Alkaloids, Comprehensive Natural Products II - Vol. 1: Natural Products Structural Diversity-I Secondary Metabolites: Organization and Biosynthesis, Elsevier, Oxford, 2014, p. 977—1007.

[102] S Deng, MK Shanmugam, AP Kumar, CT Yap, G Sethi, A Bishayee, Targeting autophagy using natural compounds for cancer prevention and therapy, Cancer 125 (2019) 1228–1246.

[103] LM Ortiz, P Lombardi, M Tillhon, AI Scovassi, Berberine, an epiphany against cancer, Molecules 19 (2014) 12349–12367.

[104] HC Chou, YC Lu, CS Cheng, YW Chen, PC Lyu, CW Lin, et al., Proteomic and redox-proteomic analysis of berberine-induced cytotoxicity in breast cancer cells, J Proteomics 75 (2012) 3158–3176.

[105] ZC Li, HG Li, YX Lu, P Yang, ZY Li, Berberine inhibited the proliferation of cancer cells by suppressing the activity of tumor pyruvate kinase M2, Nat Prod Commun 12 (2017) 1415—1418.

[106] LX Fan, CM Liu, AH Gao, YB Zhou, J Li, Berberine combined with 2-deoxy-d-glucose synergistically enhances cancer cell proliferation inhibition via energy depletion and unfolded protein response disruption, Biochim Biophys Acta 1830 (2013) 5175–5183.

[107] H Impheng, S Pongcharoen, L Richert, D Pekthong, P Srisawang, The selective target of capsaicin on FASN expression and de novo fatty acid synthesis mediated through ROS generation triggers apoptosis in HepG2 cells, PLoS One 9 (2014) e107842.

[108] M Cruz-Correa, DA Shoskes, P Sanchez, R Zhao, LM Hylind, SD Wexner, et al., Combination treatment with curcumin and Quercetin of adenomas in familial adenomatous polyposis, Clin Gastroenterol Hepatol 4 (2006) 1035–1038.

[109] RE Carroll, RV Benya, DK Turgeon, S Vareed, M Neuman, L Rodriguez, et al., Phase IIa clinical trial of curcumin for the prevention of colorectal neoplasia, Cancer Prev Res (Phila) 4 (2011) 354–364.

[110] GB Maru, RR Hudlikar, G Kumar, K Gandhi, MB Mahimkar, Understanding the molecular mechanisms of cancer prevention by dietary phytochemicals: From experimental models to clinical trials, World J Biol Chem 7 (2016) 88–99.

[111] F Levi, C Pasche, F Lucchini, R Ghidoni, M Ferraroni, C La Vecchia, Resveratrol and breast cancer risk, Eur J Cancer Prev 14 (2005) 139–142.

[112] KR Patel, E Scott, VA Brown, AJ Gescher, WP Steward, K Brown, Clinical trials of Resveratrol, Ann N Y Acad Sci 1215 (2011) 161–169.

[113] W Tan, N Li, R Tan, Z Zhong, Z Suo, X Yang, et al., Berberine interfered with breast cancer cells metabolism, balancing energy homeostasis, Anti-cancer Agents Med Chem 15 (2015) 66–78.

[114] M Ali-Seyed, I Jantan, K Vijayaraghavan, SN Bukhari, Betulinic Acid: Recent Advances in Chemical Modifications, Effective Delivery, and Molecular Mechanisms of a Promising Anticancer Therapy, Chem Biol Drug Des 87 (2016) 517–536.

[115] MA Khan, D Singh, A Ahmad, HR Siddique, Revisiting inorganic nanoparticles as promising therapeutic agents: A paradigm shift in oncological theranostics, Eur J Pharm Sci 164 (2021) 105892.

CHAPTER 26

An introduction to herbal medicines and palliative care of cancer patients and related diseases through times in Turkey

Mahsa Pourali Kahriz[a], Parisa Pourali Kahriz[a], Fahad Ahmed[b] and Khalid Mahmood Khawar[a]

[a]*Department of Field Crops, Facultury of Agriculture, Ankara University, Ankara* [b]*Department of Basic Health, Yidirim Beyazit University, Üniversiteler Mah. İhsan Doğramacı Bulvarı Ankara Atatürk Eğitim Araştırma Hastanesi Yanı Bilkent Çankaya/Ankara, Turkey*

26.1 Introduction

Starting from the Eastern borders of Artvin, Agrı, to the western borders of Edirne, the Turkish people are very fond of using medicinal plants in folk medicines and palliative care (PC) systems to treat various diseases and infections. They use plants in every sphere of social activities, religious rituals, and daily life. Most common uses include making pesticides, foods, fiber for clothing, cosmetics, construction, firewood, construction industry, and the making of musical instruments [1,2]. Present-day Turkey is the center of diversity or origin of many economic and medicinal plants. It hosted Mesopotamian civilization that is counted as one among the three oldest civilizations (Nile, Indus, and Mesopotamia) of the World. Nine other important civilizations (Hattis, Hittites, Urartus, Phrygians, Lydians, Ionians, Carians, Lycians, and Hellenics) also lived on this map (Turkey). This land has also seen the grandeur of the culturally rich Romans, Byzantines, Arab Caliphates, Seljuks, and Ottomans empires. Thus, the people living in present-day Turkey have accumulated a huge experience and knowledge through centuries for the use of many medicinal plant species in the local medical and PC systems [3–5]. This knowledge has passed from generation to generation either in verbal or written form. It is understood that the local knowledge of using medicinal plants among the old and young generations is under erosion, due to the introduction of modern living, gradual breaking up of human relationships with plants and soil [6,7].

Herbal Medicines: A Boon for Healthy Human Life.
DOI: https://doi.org/10.1016/B978-0-323-90572-5.00004-4

26.2 The pattern of prescribing medicines through time

The patterns of prescription adapted during the Hittite civilization are accepted as the oldest known prescription in the written history. Similarly, Materia Medica, of Dioscorides, could rightly be called the first pharmacopeia of the world. This book put forward about 600 plants along with drug preparation methodologies, the majority of these grow in the natural flora of Turkey. [8,9]. Hippocrates (460-377 BC) also known as the founder of modern medicine, has described 236 plant species, and their healing properties [8]. It is said that ~4000 plants were used during the Arab Moroccan period on this land by the medical doctors (physicians) called tabibs in the treatment of several diseases and in the PC systems [9].

Starting from the 7th century, they translated a lot of literatüre from Roman, Greeks and Romans, South Asian (Indian, Pakistani), Iranian, and other available sources making significant contributions to the medical sciences.

26.3 Some important physicians during the Arab period

Ebû Bekir Muhammed bin Zekeriyyâ er-Râzî (850–923), Ebu Hanife Dineverî (895–992), Ebu'l Kasım Halef ibn Abbas ez-Zehravi (936–1013). Al-Biruni (973–1051) and ibn-i-Sina (Avecena) (980–1037) were the most important scientists of that era. These were followed by many others like Ebû Mervân Abdülmelik b. Muhammed b. Mervân b. Zühr el-İşbîlî (1094–1162), Abdurrahman el-Ghâfikī. (? –1165), Ebû'l-Velīd Muḥammed ibn Aḥmed ibn Muḥammed ibn Rüshd (1128–1198), (1197–1248), Ahmed b. Abdulvehhâb en-Nuveyrî (1279–1332) and Davud bin Omer el-Antaki (1541–1599) [8].

26.4 Turkish medical systems in Anatolia

The Turks practiced treatments using herbal medicines in Central Asia. During migration from Central Asia to Anatolia the Turkish people brought their folk medicines with them and hybridized these with the traditional Anatolian practices. During this period, the Seljuks established many hospitals in various areas. The most important of these was Gevher Nesibe Dârushifâ ve Tıp medresesi (Gevher Nesibe Hospital and Medical College) established at Kayseri. The Ottomans also established many new hospitals and dispensaries in the Anatolian provinces of Bursa, Edirne, Manisa, Istanbul, etc [8].

26.5 Hospitals and teaching of medicine

The most famous medical doctors (Tabibs) and surgeons (Jerrahs) of the Ottoman Empire, include Sherafeddin Sabuncuoğlu (1386–1470) and Merkez Efendi. Famous hospitals and schools of medicine and surgery of the Ottoman period includeTibkhâne-i Amire that was established with the efforts of Shanizade Mehmed Ataullah Efendi (1771–1826) and Behchet Efendi (1774–1834) [8].

26.6 Post World War I period and democracy

Prof Dr. Turhan Baytop (1920–2002) P. Belon (1517–1567), L. Rauwolff (1535–1596), J.P. Tournefort (1656–1708) G.A. Olivier (1756–1814), P.M.R Aucher-Eloy (1793–1838), K.H.E. Koch (1809–1879), E. Boiser (1810–1885), G.T. Kotschy (1813–1866), E. Bourgeau (1813–1877), P. Tchihatcheff (1818–1890), B. Balansa (1825–1891), L. Charrel (1839–1924), P. Sintenis (1847–1907)), W. Siehe (1859–1928), J.F.N. Bornmüller (1862–1948), K. Krause (1883–1963), P.M. Zhukovsky (1888–1975), O. Schwarz (1900–1983), A. Huber-Morath (1901–1990), and Peter Handland Davis (1918–1983) [8] are the most important scientists, medical doctors, and botanist of the post democracy period.

Thereafter, Turkey passed to democracy in 1923. Based on modern medical practices, the government enacted a Medical law in 1923 with the opening of several faculties of medicine, dentistry, and pharmacy in Istanbul, Ankara, and other cities to date. These faculties are contributing positively to the treatment and diagnosis of humans and diseases as per modern standards in parallel to the Scientific and technological education in the modern world [8]. These days medical scientists in collaboration with eminent plant scientists (ethnobotanists), taxonomists agricultural scientists (especially plant breeders and agronomists), and biochemists continue to study the beneficial local flora and have discovered many active compounds desired for human and animal health. For efficient use of the many medicinal plants, the plant breeders and agronomists have developed many new cultivars and varieties to obtain active compounds uniformly. K. Husnu Can Beşer, Neşet Arslan and Abdul Razak Memon, Sebahattin Özcan, Khalid Mahmood Khawar, Hatice Dumanoglu, Ercument Osman Sarihan, Reyhan Bahtiyarca Bagdat are some important famous plant scientists, who have contributed actively to the extraction of secondary metabolites, multiplication, and understanding of medicinal plants under *in vitro* and *ex vitro* conditions during the last quarter of 20th and first quarter of 21st century in the or Common Era. Biopharming is a new concept in Agriculture [10].

The most important work during 1998 is "Turkish ethnobotanical research archives during the Republican era" in the form of an MSc thesis by Narin Sadikoglu under the supervision of Prof. Dr. Kerim Alpınar. The thesis is available in the archive of the Turkish Higher education Commission Library [11] and Istanbul University Faculty of Pharmacy, Department of Pharmaceutical Botany. This study includes uses of plants belonging to Sivas, Istanbul, and Konya provinces; mostly used in human health, beliefs, and they are used as food. The other studies of significant importance were done by Karaman, and Kocabas [12], and Ozturk et al. [13]. A selected list of the most important local plant species used in palliative and health care systems is given in Table 26.1.

Table 26.1: Some selected plant species used in palliative care of general and Cancer patients in Turkey.

S. No	Bimomial name of species and family	Parts used	Usage form	References
Stomach wounds				
1.	*Ferula orientalis* L(Apiaceae)	Roots	Decoction	[50,51]
2.	*Foeniculum vulgare* Miller. (Apiaceae)	Roots, fruits and herbs	Decoction	[52]
3.	*Glycyrrhiza glabra* L. var. glandulifera (Waldst et Kit), Boiss. (Fabaceae)	Roots	Decoction	[53–55]
4.	*Hypericum atomarium* Boiss(Hypericaceae (Guttiferae)	Roots		[53,54,56]
5.	*Hypericum perforatum* L. (Hypericaceae(Guttiferae)	Roots		[57,58]
Cancer				
6.	*Urtica dioica* L.(Urticaceae)	Roots andLeaves	Decoction, infusion, tea, masage,	[59,60]
7.	*Rubus sanctus* Schreber (Rosaceae)	Roots; flowers leaves and Fruits	Infusion, tea, decoction, öintment	[61–65]
8.	*Malva neglecta* Wallr. (Malvaceae)	Roots	Decoction	[54,57]
9.	*Papaver somniferum* (Papaveraceae)	Mature flowers	Latex	[66]
10.	*Glycyrrhiza glabra* L. (Fabaceae)	Roots	Decoction, Maceration	[53–55]
11.	*Eryngium campestre* L. var. virens Link.(Apiaceae)	Roots	Raw, fresh Decoction	[56,67]
Rheumatism				
12.	*Apropyron repens* (L) P. Beauv. (Poaceae (Graminae)	Roots	Tea, ointment	[54]
13.	*Arum rupicola* var. *rupicola* Boiss.(Endemic) (Araceae)	Roots	Raw, fresh Infusion	[7,69]
14.	*Astragalus latexmifer* Lab. (Fabacea)	root latex androots	Latex, glue	[60,70]
15.	*Dracunculus vulgaris* Schott (Araceae)	Roots	Fresh, Dried, Decoction	[71,72]
16.	*Ecbalium elaterium* (L) A. Rich. (Cucurbitaceae)	Roots, Leaves, and Fruits	Raw, fresh sliceed, decoction, ointment,	[50,59,71,73]
17.	*Elytrigia repens* (L.) Desv. ex Nevski (Poaceae (Graminae)	Roots	Tea, ointment	[74]
18.	*Rubia tinctorium* L. (Rubiaceae)	roots	Dye, tea	[55,59,60 68,75,76]
19.	*Tamus communis* L. (Dioscoreaceae)	Roots	Latex,	[73]
20.	*Tamus communis* L. subsp. *communis* (Dioscoreaceae)	Roots	latex	[77]

26.7 Palliative care system

Natural limits of death are blurred [14]. Therefore present situation offers enormous contradictory and confusing and unexpected possibilities and information provided by specialists and nonspecialists about life [15] and diseases [16].

PC is defined as "a methodology that improves the personal satisfaction or quality of life of patients and their families dealing with the issues related to perilous sickness. Furthermore, PC also acts through the counteraction and help of sufferings by PC methods for early identification, perfect evaluation, and therapy of pain and other issues, related to physical, psychosocial, and spiritual problems" [17].

PC focus on maximizing the quality of life of a patient by effective control of disease symptoms by provision of meaningful spiritual, psychological and social support, at a very difficult or difficult time in life using all treatment and care options [18].

Therefore PC is an important organ of the treatment systems and recognizes the need for continuous care, which begins from the time, any disease starts including cancer treatment or the diseases after diagnosis until the patient is cured, or moves gradually to death.

PC recognizes and encourages multidisciplinary approaches with the help of physicians, community medicine practitioners, nurses, psycho-oncologists, paramedical professionals with an integrative approach.

26.8 The general and oncologists approaches

The PC system offers support to challenges to conventional medical care including the use of herbal medicines, acupuncture, homeopathy, mind-body, and other techniques to better understand and cure the patients. A resolution was published by World Health Assembly in 2014, which is recognized as the first-ever global proclamation on PC. This resolution includes PC in the definition of universal health coverage. It emphasizes member states to strengthen PC as a core component in healthcare systems with a stress on primary health care and community/home-based care World over [19].

The recognition of and development of PC in Turkey started in the early 1990s with the provision of basic services that focussed on developing pain control strategies in cancer patients [20]. Thereafter Turkish Republic Ministry for Health through technical support from the Middle East Cancer Consortium after shaped several dimensions of cancer control programs during the 2000–210 period [21]; considering PC for cancer patients at the grassroots level. The main focusing points were the establishment of the Turkish Republic National Cancer Control Program during 2009, and the developing Pallia Turk Project policy document called during 2010 [22]. The Pallia–Turk project planned the establishment of primary, secondary, and tertiary levels PC centers; to provide the related services. with major involvement of home care teams and family physicians with the integration of economic, physiological, physical, and social stakeholders to meet the and needs of the patients [23].

Despite these policy changes, still in 2017, however, GAPCEL (The Global Atlas of PC at the End of Life) reports patchy work on the development of PC activism in Turkey when compared to the population size [23]. International Agency for Research on Cancer, Globocan 2020 also confirms that out of 233,834 new diagnosed cancer cases in 2020; approximately 126,335 Turkish citizens died due to cancer [24]. Besides, one in every four deaths in Turkey is due to cancer [17]. The size of the cancer patients' population is increasing significantly as people over the age of 65 are expected to double by 2040 [25]. Therefore, there is a need to improve the PC of the patients with appropriate planning at the national level to improve their quality of life by improved PC programs at the national level.

26.9 Concern about erosion of ethnomedicinal knowledge about plants

The rich cultural history of Turkey is continuously playing a distinct role in the plant–human relationship, both in verbal and written form passing down from generation to overtime. In line with advances in technology in recent years, there is increased economic migration from rural areas to cities. This has resulted in reduced understanding and recognition of these plant species, loss of habitat of many plants from the Turkish flora. There is a need to protect and guaranty the survival of this knowledge by securing it through transcription for benefit of future generations [1–5]. Therefore, documenting and conservation of "medicinal plants used in PC systems is significantly important [26,27].

26.10 Need of documentation with mutidisciplinery approach

Documenting available information, researching, and confirming the correct use of plants through modern research is very important [27]. In line with this modern researchers of Turkey are continuously studying and evaluating local flora using different life sciences including medicinal, pharmaceutical, agricultural, biochemical, ethnomedicinal, and sciences to understand the medicinal characteristics of the local flora.

Farmers, gardeners, botanists, and all those taking interest in plants (professionals or amateur) know that plants are important to provide an abode to chirping birds, good natural aroma all around with developöemt of green atmosphere around with soothing, tranquil atmosphere that could play a very powerful role in improving quality of life and mental health [28–30]. It is established that the plants reduce stress levels, boost human moods, source of joy with improvement in overall happiness. This alleviates depression and loneliness and stress due to chronic cancer dying patients [28,29]. Moreover, there are some hospitals that encourage patients to work with plants and dirt to take their minds off their current problems providing them with moments of happiness and joy [31]. Plants improve the quality of air helping patients to live by themselves. Therefore plants growing appropriate to local climate act as a gift for everyone and anyone. Having plants around are known to lower stress and pain levels [30–33].

26.11 Therapeutic value

Moreover, the concept of introducing plants in life of the people to encourage patients and their relatives or families with easy methods to come closer to increasing number of plants and use them in any form to improve their quality of life through houseplants or garden plants. used as part of hospice care until recently. Not only do plants clean and oxygenate the air, but they also eliminate harmful toxins [34–36]. They purify and oxygenate the air, with improved respiratory comfort by filtering most of the pollutants in the air [34,35].

The plants lower stress and anxiety among people especially those who have the problem of blood pressure and heart diseases [37].

The plants improve cognitive functions like improving the ability to focus and remember and improve self-esteem and confidence with emotional and visual gratification [34,36,37].

Frequent interaction with environments helps cancer patients to forget the problem. Therefore, cancer patients are generally encouraged for inter-disciplinary treatment with interdisciplinary palliative medicine care [38–41] to cancer and other diseases with the help of integrated technologies including herbal medicines and foods within a single medical institution [42].

Complementary medicines especially herbal medicines used in PC, focus on a patient-centered treatment to improve life quality and reduce the related problems in cancer treatments [43]. Therefore integration of medicinal herbs as a part of complementary medicine treatment carries weight and in minimizing the toxic effects of the cancer treatments [43,44]. The integration of traditional medicines in PC treatments could provide an opportunity for addressing the bio, psycho, social, cultural, and spiritual aid to the patients in many oncology hospitals [45]. Medical doctors are convinced to shift to the use of medicinal herbs for centuries; to deliver chances of improved quality of life to chronically ill people including cancer patients [45–47].

PC is offered in most of the oncology hospitals in Turkey, which care the significance of concerns the quality of life of the patients and related survival or disease-focused parameters [48].

Development of medicinal, pharmaceutical, agricultural, botanic (taxonomy and ethnobotanical biotechnological and biochemical sciences), and technologies are continuously participating in the understanding of features of these and many new medicinal plant species for appropriate use in palliative and health care systems [1,3,15,21,38,45,46,49]. Turkish government encourages the reproduction and dissemination of medicinal plants through scientific projects at the National level through TÜBITAK. The state is expected to further encourage, the researchers and universities in their pursuits to reproduce, multiply and disseminate these enormously important plant species and taxons through the development of *in vitro* and *ex vitro* techniques.

26.12 Conclusions

It is understood that there is a need to develop more PC (and supportive/end-of-life care) approaches for everyone with strong social networks. This review describes briefly the popular knowledge about the plants used in PC among Turkish incurables including cancer patients. It will be better to call it transitional knowledge needing more studies. This is an exploratory descriptive study that describes some of the many plants used by Turkish people with cancer and other diseases used in normal and PC in parallel to the treatment procedures carried out in hospitals. The information provided in this study will expand the sphere of effective PC and plant-based studies used in treating all types of patients.

References

[1]   I Ugulu, Traditional ethnobotanical knowledge about medicinal plants used for external therapies in Alasehir, Turkey, Int J Med Arom Plants 1 (2) (2011) 101–106.

[2]   ME Bozyel, E Bozyel-Merdamert, K Canli, EM Altuner, Anticancer uses of medicinal plants in Turkish traditional medicine, Int J Acad Appl Res 22 (2) (2019) 465–484.

[3]   M Kargıoğlu, S Cenkci, A Serteser, N Evliyaoğlu, M Konuk, M.Ş Kök, et al., An ethnobotanical survey of inner-West Anatolia, Turkey, Hum Ecol 36 (2008) 763–777.

[4]   T Baytop, Turkiye'de Bitkiler ile Tedavi, Nobel Tip Kitabevleri; Bitkibilim, Sifalı Bitkiler, II. baskı., Nobel Tıp Kitapevleri Ltd. Şti., Tayf Ofset Baskı, İstanbul, 1999, p. 480.

[5]   KHC Başer, et al., Chapter 46: Most widely traded plant drugs of Turkey, in: Tuley De Silva, et al. (Eds.), Traditional and Alternative Medicine-Research 3& Policy Perspectives, NAM-Daya Publ. House, Delhi, 2009, pp. 443–454.

[6]   M Öztürk, Ethnobotany-time for a new relationship-case study from Turkey, in: Ikram ul-haq (Ed.), 11th National Meeting of Plant Scientists (NMPS) and 2nd Intern. Conf. of Plant Scientists, GC University, Lahore, 2011.

[7]   E Altundag, M Ozturk, Ethnomedicinal studies on the plant resources of east Anatolia, Turkey, Procedia-Social and Behavioral Sciences 19 (2011) 756–777.

[8]   I Ugulu, Traditional ethnobotanical knowledge about medicinal plants used for external therapies in Alasehir, Turkey, Int J Med Arom Plants 1 (2) (2011) 101–106.

[9]   ME Bozyel, E Bozyel-Merdamert, A Benek, D Turu, MA Yakan, K Canlı, Ethnomedicinal uses of araceae taxa in Turkish traditional medicine, Int J Acad Appl Res 4 (5) (2020) 78–87.

[10] Ç Orçun, M Gölükçü, in: Ç Orçun, M Gölükcü (Eds.), III. Tıbbi ve Aromatik Bitkiler Sempozyumu Tam Metin Bildirileri Kitabı, 51, Süleyman Demirel Üniversitesi Ziraat Fakültesi 4, Baskı, 2013, pp. 1–189. ISBN:975-7929-79-4.

[11] B Keykubat, Tıbbi Aromatik Bitkiler ve İyi Yasam, İzmir Ticaret Borsası, Ar-Ge Müdürlüğü yayını, İzmir, 2016.

[12] A Aksoy, HÇ Kaymak, Türkiye Domates Sektörüne Genel Bakış, Iğdır Üniversitesi Fen Bilimleri Enstitüsü Dergisi 6 (2) (2016) 121–129.
[13] N Sadikoglu, Cumhuriyet Dönemi Türk Etnobotanik Araştırmalar Arşivi, Yüksek Lisans Tezi, İstanbul Üniversitesi, Sağlık Bilimleri Enstitüsü, İstanbul, 1998. https://tez.yok.gov.tr/UlusalTezMerkezi/tezSorguSonucYeni.jsp.
[14] S Karaman, YZ Kocabas, Traditional medicinal plants of K. Maras (Turkey), Sciences (New York) 1 (3) (2001) 125–128.
[15] F Oztürk, M Dölarslan, G Ebru, Etnobotanik ve Tarihsel Gelişimi, Türk Bilimsel Derlemeler Dergisi 9 (2) (2016) 11–13.
[16] DGP. 2019. Deutsche Gesellschaft für Palliativmedizin, https://www.dgpalliativmedizin.de/ (accessed March 31, 2021)
[17] A Coulter, Measuring what matters to patients, BMJ 356 (2017) 816, doi:10.1136/bmj.j816.
[18] P Teillhard de Chardin, The Phenomenon of Man, with an Introduction by Sir Julian Huxley, Collins, London, 1963.
[19] WHO (2018). Palliative care. http://www.who.int/news-room/fact-sheets/detail/palliative-care, https://www.who.int/nmh/countries/tur_en.pdf. (Accessed 01.06.2021)
[20] D Doyle, Editorial palliative medicine, Palliat Med 7 (1993) 253–255.
[21] N Ezer, K Avcı, Çerkeş (Çankırı) yöresinde kullanılan halk ilaçları, Hacettepe Üniversitesi Eczacılık Fakültesi Dergisi 24 (2) (2004) 67–80.
[22] MZ Öztürk, G Çetinkaya, S Aydın, Köppen-Geiger iklim sınıflandırmasına göre Türkiye'nin iklim tipleri, Cografya dergisi 35 (2017) 17–27.
[23] PH Davis, Flora of Turkey and The East Aegean Islands, 1-9, Edinburgh University Press, Edinburgh, 1965–1985.
[24] ES Utku, E Hacikamiloglu, M Gultekin, Experience associated with the developing nationwide palliatve care services in the comunity: what can one learn from it for the future? in: M Silbermann (Ed.), Palliative Care: Perspectives, Practices and Impact on Quality of Life. A Global View, Nova Science Pub Inc., 415 Oser Avenue, Suite N Hauppauge, NY, 11788 USA, 2017, pp. 287–292.
[25] SR Connor, Global atlas of palliative care at the end of life, World Palliative Care Alliance, World Health Organization, London, 2020, pp. 12–16.
[26] IARC. Global cancer observatory. Secondary Global Cancer Observatory 2020. https://gco.iarc.fr/.
[27] AARPThe aging readiness & competitiveness initiative: Turkey, Secondary The Aging Readiness & Competitiveness Initiative, Turkey, 2019. https://arc.aarpinternational.org/countries/turkey.
[28] G Kendir, A Güvenç, Etnobotanik ve Türkiye'de yapılmış etnobotanik çalışmalara genel bir bakış, Hacettepe Üniversitesi Eczacılık Fakültesi Dergisi 30 (1) (2010) 49–58.
[29] C Muthu, M Ayyanar, N Raja, S Ignacimuthu, Medicinal plants used by traditional healers in Kancheepuram District of Tamil Nadu, India, J Ethnobiol Ethnomed 2 (1) (2006) 1–10.
[30] FW Telewski, A unified hypothesis of mechanoperception in plants, Am J Bot 93 (10) (2006) 1466–1476.
[31] WH Lewis, MP Elvin-Lewis, Medical Botany: Plants Affecting Human Health, John Wiley & Sons, New York, 2003.
[32] K Choudhary, M Singh, U Pillai, Ethnobotanical survey of Rajasthan—an update, American-Eurasian Journal of Botany 1 (2) (2008) 38–45.
[33] D Conquergood, Health theatre in a Hmong refugee camp: performance, communication, and culture, TDR 32 (3) (1988) 174–208.
[34] WHO Air Pollution. WHO. http://www.who.int/airpollution/en/ (accessed March 3, 2021).
[35] I Parajuli, H Lee, KR Shrestha, Indoor air quality and ventilation assessment of rural mountainous households of Nepal, Int J Sust Built Env 5 (2016) 301–311 10.1016.
[36] C Hall, M Knuth, An update of the literature supporting the well-being benefits of plants: a review of the emotional and mental health benefits of plants, Journal of Environmental Horticulture 37 (1) (2019) 30–38.
[37] CR Hall, MW Dickson, Economic, environmental, and health/well-being benefits associated with green industry products and services: a review, Journal of Environmental Horticulture 29 (2) (2011) 96–103.

[38]   A Smith, M Pitt, Sustainable workplaces: improving staff health and well-being using plants, Journal of Corporate Real Estate 11 (1) (2009) 52–63.

[39]   SH Park, RH Mattson, Effects of flowering and foliage plants in hospital rooms on patients recovering from abdominal surgery, HortTechnology 18 (4) (2008) 563–568.

[40]   AB Hamric, LJ Blackhall, Nurse-physician perspectives on the care of dying patients in intensive care units: collaboration, moral distress, and ethical climate, Crit Care Med 35 (2) (2007) 422–429.

[41]   B Kane, S Luz, Achieving diagnosis by consensus, Computer Supported Cooperative Work 18 (4) (2009) 357–392.

[42]   BR Ferrell, JS Temel, S Temin, ER Alesi, TA Balboni, E.M Basch, T.J Smith, Integration of palliative care into standard oncology care: American Society of Clinical Oncology clinical practice guideline update, J Clin Oncol 35 (1) (2017) 96–112.

[43]   TJ Smith, S Temin, ER Alesi, AP Abernethy, TA Balboni, EM Basch, et al., American Society of Clinical Oncology provisional clinical opinion: the integration of palliative care into standard oncology care, J Clin Oncol 30 (8) (2012) 880–887.

[44]   C Bron, M Wensing, JL Franssen, RA Oostendorp, Tratment of myofascial trigger points in common shoulder disorders by physical therapy: a randomized controlled trial [ISRCTN75722066], BMC Musculoskeletal Disorders 8 (1) (2007) 1–8.

[45]   R Muecke, M Paul, C Conrad, C Stoll, K Muenstedt, O Micke, …, PRIO (Working Group Prevention and Integrative Oncology of the German Cancer Society), Complementary and alternative medicine in palliative care: a comparison of data from surveys among patients and professionals, Integr Cancer Ther 15 (1) (2016) 10–16.

[46]   DB Graves, The emerging role of reactive oxygen and nitrogen species in redox biology and some implications for plasma applications to medicine and biology, J Phys D: Appl Phys 45 (26) (2012) 263001.

[47]   HM Chochinov, BJ Cann, Interventions to enhance the spiritual aspects of dying, J Palliat Med 8 (S1) (2005) s103.

[48]   DA Hussain, MM Hussain, Nigella sativa (black seed) is an effective herbal remedy for every disease except death—a prophetic statement which modern scientists confirm unanimously: a review, Adv Med Plant Res 4 (2) (2016) 27–57.

[49]   JW Sheldon, MJ Balick, SA Laird, GM Milne, Medicinal plants: can utilization and conservation coexist? Advances in Economic Botany 12 (1997) i–104.

[50]   K Gallacher, B Jani, D Morrison, S Macdonald, D Blane, PJ Erwin, et al., Qualitative systematic reviews of treatment burden in stroke, heart failure and diabetes - methodological challenges and solutions, BMC Med Res Method, 2013, pp. 10–13.

[51]   E Sezik, M Tabata, E Yesilada, G Honda, K Goto, Y Ikeshiro, Traditional medicine in Turkey in folk medicine in north-east Anatolia, J Ethnopharmacol 35 (1991) 191–196.

[52]   G Bulut, E Tuzlacı, A Doğan, İ Şenkardeş, An ethnopharmacological review on the Turkish Apiaceae species, İstanbul Ecz Fak Dergi /J Fac Pharm Istanbul 44 (2) (2014) 163–179.

[53]   I Kaval, L Behçet, U Cakilcioglu, Ethnobotanical study on medicinal plants in Geçitli and its surrounding (Hakkari-Turkey), J Ethnopharmacol 155 (2014) 171–184.

[54]   İ Şenkardeş, E Tuzlacı, Some Ethnobotanical Notes from Gündoğmuş District (Antalya/Turkey), MÜSBED 4 (2) (2014) 63–75.

[55]   S Kizil, Genetic diversity of medicinal plants and their growing potential in the southeastern Anatolia region of Turkey. Acta Hortic 826 (2009) 241–254. https://doi.org/10.17660/ActaHortic.2009.826.33.

[56]   S Tekin, Üzümlü (Erzincan) ilçesinin etnobotanik özellikleri, Erzincan Üniversitesi Fen Bilimleri Enstitsü, Biyoloji Anabilim Dalı, Yüksek Lisans tezi, 2011, p. 126.

[57]   R Polat, Ethnobotanical study on medicinal plants in Bingol (City center) (Turkey), Journal of Herbal Medicine 16 (2019) 1–11.

[58]   M Korkmaz, Z Alpaslan, Ergan Dağı (Erzincan-Türkiye)'nın Etnobotanik özellikleri, Bağbahçe Bilim Dergisi 1 (3) (2014) 1–31 2014.

[59]   G Kendir, A Güvenç, Etnobotanik ve Türkiye'de yapılmış etnobotanik çalışmalara genel bir bakış, Hacettepe Univ J Faculty Pharm 1 (2010) 49–80.

[60]   Y Bağcı, R Erdoğan, S Doğu, Sarıveliler (Karaman) ve Çevresinde Yetişen Bitkilerin Etnobotanik Özellikleri Selçuk, Üniversitesi Fen Fakültesi Fen Dergisi 42 (1) (2016) 84–107.

[61]   DA Güler, A Aydin, M Koyuncu, İ Parmaksiz, Anticancer activity of papaver somniferum, Journal of the Turkish Chemical Society Section A: Chemistry 3 (3) (2016) 349–366.

[62]   S Karakaya, A Polat, O Aksakal, YZ Sumbullu, U İncekara, An ethnobotanical investigation on medicinal plants in South of Erzurum (Turkey), Ethnobotany Research & Applications 18 (13) (2019) 1–18.

[63]   A Ünver, Lamas çayı çevresindeki köylerde (Erdemli, Silike/Mersin) etnobotanik araştırmalar, Etnobatany reseach in the villages around Lamas rtream (Erdemli/Silifke-Mersin), Balıkesir Üniversity, Instute of Science, Deparment of Biology. Msc Thesis, Balıkesir, 2019, p. 195. 2019.

[64]   G Dogan, E Bagci, Elazığ'ın Bazı Yerleşim alanlarında halkın geleneksel ekolojik bilgisine dayanarak kullandığı bitkiler ve etnobotanik ozellikleri, Firat University Journal of Science 23 (2) (2011) 77–86.

[65]   AD Koca, Ş Yıldırımlı, Ethnobotanical properties of Akçaoca district in Düzce (Turkey), Hacettepe Journal of Biology and Chemistry 38 (1) (2010) 63–69.

[66]   Bülent Olcay, Ş Kültür, Medicinal Plants Used In Traditional Treatment of Hypertension in Turkey, 6, 3rd ed., Adıyaman University, Graduate School of Natural and Applied Sciences, Department of Biology, 2020, pp. 80–95.

[67]   MM Hürkul, A Köroğlu, Etnobotanik Bir Derleme: Amygdaloideae (Rosaceae) Alt Familyası, FABAD Journal of Pharmaceutical Sciences 44 (1) (2019) 35–46.

[68]   R Polat, Medicinal Plants Used In Traditional Treatment of Hypertension in Turkey, 16, J Herb Med, 2019, pp. 1–11.

[69]   L Behçet, M Arık, An ethnobotanical investigation in East Anatolia (Turkey) Tr. Doğa ve Fen Derg. – Tr, J Nature Sci 2 (1) (2013) 1–14.

[70]   U Özgen, Y Kaya, M Coşkun, Ethnobotanical studies in the villages of the District of Ilıca (Province Erzurum) Turkey, Econ Bot 58 (4) (2004) 691–696.

[71]   AM Gençler Özkan, M Koyuncu, Traditional medicinal plants used in Pınarbaşı area (Kayseri-Turkey), Turkish Journal of Pharmaceutical Sciences 2 (2) (2005) 63–82.

[72]   E Deveci, TÇ Gulsen, S Karakurt, ME Duru, Anti-Colorectal Cancer Effects of Medicinal Plants: Euphorbia helioscopia, Ferula elaeochytris, and Sideritis albiflora., 5, 1st ed., Commag J Biol, 2021, pp. 73–77.

[73]   A Aras, F Türkan, U Yildiko, MN Atalar, O Kılıç, MH Alma, et al., Biochemical constituent, enzyme inhibitory activity, and molecular docking analysis of an endemic plant species. Thymus migricus, 75, Chemical Papers, 2021, pp. 1133–1146. https://doi.org/10.1007/s11696-020-01375-z.

[74]   SA Sargin, Ph.D. Thesis, Balıkesir University, Institute of Science, Biology, Balıkesir, 2013, p. 461.

[75]   R Polat, PhD. Thesis, Balıkesir, Balıkesir University, Institute of Sciences, Department of Biology, 2010, p. 334.

[76]   H Akan, SY Bakır, Investigation of The Etnobotanical Aspects The Town Kâhta and Village of Narince (Kâhta (Adıyaman) Merkezi ve Narince Köyü'nün Etnobotanik Açıdan Araştırılması), BEU Journal of Science 4 (2) (2015) 219–248.

[77]   E Sezik, E Yeşilada, G Honda, Y Takaishi, T Takeda, Traditional medicine in Turkey X. Folk medicine in central Anatolia, 72, 2nd–3rd ed., J Ethnopharm, 2001, pp. 95–115.

CHAPTER 27

Effects of Amygdalin on prostate cancer

Charles E. Dowling II, Berthon Eliche and Omar Bagasra
Claflin University, 400 Magnolia Street, Orangeburg, SC 29115, USA

27.1 Introduction

Cancer is one of the most dreaded diseases of the 20th century and it is spreading further with persistence and increasing prevalence in the 21st century. Following heart disease, cancer is the second leading cause of death in the United States [1]. It was predicted that from 1969 through 2020, the number of heart disease deaths would decrease 21.3% among men (−73.9% risk, 17.9% growth, 34.7% aging) and 13.4% among women (−73.3% risk, 17.1% growth, 42.8% aging) while the number of cancer deaths would increase 91.1% among men (−33.5% risk, 45.6% growth, 79.0% aging) and 101.1% among women (−23.8% risk, 48.8% growth, 76.0% aging) [2]. If this trend continues, then cancer will replace heart disease as the leading cause of death by the year 2020. The cancer phenotype has four major characteristics: uncontrolled cell proliferation, genomic instability, immortality, and the ability to disrupt local and distant tissues.

In men, the most common malignancies are lung, prostate, colorectal, stomach, and liver cancers, whereas, in females' breast, colorectal, lung, cervix, and thyroid cancers are most prevalent.

Prostate cancer (PCa) is a disease of increasing significance worldwide [1]. In many industrialized nations including the United States, it is one of the most common cancers and among the leading causes of cancer deaths. PCa is the most frequently diagnosed nonskin cancer in the United States and the third leading cause of cancer deaths. The United States has one of the most active PCa early detection programs in the world, and also the highest incidence of PCa [1,2]. Once prostate-specific antigen (PSA) tests were approved by the FDA in 1986 and became available for PCa screening, the United States has experienced a huge increase in PCa incidence [3]. The American Cancer Society estimates that in 2021, 248,530 men will be diagnosed with PCa and 34,130 men will die from it. About one man in six will be diagnosed with PCa during his lifetime. PCa occurs mainly in older men. About 6 cases in 10 are diagnosed in men >65, and it is rare before age 40 [3]. Recently, large-scale epidemiological analyses on PSA screening for PCa screening have challenged the utility pf PSA as the main diagnostic test since it is not specific to PCa; common conditions, such as benign prostatic hyperplasia and prostatitis, also increase PSA levels. Approximately 1.5 million US men age 40–69 years have a PSA level greater than 4.0 µg/L (a widely used cutoff value for a potential positive screening for PCa). Refinements designed to improve the PSA test's sensitivity and specificity for PCa include the determination of PSA density, PSA velocity, PSA doubling time, and percentage of free PSA [4]. Potential harms from PSA screening include additional medical visits, adverse effects of prostate biopsies, anxiety, and most importantly the over-diagnoses of PCa (the identification of PCa that would never have caused symptoms in the patient's lifetime, leading to unnecessary treatment and associated adverse effects) [4].

Very early detection is the key to effective treatment of PCa and to the prevention of deaths due to the progression of untreatable advanced stages of cancer. Mitigating factors, especially benign prostatic hyperplasia and prostatitis result in a low accuracy (about 60%) of prostate-specific antigen (PSA) testing [3,4]. Thus, there is an urgent need for a more reliable biomarker to identify PCa at a very early stage and to identify "at-risk" individuals. Currently, there are no satisfactory ways to differentiate between Stages I/II indolent and lethal (aggressive) PCa at diagnosis. Both early identification and indolent/lethal differentiation are critical because PCa, if identified while confined to the prostate, (1) is "curable" for aggressive tumors by surgery and subsequent treatment and (2) "watchful waiting" would be appropriate for indolent tumors [3,4].

As mentioned in the proceeding section that cancer cells consume 200 times more glucose than normal cells. A significance amount of the energy comes from aerobic glycolysis; about 95%. Rest 5% is acquired from the Krebs cycle. Due to this increased energy requirement these cancer cells are sensitive to glucose deprivation-induced cytotoxicity. This can be exploited by inhibitors that interrupt glycolytic pathways, glucose cellular uptake and oxidative metabolism [5]. We have explored two natural compounds-Amygdalin and Lycopene, well known for their therapeutic effects. We evaluated Amygdalin-a natural product in combination with a FDA-approved drug dichloroacetate (DCA) for an additive or synergistic effects [5–10].

Amygdalin is a naturally occurring chemical compound found in plant seeds like apples, pears, and members of the *Prunus* species like apricots, plums, peaches, etc. The man-made version of this naturally compound is laevo-mandelonitrile-beta-glucuronoside which is commonly known as Laetrile (or Vitamin B-17). Amygdalin consists of two glucose molecules connected by a glyosidic bond that is attached to a cyanide group and benzene ring. When Amygdalin is orally ingested, by the action of β-galactosidase it is hydrolyzed into two molecules of glucose and a molecule of mandelonitrile [6–10]. Mandelonitrile is unstable and subsequently breaks into benzaldehyde and hydrogen cyanide (HCN). HCN is toxic since it inhibits the function of cytochrome oxidase in electron transport chain. The enzyme

Rhodanese help protect tissues from cyanide poisoning. This enzyme donates sulfur and forms a neutral thiosulphate from a highly toxic HCN. This is then easily filtered out renally. Amygdalin has been used for cancer treatment both as a single agent and in combination with a metabolic therapy program that consists of a specialized diet, high-dose vitamin supplements, and pancreatic enzymes for over a century [7]. HCN appears to be the main compound responsible for its health benefits, it is also believed to be its primary anticancer ingredient. Since unlike healthy cells cancer cells generate most of their energy from aerobic glycolysis, the HCN acts as an antimetabolite for these cells. Malignant cells also have higher than normal levels of beta-glucosidase while being deficient in rhodanese [8]. When Amygdalin reaches cancer cells through IV administration, β-glucosidase breaks it down to releases benzaldehyde and hydrocyanic acid. This creates a toxic synergy that causes apoptosis in cancer cell. However, rhodanese in normal cells neutralizes free cyanide molecules and renders them harmless [9–13].

Dichloroacetic acid is a colorless, odorless, relatively nontoxic compound. Sometimes referred to as bichloroacetic acid, the salt and esters of dichloroacetic acid are called DCAs. Having the chemical formula $CHCl_2COOH$, it can be prepared through various means; DCA is typically prepared by the reduction of trichloroacetic acid (TCA), It can also be prepared by the reaction of calcium carbonate and sodium cyanide in water followed by acidifying with hydrochloric acid also, it can be made by passing acetylene through solutions of hypochlorous acid. DCA has several therapeutic applications based on its pharmacological property of inhibiting pyruvate dehydrogenase kinase [10,14,15] which thus enhances the flux of pyruvate into the mitochondria by indirect activation of pyruvate dehydrogenase.

DCA's anticancer properties were discovered in 2007 by Evangelos Michelakis of the University of Alberta, Canada. In his experiments on rats, he experienced a "eureka moment" when he saw the tumors of various types had been dramatically reduced over the course of a few weeks. However, it showed very minimal disruption to healthy cells. DCA has been used to treat inherited mitochondrial disorders that result in lactic acidosis, as well as pulmonary hypertension and several different solid tumors, the latter through its ability to reverse the Warburg effect in cancer cells and restore aerobic glycolysis. The Warburg effect characterized in the early 1920s by Otto Warburg describes the metabolic energy production of most cancer cells that rely on aerobic glycolysis in the presence of oxygen [10,15]. Glycolysis contains several advantages for cancer progression such as lactic acidosis allowing tumor growth due to damage of extracellular matrix and increased cell mobility [14], decreased cell respiration leads to lower production of reactive oxygen species in mitochondria as well as decreased DNA damage and enables apoptosis resistance [15].

The first step in DCA metabolism is conversion to glyoxylate catalyzed by glutathione transferase zeta 1 (GSTZ1) expressed in the liver cytosol and primarily in the mitochondria. This glyoxylate arising from DCA is converted to secondary metabolites, including glycine, carbon

Table 27.1: Compares LNCaP and DU145 (LNCap cell line).

	LNCaP	Du145
Origin	50-year-old adult male with blood type B+	69-year-old adult male of blood group O, Rh+
Media	Grows in RPMI-1640 media	Grows in EMEM and can also be grown in RPMI-1640 media
Metastatic site	Derived from the left supraclavicular lymph node	Its metastatic site was the brain
Androgen sensitivity	Androgen dependent	Androgen independent
Tumorigenicity	Very low	Moderate
Intracellular protein content	Extremely high	Much lesser compared to LNCaP

dioxide, and oxalate, as documented in studies in rats and mice [15]. Some of the glycine was excreted in urine in the form of hippuric acid, phenylacetyl glycine, and other glycine conjugates [14,15].

Our studies were based on the hypothesis that if we combined the three chemicals two phytochemicals (i.e., Amygdalin and lycopene) and one FDA-approved drug (DCA) then the oncolytic effects of the combination will be much greater at a low concentration of Amygdalin.

27.2 Materials and methods

27.2.1 Materials

Cell lines: LNCaP and Du145.
Media: Roswell Park Memorial Institute-1640 Media (RPMI), SILAC RPMI-1640 Flex Media (Sugar free), Phosphate Buffered Solution (PBS), Fetal Bovine Serum (FBS), Trypsin.
Antimetabolites: Sodium DCA and Amygdalin.
Instruments: Elisa Multiskan FC, Redox dye, hemocytometer, PM1 96 well plates.
The Du145 cell line is different from LNCaP in certain areas. Table 27.1 highlights some of these differences.

27.2.2 Methods

27.2.2.1 Cell culturing

The PCa cell lines used are DU145 and LNCaP and were obtained from the American Type Culture Collection. Both Du145 and LNCaP are from a 69 and 50-year-old Caucasian males, respectively, and they thrive in a complete growth medium of RPMI 1640 media, heat-inactivated FBS to a final concentration of 10% and 1% L-glutamine-penicillin-streptomycin solution. Cell lines were cultured in a T-25 flask containing 5 mL of RPMI 1640 complete

Table 27.2: Indicates the mean values as well as the P values for the control and the compounds mostly utilized.

	D-Ribose	Sucrose	D-Malic acid	L-Lyxose
Control	0.4081	0.5798	0.8069	0.4053
Mean: 0.3450	**P = 0.0388**	**P = 0.0123**	**P = 0.0019**	**P = 0.0357**

The **bold** highlights a "statistically significant P value."

Table 27.3: Indicates the mean values as well as the P values for the control and the compounds mostly utilized.

	D-Sorbitol	D-Ribose	Inosine
Control	0.4066	0.3983	0.7102
Mean: 0.2968	**P = 0.0204**	**P = 0.0207**	**P = 0.0125**

The **bold** highlights a "statistically significant P value."

Table 27.4: Indicates the mean values as well as the P values for the control and the compounds utilized.

	D-Ribose	Sucrose	D-Malic acid	L-Lyxose
LNCaP ONLY	0.4081	0.5798	0.8069	0.4054
LNCaP + Amyg.	0.3983	0.3378	0.2936	0.3783
P value	*0.5665*	**0.0122**	**0.0012**	*0.1363*

The **bold** highlights a "statistically significant P value" while the *italic* highlights a "not statistically significant P value."

Table 27.5: Indicates the mean values as well as the P values for the control and the compounds utilized.

	N-Acetyl-D-Glucosamine	L-Asparagine	Citric acid	D-Psicose
Control	3.2472	2.3468	1.3909	1.6718
Mean: 0.4751	**P = 0.0002**	**P = 0.0389**	*P = 0.0902*	**P = 0.0220**

The **bold** highlights a "statistically significant P value" while the *italic* highlights a "not statistically significant P value."

media and subsequently incubated at 37°C, media was changed promptly, and cells split into a bigger T-75 flask was done when fully grown in the small flask (Tables 27.2–27.15).

27.2.2.2 Cell splitting

First, the media was removed from the respective flasks and subsequently the flasks were washed with 1mL of PBS. After washing, PBS is removed, and trypsin is added in drops to the flask. In cell culturing, trypsin functions to dissociate adherent cells from the vessels they are being cultured in by breaking the disulfide and peptide bonds between proteins which enable the cells adhere to the vessels. On complete dissociation from the flask, 1 mL of FBS was added to the flask to deactivate the trypsin.

Table 27.6: Indicates the mean values as well as the *P* values for the control and the compounds utilized.

	D-Ribose	Sucrose	D-Malic acid	L-Lyxose
LNCaP ONLY	0.4081	0.5798	0.8069	0.4053
LNCaP + DCA	0.4636	0.5212	0.3946	0.4962
P value	*0.0720*	*0.3214*	**0.0031**	**0.0202**

The **bold** highlights a "statistically significant *P* value" while the *italic* highlights a "not statistically significant *P* value."

Table 27.7: Indicates the mean values as well as the *P* values for the control and the compounds utilized.

	D-Mannitol	Adonitol	Adenosine	Glycyl-L-Aspartic acid	Phenylethylamine
Control	0.6820	0.6574	1.9623	1.6498	2.2303
Mean: 0.5370	*P = 0.1511*	*P = 0.1681*	*P = 0.0755*	*P = 0.0896*	**P = 0.0281**

The **bold** highlights a "statistically significant *P* value" while the *italic* highlights a "not statistically significant *P* value."

Table 27.8: Indicates the mean values as well as the *P* values for the control and the compounds utilized (DU145).

	D-Ribose	Sucrose	D-Malic acid	L-Lyxose
LNCaP ONLY	0.4081	0.5798	0.8069	0.4053
LNCaP + DCA + Amyg.	0.5422	0.6027	0.5073	0.5256
P value	**0.0344**	*0.6649*	**0.0103**	**0.0125**

The **bold** highlights a "statistically significant *P* value" while the *italic* highlights a "not statistically significant *P* value."

Table 27.9: Indicates the mean values as well as the *P* values for the control and the compounds utilized.

	D-Mannose	a-D-Glucose	Maltose	a-Keto-Butyric acid
Control	0.7426	0.8092	0.8225	0.7235
Mean: 0.4454	*P = 0.0874*	*P = 0.0774*	*P = 0.0804*	*P = 0.0795*
	Uridine	Maltotriose	Methyl Pyruvate	
Control	0.6730	0.8455	0.6699	
Mean: 0.4454	*P = 0.0868*	*P = 0.0757*	*P = 0.0753*	

The *italic* highlights a "not statistically significant *P* value."

Table 27.10: Indicates the mean values as well as the *P* values for the control and the compounds utilized.

	N-Acetyl-D-Glucosamine	D-Trehalose	D-Mannose	a-D-Glucose
Control	0.7678	0.6712	0.8189	0.8838
Mean: 0.4763	*P = 0.0838*	**P = 0.0373**	*P = 0.0639*	*P = 0.0846*
	Maltose	Lactulose	Maltotriose	
Control	0.8576	0.6588	0.9082	
Mean: 0.4763	*P = 0.0941*	*P = 0.0991*	*P = 0.0700*	

The **bold** highlights a "statistically significant *P* value" while the *italic* highlights a "not statistically significant *P* value."

Table 27.11: Indicates the mean values as well as the *P* values for the control and the compounds utilized.

	D-Mannose	a-D-Glucose	Maltose	a-Keto-Butyric acid	Uridine	Maltotriose	Methyl Pyruvate
DU145 ONLY	0.7426	0.8092	0.8225	0.7235	0.6730	0.8455	0.6699
DU145 + Amg.	0.8189	0.8838	0.8576	0.7615	0.6121	0.9082	0.5371
P value	*0.6052*	*0.6905*	*0.8522*	*0.7987*	*0.5578*	*0.7388*	*0.1937*

The *italic* highlights a "not statistically significant P value."

Table 27.12: Indicates the mean values as well as the *P* values for the control and the compounds utilized.

	a-D-Glucose	Maltose	Uridine	Maltotriose	Methyl Pyruvate	D-Psicose
Control	0.7566	0.7528	0.6818	0.7583	0.6783	2.3574
Mean: 0.4413	*P = 0.0933*	*P = 0.0904*	*P = 0.1144*	*P = 0.0927*	*P = 0.0827*	**P = 0.0067**

The **bold** highlights a "statistically significant P value" while the *italic* highlights a "not statistically significant P value."

Table 27.13: Indicates the mean values as well as the *P* values for the control and the compounds utilized.

	D-Mannose	a-D-Glucose	Maltose	a-Keto-Butyric acid	Uridine	Maltotriose	Methyl Pyruvate
DU145 ONLY	0.7426	0.8092	0.8225	0.7235	0.6730	0.8455	0.6699
DU145 + DCA	0.6339	0.7566	0.7528	0.6221	0.6818	0.7583	0.6783
P value	*0.4407*	*0.7505*	*0.6821*	*0.4235*	*0.9421*	*0.6229*	*0.9337*

The *italic* highlights a "not statistically significant P value."

Table 27.14: Indicates the mean values as well as the *P* values for the control and the compounds utilized.

	N-Acetyl-D-Glucosamine	D, L-Malic acid	Uridine	Adonitol	Bromo Succinic acid
Control	3.1639	2.0309	0.9206	2.0464	1.7499
Mean: 0.7089	**P = 0.0006**	*P = 0.0676*	*P = 0.1342*	*P = 0.0661*	*P = 0.0900*

The **bold** highlights a "statistically significant P value" while the *italic* highlights a "not statistically significant P value."

Table 27.15: Indicates the mean values as well as the *P* values for the control and the compounds utilized.

	D-Mannose	a-D-Glucose	Maltose	a-Keto-Butyric acid	Uridine	Maltotriose	Methyl Pyruvate
DU145 ONLY	0.7426	0.8092	0.8225	0.7235	0.6730	0.8455	0.6699
DU145 + DCA	0.8886	0.9114	0.9112	0.7554	0.9161	0.8952	0.7547
p-Value	*0.3461*	*0.5183*	*0.5802*	*0.7733*	*0.1312*	*0.7524*	*0.3708*

The *italic* highlights a "not statistically significant P value."

27.2.2.3 Treatment with amygdalin and dichloroacetate

We require 20,000 cells/100 uL which provides us with 2,000,000 cells in 10 mL for each PM1 plate thus, when cells are fully grown in T-75 flask and after trypsinization, using the hemocytometer cells are counted and ensured they meet the requirements. Furthermore, 100 mM Amygdalin is added to the cell suspension of both LNCaP and Du145 separately, containing sugar-free RPMI-1640 media to a final concentration of 10 mM/well. Also, in a different cell suspension 50mM DCA is added to a final concentration of 5 mM/well. This is done for amygdalin only, DCA only, amygdalin and DCA together for both LNCaP and DU145 cell lines.

27.2.2.4 Plating the cells in a PM-M1 96 well plate

Biolog metabolic Phenotype Microarray Microplate (PM-M) provides an easy-to-use technology for scanning and measuring the energy metabolism pathways present in a wide range of mammalian cell types from *in vitro* cultured cells to primary cells, the PM1 specifically measures the carbon-energy utilization in these cells. Using a 100 uL multichannel pipette, cell solutions containing our desired compound are transferred into a PM1 plate and subsequently incubated at 37°C in a humidified atmosphere with 95% air, 5% CO_2 for 48 hours. Ten microliter per well Biolog Redox dye is added to all wells after the 48-hour time frame using a multichannel pipette and allowed to stand for 1–6 hours before being read with the Elisa Multiskan FC reader at an absorbance of 590 nm. We collected a 2-hour reading.

27.2.3 Results

From the results of our carbohydrate metabolic array (PM-1) plate, we observe that LNCaP in the absence of any antimetabolites utilized highly; D-ribose, sucrose, D-malic acid, and L-lyxose (Fig. 27.1). D-ribose, a pentose monosaccharide naturally produced by the human body and not found in food sources, L-lyxose also is a pentose monosaccharide, a C-2 carbon epimer of the sugar xylose. Sucrose generally known as table sugar, a disaccharide composed of two monosaccharides glucose and fructose while malic acid, an organic compound, a dicarboxylic acid made by all living compound. All these compounds except for L-lyxose are readily produced/used up by the human body in different metabolic pathways.

In addition, the results of our carbohydrate metabolic array (PM-1) plate, we observe that LNCaP in the absence of any antimetabolites utilized highly; D-ribose, sucrose, D-malic acid, and L-lyxose (Figs. 27.1–27.7). D-ribose, a pentose monosaccharide naturally produced by the human body and not found in food sources, L-lyxose also is a pentose monosaccharide, a C-2 carbon epimer of the sugar xylose. Sucrose generally known as table sugar, a disaccharide composed of two monosaccharides glucose and fructose while malic acid, an organic compound, a dicarboxylic acid made by all living compounds. All these compounds except for L-lyxose are readily produced/used up by the human body in different metabolic pathways.

LNCaP cell line

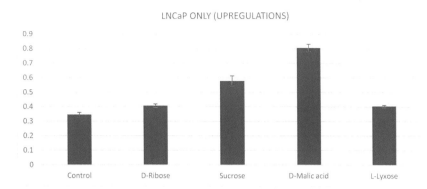

Figure 27.1
The compounds mostly utilized by LNCaP in sugar free media without any antimetabolite.

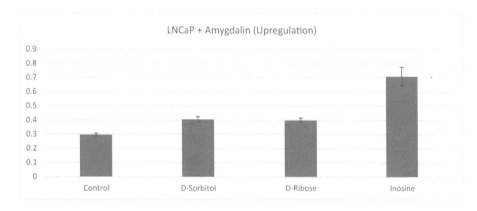

Figure 27.2
The compounds mostly utilized by LNCaP in the presence of amygdalin and sugar free media.

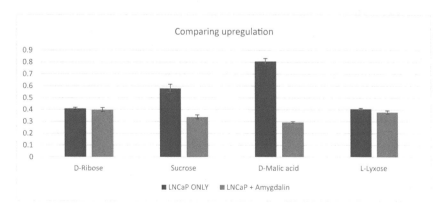

Figure 27.3
Compares the mostly used compounds for LNCaP only and LNCaP in the presence of amygdalin.

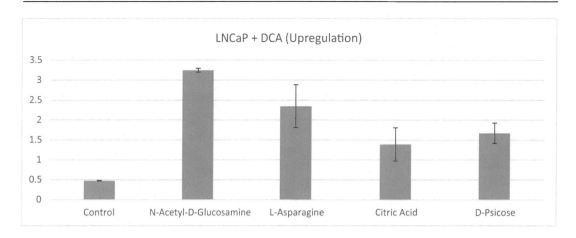

Figure 27.4
The compounds mostly utilized by LNCaP in the presence of dichloroacetate and sugar free media.

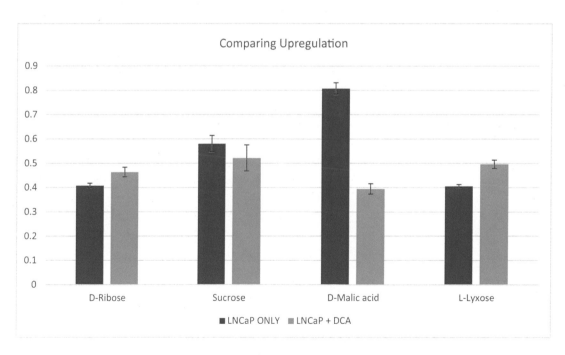

Figure 27.5
Compares the mostly used compounds for LNCaP only and LNCaP in the presence of dichloroacetate.

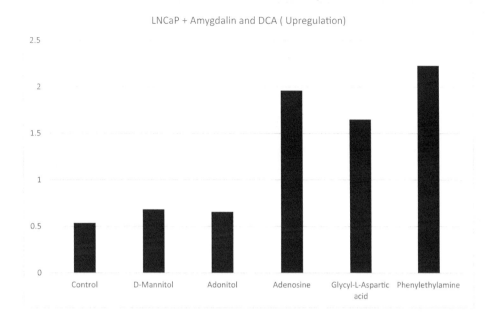

Figure 27.6
The compounds mostly utilized by LNCaP in the presence of amygdalin, dichloroacetate, and sugar free media.

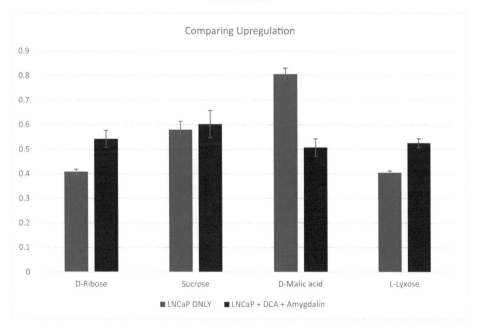

Figure 27.7
Compares the mostly used compounds for LNCaP only and LNCaP in the presence of both amygdalin and dichloroacetate.

27.3 Discussion

Amygdalin is a natural compound primarily found in seeds of numerous fruits. In the human body, it is hydrolyzed to benzaldehyde, glucose, and cyanide, which is considered the active component of amygdalin. Since the paradigm shift toward allopathic medicine in the 18th–19th century, in many cultures natural or herbal medicine was the main mode of therapy. Many of the current conventional medicine products are derived from homeopathic and shaman medicine which was practiced for many millennia. Even today alternate medicine is popularly used due to side-effects and failure of approved methods, especially for treating cancer. Amygdalin, a cyanogenic diglycoside is commonly administered for cancer with other allopathic therapies like vitamins and seeds of fruits like apricots and bitter almonds, due to its ability to hydrolyze to HCN, benzaldehyde, and glucose. As a result of potential cyanide toxicity, use of amygdalin have been cautiously used. *In-vitro* and *in-vivo* studies using various doses and modes of administration, like IV administration studies that showed no HCN formation, point to the role played by the gut microbiota for the commonly seen poisoning on consumption. The common gut bacteria anaerobic Bacteriodetes has a high β-glucosidase activity needed for amygdalin hydrolysis to HCN. However, there are certain conditions under which these HCN levels rise to cause toxicity.

In the presence of oxygen, normal cells primarily use the mitochondrial tricarboxylic acid (TCA) cycle and oxidative phosphorylation to produce energy and rely on glycolysis only when their oxygen supply is limited. In contrast, cancer cells frequently utilize glycolysis even in the presence of enough amounts of oxygen [16]. Thus, their reduced dependence on mitochondrial oxidative phosphorylation and more reliance on glycolysis provides a wide range of potential targets for therapy. Targeting aerobic glycolysis is a promising strategy to preferentially kill cancer cells that are dependent on this pathway and in recent years multiple glycolytic inhibitors have been developed [17].

All the enzymes within the glycolytic pathway, electron transport chain, and TCA cycle potentially represent targets for anticancer treatment and inhibitors have been developed that target molecular components of these pathways. Amygdalin is one of the antimetabolites used in my research which belongs to a family of compounds called cyanogenic glycosides. Cyanogenic glycosides are considered nontoxic until cyanide is released, and this cyanide is believed to be the active cytotoxic ingredient in amygdalin. Cyanide acts through the inhibition of cytochrome-c oxidase in the respiratory electron transport chain of the mitochondria, impairing both oxidative metabolism and the associated process of oxidative phosphorylation, thereby causing death through energy deprivation and oxygen uptake [18], studies by Ruan et al. [19] have shown that apricot and peach kernels dispensed in its raw form, induce apoptosis and differentiation of tumor cells.

We observe that LNCaP in the absence of any antimetabolites highly utilize; D-ribose, sucrose, D-malic acid and L-lyxose, however in the presence of amygdalin, malic acid an

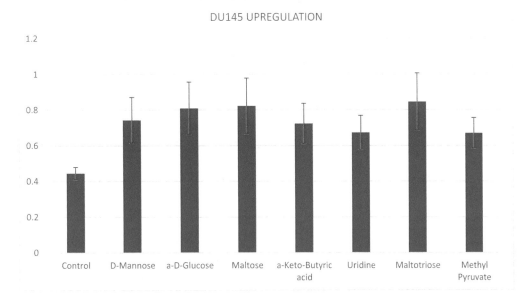

Figure 27.8
The compounds mostly utilized by DU145 in sugar free media without any antimetabolite.

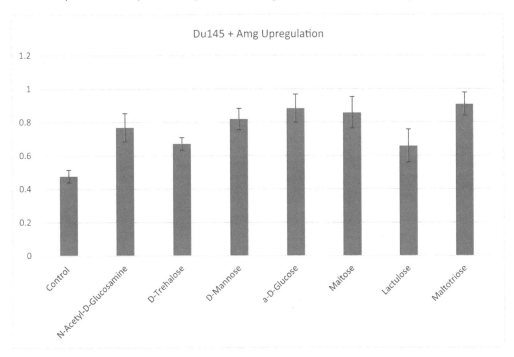

Figure 27.9
The compounds mostly utilized by DU145 in the presence of amygdalin and sugar free media.

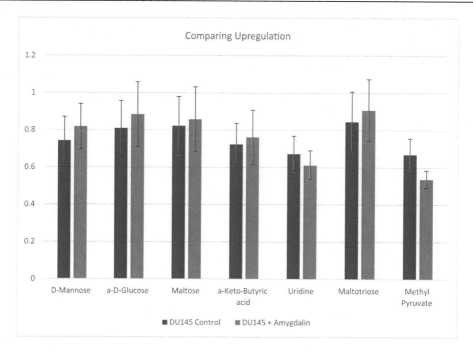

Figure 27.10

Compares the mostly used compounds for DU145 only and DU145 in the presence of amygdalin.

intermediate of the TCA cycle and sucrose which is broken down through the glycolytic pathway are inhibited as well as all pathways that has them as an intermediate. This is in line with the results recorded by Hye and Aree [20] on "amygdalin regulates apoptosis and adhesion in Hs578T triple-negative breast cancer cells," amygdalin was previously demonstrated to induce apoptosis by increasing expression of Bax and decreasing expression of Bcl-2 and procaspase-3 in DU145 and LNCaP PCa cells [21]. Their results indicated that amygdalin induces apoptosis by increasing the expression of Bax and decreasing the expression of Bcl-2 in Hs578T breast cancer cell comparisons.

Furthermore, DCA, a pyruvate dehydrogenase kinase inhibitor, which activates pyruvate dehydrogenase (PDH), and increases glucose oxidation by promoting influx of pyruvate into the Krebs cycle has been shown to cause production of reactive oxygen species [22]. Its antitumor properties induce cell-cycle arrest and apoptosis in prostate [10], colorectal [23], and breast cancer [24]. Our experiment when DCA was added, showed alternative compounds of N-acetyl-D-glucosamine (a monosaccharide derivative of glucose), L-asparagine (a nonessential amino acid) that enters the TCA cycle as oxaloacetate, citric acid (a weak organic acid naturally occurring in citrus fruit) an intermediate of the TCA cycle as well as D-psicose a low energy monosaccharide, a C-3 epimer of D-fructose were used by the cancer cells to sustain its growth as opposed to those used in the control. Thus, DCA inhibited the use of compounds used up in the control, this is consistent with reports by Lin et al. [25] on *"dichloroacetate*

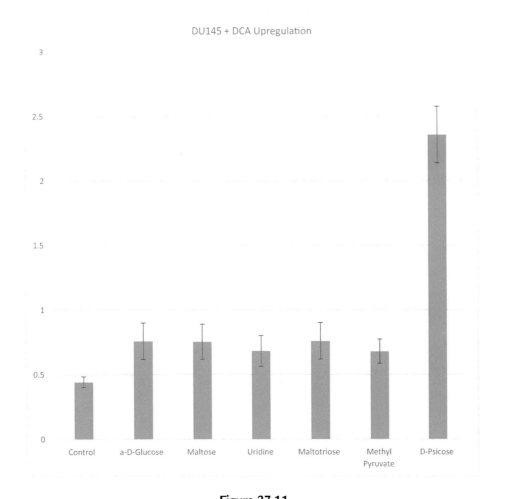

Figure 27.11
The compounds mostly utilized by DU145 in the presence of dichloroacetate and sugar free media.

induces autography in colorectal cancer cells and tumors" which shows that DCA induces autophagy in human colorectal and PCa cells with minimal apoptosis and necrosis as well as causing a significantly decreased lactate excretion in all DCA-treated cells with no change in glucose uptake, suggesting a reduction in the Warburg effect [12–13].

Combining both antimetabolites, amygdalin and DCA can potentially create a very effective alternative cancer therapy which will bring about significant synergistic effects and will exhibit selective anticancer effects that can be used to treat patients. Our data recorded the use of compounds mostly comprised of sugar alcohols and amino acids by the cells (Figs. 27.8–27.12) to survive. This we largely believe to be as a result of a phenomenon known as "gluconeogenesis," a process of the body creating sugars from noncarbohydrate precursors. The presence of the two antimetabolites inhibited the glycolytic pathway causing the cells to adopt an alternative mechanism for sustenance (Figs. 27.13 and 27.14).

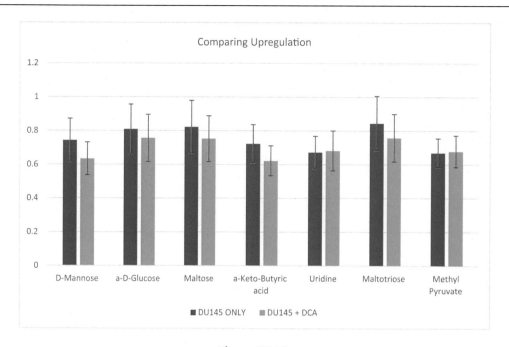

Figure 27.12
Compares the mostly used compounds for DU145 only and DU145 in the presence of dichloroacetate.

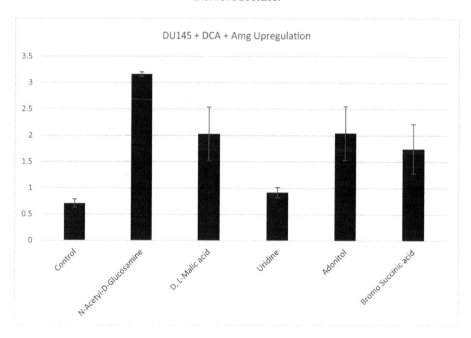

Figure 27.13
The compounds mostly utilized by DU145 in the presence of amygdalin, dichloroacetate and sugar free media.

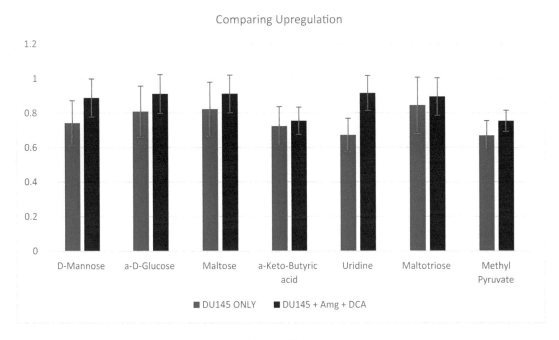

Figure 27.14

Compares the mostly used compounds for DU145 only and DU145 in the presence of both amygdalin and dichloroacetate.

In summary, a combining both antimetabolites, amygdalin and DCA can potentially create a very effective alternative cancer therapy which will bring about significant synergistic effects and will exhibit selective anticancer effects that can be used to treat patients. Our data recorded the use of compounds mostly comprised of sugar alcohols and amino acids by the cells to survive. This we largely believe to be the result of a phenomenon known as "gluco-neogenesis," a process of the body creating sugars from noncarbohydrate precursors. The presence of the two antimetabolites inhibited the glycolytic pathway causing the cells to adopt an alternative mechanism for sustenance. We hypothesized that such an adaptive pathway may be the limiting step for the cancer cells *in vivo*.

References

[1]   C Pernar, E Ebot, K Wilson, L Mucci, The epidemiology of prostate cancer, Cold Spring Harb Perspect Med 8 (12) (2018) a030361.
[2]   R Siegel, K Miller, A Jemal, Cancer statistics. CA: A Cancer J Clin. 69(1) (2019) 7–34.
[3]   American Cancer Society. Prostate cancer, prostate cancer information and overview 2019.
[4]   B Djavan, A Zlotta, C Kratzik, M Remzi, C Seitz, C Schulman, et al., PSA, PSA density, PSA density of transition zone, free/total PSA ratio, and PSA velocity for early detection of prostate cancer in men with serum PSA 2.5 to 4.0 ng/mL, Urology 54 (3) (1999) 517–522.

[5]   A Třísková, J Rudá Kučerová, Can amygdalin provide any benefit in integrative anticancer treatment?
       Klinicka Onkol 32 (5) (2019) 360–366.

[6]   J Shi, Q Chen, M Xu, Q Xia, T Zheng, J Teng, et al., Recent updates and future perspectives about amyg-
       dalin as a potential anticancer agent: A review, Cancer Med 8 (6) (2019) 3004–3011.

[7]   P Liczbiński, B Bukowska, Molecular mechanism of amygdalin action in vitro: review of the latest re-
       search, Immunopharmacol Immunotoxicol 40 (3) (2018) 212–218.

[8]   M Saleem, J Asif, M Asif, U Saleem, Amygdalin from apricot kernels induces apoptosis and causes cell
       cycle arrest in cancer cells: an updated review, Anti-Cancer Agents Med Chem 18 (12) (2019) 1650–1655.

[9]   G Sutendra, E Michelakis, Pyruvate dehydrogenase kinase as a novel therapeutic target in oncology, Front
       Oncol 3 (38) (2013).

[10]   W Cao, S Yacoub, K Shiverick, K Namiki, Y Sakai, S Porvasnik, et al., Dichloroacetate (DCA) sensitizes
       both wild-type and over expressingBcl-2prostate cancer cells in vitro to radiation, Prostate 68 (11) (2008)
       1223–1231.

[11]   H Chang, M Shin, H Yang, J Lee, Y Kim, M Lee, et al., Amygdalin induces apoptosis through regulation
       of Bax and Bcl-2 expressions in human DU145 and LNCaP prostate cancer cells, Biol Pharm Bull 29 (8)
       (2006) 1597–1602.

[12]   RA Gatenby, RJ Gillies, Why do cancers have high aerobic glycolysis? Nat Rev Cancer 4 (11) (2004) 891–
       899.

[13]   DR Plas, CB Thompson, Cell metabolism in the regulation of programmed cell death, Trends Endocrinol
       Metab 13 (2) (2002) 75–78.

[14]   A Gonzalez-Leon, IR Schultz, G Xu, RJ Bull, Pharmacokinetics and metabolism of dichloroacetate in the
       F344 rat after prior administration in drinking water, Toxicol Appl Pharmacol 146 (1997) 189–195.

[15]   MO James, Z Yan, R Cornett, VM Jayanti, GN Henderson, N Davydova, et al., Pharmacokinetics and
       metabolism of [14C] dichloroacetate in male Sprague-Dawley rats. Identification of glycine conjugates,
       including hippurate, as urinary metabolites of dichloroacetate, Drug Metabol Dispos 26 (1998) 1134–1143.

[16]   L Leal-Esteban, L Fajas, Cell cycle regulators in cancer cell metabolism, Biochimica et Biophysica Acta
       (BBA)—Mol Basis Dis 1866 (5) (2020) 165715.

[17]   AF Abdel-Wahab, W Mahmoud, RM Al-Harizy, Targeting glucose metabolism to suppress cancer progres-
       sion: prospective of anti-glycolytic cancer therapy, Pharmacol Res 150 (3) (2019) 104511.

[18]   N Chaouali, I Gana, A Dorra, F Khelifi, A Nouioui, W Masri, et al., Potential toxic levels of cyanide in
       almonds (Prunus amygdalus), Apricot Kernels (Prunus armeniaca), Almond Syrup 2013 (2013) 1–7. ISRN
       Toxicol 2013, 610648. Published online 2013 Sep 19. doi:10.1155/2013/610648.

[19]   W Ruan, M Lai, J Zhou, Anticancer effects of Chinese herbal medicine, science or myth? J Zhejiang Univ
       Sci B 7 (12) (2006) 1006–1014.

[20]   H Lee, A Moon, Amygdalin regulates apoptosis and adhesion in Hs578T triple-negative breast cancer cells,
       Biomol Therap 24 (1) (2016) 62–66.

[21]   HK Chang, MS Shin, HY Yang, JW Lee, YS Kim, MH Lee, et al., Amygdalin induces apoptosis through
       regulation of Bax and Bcl-2 expressions in human DU145 and LNCaP prostate cancer cells, Biol Pharm
       Bull 29 (2006) 1597–1602.

[22]   S Bonnet, SL Archer, J Allalunis-Turner, A Haromy, C Beaulieu, R Thompson, et al., A mitochondria-K+
       channel axis is suppressed in cancer and its normalization promotes apoptosis and inhibits cancer growth,
       Cancer Cell 11 (1) (2007) 37–51.

[23]   BM Madhok, S Yeluri, SL Perry, TA Hughes, DG Jayne, Dichloroacetate induces apoptosis and cell-cycle
       arrest in colorectal cancer cells, Brit J Cancer 102 (12) (2010) 1746–1752.

[24]   RC Sun, M Fadia, JE Dahlstrom, CR Parish, PG Board, AC Blackburn, Reversal of the glycolytic pheno-
       type by dichloroacetate inhibits metastatic breast cancer cell growth *in vitro* and *in vivo*, Breast Cancer Res
       Treat 120 (1) (2010) 253–260.

[25]   G Lin, DK Hill, G Andrejeva, JK Boult, H Troy, AC Fong, et al., Dichloroacetate induces autophagy in
       colorectal cancer cells and tumors, Brit J Cancer 111 (2) (2014) 375–385.

CHAPTER 28

Protective effects of plant-derived natural products against hepatocellular carcinoma

Meenakshi Gupta and Maryam Sarwat

Amity Institute of Pharmacy, Amity University, Noida, Uttar Pradesh, India

28.1 Introduction

Of all the liver cancers, hepatocellular carcinoma (HCC) contributes to about 70%–80% of its share and is reported to be the third most common cause of cancer mortality in India during the past few decades. Risk factors that contribute directly or indirectly in hepatotumourigenesis include alcoholism, aflatoxin B1, diabetes, hepatitis B virus (HBV), and hepatitis C virus (HCV) infection, iron accumulation, non-alcoholic fatty liver disease and obesity [1,2]. Management protocols for the treatment of the HCC include radiotherapy, surgical resection, embolization, ablation, and chemotherapy [3]. Clinical therapy with sorafenib is found to be effective but it is only able to prolong the patient's survival for 8–9 months [4]. The use of current therapy is limited because of poor prognosis, recurrence, various side effects, and complexities associated with them. These limitations call for the development of newer and additional therapies for the treatment of HCC and associated risks.

Complementary and alternative medicine approaches can improve the quality of life and may prolong the survival of patients [5]. Way back more than 4000 years, the herbal system of medicine is being practiced in India, China, and Egypt. Over two decades, various herbal products are being introduced to treat chronic liver diseases which gained popularity because of fewer side effects, long curative effects, cost-effectiveness, and higher safety margins [6]. Various studies have reported the beneficial effects of herbal species against liver diseases by targeting several mechanisms like inhibiting apoptosis and tissue inflammation, preventing oxidative injury, blocking fibrogenesis, and suppressing tumorigenesis (Table 28.1). This chapter provides an insight into understanding the mechanism of action of various herbal drugs used in the management of HCC by attenuating the risk factors.

Herbal Medicines: A Boon for Healthy Human Life.
DOI: https://doi.org/10.1016/B978-0-323-90572-5.00009-3

Table 28.1: Role of various plants and active constituents in the treatment of hepatocellular carcinoma.

S. No.	Herbal plant	Activity
1.	Sulforaphane	Anti-inflammatory, antioxidant, and apoptosis promoter
2.	Green tea	Antioxidant, induction of apoptosis, and anti-inflammatory
3.	Ginger	Anti-inflammatory, antioxidant, and apoptosis promoter
4.	Wolfberry	Antioxidant, immunoregulation
5.	Berberine	Anti-inflammatory effect, Apoptosis inducer, and inhibition of tumor cell proliferation
6.	Capillary wormwood	Apoptosis inducer
7.	*Andrographis paniculata*	Apoptosis inducer, necrosis of cancer cells, and cytotoxic activity antiviral activity
8.	*Alpinia officinarum*	Anti-inflammatory, antioxidant, and cytotoxic properties
9.	*Glycyrrhiza glabra*	Antiviral activity, anti-inflammatory, antioxidant, and immunomodulatory activities
10.	*Illicium verum*	Cytotoxic activity, antioxidant, anti-inflammatory, apoptosis inducer, and inhibition of tumor metastasis
11.	*Nigella sativa*	Anti-inflammatory, antioxidant, cytotoxic, and immunomodulatory

28.2 Oxidative stress and capturing it to prevent hepatocellular carcinoma

28.2.1 Sulforaphane

Sulforaphane (SFN) [1-isothiocyanate-(4R)-(methylsulfinyl) butane] (Fig. 28.1A) is a naturally occurring isothiocyanate present in cruciferous vegetables such as broccoli, brussel sprouts, and cabbage. Researchers are showing increasing interest in this compound because of its cytoprotective and anticancer activities, which are directly linked to its antioxidant potential. The protective effect of SFN has been observed both in *in vitro* and *in vivo* studies. The protective effect of sulfur-radish extract and SFN is demonstrated by Baek and team members on carbon tetrachloride (CCl_4) induced liver injury in mice. CCl_4 is known for increasing the serum levels of alanine aminotransferase (ALT), lipid peroxidation, and necrosis. These effects are ameliorated by sulfur-radish extract and SFN and the hepatoprotective activity is observed [7]. The protective activity of SFN against cisplatin-induced hepatic damage, oxidative stress, and mitochondrial dysfunction is also been demonstrated. Upon administration of SFN, the level of antioxidant enzymes and mitochondrial function is preserved [8]. SFN (3 mg/kg), against D-galactosamine (GalN;300mg/kg) and lipopolysaccharide (LPS;30μg/kg) induced fulminant hepatic failure in rats, caused reduced mortality, and alleviation in pathological liver injury. Signification decrease in the serum enzymes and lipid peroxidation is also observed. Reduction in the serum tumor necrosis factor-α (TNF-α), interleukin-6 (IL-6), and interleukin-10 (IL-10) is also noted in the GalN/LPS-treated rats.

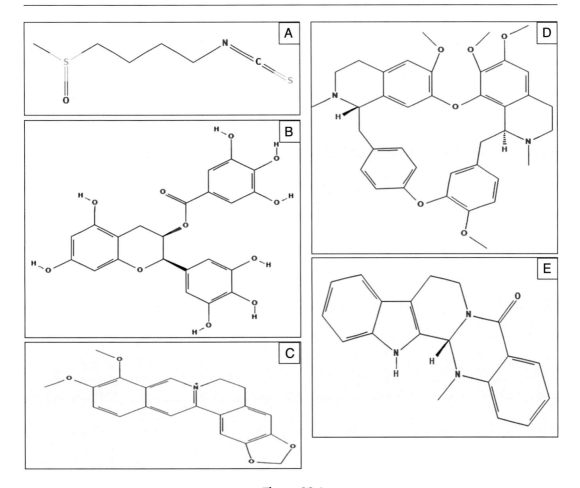

Figure 28.1
Chemical structure of the active constituents: (A) sulforaphane; (B) Epigallocatechin gallate; (C) Berberine; (D) Tetrandrine; (E) Evodiamine.

These findings suggest the antioxidant, anti-inflammatory, and antiapoptotic activity of SFN in GalN/LPS-induced liver injury [9].

28.2.2 Green tea

Green tea is an unfermented product obtained from the leaves of *Camellia sinensis* and is a famous drink especially in the Southern-Eastern countries of Asia. The major chemical constituents of green tea are epigallocatechin gallate (EGCG), epigallocatechin, epicatechin gallate, and epicatechin. Out of these, EGCG (Fig. 28.1B) is the most studied and explored active constituent due to its antiaging, antibacterial, anticaries, antidiabetic, antidiarrheal, antifibrotic, anti-inflammatory, antioxidative, antiparkinsonism, antistroke, antiatherosclerotic, and anticancer activities [2]. The anticancer potential of EGCG has been revealed by many *in-vitro* studies. The effect of EGCG on proliferation and cell cycle progression of HepG2 cell is examined and it is observed that EGCG induces apoptosis and arrests cell cycle in the G1 phase. Significant increase in the expression profile of p53 and p21/WAF1 protein is also observed, which further contributes to the cell cycle arrest. The apoptotic effect of EGCG may be possibly due to an increase in the levels of Bax protein and Fas/APO-1 and its two form ligands, membrane-bound Fas ligand (mFasL) and soluble Fas ligand (sFasL) [10]. EGCG is also been reported to inhibit the phosphorylation of IGF-1R, followed by decrease in the activation of its downstream signaling molecules like Akt, ERK, GSK-3β, and STAT-3. This leads to suppression in the growth of HepG2 human HCC cells [11].

28.2.3 Zingiber officinale

Ginger is a familiar dietary spice known for its medicinal properties. The rhizomes of ginger contain various active phenolic constituents such as gingerol, paradol, and shogoal. This spice has a long history in the treatment of cardiovascular, respiratory, and rheumatic disorders. It is also found effective in conditions like cough, cold, nausea, vomiting, and motion sickness [12,13]. The anticancer activity of ginger and its phyoconstituents have been reported in the literature. Abdel-Azeem and teammates investigated the efficacy of ginger and vitamin E against acetaminophen (APAP) induced acute hepatotoxicity in rats. APAP significantly increase the levels of serum biomarkers like ALT, aspartate aminotransferase (AST), alkaline phosphatase (ALP), arginase activities, and total bilirubin concentration, inducing significant liver injury. Significant reduction in the levels of total proteins and albumin is also observed by the authors upon administration of APAP. An increase in the malondialdhyde (MDA) levels, plasma triacylglycerols, and total antioxidant capacity is also noted in APAP control group. Morphometric assessment in APAP group showed alterations in the normal architecture of liver cells along with their necrosis and vacuolization. Pretreatment with ginger or vitamin E restore all these histo-architectural alterations. Significant lowering in the levels of AST, ALT, ALP, arginase, and total bilirubin in plasma was observed upon treatment. The oxidative stress was also significantly ameliorated by ginger and vitamin E treatment [14]. The apoptotic effect of 6-shogaol (6-SG), an alkanone obtained from ginger rhizomes, is reported in mahlavu cells via an oxidative stress-mediated caspases-dependent mechanism. An initial overproduction of reactive oxygen species (ROS) followed by a severe depletion of intracellular glutathione (GSH) contents is observed in cells upon treatment with 6-SG. Both these events consequently decreased mitochondrial transmembrane potential, followed by activation of caspases 3/7 resulting in DNA fragmentation [15]. Ginger shows direct interaction with cellular microtubules, disrupting their structure and inducing apoptosis in cancer cells [12]. In ethionine-induced hepatoma in rats, ginger extract significantly reduces the elevated expression of NFκB and TNF-α. It acts as an anticancer and anti-inflammatory agent, inactivating NFκB via suppression of pro-inflammatory TNF-α [16]. In hepatic carcinogenesis induced by diethylnitrosoamine (DEN) and promoted by CCl_4, treatment with ginger extract 2 weeks before induction and throughout the experimental schedule prevents the reduction in the levels of metallothionein and endostatatin in the hepatic content. The extract normalizes the serum markers and histopathological studies also reflect the restoration of the architecture of liver cells [13]. In an *in vitro* study, the saline extract of ginger is reported to show cytotoxic activity against HepG2 cells. Dose-dependent suppression of cell proliferation is observed, when cells are treated with this extract. The IC_{50} value observed is 900 µg/mL. However, marked morphological changes including cell shrinkage and condensation of chromosomes are observed at a dose of 250 µg/mL. Further, decreased nitrate formation, depletion of GSH, and increased superoxide dismutase levels in extract-treated cells suggest its antitumor activity [14].

28.2.4 Lycium barbarum

Lycium barbarum (Wolfberry, Goji berry) is a fruit of Solanaceous defoliated shrubbery, famous in Asian countries. It is found to have bioactives responsible for its anticancer, antioxidant, hypoglycemic, and immunological activities and is reported beneficial in both eyes and liver diseases and is generally used as an antiaging remedy in Western countries [17,18]. The Lycium polysaccharide portion of wolfberry is considered most important due to its therapeutic actions like antioxidant effect, neuroprotection, antitumor activities, control of glucose metabolism and immunoregulation [19]. A randomized, double-blind, placebo-controlled clinical study consisting of 60 older healthy adults showed an increase in lymphocytes count and interleukin-2 and immunoglobulin G levels when treated with *L. barbarum* fruit juice (GoChi). A significant increase in the general feeling of well-being, such as fatigue and sleep, and memory and focus is also seen in GoChi group [18]. The GoChi group showed a significant increase in the levels of superoxide dismutase and GSH-Px between the pre-intervention and postintervention measurements and the MDA levels are reported to decrease, indicating its antioxidant efficacy [17]. The protective effect of *L. barbarum* polysaccharide (LBP) is observed on alcohol-induced liver damage in rats. Significant amelioration in the liver injury and an increase in the antioxidant function is observed upon its administration. Damage induced by ethanol is restored by LBP, when examined histopathologically [19]. The protective effect of hot water-extracted *L. barbarum* and *Rehmannia glutinosa* via inhibition of cell proliferation and stimulation of p53-mediated apoptosis is also reported in rats and/or HCC cells [20]. The antioxidant status of mice fed with high-fat diet for 2 months is studied upon administration of LBP and significant increase in levels of antioxidant enzymes and decrease in MDA level is observed, as compared to induction group [21]. Pretreatment with LBP in CCl_4-induced acute liver injury reduces the hepatic necrosis and the serum ALT level induced by CCl_4 intoxication. LBP inhibits the expression of cytochrome P450 2E1 and restored the levels of antioxidant enzymes. The nitric oxide levels and lipid peroxidation induced by CCl_4 is also decreased by LBP treatment. The hepatic inflammation is also attenuated via downregulation of proinflammatory mediators and chemokines. It also partly shows protective effect by downregulating NF-κB activity [22]. Treatment with LBP causes inhibition of human hepatoma QGY7703 cell growth and arrest the cell cycle in S phase and induces apoptosis. Alteration in the amount of RNA and concentration and distribution of Ca^{2+} is also noted in these cells [23].

28.3 Apoptosis and antiapoptotic treatment to prevent hepatocellular carcinoma

28.3.1 Andrographis paniculata

Andrographis paniculata is a familiar herb in the South East Asian traditional medicine system. It is widely used because of its distinguished therapeutic properties such as anti-hyperglycaemic [24], antiplatelet aggregation [25], antithrombotic [26], antimalarial [27], antifungal [28], antioxidant [24], anti-inflammatory [29], immuno-stimulant [30], antifertility [31], hepatoprotective, and antihepatotoxic [32,33]. Various studies have shown the sensitizing effect of andrographolide and other constituents of *A. paniculata* against cancer cells. Andrographolide stimulate the apoptotic signaling pathway and activates the initiator caspases for the extrinsic death receptor pathway and mitochondrial pathway, which could be the possible mechanism behind its antiapoptotic activity [34]. These researchers in the year 2008 mentioned that andrographolide sensitizes tumor necrosis factor-related apoptosis-inducing ligand (TRAIL) induced apoptosis in TRAIL resistant human cancer cells. The possible mechanism is p53-dependent transcriptional up-regulation of death receptor (DR-4), a process mediated by several chronological events including ROS production, C-Jun NH2-terminal kinase (JNK) activation, p53 phosphorylation, and stabilization [35]. The antiproliferative effect of andrographolide is also reported on human colorectal carcinoma Lovo cells, arresting the cell cycle at G1–S phase and inducing the expression of p53, p21, and p16, followed by repressed activity of cyclin D1/Cdk4 and/or cyclin A/Cdk2, as well as Rb phosphorylation [36]. Andrographolide exerts its anti-HCV activity by upregulating haeme oxygenase-1 and modulating p38 MAPK/Nrf2 pathway in human hepatoma cells, further increasing the amount of biliverdin and suppressing the replication of HCV [37]. Geethangili and team isolated 14 compounds including flavonoids and labdane diterpenoids and evaluated their cytotoxic activities against Jurkat, PC-3, HepG2, Colon 205 tumor cells, and normal cells. Few compounds are moderately cytotoxic while others are found to arrest cell cycle at G0/G1 and G2/M phase [38].

28.3.2 Artemisia capillaris

Artemisia capillaris (AC), commonly known as capillary wormwood, is a widely used traditional medicine and culinary agent. It is generally used in liver cholestasis, jaundice, gallbladder dysfunction, hepatitis, digestive system diseases and in microbial infections. It is used in dermatitis, chronic HBV infection, and liver cirrhosis is because of its anti-inflammatory property [39,40]. The aqueous extract inhibits ethanol-induced apoptosis of liver cells and IL-1 and TNF-α induced cytotoxicity [41]. Additionally, it inhibits inflammatory response by preventing NF-κB activation in HCC cells. It inhibits cell growth and induces apoptosis in breast cancer and leukemia cell lines [42]. Capillin and scoparone, the major constituents of AC, display anticancer effects in breast, prostate, lung, and liver cancers. The ethyl acetate fraction of AC induces apoptosis and antiangiogenesis effect by inhibiting PI3K/AkT pathway in HCC. The researchers deduced the increased expression of cleaved caspase-3 and PARP, a major indicator of apoptosis [40]. The ethanolic extract of AC reduces cell growth and prompts mitochondria-mediated apoptosis by impeding the PI3K/AkT pathway in *in vitro* and *in vivo* conditions [42].

28.3.3 Berberine

Berberine (Fig. 28.1C), an isoquinoline alkaloid, is extracted from *Coptidis rhizoma* (Huanglian in Chinese), *Berberis aquifolium, Berberis aristata, Berberis vulgaris* (barberry), and *Hydrastis canadensis* (goldenseal). It possesses a wide range of therapeutic activities such as antibacterial, antiprotozoal, anticholinergic, antihypertensive, anti-inflammatory, antidiabetic, antihyperlipidemic, antidiarrheal, and antiarrhythmic effects [43]. Recently researchers have reported the antineoplastic activity of berberine against various types of cancer, including liver cancer. Berberine induces apoptosis in HepG2, SMMC–7721, Bel–7402, LNCaP, PC-3, HONE1, and HK1 cell lines through various cutting mechanisms [43–47]. Hou and coworkers showed the inhibition of tumor cells growth by berberine in a dose- and time-dependent manner. Berberine significantly inhibits the expression of CD147 in a dose-dependent manner followed by induction of cell death by autophagy and apoptosis [44]. Berberine inhibits the growth of human hepatoma cells by prompting AMPK–mediated caspase–dependent mitochondrial pathway cell apoptosis, without affecting the viability of normal cells [45]. Berberine significantly supresses pro-inflammatory response in macrophages by inhibiting MAPK signaling and cellular ROS via AMPK activation [48]. To understand the underlying mechanism behind berberine induced cell death, the authors examined autophagy and apoptosis in HepG2 and MHCC97-L cell lines. The results reflect increase in the Bax expression, permeable transition pore formation, release of cytochrome C to cytosol, followed by activation of caspases 3 and 9. The possible mechanism behind autophagic cell death in cells is the beclin-1 activation and suppression of mTOR pathway by inhibiting Akt and upregulating p38 MAPK signaling [47]. Cotreatment with vincristine and berberine shows significant cytotoxicity and increased apoptotic cell death against liver cancer cell lines in comparison to the individual drugs. The possible mechanism is the strengthened mitochondrial damage, increased ROS generation, and decreased MMP [46].

28.4 Inflammation and anti-inflammatory herbal drugs to prevent hepatocellular carcinoma development

28.4.1 Alpinia officinarum

Alpinia officinarum is a perennial medicinal plant extensively cultivated in south-east Asia. It is a member of Zingiberaceae family and is also known as lesser galangal [49]. The major constituent includes diterpenoids, diarylheptanoids (linear, cyclic, and dimeric), flavonoids, lignin, phenylpropanoid, and volatile oils [50]. Various reports have stated the antioxidant, anti-inflammatory, antiapoptotic, analgesic, antihyperlipidemic, antimicrobial, antiemetic antiphlogistic, stomachic, carminative, antispasmodic, and cytotoxic properties of *A. officinarum* [51]. The preventive and therapeutic potential of *A. officinarum* rhizome extract (AORE) alone and in combination with cisplatin in DEN and phenobaritone (PB) induced rat is studied. The results of the study reveal restoration of serum alpha-fetoprotein level to normal and reversal to normal architecture of liver cells. The possible mechanism behind the hepato-protective effect of AORE may be the antiapoptotic and free radical scavenging activity of the flavonoid (galangin), which modulates the activities of enzymes and thus, protects the cells from the genotoxicity of chemicals [52]. An *in silico* and *in vitro* study conducted by Elgazar and collaborators indicates the benefits of various compounds isolated from *A. officinarum* in inflammatory disease. *In silico* studies showed the interaction with p38 alpha MAPK. When the activities of these compounds were tested against HepG2 cells stimulated by LPS, down-regulation of TNFα mRNA, Il-1β mRNA, and IL-6 mRNA expression level was observed in cells [53].

28.5 Hepatitis B, C infection and treatment to prevent hepatocellular carcinoma development

28.5.1 Glycyrrhiza glabra

Glycyrrhiza glabra is a native herb of central and South-Western Asia, and Mediterranean region and is cultivated in temperate and subtropical regions of the world. Its dried roots have a characteristic odor and are sweet in taste. They are known for their antioxidant, anti-inflammatory, and immunomodulatory activities. The active constituent of roots of *G. glabra,* glycyrrhizin, is a glycosylated saponin. Glycyrrhizin is known for its antiviral, anti-inflammatory, antitumor, and hepatoprotective activities. It inhibits HCV titer dose dependently, reduces the virus by 50% and synergizes its activity with interferon. Transient transfection of liver cells with HCV3a core plasmid verified these results as the expression of this gene was inhibited in a dose-dependent manner both at mRNA and protein level, keeping the GAPDH constant [54]. Liquorice shows a protective effect against DEN/CCL$_4$ induced liver cancer in rats and significantly alleviates the deteriorated conditions caused by DEN/ CCL$_4$ in presence/absence of cisplatin in all the parameters assessed [55]. SNMC, a glycyrrhizin-based preparation, when investigated for its long-term potential against treatment of hepatic steatosis in iron-overloaded HCV transgenic mice, increased the expression of carnitine palmitoyl transferase I, activating mitochondrial b-oxidation and decreasing the ROS load in liver cells [56]. Glycyrrhetinic acid reduces cell viability in HCC cells, increasing lactate dehydrogenase release and enhancing the expression of Bax, cleaved caspase-3, and LC3-II [57].

28.6 Cytotoxic agents, apoptosis, and hepatocellular carcinoma

28.6.1 Illicium verum

Illicium verum (Illiciaceae) has been classified as "both food and medicine" by the Ministry of Health, the People's Republic of China, denoting its low toxicity to humans [58]. It is an aromatic evergreen tree commonly called as star anise or Chinese star anise. It is widely distributed in China, Pakistan, and other Asian countries [59]. The active constituents include monoterpenoids, sesquiterpenoids, phenylpropanoids, lignans, flavonoids, and volatile compounds. It also contains tannins, bitter principles and essential oils like transanethole, limone, α-pinene, β-phellandrene, farnesol, safrol, and α-terpineol [2,59,60]. It has its use in phytotherapy and aromatization of foods, cosmetics, and pharmaceutical products. The analgesic, antioxidant, anti-inflammatory, anticonvulsive, insecticidal, antimicrobial, antifungal, and sedative activity of star anise has been reported in several studies. The cytotoxic potential of *I. verum* is also reflected in the literature [61]. Treatment with *I. verum* against colon cancer shows a dose-dependent change in the nuclear morphology and decrease in the mitochondrial membrane potential. Inhibition in the cell migration, invasion, and colony formation is also observed when cells are treated with various doses of *I. verum* [59]. Assessment of anticarcinogenic potential of *I. verum* in DEN initiated and PB promoted hepatocarcinogenesis showed decrease in the tumor burden, reduced oxidative stress and increase in the level of phase II enzymes [62].

28.6.2 Nigella sativa

Nigella sativa (Ranunculaceae) is annual flowering plant native to South and Southwest Asia and commonly found in Northern Africa, the Middle East, and Southern Europe. It is commonly called as blackseed, black cumin, black caraway, fennel flower, Hak Jung Chou (in China), habbat al-barakah (in the Middle East), kalonji (in India), nigella, nutmeg flower and Kalo jeera (in Bangladesh). It has delicate flowers (having 5–10 petals), white, pale blue, or pale purple in color. It is generally used as food additive and is known for its antiasthmatic, immunomodulatory antihypertensive, hypoglycemic, anti-inflammatory, antioxidant, antimicrobial, antiparasitic, anticancer property [2]. Literature has reflected wide usage of nigella against HCC. Polyherbal formulation containing *N. sativa, H. indicus*, and *S. glabra,* when orally administered in dose 6g/kg/day show long-term protection against DEN-induced hepatic adenoma in wistar rats [63,64]. Topical application of *N. sativa* extract (100 mg/kg) inhibits two-stage initiation/promotion of skin cancer and deferred the inception of skin papilloma in mice, challenged with 7,12-dimethylbenzanthracene (DMBA)/croton oil. Intraperitonial administration of this extract reduces the methylcholanthrene (MCA)-induced soft tissue sarcomas by 70% [65]. In human breast cancer cell lines, MCF-7, aqueous, and ethanolic extract of *N. sativa* seeds show antiproliferative activity, both separately and in combination [66]. The protective effect of *N. Sativa* oil (4g/kg/day; i.g.) on female sprague dawley rats challenged with DMBA to induce mammary cancer is studied by Aziz and team. It is observed that the frequency of mammary papillary, comedo, and cribriform carcinoma reduces in rats treated with this oil and this count is even more recused in rats pre-treated with this oil. Reduction in the levels of serum biomarkers like total sialic acid and lipid-bound sialic acid, serum endocrine derangement markers (prolactin, estradiol, and progesterone), and apoptotic markers (serum TNFα, tissue caspase-3 activity, and DNA fragmentation) is observed in the tissues isolated from treatment group [67]. *In vitro* treatment with aqueous extract of *N. sativa* (0.1–1.0% concentration) causes a noteworthy reduction in the cell proliferation and preserved the morphological architecture of HepG2 cells, including cell shrinkage, membrane damage, DNA damage and cell death [68].

28.7 Other herbal management

28.7.1 Compound Astragalus *and* Salvia miltiorrhiza *extract (CASE)*

Astragalus membranaceus (Leguminosae) and *Salvia miltiorrhiza* (Lamiaceae) are used in the treatment of liver disease as a single or composite formulation. This synergistic composite extract targets TGF-β/Smad pathway for the amelioration of liver fibrosis and HCC [69]. For the overall oncogenic role of TGF-β/Smad signaling in HCC, activation of MAPK and its linked phosphorylation of Smad2/3 and their preferential nuclear import is considered necessary. CASE is demonstrated to block the MAPK-activation and its dependent linker phosphorylation of Smad2/3 and its related cascades. This blocking leads to significant downregulation of PAI-1gene expression [70].

28.7.2 Chinese herbal medicines

For over thousands of years, Chinese traditional medicines are being used as a treatment option in lung, breast, gastric and pancreatic cancer. Many clinical studies have shown the potential of Chinese herbal medicines against various cancers [71–73]. Songyou Yin (SYY) downregulates cytokine secretion controlled by PI3K/AKT signaling and attenuates the invasion and metastasis of LX2, a hepatic stellate cell line [74]. This herbal drug is useful in HCC treatment as it significantly inhibits tumor growth and prolongs survival by induction of apoptosis and downregulation of metalloproteinase-2 (MMP2) and vascular endothelial growth factor [75]. Buffalin, a component of a Chinese medicine, Chansu, when tested on Huh-7 and HepG-2, demonstrates marked inhibition in cell proliferation. It promotes apoptosis by inducing endoplasmic reticulum stress by modulating IRE1–JNK pathway. The autophagy pathway, marked by the conversion of LC3-I to LC3-II, is activated, leading to increased level of Beclin-1 protein, decreased p62 expression and stimulation of autophagic flux [76]. Roots of Radix *Stephaniae tetrandrae* contains a bisbenzylisoquinoline alkaloid, tetrandrine (Fig. 28.1D), widely used in traditional Chinese medicines. It induces apoptosis in liver cancer cells dose- and time-dependently, followed by alterations in the architecture of cells, fragmentation of chromatin, and activation of caspases. Tetrandrine generates ROS, which is an important player in induction of apoptosis [77]. A triterpene-saponin, Ardipusilloside I, isolated from a Chinese medicine, *Ardisia pusilla*, inhibits the invasion and metastasis of HepG2 and SMMC-7721 cells. It acts by reducing the expression of MMP-9 and MMP-2 proteins. It activates Rac1, which enhances the activity of E-cadherin and decrease the metastatic potential [78]. Cinobufacini has been used for the treatment of sarcoma, leukaemia and HCC in Chinese system of medicine. Cotreatment with doxorubicin significantly inhibits the growth of HCC cells and induces significant apoptosis by targeting Bcl-2, Bax, Bid, and cytochrome c proteins and RNAs [79]. Known as potential antibacterial, antiviral, anti-inflammatory, and antitumor drug in Chinese medicine system, Shufeng Jiedu Capsule is tested against HCC cells along with doxorubicin. It is observed that the combination induces significant apoptosis and inhibits migration and invasion of HCC cells by regulating the proteins and mRNA linked to apoptosis and migration cascades. The drug, as a whole, has stronger activity than its individual component or mixture of some components [80]. In Taiwan, *Gynura divaricata* subsp. Formosana is given to patients with liver disorders such as hepatitis and liver cancer. The anticancer and cancer-stabilization effect of aqueous extract of the aerial part of *G. divaricata* has been studied by Yen and coworkers both in *in vitro* and *in vivo* environments. Moderate cytotoxicity in Huh7 cell, inhibition in the cancer sphere formation, and reduction in the CSC marker expression is observed. The extract inhibits the Huh7 tumor growth, Ki-67expression and extended the anti-HCC effect of cisplatin in *in vivo* conditions. The Wnt reporter activity and expression of Wnt target 1 genes are also reduced by this extract [81]. A large number of studies have been conducted on evodiamine (Fig. 28.1E), a quinazolinocarboline alkaloid, isolated from traditional Chinese medicine, Wu-Chu-Yu (*Evodia rutaecarpa*). It is known for its anti-inflammatory, cognitive enhancing, vasorelaxant, and cardiotonic properties. Recently, it is revealed to have an antitumor effect against gastrointestinal and genitourinary cancers. Evodiamine induces cell cycle arrest at the G2/M phase, followed by induction of apoptosis. Hu and team provide evidence for the time- and dose-dependent inhibition of mice and human HCC cells (Fig. 28.2). The expression profile of WW domain-containing oxidoreductase (WWOX), after treatment with evodiamine, is increased in HCC cell lines and Hepa1-6 hepatoma-bearing mice in a dose-dependent manner [81].

Various parts of the plants that can be useful in the treatment of HCC are depicted in Fig. 28.3.

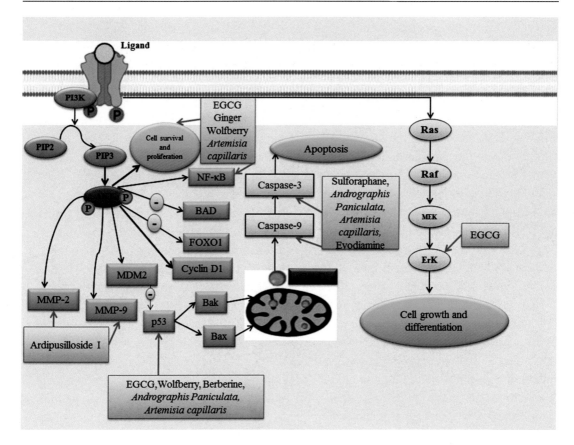

Figure 28.2

The signaling pathways involved in the anti-HCC potential of natural compounds. The PI3K/AKT cascade is known to inhibit apoptosis via phosphorylation of the apoptotic proteins, which results in the cytoplasmic sequestration. This inhibits the release of cytochrome-C from mitochondria. This also inactivates FOXO1 (a transcription factor) by phosphorylation, leading to inactivation of the pro-apoptotic proteins and cell survival. Natural products like EGCG attenuates both the Ras/Raf and PI3K/AKT signaling pathways leading to the inhibition of cell proliferation and induction of apoptosis. Sulforaphane, *Andrographis paniculata, Artemisia capillaris,* and evodiamine arrests the cell cycle at various stages, inducing cell apoptosis. Ardipusilloside reduces the expression of MMP-9 and MMP-2 proteins, decreasing the metastatic potential. Ginger, wolfberry, *Artemisia capillaris* acts as anticancer agents by inactivating NFκB via suppression of pro-inflammatory TNF-α. PI3K:Phosphoinositide 3-kinase, PIP2: Phosphatidylinositol-4,5-bisphosphate, PIP3: Phosphatidylinositol-3,4,5-trisphosphate, MDM2: Mouse double minute 2 homolog, NFκB: Nuclear factor kappa-light-chain-enhancer of activated B cells, Cyt C: Cytochrome C, MEK: Mitogen-activated protein kinase kinase, FOXO1:Forkhead box protein O1, ErK: Extracellular-signal-regulated kinase.

Figure 28.3

Medicinal plants having anti-HCC potential: (A) Rhizomes of *Zingiber officinale*; (B) fruits of *Lycium barbarum*; (C) stem and leaves of *Andrographis paniculate*; (D) *Artemisia capillaris* plant; (E) parts of *Alpinia officinarum*; (F) roots of *Glycyrrhiza glabra;* (G) *Illicium verum*; (H) *Nigella sativa* seeds and flower.

28.8 Conclusions and future perspectives

Of all forms of liver cancer, HCC accounts for about 80%–90% share and is highly related to the increase in mortality caused by cancer. It is a multistage process that converts a normal cell into a malignant one. Several factors contribute to the pathogenesis of HCC including alcohol consumption, improper dietary intake, metabolic disorders, some drugs, HBV, and HCV infection. This chapter summarizes the potential of various herbal drugs and their active constituents for the prevention and treatment of liver cancer. These herbal drugs act through various mechanisms, inhibiting the development and progression of HCC. Therefore, we recommend the use of these herbs as an adjuvant and maintenance therapy for HCC. These herbal drugs can also be added to the different recipes in the food industries to reduce the occurrence of liver diseases.

Conflict of interest

The authors declare no conflict of interest.

Acknowledgment

This work was funded by Central Council for Research in Unani Medicine (F.No. 3-31/2014-CCRUM/Tech) Ministry of AYUSH, Government of India.

References

[1]   M Gupta, K Chandan, M Sarwat, Role of microRNA and long non-coding RNA in hepatocellular carcinoma, Curr Pharm Des 26 (2020) 415–428.

[2]   NM Abdel-Hamid, SA Abass, AA Mohamed, DM Hamid, Herbal management of hepatocellular carcinoma through cutting the pathways of the common risk factors, Biomed Pharmacother 107 (2018) 1246–1258.

[3]   S Patel, M Sarwat, TH Khan, Mechanism behind the anti-tumour potential of saffron (Crocus sativus L.): the molecular perspective, Crit Rev Oncol Hematol 115 (2017) 27–35.

[4]   Z Wang, J Li, Y Ji, P An, S Zhang, Z Li, Traditional herbal medicine: a review of potential of inhibitory hepatocellular carcinoma in basic research and clinical trial, Evidence-Based Complementary and Alternative Medicine 2013 (2013) 1–7.

[5]   M Hong, S Li, HY Tan, N Wang, SW Tsao, Y Feng, Current status of herbal medicines in chronic liver disease therapy: the biological effects, molecular targets and future prospects, Int J Mol Sci 16 (2015) 28705–28745.

[6]   F Stickel, D Schuppan, Herbal medicine in the treatment of liver diseases, Dig Liver Dis 39 (2007) 293–304.

[7]   SH Baek, M Park, JH Suh, HS Choi, Protective effects of an extract of young radish (Raphanus sativus L) cultivated with sulfur (sulfur-radish extract) and of sulforaphane on carbon tetrachloride-induced hepatotoxicity, Biosci Biotechnol Biochem 72 (2008) 1176–1182.

[8]   L Gaona-Gaona, E Molina-Jijón, E Tapia, C Zazueta, R Hernández-Pando, M Calderón-Oliver, et al., Protective effect of sulforaphane pretreatment against cisplatin-induced liver and mitochondrial oxidant damage in rats, Toxicology 286 (2011) 20–27.

[9]   RH Sayed, WK Khalil, HA Salem, BM El-Sayeh, Sulforaphane increases the survival rate in rats with fulminant hepatic failure induced by D-galactosamine and lipopolysaccharide, Nutr Res 34 (2014) 982–989.

[10]   PL Kuo, CC Lin, Green tea constituent (−)-epigallocatechin-3-gallate inhibits Hep G2 cell proliferation and induces apoptosis through p53-dependent and Fas-mediated pathways, J Biomed Sci 10 (2003) 219–227.

[11] M Shimizu, Y Shirakami, H Sakai, H Tatebe, T Nakagawa, Y Hara, et al., EGCG inhibits activation of the insulin-like growth factor (IGF)/IGF-1 receptor axis in human hepatocellular carcinoma cells, Cancer Lett 262 (2008) 10–18.

[12] D Choudhury, A Das, A Bhattacharya, G Chakrabarti, Aqueous extract of ginger shows antiproliferative activity through disruption of microtubule network of cancer cells, Food Chem Toxicol 48 (2010) 2872–2880.

[13] MA Mansour, SA Bekheet, SS Al-Rejaie, OA Al-Shabanah, TA Al-Howiriny, AC Al-Rikabi, et al., Ginger ingredients inhibit the development of diethylnitrosoamine induced premalignant phenotype in rat chemical hepatocarcinogenesis model, Biofactors 36 (2010) 483–490.

[14] AS Abdel-Azeem, AM Hegazy, KS Ibrahim, AR Farrag, EM El-Sayed, Hepatoprotective, antioxidant, and ameliorative effects of ginger (Zingiber officinale Roscoe) and vitamin E in acetaminophen treated rats, Journal of Dietary Supplements 10 (2013) 195–209.

[15] CY Chen, TZ Liu, YW Liu, WC Tseng, RH Liu, FJ Lu, et al., 6-shogaol (alkanone from ginger) induces apoptotic cell death of human hepatoma p53 mutant Mahlavu subline via an oxidative stress-mediated caspase-dependent mechanism, J Agric Food Chem 55 (2007) 948–954.

[16] SH Habib, S Makpol, NA Hamid, S Das, WZ Ngah, YA Yusof, Ginger extract (Zingiber officinale) has anti-cancer and anti-inflammatory effects on ethionine-induced hepatoma rats, Clinics 63 (2008) 807–813.

[17] H Amagase, B Sun, C Borek, *Lycium barbarum* (goji) juice improves in vivo antioxidant biomarkers in serum of healthy adults, Nutr Res 29 (2009) 19–25.

[18] H Amagase, B Sun, DM Nance, Immunomodulatory effects of a standardized *Lycium barbarum* fruit juice in Chinese older healthy human subjects, J Med Food 12 (2009) 1159–1165.

[19] D Cheng, H Kong, The effect of *Lycium barbarum* polysaccharide on alcohol-induced oxidative stress in rats, Molecules 16 (2011) 2542–2550.

[20] JC Chao, SW Chiang, CC Wang, YH Tsai, MS Wu, Hot water-extracted *Lycium barbarum* and Rehmannia glutinosa inhibit proliferation and induce apoptosis of hepatocellular carcinoma cells, World J Gastroenterol 12 (2006) 4478–4484.

[21] HT Wu, XJ He, YK Hong, T Ma, YP Xu, HH Li, Chemical characterization of *Lycium barbarum* polysaccharides and its inhibition against liver oxidative injury of high-fat mice, Int J Biol Macromol 46 (2010) 540–553.

[22] J Xiao, EC Liong, YP Ching, RC Chang, KF So, ML Fung, et al., *Lycium barbarum* polysaccharides protect mice liver from carbon tetrachloride-induced oxidative stress and necroinflammation, J Ethnopharmacol 139 (2012) 462–470.

[23] M Zhang, H Chen, J Huang, Z Li, C Zhu, S Zhang, Effect of *lycium barbarum* polysaccharide on human hepatoma QGY7703 cells: inhibition of proliferation and induction of apoptosis, Life Sci 76 (2005) 2115–2124.

[24] XF Zhang, BK Tan, Antihyperglycaemic and anti-oxidant properties of *andrographis paniculata* in normal and diabetic rats, Clin Exp Pharmacol Physiol 27 (2000) 358–363.

[25] RA Burgos, MA Hidalgo, J Monsalve, TP LaBranche, P Eyre, JL Hancke, 14-deoxyandrographolide as a platelet activating factor antagonist in bovine neutrophils, Planta Med 71 (2005) 604–618.

[26] HY Zhao, WY Fang, Antithrombotic effects of *Andrographis paniculata* nees in preventing myocardial infarction, Chin Med J 104 (1991) 770–785.

[27] VK Dua, VP Ojha, R Roy, BC Joshi, N Valecha, CU Devi, et al., Anti-malarial activity of some xanthones isolated from the roots of *Andrographis paniculata*, J Ethnopharmacol 95 (2004) 247–251.

[28] A Sule, QU Ahmed, J Latip, OA Samah, MN Omar, A Umar, et al., Antifungal activity of *Andrographis paniculata* extracts and active principles against skin pathogenic fungal strains in vitro, Pharm Biol 50 (2012) 850–866.

[29] LL Ji, Z Wang, F Dong, WB Zhang, ZT Wang, Andrograpanin, a compound isolated from anti-Inflammatory traditional Chinese medicine *Andrographis paniculata*, enhances chemokine SDF-1α-induced leukocytes chemotaxis, J Cell Biochem 95 (2005) 970–978.

[30] A Puri, R Saxena, RP Saxena, KC Saxena, V Srivastava, JS Tandon, Immunostimulant agents from *Andrographis paniculata*, J Nat Prod 56 (1993) 995–999.

[31]   MA Akbarsha, B Manivannan, KS Hamid, B Vijayan, Antifertility effect of *Andrographis paniculata* (Nees) in male albino rat, Indian J Exp Biol 28 (1990) 421–426.

[32]   SS Handa, A Sharma, Hepatoprotective activity of andrographolide against galactosamine & paracetamol intoxication in rats, Indian J Med Res 92 (1990) 284–292.

[33]   A Kapil, IB Koul, SK Banerjee, BD Gupta, Anti-hepatotoxic effects of major diterpenoid constituents of *Andrographis paniculata*, Biochem Pharmacol 46 (1993) 182–185.

[34]   J Zhou, S Zhang, O Choon-Nam, HM Shen, Critical role of pro-apoptotic Bcl-2 family members in andrographolide-induced apoptosis in human cancer cells, Biochem Pharmacol 72 (2006) 132–144.

[35]   J Zhou, GD Lu, CS Ong, CN Ong, HM Shen, Andrographolide sensitizes cancer cells to TRAIL-induced apoptosis via p53-mediated death receptor 4 up-regulation, Mol Cancer Ther 7 (2008) 2170–2180.

[36]   MD Shi, HH Lin, YC Lee, JK Chao, RA Lin, JH Chen, Inhibition of cell-cycle progression in human colorectal carcinoma Lovo cells by andrographolide, Chem Biol Interact 174 (2008) 201–210.

[37]   JC Lee, CK Tseng, KC Young, HY Sun, SW Wang, WC Chen, et al., Andrographolide exerts anti-hepatitis C virus activity by up-regulating haeme oxygenase-1 via the p38 MAPK/N rf2 pathway in human hepatoma cells, Br J Pharmacol 171 (2014) 237–252.

[38]   M Geethangili, YK Rao, SH Fang, YM Tzeng, Cytotoxic constituents from *Andrographis paniculata* induce cell cycle arrest in jurkat cells, Phytother Res 22 (2008) 1336–1341.

[39]   E Jang, BJ Kim, KT Lee, KS Inn, JH Lee, A survey of therapeutic effects of Artemisia capillaris in liver diseases, Evidence-Based Complementary and Alternative Medicine 2015 (2015) 1–10.

[40]   KH Jung, M Rumman, H Yan, MJ Cheon, JG Choi, X Jin, et al., An ethyl acetate fraction of Artemisia capillaris (ACE-63) induced apoptosis and anti-angiogenesis via inhibition of PI3K/AKT signaling in hepatocellular carcinoma, Phytother Res 32 (2018) 2034–2046.

[41]   HN Koo, SH Hong, HJ Jeong, EH Lee, NG Kim, SD Choi, et al., Inhibitory effect of Artemisia capillaris on ethanol-induced cytokines (TNF-α, IL-1 α) secretion in HEPG2 cells, Immunopharmacol Immunotoxicol 24 (2002) 441–453.

[42]   J Kim, KH Jung, HH Yan, MJ Cheon, S Kang, X Jin, et al., Artemisia Capillaris leaves inhibit cell proliferation and induce apoptosis in hepatocellular carcinoma, BMC Complementary and Alternative Medicine 18 (2018) 1–10.

[43]   MS Choi, JH Oh, SM Kim, HY Jung, HS Yoo, YM Lee, et al., Berberine inhibits p53-dependent cell growth through induction of apoptosis of prostate cancer cells, Int J Oncol 34 (2009) 1221–1230.

[44]   Q Hou, X Tang, H Liu, J Tang, Y Yang, X Jing, et al., Berberine induces cell death in human hepatoma cells in vitro by downregulating CD147, Cancer Sci 102 (2011) 1287–1292.

[45]   X Yang, N Huang, Berberine induces selective apoptosis through the AMPK–mediated mitochondrial/caspase pathway in hepatocellular carcinoma, Molecular Medicine Reports 8 (2013) 505–510.

[46]   L Wang, D Wei, X Han, W Zhang, C Fan, J Zhang, et al., The combinational effect of vincristine and berberine on growth inhibition and apoptosis induction in hepatoma cells, J Cell Biochem 115 (2014) 721–730.

[47]   N Wang, Y Feng, M Zhu, CM Tsang, K Man, Y Tong, et al., Berberine induces autophagic cell death and mitochondrial apoptosis in liver cancer cells: the cellular mechanism, J Cell Biochem 111 (2010) 1426–1436.

[48]   HW Jeong, KC Hsu, JW Lee, M Ham, JY Huh, HJ Shin, et al., Berberine suppresses proinflammatory responses through AMPK activation in macrophages, American Journal of Physiology-Endocrinology and Metabolism 296 (2009) E955–E964.

[49]   WJ Kress, AZ Liu, M Newman, QJ Li, The molecular phylogeny of Alpinia (Zingiberaceae): a complex and polyphyletic genus of gingers, Am J Bot 92 (2005) 167–178.

[50]   WJ Zhang, JG Luo, LY Kong, The genus Alpinia: a review of its phytochemistry and pharmacology, World Journal of Traditional Chinese Medicine 2 (2016) 26–41.

[51]   S Ghosh, L Rangan, Alpinia: the gold mine of future therapeutics, 3 Biotech 3 (2013) 173–185.

[52]   SA Abass, NM Abdel-Hamid, TK Abouzed, MM El-Shishtawy, Chemosensitizing effect of Alpinia officinarum rhizome extract in cisplatin-treated rats with hepatocellular carcinoma, Biomed Pharmacother 101 (2018) 710–718.

[53]   AA Elgazar, NM Selim, NM Abdel-Hamid, MA El-Magd, HM El Hefnawy, Isolates from Alpinia offici-narum Hance attenuate LPS-induced inflammation in HepG2: Evidence from in silico and in vitro studies, Phytother Res 32 (2018) 1273–1288.

[54]   UA Ashfaq, MS Masoud, Z Nawaz, S Riazuddin, Glycyrrhizin as antiviral agent against Hepatitis C Virus, J Transl Med 9 (2011) 1–7.

[55]   F Hemieda, HM Serag, E El-Baz, S Ramadan, AE Faried, S Hemieda, et al., Therapeutic efficacy of licorice and/or cisplatin against diethylnitrosamine and carbon tetrachloride-induced hepatocellular carcinoma in rats, Journal of American Science 12 (2016) 10–19.

[56]   M Korenaga, I Hidaka, S Nishina, A Sakai, A Shinozaki, T Gondo, et al., A glycyrrhizin-containing preparation reduces hepatic steatosis induced by hepatitis C virus protein and iron in mice, Liver Int 31 (2011) 552–560.

[57]   ZH Tang, T Li, LL Chang, H Zhu, YG Tong, XP Chen, et al., Glycyrrhetinic acid triggers a protective autophagy by activation of extracellular regulated protein kinases in hepatocellular carcinoma cells, J Agric Food Chem 62 (2014) 11910–11916.

[58]   L Wei, R Hua, M Li, Y Huang, S Li, Y He, et al., Chemical composition and biological activity of star anise *Illicium verum* extracts against maize weevil, Sitophilus zeamais adults, Journal of Insect Science 14 (2014) 1–13.

[59]   M Asif, AH Yehya, MA Al-Mansoub, V Revadigar, MO Ezzat, MB Ahamed, et al., Anticancer attributes of *Illicium verum* essential oils against colon cancer, S Afr J Bot 103 (2016) 156–161.

[60]   LD Wu, CL Xiong, ZZ Chen, RJ He, YJ Zhang, Y Huang, et al., A new flavane acid from the fruits of *Illicium verum*, Nat Prod Res 30 (2016) 1585–1590.

[61]   JY Kim, SS Kim, TH Oh, J Baik, G Song, N Lee, et al., Chemical composition, antioxidant, anti-elastase, and anti-inflammatory activities of Illicium anisatum essential oil, Acta Pharm 59 (2009) 289–300.

[62]   AS Yadav, D Bhatnagar, Chemo-preventive effect of Star anise in N-nitrosodiethylamine initiated and phenobarbital promoted hepato-carcinogenesis, Chem Biol Interact 169 (2007) 207–214.

[63]   SR Samarakoon, I Thabrew, PB Galhena, D De Silva, KH Tennekoon, A comparison of the cytotoxic potential of standardized aqueous and ethanolic extracts of a polyherbal mixture comprised of Nigella sativa (seeds), Hemidesmus indicus (roots) and Smilax glabra (rhizome), Pharmacognosy Research 2 (2010) 335–342.

[64]   SS Iddamaldeniya, MI Thabrew, SM Wickramasinghe, N Ratnatunge, MG Thammitiyagodage, A long-term investigation of the anti-hepatocarcinogenic potential of an indigenous medicine comprised of Nigella sativa, Hemidesmus indicus and Smilax glabra, Journal of carcinogenesis 5 (2006) 1–7.

[65]   MJ Salomi, SC Nair, KR Panikkar, Inhibitory effects of Nigella sativa and saffron (Crocus sativus) on chemical carcinogenesis in mice, Nutr Cancer 1991 (1991) 67–72.

[66]   IO Farah, RA Begum, Effect of Nigella sativa (N. sativa L.) and oxidative stress on the survival pattern of MCF-7 breast cancer cells, Biomed Sci Instrum 39 (2003) 359–364.

[67]   MA El-Aziz, HA Hassan, MH Mohamed, AR Meki, SK Abdel-Ghaffar, MR Hussein, The biochemical and morphological alterations following administration of melatonin, retinoic acid and Nigella sativa in mammary carcinoma: an animal model, Int J Exp Pathol 86 (2005) 383–396.

[68]   F Khan, G Kalamegam, M Gari, A Abuzenadah, A Chaudhary, M Al Qahtani, et al., Evaluation of the effect of Nigella sativa extract on human hepatocellular adenocarcinoma cell line (HepG2) in vitro, BMC Genomics 15 (2014) 1–2.

[69]   X Liu, Y Yang, X Zhang, S Xu, S He, W Huang, et al., Compound Astragalus and Salvia miltiorrhiza extract inhibits cell invasion by modulating transforming growth factor-β/Smad in HepG2 cell, J Gastroen-terol Hepatol 25 (2010) 420–426.

[70]   A Boye, C Wu, Y Jiang, J Wang, J Wu, X Yang, et al., Compound Astragalus and Salvia miltiorrhiza extracts modulate MAPK-regulated TGF-β/Smad signaling in hepatocellular carcinoma by multi-target mechanism, J Ethnopharmacol 169 (2015) 219–228.

[71]   YW Lee, TL Chen, YR Shih, CL Tsai, CC Chang, HH Liang, et al., Adjunctive traditional Chinese medicine therapy improves survival in patients with advanced breast cancer: A population-based study, Cancer 120 (2014) 1338–1344.

[72] H Guo, JX Liu, L Xu, T Madebo, JP Baak, Traditional Chinese medicine herbal treatment may have a relevant impact on the prognosis of patients with stage IV adenocarcinoma of the lung treated with platinum-based chemotherapy or combined targeted therapy and chemotherapy, Integr Cancer Ther 10 (2011) 127–137.

[73] Y Xu, AG Zhao, ZY Li, G Zhao, Y Cai, XH Zhu, et al., Survival benefit of traditional Chinese herbal medicine (a herbal formula for invigorating spleen) for patients with advanced gastric cancer, Integr Cancer Ther 12 (2013) 414–422.

[74] Y Bu, QA Jia, ZG Ren, TC Xue, QB Zhang, KZ Zhang, et al., The herbal compound Songyou Yin (SYY) inhibits hepatocellular carcinoma growth and improves survival in models of chronic fibrosis via paracrine inhibition of activated hepatic stellate cells, Oncotarget 6 (2015) 40068–40080.

[75] XY Huang, L Wang, ZL Huang, Q Zheng, QS Li, ZY Tang, Herbal extract "Songyou Yin" inhibits tumor growth and prolongs survival in nude mice bearing human hepatocellular carcinoma xenograft with high metastatic potential, J Cancer Res Clin Oncol 135 (2009) 1245–1255.

[76] F Hu, J Han, B Zhai, X Ming, L Zhuang, Y Liu, et al., Blocking autophagy enhances the apoptosis effect of bufalin on human hepatocellular carcinoma cells through endoplasmic reticulum stress and JNK activation, Apoptosis 19 (2014) 210–223.

[77] C Liu, K Gong, X Mao, W Li, Tetrandrine induces apoptosis by activating reactive oxygen species and repressing Akt activity in human hepatocellular carcinoma, Int J Cancer 129 (2011) 1519–1531.

[78] J Xia, Y Inagaki, J Gao, F Qi, P Song, G Han, et al., Combination of cinobufacini and doxorubicin increases apoptosis of hepatocellular carcinoma cells through the Fas-and mitochondria-mediated pathways, Am J Chin Med 45 (2017) 1537–1556.

[79] J Xia, L Rong, T Sawakami, Y Inagaki, P Song, K Hasegawa, et al., Shufeng Jiedu Capsule and its active ingredients induce apoptosis, inhibit migration and invasion, and enhances doxorubicin therapeutic efficacy in hepatocellular carcinoma, Biomed Pharmacother 99 (2018) 921–930.

[80] CH Yen, CC Lai, TH Shia, M Chen, HC Yu, YP Liu, et al., Gynura divaricata attenuates tumor growth and tumor relapse after cisplatin therapy in HCC xenograft model through suppression of cancer stem cell growth and Wnt/β-catenin signalling, J Ethnopharmacol 213 (2018) 366–375.

[81] CY Hu, HT Wu, YC Su, CH Lin, CJ Chang, CL Wu, Evodiamine exerts an anti-hepatocellular carcinoma activity through a WWOX-dependent pathway, Molecules 22 (2017) 1175–1182.

CHAPTER 29

Herbal medicines and bladder cancer

Neeraj Agarwal

Department of Medicine, Cedars Sinai Medical Center, Los Angeles, CA, United States

29.1 Bladder cancer

Bladder cancer is among the most prevalent cancers worldwide, with 549,393 new cases reported in 2018 [1]. In the United States, bladder cancer is the fourth most common type of cancer in men and the ninth most common cancer in women. More than 50,000 patients are diagnosed with urothelial bladder cancer (UBC) and ~15,000 die of this devastating disease every year. Most cases of UC are caused by chemical carcinogens found in tobacco and industrial products. Half a million Americans live with UC, with a cost per patient from diagnosis to death of ~$150K, the greatest of any cancer in Medicare. Furthermore, while the death rate from metastatic disease is improving primarily due to immunotherapy, most patients still die. Bladder cancer is classified as non-muscle invasive bladder cancer (NMIBC), muscle-invasive bladder cancer (MIBC), or as a metastatic form of the disease.

29.2 Therapeutic advances in bladder cancer

For NMIBC, intravesical chemotherapy along with transurethral resection of the bladder is the first choice of treatment and result in a significant reduction of disease and its recurrence [2]. Mitomycin C, epirubicin, thiotepa, gemcitabine, and doxorubicin are the commonly used chemotherapy agents for this purpose. Number of clinical trials are also ongoing for the better and more efficacious intravesical delivery of chemical agents. Administration of Bacillus Calmette–Guerin (BCG) in the bladder for NMIBC patients with a high risk of recurrence is effective [3–5]. Patients' refractory to BCG-therapy has been given other immunomodulatory agents in combination. An immune checkpoint inhibitor, PD1-blocking antibody, pembrolizumab has been approved recently by FDA for NMIBC patients refractory to BCG [6].

For MIBC patients, the current standard of care is neoadjuvant platinum-based chemotherapy followed by radical cystectomy [7,8]. MIBC patients who present with genetic alterations in FGFR gene have been FDA-approved to be treated with FGFR inhibitor erdafitinib [9].

Herbal Medicines: A Boon for Healthy Human Life.
DOI: https://doi.org/10.1016/B978-0-323-90572-5.00006-8

29.3 Complementary and alternative medicine for bladder cancer

Complementary and alternative medicine (CAM) is defined as a group of diverse medical and healthcare systems, practices, and products that are not considered as conventional medicine. The use of CAM has been gaining popularity among cancer patients. Reasons for the use of CAM by patients are decreasing faith in conventional therapy and perceived safety of natural products. CAM use is often not disclosed by patients to their physicians because of lack of enquiry by doctor, fear of doctor's disapproval, and general perception that CAM use is independent of conventional care [10]. The safety, efficacy, cost-effectiveness, and mechanism are not well-known for CAM. Good quality clinical research work will help patients and practitioners make informed decisions. We will discuss below in detail about use of herbal remedies as part of CAM therapy for urological malignancies supported by some scientific research and clinical reports.

29.4 Korean medicine against bladder cancer

The Korean herbal therapy is based on the textbook Donguibogam. The herbal remedy mentioned in Korean medicine is Saeng-Maek-San which is comprised of *Liriopis Tuber* (Tuber of *Liriope platyphylla, Liliaceae*), *Ginseng Radix* (root of *Panax ginseng*), and *Schisandrae Fructus* (fruit of *Schisandra chinensis*) [11]. Saeng-Maek-San extract is the major herbal remedy for UBC patients. Other herbs such as *Astragali Radix* (root of *Astragalus membranaceus*) and *Oldenlandia diffusa* are also added to increase energy levels and anticancer effects. Patients with lower urinary tract symptoms are also administered with Jeoryengtang or Paljeongsan. *Rehmanniae Radix* (root of *Rehmannia glutinosa*), node of Lotus rhizome, and Typhae pollen are added for patients with hematuria. The decoction is prepared by boiling mixture of chopped herbs in water at 100°C for 4 hours and extracted two times. The quality of the herbs is tested according to Korea Food & Drug Administration (K-FDA). Oral administration of 100 mL decoction is prescribed three times a day.

29.5 Case reports of urothelial bladder cancer patients treated with Korean medicine

1. A 59-year-old Asian man with hematuria has two masses at the distal right ureter and bladder in May 2015. While waiting for surgical resection, he received Korean medicine against bladder cancer (KMBC) for 38 days. He underwent Ureteroneocystostomy in June 2015 and papillary urothelial carcinoma was confirmed with invasion into subepithelial connective tissue (pT1) and high grade (Gr2). The other mass in the bladder scheduled to be removed transurethrally could not be found. UBC was undetectable up to the last follow-up in October 2015.

2. A 37-year-old Asian man initially diagnosed with papillary urothelial carcinoma in June 2013 developed multiple masses at the anterior and posterior wall of bladder in January 2015. Patient refused to perform transurethral resection of bladder tumor (TURBT) with the fear of frequent UBC recurrence. KMBC treatment alone was started as alternative from January 2015. In June 2015, cystoscopy was carried out revealing disappearance of smaller masses and shrinkage of bigger masses as compared to 5 months ago.

3. A 71-year-old Asian woman with history of hypertension and diabetes diagnosed with urothelial carcinoma (pT1) in May 2013. TURBT was performed followed with BCG vaccine treatment. UBC keep recurring and was resected with TURBT three times in August 2013, November 2013, and March 2014. The tumor advanced to T2 stage and high grade (Gr 3) with invasion into subepithelial connective tissue and muscularis propria. Cystectomy was recommended but considering significant alteration in the quality of life and age, she decided to start KMBC alone in March 2014. Thereafter, cystoscopy was carried out every 3 months and recurrent UBC was seen till the last follow-up in September 2015. Her period of disease-free survival (DFS) was exceptionally long over 18 months considering the history of recurring UBC.

4. A 58-year-old Asian woman working as a painter in shipbuilding yard was presented with hematuria. She was diagnosed with adenocarcinoma with pT1N0M0 in March 2011 and undergone TURBT. Her UBC recurred after 4 months with pT1 stage and removed with TURBT. With hope to avoid tumor recurrence, she decided to start KMBC from October 2011. After KMBC treatment, cystoscopy was done every 3 months, and no UBC was detected until the last follow-up in November 2014. Her DFS was very long over 3 years.

5. A 53-year-old Asian man presented with urinary frequency, urgency, and nocturia. He was diagnosed with papillary urothelial carcinoma with pTa and low grade (Gr 1) in January 2010. TURBT was performed followed by BCG vaccine treatment. He then received TURBT eight times because of recurring UBC until October 2014. Eventually the tumor advanced to Gr 2 and his UBC relapsed three times within 6 months. He decided to start on KMBC treatment in October 2014. After KMBC treatment, cystoscopy was performed every 3 months and no recurrent UBC was seen up to the last follow-up in September 2015. His DFS was 12 months with improvement in other urinary symptoms like painful and frequent urination.

6. A 39-year-old Asian man presented with hematuria, received TURBT, and diagnosed with papillary urothelial carcinoma with pTa and Gr 2 in April 2013. UBC recurred and he received TURBT twice in January 2014 and May 2014. His disease progressed and the region of UBC expanded to the whole bladder. He decided to start on KMBC treatment in June 2014. Cystoscopy was carried out every 3 months and no recurrent UBC was detected up to the last follow-up in October 2015. His DFS was around 17 months.

29.6 Chinese herbal therapy for bladder cancer

Guizhi Fuling Wan (GFW) is a popular Chinese herbal medicine used throughout Asia as a treatment for blood stasis [12–14]. It is comprised of five different herbs namely *Cinnamomi Ramulus, Poria Cocos, Paeoniae Radix Rubra, Persicae Semen*, and *Moutan Cortex* [15]. GFW has sedative, analgesic, and anti-inflammatory properties. It has been shown to inhibit the growth of hepatocellular carcinoma [16,17] and cervical cancer [18]. GFW treatment inhibits the growth of bladder cancer cells isolated from patients. It suppressed the proliferation of BFTC 905 cells isolated from a female grade III stage C UBC [19] and TSGH 8301 cells isolated from a male grade II stage A UBC [20]. GFW has also shown to be as effective as chemotherapeutic agents, mitomycin C, epirubicin, and cisplatin, in treating UBC. GFW further been shown to have selective toxicity for cancer cells but not normal bladder urothelium [21].

Fruiting bodies of a medicinal mushroom (*Ganoderma lucidum*) have been used in Chinese medicine for over 2000 years. It is commonly known as "Lingzhi" in China. Ethanol extracts of fruiting bodies and spores of Lingzhi showed growth inhibition of human bladder cancer cell lines by arresting cells in G2/M phase of cell cycle [22].

Justicia procumbens is a traditional herbal remedy in south-western and southern provinces of China and Taiwan for treatment of fever, pain, and cancer. Compound 6'-hydroxyjusticidin A (JR6) present in *J. procumbens* can inhibit growth of human bladder cancer cell line "EJ." JR6 induces the generation of reactive oxygen species and also induced apoptosis by activating caspase-dependent pathway [23].

A flavonoid "Baicalein" derived from the roots of *Scutellaria baicalensis*, a plant widely used in Chinese herbal medicine. Baicalein inhibited the growth of T24 human bladder cancer cells by arresting them in G1/S phase of cell cycle. It induces apoptosis via loss of mitochondrial transmembrane potential, release of cytochrome c and activation of caspases [24].

Angelica sinensis (Oliv.) Diels (Umbelliferae) pronounced as "Danggui" in Mandarin is one of the most common herbs in traditional Chinese medicine. It is used to replenish blood and treat several gynecological symptoms. N-butylidenephthalise (BP) isolated from Danggui causes bladder cancer cells death and induced apoptosis via activating caspase-dependent pathway. It also suppressed the migration and invasion of cancer cells by modulating epithelial-to-mesenchymal transition [25].

Erianin is a natural product extracted from traditional Chinese herbal medicine *Dendrobium chrysotoxum*. Erianin suppressed the growth of bladder cancer cell lines EJ and T24 by inducing G2/M cell cycle phase arrest. It also induced apoptosis via JNK signaling pathway [26].

Licochalcone A (LCA) is a phenolic chalcone found in the root of *Glycyrrhiza glabra* and is widely used in herbal medicines in Asia. LCA treatment has antiproliferative effect on human bladder cancer cells. It induces ROS production which in turn cause G2/M phase arrest and apoptosis [27].

Herbal plant *Solanum nigrum* L. is indigenous to southeast Asia and is frequently used as valuable ingredient in traditional Chinese medicine for cancer therapy. It is frequently used to prevent the postoperative recurrence of bladder cancer. Quercetin is the main ingredient affecting various cancer-associated genes. *S. nigrum* affects multiple pathways including HIF-1, TNF, p53, MAPK, PI3K/Akt, apoptosis, and bladder cancer pathway [28].

29.7 Frankincense oil in bladder cancer treatment

Frankincense oil is extracted from milk like gum resin obtained from Boswelia trees (family *Burseraceae*). It is one of the most common oils used in aromatherapy practices. There are several varieties of Boswelia trees found in India, East Africa, China, Somalia, and Arabia. Type of soil and climate can affect its qualities. It is valued for its healing properties and also used for religious rituals [29]. It has been used in incense, fumigans, and as fixative of perfumes since thousands of years. Boswellic acids are the main ingredients responsible for its therapeutic effects.

Frankincense resin have anti-inflammatory properties attributed to its ability to regulate immune cytokines production [30] and leukocyte infiltration [31,32] as well as anticarcinogenic properties. It is used in the treatment of rheumatoid arthritis and other inflammatory diseases such as Crohn's disease [33,34]. It also has antibacterial and antifungal activity [35]. It has also been demonstrated to possess antiproliferative and proapoptotic activities against rat astrocytoma [36] and human leukemia cell lines [37]. In patients, frankincense oil reduces peritumoral edema in glioblastoma [36] and reverses brain metastases in breast cancer patients [38]. In bladder cancer, Frank et al. (2009) reported that frankincense oil suppressed the growth of human bladder cancer cell line J82 significantly more effectively compared to immortalized normal urothelium cell line UROtsa [39]. This could be because of antiproliferative and pro-apoptotic properties of frankincense oil. Frankincense oil induced the cell death and antiproliferative pathways in J82 cell line.

29.8 Other herbal remedies used in bladder cancer treatment

• A clinical trial was conducted in China with the use of medicinal mushroom *Grifola umbellata* (Zhuling) in 22 bladder cancer patients with transurethral resections or cystectomies for recurrent bladder cancer. Patients were given oral Zhuling and monitored for an average period of 26.5 months. Fifteen patients had no recurrence and seven had an average recurrence time of 19.2 months significantly lower than their first recurrence. Zhuling is like *Grifola frondosa* (maitake) which is frequently used in Japan and in the West by cancer patients because of its immunomodulating properties [40]. *Astragalus membranaceus* (astragalus) root has shown a favorable response for bladder cancer in animal models [41]. Other botanical immunomodulators shown to be beneficial in other cancer types are *Trametes versicolor* (yun zhi, cloud mushroom), *P. ginseng* (Asian ginseng) root, *Eleutherococcus senticosus* (eleuthero) root, and *Lentinula edodes* (shiitake mushroom) [42–44].

- Propolis, a resinous mixture collected by honey bees (*Apis mellifera*) from various plant sources, has been used as medicine for centuries because of its antimicrobial and anti-inflammatory activities [45]. Brazil has the widest range of propolis types based on the presence of chemical differences [46]. Brazilian red propolis (BRP) is the newly identified variety and is a source of bioactive compounds like charcones, pterocarpans, isoflavonoids, and polyphenols [47,48]. BRP ethanol extract treatment in human bladder cancer cell line 5637 showed decreased cell viability in a dose-dependent manner. BRP treatment also induced both early and late apoptosis in cancer cell lines. It also inhibited the cell migration potential of cancer cells [49].
- Water-soluble derivative of propolis (WSDP) is prepared by extracting green propolis resin in water using 8% L-lysine [50]. WSDP is reported to have anti-infectious [51], anti-inflammatory [52], immunomodulatory [53,54], antitumor [55], antioxidant [56–58], and antimetastatic [53] effects. In BBN-carcinogen induced bladder cancer model in rats, WSDP treatment reduced the angiogenesis within tumors probably due to its antioxidant property [59].
- Components of magnolia bark extract, honokiol (Hono), and magnolol (Mag) are widely reported as antineoplastic agents. Hono and Mag synergistically inhibit the growth of human bladder cancer cells by causing cell cycle arrest as well as inducing apoptosis and autophagy. They also inhibit the migratory and invasive potential of cancer cells [60].
- Curcumin is a well-known plant-based polyphenol derivative found in turmeric. It has several pharmacological effects such as antioxidant, anti-inflammatory, antidiabetic, and antitumor properties. It has the potential to suppress the growth of various cancer cells including bladder cancer by affecting molecular signaling pathways [61].

References

[1]   F Bray, J Ferlay, I Soerjomataram, RL Siegel, LA Torre, A Jemal, Global cancer statistics 2018: GLOBO-CAN estimates of incidence and mortality worldwide for 36 cancers in 185 countries, CA Cancer J Clin 68 (6) (2018) 394–424.

[2]   M Babjuk, M Burger, EM Comperat, P Gontero, AH Mostafid, J Palou, et al., European association of urology guidelines on non-muscle-invasive bladder cancer (TaT1 and carcinoma in situ)—2019 Update, Eur Urol 76 (5) (2019) 639–657.

[3]   S Monro, KL Colon, H Yin, J Roque 3rd, P Konda, S Gujar, et al., Transition metal complexes and photodynamic therapy from a tumor-centered approach: challenges, opportunities, and highlights from the development of TLD1433, Chem Rev 119 (2) (2019) 797–828.

[4]   RJ Sylvester, MA van der, DL Lamm, Intravesical bacillus Calmette-Guerin reduces the risk of progression in patients with superficial bladder cancer: a meta-analysis of the published results of randomized clinical trials, J Urol 168 (5) (2002) 1964–1970.

[5]   DL Lamm, BA Blumenstein, JD Crissman, JE Montie, JE Gottesman, BA Lowe, et al., Maintenance bacillus Calmette-Guerin immunotherapy for recurrent TA, T1 and carcinoma in situ transitional cell carcinoma of the bladder: a randomized Southwest Oncology Group Study, J Urol 163 (4) (2000) 1124–1129.

[6]   KM Wright, FDA Approves Pembrolizumab for BCG-unresponsive NMIBC. Oncology (Williston Park) 34 (2) (2020) 44.

[7]   JA Witjes, HM Bruins, R Cathomas, EM Comperat, NC Cowan, G Gakis, et al., European association of urology guidelines on muscle-invasive and metastatic bladder cancer: summary of the 2020 guidelines, Eur Urol 79 (1) (2021) 82–104.

[8]   TJN Hermans, CS Voskuilen, MS van der Heijden, BJ Schmitz-Drager, W Kassouf, R Seiler, et al., Neoadjuvant treatment for muscle-invasive bladder cancer: the past, the present, and the future, Urol Oncol 36 (9) (2018) 413–422.

[9]   JM Broderick, FDA approves Erdafitinib for bladder cancer. Onclive, April 12, 2019.

[10]   FL Bishop, A Rea, H Lewith, YK Chan, J Saville, P Prescott, et al., Complementary medicine use by men with prostate cancer: a systematic review of prevalence studies, Prostate Cancer Prostatic Dis 14 (1) (2011) 1–13.

[11]   Heo J Donguibogam. 2013 (Ministry of Health &Welfare, Seoul).

[12]   CC Tsai, ST Kao, CT Hsu, CC Lin, JS Lai, JG Lin, Ameliorative effect of traditional Chinese medicine prescriptions on alpha-naphthylisothiocyanate and carbon-tetrachloride induced toxicity in rats, Am J Chin Med 25 (2) (1997) 185–196.

[13]   KJ Lin, JC Chen, W Tsauer, CC Lin, JG Lin, CC Tsai, Prophylactic effect of four prescriptions of traditional Chinese medicine on alpha-naphthylisothiocyanate and carbon tetrachloride induced toxicity in rats, Acta Pharmacol Sin 22 (12) (2001) 1159–1167.

[14]   Y Hiyama, T Itoh, Y Shimada, T Shimada, K Terasawa, A case report of Moyamoya disease: successfully treated with Chinese medicine, Am J Chin Med 20 (3-4) (1992) 319–324.

[15]   YH Lin, KK Chen, JH Chiu, Coprescription of Chinese Herbal Medicine and Western Medications among prostate cancer patients: a population-based study in Taiwan, Evid Based Complement Alternat Med 2012 (2012) 147015.

[16]   WH Park, ST Joo, KK Park, YC Chang, CH Kim, Effects of the Geiji-Bokryung-Hwan on carrageenan-induced inflammation in mice and cyclooxygenase-2 in hepatoma cells of HepG2 and Hep3B, Immunopharmacol Immunotoxicol 26 (1) (2004) 103–112.

[17]   WH Park, SK Lee, HK Oh, JY Bae, CH Kim, Tumor initiation inhibition through inhibition COX-1 activity of a traditional Korean herbal prescription, Geiji-Bokryung-Hwan, in human hepatocarcinoma cells, Immunopharmacol Immunotoxicol 27 (3) (2005) 473–483.

[18]   Z Yao, Z Shulan, Inhibition effect of Guizhi-Fuling-decoction on the invasion of human cervical cancer, J Ethnopharmacol 120 (1) (2008) 25–35.

[19]   YT Cheng, YL Li, JD Wu, SB Long, TS Tzai, CC Tzeng, et al., Overexpression of MDM-2 mRNA and mutation of the p53 tumor suppressor gene in bladder carcinoma cell lines, Mol Carcinog 13 (3) (1995) 173–181.

[20]   MY Yeh, DS Yu, SC Chen, MS Lin, SY Chang, CP Ma, et al., Establishment and characterization of a human urinary bladder carcinoma cell line (TSGH-8301), J Surg Oncol 37 (3) (1988) 177–184.

[21]   CC Lu, MY Lin, SY Chen, CH Shen, LG Chen, HY Hsieh, et al., The investigation of a traditional Chinese medicine, Guizhi Fuling Wan (GFW) as an intravesical therapeutic agent for urothelial carcinoma of the bladder, BMC Complement Altern Med 13 (2013) 44.

[22]   QY Lu, YS Jin, Q Zhang, Z Zhang, D Heber, VL Go, et al., Ganoderma lucidum extracts inhibit growth and induce actin polymerization in bladder cancer cells in vitro, Cancer Lett 216 (1) (2004) 9–20.

[23] XL He, P Zhang, XZ Dong, MH Yang, SL Chen, MG Bi, JR6, a new compound isolated from *Justicia procumbens*, induces apoptosis in human bladder cancer EJ cells through caspase-dependent pathway, J Ethnopharmacol 144 (2) (2012) 284–292.

[24] HL Li, S Zhang, Y Wang, RR Liang, J Li, P An, et al., Baicalein induces apoptosis via a mitochondrial-dependent caspase activation pathway in T24 bladder cancer cells, Mol Med Rep 7 (1) (2013) 266–270.

[25] SC Chiu, TL Chiu, SY Huang, SF Chang, SP Chen, CY Pang, et al., Potential therapeutic effects of N-butylidenephthalide from Radix Angelica Sinensis (Danggui) in human bladder cancer cells, BMC Complement Altern Med 17 (1) (2017) 523.

[26] Q Zhu, Y Sheng, W Li, J Wang, Y Ma, B Du, et al., Erianin, a novel dibenzyl compound in Dendrobium extract, inhibits bladder cancer cell growth via the mitochondrial apoptosis and JNK pathways, Toxicol Appl Pharmacol 371 (2019) 41–54.

[27] SH Hong, HJ Cha, H Hwang-Bo, MY Kim, SY Kim, Ji SY, et al., Anti-Proliferative and Pro-Apoptotic Effects of Licochalcone A through ROS-Mediated Cell Cycle Arrest and Apoptosis in Human Bladder Cancer Cells, Int J Mol Sci 20 (15) (2019) 3820.

[28] Y Dong, L Hao, K Fang, XX Han, H Yu, JJ Zhang, et al., A network pharmacology perspective for deciphering potential mechanisms of action of Solanum nigrum L. in bladder cancer, BMC Complement Med Ther 21 (1) (2021) 45.

[29] GAM Gold, Frankincense, and Myrrh: An Introduction to Eastern Christian Spirituality, Crossroads Pub Co, New York, NY, 1997.

[30] MR Chevrier, AE Ryan, DY Lee, M Zhongze, Z Wu-Yan, CS Via, Boswellia carterii extract inhibits TH1 cytokines and promotes TH2 cytokines in vitro, Clin Diagn Lab Immunol 12 (5) (2005) 575–580.

[31] ML Sharma, A Khajuria, A Kaul, S Singh, GB Singh, CK Atal, Effect of salai guggal ex-Boswellia serrata on cellular and humoral immune responses and leucocyte migration, Agents Actions 24 (1-2) (1988) 161–164.

[32] GB Singh, CK Atal, Pharmacology of an extract of Salai Guggal ex-Boswellia serrata, a new non-steroidal anti-inflammatory agent, Agents Actions 18 (3-4) (1986) 407–412.

[33] N Banno, T Akihisa, K Yasukawa, H Tokuda, K Tabata, Y Nakamura, et al., Anti-inflammatory activities of the triterpene acids from the resin of Boswellia carteri, J Ethnopharmacol 107 (2) (2006) 249–253.

[34] L Langmead, DS Rampton, Review article: complementary and alternative therapies for inflammatory bowel disease, Aliment Pharmacol Ther 23 (3) (2006) 341–349.

[35] S Weckesser, K Engel, B Simon-Haarhaus, A Wittmer, K Pelz, CM Schempp, Screening of plant extracts for antimicrobial activity against bacteria and yeasts with dermatological relevance, Phytomedicine 14 (7-8) (2007) 508–516.

[36] M Winking, S Sarikaya, A Rahmanian, A Jodicke, DK Boker, Boswellic acids inhibit glioma growth: a new treatment option? J Neurooncol 46 (2) (2000) 97–103.

[37] K Hostanska, G Daum, R Saller, Cytostatic and apoptosis-inducing activity of boswellic acids toward malignant cell lines in vitro, Anticancer Res 22 (5) (2002) 2853–2862.

[38] DF Flavin, A lipoxygenase inhibitor in breast cancer brain metastases, J Neurooncol 82 (1) (2007) 91–93.

[39] MB Frank, Q Yang, J Osban, JT Azzarello, MR Saban, R Saban, et al., Frankincense oil derived from Boswellia carteri induces tumor cell specific cytotoxicity, BMC Complement Altern Med 9 (2009) 6.

[40] DA Yang, Inhibitory effect of Chinese herb medicine Zhuling on urinary bladder cancer. An experimental and clinical study, Zhonghua Wai Ke Za Zhi 29 (6) (1991) 393–395 9.

[41] S Kurashige, Y Akuzawa, F Endo, Effects of astragali radix extract on carcinogenesis, cytokine production, and cytotoxicity in mice treated with a carcinogen, N-butyl-N'-butanolnitrosoamine, Cancer Invest 17 (1) (1999) 30–35.

[42] PM Kidd, The use of mushroom glucans and proteoglycans in cancer treatment, Altern Med Rev 5 (1) (2000) 4–27.

[43] RJ Cha, DW Zeng, QS Chang, Non-surgical treatment of small cell lung cancer with chemo-radio-immunotherapy and traditional Chinese medicine, Zhonghua Nei Ke Za Zhi 33 (7) (1994) 462–466.

[44] VI Kupin, EB Polevaia, Stimulation of the immunological reactivity of cancer patients by Eleutherococcus extract, Vopr Onkol 32 (7) (1986) 21–26.

[45] A Daugsch, CS Moraes, P Fort, YK Park, Brazilian red propolis–chemical composition and botanical origin, Evid Based Complement Alternat Med 5 (4) (2008) 435–441.

[46] AA Righi, G Negri, A Salatino, Comparative chemistry of propolis from eight Brazilian localities, Evid Based Complement Alternat Med 2013 (2013) 267878.

[47] SM Alencar, TL Oldoni, ML Castro, IS Cabral, CM Costa-Neto, JA Cury, et al., Chemical composition and biological activity of a new type of Brazilian propolis: red propolis, J Ethnopharmacol 113 (2) (2007) 278–283.

[48] B Trusheva, M Popova, V Bankova, S Simova, MC Marcucci, PL Miorin, et al., Bioactive constituents of Brazilian red propolis, Evid Based Complement Alternat Med 3 (2) (2006) 249–254.

[49] KR Begnini, PM Moura de Leon, H Thurow, E Schultze, VF Campos, F Martins Rodrigues, et al., Brazilian red propolis induces apoptosis-like cell death and decreases migration potential in bladder cancer cells, Evid Based Complement Alternat Med 2014 (2014) 639856.

[50] Nicolov N MN, Bancova V, Popov S, Ignatov R, Valdimirova I, Inventormethod for the preparation of water-soluble derivative of propolis. Bulgaria Patent 79903. 1987.

[51] V Dimova NI, N Manolovab, V Bankovac, N Nikolovd, S Popov, Immunomodulatory action of propolis. Influence on anti-infectious protection and macrophage function, Apidologie 22 (1991) 155–162.

[52] V Dimov, N Ivanovska, V Bankova, S Popov, Immunomodulatory action of propolis: IV. Prophylactic activity against gram-negative infections and adjuvant effect of the water-soluble derivative, Vaccine 10 (12) (1992) 817–823.

[53] N Orsolic, I Basic, Immunomodulation by water-soluble derivative of propolis: a factor of antitumor reactivity, J Ethnopharmacol 84 (2-3) (2003) 265–273.

[54] N Orsolic, AH Knezevic, L Sver, S Terzic, Basic I. Immunomodulatory and antimetastatic action of propolis and related polyphenolic compounds, J Ethnopharmacol 94 (2-3) (2004) 307–315.

[55] N Orsolic, I Kosalec, Basic I. Synergistic antitumor effect of polyphenolic components of water soluble derivative of propolis against Ehrlich ascites tumour, Biol Pharm Bull 28 (4) (2005) 694–700.

[56] N Orsolic, S Terzic, Z Mihaljevic, L Sver, I Basic, Effects of local administration of propolis and its polyphenolic compounds on tumor formation and growth, Biol Pharm Bull 28 (10) (2005) 1928–1933.

[57] N Orsolic, I Basic, Water-soluble derivative of propolis and its polyphenolic compounds enhance tumoricidal activity of macrophages, J Ethnopharmacol 102 (1) (2005) 37–45.

[58] N Orsolic, V Benkovic, A Horvat-Knezevic, N Kopjar, I Kosalec, M Bakmaz, et al., Assessment by survival analysis of the radioprotective properties of propolis and its polyphenolic compounds, Biol Pharm Bull 30 (5) (2007) 946–951.

[59] CA Dornelas, FV Fechine-Jamacaru, IL Albuquerque, HI Magalhaes, TA Dias, MH Faria, et al., Angiogenesis inhibition by green propolis and the angiogenic effect of L-lysine on bladder cancer in rats, Acta Cir Bras 27 (8) (2012) 529–536.

[60] HH Wang, Y Chen, CY Changchien, HH Chang, PJ Lu, H Mariadas, et al., Pharmaceutical evaluation of Honokiol and Magnolol on Apoptosis and Migration Inhibition in human bladder cancer cells, Front Pharmacol 11 (2020) 549338.

[61] M Ashrafizadeh, H Yaribeygi, A Sahebkar, Therapeutic effects of curcumin against bladder cancer: a review of possible molecular pathways, Anticancer Agents Med Chem 20 (6) (2020) 667–677.

CHAPTER 30

Mixed effects and mechanisms of cannabinoids for triple-negative breast cancer treatment

Khanh Tran

Department of Molecular Medicine and Pathology, University of Auckland, Auckland, New Zealand

30.1 Introduction

Breast cancer is the most commonly diagnosed cancer in women worldwide. In 2019, the total number of female patients diagnosed with breast cancer for the first time worldwide was approximately 2 million, which accounted for 18.9% of all female cancer incidence [1]. The incidence of breast cancer in developed countries is 66.4/100,000 people, which is twice as high as that in developing countries [2]. In New Zealand, the rate of breast cancer in 2019 was approximately 80 per 100,000 [1]. Geographical regions that have high incidences of breast cancer include West and North Europe, Australia/New Zealand, and North America [1]. It has been estimated that the possibility for a woman to develop breast cancer in her lifetime is approximately 1 in 10 [3]. Although mammography screening has decreased the breast cancer death rates [4,5], breast cancer remains the leading cause of cancer mortality in women. In 2019, there were approximately 689,000 breast cancer deaths for females globally, accounting for 15.9% of total female cancer mortality [1]. Breast cancer is a burden for patients and society, and like with any other malignant disease, being diagnosed with breast cancer is always an acute emotional shock, which may permanently and substantially affect the physical and intellectual capacity of the patients [6,7]. In Sweden, the total annual cost required from a patient younger than 50 years of age with metastatic breast cancer was estimated to be $43,565 USD [8]. In France, trastuzumab accounted for 44% of total treatment cost (Poncet, 2009), and in Canada, the annual cost for trastuzumab required for a metastatic breast cancer patient was $28,350 USD [9]. Therefore, it is imperative that breast cancer be effectively managed, and for that purpose, searching for novel treatments is crucial.

30.2 Classification of breast cancer

It has been established that the expression status of estrogen receptor (ER), progesterone receptor (PR), and HER-2 imposes significant influences on the development, prognosis, and treatment outcome of breast cancer [10]. Therefore, breast cancer has been classified according to differential expression of ER, PR, and HER-2. Based on the presence of ER in cancer cells, breast cancer is classified into ER-positive (ER+), which expresses the ER α gene, and ER-negative, which lacks the estrogen α gene. The survival and growth of ER+ breast cancer cells are dependent on estrogen binding, whereas ER− cells rely on other growth factors, including epidermal growth factor (EGF) and endothelial vascular growth factor (VEGF) [11]. ER− breast cancer accounts for approximately 75% of all breast cancer. Compared to ER+, ER− breast cancer is often associated with a poorer prognosis and more aggressive progression. In contrast to ER+, ER− is unlikely to respond to hormonal treatments [12]. The presence of PR in the cancer cells, along with ER, is more predictive of a hormonal responsive tumor than the presence of ER alone [13,14]. Tumors with ER+/PR+ were reported to exhibit a response rate of 75% to hormonal therapies, compared to the response rate of 33% in ER+/PR− tumors [15]. Furthermore, approximately one in five breast cancer patients exhibit amplification of the HER-2 gene and overexpression of the receptor. HER-2 positive breast cancer is associated with more aggressive behaviors of the tumors [16]. In addition, HER-2 positive breast cancer is also unlikely to respond to hormonal therapies. However, they respond to anti-HER-2 monoclonal antibodies such as trastuzumab, a specific targeted treatment for HER-2 positive breast cancer [12].

30.3 Triple-negative breast cancer

Triple-negative breast cancer (TNBC) is a subtype of breast cancer characterized by the lack of ER, PR, and HER-2. TNBC accounts for 10%–20% of all breast cancer [11]. Compared to other subtypes of breast cancer, clinical features of TNBC include poor outcome, aggressive progression of the primary tumors and metastasis, shorter survival, and high mortality rate. In a cohort study of 1601 breast cancer patients, Dent (2007) found that 11.2% of the patients were of TNBC subtype [17]. In addition, the results showed that TNBC has an increased risk of distant recurrence and death within 5 years, but not after that. Similarly, it has been noted that there is a sharp decrease in survival during the first 5 years after diagnosis of TNBC. A study on 496 patients with invasive breast cancer from the Carolina Breast Cancer Study showed that TNBC accounted for 26% of the total breast cancer patients in the study [18]. TNBC tumors were mainly at the grade 3 with high mitotic indices. In addition, TNBC was more frequent in premenopausal patients and often associated with overexpression of epidermal growth factor receptor (EGFR) and p53 [19]. TNBC was also reported to relate to BCRA1 mutation closely. It was reported that 75% of breast cancer tumors with BCRA1 mutation have a TNBC phenotype [10]. TNBC lacks ER and HER-2; therefore, TNBC does not benefit from endocrine therapy or trastuzumab [12]. The primary treatment options for TNBC are surgery, chemotherapy, and radiation. In addition to these conventional treatment options, several targeted therapies, including EGFR inhibitors, PARP inhibitors, and angiogenesis inhibitors, are in clinical trials [20,21]. As there are currently no targeted treatments for TNBC, there is an urgent need to develop novel treatments for this subtype of cancer.

30.4 Marijuanna-derived compounds

Cannabinoids can be classified into three main groups, including natural cannabinoids that are derived from the plant *Cannabis sativa* (phytocannabinoids), endogenous cannabinoids that are originated within the body (endocannabinoids), and cannabinoid analogues that are artificially synthesized (synthetic cannabinoids) [22,23]. Phytocannabinoids are 21-carbon-containing terpenophenolic compounds derived from the plants *C. sativa* and *Cannabis indica*. At least 85 phytocannabinoids has been isolated and chemically characterized. Based on their chemical structures, phytocannabinoids are further categorized into different groups. Of these, the tetrahydrocannabinol (THC) group contains highly potent cannabinoids, including delta-9-tetrahydrocannabinol (Δ9-THC) and delta-8-tetrahydrocannabinol (Δ8-THC), which are the most abundant and largely responsible for the psychological and physiological effects of marijuana [23]. After the presence of CB1 and CB2 cannabinoid receptors in the brain and the immune system was recognized, another group of cannabinoids has been found. Endocannabinoids are a group of ligands for cannabinoid receptors that are endogenously biosynthesized within the body. Endocannabinoids are all arachidonic acid derivatives, including arachidonoylethanolamine (anandamide or AEA), 2- arachidonoyl glycerol (2-AG), 2-arachidonoyl glyceryl ether (noladin ether), N-arachidonoyl-dopamine (Gupta GP), and *O*-arachidonoyl-ethanolamine (Virodhamine or OAE) [24,25]. Endocannabinoids and cannabinoids receptors present within the body are two major components of the endo-cannabinoid system, which has now been revealed to have important roles in modulation of neurotransmitter release; control of cell survival, transformation, proliferation, and metabolism; regulation of pain perception, cardiovascular, gastrointestinal, and respiratory functions [26].

Synthetic cannabinoids are a large group of structurally diverse compounds that can bind to the cannabinoid receptors and have cannabimimetic activities. Classical synthetic cannabinoids refer to compounds that retain parts of the dibenzopyran ring of THC [27]. The first generation of classical synthetic cannabinoids includes HU-210 and Nabilone (Cesamet, Lilly), the latter of which was approved by the US Food and Drug Administration for treatment of chemotherapy-induced nausea and vomiting [28]. The second generation of classical synthetic cannabinoids synthesized by J.W. Huffman's group has a variety of core ring structures. The compounds of this group are named JWH- compounds to honor JW Huffman, the creator of JWH-018, and other JWH- compounds, including JWH-133, JWH-018, JWH-075, and JWH-019 [29]. In contrast to classical synthetic cannabinoids, non-classical compounds lack the dibenzopyran in their molecular structures, including cyclohexylphenol compounds and aminoalkylindole (AAI) compounds. CP cannabinoids were developed by Pfizer, including CP-47,497 and CP-55,940 [30]. Aminoalkylindole (AAI) cannabinoids were developed at Sterling Winthrop, with WIN55,212-2 being the most potent compound in this series [31].

30.5 Cannabinoid receptors

To date, two specific receptors for cannabinoid ligands have been cloned from mammalian cells. The first cannabinoid receptor, the CB1 receptor, was cloned in 1990 from a rat cerebral cortex cDNA library [32]. The second cannabinoid receptor, the CB2 receptor, was cloned in 1993 by Munro et al. from human promyelocytic leukemic HL-60 cells [33]. Although the two receptors have many common cannabinoid ligands, they differ substantially from each other in many aspects, including amino acid sequence, distribution location within the body, and downstream signaling cascades. Human and rat CB1 receptor proteins consist of 427 and 473 amino acids, respectively, with 97%–99% amino acid sequence identity across species.

The CB2 receptor was cloned from human leukemic HL-60 cells as a cDNA fragment that encodes a protein of 360 amino acids, with 82% and 81% amino acid sequence identity to mouse and rat CB2 receptors, respectively. CB1 and CB2 receptor proteins show 44% identity in general and 68% for the transmembrane residues, which are considered to determine ligand specificity for the receptors [23]. The distribution location of CB1 and CB2 cannabinoid receptors is also different. CB1 receptors are found primarily in the brain and are the most abundant G protein-coupled receptor in the brain. Particularly, the CB1 receptor is highly expressed in basal ganglia and cerebellum, cortex and hippocampus, amygdala, thalamus, hypothalamus, pons, and medulla [23]. In addition, the CB1 receptor is also expressed in peripheral nerve terminals and extraneural tissues such as testes, eyes, vascular endothelium, and spleen [34]. CB2 receptors are found primarily in the immune cells (B and T cells and macrophages) and tissues (spleen, tonsils, and lymph nodes) [35].

Cannabinoid receptors belong to the superfamily of G protein-coupled receptors (GPCRs), also known as seven-transmembrane domain (7TM) receptors [36]. It has been established that CB1 receptors are coupled to $G_{i/o}$ and G_s proteins, while CB2 receptors are coupled to $G_{i/o}$ proteins [37,38]. Activation of CB1 and CB2 receptors leads to inhibition of adenylyl cyclase. In addition, activated cannabinoid receptors can also modulate signaling cascades that are involved in the regulation of cell survival, growth, and proliferation [39]. Major downstream signaling cascades of cannabinoid receptors include extracellular signal-regulated kinase, c-Jun N-terminal kinase (JNK), p38 mitogen-activated protein kinase, phosphatidylinositol 3-kinase/Akt and focal adhesion kinase [40]. In addition, the CB1 receptor can modulate certain types of ion channels, which plays a crucial role in the neuromodulatory actions of the endocannabinoids. CB1 receptor can inhibit N-, L- and P- or Q-type voltage-sensitive Ca^{2+} channels [41,42] and activate G protein-activated inwardly rectifying K^+ channels [43].

30.6 Anticancer effects of cannabinoids in triple-negative breast cancer

TNBC lacks ER and HER2, the targets for selective estrogen receptor modulators and anti-HER2 monoclonal antibodies. Therefore, it is imperative that alternative treatment for TNBC be developed in order to effectively manage this poor prognosis and highly aggressive subtype of breast cancer. Due to the lack of specific treatment for TNBC and that cannabinoids produced antitumoral effects in some other cancers, recent attention has also been drawn to the possibility for cannabinoids to be used as a treatment for TNBC [44].

It has been reported that cannabinoid receptors are overexpressed in primary human breast tumors compared to normal breast tissue [44]. In breast cancer tissue, Caffarel (2006) showed that CB2 expression was higher than CB1 expression in the same tumor [45]. In addition, CB2 expression appeared to correlate positively with the grades of the tumors. Compared to ER(+), PR(+), and HER2(+) breast cancer tumors, CB2 mRNA was found to be higher than in ER(-), PR(-), and HER2(-), respectively [45]. Studies have also confirmed the presence of CB1 and CB2 receptors in TNBC cells using reverse transcriptase and real-time PCR coupled with confocal microscopy [44,46].

In vitro, treatment with cannabinoids has been carried out in various TNBC cells of which the most commonly used cell lines are MDA-MB-231 and MDA-MB-468. Studies demonstrated that cannabinoids inhibit the survival and proliferation of TNBC cells in a dose-dependent and time-dependent manner [45]. In addition, using CB1 and CB2 receptor antagonists was found to prevent cannabinoid-induced cell death, suggesting that the inhibitory effects of cannabinoids against TNBC cells were mediated via cannabinoid receptors. Antiproliferative effects of cannabinoids *in vitro* can be attributed to induction of apoptosis and cell cycle arrest [44,47].

In vivo, cannabinoids have been reported to suppress the growth of TNBC xenografts. In a study conducted by Ligresti *et al.* (2006), MDA-MB-231 cells were subcutaneously inoculated into the dorsal right side of male athymic mice [47]. The mice were intratumorally treated with Δ9-THC or cannabidiol (5mg/kg) twice a week for 16 days. It was found that Δ9-THC and cannabidiol significantly reduced the volume of the xenografts. Using SCID mice, Qamri (2009) reported that both JWH-133 and WIN55,212-2 suppressed the growth of MDA-MB-231 xenografts, and the effects were mediated via CB1 and CB2 receptors. Cannabinoids have also been reported to inhibit the metastasis of TNBC tumors [44]. Treatment with cannabidiol (i.p. injection of 5mg/kg every 72 h for 21 days) significantly reduced metastatic lung infiltration from the primary tumors induced by injection of MDA-MB-231 cells into the left paw of the mice [47]. Similarly, treatment with JWH-133 and WIN55,212-2 was shown to reduce lung metastasis by 65% to 80%, respectively [44].

The involvement of CB1 and CB2 receptors has been reported both *in vitro* and *in vivo*. CB1 and CB2 receptor antagonists have been demonstrated to reverse the inhibitory effects of cannabinoids on cultured TNBC cells and on TNBC xenografts, suggesting cannabinoid-induced cell death *in vitro* and tumor growth suppression *in vivo* were mediated via CB1 and CB2 receptors. Furthermore, the mediative role of the CB2 receptor has been confirmed using CB2 targeting siRNA that was transfected into MDA-MB-231 cells. The results showed that CB2 siRNA was able to block JWH-133 and WIN55,212-2 effects by decreasing level of CB2 expression in the transfected [44]. Taken together, studies indicate that the anticancer effects of cannabinoids in TNBC were mediated via CB1 and CB2 receptors.

30.7 Cellular mechanisms of anticancer effects of cannabinoids

So far, several mechanisms for anticancer effects of cannabinoids have been identified, including apoptosis induction, cell cycle arrest, antiangiogenesis, and inhibition of migration and invasion. Cannabinoids were found to induce apoptosis in glioma cells and other cancer cells in culture [48–50]. Cannabinoids also increased apoptotic activities in tumors treated with cannabinoids, which was associated with the suppression of tumor growth [51,52]. Existing evidence indicated that cannabinoid-induced apoptosis was mediated via the activation of cannabinoid receptors, which in turn triggers the proapoptotic mitochondrial intrinsic pathway. In glioma cells and pancreatic cancer cells, activation of cannabinoid receptors resulted in two peaks of ceramide generation by the mechanisms of sphingomyelin hydrolysis and *de novo* synthesis, respectively [53,54]. Studies showed that the second peak of ceramide accumulation accounted for the apoptosis induced by cannabinoids [55]. The mechanism by which accumulation of the sphingolipid ceramide leads to apoptosis has been reported to be mediated by the stress-regulated protein p8, which is upregulated by ceramide accumulation. p8 upregulation leads to the upregulation of the activating transcription factor 4 (ATF-4) and the C/EBP-homologous protein and through which induce apoptosis [56].

Cannabinoids also cause cell cycle arrest in cancer cells of prostate carcinoma [56], thyroid epithelioma [57], breast carcinoma [44], lung carcinoma [51], and gastric carcinoma [58]. It has been suggested that activation of cannabinoid receptors lead to the inhibition of adenylyl cyclase and the cAMP/protein kinase A (PKA) pathway. As PKA inhibits Raf-1, cannabinoids prevent the inhibition of Raf-s and consequently result in prolonged activation of Raf-1/MEK/ERK signaling cascade. The activation of Raf-1/MEK/ERK has been found to be associated with cell arrest in various cancer cells. Also, prolonged activation of Raf-1/MEK/ERK may induce cell cycle arrest by modulating the expression of molecules that involve in the cell cycle regulation, including $p16^{Ink4a}$, $p15^{Ink4b}$, and $p21^{Cip1}$, which can lead to the cell cycle arrest at the G1 phase [59,60]. In addition, cannabinoid-induced cell cycle arrest in thyroid epithelioma was also reported to be mediated via the induction of the cyclin-dependent kinase inhibitor p27^{kip1} [61].

For a tumor to grow beyond the minimal size, it must recruit new blood vessels by producing proangiogenic factors that promote the formation of new vessels for nutrition, gas exchange, and waste disposal. Targeting the formation of new blood vessels in the tumor, therefore, is an important approach in anticancer drug development. Studies indicated that cannabinoid treatment could change the blood vessel pattern in xenograft tumors from a hyperplasic network of dilated vessels to a pattern of blood vessels characterized by narrow, differentiated and impermeable capillaries [62]. The antiangiogenesis effects of cannabinoids are associated with inhibition of the expression of vascular epidermal growth factor (VEGF) and other proangiogenic factors such as placental growth factor and angiopoietin 2 [61]. In addition, cannabinoids can also inhibit the formation of new blood vessels by downregulating the expression of VEGF receptors [61]. It has been suggested that ceramide synthesis *de novo* is involved in the mechanism of antiangiogenesis of cannabinoids. Inhibition of ceramide synthesis *de novo* prevented the cannabinoid-induced inhibition of VEGF production *in vitro* and *in vivo* [39].

Recently, increasing evidence shows that cannabinoids can also inhibit the invasiveness of cancer tumors. WIN55,212-2 and JWH-015 significantly decreased *in vitro* chemotaxis and chemoinvasion of lung cancer cells and inhibited *in vivo* metastasis from the xenografts to the lungs [51]. Similarly, 2-methyl-arachidonyl-2′-fluoro-ethylamide (Met-F-AEA) significantly reduced the number of metastatic nodes following injection of Lewis lung cancer cells into the paw [61]. Mechanisms for the antimigration and anti-invasion effects of cannabinoids may be related to the inhibition of Akt, which is involved in the regulation of migration. In addition, cannabinoids were also found to downregulate the expression and activity of matrix metalloproteinase-2 (MMP2), which plays an important role in tissue remodeling and is closely associated with angiogenesis, tissue repair, and metastasis [44,62].

30.8 Molecular mechanisms of cannabinoids

30.8.1 Modulation of receptor expression by synthetic cannabinoids

In normal noncancerous human breast tissue, cannabinoid receptors are expressed at lower levels than breast cancer tissue [44]. In breast cancer tissue, it was reported that CB2 expression was higher than the CB1 expression in the same tumors and CB2 expression seemed to correlate with the grades of the tumors [45]. Compared to ER(+) and PR(+) breast cancer tumors, CB2 mRNA was found to be 3.6- and 2.3-fold higher in ER(-) and PR(-) tumors, respectively [45].

Alteration in cannabinoid receptor expression was also associated with cannabinoid exposure. Chronic treatment with cannabinoids led to downregulation of cannabinoid receptors [63–65]. Daily treatment with Δ9-THC for 14 days caused a 30% reduction in cannabinoid receptor binding [66]. Downregulation of cannabinoid receptors following long-term treatment with cannabinoid is attributed to the internalization and degradation of the receptors. Cannabinoid receptors belong to GPCRs; and like many other GPCRs, cannabinoid receptors undergo agonist-induced or constitutive internalization from the cell membrane to low pH endosomes, where a certain amount of the receptors is recycled back to the cell surface, another part stays in the cytoplasm and contributes to the intracellular pool of the receptors, and the rest of the endocytosed receptors are sent to lysosomes for degradation [67,68].

The intracellular reservoir of the CB1 receptor is reported to account for 85% of the total cellular CB1 cannabinoid receptor; however, functions of the intracellular pool remain to be elucidated. As reviewed by Rozenfeld (2011), the intracellular pool of cannabinoid receptors serves as a source from which surface CB1 receptors are replenished for endocytosed receptors [69]. However, Grimsey (2010) reported that the intracellular pool of the cannabinoid receptors does not contribute to the recycling of the cell surface receptor population [70]. Mechanisms for cannabinoid receptor internalization and recycle are not fully established. HU308-induced endocytosis of CB2 cannabinoid receptors was found to be mediated via Rab5, a small GTPase localized to early endosomes, while CB2 recycle was mediated via a recycling endosome-Rab11-dependent pathway [71]. WIN55,212-2 induced CB1 and CB2 downregulation, and CB1 receptors were sorted in lysosomal compartments for degradation by a G-protein-associated sorting protein (GASP-1) and an adaptor protein 3 (AP-3) [65,72].

In contrast to ER(+) breast cancer, where estrogen-ER-ERE pathway plays the central role in cellular processes, ER(-) breast cancer is under the regulation from other signaling pathways that are independent of ER. Previous findings have demonstrated that overexpression of EGFR is common in ER(-) breast cancer [73], suggesting EGFR signaling may play important roles in regulating the cancer cell fate. Dominant-negative EGFR tumors present a strong reduction of VEGF expression and an increase in the apoptotic rate. And EGFR-dependent Ha-ras activation has a crucial role in VEGF expression, tumor angiogenesis, and growth [74]. The vital importance of EGFR in various cancer types has made it a key target for cancer treatment therapies, with the general goal being EGFR downregulation. Downregulation in EGFR expression after cannabinoid treatment has been previously reported in various cancer types. Anandamide induced cell death and decreased EGFR levels on LNCaP, DU145, and PC3 prostate cancer cells and inhibited the EGF-stimulated growth of the cells (Mimeault, 2003). Another cannabinoid, WIN55,212-2, was reported to induce growth inhibition of PDV.C57 epidermal tumor xenografts and decrease EGFR and phosphorylated EGFR [62].

30.8.2 Modulation of downstream signaling of EGFR and cannabinoid receptors

An increasing number of studies have demonstrated that p38-MAPK plays a crucial role in cannabinoid receptor downstream signaling, in which activation of p38-MAPK is associated with apoptosis induction. Cannabinoid-induced activation of p38 MAPK has been reported *in vitro* cell lines and in a number of other cancer types. Also, p38-MAPK-related apoptosis following cannabinoid treatment in various cancer cell types. In Jurkat human leukemia cells, Δ9-THC and JWH-133 treatment-induced CB2-mediated cell death and activation of p38-MAPK (Herrera, 2005). In mantle cell lymphoma cells, R(+)-methanandamide and WIN55,212-2 were reported to induce apoptosis via a sequence of events, including accumulation of *de novo* synthesized ceramide, activation of p38 MAPK and depolarization of the mitochondrial membrane [75]. Not only in cancer cells, cannabinoid-induced activation of p38 MAPK also occurs in central nervous system cells; however, the effect is mediated via CB1 cannabinoid receptor. In rat and mouse hippocampal slices, anandamide, 2-arachidonoylglycerol, WIN55,212-2, and Δ9-THC activated p38 MAPK via CB1 receptor, but the cannabinoids did not activate c-JNK, another mitogen-activated protein kinase [76].

The mechanisms on how activation of cannabinoid receptors leads to activation of p38 MAPK is still not fully understood. In some cell types, G protein-coupled receptors can stimulate p38 MAPK and JNK activities via the protein kinase C and the tyrosine kinase Src [76]. In other cell types in the hippocampus, using a specific inhibitor of the Src-family kinase, PP2, did not prevent the cannabinoid-induced activation of p38 MAPK, suggesting that activation of p38 MAPK by cannabinoids is independent of Src-family kinases in these cells [76]. Another mechanism for p38 MAPK activation is that CB1 activation stimulates PI3K [77], which in turn can be upstream of p38 MAPK in some certain cell types [78]. After being activated, the following targets of p38 MAPK are of Bcl-2 family proteins. Specifically, Cai (2006) reported that apoptosis of PC12 pheochromocytoma cells by sodium arsenite treatment may be due to direct phosphorylation of a Bcl-2 family protein, Bim, at Ser-65 by p38 MAPK [79]. Alternatively, p38 MAPK can also be upstream of caspase activation [80].

It is commonly implicated that ERK activation leads to cell proliferation. However, increasing evidence indicated that the biological effects of ERK activation depend on several factors, of which the duration of ERK activation acts as a key factor. Prolonged activation of ERK induces cell death via apoptosis rather than proliferation [40,81]. Mechanisms for cannabinoid-stimulated activation of ERK include prolonged accumulation of ceramide and/or inhibition of the adenylyl cyclase (AC)—PKA pathway. Activation of cannabinoid receptors led to two peaks of ceramide generation; the short-term peak is associated with sphingomyelin hydrolysis via sphingomyelinase, while the long-term is associated with palmitoyltransferase induction and enhanced ceramide synthesis *de novo* [77,82]. The second peak of ceramide accumulation is related to ERK activation and apoptosis through a mechanism, in which ceramide directly binds to the ceramide-binding motif of Raf-1 and results in Raf-1 activation. Another mechanism is that cannabinoids inhibit the AC/PKA pathway, which inhibits indirectly Raf-1. By preventing inhibitory effects of Raf-1, cannabinoids can indirectly activate MEK/ERK and consequently induce apoptosis [83]. Taken together, increasing evidence suggest that activation of MAPKs pathways is likely to play a crucial role in the tumor suppression effect of cannabinoids.

NF-κB is a key regulator for the genes involving in immune responses and antiapoptotic activities [84]. Importantly, NK-κB belongs to downstream signaling pathways of both cannabinoid receptors and EGFR. Activated NF-κB is found in many cancer types, and studies have shown that activation of NF-κB can initiate the expression of genes encoding antiapoptotic, angiogenesis, cell cycle regulatory, and growth factors, which promote the formation and development of malignant tumors. Conversely, inhibition of NF-κB results in increased apoptotic activities and cell cycle arrest, consequently inducing tumor regression. Therefore, NF-κB is an important target for cancer treatment therapies. NF-κB inhibition by cannabinoids and other compounds has been reported in a number of *in vitro* studies and with other cancer types. The inhibition of NF-κB in the tumors may be partly responsible for the tumor regression observed in mice treated with WIN55,212-2. Similarly, AS602868, a specific inhibitor of IKK2, blocked NF-κB activation and led to apoptosis in human primary acute myeloid leukemia cells [85]. Nonspecific inhibitors of NF-κB such as anti-inflammatory agents and non-steroidal anti-inflammatory drugs mediate the regression of adenomatous polyps of the colon, prevent the development of colon cancer, and increase cancer cell apoptosis in existing malignant tumors [86].

In A549 lung adenocarcinoma epithelial cells, anandamide was found to inhibit TNFα-induced NF-κB activation by direct inhibition of the IκB kinase (IKK) β and the IKKα subunits of κB inhibitor (IκB) kinase complex [87]. To determine the involvement of cannabinoid receptors, Sancho (2003) utilized A549 and 5.1 cell lines expressing only CB1 and CB2 cannabinoid receptors, respectively. The results showed that anandamide-induced NF-κB inhibition was independent of both CB1 and CB2 cannabinoid receptors. WIN,55212-induced downregulation of NF-κB expression suggests a potential of the compound to be used as an adjuvant treatment as NF-κB activation may induce the expression of the multidrug resistance P-glycoprotein, and inhibition of NF-κB has shown to increase the apoptotic response to chemotherapy and radiation therapy [88]. Also, mere inhibition of NF-κB may be insufficient for a pronounced apoptotic response. Therefore, combinations of NF-κB inhibitors and conventional chemoradio therapies may enhance the effectiveness of the treatment and reduce the risk for drug resistance. WIN55,212-2 inhibited the expression of NF-κB *in vivo*. Therefore, it would be interesting to further evaluate WIN55,212-2 as a potential adjuvant treatment for TNBC by combining WIN55,212-2 treatment in conjunction with cytotoxic agents such as anthracyclines or platinum compounds, or with radiation therapy to determine whether co-treatment with WIN55,212-2 can increase the sensitivity of TNBC cancer cells to chemoradio therapies, thereby enabling decreased doses.

Besides the MAPKs pathways, the survival PI3K/Akt/mTOR pathway is also involved in downstream signaling of both cannabinoid receptors and EGFR. It has been well established that the PI3K/Akt/mTOR pathway plays a pivotal role in cellular survival processes, including cell growth, proliferation, invasion, and migration. Common results from previous studies indicated that the inhibitory effects of cannabinoids in cancer cells were frequently associated with Akt inhibition. Conversely, sustained activation of Akt was related to the protective effects of cannabinoids in neuronal cells. Therefore, PI3K/Akt/mTOR pathway is considered an important target for novel treatments. Cannabinoid-induced inhibition of Akt was found to result from *de novo* synthesis of ceramide. In glioma cells, cannabinoids induce intracellular ceramide accumulation by mechanisms of sphingomyelin hydrolysis and ceramide synthesis *de novo* [54,89]. In turn, *de novo* synthesized ceramide leads to activation of ERK and inhibition of Akt. Blockade of the synthesis *de novo* of ceramide by L-cycloserine prevented THC-induced ERK activation and THC-induced Akt inhibition [77]. In addition to Akt inhibition, Akt activation by cannabinoid treatment was also reported. Lung carcinoma and glioblastoma cells treated with either THC, WIN55,212-2, or HU-210 showed activation of both ERK and Akt. The activating effect was abolished by blockade of EGFR signal transactivation with the selective EGFR inhibitor AG1478 or the metalloprotease inhibitor BB94. This suggested the cannabinoid-induced activation of ERK and Akt was dependent on EGFR function [90]. Co-activation of ERK and Akt following cannabinoid treatment was reported to protect astrocytes from ceramide-induced apoptosis [77].

WIN55,212-2 treatment of LNCaP prostate cancer cells resulted in an induction of p27 and down-regulation of cyclins D1, D2, E [49]. In a study by Portella et al. (2003), nude mice inoculated with K-ras-transformed FRTL-5 cells and peritumorally treated with 2-methyl-arachidonyl-2-fluoro-ethylamide (Met-F-AEA) showed a 50% reduction in tumor volume compared to vehicle-treated mice [61]. Met-F-AEA also resulted in upregulation of p27(kip1) by 1.6-fold, and this upregulation was attenuated by the selective CB1 receptor antagonist, SR141716A, suggesting the cannabinoid-induced upregulation of p27(kip1) was mediated at least in part via the CB1 receptor.

30.9 Mixed actions of cannabinoids in other cancer types

A number of studies also have reported that cannabinoids increased the proliferation of several cancer cell types. One possible reason could be that cannabinoids may develop different biological effects on different cell types and their expression levels of cannabinoid receptors. Cannabinoids were found to have protective effects toward cultured neurons against excitotoxicity. In contrast, studies also demonstrated cytotoxicity effects of cannabinoids on various cancer cells. Furthermore, even to the same cell type, cannabinoids have also been reported to develop different effects. Delta-9-THC induced hippocampal neuron death through neuronal apoptosis mechanism [91]; the compound was also reported to protect spinal neurons from excitotoxicity produced by kainate [92]. Not only depending on cell types, there also exists evidence that cannabinoids may have biological actions in different directions at different ranges of concentrations. Hart et al. reported that THC only induced apoptosis in cancer cells at relatively high concentrations. In contrast, nanomolecular concentrations of THC accelerate the proliferation of the cancer cells in an EGFR- and metalloprotease-dependent mechanism [93]. The difference is important in regard to clinical relevance as after oral or rectal administration of THC or its derivatives, the maximum serum concentrations of THC were only 35–350 nM [94,95].

Evidences suggest that some cannabinoids can trigger the proliferation of cancer cells at suitably low concentrations. In a study by Sanchez *et al.*, THC and R-/(+)-methanandamide (MET) at nanomolar concentrations induced accelerated proliferation of PC-3 prostate cancer cells. In addition, the stimulation is associated with cannabinoid-induced activation of PI3K cascade and NGF synthesis, a neurotrophic factor previously reported to involve in prostate cells' proliferation [96]. In other cell types of glioblastoma and lung carcinoma, THC at nanomolar concentrations also accelerate the proliferation of cancer cells and this effect was mediated by EGFR activity in a mechanism where cannabinoid-induced EGFR transactivation was mediated via metalloprotease and tumor necrosis factor α-converting enzyme (TACE/ADAM17) [93]. JWH133 with the chemical name 3-(1′,1′-Dimethylbutyl)-1-deoxy-Δ8-THC possesses a molecular structure that is remarkably similar to that of Δ9-THC [97,98]. It, therefore, may potentially possess the similar ability to induce accelerated proliferation in cancer cells, which was observed with Δ9-THC. Another possible mechanism for cannabinoid stimulation effect on tumor growth is that cannabinoids promote the formation of new blood vessels, which better nourish tumor cells and make the tumor grow faster. It was previously reported that THC induced activation of NGF synthesis [96], and in a study by Romon, recombinant NGF and NGF produced by breast cancer cells promote breast cancer angiogenesis and endothelial cell invasion [99]. NGF increased the secretion of VEGF in both endothelial and breast cancer cell. NGF also leads to the activation of the PI3K/Akt pathway, which was previously reported to be also activated by THC [96,100]. In another aspect, it has been established that EGFR activation may stimulate invasion, angiogenesis, and metastasis of cancer tumors [as reviewed by 101]. This may explain why cetuximab, an EGFR monoclonal antibody, is found more effective *in vivo*, where the drug can develop its effects toward invasion, angiogenesis, and metastasis, than *in vitro* [102]. THC was found to lead to EGFR transactivation and activation of ERK and Akt/PKB survival pathways in an EGFR-dependent manner [93]. Therefore, further models for assessing the possibility of cannabinoids to promote cancer growth are needed.

References

[1]   C Fitzmaurice, et al., Global, regional, and national cancer incidence, mortality, years of life lost, years lived with disability, and disability-adjusted life-years for 29 cancer groups, 1990 to 2017: a systematic analysis for the global burden of disease study, JAMA Oncol 5 (12) (2019) 1749–1768.

[2]   H Sung, et al., Global cancer statistics 2020: GLOBOCAN estimates of incidence and mortality world-wide for 36 cancers in 185 countries, CA: Cancer Journal for Clinicians 71 (3) (2021) 209–249.

[3]   A Jemal, et al., Cancer statistics, 2007, CA Cancer J Clin 57 (1) (2007) 43–66.

[4]   A Coldman, et al., Pan-Canadian study of mammography screening and mortality from breast cancer, J Natl Cancer Inst 106 (11) (2014) 1–7.

[5]   D Roder, et al., Population screening and intensity of screening are associated with reduced breast cancer mortality: evidence of efficacy of mammography screening in Australia, Breast Cancer Res Treat 108 (3) (2008) 409–416.

[6]   J Giese-Davis, et al., Decrease in depression symptoms is associated with longer survival in patients with metastatic breast cancer: a secondary analysis, J Clin Oncol 29 (4) (2011) 413–420.

[7]   A Mehnert, U Koch, Prevalence of acute and post-traumatic stress disorder and comorbid mental disorders in breast cancer patients during primary cancer care: a prospective study, Psychooncology 16 (3) (2007) 181–188.

[8]   M Lidgren, et al., Health related quality of life in different states of breast cancer, Qual Life Res 16 (6) (2007) 1073–1081.

[9]   TS Foster, et al., The economic burden of metastatic breast cancer: a systematic review of literature from developed countries, Cancer Treat Rev 37 (6) (2011) 405–415.

[10]   WD Foulkes, IE Smith, JS Reis-Filho, Triple-negative breast cancer, N Engl J Med 363 (20) (2010) 1938–1948.

[11]   S Cleator, W Heller, RC Coombes, Triple-negative breast cancer: therapeutic options, Lancet Oncol 8 (3) (2007) 235–244.

[12]   KR Bauer, et al., Descriptive analysis of estrogen receptor (ER)-negative, progesterone receptor (PR)-negative, and HER2-negative invasive breast cancer, the so-called triple-negative phenotype, Cancer 109 (9) (2007) 1721–1728.

[13]   VW Setiawan, et al., Breast cancer risk factors defined by estrogen and progesterone receptor status: the multiethnic cohort study, Am J Epidemiol 169 (10) (2009) 1251–1259.

[14]   M Dowsett, AK Dunbier, Emerging biomarkers and new understanding of traditional markers in personalized therapy for breast cancer, Clin Cancer Res 14 (24) (2008) 8019–8026.

[15]   WL McGuire, Steroid receptors in human breast cancer, Cancer Res 38 (11 Part 2) (1978) 4289–4291.

[16]   GDL Phillips, et al., Targeting HER2-positive breast cancer with trastuzumab-DM1, an antibody–cytotoxic drug conjugate, Cancer Res 68 (22) (2008) 9280–9290.

[17]   R Dent, et al., Triple-negative breast cancer: clinical features and patterns of recurrence, Clin Cancer Res 13 (15) (2007) 4429–4434.

[18]   LA Carey, et al., Race, breast cancer subtypes, and survival in the Carolina Breast Cancer Study, JAMA 295 (21) (2006) 2492–2502.

[19]   EA Rakha, et al., Prognostic markers in triple-negative breast cancer, Cancer 109 (1) (2007) 25–32.

[20]   T Asmis, et al., Comorbidity, age and overall survival in cetuximab-treated patients with advanced colorectal cancer (ACRC)—results from NCIC CTG CO. 17: a phase III trial of cetuximab versus best supportive care, Ann Oncol 22 (1) (2011) 118–126.

[21]   B Gerber, et al., Effect of luteinizing hormone–releasing hormone agonist on ovarian function after modern adjuvant breast cancer chemotherapy: the GBG 37 ZORO study, J Clin Oncol 29 (17) (2011) 2334–2341.

[22]   J Gertsch, RG Pertwee, V Di Marzo, Phytocannabinoids beyond the Cannabis plant–do they exist? Br J Pharmacol 160 (3) (2010) 523–529.

[23]   A Howlett, et al., International union of pharmacology. XXVII. Classification of cannabinoid receptors, Pharmacol Rev 54 (2) (2002) 161–202.

[24]   D Piomelli, The molecular logic of endocannabinoid signalling, Nat Rev Neurosci 4 (11) (2003) 873–884.

[25]   P Morales, DP Hurst, PH Reggio, Molecular targets of the phytocannabinoids: a complex picture, Phyto-cannabinoids (2017) 103–131.

[26]   V Di Marzo, The endocannabinoid system in obesity and type 2 diabetes, Diabetologia 51 (8) (2008) 1356–1367.

[27]   DR Compton, et al., Pharmacological profile of a series of bicyclic cannabinoid analogs: classification as cannabimimetic agents, J Pharmacol Exp Ther 260 (1) (1992) 201–209.

[28]   MP Davis, Oral nabilone capsules in the treatment of chemotherapy-induced nausea and vomiting and pain, Expert Opin Investig Drugs 17 (1) (2008) 85–95.

[29]   JW Huffman, et al., Structure–activity relationships for 1-alkyl-3-(1-naphthoyl) indoles at the cannabinoid CB1 and CB2 receptors: steric and electronic effects of naphthoyl substituents. New highly selective CB2 receptor agonists, Bioorg Med Chem 13 (1) (2005) 89–112.

[30]   JW Huffman, et al., Synthesis and pharmacology of 1-deoxy analogs of CP-47,497 and CP-55,940, Bioorg Med Chem 16 (1) (2008) 322–335.

[31]   GA Thakur, et al., Methods for the Synthesis of Cannabinergic Ligands, Marijuana and Cannabinoid Research, Springer, New York, 2006, pp. 113–148.

[32]   LA Matsuda, et al., Structure of a cannabinoid receptor and functional expression of the cloned cDNA, Nature 346 (6284) (1990) 561–564.

[33]   S Munro, KL Thomas, M Abu-Shaar, Molecular characterization of a peripheral receptor for cannabinoids, Nature 365 (6441) (1993) 61–65.

[34]   K Tsou, et al., Cannabinoid CB1 receptors are localized primarily on cholecystokinin-containing GABAergic interneurons in the rat hippocampal formation, Neuroscience 93 (3) (1999) 969–975.

[35]   D Shire, et al., Molecular cloning, expression and function of the murine CB2 peripheral cannabinoid receptor, Biochimica et Biophysica Acta (BBA)-Gene Structure and Expression 1307 (2) (1996) 132–136.

[36]   S Galiègue, et al., Expression of central and peripheral cannabinoid receptors in human immune tissues and leukocyte subpopulations, Eur J Biochem 232 (1) (1995) 54–61.

[37]   M Holland, et al., Cannabinoid CB1 receptors fail to cause relaxation, but couple via Gi/Go to the inhibition of adenylyl cyclase in carotid artery smooth muscle, Br J Pharmacol 128 (3) (1999) 597–604.

[38]   S Oesch, et al., Cannabinoid receptor 1 is a potential drug target for treatment of translocation-positive rhabdomyosarcoma, Mol Cancer Ther 8 (7) (2009) 1838–1845.

[39]   M Guzman, Effects on cell viability, Cannabinoids (2005) 627–642.

[40]   G Velasco, C Sánchez, M Guzmán, Towards the use of cannabinoids as antitumour agents, Nat Rev Cancer 12 (6) (2012) 436–444.

[41]   CC Felder, et al., Comparison of the pharmacology and signal transduction of the human cannabinoid CB1 and CB2 receptors, Mol Pharmacol 48 (3) (1995) 443–450.

[42]   W Twitchell, S Brown, K Mackie, Cannabinoids inhibit N-and P/Q-type calcium channels in cultured rat hippocampal neurons, J Neurophysiol 78 (1) (1997) 43–50.

[43]   DJ Henry, C Chavkin, Activation of inwardly rectifying potassium channels (GIRK1) by co-expressed rat brain cannabinoid receptors in Xenopus oocytes, Neurosci Lett 186 (2-3) (1995) 91–94.

[44]   Z Qamri, et al., Synthetic cannabinoid receptor agonists inhibit tumor growth and metastasis of breast cancer, Mol Cancer Ther 8 (11) (2009) 3117–3129.

[45]   MM Caffarel, et al., Δ9-tetrahydrocannabinol inhibits cell cycle progression in human breast cancer cells through Cdc2 regulation, Cancer Res 66 (13) (2006) 6615–6621.

[46]   A Ligresti, et al., New potent and selective inhibitors of anandamide reuptake with antispastic activity in a mouse model of multiple sclerosis, Br J Pharmacol 147 (1) (2006) 83–91.

[47]   A Ligresti, et al., Antitumor activity of plant cannabinoids with emphasis on the effect of cannabidiol on human breast carcinoma, J Pharmacol Exp Ther 318 (3) (2006) 1375–1387.

[48]   I Galve-Roperh, et al., Mechanism of extracellular signal-regulated kinase activation by the CB1 cannabinoid receptor, Mol Pharmacol 62 (6) (2002) 1385–1392.

[49]   S Sarfaraz, et al., Cannabinoid receptor agonist-induced apoptosis of human prostate cancer cells LNCaP proceeds through sustained activation of ERK1/2 leading to G1 cell cycle arrest, J Biol Chem 281 (51) (2006) 39480–39491.

[50]   H Mukhtar, F Afaq, S Sarfaraz, Cannabinoid Receptors: A Novel Target for Therapy for Prostate Cancer, University of Wisconsin-Madison, 2008.

[51]   A Preet, et al., Cannabinoid receptors, CB1 and CB2, as novel targets for inhibition of non–small cell lung cancer growth and metastasis, Cancer Prev Res 4 (1) (2011) 65–75.

[52]   N Olea-Herrero, et al., The cannabinoid JWH-015 activates NF:B in prostate cancer PC-3 cells: involvement of CB2 and PI3K/Akt, F1000 Research 2 (2011) 1.

[53]   K Kitatani, J Idkowiak-Baldys, YA Hannun, The sphingolipid salvage pathway in ceramide metabolism and signaling, Cell Signal 20 (6) (2008) 1010–1018.

[54]   G Velasco, et al., Cannabinoids and ceramide: two lipids acting hand-by-hand, Life Sci 77 (14) (2005) 1723–1731.

[55]   M Guzmán, C Sánchez, I Galve-Roperh, Control of the cell survival/death decision by cannabinoids, J Mol Med 78 (11) (2001) 613–625.

[56]   A Carracedo, et al., Cannabinoids induce apoptosis of pancreatic tumor cells via endoplasmic reticulum stress–related genes, Cancer Res 66 (13) (2006) 6748–6755.

[57]   M Bifulco, et al., Control by the endogenous cannabinoid system of RAS oncogene-dependent tumor growth, FASEB J 15 (14) (2001) 1–17.

[58]   JM Park, et al., Antiproliferative mechanism of a cannabinoid agonist by cell cycle arrest in human gastric cancer cells, J Cell Biochem 112 (4) (2011) 1192–1205.

[59]   JA McCubrey, et al., Roles of the Raf/MEK/ERK pathway in cell growth, malignant transformation and drug resistance, Biochim Biophys Acta 1773 (8) (2007) 1263–1284.

[60]   J-i Okano, AK Rustgi, Paclitaxel induces prolonged activation of the Ras/MEK/ERK pathway independently of activating the programmed cell death machinery, J Biol Chem 276 (22) (2001) 19555–19564.

[61]   G Portella, et al., Inhibitory effects of cannabinoid CB1 receptor stimulation on tumor growth and metastatic spreading: actions on signals involved in angiogenesis and metastasis, FASEB J 17 (12) (2003) 1771–1773.

[62]   ML Casanova, et al., Inhibition of skin tumor growth and angiogenesis in vivo by activation of cannabinoid receptors, J Clin Invest 111 (1) (2003) 43–50.

[63]   F Fan, et al., Cannabinoid receptor down-regulation without alteration of the inhibitory effect of CP 55,940 on adenylyl cyclase in the cerebellum of CP 55,940-tolerant mice, Brain Res 706 (1) (1996) 13–20.

[64]   L Nong, et al., Altered cannabinoid receptor mRNA expression in peripheral blood mononuclear cells from marijuana smokers, J Neuroimmunol 127 (1-2) (2002) 169–176.

[65]   L Martini, et al., Ligand-induced down-regulation of the cannabinoid 1 receptor is mediated by the G-protein-coupled receptor-associated sorting protein GASP1, FASEB J 21 (3) (2007) 802–811.

[66]   J Romero, et al., Time-course of the cannabinoid receptor down-regulation in the adult rat brain caused by repeated exposure to Δ9-tetrahydrocannabinol, Synapse 30 (3) (1998) 298–308.

[67]   C Leterrier, et al., Constitutive endocytic cycle of the CB1 cannabinoid receptor, J Biol Chem 279 (34) (2004) 36013–36021.

[68]   DF Wu, et al., Role of receptor internalization in the agonist-induced desensitization of cannabinoid type 1 receptors, J Neurochem 104 (4) (2008) 1132–1143.

[69]   R Rozenfeld, Type I cannabinoid receptor trafficking: all roads lead to lysosome, Traffic 12 (1) (2011) 12–18.

[70]   NL Grimsey, et al., Cannabinoid receptor 1 trafficking and the role of the intracellular pool: implications for therapeutics, Biochem Pharmacol 80 (7) (2010) 1050–1062.

[71]   NL Grimsey, et al., Cannabinoid receptor 2 undergoes Rab5-mediated internalization and recycles via a Rab11-dependent pathway, Biochim Biophys Acta 1813 (8) (2011) 1554–1560.

[72]   A Tappe-Theodor, et al., A molecular basis of analgesic tolerance to cannabinoids, J Neurosci 27 (15) (2007) 4165–4177.

[73] DK Biswas, et al., Epidermal growth factor-induced nuclear factor κB activation: a major pathway of cell-cycle progression in estrogen-receptor negative breast cancer cells, Proc Natl Acad Sci 97 (15) (2000) 8542–8547.

[74] F Roch, G Jiménez, J Casanova, EGFR signalling inhibits Capicua-dependent repression during specification of Drosophila wing veins, Development 129 (4) (2002) 993–1002.

[75] K Gustafsson, et al., Cannabinoid receptor-mediated apoptosis induced by R (+)-methanandamide and Win55, 212-2 is associated with ceramide accumulation and p38 activation in mantle cell lymphoma, Mol Pharmacol 70 (5) (2006) 1612–1620.

[76] P Derkinderen, et al., Cannabinoids activate p38 mitogen-activated protein kinases through CB1 receptors in hippocampus, J Neurochem 77 (3) (2001) 957–960.

[77] T Gómez del Pulgar, G Velasco, M Guzman, The CB1 cannabinoid receptor is coupled to the activation of protein kinase B/Akt, Biochem J 347 (2) (2000) 369–373.

[78] IA Yamboliev, et al., Evidence for modulation of smooth muscle force by the p38 MAP kinase/HSP27 pathway, American Journal of Physiology-Heart and Circulatory Physiology 278 (6) (2000) H1899–H1907.

[79] T Cai, et al., Manganese induces the overexpression of α-synuclein in PC12 cells via ERK activation, Brain Res 1359 (2010) 201–207.

[80] J Kralova, et al., p38 MAPK plays an essential role in apoptosis induced by photoactivation of a novel ethylene glycol porphyrin derivative, Oncogene 27 (21) (2008) 3010–3020.

[81] M Guzman, Cannabinoids: potential anticancer agents, Nat Rev Cancer 3 (10) (2003) 745–755.

[82] RN Kolesnick, M Krönke, Regulation of ceramide production and apoptosis, Annu Rev Physiol 60 (1) (1998) 643–665.

[83] I Galve-Roperh, et al., Anti-tumoral action of cannabinoids: involvement of sustained ceramide accumulation and extracellular signal-regulated kinase activation, Nat Med 6 (3) (2000) 313–319.

[84] N Li, M Karin, Signaling pathways leading to nuclear factor-κB activation, Methods Enzymol 319 (2000) 273–279.

[85] C Frelin, et al., Targeting NF-κB activation via pharmacologic inhibition of IKK2-induced apoptosis of human acute myeloid leukemia cells, Blood 105 (2) (2005) 804–811.

[86] Y Yamamoto, RB Gaynor, Role of the NF-kB pathway in the pathogenesis of human disease states, Curr Mol Med 1 (3) (2001) 287–296.

[87] Ro Sancho, et al., Anandamide inhibits nuclear factor-κB activation through a cannabinoid receptor-independent pathway, Mol Pharmacol 63 (2) (2003) 429–438.

[88] X Dolcet, et al., NF-kB in development and progression of human cancer, Virchows Arch 446 (5) (2005) 475–482.

[89] A Ellert-Miklaszewska, B Kaminska, L Konarska, Cannabinoids down-regulate PI3K/Akt and Erk signalling pathways and activate proapoptotic function of Bad protein, Cell Signal 17 (1) (2005) 25–37.

[90] S Hart, OM Fischer, A Ullrich, Cannabinoids induce cancer cell proliferation via tumor necrosis factor α-converting enzyme (TACE/ADAM17)-mediated transactivation of the epidermal growth factor receptor, Cancer Res 64 (6) (2004) 1943–1950.

[91] GC-K Chan, TR Hinds, S Impey, DR Storm, Hippocampal neurotoxicity of Δ9-tetrahydrocannabinol, J Neurosci 18 (14) (1998) 5322–5332.

[92] ME Abood, G Rizvi, N Sallapudi, SD McAllister, Activation of the CB1 cannabinoid receptor protects cultured mouse spinal neurons against excitotoxicity, Neurosci Lett 309 (3) (2001) 197–201.

[93] OMF Stefan Hart, A Ullrich, Cannabinoids induce cancer cell proliferation via tumor necrosis factor α-converting enzyme (TACE/ADAM17)-mediated transactivation of the epidermal growth factor receptor, Cancer Res 64 (6) (2004) 1943–1950.

[94] R Brenneisen, A Egli, MA Elsohly, V Henn, Y Spiess, The effect of orally and rectally administered delta 9-tetrahydrocannabinol on spasticity: a pilot study with 2 patients, Int J Clin Pharmacol Ther 34 (10) (1996) 446–452.

[95]   SJ Heishman, HJE H.M., EJ Cone, Acute and residual effects of marijuana: profiles of plasma THC levels, physiological, subjective, and performance measures, Pharmacol Biochem Behav 37 (3) (1990) 561–565.

[96]   G Maria, LR-L Sanchez, AM Sánchez, Inés Diáz-Laviada, activation of phosphoinositide 3-kinase/PKB pathway by CB1 and CB2 cannabinoid receptors expressed in prostate PC-3 cells. Involvement in Raf-1 stimulation and NGF induction, Cell Signal 15 (9) (2003) 851–859.

[97]   JW Huffman, CB2 receptor ligands, Mini Rev Med Chem 5 (7) (2005) 641–649.

[98]   KS Marriott, JW Huffman, Recent advances in the development of selective ligands for the cannabinoid CB(2) receptor, Curr Top Med Chem 8 (3) (2008) 187–204.

[99]   R Romon, et al., Nerve growth factor promotes breast cancer angiogenesis by activating multiple pathways, Mol Cancer 9 (1) (2010) 157.

[100] LF Reichardt, Neurotrophin-regulated signalling pathways, Philosophical Transactions of the Royal Society B: Biological Sciences 361 (1473) (2006) 1545–1564.

[101] DO Annette K Larsen, KEl Ouadrani, A Petitpre, Targeting EGFR and VEGF(R) pathway cross-talk in tumor survival and angiogenesis, Pharmacol Ther 131 (1) (2011) 80–90.

[102] Baselga, The EGFR as a target for anticancer therapy—focus on cetuximab, Eur J Cancer 37 (Supplement 4(0)) (2001) 16–22.

PART 9

Concern

A sound comprehension of molecular biology and relevant biotechnology is a prerequisite for research on the molecular mechanisms of traditional Chinese medicine, mainly medicinal herbs

Zhengqi Liu[a], Tao Xu[b], Li Peng[b], Zhaozhao Hua[c], Jian Li[d], Zhigang Jiang[e] and Dezhong Joshua Liao[a]

[a]Office of Research and Education Administration, The Second Hospital, Guizhou University of Traditional Chinese Medicine, Guiyang, Guizhou Province, China [b]Department of Cardiology, The Second Hospital, Guizhou University of Traditional Chinese Medicine, Guiyang, Guizhou Province, China [c]Department of Obstetrics, The Second Hospital, Guizhou University of Traditional Chinese Medicine, Guiyang, Guizhou Province, China [d]Department of Oncology, The Second Hospital, Guizhou University of Traditional Chinese Medicine, Guiyang, Guizhou Province, China [e]School of Public Health Sciences, Zunyi Medical University, Zunyi City, Guizhou Province, China

31.1 Introduction

Medical practice basically consists of two parts, that is, diagnosis and treatment. Either of the two, especially the diagnosis part, is established on complex theories with the accumulation of not only profuse but also profound knowledge. Virtually, in all schools of medicine in the world, the theory about disease treatments was initially derived from the lore of common people and a long folk experience in using herbs to treat various illnesses, as centuries ago chemistry was not sophisticated enough for isolation and purification of chemicals from plants or other raw materials, and certainly not for chemical synthesis. Therefore, centuries ago, medicines used for disease treatments differed not a whit among China, Japan, India, and western countries. Traditional Chinese medicine (TCM) and western medicine (WM) split mainly over the theory on how diseases are developed, which is the bedrock of diagnosis. Ever since its inception over 2000 years ago, TCM has had its school of theory centralizing on "meridians" and "five visceral organs plus six hollow organs"; even now, it is still impossible to correctly describe this theory in the WM language. The "meridians" are not those marking geographical locations on the earth but are somewhat a system of dozens of channels that are somewhat related to, but still quite separate from, distributions of nerves and circulations of blood and lymphatic vessels. The "five visceral organs," although named as the liver, spleen, heart, lung, and kidney as in the WM, differ greatly from the same organs described in the textbooks of anatomy and physiology. For instance, the "spleen" in the school of TCM theory controls largely food digestion, whereas in WM food digestion is carried out by the digestive system that includes the liver while the spleen is mainly involved in immune functions. The "spleen" defined by TCM is indispensable for a person's life, but chirurgical removal of the spleen defined by WM is not life-threatening. Probably few, if any, physicians other than TCM practitioners can comprehend the reference listed in the Pubmed entitled "Narrative review of the mechanisms of action of dachengqi decoction in the treatment of hyperlipidemic pancreatitis on six-hollow-organs to be unblocked theory" [1], because the "six-hollow-organs to be unblocked theory" is unfathomable or opaque to them. This exemplifies the insuperable difficulties that interdict WM practitioners to learn and practice TCM.

Two basic facts about TCM have been well accepted by most WM practitioners in the world. One is that TCM is indeed effective and thus useful in treating many medical problems, and the other is that the effectiveness and adversity of many, if not the most, TCM remedies or therapeutical approaches are largely empirical and have not yet received unerring evidence from scrupulously designed laboratory studies and clinical trials, as has been summarized by Fung et al. [2]. Actually, there is not one iota of tenable evidence from clinical trials for many herb remedies. Realizing these deficiencies of TCM, the Chinese government has in recent decades injected a colossal amount of funding into TCM research and clinical trials. In the research quarter, a significant portion of the funding goes into studies on scouting out the molecular mechanisms behind remedies or other therapeutic approaches of TCM, such as acupuncture, in an attempt to translate the TCM theory into the language of WM or to find its WM equivalent [3]. This "translational research" is an obvious imperative for global promulgation of TCM so that WM practitioners can fathom how and why a particular TCM remedy or approach works for a particular disease. However, the current status of this line of "translational research" has multiple and various problems. A particular one is germane to the studies on delving into the molecular mechanisms of how TCM remedies or approaches work via certain particular cascades of molecular changes, which is called "molecular signaling pathway or transduction" in a jargon of molecular medicine and molecular biology that are key elements of modern WM.

31.2 There are unsolved problems in translation of traditional Chinese medicine theory into the western medicine language

In TCM, diagnoses of diseases are made based on information garnered using four approaches dubbed as "look, smell, ask, and feel," respectively. "Look" means looking at the appearance of the patient, including his tongue. "Smell" usually means to determine whether the patient releases any weird odor or sound since the Chinese word "smell" means not only "smelling" but also "listening." "Ask" is to enquire about the patient's symptoms and history of illness. "Feel" is to feel the patient's pulse, as TCM divides pulse style into quite a few types that are fathomable only to TCM practitioners. A conflation of these four lines of information leads to the establishment of TCM theory that centralizes on the aforementioned "meridians" and "five visceral organs plus six hollow organs." However, TCM so established faces a question as to whether TCM, both in its theory and in its clinical practice, needs to be advanced along with the advancement of science and technology achieved in the past thousand years. Many, probably most, of today's TCM practitioners answer this question with yes by making diagnoses heavily based on data from various clinical laboratories and sophisticated equipment, including computed tomography and nuclear magnetic resonance. As the sequel, they make two separate sets of diagnosis and associated description, one in TCM language and appellation and the other in WM language and terminology, such as "prostate cancer," although there is neither "prostate" nor "cancer" in the lexicon of TCM. It resembles a situation of speaking in Chinese and then in English anew, leaving the listeners wondering whether the version in their mother tongue really matches the one in the foreign language. Today, probably very few TCM physicians in comprehensive hospitals in China make diagnoses solely based on the aforementioned four lines of information without assistance from data from clinical laboratories and sophisticated instruments. Coupling modern laboratory data with TCM theory and making diagnoses in WM terminology are obviously important for translating TCM into the WM language so that all medical practitioners in the world can comprehend. However, these efforts have encountered variegated problems, with those listed below as examples:

1. Translation of TCM into the WM language requires the availability of molecular mechanisms of TCM, and thus requires research into TCM mechanisms using the same approaches as used for WM research. Research in WM employs various animal species at different evolutionary levels, some being mammals and some others being invertebrates. However, most animal models still more or less differ from the corresponding diseases in the human, although researchers often tout the strengths of their models. These animal models are frequently used in TCM research as well, but TCM is not built on the anatomy, physiology, anatomic pathology, and pathophysiology that are the bedrock of WM. Actually, even a fundamental question remains unanswered as to whether these animal species have "meridians" and, if they have, whether the "meridians" in these evolutionarily lower species are selfsame to their human counterparts. For instance, some meridians are somewhat hand specific (such as "shou-shao-ying-xin" meridian) while some others are foot specific (such as the "zu-yang-ming-wei" meridian). Has this issue of hand or foot specificity been solved in mice and rats that have no hands but four feet? This and other species disparities alienate animal models from human diseases diagnosed based on TCM.

2. The aforementioned four lines of information lack objective criteria, cannot be quantitatively measured for statistical comparison and, even worse, are collected mainly based on the personal experience of individual physicians. Therefore, unlike WM that can be quickly promulgated by medical school education, TCM meets insurmountable hurdles in standardization and dissemination even via schooling, because personal experience is hard to pass from teachers to students.

3. While translating TCM theory into the WM language is hard, matching laboratory data, such as blood cell counts, ultrasound data, computed tomography or nuclear magnetic resonance images, etc., with TCM concepts is even harder, if not insurmountable.

4. Studies on WM mechanisms, either for how a disease is developed or for how a treatment works or causes side-effects, have gone into the molecular level that is overarched by molecular biology, mainly gene-related knowledge with some of the aspects enlarged below. However, TCM goes only to the "organ" (such as the liver, heart, or kidney by its own definition) level without counterparts at the gene level, which makes it difficult to integrate molecular data to TCM theory, as elaborated below.

31.3 One gene has many RNA and protein products

Traditional biology has a concept that nuclear DNA harbors genes that are transcribed to mRNAs. The middle region of an mRNA, incepting from a translation start codon and ending at a translation stop codon, is coined as an open reading frame or a cistron for translation to a protein. However, it is now known that less than 2% of the human genome is assigned to open reading frames coding for proteins, although virtually the whole genome in a human cell, which is composed of 3.2–3.5 billion nucleotides, is transcribed. The vast majority of the remaining 98% of the genomic sequence that is noncoding harbors transcriptional promoters or enhancers or encodes a multitudinous number of regulatory RNAs, all of which regulate directly or indirectly the production of proteins. The human genome is currently considered to contain only about 20,000 protein-coding genes [4–7]; this number is obviously too small to explain so diversified an assortment of cellular functions as are in a human body. The explanation for this discrepancy, in our opinion, is that "gene" should not be defined at the DNA level, but should be at the RNA level, meaning that each mRNA or long noncoding regulatory RNA should be regarded as one gene [8–11]. This idea is derived from the finding that one genomic DNA locus that is canonically defined as one gene can produce different mRNA variants, long noncoding RNA variants, PIWI-interacting RNAs (piRNA), extracellular RNAs (exRNA), antisense RNAs, etc., besides many small regulatory RNAs like microRNAs (miRNA) and small interfering RNAs (siRNAs), as illustrated in Fig. 31.1 [8,9]. The Titin gene expressed in the skeletal muscle is the best example, expressed in the human to over one-million mRNA variants mainly via complicated alternative-splicing of its 363 exons [12]. This new definition of "gene," which awaits consensus among molecular

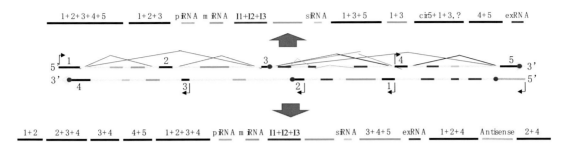

Figure 31.1

An oversimplified graphic illustration of RNA production from a canonically defined gene. A gene is hypothetically located on the plus (Watson) strand (the long light-grey line with the 5'-end at the left) of the DNA double helix. Its transcription may produce a long transcript with five exons (the five short thick black lines that are numbered) or, alternatively, may stop at the exon 3 to produce a short transcript. Red dots indicate transcription termination sites. The long transcript may be spliced to an mRNA with all five exons (usually annotated as the wild type), an mRNA with only exons 1, 3, and 5, a noncoding RNA with exons 1 and 3 (the shorter green line on the top), and/or probably even a circular RNA with exons 5, 1, and 3. The shorter transcript may be spliced to an mRNA with all three exons. The intron 3 of this gene harbors another gene that contains three exons dubbed as I1, I2, and I3 (the three thick red lines between exons 3 and 4), the transcript of which is spliced to an mRNA with all three exons (red line on the top). Intron 1 encodes a piRNA (the thick blue line) and a miRNA (the thick dark-yellow line), intron 2 encodes a single-exon long noncoding RNA (the thick green line between exons 2 and 3), and intron 4 encodes an siRNA (the thick light-yellow line) and an exRNA (the thick pink line). Moreover, transcription of this gene can also be alternatively initiated at exon 4 and the resulting transcript is spliced to a short mRNA with only exons 4 and 5. The minor (Crick) strand (the long light-grey line with the 5'-end at the right) harbors another gene. Transcription of this gene may produce a long transcript containing four exons, with the exon 4 partly reverse-complementary to the exon 1 of the gene on the plus DNA strand. This long transcript may be spliced to an mRNA with all four exons (the wild type), an mRNA with exons 4 and 5, and/or an mRNA with exons 1, 2, and 4. Transcription of this minor-strand gene has two alternative initiation sites at exon 2 and exon 3, respectively, and has an alternative stop site at the exon 3. The transcripts initiated from exon 2 or 3 or terminated at exon 3 may be alternatively spliced as well, engendering different shorter RNAs that may be coding (mRNAs) or long-noncoding, such as the one with exons 1, 2, and 3, the one with exons 2, 3, and 4, the one with exons 2 and 4, and/or the one with exons 3 and 4. The 5'-end of the minus strand is also transcribed to an antisense RNA (the light-grey line) that is partly reverse-complementary to the exon 5 of the gene on the plus DNA strand. In addition, intron 1 of the minor strand gene harbors an exRNA and a single-exon noncoding RNA. The mature RNAs produced by the plus and minor strands are shown, respectively, in the top and the bottom parts of the graph. Note that different combinations of alternative uses of transcriptional initiation sites, alternative uses of transcription termination sites, and alternative uses of splicing sites can still produce many other mRNAs or long-noncoding RNAs that are not illustrated herein to avoid overwhelming the graph.

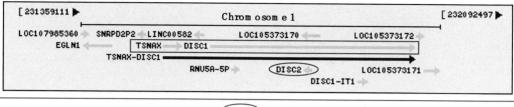

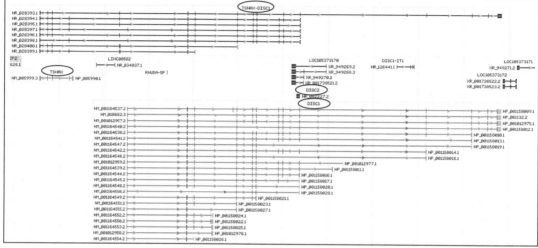

Figure 31.2

Illustration of how one gene can produce many different types of RNAs, using as an example images copied from the NCBI database about the human TSNAX-DISC1 transcriptional-readthrough gene. Top panel: as shown in the red box, the canonically defined gene TSNAX and its down-stream gene DISC1 locate on the plus strand (arrows point from the left to the right) of the DNA double helix in the chromosome 1, whereas the DISC2 gene (in the red circle) locates on the minus strand (arrow points from the right to the left) within the DISC1. Transcription of the TSNAX sometimes does not stop at its end but instead goes into the DISC1 gene, producing an RNA transcript containing both TSNAX and DISC1, which is defined as a read-through gene (the long red arrow). Within this TSNAX-DISC1 gene, there are the DISC1-IT1, LINC00582, and RNU5A-5P genes, besides the unannotated gene LOC105373170. Bottom panel: the TSNAX has only one mRNA (indicated as a green line), whereas the DISC1 has 23 mRNA variants due to both alternative uses of transcription termination sites and alternative uses of splicing sites. However, all eight RNA variants of the TSNAX-DISC1 gene are long noncoding (indicated as blue lines).

biologists, greatly enlarges the number of genes, thus making it easier to explain the great variety of cellular functions in the human body. This new definition basically says that a canonically defined gene actually contains quite a few other genes (Fig. 31.1 and 31.2), which requires researchers' attention to a number of technical details in molecular biology research. A major one pertains to how to avoid mistaken detection of other RNA variants or protein isoforms of the same gene or even RNA or proteins of other gene(s) within the gene (more correctly, the genomic locus) of interest. For example, detecting RNA of either the TSNAX gene (Gene ID: 7257) or the DISC1 gene (Gene ID: 27185) will also detect the RNA(s) of the read-through gene TSNAX-DISC1 (Gene ID: 100303453) if read-through occurs (i.e., if this gene is expressed), as illustrated in Fig. 31.2, and currently we still have no technical strategy to surmount this weakness [8].

Multiplicity at the protein level can occur to a single mRNA as well. Translation of an mRNA may have alternative initiation sites, that is, uses of different start codons, and alternative termination sites, that is, uses of different stop codons [13–16]. These different start or stop codons may be within the same open reading frame but may also belong to different open reading frames, making the RNA "bi- or poly-cistronic," that is, encoding two or multiple unrelated proteins [17]. Reiterated, one mRNA may encode different protein isoforms of the same gene or may encode unrelated proteins. In the latter case, a single RNA may be regarded as two or multiple genes. For instance, the CDKN2A gene is transcribed to several mRNA variants besides to a noncoding RNA; one of the mRNAs not only has an open reading frame for a protein isoform related to the proteins encoded by the other mRNA variants but also has an alternative open reading frame for an unrelated protein [18–23]. In summation, alternative uses of different transcription initiation or termination sites of a gene, alternative uses of different splice sites of an RNA transcript, alternative uses of different open reading frames of the same mRNA, and alternative uses of different start or stop codons of the same open reading frame can, in different combinations, magically produce a huge number of protein species from a single genomic DNA locus that is canonically defined as a gene.

31.4 Most published traditional Chinese medicine studies do not address the issue of RNA or protein multiplicity

The super intricacy of a genomic locus and its RNA and protein products described above delivers us a clear message that a sound comprehension of modern molecular biology is a prerequisite for correctly studying molecular mechanisms of how a disease is developed and how a therapy works. Since one gene can produce multiple mRNAs and long noncoding RNAs, it is inappropriate to state that expression of the gene studied is increased, decreased, or unchanged. Instead, it should be specific on which one or ones of the RNA variants are analyzed. If reverse transcription (RT) and polymerase chain reactions (PCR) are used to determine RNA expression, locations, and sequences of the primers used should be provided with an illustrational figure to help readers in understanding which RNA variant(s) are targeted and which others are avoided. RNA multiplicity is likely associated with multiplicity of protein isoforms, as we have shown using a simple top-down tactic of relatively high throughput proteomics [24–27]. Hence, when western blotting (WB) is used to detect protein expression, it is inappropriate to state that the protein level of the gene studied is changed or unchanged but in lieu which protein isoform(s) that are analyzed should be mentioned, with a graph to

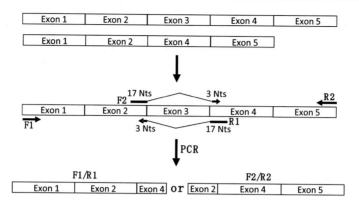

Figure 31.3
Depiction of one simple primer-design strategy for RT-PCR often used by us to specifically amplify an RNA variant without mis-amplifying the other(s). Say the gene in question has two RNA variants with one lacking exon 3 due to alternative splicing. A 20-mer forward (F) or reverse (R) primer is designed to skewedly span exons 2 and 4, with its last three nucleotides (Nts) on the exon 2 (R1 primer) or exon 4 (F2 primer) as illustrated. PCR with the F1/R1 or F2/R2 primer pair at a high annealing temperature will result in an amplicon containing exons 1 and 2 and part of exon 4 or containing part of exon 2 and exons 4 and 5, respectively, because the 3-nt annealing will melt at a high (over 57°C) annealing temperature.

clearly illustrate the relationship among different isoforms and the location(s) of the antibody epitope(s) on the protein sequence(s).

In the reality of the current biomedical research, the multiplicities of RNAs and proteins of individual genes have unfortunately been largely neglected [10,28], since probably most published studies reporting RNA or protein expression of genes do not provide information of how many RNA variants and protein isoforms of the gene in question have been listed in the National Centre for Bioinformation of the United States (NCBI) database. Actually, many genes have some RNA variants that have not been listed in the NCBI but have been reported in the literature, including some identified by us [29–32]. Accordingly, the strategy and technical details for specific detection of certain particular RNA variants while skipping some others, such as the one depicted in Fig. 31.3 that we often used, are not provided in most publications. Germaine to the determination of protein expression with WB, many published studies only mention that an N- or C-terminal antibody is used, which is not sufficient enough, although these reports are better than many others that do not even provide this information. All these deficiencies occur more often in the published studies on TCM mechanisms. Actually, few, if any, TCM studies provide information of the multiplicity of RNAs and proteins of the gene in question, to our knowledge.

In our experience in using RT-PCR and WB techniques, we often encounter that it is onerous and in rare cases impossible to specifically amplify a particular RNA variant without mis-amplifying one or more others of the same gene, and that it is even more insurmountable to specifically detect one particular protein isoform with commercially available antibodies in WB. When obtaining additional band(s) on a WB membrane, most researchers choose to cut away the unwanted band(s), assuming that they are spuriousness produced by nonspecific proteins without considering the possibility that they might actually be additional isoforms of the gene in question. This practice, which is likely misleading and biased [28], is more frequently seen in the mechanistic research of TCM.

31.5 Many traditional Chinese medicine studies do not consider technical pitfalls in common methods for DNA and RNA analyses

Molecular biology has swiftly advanced in recent decades, mainly because of the swift advancement of biotechnology starting from the establishment and wide dispersion of the PCR technique nearly three decades ago. The ensuing meteoric success in developing modern DNA sequencing techniques and associated bioinformatic implements has expedited DNA and RNA research, making whole genome sequencing feasible and affordable to many investigators. Nowadays, the second and third generations of DNA sequencing technology have been widely utilized by many biomedical workers and have resulted in a stupendous amount of sequence data deposited in public databases or investigators' own computers. However, because these modern sequencing techniques as well as associated sample-processing methods and bioinformatic implements are established on complicated principles and require sophisticated instruments, they are usually performed by companies or core facilities and rarely by individual investigators themselves. A flaw thus ensues that few researchers who produce data of DNA or RNA sequences or expression levels in a high throughput manner really know well the pitfalls and weaknesses of the techniques involved. Actually, even RT and PCR methods, which are daily used in most laboratories and are the basic elements of most sequencing techniques, have variegated weaknesses and technical pitfalls, as we have pointed out before [11,33–35]. For instance, most techniques for DNA and RNA analyses employ the strategy of hybridization between A and T or C and G bases in a DNA, and are based on a principle that DNA polymerization is primed by a short DNA or RNA primer. Since the whole genome of 3.2-3.5 billion nucleotides is constituted by these four bases, the genome has a heap of identical sequences, especially short ones. Since each primer is very short, usually about several to dozens of nucleotides, mispriming, that is, annealing the primer to one or more additional regions of the same gene or other genes due to sequence homologue, becomes a common and hitherto-intractable problem. Actually, one or two mismatches between the short primer and the template can still prime DNA polymerization in a slew of occasions, which further increases the chance for mispriming.

PCR results are often presented qualitatively as a band in an agarose gel or quantitatively as a number given by a real-time PCR machine. Because either routine RT-PCR or quantitative

RT-PCR has its strengths and weaknesses, both methods should be used to compensate for each other's deficiency and to ensure the specificity of the resulting amplicon. Unfortunately, almost all TCM researchers publish only quantitative RT-PCR results, and few present also an image of the PCR amplicon in an agarose gel to corroborate that the amplicon is the only one engendered and has the correct molecular weight.

31.6 Signaling transduction networking is even more intricate and its studies are more problematic

Cellular functions are largely elicited by proteins, as many nonprotein cellular components, such as various fatty acids and chemicals like iron, copper, zinc, etc., function via affecting the activity of proteins (enzymes in many cases). Various regulatory RNAs also function by directly or indirectly regulating protein production. Proteins usually exert their effects by eliciting different cascades of biochemical reactions, which is called "signal transduction" or "molecular pathway" in an idiom of molecular biology. Therefore, the aforementioned multiplicity of RNAs or proteins of individual genes is just the tip of the iceberg, meaning that the actual complexity of mechanisms behind even a very simple cellular function is much greater and more recondite: a protein, once translated from an mRNA, may be cleaved to shorter peptides before it has functions, as exemplified by the pro-insulin and many caspases that need to be cleaved to be the effective or mature protein. Usually, mature proteins still need to be subjected to different chemical modifications, such as phosphorylation or acetylation, to be effective or more effective. A protein having experienced such modifications may need to be transferred, likely by one or more other proteins or through a channel formed by certain proteins, to another intracellular compartment, such as the nucleus or even the nucleolus. At the right compartment, the protein may bind to another protein, such as a receptor, or to a complex of cofactors consisting of many, even over one hundred, other proteins, with formation of spliceosome, an enzymatic complex that splices RNAs, as an example. This protein dimer or complex may function as a regulator (activator or suppressor) to regulate the RNA transcription, RNA splicing, RNA stability, or RNA translation of another gene, or as an enzyme to chemically modify another gene, RNA, or protein. The protein product of this other gene may have quite different types of function, or may just repeat some or all of the steps described above, collectively forming a chain of biochemical reactions, so that the order from the cell's nucleus or from the organism (the human body) as a whole is delivered and executed. A caveat that needs to be given is that once a signaling pathway goes awry due to whatever reason, the cell can easily establish a bypass to circumvent it, which makes signal transduction much more convoluted.

Currently, most mechanistic studies on TCM-involved signal transduction stay superficial, as they end at the determination of the RNA and protein levels without analyzing the formation of protein complex and the activities of the complex. For instance, many investigators claim

Table 31.1: Projects on signal transduction elicited by medicinal herbs funded by the National Natural Science Foundation of China during 2010–2019.

	2010	2011	2012	2013	2014	2015	2016	2017	2018	2019
Number of projects	1118	1562	1845	1997	1965	2232	2206	2446	2492	2488
Funds (M $)[a]	51.06	104.57	148.69	151.29	156.79	147.31	132.93	150.59	160.32	152.55

[a] The funds are in Chinese Yuan but herein converted to US dollars based on exchange rate of 7.0 Yuan to 1.0 dollar.

that they study the role of the NF-™B, but they determine only the RNA or protein levels of only one or two components of the NF-™B protein complex without analyzing its transcriptional activity using such as a gel-shift assay that is often used by us [36] and many others, as exemplified by the work of Li et al. [37].

31.7 Traditional Chinese medicine professionals have an innate disadvantage in translating traditional Chinese medicine theory into the western medicine language

From the WM point of view, a disease occurs because one or more steps of one or more molecular cascades of the body's order-delivering (signaling) system go awry, while correction of the slip(s) is a job for the therapy. Therefore, WM puts a tremendous effort (1) on research on molecular signaling transduction to obtain mechanistic insights into how diseases are developed and (2) on the development of new therapeutic drugs or approaches that target certain step(s) of the relevant pathway(s). These lines of knowledge are reaped through a tight collaboration among workers specializing in assorted disciplines of natural science, including chemistry, physics, mathematics, biology, molecular biology, genetics, etc. This motley crew of researchers works closely with various medical practitioners who see patients or handle specimens from patients; together they tackle the two tasks mentioned above. This integration of various disciplines helps significantly in untangling the super convolution of modern biomedicine.

Translation of TCM theory and concepts into the WM language requires a solid comprehension of how WM expound on how diseases are developed and how treatments work at the molecular level. Therefore, in China, an enormous amount of funding has been awarded to research on the molecular signaling pathways involved by therapeutic remedies or approaches of TCM. We rummaged through the public database of the National Natural Science Foundation of China, the Chinese counterpart of National Institute of Health in the United States pertaining to governmental research funding, for the projects funded by it on the research of signaling transduction elicited by medicinal herbs. The result, although obviously very preliminary and not very accurate, shows that in the past ten years the Chinese government has injected nearly an equivalent of 1.4-billion US dollars to over 20,000 projects on this research area alone (Table 31.1), which constitutes the major portion of the funding on mech-

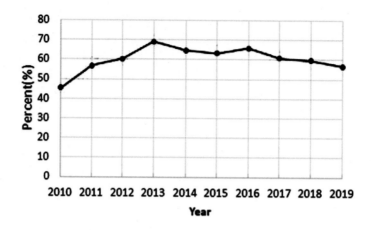

Figure 31.4

Percentage of the projects on molecular signaling pathways in mechanistic research of TCM funded by the National Natural Science Foundation of China during 2010–2019. *TCM*, Traditional Chinese medicine.

anistic research of TCM (Fig. 31.4) with the funding on many other TCM quarters excluded. A vexing flaw is that the principal investigators in these projects are TCM physicians who are engrossed by clinical work with little time in research laboratories. Few of these awarded projects have wizards with specialties in molecular biology as key participants, presumably because it is too hard for the non-TCM professionals to comprehend the TCM language. A caveat is that many chemists and biologists do work on medicinal herbs, but they focus on isolating effective chemicals from herbs and determining the therapeutic functions and possible adversities of the isolated chemicals. This is not TCM research but is a traditional research of WM that has led to the discovery and ensuing production of many medicinal drugs in the past century, exemplified by many traditional painkillers such as opium and morphine as well as their analogs and chemical derivatives.

In China, the schooling of most TCM practitioners includes little molecular biology, molecular medicine, and modern biotechnology, with the aforementioned intricate genomic organization and technical tactics hardly mentioned. As a repercussion, most TCM professionals lack sufficient theoretical knowledge of biotechnology and lack bench experience in the techniques routinely used in biomedical research, which makes it hard for them to avoid or circumvent the relevant technical pitfalls. Moreover, most TCM practitioners read little scientific literature in English because most literature relevant to TCM is written in Chinese, which collectively makes most TCM professionals putting little effort on English education. Insufficient reading of English literature on modern biomedicine renders them poor in English reading skills, especially in scientific and technological vocabulary, which in turn confers on them more difficulty in reading the relevant literature in the English language, thus getting into a vicious circle.

31.8 How should the above mentioned constraints be vanquished to advance traditional Chinese medicine?

It is hard for WM workers and even much harder for TCM professionals to grasp all new developments in relevant disciplines of science. Therefore, a feasible tactic to surmount all the aforementioned constraints that impede the TCM advancement and its translation into the WM language is to establish a tight collaboration among TCM professionals, WM practitioners, and pundits with different expertise. Moreover, WM physicians should be inspired to allocate much more time to TCM learning, so that they can better communicate with TCM professionals. However, more critically, TCM workers should be spurred to put much more effort on learning molecular biomedicine, including molecular biology and associated bench technology, with raising their English level as a prerequisite and essential. This requires Chinese medical schools of WM to assign more hours in TCM teaching while requiring TCM school to add molecular biology and biotechnology into their curricula or, if these courses are already in the existing curricula, to allocate more teaching hours on these subjects, besides more hours on other WM subjects such as physiology and pathology. In addition, savants in many realms of natural science, especially biology and molecular biology, should be actuated to learn TCM theory and collaborate with TCM wizards in TCM research.

31.9 Conclusions

TCM has in the past centuries shown its marvelous effectiveness in multifarious medical problems. Actually, there are some illnesses to which WM can do little good and TCM remedies become irreplaceable, which has been showcased by the ongoing subjugation of the SARS-Cov-2 virus by the Chinese [38–43]. Notwithstanding, most herb recipes and other therapeutic approaches of TCM still lack indubitable proofs from laboratory studies and clinical trials. Most published studies on the molecular mechanisms of TCM have more or less deficiencies or are somewhat superficial. TCM needs to correct these deficiencies and to advance and promulgate itself, keeping abreast of the meteoric progress in WM and other quarters of science and technology. A promising strategy to achieve these goals is to form a new school of medicine by conflating TCM and WM, which is an attempt both medical professionals and governments in China have long been making. A cornerstone of this coalescence is to assimilate TCM theory and concepts into the school of WM, as the latter has already been much more advanced and more globally disseminated and used. This integration requires TCM professionals, especially those who perform mechanistic research, to equip themselves with a sound edifice of knowledge of modern biomedicine, including the organization of chromosomal DNA and associated biotechnology described herein, before their ingress into the research of molecular mechanisms behind TCM.

References

[1]   G Liu, F Liu, L Xiao, Q Kuang, X He, Y Wang, Narrative review of the mechanisms of action of dachengqi decoction in the treatment of hyperlipidemic pancreatitis on six-hollow-organs to be unblocked theory, Ann Palliat Med 9 (4) (2020) 2323–2329.

[2]   FY Fung, YC Linn, Developing traditional Chinese medicine in the era of evidence-based medicine: current evidences and challenges, Evid Based Complement Alternat Med 2015 (2015) 425037.

[3]   R Xue, Z Fang, M Zhang, Z Yi, C Wen, T Shi, TCMID: traditional Chinese medicine integrative database for herb molecular mechanism analysis, Nucleic Acids Res 41 (Database issue) (2013) D1089–D1095.

[4]   MJ Bamshad, SB Ng, AW Bigham, HK Tabor, MJ Emond, DA Nickerson, Exome sequencing as a tool for Mendelian disease gene discovery, Nat Rev Genet 12 (11) (2011) 745–755.

[5]   JE Belizario, The humankind genome: from genetic diversity to the origin of human diseases, Genome 56 (12) (2013) 705–716.

[6]   RC Pink, K Wicks, DP Caley, EK Punch, L Jacobs, DR Carter, Pseudogenes: pseudo-functional or key regulators in health and disease? RNA 17 (5) (2011) 792–798.

[7]   KD Pruitt, J Harrow, RA Harte, C Wallin, M Diekhans, DR Maglott, et al., The consensus coding sequence (CCDS) project: Identifying a common protein-coding gene set for the human and mouse genomes, Genome Res 19 (7) (2009) 1316–1323.

[8]   Y He, C Yuan, L Chen, M Lei, L Zellmer, H Huang, Transcriptional-readthrough rnas reflect the phenomenon of "A Gene Contains Gene(s)" or "Gene(s) within a Gene" in the human genome, and thus are not chimeric RNAs, Genes (Basel) 9 (1) (2018) pii E40, doi:10.3390/genes9010040.

[9]   Y Jia, L Chen, Y Ma, J Zhang, N Xu, DJ Liao, To know how a gene works, we need to redefine it first but then, more importantly, to let the cell itself decide how to transcribe and process Its RNAs, Int J Biol Sci 11 (12) (2015) 1413–1423.

[10] X Liu, Y Wang, W Yang, Z Guan, W Yu, DJ Liao, Protein multiplicity can lead to misconduct in western blotting and misinterpretation of immunohistochemical staining results, creating much conflicting data, Prog Histochem Cytochem 51 (3-4) (2016) 51–58.

[11] C Yuan, Y Han, L Zellmer, W Yang, Z Guan, W Yu, et al., It is imperative to establish a pellucid definition of chimeric RNA and to clear up a lot of confusion in the relevant research, Int J Mol Sci 18 (4) (2017) pii E714, doi:10.3390/ijms18040714.

[12] W Guo, SJ Bharmal, K Esbona, ML Greaser, Titin diversity—alternative splicing gone wild, J Biomed Biotechnol 2010 (2010) 753675, doi:10.1155/2010/753675.

[13] C Touriol, S Bornes, S Bonnal, S Audigier, H Prats, AC Prats, et al., Generation of protein isoform diversity by alternative initiation of translation at non-AUG codons, Biol Cell 95 (3-4) (2003) 169–178.

[14] X Cao, SA Slavoff, Non-AUG start codons: expanding and regulating the small and alternative ORFeome, Exp Cell Res 391 (1) (2020) 111973.

[15] MV Rodnina, N Korniy, M Klimova, P Karki, BZ Peng, T Senyushkina, et al., Translational recoding: canonical translation mechanisms reinterpreted, Nucleic Acids Res 48 (3) (2020) 1056–1067.

[16] AV Kochetov, Alternative translation start sites and hidden coding potential of eukaryotic mRNAs, Bioessays 30 (7) (2008) 683–691.

[17] TA Karginov, DPH Pastor, BL Semler, CM Gomez, Mammalian polycistronic mRNAs and disease, Trends Genet 33 (2) (2017) 129–142.

[18] CJ Sherr, Divorcing ARF and p53: an unsettled case, Nat Rev Cancer 6 (9) (2006) 663–673.

[19] X Tian, J Azpurua, Z Ke, A Augereau, ZD Zhang, J Vijg, et al., INK4 locus of the tumor-resistant rodent, the naked mole rat, expresses a functional p15/p16 hybrid isoform, Proc Natl Acad Sci USA 112 (4) (2015) 1053–1058.

[20] DE Quelle, F Zindy, RA Ashmun, CJ Sherr, Alternative reading frames of the INK4a tumor suppressor gene encode two unrelated proteins capable of inducing cell cycle arrest, Cell 83 (6) (1995) 993–1000.

[21] D Duro, O Bernard, V Della, BR V, CJ Larsen, A new type of p16INK4/MTS1 gene transcript expressed in B-cell malignancies, Oncogene 11 (1) (1995) 21–29.

[22] L Mao, A Merlo, G Bedi, GI Shapiro, CD Edwards, BJ Rollins, et al., A novel p16INK4A transcript, Cancer Res 55 (14) (1995) 2995–2997.

[23] S Stone, P Jiang, P Dayananth, SV Tavtigian, H Katcher, D Parry, et al., Complex structure and regulation of the P16 (MTS1) locus, Cancer Res 55 (14) (1995) 2988–2994.

[24] J Qu, J Zhang, L Zellmer, Y He, S Liu, C Wang, et al., About three-fourths of mouse proteins unexpectedly appear at a low position of SDS-PAGE, often as additional isoforms, questioning whether all protein isoforms have been eliminated in gene-knockout cells or organisms, Protein Sci 29 (4) (2020) 978–990.

[25] Y Ren, J Savill, Apoptosis: the importance of being eaten, Cell Death Differ 5 (7) (1998) 563–568.

[26] R Yan, J Zhang, L Zellmer, L Chen, D Wu, S Liu, Probably less than one-tenth of the genes produce only the wild type protein without at least one additional protein isoform in some human cancer cell lines, Oncotarget 8 (47) (2017) 82714–82727.

[27] J Zhang, X Lou, H Shen, L Zellmer, Y Sun, S Liu, Isoforms of wild type proteins often appear as low molecular weight bands on SDS-PAGE, Biotechnol J 9 (8) (2014) 1044–1054.

[28] Y He, C Yuan, L Chen, Y Liu, H Zhou, N Xu, et al., While it is not deliberate, much of today's biomedical research contains logical and technical flaws, showing a need for corrective action, Int J Med Sci 15 (4) (2018) 309–322.

[29] Y Sun, S Cao, M Yang, S Wu, Z Wang, X Lin, et al., Basic anatomy and tumor biology of the RPS6KA6 gene that encodes the p90 ribosomal S6 kinase-4, Oncogene 32 (14) (2013) 1794–1810.

[30] Y Sun, X Lou, M Yang, C Yuan, L Ma, BK Xie, et al., Cyclin-dependent kinase 4 may be expressed as multiple proteins and have functions that are independent of binding to CCND and RB and occur at the S and G 2/M phases of the cell cycle, Cell Cycle 12 (22) (2013) 3512–3525.

[31] M Yang, Y Sun, L Ma, C Wang, JM Wu, A Bi, et al., Complex alternative splicing of the smarca2 gene suggests the importance of smarca2-B variants, J Cancer 2 (2011) 386–400.

[32] M Yang, J Wu, SH Wu, AD Bi, DJ Liao, Splicing of mouse p53 pre-mRNA does not always follow the "first come, first served" principle and may be influenced by cisplatin treatment and serum starvation, Mol Biol Rep 39 (9) (2012) 9247–9256.

[33] Z Peng, C Yuan, L Zellmer, S Liu, N Xu, DJ Liao, Hypothesis: artifacts, including spurious Chimeric RNAs with a short homologous sequence, caused by consecutive reverse transcriptions and endogenous random primers, J Cancer 6 (6) (2015) 555–567.

[34] B Xie, W Yang, Y Ouyang, L Chen, H Jiang, Y Liao, et al., Two RNAs or DNAs May artificially fuse together at a short homologous sequence (SHS) during reverse transcription or polymerase chain reactions, and thus reporting an SHS-containing Chimeric RNA requires extra caution, PLoS One 11 (5) (2016) e0154855, doi:10.1371/journal.pone.0154855.

[35] C Yuan, Y Liu, M Yang, DJ Liao, New methods as alternative or corrective measures for the pitfalls and artifacts of reverse transcription and polymerase chain reactions (RT-PCR) in cloning chimeric or antisense-accompanied RNA, RNA Biol 10 (6) (2013) 958–967.

[36] H Biliran Jr., S Banerjee, A Thakur, FH Sarkar, A Bollig, F Ahmed, et al., c-Myc-induced chemosensitization is mediated by suppression of cyclin D1 expression and nuclear factor-kappa B activity in pancreatic cancer cells, Clin Cancer Res 13 (9) (2007) 2811–2821.

[37] Q Li, J Gao, X Pang, A Chen, Y Wang, Molecular mechanisms of action of emodin: as an anti-cardiovascular disease drug, Front Pharmacol 11 (2020) 559607.

[38] N Shi, L Guo, B Liu, Y Bian, R Chen, S Chen, et al., Efficacy and safety of Chinese herbal medicine versus Lopinavir-Ritonavir in adult patients with coronavirus disease 2019: A non-randomized controlled trial, Phytomedicine 81 (2020) 153367.

[39] N Liang, Y Ma, J Wang, H Li, X Wang, L Jiao, et al., Traditional Chinese medicine guidelines for coronavirus disease 2019, J Tradit Chin Med 40 (6) (2020) 891–896.

[40] DYW Lee, QY Li, J Liu, T Efferth, Traditional Chinese herbal medicine at the forefront battle against COVID-19: clinical experience and scientific basis, Phytomedicine 80 (2021) 153337.

[41] A Al-Romaima, Y Liao, J Feng, X Qin, G Qin, Advances in the treatment of novel coronavirus disease (COVID-19) with Western medicine and traditional Chinese medicine: a narrative review, J Thorac Dis 12 (10) (2020) 6054–6069.

[42] S Xi, Y Li, L Yue, Y Gong, L Qian, T Liang, et al., Role of traditional chinese medicine in the management of viral pneumonia, Front Pharmacol 11 (2020) 582322.

[43] J Huang, G Tao, J Liu, J Cai, Z Huang, JX Chen, Current prevention of COVID-19: natural products and herbal medicine, Front Pharmacol 11 (2020) 588508.

Index

Page numbers followed by "*f*" and "*t*" indicate, figures and tables respectively.

Printed in the United States
by Baker & Taylor Publisher Services